Routledge Revivals

The History of the Study of Landforms
or
The Development of Geomorphology

Volume 2

This reissue, originally published in 1973, is entirely devoted to the life and work of the world's most famous geomorphologist, William Morris Davis (1850-1934). It contains an in depth treatment of Davis' many contributions to the study of landforms, including: the cycle of erosion; denudation chronology; arid and karst geomorphology; and the coral reef problem.

The History of the Study of Landforms
or
The Development of Geomorphology

Volume 2: The Life and Work of
William Morris Davis

Richard J. Chorley,
Robert P. Beckinsale
and
Antony J. Dunn

Routledge
Taylor & Francis Group

First published in 1973
by Methuen

This edition first published in 2009 by Routledge
2 Park Square, Milton Park, Abingdon, Oxon, OX14 4RN

Simultaneously published in the USA and Canada
by Routledge
270 Madison Avenue, New York, NY 10016

Routledge is an imprint of the Taylor & Francis Group, an informa business

Publisher's Note
The publisher has gone to great lengths to ensure the quality of this reprint but points
out that some imperfections in the original copies may be apparent.

Disclaimer
The publisher has made every effort to trace copyright holders and welcomes
correspondence from those they have been unable to contact.

ISBN 13: 978-0-415-56369-7 (set)
ISBN 13: 978-0-416-26890-4 (hbk)
ISBN 13: 978-0-415-56795-4 (pbk)
ISBN 13: 978-0-203-47253-8 (ebk)

ISBN 10: 0-415-56369-0 (set)
ISBN 10: 0-415-56795-5 (pbk)
ISBN 10: 0-416-26890-0 (hbk)
ISBN 10: 0-203-47253-5 (ebk)

THE HISTORY OF THE STUDY OF LANDFORMS OR THE DEVELOPMENT OF GEOMORPHOLOGY

VOLUME TWO

Davis in a characteristic pose studying a marine terrace at Beverly Boulevard, Los Angeles on 27 February 1931 (Courtesy J. H. Maxson)

THE HISTORY OF THE STUDY OF
LANDFORMS

OR
THE DEVELOPMENT OF
GEOMORPHOLOGY

VOLUME TWO: THE LIFE AND WORK
OF WILLIAM MORRIS DAVIS

BY

RICHARD J. CHORLEY
ROBERT P. BECKINSALE
ANTONY J. DUNN

Routledge
Taylor & Francis Group
LONDON AND NEW YORK

First published in 1973
by
Routledge
2 Park Square, Milton Park,
Abingdon, Oxon, OX14 4RN

270 Madison Ave, New York NY 10016

Transferred to Digital Printing 2006

© *1973 R. J. Chorley & R. P. Beckinsale*
Filmset in Photon Imprint 11 on 12 pt. by
Richard Clay (The Chaucer Press), Ltd, and

ISBN 0–416–26890–0 (hbk)

Distributed in the U S A by
HARPER & ROW PUBLISHERS, INC.
BARNES & NOBLE IMPORT DIVISION

Contents

CONTENTS

List of Illustrations

Davis in a characteristic pose studying a marine terrace at
Beverly Boulevard, Los Angeles on 27 February 1931   *frontispiece*

Preface

Volume Two of *The History of the Study of Landforms* needs no further preface than that which was provided by the publication of Volume One some nine years ago. In the latter we showed how, towards the end of the nineteenth century, many of the hitherto disparate themes in geomorphology were beginning to converge, and the present volume attempts to evaluate the central role of William Morris Davis in this synthesis. This task has been neither a rapid nor an easy one.

The lack of any previous substantive biography of Davis has meant that much labour and space has had to be devoted to assembling and presenting the mass of personal information relating to his long and varied life, which has up to now remained scattered and unpublished. If we have been generous in the inclusion of this original material it is in an attempt to redress the surprising lack of biographical interest in Davis since his death almost forty years ago. Davis, who provided a massive biography of G. K. Gilbert for the National Academy of Sciences, clearly expected a similarly elaborate treatment for himself. The obvious scholar to produce this was his loyal student Douglas Johnson, but only eight years after the death of Davis he himself suffered a heart attack and died two years later, leaving undone the task which undoubtedly he had planned for his retirement. Ironically, the preparation of Davis' official biography fell to R. A. Daly who, although a colleague at Harvard, did not share Johnson's undiluted admiration for the Master. The biography, which appeared in 1945, was perfunctory and made little use of the considerable personal material which was then available. The present authors have been fortunate in securing much of this material from a variety of sources, to which the length of our acknowledgements attests. Richard Chorley was particularly fortunate in being able to interview Davis' eldest son only three months before the latter died in his eighty-first year.

A further difficulty facing anyone attempting an analysis of the work of W. M. Davis stems from the very magnitude of his achievement. Not only is the sheer productivity of his scholarship, which was sustained for over half a century, daunting to the would-be analyst, but the strong and extreme views which are held by contemporary workers regarding Davis' contributions constantly force the assessor back to the source material. We have tried in this volume to judge Davis' work within the temporal context of its original production, without, we hope, succumbing to either the attacks of his hostile

critics or to the unthinking praise of some of his personal followers. In
Volume Three we will deal with the general history of geomorphology
during the Davis era, to which the work of the Master provides a background, and will return there to the major features of Davis' work; and in
Volume Four a treatment of the 'geomorphological revolution' which followed the Second World War will naturally include recent reactions to
Davisian geomorphology.

Although for us the proverb 'To understand all is to forgive all' presumes
an unattainable generosity, a decade of living with William Morris Davis has
left three not uncritical biographers with an increasing respect for the
world's most influential geomorphologist.

R. J. CHORLEY   R. P. BECKINSALE   A. J. DUNN
University of Cambridge   *University of Oxford*   *Northumberland*

Acknowledgements

The authors and publisher would like to record their gratitude to the following editors, publishers and individuals for permission to reproduce quotations and illustrations, for the provision of source material and for other assistance.

ORIGINAL DOCUMENTS BY W. M. DAVIS

Mrs J. P. Buwalda, Pasadena, California; Professor E. Mott Davis Jr, Austin, Texas; Mr and Mrs R. Mott Davis, Silver Spring, Maryland; William Morris Davis II, Bass River, Massachusetts; The Secretary, Harvard Corporation and the Harvard College Library; Dr P. B. King, Menlo Park, California; Professor G. J. Martin, New Haven, Connecticut; Dr John H. Maxson, Denver, Colorado; Archives of the National Academy of Sciences, Washington, D.C.; Colonel and Mrs Charles D. Osbourne, Auburn, New York; Professor H. F. Raup, Kent, Ohio; The Director, Royal Geographical Society, London; Professor R. J. Russell, Baton Rouge, Louisiana; Professor S. S. Visher, Bloomington, Indiana; Charles J. Wellner, *The Citizen-Advertiser*, Auburn, New York; Professor A. O. Woodford, Claremont, California; Dr J. K. Wright, Lyme, New Hampshire.

JOURNALS

Annals of the Association of American Geographers, Vol. 1, pp. 23–4, 30, 35–6, 40, 43, 47, 55–6, 64–5 and 69–70; Vol. 2, pp. 74–6, 78–81, 83–5, 88–9, 93–8; Vol. 5, pp. 62, 71–4, 77–8, 86–7, 90–1, 98–9; Vol. 6, pp. 268–9, 284–5 and 288; Vol. 10, pp. 14 and 144–6; Vol. 14, pp. 190 and 200–1; Vol. 19, p. 209; Vol. 22, pp. 217 and 230; Vol. 25, pp. 24 and 31; Vol. 30, pp. 225, 231, 247 and 254; Vol. 40, pp. 174–9, 188, 191–2, 195, 197, 209–10 and 213; Vol. 45, pp. 312–13.
American Geologist, Vol. 23, pp. 212–13, 215–16, 223, 226 and 232–3; Vol. 33, pp. 166 and 167.
American Journal of Science, 2nd Ser., Vol. 42, p. 82; 3rd Ser., Vol. 28, pp. 123 and 131–2; 4th Ser., Vol. 35, pp. 176–8, 180–1 and 188; 4th Ser., Vol. 40, pp. 231–2, 239–43, 247 and 266–7; 4th Ser., Vol. 44, pp. 346–7 and 349–50.
Annales de Géographie, Vol. 21, pp. 2, 13 and 14.
Bulletin of the Geographical Society of Philadelphia, Vol. 10, pp. 32–3.

Bulletin of the Geological Society of America, Vol. 11, p. 600; Vol. 23, pp. 94–5, 100, 105, 106, 114 and 121; Vol. 24, pp. 686–7; Vol. 33, pp. 94, 587–90 and 592–5; Vol. 41, pp. 295, 486, 503, 509–10, 551, 560, 562, 603 and 623; Vol. 42, pp. 308–9; Vol. 43, pp. 406, 412–15 and 439–40; Vol. 44, pp. 1046, 1048, 1051 and 1062; Vol. 49, pp. 1339–41, 1343, 1346, 1360–1, 1372, 1387, 1391–2 and 1414.

Educational Review, Vol. 13, pp. 437–8.

Geographen Kalender, Vol. 7, pp. 1–2, 4, 6, 8, 10–12, 14, 18–19, 24, 26–8, 36, 40–4, 50, 52–4, 56–8, 62–4, 66 and 68–72.

Geographical Journal, Vol. 5, pp. 127, 130 and 141; Vol. 14, pp. 481–5, 487–9, 491–7, 499–500 and 503; Vol. 22, pp. 416 and 420; Vol. 34, pp. 301–3, 305, 307–8 and 312–14; Vol. 55, pp. 212 and 216–17; Vol. 121, pp. 89–90.

Geographical Review, Vol. 7, pp. 179–80; Vol. 8, pp. 266–7 and 271–3; Vol. 10, pp. 249–51, 254, 257 and 260; Vol. 13, pp. 318–22, 601 and 604–7; Vol. 16, pp. 128 and 350–1; Vol. 24, pp. 11, 177–8 and 297.

Geographical Teacher, Vol. 10, pp. 289–90.

Geological Magazine, Vol. 10, N.S. Dec. 4, pp. 146–7.

Harvard Graduates' Magazine, Vol. 11, pp. 364–5; Vol. 38, pp. 395–7 and 402–4.

Journal of Geology, Vol. 2, pp. 68 and 91–5; Vol. 10, pp. 84–7, 89, 92–6 and 100; Vol. 12, pp. 675, 679, 681 and 684; Vol. 13, pp. 382–4, 386–98 and 402; Vol. 26, pp. 293, 297–8, 302 and 307–8; Vol. 31, pp. 2, 3, 5, 7–10, 24–7 and 29–30; Vol. 38, pp. 1–7, 14, 140–1, 150 and 158.

Journal of Geomorphology, Vol. 2, pp. 303 and 372; Vol. 3, pp. 59, 63–4, 156 and 261–2.

Proceedings of the Geologists' Association of London, Vol. 16, p. 93; Vol. 21, p. 152.

Professional Geographer, Vol. 15, pp. 2–3.

Scottish Geographical Magazine, Vol. 22, pp. 76–9, 82 and 89; Vol. 26, pp. 563–9, 575–7 and 582–4; Vol. 41, pp. 65, 67, 70 and 73–4.

Transactions of the Institute of British Geographers, Vol. 31, pp. 12–13.

American Naturalist, Vol. 21, pp. 812–13.

American Meteorological Journal, Vol. 6, p. 462.

Geologische Rundschau, Vol. 16, p. 314.

Journal of Geography, Vol. 4, pp. 3–4.

National Geographic Magazine, Vol. 1, pp. 2–3, 8, 11, 16, 21–4, 220–7, 229–36, 239–44 and 251; Vol. 2, pp. 89 and 91–7; Vol. 7, p. 195–6.

Popular Science Monthly, Vol. 57, pp. 70, 85, 168 and 515–16; Vol. 78, pp. 228–9.

Proceedings of the Boston Society of Natural History, Vol. 24, pp. 195–6, 367 and 375; Vol. 29, pp. 275, 277–8, 284, 295–6, 298, 300, 303, 305, 307 and 309–10.

Proceedings of the National Academy of Sciences, Vol. 1, p. 149; Vol. 2, pp. 467–8 and 473–5; Vol. 3, pp. 478–9 and 652–4; Vol. 4, p. 203; Vol. 10, pp. 206–9.

Proceedings of the American Association for the Advancement of Science, Vol. 33, pp. 429–32.

Quarterly Journal of the Geological Society of London, Vol. 65, pp. 281–2, 287, 292–3, 295, 303–4, 311–12, 318–20 and 346.

School Review, Vol. 1, p. 338.

Science, Vol. 1, p. 356; Vol. 21, pp. 825–8; Vol. 56, pp. 123 and 127; Vol. 63, p. 446; Vol. 83, p. 90.

Scientific Monthly, Vol. 15 (September 1922), pp. 196–9, 201–2, 204, 206, 210 and 212–13; Vol. 38 (May 1934), pp. 395–421.

Transactions and Proceedings of the New Zealand Institute, Vol. 51, pp. 22 and 27–9.

Transactions of the San Diego Society of Natural History, Vol. 7, pp. 216–17 and 219–20.

BOOKS AND NEWSPAPERS

American Geographical Society, New York: A. P. Brigham, *Memorial Volume of the Transcontinental Excursion of 1912* (1915), pp. 10–25; W. M. Davis, *The Lesser Antilles* (1926), pp. 12–16, 44, 164–5, 184–5 and 194–7; W. M. Davis, *The Coral Reef Problem* (1928), pp. 2, 3, 18, 20–1, 39–40, 51–2, 112, 118–20, 127, 139–41, 143–5, 218–19, 307, 333–4, 373–4 and 546–8.

Boston Evening Transcript: W. M. Davis, *The British Association Meeting in South Africa* (1905); W. M. Davis, *A Year Abroad* (1909).

Boston Herald: W. M. Davis, *Oration at Phi Beta Kappa Harvard Chapter* (1922).

Carnegie Institute, Washington, D.C.: R. Pumpelly (Ed.), *Explorations in Turkestan* (1905), pp. 66–70.

Collins, London: N. Barlow (Ed.), *The Autobiography of Charles Darwin 1809–1882* (1958), pp. 98–9.

Columbia University Press, New York: D. W. Johnson, *Stream Sculpture on the Atlantic Slope* (1931), pp. vii–viii and 56.

Constable, London: O. W. Villard, *John Brown 1800–1859* (1910), pp. 509–10.

Eastern Michigan University Press, Ypsilanti, Michigan: G. J. Martin, *Mark Jefferson: Geographer* (1968), pp. 40–1 and 140.

Ginn & Co., Boston: W. M. Davis, *Geographical Essays* (1909), pp. 32–4, 36, 46, 57–8, 127 and 150.

Harpers, New York: W. A. Neilson (Ed.), *Charles W. Eliot: The Man and his Beliefs* (1926), p. 225.

Harvard University Press, Cambridge, Massachusetts: S. E. Morrison (Ed.),

The Development of Harvard University Since the Inauguration of President Eliot (1930), pp. 206, 315 and 321–2.

Houghton Mifflin Co., Boston and New York: A. D. Hallowell, *James and Lucretia Mott: Life and Letters* (1884), pp. 22–5, 128–30, 295–6, 326, 328 and 331–2.

International Geographical Union: *Report of the Eighth International Geographical Congress* (1904), pp. 150–8 and 160–1.

Longmans, London: H. R. Mill, *An Autobiography* (1951), p. 90.

Macmillan and Co., London: W. Penck, *Morphological Analysis of Landforms* (1953), pp. 7–9 and 121.

The Nation: W. M. Davis, *The British Association Meeting in South Africa* (1905).

National Academy of Sciences, Washington, D.C.: G. A. Comstock, *Biographical Memoir of B. A. Gould* (1924), pp. 159–60 and 167; W. M. Davis, *Biographical Memoir of G. K. Gilbert* (1927), pp. 1, 98, 107–8, 203 and 303; R. A. Daly, *Biographical Memoir of W. M. Davis* (1945), pp. 265, 272, 274–5 and 277–80.

University of Nebraska, Lincoln, Nebraska: V. E. Rigdon, *Contributions of William Morris Davis to Geography in America* (1934), pp. viii, 61, 65–6 and 107.

University of Oklahoma Press, Norman, Oklahoma: C. N. Gould, *Covered Wagon Geologist* (1959), p. 133.

Private Printing, A. M. Johnson, Washington, D.C.: A. M. Johnson, *William Morris Davis: 1815–1891* (1951), p. 37.

Smithsonian Institute, Washington, D.C.; G. P. Merrill, *Bulletin 109* (1920), p. 927.

B. G. Teubner, Stuttgart: A. Hettner, *Die Oberflächenformen des Festlandes* (1928). English translation by Macmillan, *The Surface Features of the Land* (1972) pp. 98, 103–6 and 135–6.

John Wiley, New York: W. D. Thornbury, *Principles of Geomorphology* (1954), pp. 10–11.

ILLUSTRATIONS

Advertiser Publishing Co. Ltd., Honolulu for Fig. 51; American Geographical Society for Figs. 89, 90, 106, 107, 111, 120, 121, 122, 123 and 125; *American Journal of Science* for Figs. 25, 110, 112, 113, 114, 115, 116, 119, 129, 131, 132 and 161; *Annals of the Association of American Geographers* for Figs. 68, 69, 70, 71, 72, 73, 74, 75, 76 and 77; *Bulletin of the Museum of Comparative Zoology*, Harvard College for Figs. 26, 32, 34 and 40; *Bulletin of the Geological Society of America* for Figs. 127, 128, 130, 133, 134, 141, 142, 143, 144, 145, 146, 147, 148, 149, 150, 151A, 151B, 152, 153, 154, 155 and 156; Columbia University Press for Figs. 35 and 36; *Geographical Journal* for Figs. 29, 31, 46, 124 and 160; *Geographical Review* for Fig. 98;

Ginn and Co. Boston for Figs. 22, 23, 24, 30, 33, 45 and 82; Houghton Mifflin Co., Boston and New York for Figs. 2, 3, 4 and 7; *Journal of Geology* (The University of Chicago Press) for Figs. 138, 139 and 140; Museum of Comparative Zoology, Harvard College for Figs. 10, 11, 12 and 13; *National Geographic Magazine* for Figs. 37, 38, 39, 41, 42, 43, 47 and 48; *Proceedings of the Boston Society of Natural History* for Figs. 58 and 59; Dr Tuan and the Indiana Academy of Science (Proceedings) for Fig. 151C; *Proceedings of the National Academy of Sciences* for Figs. 117 and 118; *Quarterly Journal of the Geological Society of London* for Figs. 63, 64 and 65; *Scientific Monthly* for Figs. 99, 100 and 101; *Scottish Geographical Magazine* for Figs. 60, 104 and 105; Secretary of the Harvard Corporation and Harvard College Library for Figs. 44, 81, 83 and 96; Teubner, Leipzig for Figs. 61, 78, 79, 80, 86, 87 and 93; *Transactions of the San Diego Society of Natural History* for Figs. 135, 136 and 137; Professor E. Mott Davis Jr. of Austin, Texas for Figs. 6 and 18; the late R. Mott Davis of Silver Spring, Maryland for Figs. 8, 19, 20, 21, 27, 52, 53, 67, 85, 88, 91 and 159; William Morris Davis II of Bass River, Massachusetts for Figs. 1, 94 and 97; Dr P. B. King of Menlo Park, California for Fig. 109; J. S. Diller U.S. Geological Survey for Figs. 49 and 50; Professor G. J. Martin of New Haven, Connecticut for Figs. 62 and 66; Dr John H. Maxson of Denver, Colorado for Figs. Frontispiece and 158; Professor H. F. Raup of Kent, Ohio for Fig. 103; Dr E. Sampson of Princeton, New Jersey for Fig. 126; Dr A. N. Strahler of Cape Cod for Fig. 157; Dr Howard M. Turner of Boston, Massachusetts for Fig. 28; and the late Dr John K. Wright of Lyme, New Hampshire for Fig. 163.

INFORMATION

Dr Tessana S. Adkins, Cambridge, England; Professor J. E. Allen, Portland, Oregon; Professor G. W. Bain, Amherst, Massachusetts; Professor Walter H. Bucher, Houston, Texas; Professor C. H. Behre, New York City, New York; Dr Walter L. Berg, Washington State University; Professor M. P. Billings, Cambridge, Massachusetts; Miss Ann Billingsley, Burton, Washington; Professor Eliot Blackwelder, Stanford, California; Professor A. L. Bloom, Ithaca, New York; Mrs Rose Blyth, News Bureau, California Institute of Technology; Professor Baylor Brooks, San Diego, California; Anna Hartshorne Brown, Westtown, Pennsylvania; Mrs Kirk Bryan, Santa Fé, New Mexico; Mrs J. P. Buwalda, Pasadena, California; Professor André Cailleux, Sorbonne, Paris; Dr Ian Campbell, San Francisco, California; Dr T. H. Clark, Montreal, Quebec; Dr Francis J. Dallett, The Athenaeum, Philadelphia; Edward L. Davis Esq., Philadelphia, Pennsylvania; Professor E. Mott Davis Jr, Austin, Texas; Mrs N. Burt Davis, Higham, Massachusetts; Mr and Mrs R. Mott Davis, Silver Spring, Maryland; C. King Davis Esq., Wayland, Massachusetts; W. M. Davis II, Bass River, Massachusetts; Dr F. Dixey, Steyning, England; Professor Stanley D. Dodge, Ann Arbor,

Michigan; Professor R. C. Emmons, Madison, Wisconsin; Dr Kimball C. Elkins, Cambridge, Massachusetts; Dr R. L. Engel, Wofford Heights, California; P. Evans Esq., Berkhampstead, England; Dr Richard S. Eustis, Hancock, New Hampshire; Miss N. Felland, American Geographical Society, New York City; Dr J. Gilluly, Lakewood, Colorado; Professor Waldo S. Glock, Boulder, Colorado; Dr Winifred Goldring, Slingerlands, New York; Professor Richard P. Goldthwait, Columbus, Ohio; Mrs J. Mott Hallowell, Wianno, Massachusetts; Mrs Barbara Harrington, Royal Astronomical Society, London; Dr G. F. Hanson, Madison, Wisconsin; Professor Richard Hartshorne, Madison, Wisconsin; Professor Arthur D. Howard, Stanford, California; Dr B. F. Howell, Princeton, New Jersey; Professor Preston E. James, Syracuse, New York; Professor Etienne Juillard, Strasbourg; Dr P. B. King, Menlo Park, California; Professor F. Kniffen, Baton Rouge, Louisiana; Mrs Eleanora Bliss Knopf, Stanford, California; Dr Clarence E. Koeppe, La Jolla, California; Dr Earl E. Lackey, Natick, Massachusetts; Dr and Mrs Frederic H. Lahee, Dallas, Texas; Professor John Leighly, Berkeley, California; A. Locke Esq., Menlo Park, California; Dr S. W. Lohman, Denver, Colorado; Professor Chester R. Longwell, Palo Alto, California; Dr T. S. Lovering, Denver, Colorado; Dr David Lowenthal, American Geographical Society, New York City; Esther Ann Manion, National Geographic Society, Washington, D.C.; Professor G. J. Martin, New Haven, Connecticut; Dr John H. Maxson, Denver, Colorado; Mrs Louisa Davis McCall, Radnor, Pennsylvania; Dr Paul K. McClure, National Academy of Sciences, Washington, D.C.; D. H. McLaughlin Esq., San Francisco, California; the late Dr Levi F. Noble, Valyermo, California; Dr T. B. Nolan, Washington, D.C.; Professor Robert S. Platt, Chicago, Illinois; Dr W. W. Rubey, Houston, Texas; Dr Erwin Raisz, Cambridge, Massachusetts; Professor H. F. Raup, Kent, Ohio; Leonora S. Reno, Pasadena, California; Professor Richard J. Russell, Baton Rouge, Louisiana; Dr E. Sampson, Princeton, New Jersey; Dr G. Austin Schroter, Los Angeles, California; Professor Earl B. Shaw, Worcester, Massachusetts; Professor Carl O. Sauer, Berkeley, California; Professor Robert P. Sharp, Pasadena, California; Martin Simons Esq., Adelaide, Australia; Dr Richard F. Sivel, American Entomological Society, Philadelphia; Professor Guy-Harold Smith, Columbus, Ohio; Professor J. Russell Smith, Swarthmore, Pennsylvania; Dr James G. Stephens, Southern Methodist University; Dr Alexander Stoyanow, Los Angeles, California; Dr Frederick B. Tolles, Swarthmore, Pennsylvania; Dr Walter S. Tower, Carmel, California; Dr E. L. Troxell, Winter Park, Florida; Dr Ricardo H. Tschamler, Cordoba, Argentina; Dr Howard M. Turner, Boston, Massachusetts; Professor E. Van Cleef, Columbus, Ohio; Professor O. D. Von Engeln, Ithaca, New York; Professor S. S. Visher, Bloomington, Indiana; Dr Nicholas B. Wainwright, Historical Society of Pennsylvania, Philadelphia; Professor Robert W. Webb, Santa

Barbara, California; Dr J. Marvine Weller, Chicago, Illinois; Professor George W. White, Urbana, Illinois; Dr R. N. Williams, Historical Society of Pennsylvania, Philadelphia; Dr Edwin B. Wilson, Brookline, Massachusetts; Dr Eldred D. Wilson, Tucson, Arizona; W. H. Wilcockson Esq., Sheffield, England; Dr Michael J. Woldenberg, Cambridge, Massachusetts; Professor A. O. Woodford, Claremont, California; Dr John K. Wright, Lyme, New Hampshire.

PERSONAL ASSISTANCE

Thanks are due to Mr R. J. Bennett, Mr R. S. Jarvis and Mrs Celia Kamau for rendering valuable assistance in the analysis of some of Davis' publications.

The authors, once again, unite in thanking Monica Beckinsale whose literary sense and skill are in evidence in the comprehensive and authoritative Index, which contributes greatly to the value of this Volume.

The authors are particularly grateful for the invaluable assistance and encouragement given by the following members of the Davis family, without which the completion of this volume would have been impossible: the late Mr R. Mott Davis and Mrs R. Mott Davis of Silver Spring, Maryland; Professor E. Mott Davis Jr of Austin, Texas; William Morris Davis II of Bass River, Massachusetts; Mrs J. Mott Hallowell of Wianno, Massachusetts; Mrs N. Burt Davis of Higham, Massachusetts; and C. King Davis Esq., of Wayland, Massachusetts.

CHAPTER ONE

Introduction

In writing our *History of the Study of Landforms* we have always been aware that there loomed ahead a precipice in the form of William Morris Davis. Our course, so far, has been beset by numerous pitfalls and obstacles many of which have proved exciting and hindering as the trail was unblazed and there were few major landmarks to guide us. Yet we knew where we wanted to go even if the details of the route demanded careful reconnaissance. But the fact is that Davis has proved to be a rock-wall that could be tackled only with modern equipment. In our ignorance we assumed that many Americans with psychological crampons had climbed the Davisian precipice, that a path already existed up which we could safely scale the heights to assess the summit panorama and the country ahead. Instead we found ourselves with numerous scattered writings, like tortuous paths up the scree at the foot of a rock-wall which had barely been reconnoitred. We had no option but to climb our Everest and write a biography of Davis. It seemed astonishing to us that whereas, for example, geologists have several books, many quite recent, on Lyell, geographers have no sizeable book on Davis. But we can assure the reader that Davis is the *only* man to whom in our history of landforms we would devote a whole volume.

To those who think that Davis was a geographer rather than a geomorphologist, we recommend an incident on the International Transcontinental Excursion across the United States in 1912 when, according to Mark Jefferson: 'A very lively discussion . . . ended in Dr. Niermeyer telling Davis that he was no geographer but a geomorphologist':

> As usual he (Davis) spoke only of physical geography. He is too fair to deny this when accused of it and admits that he is really a physical geographer, but when Dr. Niermeyer of Utrecht rose and called attention to the point, everyone got excited and seemed to feel that Niermeyer was out of order. (*Letter: Mark Jefferson to his wife, 5 September 1912; from Martin, 1968, p. 140*)

We can recommend too the views of Douglas Wilson Johnson, one of Davis' most famous geomorphological students. Apparently Davis' geography was always dominated by geomorphology and at best was a rather crude form of environmental determinism. Most of his writings were not strictly 'geographical' in character, at least in any modern sense:

The reader who studies his *Geographical Essays* in the chronological order of their publication (indicated in the Preface of the work) will be impressed by the fact that in his earlier writings he makes *present form* the criterion by which things geographical are to be distinguished from things geological. Of the trilogy 'structure, process and stage', structure and process are definitely excluded as geological, and the portion of geological time occupied in the evolution of a landform to a given stage of development seems also to be excluded. At this time there was no question of the relation of form to organic life. Later Davis makes this *relationship* the unifying principle of geography, and seems to exclude from geography all physical studies not specifically directed to the elucidation of the relationship between physical environment and organic life.

Thus while Professor Davis's writings, considered as a whole, may appear to support the view that geomorphology is geography rather than geology, he has in later years moved much closer to the conception of geography advocated in the present essay. I understand that he would not now consider most of his writings as strictly geographic in character, and that he would classify himself as a physiographer or geomorphologist rather than as a geographer. (*Johnson, 1929, p. 209, footnote*)

There probably never was any doubt that the biography of Davis belongs quite properly to any history of the study of landforms. For that reason we have already outlined his early ideas in Volume One where he appears at the end of an epoch. In the present volume we expand in detail upon his geomorphological and geographical ideas and in subsequent volumes we intend to show their influence on his contemporaries and on modern authors. However, in Volume One and in Volume Three we deal mainly with published works which can be readily assessed on their relative merits. In the present volume we have perforce dealt also with his private influence – with his copious letters which may be called his 'hidden persuaders', and with his personal impact or what might be termed his psycho-physical persuasion.

However, to assess the influence of the abundant letters, apart from the herculean task of obtaining and sifting them, was less difficult than the assessment of Davis' character and personal influence. We felt that our three-pronged attack – by an oldish don, a youngish don and an active civil servant – needed reinforcement, so we consulted psychologists who left us in no doubt that we were fifty years too late, that our posthumous method needed more than psychological skill, that what we should do was to let Davis reveal himself and let his host of acquaintances reveal both themselves and him. Consequently we have refused to dogmatize and have, like Chesterton's Father Brown, set out not to prove anything but merely to display what will prove itself.

It is, of course, impossible to divorce the person from his ideas and personal achievements, and in his lifetime Davis was too important and too prolific to be ignored. Not surprisingly his contemporaries fell mainly into two categories: those who disagreed with his theories and clearly resented his dominance of geographic thought, and those who supported his ideas and regarded him with genuine respect. There should be a middle choice but it is hard to find and now that the influence of Davis is waning it would be all too easy to take the side of the less friendly critics. However, although this biography is not an exercise in adulation, the evidence available would not allow it to become a massive project of demolition. Its main aims are to sketch the social and scientific climate and the domestic and scholastic environment that moulded the man and his ideas, and to assess his value in the history of geomorphological progress.

The reader will not be surprised to find that the letters quoted vary widely in quality as evidence. Davis' own letters belong to an age when wordy epistles were penned daily fresh from an experience or demand. So many of them have a coating of justification, wishful thinking, glorification, pontification and so on that they reveal a fascinating ambivalence. But once dispatched they were gone from him beyond recall except in memory and we alone see the total sequence and the final balance sheet.

The numerous assessments of Davis written by his acquaintances, some of whom happily survive, also reveal widely different opinions and facets, as the following extracts show.

I can say this, that up to the very end he maintained an amazingly high degree of activity and great enthusiasm. He was a fluent, well-organised lecturer, and this made him an effective teacher. I can remember walking into classrooms that he had just vacated and seeing typical, very beautiful Davis landscape diagrams all over the blackboards. One story oft repeated around here concerned a graduate student on his way to one of Davis' classes late one afternoon. This student was notably indolent, and on that particular day he was at a very low ebb in terms of his general outlook on life, science, education, and the state of the Union. An hour later, he came rushing down the hall with fire in his eyes, wildly gesticulating, and announcing that he was going out at once to measure slopes on alluvial fans, as this was just about the most important thing any man could do. From this we can judge that Davis had real impact on his students, and was successful in transferring his enthusiasm to them. (*Letter: R. P. Sharp to R. J. Chorley, 18 January 1962; Pasadena, California*)

I have several rather vivid memories of Davis. One was when I was an undergraduate at Harvard, and he and Daly debated the coral reef theories at a meeting of the Harvard Geological Club. The meeting went on until a late hour with no one noticing the clock. It was terrific. Both men were vehement, but completely impersonal. I was greatly stimulated at the idea that two scholars could argue so heatedly without letting personal remarks mar the encounter.

Davis was a severe critic, and spared no one's feelings. I remember a graduate seminar in which there was one Radcliffe girl who had just returned from Peru. She gave a paper describing certain landforms in the Andes, and Davis really gave her a bad time with criticisms that were quite just, but a little vigorous. She started to cry. Whereupon a male student at the seminar shook his finger at Davis and said, 'You ought to be ashamed of yourself.' I never saw Davis so completely speechless. (*Letter: Preston E. James to R. J. Chorley, 11 January 1962; Syracuse, New York*)

Both descriptions tell us something different: the first emphasizes Davis' artistic skill and powers of persuasion; the other, in a more intimate way, shows not a benign old man exercising a kindly tolerance acquired from a half century of understanding but an intellectual ice-pick.

Two further quotations must for the moment suffice to exemplify the adulation of a devoted protegé and the adverse criticism of a specialist who immediately saw a weakness in Davis' landform studies:

> The more you checked his teaching against the out-of-doors, the sounder you found it. Surely Davis read God's thoughts about the surface of the earth. (*Mark Jefferson, autobiographical notes, 1947; in Martin, 1968, p. 41*)

> Davis was simply not concerned at all with the chemistry of weathering and physics of slope formation. As he grew older, he drifted away more and more from any concern with the geological facts behind the landscape. This is illustrated by an incident that occurred when I last saw Davis, only a few years before his death. He visited Prof. Fenneman, one of his most devoted admirers and followers, while he was returning from studies in the Great Basin. He spoke before the faculty and advanced students in our department of geology at Cincinnati. He drew diagrams on the blackboard and showed photographs to illustrate his contention that many of the gentle slopes exhibited by crystalline mountain ranges in the Great Basin, were scarps of low-angle normal faults, as low as $11°$ to $17°$, or that order of magnitude. In the discussion that followed his address, I commented that I considered an alternative explanation possible, namely, that the slopes reflected the gentle dip of foliation in metamorphic schists. I asked if he had looked at the nature of the rocks on those slopes. Davis turned to me and said in a stern tone, something to this effect: 'Young man, you have yet to learn that the essential facts of geomorphology are best seen at a great distance.' There was no answer to that comment, of course. (*Letter: W. H. Bucher to R. J. Chorley, 22 February 1962; Houston, Texas*)

It soon becomes obvious that no quotation is in itself conclusive; each is representative – no more; each illustrative of the varied impressions generated by the currents of the man's complex personality and ideas. The result is a bewildering palimpsest of outlines, some coincidental and others totally dissimilar. Our biography thus becomes largely the coincidence of the maximum occurrences of certain impressions and actions.

In this connection, it seems to us that biographers of scientists often deal too exclusively with the daily round and progressive outcome of their subjects' productive years. We do not intend to assume, as Davis himself often did in landform study, that the basic origin matters little. For man, as for physical landscapes, there is a stage before youth which is of great significance, and we shall start with the young Davis' parental environment.

Davis himself clearly envisaged that his biographical memoir for the

National Academy of Sciences would be as exhaustive as his own memoir of
G. K. Gilbert (1927F) had been.

> When you get back to Shirley, look in the package of my own letters – if you
> havent thrown them away or burnd them up – and let me know if any are found
> as far back as my Rocky Mtn trip of 1869. A request comes from Colorado to send
> an article about my adventures of that summer, especially about the ascents of
> Mts. Harvard and Yale.
>
> Please take note also that these letters are, after my death, to be placed at the
> disposition of the National Academy of Sciences in Washington, for their use in
> preparing my memoir; but to be returned to Billy when they are finisht. When
> that time comes, write first to the Secy. N.A.S. and say what you have been told to
> do, and ask his instructions as to when and where. (*Letter: W. M. Davis to his son
> Edward, 2 March 1932; Pasadena, California*)

The short memoir prepared by Daly (1945) hardly measures up to Davis'
expectations, and it is hoped that the present volume will remedy this
deficiency.

This book is richly illustrated with Davis' landscape drawings, but it is
important to remember that he himself used them for more than illustrative
purposes. Davis' drawings were not mere depictions of terrain, but interpretations of it – subtle idealizations in which features significant to his
theories were cleverly accentuated to provide what amounted to *models* of
terrain types. Sometimes he would carry this method to the extreme of
producing idealized composite views. At a time when photography was
becoming both popular and inexpensive, it is significant that Davis did not
exploit this technique, for he required not the objective camera eye but the
trained eye of the scientist. Just as later geomorphologists recorded their
observational data in tabular or other form, Davis used his field sketches as
data storage systems which were made more explicit and internally logical as
they were recollected in tranquillity. More than this, as we shall see particularly in his work on coral reefs, Davis' drawings were often *vehicles for his
research*, wherein initial forms were postulated and the consequences of
possible sequences of events logically deduced by careful geometrical drawings, and the end products compared with the real world according to the
dictates of the method of multiple working hypothesis.

Thus we make no apology for devoting the whole of the second volume of
this *History of the Study of Landforms* to William Morris Davis, nor for stressing his biographical details. This remarkable scholar found the study of landforms a fortuitous assemblage of peripheral principles developed as the by-
products of geology, geography, engineering and other sciences; he left it a
coherent discipline. No matter whether, as his detractors assert, he created
a Frankenstein whose limited genetic composition was to restrict the development of the science for the next half century, or whether his conceptual
model was as capable of adaptation and proliferation as his supporters aver,

Davis' genius resided in his own peculiar brand of alchemy which assembled, and breathed life into, a viable, intellectual entity.

The details of his career and personality find their natural place in this volume, for the fundamentally deductive and evolutionary basis of his work sprang as much from his own mental make-up and the intellectual environment of his day as from requirements inherent in the nature of landforms themselves. The ebb and flow of his scholarship were largely dictated by the circumstances of his life. He was first and last a Victorian gentleman; proud, austere, disciplined and aloof. Yet within him there was a sustained flame which produced some of his most imaginative and benign writings long after his allotted three-score years and ten. By modern standards he was simple and direct to the point of naïvety, by any absolute standards he embodied the traditional integrity of his Quaker stock. He adored his parents and his own family but, at least during his early and mature years, was often unable to establish warm personal understanding with others. This inability, combined with an almost religious egocentricity which characterized most of his life, led not only to much loneliness but also to significant dogmatism and to many misunderstandings with his academic contemporaries. His Quaker values, emotional reticence and love of nature, evolution and logical deduction are inexorably intertwined in his writings.

Davis' huge literary output was at one and the same time wide-ranging and repetitive; concerned with limited themes but subjected to almost infinite variations; sustained by world-wide excursions yet prompted by mental deduction; rigidly constructed yet flexible; generated with scholarly aloofness yet popularized with uncanny insight. By devoting the whole of one large volume to him we have been able to identify and examine his main contributions to the study of landforms, or geomorphology, as well as to the natural sciences related to it. We have not neglected his writings on meteorology, geography, geology and the philosophy of science but they seemed to us to lack the outstanding importance of his studies of landforms. We happen to know that he himself would have championed our view. He wished to be remembered as 'a founder of physiographic geology', by which is meant the study of landforms, and as 'a foremost investigator and an inspiring teacher'.

Inevitably some details of his geomorphic work will be mentioned again in Volume Three of our *History of the Study of Landforms* where we shall discuss the contributions to geomorphology made by his fellow-scientists. But, as we have said, it was necessary to deal first with the work of William Morris Davis because he to a great extent provided the backcloth against which the development of geomorphology between the 1880s and 1930s was played out. His was the yardstick with which the success of others was judged; his the conception which evoked both the action and reaction of his contemporaries.

PART ONE

Youth

F IG. 1. Davis at the age of twelve (Courtesy W. M. Davis II, Bass River, Mass.)

CHAPTER TWO

Early Environment

William Morris Davis was born in Philadelphia on 12 February 1850, eleven years before the outbreak of the American Civil War. Though not taking any direct part in the war, its background of common tragedy and hysterical clamour must have affected him as it did other Americans. Not only did it colour his life in this general way but it touched him directly through his father, who served on General Fremont's staff, and through his grandmother, Lucretia Mott. Much of these early years was spent within an atmosphere subject to the powerful influence of his grandmother. He could hardly have

FIG. 2. James and Lucretia Mott, from a daguerreotype by Langenheim about 1842 (From Hallowell, 1884)

9

followed a more potent guide-star. She had a personality as tensile as steel, and ideas that were as strong and sharp-edged. From 1850 until 1868, when the family went on a trip to Europe, they lived in daily contact with Mrs Mott. Part of the time they shared the same home, and during the remainder they occupied the next-door property. The influence of this woman is an almost incontrovertible assumption because so many of his adult characteristics and attitudes, like his extreme seriousness and his opposition to servility, closely resembled hers.

In the persons of his grandmother, father and uncle (i.e. William Morris Davis (1815–91), the Congressman) Davis was presented from his early days with examples of above-ordinary success in differing spheres. The effect on an active, intelligent young mind must have been like a constant goad. From his earliest youth there is a purposefulness about all his activities and a deliberate daily routine designed to waste as little time as possible. To these environmental influences was added that inborn feeling of distinction and aristocracy, said to be a characteristic of many Aquarians, which was to make William Morris Davis such a formidable personality in later life.

GRANDMOTHER MOTT, 1793–1880

For Lucretia Coffin Mott her upbringing within the simple Nantucket seafaring economy and Quaker community acted like an enzyme within her small frame bursting forth at maturity as an irresistible mixture of intellectual enthusiasm and strong liberal principles. The very austerity and plainness of Quaker life, instead of stunting her development, provided a testing ground that turned her out as the inflexible and tireless opponent of all forms of illiberalism. Quakerism in Nantucket, as described below in Davis' sister Anna's biography of their grandparents, seems to have provided a hard and chastening training.

> Their situation in life required the most unflinching self-reliance, and in that day of farming and fishing, it followed, of course, that their physical powers were sufficiently taxed for their most vigorous expansion . . . Not only the smaller fish, but the whale itself, was pursued from the shore; and at the first dawn of day the men were in readiness to leave their homes, having taken their morning meal with such parts of the families as had hastened its preparation. The men proceeded on their adventurous voyage, full of expectation and hope, and in entire confidence that the women would be no idle worshippers at home. The cows were milked, the butter was churned, the wool was carded and spun, the cloth was woven, and the unpainted floors scoured and neatly sanded; the oven had been previously heated for the rye and Indian bread, the pumpkin pies, and other substantial provisions for the table, that the father and his sons might be made double welcome on their return at nightfall. The men returned, the boats had been successful, and the joy of the family was complete. Some of the men had gigantic strength and some of the matrons would walk from fifteen to twenty miles without thinking it a

hardship. Here were fine constitutions, and a long life seemed to be the legitimate attribute. (*Hallowell, 1884, pp. 22–3*)

Mental strength may well have been an advantage to Mrs Mott. She, and to a lesser extent Davis, expended much nervous energy to keep going.

It is but seldom that vice grows on a barren soil like this, which produces nothing without extreme labor. How could the common follies of society take root in so despicable a soil? They generally thrive on its exuberant juices; here we have none but those which administer to the useful, to the necessary, and to the indispensable comforts of life ... The inhabitants abhor the very idea of expending in useless waste and vain luxuries the fruits of prosperous labor ... The simplicity of their manners shortens the catalogue of their wants ... At home the tender minds of the children must be early struck with the gravity, the serious, though cheerful deportment of their parents; they are inured to a principle of subordination, arising neither from sudden passions, nor inconsistent pleasure. They are corrected with tenderness, nursed with most affectionate care, clad with that decent plainness from which they observe their parents never to depart; in short, by the force of example, more than by precept, they learn to follow the steps of their parents, and to despise ostentatiousness as being sinful. They acquire a taste for that neatness for which their fathers are so conspicuous; they learn to be prudent and saving; the very tone of voice in which they are addressed establishes in them that softness of diction which ever after becomes habitual. If they are left with fortunes, they know how to save them, and how to enjoy them with moderation and decency; if they have none, they know how to venture, how to work and toil as their parents have done before them. At meetings they are taught the few, the simple tenets of their sect; tenets fit to render men sober, industrious, just, and merciful ... There are but two congregations in this town, and but one priest on the whole island. This lonely clergyman is the Presbyterian minister, who has a very large and respectable congregation; the other is composed of Quakers, who admit of no one particular person entitled to preach, to catechise, and to receive certain salaries for his trouble. Most of these people are continually at sea, and often have the most urgent reasons to worship the Parent of Nature in the midst of the storms which they encounter. These two sects live in perfect peace and harmony with each other. Every one goes to that place of worship which he likes best, and thinks not that his neighbor does wrong by not following him ... As the sea excursions are often very long, the wives are necessarily obliged to transact business, to settle accounts, and, in short, to rule and provide for their families. These circumstances being oft-repeated give women the ability, as well as the taste, for that kind of superintendency to which, by their prudence and good management, they seem to be in general very equal. This ripens their judgment, and justly entitles them to a rank superior to that of other wives. To this dexterity in managing their husband's business whilst he is absent, the Nantucket women unite a great deal of industry. They spin, or cause to be spun, abundance of wool and flax, and would be forever disgraced and looked upon as idlers if all the family were not clad in good, neat, and sufficient homespun cloth. First-days are the only seasons when it is lawful for both sexes to exhibit garments of English manufacture, and even these are of the most moderate price, and of the gravest colors . . .

The absence of so many men at particular seasons leaves the town quite desolate, and this mournful situation disposes the women to go to each others' homes much oftener than when their husbands are at home. The house is always cleaned before they set out, and with peculiar alacrity they pursue their intended visit, which consists of a social chat, a dish of tea, and an hearty supper . . . The young fellows easily find out which is the most convenient house, and there they assemble with the girls of the neighborhood. Instead of cards, musical instruments, or songs, they relate stories of their various sea-adventures, . . . and if anyone has lately returned from a cruise, he is generally the speaker of the night. Pyes and custards never fail to be produced on such occasions; . . . they laugh and talk together until the father and mother return, when all retire to their respective homes, the men reconducting the partner of their affections. Thus they spend many of the youthful evenings of their lives; no wonder therefore that they marry so early. (*Hallowell, 1884, pp. 23–5*)

The moral of the last conclusion is a little hard to follow but the general impression of orderly moderation has much to commend it. Mrs Mott belonged to this community and, as an ordinary mother and housewife, accepted its standards of behaviour. Yet she was exceptional, particularly in the visionary fervour with which she expressed her beliefs. This fervour drove her almost fatalistically along a course well removed from the traditional paths permissible to women. True, throughout her life she accepted the rules of conduct expected of a devout Quaker, dressed soberly, used the Quaker form of address, kept to a plain and frugal way of life, and, as a young woman, took a full part in monthly meetings. But in a very short time her exceptional talents were recognized and she was appointed a Minister. This would have been honour enough for many women but its only effect was to fortify her ambitions.

Lucretia Mott was the type of person whose feelings, once roused, grow with such intensity and speed that the pressure generated within can only be relieved by some positive outburst. She was a born rebel, a natural enemy of standard contemporary prejudices which she fought with all the arguments at her command. Problems or disputes within the Quaker community were by custom discussed and settled in open meetings without a formal vote. It was natural therefore for Mrs Mott to employ logical persuasion and emotion as her main offensive weapons – though as a woman and a Quaker there was little else permitted. The major social problems of the day included slavery, women's rights and trade unions. Mrs Mott supported the radical view on each, as indeed did her husband James. While she was attracted to many lost causes, her main energies became concentrated against slavery, a problem which had been forced into prominence by the altered view of the dignity of man. Literary notice of this was served by the publication of *Uncle Tom's Cabin* in 1852. Unrest among the intellectuals derived practical support from the resentment springing up between the American North and the South as the economic balance of power gradually began to swing in favour of the former. A war, which really had little to do with slavery but marked a contest between rival economies, brought an end to slavery even if it did little to alter the problems of Negro status.

FIG. 3. James Mott, from a photograph by Gutekunst in 1863
(From Hallowell, 1884)

Mrs Mott had been prominent in the formation and operation of the Philadelphia Female Anti-Slavery Society. A society of women in that time was rare in itself; one that campaigned against an accepted institution was not only unusual but clearly hazardous. Mrs Mott could easily have stopped here but vocal attacks did not satisfy her. Conviction obliged her to carry her opinions well beyond the confines of Philadelphia, even into the South itself. Her convictions were so strong that the very real possibility of personal assault did nothing at all to check her utterance of what she felt was right.

The young generation of this day would probably find it difficult to conceive of the savage form of opposition to the abolitionists, which prevailed during many years. In these perilous periods, Mrs. Mott proved her fidelity to her principles of non-resistance, as well as her anti-slavery faith. Self-possessed and unshrinking in the stormiest scenes, a mob howling around the house, assailing its windows with stones, or clamoring within its walls, scattering vitriol among the audience, leaping on the platform, drowning the voices of the speakers in their own mad cries, she held fast her integrity, never compromising in the slightest degree a principle, and never giving her consent that the protection of the police should be asked for the maintenance of our rights.

In the year 1838, when Pennsylvania Hall was burned by a mob, and the Mayor of Philadelphia connived at the outrage, the furious rioters marched through the streets threatening an assault upon the house of James and Lucretia Mott. Warned of the peril, and aware of the unsated wrath of the savage men, Mrs. Mott made preparation for the attack by sending her younger children and some articles of clothing out of the house, and with her husband and a few friends sat in their parlor, quietly awaiting the approach of the mob. Before it reached the house a suggestion that it should attack the shelter for Colored Orphans in another part of the city diverted its course, and the rioters proceeded to that work of destruction. During the night they passed the house of Edward and Mary Needles, prominent abolitionists, who were also serenely expecting their arrival. But they satisfied their rage by hideous yells, and passed on. (*Hallowell, 1884, pp. 128–9*)

On Friday afternoon the rumors were thick and strong that this house would be assaulted the coming night. A few light pieces of furniture, and some clothing, were removed to the next house, and in the evening we sat down to await the event, whatever it might be. Mr. and Mrs. Mott sat near the middle of the room, with many friends around them. Thomas went out into the street now and then to reconnoitre, and then return and tell us the result of his observations. Several young men came in ready for any emergency which might require their services, and at any rate, to cheer us by their presence and sympathy. About eight o'clock Thomas came running in, saying, 'They're coming!' The excited throng was pouring along up Race-street; we could hear their shouts distinctly; but they crossed Ninth-street without turning up, and for the present we were relieved from apprehension. We have heard since, that when the mob reached Ninth-street, a young man friendly to the family joined in the cry, 'On to Mott's,' at the head of the gang, and rushed on up Race-street, – they blindly following their leader, – and thus we escaped. We thought, however, they might still be down upon us, and sat in calm expectation of their advance; hearing every few minutes by some of our friends who were on the alert what points were occupied, and what movements were going on. At length, learning that the mob seemed broken and scattered, we concluded we were to escape that night at least, and retired to rest.
(*Hallowell, 1884, pp. 129–30*)

As well as being an outspoken critic, Lucretia Mott and her husband James took a practical part in the campaign against slavery. Their home was used as a refuge by the organizers of the 'Underground Railroad', which was set up to assist the safe escape of runaway slaves. In Still's *The Underground Railroad* (1872) she is listed as 'one of the 24 Station-Masters, Prominent Anti-Slavery Men, and Supporters of the UGRR'. Levi Coffin, a relative of Mrs Mott, though not it is thought a close one as he came from Cincinnati,

F I G. 4. Lucretia Coffin Mott, from a photograph by Gutekunst in 1875
(From Hallowell, 1884)

was one of 'The Railroad' organizers. John Brown's wife spent part of the last tense weeks before her husband's execution at the home of Lucretia Mott. Richard Price Hallowell, the husband of Davis' sister, was one of those who went to Harper's Ferry to escort the body of John Brown for interment at North Elba, New York. Lucretia was more than devoted to the cause, as one author put it, 'If John Brown was insane on the subject of slavery, so (was) Lucretia Mott' (Villard, 1910, pp. 509–10). In fact almost every member of the Mott–Davis family circle was an active abolitionist, either through conviction or the persuasion of Mrs Mott.

Not much of the intensity of feeling for the hardships of the under-privileged seems to have flowed in Davis' veins, if his letters and actions in later life are any guide. A general aversion to subject status was certainly present – several examples appear throughout this book in various letters – but there is no burning zeal to put the world to rights before breakfast. The feelings of his father and uncle were equally modulated. Davis was no political crusader. Yet similarity should not be discounted as it may have appeared in a different form. While Mrs Mott's energy exhausted itself in the public battle against slavery, the driving force of Davis' soul acted within the narrower, more tranquil environment of the university campus. He had the same inflexible faith in his own opinions and, like his grandmother, believed that logical argument would always prevail. His oratory too had much of her emotional appeal. If Mrs Mott had a flaw it was her failure to accept that others might have equal access to the truth.

> . . . R   B   gives forth her views both on slavery and on woman, which are in the main good, but from not attending any of our meetings, nor reading our reports, she is ignorant, and thinks the advocates of these causes are as fanatical as the papers and popular opinion represent them. Some years ago, when speaking in a meeting in Richmond, Va., she introduced some remarks on anti-slavery, – some rose and left the room. She said, 'Stop, friends! I am no abolitionist.' Later, having seen an account of this in the papers, I expressed my regret that she should pander to the pro-slavery prejudices of the people by such a disclaimer. She replied to me, 'Oh, my dear, I was not correctly reported; I said, "I am no modern abolitionist." ' Now she knew no better than to suppose that would be a satisfactory explanation to me! (*Lucretia Mott, quoted in Hallowell, 1884, p. 382*)

The present reaction among academic geographers to Davis' ideas may have received an initial impulse from a similar failure to admit flaws in his theories.

But to return to Lucretia Mott. As might be expected of a woman with a high degree of intellect and religious integrity, she strongly disapproved of any lightness of character. She abhorred alcohol, never read novels (as distinct from religious, historical or other mind-improving books) and disliked sport and gambling. Much of this strictness was repeated in Davis. Exhibitions of drunkenness shocked him profoundly and he rarely admits to reading a novel. Yet he enjoyed watching occasional games of college football and he played golf whenever he could find a partner.

> He was always willing for a game and he was fond of games. He enjoyed professional baseball and Harvard baseball and football. I recall the first baseball game which I attended with him and my mother took place in Cambridge in 1890 when Harvard beat Yale. His interest was prompted perhaps by his nephew Frank Hallowell, who was a star performer at Harvard in both baseball and football and

who was my boyhood hero. (*Interview: R. Mott Davis with R. J. Chorley, 7 September 1962*)

On gambling Davis showed himself a stern parent such that his children hid from him even the most innocent manifestations of this. His eldest son recalled how, when Davis discovered that his adolescent boys were frequenting a pool hall in Cambridge, he immediately had a pool table installed in the basement at home. However, as R. Mott Davis put it rather wistfully, it was never quite the same!

If Davis did not allow his feelings to be so deeply touched by social problems as were those of Lucretia Coffin Mott, he was nevertheless profoundly affected by the example of his grandmother's independent character, outspokeness, belief in her opinions and abilities, and sense of duty.

During 1850, the year of Davis' birth, his father purchased Oak Farm, which stood in the suburbs about eight miles north of Philadelphia. This property was shared with Thomas Mott and his family. At the same time James and Lucretia Mott bought 338 Arch Street, a large property in Philadelphia. By a somewhat curious arrangement these two properties were shared alternately.

> In 1850 James and L. Mott purchased 338 Arch Street (below Twelfth); This house being too large for their own small family, – now only themselves and their daughter Martha; – an arrangement was entered into, by which Edward and Maria Davis with their children, and Thomas and Mariana Mott, with theirs, made it their winter home; they in turn taking their parents and sister into the household at Oak Farm for the summer. (*Hallowell, 1884, p. 326*)

This close attachment as a family group persisted throughout the whole time his grandmother was alive, and probably meant that she had as much effect on his character training as his parents. 338 Arch Street was described briefly:

> It was an ordinary looking house on the outside, like many another of its size in the monotonous city, built of smooth red brick, with white marble facings and broad white marble steps. According to Philadelphia fashion, the lower shutters were heavy and solid, and were painted white. When nightfall came on, it was not the way then, however, to close and bolt these tight, as it is now; they were left open till bed-time, and passers-by could glance in at the bright, cozy parlor, with its animated circle around the evening lamp. How cheery the windows looked to those who came belated home!
>
> In the broad hall stood two roomy arm-chairs, – 'beggar's chairs', we children used to call them, they were in such constant requisition for applicants of all sorts, 'waiting to see Mrs. Mott, miss'. The two parlors, connected by folding-doors, were large, square rooms, of handsome proportion and home-like pleasant appearance. Although the furniture was old-fashioned mahogany and black hair-cloth, and ornaments were few, there was a general air of comfort and every-day use which was very attractive. The carpet was sure to be of bright colors, and of

rather striking design, for my grandmother cordially disliked what she called 'dingy carpets'. She also disliked the prevailing style of dark, heavily curtained rooms; and when she came into the parlor in the afternoon, would step quickly across to the windows, and draw up the green venetian blinds, letting the sunlight stream in. In these cheerful rooms guests of all kinds found gracious courtesy. Could the old walls speak, how many illustrious names they would recall! How many stirring sentiments ring in our forgetful ears! What echoes of laughter and merriment would they not throw back to us! (*Hallowell, 1884, p. 328*)

The general atmosphere of the house is sketched in slightly more detail.

Even pleasanter than these, were the family meetings, which have been described already; and the birthday and Christmas celebrations. Our grandmother's birthday, First mo. 3rd, called together old and young; our grandfather's was observed in a quieter way, as it occurred in midsummer, when the families were apt to be scattered. There were also the delightful Seventh-day dinners of hominy soup, – a primitive compound of Nantucket origin, – when children and grandchildren were expected to happen in without invitation. Occasionally there were solemn and stately entertainments, a bore to everybody; but such did not flourish in the every-day air of 'Three-thirty-eight'. The life there, busy even for the youngest inmate, was one of simple duties and pleasures, shared by old and young. Few servants were kept, and few were needed, where the work was so well divided. It was a lively house-hold, full of busy people going here and there, and of children running up and down, but there was no sense of discord or confusion'. (*Hallowell, 1884, p. 331*)

In 1852 Lajos Kossuth, the Hungarian politician, dined at Arch Street during his American visit. In the same year James Mott retired from his textile business and five years later the Motts moved to 'Roadside', a house just across the road from Oak Farm.

The matriarchal nature of the Mott establishment cannot be disguised. While Lucretia's granddaughter and biographer is anxious to establish that she never neglected her household duties, there is no doubt that she was greatly assisted by her numerous daughters, daughters-in-law and grandchildren, without whose help she could not possibly have kept her home going and simultaneously have gone on many campaign travels. In the background there is always a very clear impression of Mrs Mott as the centre of this private universe with her family revolving round her and because of her.

It is strange that, in actual fact, few references can be found to his grandmother in Davis' many surviving letters. One contains a very brief but telling allusion to her outspokeness:

Going down hill after dinner, we came to the Merced Group of Big Trees, and saw our first specimens of *Sequoia gigantea*. I don't know of any measurements of their height, but I should say decidedly over 200 feet, small for their kind, and scarcely overtopping the pines, cedars and spruces about them, but big trees for all, especially at their base, where they are much thicker in proportion to their height, than those about them. Around the largest, I took twenty-eight paces, the area of its section would more than equal that of our room here – not closely descriptive, as you don't know the room. We shall see and you shall hear more of them when we stop at the Mariposa Grove, on our way back to Merced.

Would you touch a needle or break the bark of these old fellows? I wouldn't much, but some scrub actually cut his initials on one of them, 'G-L-H'. I wish grandma had been with us to call him a fool. (*Tour, 5 October 1877; Yosemite Valley, California*)

There is also a remarkable lack of reference in Davis' letters to many other feminine affairs. Yet there is every reason to think that, outside the purely academic world, he preferred female to male company. His three marriages and the feeling of restlessness in between strongly support this. As his grandmother was such an exceptional person it is natural to ascribe to her the leading part in his development. Yet it might have been his mother. Certainly the male influence during the formative years was weak. The absence of his father during the war, a houseful of women, whatever it was – his shyness, prim precision all suggest an oversubjection to feminine habits during his formative years. The clearest parallel or influence clearly appears to be his grandmother whom he closely resembled in the severity of his opinions, high moral integrity, serious approach to life, appreciation of natural beauty, repeated attempts to express himself in verse and even in the way he ordered his own household when he was a parent. Yet he did not inherit or acquire her quality of personal domination for he never appears as a natural leader in a military sense; his was a domination of persuasive ideas.

THE PATERNAL SIDE

On his father's side, Davis was descended from Samuel Davis, one of three Quaker brothers who emigrated from Wales to Montgomery County, Pennsylvania, about 1730 (Johnson, 1951). The seventh son of Samuel and his wife Jane Rees was born on New Year's Day 1746 and christened after his father. Samuel the younger married Elizabeth Williams in June 1772 and took an active part in the War of Independence, carrying dispatches for General Washington and rising to the rank of Major. Elizabeth died in the great plague which swept Philadelphia in 1793, leaving a son, Evan, who served under Andrew Jackson in the War of 1812 for which he was expelled from the Society of Friends. Evan 'apparently a man of modest but adequate means' was married twice, the first time in 1808 to Elizabeth Evans, a descendant of John Evans who emigrated to Pennsylvania in 1707. It is interesting that in 1906 William Morris Davis wrote to a local genealogical expert:

I want to find out whether Lewis Evans, who wrote a remarkably good account of the 'Middle British Colonies', about 1750, was of the same family from which descended the Evans mentioned below. It happens that Lewis Evans had Benjamin Franklin print his book, and that B. F. thereby had a chance to introduce some of his own ideas in the legends on the map of the first edition (1745?) which were removed from the second edition (1755?). From this legend it has been supposed that the said Evans (who I hope is my very great grandfather) originated certain facts regarding the progressive movement of storms, but I wish to show that the credit for the statement belongs to Franklin (who is my first cousin, five times removed).

If I can find out about Lewis Evans, and if he proves to be of the desired ancestral tree, I can make a pretty story for the April meeting of your highly enterprising and meritorious Philosophical Society. (*Letter: W. M. Davis to Mr Williams, 14 February 1906; Cambridge, Mass.*)

Edward Morris Davis, Davis' father, was the second son of Evan and Elizabeth Davis, being born in a 'mansion' on Arch Street, Philadelphia, in 1811. Elizabeth died in 1814 and Evan married a second time to Rachel Hill, producing a son – William Morris Davis, the future Congressman and half-uncle of the geomorphologist. The half-brothers Edward and William both attended the Friends' School at West-town, Chester County, Pennsylvania, and their subsequent careers as merchants and financiers were very much interconnected.

In 1834 Edward became a member of the American Anti-Slavery Society, founded by James and Lucretia Mott the previous year, undoubtedly as a result of which he married their daughter Maria in 1836. This association with abolitionism adversely affected his cotton and textile trading business, although in the following year he is listed as an importer of silk goods. However, we know that in May 1838 he crossed the Atlantic in the *Sirius*, the first steamship ever to make the passage, to conduct business in Paris (he was there again in 1840) where he received a letter from his mother-in-law telling of the burning of the Pennsylvania Hall by anti-abolitionists. As we have already described, in the same riots the Mott home narrowly escaped destruction (Hallowell, 1884, pp. 129–30). For the next decade or so Edward and Maria lived at 138 North 9th Street, Philadelphia, next door to the Motts. There was later a connecting door between the houses so that they could be used as one in connection with their abolitionist activities, and Lucretia Mott wrote to a friend in 1847 'We shall open E. M. Davis' house and can accommodate abolitionists a score' (Cromwell, 1958, p. 103).

The business interests of Edward Morris Davis were many and complex. For two years he was a Director of the Pennsylvania Rail Road Company; in 1848 he was registered as a partner of Morris Longstreth Hallowell, the father of his daughter's future husband; in 1854 he was President of the Barclay Coal Company having one million dollars capital, and formed a land company to develop a new residential township in Chelten Hills, Montgomery County, Pennsylvania, where he also established a camp for escaped Negro slaves; in 1860 he resigned as Director of the Northwest

Mining Company, of which his half-brother, William Morris Davis, was a large stockholder (Johnson, 1951, p. 15). Somehow Edward managed to combine a successful business sense with participation in the abolition campaign. This involved him in no small difficulty but never blurred his view of moral responsibility.

> As early as 1834 he attached himself to the American Anti-Slavery Society and remained an enthusiastic member until he witnessed the triumph over the evil which it opposed. Often his goods lay untouched on the shelves until sold at auction, because his customers feared to offend Southern sentiment by dealing with an Abolitionist. No consideration deflected him from his determination to release the slave. (*Philadelphia Advertiser, Saturday, 27 September 1879*)

However, the moral sacrifice happened to prove beneficial in the long term.

> The family tale is that in earlier years he made a very great deal of money dealing in cotton, but when he was caught up in the abolitionist movement he threw over the cotton trade, because it was based on slavery, and turned around and made a great deal of money in wool, instead. In any case he was a very active businessman who made (and I believe also lost, at intervals) a lot of money in those early days of the American tycoons. (*Letter: E. Mott Davis to R. J. Chorley, 17 February 1962*)

Mrs Mott undoubtedly valued his support and praised his virtues on many occasions.

> Please hand the accompanying leaves to dear Edward M. Davis. I am glad that you have so worthy a son-in-law, who, I dare say, seems every day more like a son indeed. (*Hallowell, 1884, pp. 295–6*)

However, neither Edward's ability nor his varied interests saved him from serious financial loss during the business panic of 1858.

> 1858 was a bad financial year and Edward Morris Davis' firms of E. M. Davis and Co. and Davis, Burton and Co. were caught in the aftermath of the depression of 1857. They owed George O. Hovey of Boston 106,724 dollars, which Edward met partly from loans from his father-in-law and from William Morris Davis [uncle of W. M. Davis, the geomorphologist]: Edward refused to escape by a legal loophole to leave his younger partners to suffer. (*Biog. Encycl., 1874, p. 45; Johnson, 1951, p. 15*)

This set-back, the immediate reason for the family moving in to live with the Motts, was soon overcome. In 1877 Edward was listed as a stockbroker of Walnut Street, Philadelphia, and president of the Barclay Coal Company. The revival of Edward's fortune was assisted by his brother, William Morris Davis (1815–91), who had interests in sugar refining and appears to have been more successful for he was able to retire in 1858 at the early age of forty-three. This was something of an achievement for he started his career as an apprentice on a whaler. By 1861 he had been elected to Congress. At the start of the Civil War Edward joined the staff of General Fremont as assistant quartermaster, through his half-brother's influence this soon became a permanent appointment to a captaincy in the volunteer service. However, the enlistment in the Federal Army, though apparently an innocent action in itself, was a direct breach of the Quaker belief and so Edward (Davis' father) was expelled from the Society of Friends and even his relationship to

FIG. 5. Senator W. M. Davis, the half-brother of Edward Morris Davis. Photograph taken about 1890 (From Johnson, 1951)

Lucretia Mott was insufficient to permit the family's return to the fold. Though in spirit the family maintained the same habits and practices, they were never able to rejoin the group amity of the Quaker faith.

However, I recall my Aunt Lucy (that is, my grandfather's (i.e. W. M. Davis) third wife) noting in a letter to my mother or father after my grandfather's death that he had said that probably the major event of his childhood was the action of the Quakers in throwing his father out of the Society of Friends, which took place during the civil war. As you are doubtless aware, Edward Morris Davis was very close to his mother-in-law, Lucretia Mott; I think he changed his name to Edward Mott Davis because of this relationship but am not sure. There was evidently a strong feeling of family closeness among the extended family ... James and Lucretia Mott, their children, and grandchildren. (*Letter: E. Mott Davis Jr. to R. J. Chorley, 13 January 1963*)

Returning to Edward Morris Davis of the 1860's: after having been through the great struggles of the preceding decades for the cause of Abolition, he found himself on the sidelines of the war to end slavery . . . on the sidelines because of his Quaker belief. Apparently he was badly torn between his Quaker belief and his desire to participate in the war, and eventually he decided that the only way to be true to himself was to enlist, and he did; and so they threw him out of the Society of Friends. It seems to me that this action does make some sense in the light of his father's history. But at any rate the effect of this decision and action on his part, on the Lucretia Mott entourage, must have nearly been shattering. Certainly the common Quaker belief must have been one of the major bonds. My grandfather must have been around 12 years old at that time, and it is not surprising that it shook him as much as it seems to have. Since that time the Davises have not been in the Society of Friends, but it is symptomatic that the use of the 'plain language' has continued within the family down to my generation. The bonds within the family remained strong . . . as a matter of fact, my father was born in the house at 'Roadside' in 1888 . . . but the religious tie was broken. (*Letter: E. Mott Davis Jr. to R. J. Chorley, 13 January 1963*)

Fremont unfortunately fell into political disfavour and was relieved of his command in October 1861. Davis' father was closely concerned in the scandal that followed and it seems he was at fault.

On October 22, Edward had appeared before the House Committee on Government Contracts to testify about his activities as a purchasing agent in the Western Department. In all he had purchased between fifty and sixty thousand dollars worth of equipment. One item, 14,283 dollars for blankets bought from his son Henry C. Davis, a clerk in the store of H. Robinson and Co. of Philadelphia, came under the close scrutiny of the committee. It appeared that the blankets had been condemned by a board of survey, but Edward had secured a re-survey and suggested their use in pairs as being cheaper than buying good single blankets. It was on this basis that the blankets had been accepted, but the government paid 3.25 dollars for each blanket instead of that amount for a pair of blankets. Edward told the Committee that the bill had been in error, that the word 'pair' should have run throughout the bill, and that the omission had been a mistake. (*Reports of Committee of the House of Representatives made during the Second Session of the Thirty-Seventh Congress 1861–2; Washington, 1862, pp. 755–7. Reprinted in Johnson, 1951, p. 37*)

After the inquiry Edward M. Davis returned to Washington and maintained close relations with his half-brother. The misfortune can only have been temporary because the Davis family were rich enough to visit Europe by 1868. There was a rumour that Lincoln wished to appoint uncle William Morris Davis as Secretary to the U.S. Treasury at one time but this was blocked by Jay Cooke.

THE DAVIS FAMILY AND THE AMERICAN SCENE

William Morris Davis was the youngest of a family of three whose births were spread over thirty years. His sister Anna, the eldest child, married

Richard Price Hallowell, a wool commission merchant trading in Philadelphia. Richard left Philadelphia in 1857 as a personal stand against slavery because his firm dealt in products coming from slave labour. He went to West Medford, Massachusetts, where he pursued the same occupation, and eventually became a director of the National Bank of Commerce and a trustee of the Medford Savings Bank (*Dictionary of American Biography*). Davis stayed with the Hallowells for a while when he first started going to school. His elder brother, Henry Corbit Davis, helped to manage his father's business interests. Two other sons were born to Edward and Maria sometime in the 1840s but they were either still-born or did not survive infancy.

The family was small for those days and the wide difference in age between William Morris and his brother and sister must have forced him to depend on his own invention for amusement, which perhaps partly explains the personal self-containment and aloofness which he exhibited in later life.

The war came just when William Morris Davis, the subject of our biography, was entering on the first important years of his educational development and it deprived him of the companionship and guidance of his father. This responsibility was taken on by his mother.

> For several years before attending the local schools he was taught his lessons by his mother. She, like her own mother, knew well the power of words and laid much stress on their correct use; doubtless this early training had much to do with Davis's rigor in developing a scientific vocabulary for his favorite science and his insistence on precision of speech and writing by student or professional investigator. (*Daly, 1945, p. 265*)

In the summer of 1862 William was sent to stay with his sister and to attend school in West Medford, Massachusetts. This was doubtless the result of the Confederate offensive which led to Lee's costly victory at Bull Run (29–30 August) and to his crossing of the Potomac (7 September) – the first time that Union territory had been invaded. Although Lee retired soon afterwards (18 September), the subsequent Confederate offensive carried the fighting to within a hundred miles or so of Philadelphia at the battle of Gettysburg (30 June–3 July, 1863). The following letters from Davis' mother to his sister give both an insight into her concern for her son and into her political awareness. Their tone suggests that Davis was a lovable and dutiful son, possessing sufficient good sense to be trusted to look after himself when away at his first school.

> My dear Anna, I have been so unsettled all day that I can do nothing satisfactorily, and at last have laid aside everything else to have a little talk with you – it is the greatest blank to have Willy gone! and if it was not that I was expecting soon to go on to him, I scarcely think I could bear to have him away – he has been in my mind so constantly ever since we watched the train steadily bearing him away, and I have gone with him mentally over every part of the route – one minor direction

I forgot to give him – to see that all was safe in his pockets before leaving his berth, as in rolling and turning, it would be easy enough for checks or keys or knife to slip out – but I guess he thought of that, and I hope of the basket too, for it seemed so likely that he might forget it but by this time he must be safely with you, and wont the dear little May open her big eyes in wonderment and delight – I think she will know him, and I know she will love him – dear boy! how infinitely I love him, and pray that he may live, to continue to bless us as he has always done – I am glad that he did not wear on the thick suit that I first selected for him, for it has been very warm, and his lighter clothes must have been more comfortable – clouds are gathering now for a gust, which I hope may give us a cooler atmosphere in place of the sultry 'dog day' weather we have had – I suppose Henry is home again by this time, and passed Willy on the [Sound?]. It was such a pity that they could not have journeyed on together – Father and I went over to the Kims last evng. as per invite – met there a small company of young and old – Aunt Elizabeth and Emily Davis, Sally Hallowell – Ned and Lea Davis, Cary Gibbson – Jas. Wright and wife and son – Mrs. Gresson and two daughters and son – the visit was quite pleasant, and the bride was quite pretty, and the groom quite marked, but I had much rather have stayed at home.

3rd day – How old and stale the above seems after so long an interval since writing – and how long too since thee has written, but I have heard from you directly through Henry, and indirectly through Marianna and Bel. I was very glad that Henry staid over a day so as to go out to Medford, and glad too that Bel stayed till Willy got there – with them and Frank Wright, you had really a house full. Henry says that May is learning to talk nicely and continuously and incessantly – little darling! I hope Willy will write just how she did when she met him – Henry is amused at the many questions I ask him after each visit he pays you – he and Patty came out on 7th day afternoon – it happened very well – for Uncle Morris's folks came then to tea, and it was pleasant to have all together. Morris read aloud Fremont's speech in Boston, to the edification of us all – I hope Richard went to hear him – I thought it very good and very like him – Henry says that he and thee called at the [Review House and?] Mrs. Fremont for which I was glad, and regret that you did not see them – How galling it must be to the General to be quietly laid on the shelf in these perilous days, but if misery likes company, he can console himself with the knowledge that Mr. McLellan at last is almost left to share the same fate. Certainly he is 'out in the cold' in his headquarters at Alexandria, while a council of war is held at Centerville of all the other acting generals – It was grand for Morris and Hannah Hallowell to make that flying visit to the boys, and must have been a great pleasure and satisfaction to all parties, especially if they (the boys) have gone, as I suppose they have, to share in the dangers near Manassas – they seem to be very guarded in their expression of opinion of the merits of their general Cope, but it must be only the etiquette of the army, for they cannot be blind to his repeated failures and consequent incompetency – but his day is over – let him slide – we have not heard from Willy Wright since he left – though possibly thee may have done so, through Frank.

Grandpa and ma are home again – returned yesterday via Eddington – the former went . . . (*Letter: Maria Davis to Anna Davis Hallowell, 28 August 1862*)

My dear Anna, Thy letter of yesterday about Willy's school is quite unsettling, just as we supposed all was progressing in the most satisfactory manner – only one point seems clear, that his present school is not the right one for his proper advancement. Further than that I feel very doubtful, and after expressing one or two buts and ifs, I shall be entirely satisfied to leave the matter with Richd, thee and Willy for final settlement. It is evident thee inclines to the Boston Latin School – I presume this is not, as its name would seem to imply, mainly a Latin school, but one where this study is added to the ordinary branches. This is a very important question to Willy, for it should be poor policy for him to omit, now, or even to slight the ordinary studies of the grammar schools ... this is the first question – then comes minor ones – of whether there are two sessions a day, which would be objectionable, involving a dinner difficulty. Then the distance from the depot to the school for a winter's walk – then the probability of his gaining admission if it should seem decidedly preferable, after inquiry. I do not think the fact of going in and out of town a very serious objection – perhaps it would be quite as easily accomplished, as the walk down to Medford – but I think likely the latter would be pleasanter to him, as he would always have companions. I feel very sorry to be obliged to throw the responsibility of this change on to your shoulders, although I have entire confidence in the wisdom of your decision. I am only sorry to give you any further trouble. Whatever he does, and wherever he goes, the change should be immediate, for his class is gaining upon him every day's delay – I hope that he will enter upon the duties of whatever school he may go to, with the same zest as he showed a year ago, and when I am with him again, it will be very pleasant to me, to give him what help I can, until as in his arithmetic he outstrips my limited acquirements. Write as soon as you have made a decision, for we are impatient for the result.

Thy last letter came just as mine was ready to send to thee. All thee wrote of Frank Wright was very interesting. I was very glad thee was able to be of such good service to him. I was in town last 5th day, and called at 912 to give thy letter to them to read. I saw Hannah and Anna and had a very pleasant visit, mostly (. . .?) in listening to the account of their late visit to the boys – it was a most memorable time – but I think has made Hannah more anxious than ever before. She more fully realised the exposures and hardships incident to a soldiers life. It is almost more than she can bear to think of 'her boys, so tenderly raised, living out in the open field, and very scanty fare'. She says they are neither of them at all well, and both look haggard and worn but I suppose she or Anna has written full particulars to you, and I need not repeat here what she told me. She seemed to think she could not go to Medford this fall. After leaving the house I met Morris, just on his way home to dinner. He had a formidable roll of 'green backs' in his hand, which I told him looked very good, and quite contradicted his assertion that 'he was growing poor, and Dick rich'! I think Anna would have been off on this late call for additional nurses, if she had not been obliged to be on hand at the opening of the schools. (*Letter: Maria Davis to Anna Davis Hallowell, 7 September 1862*)

Through a mother's obvious anxiety for her son's welfare there is every indication of a well-educated woman equipped by nature with a moderately shrewd and sometimes humorous temperament. Also behind the gay recitals of family comings and goings there is the note of war, not strident or anguished but nevertheless commanding recognition. Son and mother were clearly close. While she talks of him affectionately in the way mothers do of young children, the suggestion of changing schools must have come from Davis himself and it is interesting to see that in the final decision his views are taken serious account of. Not many youngsters plan their educational future. He was not a mother's boy either. We know from a reference late in life that he went on a walking trip up the Susquehanna Valley when he was thirteen.

There is a vast difference between growing up in the atmosphere of war and discovering the pleasures and complexities of the adult world in a time of peace. War creates an environment where the daily scenes are full of drama, privation, heightened emotions, heroism and where there is a suspension of the normal, established way of things. The ordinary business of bringing up a family or making a career is apt to fall into second place. For a teenager it probably means a hastening of maturity. In the absence of fathers or elder brothers he is introduced to responsibilities that in peace-time he would not have to accept until much later. The gaiety and playfulness of childhood is thrust aside much earlier.

The Civil War, while exacting on human relationships and leaving bitter memories, was not in itself as important as the general stage in the historical process through which the United States was then passing. From an unimportant but rebellious colony of the Old World the United States was rapidly transforming itself into one of the most powerful nations of all time. In great strides the era of the simple, sturdy, frontier farmer was being replaced by the explosive dynamism of an industrial society. This is the period which has supplied an inexhaustible number of popular themes for the Hollywood film moguls – the mining saloons, the cattle drives, the railroad gangs and the ungodly insolence of the self-made captains of industry. Two years before William Morris Davis' birth the United States had successfully ended the war with Mexico, annexing the territories of Texas and New Mexico in the process. In 1849 the Californian goldrush began. Gold strikes absorbed the hopes of many settlers. In 1858 gold was discovered in Colorado; 1859 the Comstock lode began to produce a fortune of 306 million dollars and signalled the rise of Virginia City; between 1860 and 1864 there were gold discoveries in Idaho and Montana; and gold was the foundation of the infamous Tombstone. Yet by 1890 gold was already being replaced in importance by copper and by 1901 oil was surpassing both. When Davis was ten there were still no states west of the Mississippi. Some of the territory was settled but neither the people nor their institutions were of a sufficiently permanent nature to justify political recognition. At this time the Indians of the west were still a match for the soldiers but the death-knell of their independence was already tolling. The invention of the repeating rifle gave the military an overwhelming advantage to which the savage bravery of the Indian was no answer. By 1878 indiscriminate slaughter by the buffalo hunters had decimated the wild herds on which most tribes of the plains Indians depended for their existence. The massacre of Colonel Custer's cavalry by Crazy Horse and Sitting Bull in 1876 was almost the Indian's last success. The Dawes Act of 1881 was merely official recognition of the break-up of the tribal system and led eventually to the restriction of the Indians within the confines of state reservations.

Already the west represented more than gold and Indians. By 1857 Butterfield and Fargo had spanned the western half of the continent with their famous Wells Fargo coach – three weeks from St Louis at 200 dollars a head. By the following year William Russell's Pony Express was running mail between Missouri and San Francisco. In 1861 east and west were joined by telegraph, and by 1869 the first trans-continental railroad had been built. At the same time the big cattle herds were being driven from their grazing lands in the west and south to the railway termini that linked with the eastern markets. In 1868 the first cattle drive was made to Abilene. Later Ellsworth, Dodge City, Cheyenne and Laramie became cattle towns. The drives took about two months with five or six cowboys to a thousand head, often singing to the cattle at night, not out of romance for their outdoor life but to quieten the beasts, easily alarmed by the strange cries of the wild life about them. Conditions were primitive and law and order largely unrecognized. Dodge City, the cowboys' capital, records the killing of twenty-five men in its first year. Those who were shocked by this lawlessness formed themselves into vigilante groups. However, the processes of civilization were swift and by 1881 the need for vigilantism had vanished. The invention and mass production of barbed wire, improved methods of ploughing and reaping, hardier varieties of wheat and a growing European market all helped to make cattle breeding less important and arable farming a more attractive investment. Permanent settlement and all the normal democratic processes soon followed.

In the east of the United States cities were beginning to grow quickly. By 1860 New York had a million persons out of a national total of $31\frac{1}{2}$ million. Philadelphia was a flourishing business community with its own Stock Exchange and diverse interests in coal, oil, railways and textiles. As the ambitious and more adventurous types of American sought fortune in the west their places were taken by immigrants from Europe. Till 1882 the majority of these came from Britain and Western Europe, the maximum influx in one year being 640,000 persons. After this date an increasing proportion came from the countries of central and south-eastern Europe. As industry began to exploit the nation's tremendous natural resources on an increasing scale, the traditional business pattern of small companies began to blossom into mammoth corporations, which could only be compared in power to small imperial states. In 1903 there were 50 corporations with a capital of 50 million dollars, 17 with 100 million, 1 with 1500 million and 6 railroad empires with more than 10,000 million dollars capital. Without realizing it and certainly without wanting it, the United States was moving into a position of economic dominance which would ultimately force her to accept a leading role in world politics. By 1900 the United States was already beginning to compete in the European markets while her millionaire industrial tycoons had began their crusade and pillage of European art treasures.

With such a glittering choice of wealth and adventure William Morris Davis could easily have taken a more exciting career. He chose instead the serenity and contemplation of life within an eastern university. The choice, which it is fair to say was probably the right one, detached him almost completely from what was happening both in the west and within the domain of big business. Often it is impossible to imagine from reading his letters that any of this was going on. There are factual references which make it obvious that he knew of these events but often they had little but statistical interest for him. Generally he seems to live within a narrow world of university shoptalk and small gossip, quite undisturbed by and unconcerned with what might be happening elsewhere. Not that he was timid or untravelled. During his life he visited many parts of the world and not always in conditions of comfort. Yet for all that the nation's most important moments of history barely affected his thoughts or ambitions. He had a clearer vision than most men but seems to have exercised it within a very narrow range.

CHAPTER THREE

Educable Youth

Before he went to school most of Davis' thought, perceptions and activities were spent in the rural surroundings of the Old York Road where his parents had bought a farmhouse known as Oak Farm. The Old York Road was part of the original stage route between Philadelphia and New York. When Davis' family bought the farmhouse in 1850 the district was entirely undeveloped – 1,000 acres of woodland and pasture broken by the presence of an occasional small Quaker farm. It was not till the end of the Civil War that Davis' father joined the syndicate which developed the land as a fashionable, residential district. In the early years Davis' parents, Edward and Maria, shared the property with the family of Thomas Mott, a brother of Davis' mother. The two families used to stay at the home of the grandparents at 338 Arch Street during the winter months and the Motts used to join them at Oak Farm during the summer. This arrangement ended in 1857 when the grandparents bought 'Roadside', another small farmhouse on the opposite side of the road to Oak Farm. Its situation is described as sunny and surrounded by cherry, apple and pear trees. When the grandparents grew older Davis' father and mother went to live at Roadside in order to give them the closer care they were beginning to need (Hallowell, 1884; Hotchkin, 1892).

The withdrawn and serious nature of young William Morris Davis must be imputed partly to the fact that his sister and brother were both too old for companionship and that his meetings with children of his own age were restricted to the visits of cousins. This lack of contact must have thrown him in upon himself a great deal. The improving instructions of his grandmother can only have accentuated this tendency, and the evening readings from the Bible and the serious family discussions were all likely to promote a rather austere and intellectual attitude towards the ways of the world.

> His early recollections as mentioned to me, mention no contemporary playmates unrelated to family. There was a family of relatives who lived across Old York Road some distance back. His association then, if any, must have been on occasions when relatives were visiting. (*Interview: R. Mott Davis with R. J. Chorley, 7 September 1962*)

This early absence of companionship induced a marked shyness and gaucherie in the presence of other persons which persisted throughout his life. The denial of young companionship deprived him of that period of wasteful

F I G. 6. Davis, aged about one year, with his sister
(Courtesy E. Mott Davis Jr., Austin, Texas)

exuberance, which though unproductive, is the rightful heritage of every normal child. More important the give and take of these early relationships prepares the child for the time when he will have to sustain satisfactory adult relationships. With the natural outlet of play denied him, in compensation perhaps, or because it answered an integral part of his character, he turned to studies of nature. Even then it was not a child's approach, not an indiscriminate collection of specimens in competition with his fellows as is found among many children. Instead there was a deliberate attempt to amass and understand detail and to classify. To this end he employed his marked ability to draw.

I recall a note-book with pencil sketches of bees with their wings spread in minute detail. As time went on he took an interest in the stars, and his powers of close

F I G. 7. The Davis family home, 'Roadside', on the Old York Road north of
Philadelphia (From Hallowell, 1884)

observation led him to identify a variable star, being the first perhaps in the
United States to note its presence. (*Interview: R. Mott Davis with R. J. Chorley,
7 September 1962*)

He made good use of this talent and later used it to improve his lectures by
visual appeal and also for his private amusement.

It would be unfair to leave the impression that his educable life was one
arduous and prolonged exercise in mental improvement. There must have
been many occasions when the house was full of the laughter of relatives and
friends, with musical evenings and some of the less boisterous type of party
games. Davis on his world tour often refers to these occasions and clearly
missed them. He was able to play the piano by ear, and never lost his interest
in serious music.

Next door to my room are two graduate students Giles and Vogel. Giles has been
U.S. Consul in Batavia, Java, and is now studying law. Vogel is trying his hand,
head and ears on the psychology of music; and as an aid in that entertaining
undertaking he has a superb orthophonic victrola. Really, it is just about like an
orchestra; I sit there and ask myself, what is the difference, and cant find any.
Among his scores is the whole of the Fifth symphony from the Royal Albert Hall
in London; a perfect treat to hear it. I wonder if any of you remember that that is
the first symphony of my acquaintance; way back at Oak Farm, opposite Old
Roadside, when Isabel and Emily Mott played parts of it arranged for four piano
hands. Vogel generally puts on a record late in the evening, and I run in to listen. I

can hear fairly well from my own room, but not the finer points. (*Letter: Davis to his children, 7 October 1927; Stanford*)

Probably his ability to versify dates from this period, and seems a likely product of the social evenings.

DAVIS' FORMAL EDUCATION

We have seen how Davis was sent to the public grammar school in West Medford, Massachusetts, in the summer of 1862. This was only a temporary expedient which in all probability he, and certainly his mother, did not find satisfactory. With an improvement in the Civil War situation, Davis returned to Philadelphia in 1863 and for the next three years attended a private school in the city. We are told (Brigham, 1909, p. 2) that at this time he did not share much in the games of the other boys, being mainly occupied with mineralogy, entomology and astronomy.

When a child is starved of his proper diet of vigorous play fantasies, he tends to build complicated worlds of his own. These take several forms depending largely on the nature of the child. Sometimes in an artistically sensitive nature, the deprivation may affect the emotional growth. As a result the child may create a well-populated dream world, with father and mother figures of his invention. He will question society's traditional values by striking deliberately antagonistic attitudes or perhaps by becoming a writer of the now well-known 'angry young men' type. On the other hand, the boy with an ambitious potentiality will sublimate his energies through the medium of some activity whose dominating element is organization. It may be marshalling model soldiers in complex battle sequences, constructing a model railway system or even being the leader of a juvenile street gang. But in each case the interest for the boy lies in the power of control which he is able to exert.

Now it is immediately clear that Davis' youthful characteristics followed none of these trends. There was nothing hungry or angry about the youthful Davis. Also absent was the willing adaption to middle-class mores that is often readily evident in the youthful leaders of tomorrow. Instead his sensitive and introspective qualities led him not unnaturally towards scholarly pursuits. His choice of mineralogy, entomology and astronomy, all studies requiring exceptional powers of observation, classification and mathematical ability, point to a detached and sharply intellectual personality, cut off from the ordinary hopes and cares of the human world. He may youthfully have imagined himself making a voyage of discovery through the insect kingdoms and complex universes of the inorganic; more likely his phlegmatic nature fed on more prosaic sustenance.

While his father, as a prominent businessman, is unlikely to have favoured his son becoming a teacher his mother may have thought that one business son in the family was enough. His grandfather and grandmother, the Motts, had both been teachers during the early part of their lives and Davis' keenness for study may have received some encouragement from them. Whatever his father's views he assisted rather than discouraged his son's inclinations. When Davis reached sixteen his father paid for his entry to Harvard. This was far from automatic. William Morris was not following a long family tradition; he was in fact the first Davis to attend university.

The Lawrence Scientific School to which he went came into being in the year of his birth and must have been a relatively untried establishment when he chose to join it.

> Such tastes and beginnings in science led him away from the ordinary college course of that day, and in 1866, at the age of sixteen, he entered the Lawrence Scientific School of Harvard University, in Cambridge, Mass., where he was graduated with the degree of S.B. (*magna cum laude*) in 1869. A year later he received the degree of Mining Engineer (*summa cum laude*), then for the first time given at Harvard and chosen by Davis not so much for the opportunities to which it led, as for the variety of studies that led up to it. (*Brigham, 1909, pp. 2 and 4*)

During his time at Harvard he failed to become absorbed in the full flow of undergraduate social life.

> He lived, at least part of the time in a private home on Picken Street not far from a large scientific school. As an illustration of how he kept to himself, I recall he watercolored a drawing and painting of his room in this house. He was active during spare time at the Harvard Astronomical Observatory. (*Interview: R. Mott Davis with R. J. Chorley, 7 September 1962*)

He made a few friends among his immediate classroom colleagues and took a great interest in musical functions. Otherwise the main impact that non-academic college activities made on him was to foster an interest in collegiate baseball and football games. He may also have learnt to play golf but it is more likely that he acquired this interest later.

At this point there is no indication that Davis had planned the career ahead of him. His family had sufficient means to make an immediate choice unnecessary. It is fair to say that his choice of subjects probably coincided with his natural interests. That Davis himself had any intention of becoming a university teacher is extremely doubtful. He was still very much at the stage of following his inclinations and was making no attempt to harness his talents within the framework of a career. His approach to everything he did was serious enough but as yet he had not committed his interests to one area.

While he was a student three teachers trained him and they may have influenced his final choice of a career. They were Raphael Pumpelly, Josiah Dwight Whitney and Nathaniel Southgate Shaler. Brigham comments only on the influence of Whitney and Shaler and as his treatise on Davis is thorough his statements should be treated with some authority.

> He was fortunate in his teachers, of whom Whitney inspired him with an admiration for thorough scholarship, marked by wide reading in many languages and by an extended acquaintance with scientific men through their works; while Shaler impressed him with the value of invention, initiative, versatility, untiring activity in field study, and close personal association with colleagues and students. He was fortunate also in his surroundings at Harvard, where academic freedom allowed full, unhindered development, and where the material aids of ample laboratories and well stocked libraries placed the responsibility of success or failure on the teacher. (*Brigham, 1909, p. 62*)

Of these three University teachers, Pumpelly had had by far the most exciting life. He acquired his early geological training in Europe, where apparently he was something of a dare-devil and took part in an affray with Corsican bandits. At twenty-two he returned to the United States and obtained a job in Arizona as a mining geologist. Again his wild nature involved him in dangerous episodes with Apaches and other desperadoes. By 1861 he had gained selection as agent for the Japanese Government and his activities involved exploration throughout the Japanese Empire and the introduction of new mining methods. In 1862 he travelled to China and while going up the River Yangtse his capacity for meeting trouble brought him into conflict with the Taiping rebels. Undeterred he went north to Peking searching for coal deposits to power steamships. He returned home to the United States in 1866 by a protracted route which took him through the then undeveloped areas of Central Asia, Asiatic and European Russia, to arrive eventually at Paris. Once back in the United States he worked first on the copper deposits around Lake Superior and then went prospecting for iron ore in Upper Michigan. About this time he was appointed State Geologist for Michigan. In 1866 he became Professor of Mining at Harvard. His prospecting expeditions must have taken him away from Harvard much of the time but Davis heard some of his lectures and probably came under his tutelage during part of his studentship. Pumpelly certainly lectured at Harvard in 1869 and again during the winter of 1870/71. In the summer of 1869 he took Davis and four other students on a tour of the iron and copper districts of Lake Superior. Experiences drawn from his early career were alone sufficient to make his lectures entertaining and one wonders what effect they had on Davis. Pumpelly's nature was so completely alien that it can hardly have appealed to Davis; it may even have shocked him. In 1883 Davis worked under Pumpelly on the northern transcontinental survey and, as the following letter shows, may have been encouraged by him in the use of block diagrams.

> I find yours of the 23rd on my return from a short absence, and am glad that you have had such a successful season.

Please make your report in the form you think best, remembering always that the graphic representation in diagram and section is the essential feature in our system of showing results. (*Letter: R. Pumpelly to W. M. Davis, 27 September 1883; Newport, Rhode Island*)

Less is known of Professor Whitney, who became Harvard Professor of Geology in 1866. After writing a book on the mineral deposits of the Lake Superior region, he directed an investigation of the mineral resources within the States of Iowa and Wisconsin. He was also State Geologist for California and published several important reports on the gold-bearing gravels of the Sierra Nevada. Davis was selected as one of the small body of students whom the Professor took with him on the expedition into the Rockies to calculate the height of a number of unsurveyed peaks. Davis made several references to this trip at different times during his life but he tells us nothing at all about Whitney (Davis, 1896N and 1930K). Later in life he made the following references to his old professor which suggests a degree of opulence and formality.

During this time Whitney lived with Dr. B. A. Gould and the latin scholar George N. Lane in the old Mann House at the elbow of Follen Street, Cambridge, Mass. . . . a chapter in Whitney's life hardly to be imagined by those who know of him only in the seclusion of his latter years. . . .

Here was feast of reason and flow of soul – a reception in Roman style, the door opened by a house-servant blackened and chained like a slave and greeting the guests in Latin with a brogue; and as the edition did not go off rapidly in those earlier days, all the unsold copies were borrowed from the publishers and used to decorate the house, everything else being cleared away. (*Davis, 1930K, pp. 206–7*)

Shaler, the last of Davis' three university teachers, was not appointed Professor of Palaeontology until 1869, the year Davis gained his bachelor's degree. Nevertheless he may have had the greatest influence on him. It was at his summer school in Kentucky that Davis was persuaded to join the staff of his Geological Department and it was to Shaler that Davis wrote while at Cordoba. Initially they must have worked closely together and Davis' second published article was a joint work with him. Like Pumpelly, Shaler was entirely different in character from Davis. He was a worldly, cultivated Southerner born in Kentucky.

His lectures were garnished with good stories admirably told, and it was a perhaps unconscious habit for his eye to twinkle and his fingers to be pushed through his shock of hair when the laugh was really due. The language of his lectures was clear and forceful, with many expressive words of his own coining. (*Merrill, 1920, p. 927*)

The two men had little sympathy for or understanding of each other. Davis admits as much when he himself wrote:

> ... two persons more unlike than Shaler and Davis could hardly be found. (*Davis, 1930K, p. 314*)

Davis' encomium written at Shaler's death is much fuller and kinder.

> In every growing measure for over forty years, Nathaniel Southgate Shaler made himself part of our life and gave the service of an intensely active personality to the College and the Country.
>
> He had an unusual range of experience in contact with the world of men and work: a boy in a slave-holding community, a young officer of the Union Army in the civil war, later the director of a survey in his native state and member of various commissions in the state of his adoption, practised field geologist in many parts of this country, observant traveller abroad, expert in two bureaus of the national government, adviser of mining enterprises in the South and West, writer in many fields, orator and poet on our days of celebration, he thus gained that wide acquaintance with external affairs which made him so invaluable a Harvard man: student at 18, lecturer at 23, professor at 27, and dean at fifty.
>
> He was impatient of seclusion of his work and therefore related himself, but without a trace of self-seeking intrusion, to all phases of university life. Confident and courageous, abounding in initiative, he gave direction to work around him and turned the course of events. Inventive and independent strikingly individualized, he worked to best advantage as a leader or alone, not as one of two; if other names have occasionally been linked with his, the association showed his generosity rather than his need.
>
> He was always devoted to the best development of his own department, which grew and flourished under his leadership; he foresaw success for other departments which came into existence under his fostering care; he made the summer months educationally useful as they had never been before; he brought new life into an old school, and he inspired and guided the creation of a new and greater school to be, of which he was marked for the first officer. So wide a distribution of academic interests might well have weakened the efficiency of a less active man; but he was vigilant and faithful even to the details of his varied administrations; discerning and cautious in action where risks were great, yet ever ready to essay new methods and bold in taking risks where judgment advised a venture; unceasingly alert in his endeavors for the betterment of all our work, untiringly ingenious in the invention of new devices for enriching the opportunities of the university and for extending the influence of learning; cheerfully assured that, however great the tasks to be done, strength would always be found to do it.
>
> While he was indifferent to the conventions of fashion and to whatever seemed to him hollow or excessive in forms or ceremonies, he was sincerely courteous in manner and he therefore carefully retained some formalities in daily intercourse which others have carelessly abandoned. He was simple in his tastes, and his house and household were simply and genuinely hospitable. It was as much his courtesy as his appreciation of good business methods that made him punctilious in

keeping all appointments. He was unaffectedly, unconsciously original and picturesque in hearing and in speech; to the end a staunch Kentuckian, tho' citizen of
another commonwealth for nearly half a century; seldom waiting for others to
make advances yet unreservedly responsive if they did; valuing the enlivenment of
ideas that springs from free discussion and the mental exhilaration that comes of
hearty laughter in good company, and always finding the good in any company
that he met; possessed of a retentive memory that brought pertinent events from
the crowded past, fresh and glowing into the service of the present; fond of
reminiscence thus abundantly supplied, and of citing the bearing of former adventures on the case just now in hand, but this in the most natural manner. (*Davis,
'Minute on the Life and Service of Nathaniel Southgate Shaler'*)

Davis' references to Shaler include an oblique one made in 1927 which, being
less formal, is perhaps more genuine.

But you haven't heard of the Sunday afternoon meeting of the Monday night
Journal Club at Blackwelder's house a few days ago; when I gave a personal
account of Whitney and Shaler, with some Davis thrown in. Whitney, because he
was state geologist of California in the 60's before he became professor at
Harvard; Shaler, because two plaster busts of him are in the geol. building; one in
the museum room on first floor; the other in the library on second floor; rather a
large proportion, in relation to other geologists of greater fame outside of the

FIG. 8. Drawing made by Davis of an unknown person (his father?) in 1868
(Courtesy R. Mott Davis)

Harvard Yard. However, if Stanford wishes to pay tribute to my old teacher in
that duplicative fashion, no objection shall come from me. (*Letter: Davis to his
children, 15 December 1927; Stanford*)

Davis temporarily interrupted the studies leading to his bachelor's degree in
the summer of 1868 when he joined his parents on a trip to Europe.

In 1868 the summer was spent in Europe. This was the beginning of a long series
of travels; the young student then had his first sight of the Alps, where his
excursions included the crossing of the St. Gotthard pass from Lucerne to
Bellinzona. (*Brigham, 1909, p. 4*)

Work pressed less tightly on the student in those days and the contemporary
competitive anxieties were less evident; such diversions were still possible
without endangering the degree candidate's final chances of success. So not
surprisingly Davis was travelling again the next summer on a student's trip
to Lake Superior and the Rockies. There is no mention then or later of the
expedition to Lake Superior with Pumpelly but there are two interesting
references to the western trip.

My first western journey was made with three classmates in the summer of 1869,
the last vacation of my student years. So little was the West then developed that,
after our party reached Cheyenne on the then recently completed Union Pacific
railroad – which with the Central Pacific made the only transcontinental line of
the time – we had to travel overnight by stage to railroadless Denver. There we
outfitted under our leader J. D. Whitney, who had been director of the geological
survey of California before he came to Harvard in 1866 as professor of geology,
and who desired to discover whether any Rocky Mountain peaks really reached
altitudes of 18,000 feet, as were ambitiously asserted by some of their climbers.
During six weeks of camping, chiefly in the neighborhood of South Park and the
upper Arkansas valley of the Colorado Rockies, we ascended a number of summits
and found that none exceeded 14,500 feet. We gave the names of Harvard and Yale
to two of the highest in the Sawatch range beyond the Arkansas; and the next
peak to the south was afterward named Princeton. The three are nowadays
pointed out to travellers on the Rio Grande Western line as the 'College Peaks'.
(*Davis, 1930F, p. 395*)

The year of Eliot's inauguration has a special interest for me; for it was in the
summer of that year of 1869 that I first made acquaintance with Colorado, then
still a territory. My professor of Geology, Josiah Dwight Whitney, a graduate of
Yale, had been state geologist of California in the earlier sixties, and had there
become much interested in the measurement of mountain heights. So in the
summer of 1869 he organized a small expedition, including four of his students,
came to Cheyenne by rail, to Denver by stage – a very tiring overnight ride – there
outfitted with saddle horses and two supply wagons and headed for South Park.
The first mountains that we climbed were Silverheads and Lincoln. Lincoln was
the higher of the two; we reduced it to the reasonable and very sufficient height of
about 14,000 feet. From its summit a number of apparently higher peaks were

seen to the south-west, beyond the upper valley of the Arkansas. A detachment of the party went there, ascended two of the highest peaks we could find and named them after the colleges in which Whitney was then teaching and in which he had earlier studied. Princeton has since been added, next to the south, in honor of your college, Mr. President. Thus the College Peaks were named; although Harvard is the highest of the three, I must confess that Princeton is the most beautiful. Unfortunately, two other summits, some miles farther north, which we did not reach, have been insufficiently reduced in height by the sapping action of their ancient glaciers, and they therefore have the temerity to rise a little above the highest of the three college peaks; but they are not so high as Mt. Whitney in California.

How greatly has Colorado changed since those early times. Denver now a great railroad center, has reacht * almost metropolitan proportions. And here stands the beautiful residential city of Colorado Springs, where in 1869 lay the open and empty plains below Pike's Peak. But all these changes you know better than I do. (*Speech: 'Remarks of a Wandering Professor' by Davis at Colorado Springs, 5 December 1925*)

The year 1869 also saw Eliot's inauguration as the new President of Harvard. Davis' student memory is worthy of note because, among other things, it shows the rate at which Harvard grew during the second half of the nineteenth century.

My experiences with ceremonies of presidential inauguration is very limited. I have attended only two such ceremonies. One in 1869, when Charles William Eliot began his greater career as leader of Harvard progress; the other, 40 years later, when Lawrence Lowell followed him. There was a difference in those two ceremonies appropriate to the different dimensions of educational institution at the two dates. The inauguration of President Eliot was a simple affair; otherwise, how could my classmates and I, third-year scientific students, have found places in the not over-large church, the old First Parish Church opposite the College Yard, to hear his address. The address was as far beyond my boyish appreciation as it was beyond the educational methods of the time; it outlined a great plan of educational reform and advance, which he persistently carried forward to essential completion in the forty years of his forceful but patient administration. (*Speech: 'Remarks of a Wandering Professor' by Davis at Colorado Springs, 5 December 1925*)

Even before he completed the master's degree Davis was apparently committed to spend the next three years at Cordoba Observatory. The choice was a little strange because it meant that his recently acquired knowledge of geology was not going to be put to immediate use: instead he was going to rely on limited experience gained from a part-time hobby. His future course was still unplotted. The only restriction he made for himself was to follow matters with a scientific or mathematical content.

In his last student year Davis does seem to be taking more interest in

* Note Davis' passion for phonetic spelling in his later years.

college life though it is fair to remark that the 'Tom' with whom he is sharing a room was his cousin:

Here I am again, busy at copying notes and blowing my nose, the latter consumes about three 'hakfs' before dinner . . .

I left at 7.45 and slept pretty well in the sleeping car – arrived at Boston about 5.00 a.m. Went to Parker's till daylight should appear, eating an oyster stew in the mean time.

I arrived here at 7.45, found Tom in the study, and breakfast nearly ready.

I found Joe had returned on Saturday, and neither Garnett nor Marvine had made their appearance. Saw Mr. Pettie and Prof. Shaler, and reported that I was all ready to go ahead, but that cant be done till the others come and they are not here yet. Marvine left his notes here before going away and I am busy copying them – rather tiresome work.

My cold is rather worse today, so I shall keep rather close till it is better.

Yesterday mother's letter to me when I was at Barclay, and thanks from the Blight children came – but nothing from Nelly; so I suppose her present did not arrive. It was 'Alice's Adventures in Wonderland' which I bought at Ashmead's the Monday before Christmas, and left there to be sent to Barclay by mail. I shall write to Barclay about it.

Miss Nannie's cat is lost! and the family is inconsolable. Mrs Rotch has been round this morning inquiring for her.

My cold makes me feel very stupid (stupid – good word for anagrams) and I can hardly write my notes or this letter. Anna took the whole bundle to Medford, so I shall have to wait for my stockings, etc., till Sunday.

The whist club meets here tonight – five tables – supper in this room – so Tom and I must go up stairs – Tom is deep in history now and studies hardly anything else. He received some very handsome books on his birthday – 9 vols of Thackeray – Plutarch – some Dickens – Shakspeare and Stayden's Dic. of Dates etc. etc.

(Letter: Davis to his parents, 5 January 1870; Cambridge, Mass.)

There is a marked concern for his health, which seems a little ironic considering the age he reached but he does appear to have suffered periodically from chest complaints and stomach disorders. Otherwise the letter is what one might expect from a young student working towards his final examination. The next letter was written only eight days later:

Where is the 'little Chestnut St. Theatre of Susan Calton and Co'? I don't remember being told anything about a bathtub at York Road.

The valise stood the journey very well. Now for my news.

Did I tell you that I was invited to a small party at Mrs Richardsons for tomorrow night? dress-coat, etc! I am cramming dancing as if I was to be examined for it. More particulars next week (that is the English for 'nous verrons').

Marv has left Joe and taken a room near here, and Garnett has taken Marv's old place with Joe. Our work is well started again now, and goes smoothly. Prof. Pumpelly began his lectures last Friday (to be given on Tuesdays, Wednesdays and

Fridays) on Economic Geology – which he defined to be the 'modes of occurrence of useful ores, and their influence on vegetable life.'

Shaler is talking pretty *large*. If the college authorities agree, and other equally wonderful '*ifs*', he wants to make two long excursions at the end of the term or in the summer vacation. The first with 50–60 Juniors throu Vermont, the Adirondacks, some central N.Y. and back – the second with about half a dozen *picked* students including, of course, the 'immortal four' through Central N.Y. from N. to S. stopping at the . . . and . . . Colleges, the state geol. collection at Albany, and Cornell Univ. (next of the '*ifs*' was Prof. Whitney's going to Colorado).

I have been busy copying notes on the lectures that I lost in December and have not nearly finished them.

The Pierian and Glee Club Concert is next Tuesday; could you not be here in time for it? We are practising hard for it.

I went to Medford on Saturday, and saw all the folks. William and Elly G. came out to tea in a hard snow storm and with Nod and Sally, we had a jolly time – we tried playing letters and anagrams after supper, but there was so much disorder that we failed in both – especially anagrams. Our word was 'poniard' and Anna had the most words about forty so she read first and began with 'parchment'! She thought we were to write any words beginning with 'p'. Then every one would interrupt while reading the words, and they would not count right, and Richard copies from Elly's list and everything else bad – so it was soon given up. After William and Elly had left Anna played a lot of operas. Sunday morning, I went to Nod's to help make some molasses candy, which was very good when finished, and we had a good time making it.

Anna and Richard drove over in the afternoon to call on Mrs. . . . (née Nelson). As Mr. . . . was away I did not go with them. They stopped there for the bundles.

Profs. Agassiz and Eustis are still sick, but a little better than they were. Tom has an examination in History tomorrow and is digging for it; he leads the German at Mrs Richardsons tomorrow night. (*Letter: Davis to his parents, 13 January 1870; Cambridge, Mass.*)

His disapproval of the behaviour of the anagram players is interesting because it is characteristic of Davis as a young man. His attitude towards any activity, whether study or entertainment, was always serious, and he expected everyone else to behave in the same manner.

At this time Davis was writing to his parents about twice a week, and most of these letters survive. The first hint of his trip to the southern hemisphere came in the middle of February:

I'm twenty! and I wish I was only sixteen – but it's a consolation to think that it is as long from twenty to thirty as from ten to twenty, and that seems a long time – if I can be as happy and fortunate, and learn as much in the next decade as in the last I shall expect very bad things from thirty to forty, as retribution – no I shaln't either.

1860 opened my travelling career, and since then I have been on the go nearly every summer; from Phila. to Medford was the beginning, and Europe and

Colorado are the end of the first period. Colorado and California will probably begin the second period, *unless* I go to S—n H—e, of which mystery I will let you know more as soon as I learn it. (*Letter: Davis to his parents, 15 February 1870; Cambridge, Mass.*)

As his scheme matured, however, it was not without a great deal of heart-searching and Davis clearly decided to go to Cordoba despite distinct reservations on the part of his family and some Harvard associates:

This evening I have been to see Dr. G . . . I gave thy letter, father, and Dr. said there would be no order of superiority among the assistants except what they make themselves, and as to the 'glory', we shall all earn it, and shall all share it – Then I presented what I considered my petition – that I could go for a year and then leave or stay with him as I choose – he modified it by saying that I could leave at any time provided I gave *6 months* notice . . . I don't think it is unreasonably long – and I do think it very kind of him to make the offer in that way – and it raises me in my own opinion! . . . It makes me feel as if it was all right for me to go now . . .

There would be 5 or 6 hours work every night – in spells of $1\frac{1}{2}$ hours each – with a rest or nap between them – and calculation in the day-time . . .

Write your ideas of the new phase soon, and let me tell Dr. G. '*Yes*'.

> Your Very truely,
> Will.

(*Letter: Davis to his parents, 14 March 1870; Cambridge, Mass.*)

Dear Parents,

I saw Dr. Gould on Saturday, and said I would go on the condition I wrote you in my last letter. As your letter is now in Medford, I can't answer it very thoroughly.

I asked Dr. G. about the vessels and he said that Mr. and Mrs. Rock were going by sailing vessel in the first week in June – that any vessel arriving by the first of August would be time enough for me – so I can probably finish up here, and have a week or two with you before going. Garcia, the Arg. Con's. Minister at Washington, says that the 'passage out' is to be paid by a fixed sum of money, and then we can go as much over or under that as we choose or can. Dr. G. has written to find what that sum is, as it has an important bearing on our route. The cost of passage is $150–120 in gold on sailing packet, $300 in gold on steamer (which stops at St. Thomas, Para, Pernambuco, Bahia and Rio J. and then you have to change to a French or English steamer when it comes, and go on to Buenos Ayres – I believe $300 covers the whole trip). The times are, by packet 52 days – by steamer 42 and from England 36. Dr. G. said that on the packet there is more room, larger state rooms, and no change at Rio, but then they don't stop anywhere, which would make voyage seem monotonous. I can't decide wh. line to take till I know how much is allowed for the passage, even tho' thee is willing to pay any extra. I wish Marvine could be one of them – he would go under my conditions and thinks of making an application for a place, and getting recommends from our Profs. to back him.

I saw Prof. Winlock on Wednesday, and asked him what he thought of the trip.

Well he didn't think much of it – I could get as good or better practice here; he would give me situation next year if I wanted one – and I 'could have better instruments here than in Cordova. So I don't think much of his opinion. On Tuesday, I told Shaler, and he said it was *first-rate* – that Dr. G. was a person who had had a number of fights, and kept getting into them, but he thought it was a good chance and advised me to go – he is the only one who has said an unqualified 'go'.

On Thursday, I went to the Tech. and saw Pickering, his class, and laboratory – and had a very pleasant conversation with him about the school. Afterwards I took dinner at Parker's, and then went to the concert – which I had to leave before its end, so as to get out here in time to dress for a concert the Pierian and Glee Club were to give at Jamaica Plains. We rode over in sleighs, had a good audience and hall, and stunning spread after the playing was over. After eating as much ice cream, salad and oysters as possible, we sang for half an hour, and then rode home again.

On Saturday, I went to Medford, and found Richard on his way to town at 1 o'c. He had been doing *down* his article to 12 pages instead of 20; In the P.M. Anna and I sleighed down to Somerville – she has probably written to you about it.

Shaler said on Tuesday this (the expedition) is one of the results of the obs. Some persons would have advised you to go to the Tech. where you could not have had such chances – I told him, then, that the Obs. was instrumental in bringing me here.

Marvine obtained permission from Prof. Winlock to take his class in Surveying up to the Obs. by fours every Wednesday and Friday till they had all been there – and on Friday last they began – and Marv. and I each took two and put them throu' a condensed course of astronomy.

Medford is lonely without Sallie – I am going to celebrate her return by taking a spectroscope over next Saturday – won't they get tired of it ?!?!

Miss Nannie has gone to New Bedford and taken her little dog along. Mr. and Mrs. Frothingham are making a visit here.

I wish you were coming on again next week.

Truely,
Will.

(Letter: Davis to his parents, 21 March 1870; Cambridge, Mass.)

A few days later his father sent him some money, the report of the closing of the Anti-Slavery Society and the programme of some private theatricals. He advised his son to notice how profitable railroad building would be in Argentina and Davis, on 29 March, expressed his inexperience of track-laying but dutifully said he would of course keep his eyes open to its possibilities. He goes on to put in a plea for his college mate Marvine who 'wants thee to ask if there is any chance of his getting a place with Mr. Fulton (?) to survey the Barclay Mines' as he needed the money to pay his expenses at Harvard.

Last Tuesday there was a joint meeting of the Harvard Nat. Hist. Soc. and the Lyceum that Shaler has been getting up – it resulted in their joining – Jopy Cooke

(Prof of Mineralogy and Chemistry in college) is president – Shaler and Ben Watson vice-presidents – I am on a committee to arrange and amend the constitution, with Shaler, Ben Watson, Lincoln and Burgess (junior, a nice fellow and good entomologist) . . . (*Letter: Davis to his parents, 29 March 1870; Cambridge, Mass.*)

On 2 April we learn that Prof. Whitney had been fortunate in getting his money ($73,000) for a survey (presumably in Colorado) but that Davis had no intention of joining him and of backing out from the Gould expedition. There is a note of happiness and tenderness in some of these letters that is rarely met again in his later correspondence. He has been to a dance and talked with many persons:

Pumpelly has offered Marv. a position with him this summer. Pump. – as we call him – has $4000 from the State of Michigan to make a review of Foster & Whitney's Geol. Survey of the Copper Region and wants Marv. to do the description of the dressing of the ores . . . It is a splendid chance for him.

I called on Shaler on Thursday night, expecting to find the rest of the committee . . . but they could not come – so Mr. and Mrs. S. and I spent the evening together – talking of all sorts of things . . .

Saturday morning, Shaler took his first geology tramp of this season – about forty sophomores went with him – over College Hill and down towards Somerville. I left them at the reservoir on top of the hill, after getting into mud up to my ankles, and went to Medford. It was a lovely day, and I found Anna out of doors, getting some boxes ready for geranium cuttings. I helped her till the children and Molly Lord was there and claimed me to go to the river with them. We found G. Wasson just starting out for a row in Richard's boat and we all got in and spent an hour in going up and down the river – six of us in the boat . . . In the P.M. I thrashed R. at billiards . . . and I called on Ned and Mrs. Ned and I saw the baby for the first time.

We all passed Pump's examination satisfactorily but our percents are not yet made public.

I wish mother was going to be in New York next Sunday – Don't thee think she had better come father?

Truely,
Will.

(Letter: Davis to his parents, 4 April 1870; Cambridge, Mass.)

Davis' letters of this period thus show a crowded and cultured life, as well as his close association with his sister Anna's family circle. Socializing with relatives and friends; attendance at lectures on religion, botany and Atlantic cables; visits to a steelworks and to the manufacturers of the telescope to be taken to Cordoba; outings to the Music Hall and to a production of 'Rip Van Winkle'; work at the Observatory and on his mineral collection; study for Pumpelly's course on Economic Geology; a lecture on the spectroscope to the Natural History Society – all these activities, and more, made up his crowded life. He did not neglect attendance at Quaker meetings, including one in New York City:

> Sunday morning I went to meeting with grandmother, and was very glad as she spoke beautifully. I was very much interested. Some of the dresses of the young quakeresses were rather more ornate than I expected to see in a meeting house. I surrendered grandmother to Mr. Haydock after meeting, and walked up 5th Avenue with Mary – and had a jolly time. Here is the dinner table at Haydock's. [Fig. 9] ● mean a plate and its owner and the lines connecting these plates show how the conversation was carried on. The blackness shows the amount of the conversation . . . I think I shall patent the idea. (*Letter: Davis to his parents, 11 April 1870; Cambridge, Mass.*)

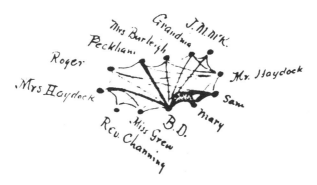

F I G. 9. Seating plan and conversation flow at the Haydocks' dinner party. B. D. denotes 'Billy Davis' and 'Grandma' Lucretia Coffin Mott (Letter: Davis to his parents, 11 April 1870; Cambridge, Mass.)

It seems that his penchant for versification had its origins at this time and that he was already rather intolerant of his inferiors:

> Have I told you about my poetry? – I am class poet! and until the last few weeks never knew I had much poetry in me – since then it has developed rapidly and now I can turn off a verse of four lines – the second and fourth lines rhyming – in quite a respectably short time – The plot of my poem is this – Prologue – describing the early life of the four principal characters – and four chapters – are for each year here. (*Letter: Davis to his parents, 25 March 1870; Cambridge, Mass.*)

> I fell in with Shaler's geolog. party in the Arlington Car. Last year there were a number of nice fellows went with him – now he has a horde of low sophomores who are a disgrace to him. They made enormous fools of themselves in the horse-car, so that I was glad to get out and leave them. (*Letter: Davis to his parents, 26 April 1870; Cambridge, Mass.*)

It seems strange that in spite of his affection for his mother he did not know the date of her birthday (30 March) and in fact admits that the only birthdays he knew were those of Lincoln, May and himself. But he thought his father's birthday gift of an atlas to his mother was 'very nice' and wondered if it were Johnston's or Colton's (*Letter: Davis to his parents, 2 April 1870; Cambridge, Mass.*).

By May his plans to go to Cordoba were already maturing.

Now for Medford – I went there in time for dinner on Saturday and dined with Sally and the children at Anna's and in the p.m. had a most jolly time with Sally getting columbines from the woods and planting them. After the company had gone, we talked about my route to the South and Harry was quite in favor of my going via England as thus I could see our Patty and could keep sending letters from different places at which the vessels touched – which is an important point. It was very pleasant to have the benefactor back again – he gave me some very good advice – and was interesting altogether, on anything about astronomy, so I certainly shan't be at the foot of the class. Hathaway is not at all vulgar but common and very *inoffensive*. Mrs. G. asked if I knew anything of Mrs. Rock, as she was anxious to hear what sort of person she was. Do you know at all? Directly after dinner which was very pleasant, – not all ceremonious – Mrs. G. hurried off to catch a car, and in a minute came running back after a Spanish Grammar she had forgotten – she said she had been studying it a week, and yet could not talk! ... One Monday, I was invited to dine with Dr. G. and meet another assistant, whom I saw, and thought rather scrubby. Mrs. G. was pleasant, and spoke feelingly of her intense desire to call on Anna before leaving and asked particularly about the Lowell R.R. trains. The scrub, Hathaway, was a divinity student not long ago, but got over that and is now working for Dr. G. in some way – getting something ready to be printed I think, he doesn't seem to know.
(*Letter: Davis to his parents, 3 May 1870; Cambridge, Mass.*)

In his reference to Hathaway, one of the other assistants chosen to go to Cordoba, there is a distinct underlining of snobbery. We see it less in the Davis of middle and senior years, although his intolerance of persons with intellects inferior to his own always persisted. The remark on Mrs Gould's slow mastery of Spanish shows that his quick brain and confidence made him contemptuous of the lesser achievements of others.

The master's examination and Davis' arrangements for his departure to Cordoba seem to have run very close. Youthful anticipation of adventures ahead is mingled with an anxiety to enjoy a last contact within the consoling intimacy of the family circle.

I never was so 'drove' before – tonight, tomorrow night and next night, the theatricals came off at Horticultural Hall in Town, and I go to all as orchestra, next week. I have had my examinations changed to Monday, Tuesday and Wednesday – Friday's work being put on Tuesday for me alone, the other three doing as I wrote last week. Besides digging up these examinations, I have all my things to pack and books, minerals, etc., to take to Medford. I don't expect to pass very well, cause I have not time to prepare on everything, but I shall have everything finished by Wednesday afternoon, May 18th, and shall try to start for home that evening if possible. A note from Dr. Gould, yesterday makes me think that perhaps I shall have to spend part of my examination days down at Clark's to see the telescope being put together. He said also that if I intended to stay here after he left (the 24th) he might ask me to take charge of the shipment of whatever was to go from this port, but as I think he knew very well that I was going home as soon as I could, the offer didn't mean anything – but, in the formal letter he sent home – a sort of credential he fixed my salary at $1000 in gold instead of $800! to commence when I sail from the U.S. the passage money is also given – Isn't that fine?

In order to close up here, I shall need some money – $100 for term bill, $241.075 for board from Jan. 4 to May 18 = 135 day = 19·286 weeks – $12.00 for washings. $10 to get home and $15 for extras – such as charity – expenses etc. – total $378 quite considerable, isnt it.

I had a jolly time at Medford and hope to have one more Sunday there before I leave and spend Wednesday afternoon or night there.

I am sorry to write such a short letter but I must study. I shall soon be home, and can stay 2–3 weeks. I hope all the young folks will be out next week. (*Letter: Davis to his parents, 10 May 1870; Cambridge, Mass.*)

His examination results were a good deal better than he forecast.

CHAPTER FOUR

Argentinian and American Interludes

As we have already noticed, Davis during his final student year at Harvard obtained a post under Dr Gould at the new Observatory at Cordoba in the Argentine. An interesting note in a paper by Gould on the discovery of a new variable star *T Coronae* four years before shows not only that Davis had come to his attention previously but that the interest of the latter in astronomy dates from his early years:

> Mr. Wm. M. Davis, Jr., of Philadelphia, saw the star on the evening of May 12 (1866), called the attention of his family and friends to the phenomenon, and noted in his journal that the star was as bright as α *Coronae*. (*Gould, 1866, p. 82*)

At the time of Davis' appointment Dr Benjamin Apthorp Gould was forty-six years old. He had graduated from Harvard in 1844 and had subsequently studied at Berlin and gained a doctorate at Göttingen. On his return to the United States he tutored privately for a short time before working, from 1852 to 1867, with the United States Coast and Geodetic Survey. In 1855 he got his first senior appointment as director of the Dudley Observatory where he was decidedly unlucky because the general business depression of 1857 weakened the Observatory's finances and a hostile newspaper campaign forced some of the trustees to bring charges of 'incompetence, disloyalty and sloth' against him (*Defence of Dr. Gould, 1858*). He was officially exonerated but the opposition persisted and he described the outcome himself:

> I was driven from my dwelling by a hired band of rioters, acting without form or pretence of law – a mere brute force. (*In Comstock, 1924, pp. 159–60*)

Apparently neither his reputation nor his enthusiasm suffered for long. In 1866 he established on the south-west coast of Ireland the famous meteorological observatory at Valentia which is still in existence. Soon after he had the idea of mapping the stars of the low latitude skies. He imagined that the absence of cloud cover would make this easier in Argentina than in the more gloomy atmosphere of north-eastern North America. Through a friendship with the Argentinian Minister in Washington he managed to interest the Argentine Government in his scheme to establish an observatory at Cordoba. The reward for his endeavours was the directorship from 1868 to 1885.

Dr Gould selected four assistants to accompany him to Cordoba, among them William Morris Davis.

Four young men had followed me from the United States to aid in the observations, and arrived in Cordoba at the close of September [Gould apparently refers here to 1870, the year of his own arrival]: Messrs. Miles Rock, John M. Thome, William M. Davis and Clarence L. Hathaway. All of them had received a collegiate education, and were earnest in their desire to contribute to the advancement of the proposed undertaking; but none of them yet possessed special astronomical training or experience. (*Chapter 1, 'Uranometria Argentina': quoted in a letter from R. H. Tschamler to R. J. Chorley, 23 August 1960*)

Davis travelled to Cordoba on the barque *Ella* and the journey between the ports of Portland and Buenos Aires lasted sixty-three days. Some of Davis' time seems to have been passed in close observation of the ship's main features. His son remembered sketch books containing drawings with minute details of the ship's rigging, even down to the various types of knots. We have Davis' description of their arrival and the first weeks at Cordoba.

After staying in B.A. [Buenos Aires] a week, we came here, and enjoyed the ride up the river and over the pampas very much – The latter strongly reminded me of the prairies in our West.

Dr. Gould arrived here about Sept 13 and was very warmly welcomed by the authorities and people and everything is being done to assist and hasten the building of the observatory – It is to be on the plain, $\frac{1}{4}$ mile S.W. of the city, and about eighty feet above the Rio Primero. There will be nothing to interfere with the horizon – The chief difficulty is to obtain water as the ground is very dry – there has been very little rain for the last six months. The time for opening the National Exposition has been postponed till next March, as the ground could not be made ready before – Their present condition tho' very unfinished shows that they will be very attractive next fall. The people here show their appreciation of The Fair by doubling the prices for board and lodging already. Three of us have taken rooms in the house of Dr. Oster, a German, who seems interested in Natural History, as he has various bottles filled with snakes, spiders etc – I dare say he will be able to assist me in collecting specimens for the Museum, but am afraid that even then my contributions will be small.

Dr. Burneister of the Museum in B.A. told me that the sierras near here are all composed of metamorphic rock, and that everything else is tertiary or recent – So that for fossils, I shall have a poor chance. (*Letter: Davis to N. S. Shaler* (?), *9 October 1870; Cordoba, Argentina*)

In the first two years accurate observation of the stars was hampered by the lack of instruments. Readings had to be made with the naked eye or with the assistance of opera glasses. The instruments needed for more accurate records were not received until 1872. Actually, Dr Gould's original assumptions proved incorrect. Cordoba was not a favourable climatic choice; the frequency of cloud cover was little less than that experienced at Boston. This may have persuaded him to establish a weather bureau and incidentally provided Davis with his first introduction to practical meteorology.

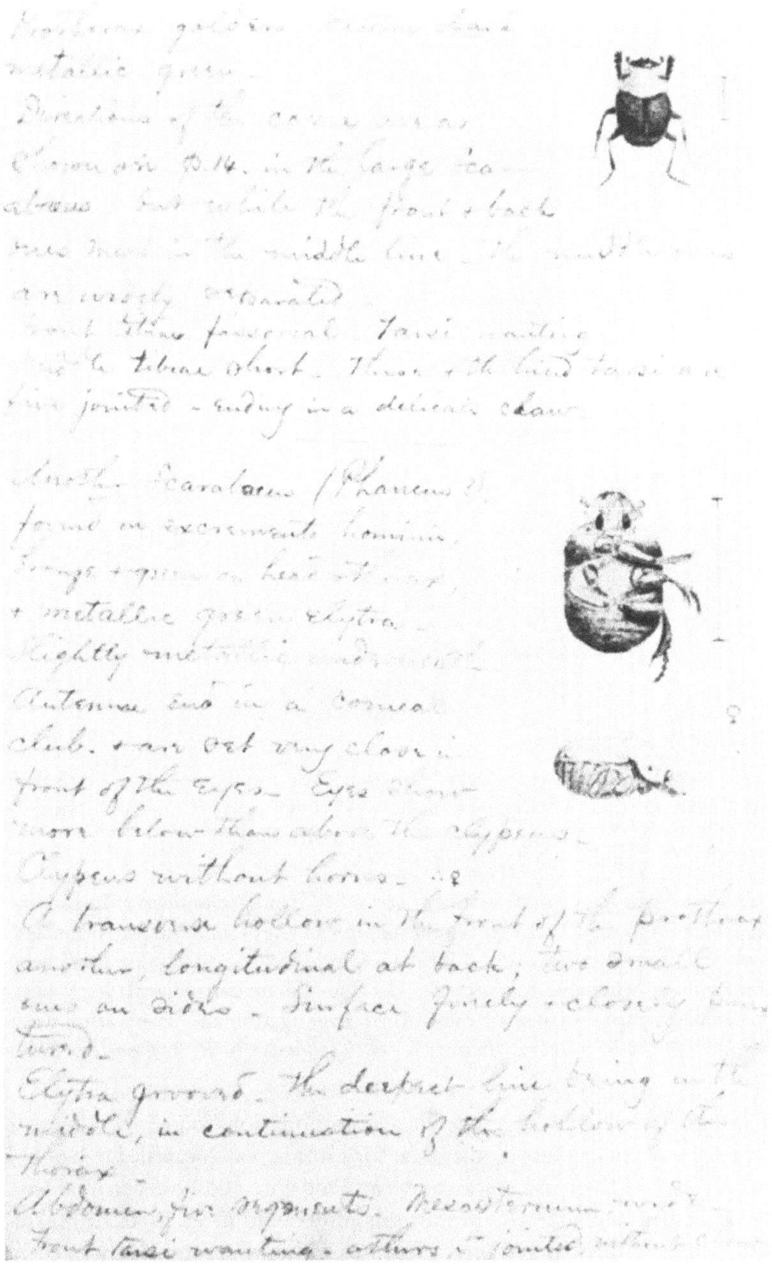

FIG. 10. Page of Davis' entomological notebook kept in Cordoba, 1871 (From *Entomological Notes, Cordoba*, 1871, p. 18) (Courtesy Museum of Comparative Zoology, Harvard University)

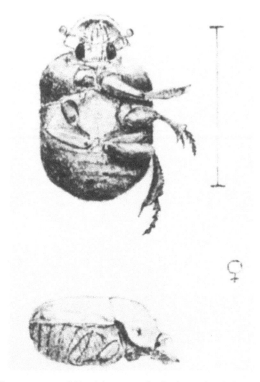

F I G. 11. Detail from a page of Davis' entomological notebook kept in Cordoba, 1871 (From *Entomological Notes, Cordoba*, 1871, p. 18) (Courtesy Museum of Comparative Zoology, Harvard University)

Davis belonged to the staff of the observatory from September 1870 to February 1873 as an assistant to B. Ą. Gould helping in the 'Uranometria Argentina' and after the installation of the Meridian Circle belonging to the staff who did observations with this instrument. He also did meteorological work but on a small scale as at that time meteorological investigation and observation was at its beginning in this country. (*Letter: R. H. Tschamler to R. J. Chorley, 23 August 1960*)

The assistants lived together in accommodation found for them. They moved once or twice before a satisfactory home was located. Dr Gould kept a friendly eye on their interests and he and his wife allowed their residence to serve as a cultural centre where the assistants could meet. Sharp sarcasm was probably one of the less attractive features of his character. He was certainly not easy to get on with. One of his assistants wrote:

He was a difficult master to serve; his methods were often indirect. He did not develop a loyal feeling among us, while I was there, rather the reverse. (*Comstock, 1924, p. 167*)

Davis left Cordoba after two and a half years there and, although we have no direct record of the circumstances surrounding his departure, the following extract from a letter written by his mother is very illuminating:

I am very glad that thee can feel that the whole thing has *not* been a failure. *I* do not feel so at all for there is always much to be learned in so varied an experience as thine has been, and though you may not have accomplished all that you hoped for, you must have done much good work. I am very sorry to find thee has so low an estimate of Dr. G.'s abilities and knowledge. I supposed he was thorough in the line of his profession and that whatever faults of disposition or temperament he had, he was a true scientist. We might have been sure that where there was so much smoke, there was some fire, and it was universal when we spoke of him to others before thy going away, that there was a shake of the head, or shrug of the shoulders, implying more than one would like to say. Aunt Marianna has always pitied Mrs. Gould and thought her marriage very unfortunate – poor soul! Perhaps the only way to keep peace in the family has been for her to be 'completely under his orders'. How glad I shall be when thee is free once again from his annoying, suspicious supervision. I cannot help a feeling of pity for him though for I suppose his heart is really in this work and what will he do if *all* his assistants leave him! However, my pity is quite a secondary feeling for just now I am only too happy to think thee has fixed an earlier time for leaving. I like thy plans very much – with one exception – the month's stay in Rio Janeiro – is not March just the season for yellow-fever there? If there was anything approaching to an epidemic I should be very unwilling for thee to go there – or even to make any stay in passing through Buenos Ayres. Thomas Baldwin called here lately and inquired after thee as he always does – he hopes that when the time comes for thy leaving Cordoba, that thee will cross the Andes, and sail up the western coast to Panama or San Francisco. I told him that that was thy plan when thee left home – but I like the idea of thy going by English steamer to Lisbon much better, as I have before suggested, and I shall be very glad for thee to travel awhile in Europe before settling down to 'teaching' or any other business. I think thee has fairly earned some relaxation. (*Letter: Davis' mother to W. M. Davis, 26 October 1872; Philadelphia*)

At this time mother and son were exchanging a voluminous correspondence and the following extracts from his mother's letters give interesting insights into the family commercial and financial fortunes:

. . . poor Richard is in the depths again over his business! the wool market has declined so much that all his last spring's profits are swept away – all, and more than all, till he is reduced pretty low and very discouraging it is. (*Letter: Davis' mother to W. M. Davis, 26 October 1872; Philadelphia*)

I was surprised to find from thy letter that in thy journey west three years ago, thee visited the copper region. I had forgotten that you went so far to the north. You must have been very near to *our* elephant and I think thee must take a greater interest in what father writes about his venture, from having visited neighboring mines – just now he has two pretty fair nibbles for buying him out – fair enough to keep us on the qui vive every day for news from one or other party, though we know well how many slips there are between the cup and the lip. I shall rejoice when it is *profitably* off his hands. For although it pays handsomely, it is too far off and too great a responsibility to continue to run it ... (*Letter: Davis' mother to W. M. Davis, 26 October 1872; Philadelphia*)

Thee will notice in the newspapers that go to thee by this mail accounts of a very singular horse epidemic prevailing throughout the country. It was so slow in reaching Phil., that we had some hopes of escaping it, but we were not to be so favored and now it is upon us in full blast. The sickness is absolutely universal. It comes on with sneezing and coughing and a profuse discharge of offensive, green matter from the nostril. Our own three horses share the fate, but seem to have the disease in a mild form. There are not many fatal cases. No-one could have realized how dependent we are on this useful animal and what a complete stagnation of business it caused. For some days last week scarcely a vehicle was seen in the streets. No street cars. In this emergency various expedients were resorted to. The dummies were allowed to go through the streets but they were not success-ful. Father got a permit from 'Councils' for a trial in Market St. of that road-engine that Henry is interested in and it proved a capital advertisement of the thing.

This disease has spread through all parts of the United States. It was very severe in Boston and lasted longer than in some other places, and possibly it may have been owing to the scarcity of horses that the late awful fire gained such head-way there. (*Letter: Davis' mother to W. M. Davis, 10 November 1872; Philadelphia*)

The calamity is wide-spread – causing a panic in stocks in other cities and distrust everywhere. Most likely it will come still nearer home to us than ever through our children – for a sale of part of our copper mines was on the eve of being consummated, and it is not at all improbable that the whole thing may fall through, or at least be postponed. It will be a sore disappointment, especially just now, when we may need all our available means to help poor Richard and Anna. Even Henry comes in for his share in the common misfortune, for he had some wool there. If it was insured, and the insurance proves good, it may be no loss and as he seems to have been born under a lucky star, perhaps it may serve him now. Has thee ever found out with your telescopes and your opera-glasses which are the lucky stars? (*Letter: Davis' mother to W. M. Davis, 10 November 1872; Philadelphia*)

A sad letter came from Anna yesterday, confirming our fears that the loss is ruinous to them. She asks to have the letter sent to Henry and it had to go into town this morning. *He* had about 8,000 dollars or pounds of wool in Norwoods store and as yet no-one can tell about insurance. I will write from memory as much of Anna's letter as I can. Mr. Collyer took tea with them on Saturday evng – after which they went 'down town' with him to his lecture. They saw the light of the fire and it was so bright they thought it must be the large lumberyards in East Cambridge. Between 11 and 12 o'clock, word came to them (she does not say how) that the fire was in the neighborhood of Federal and Congress Str. If their horse had been well they would have driven right into town but they felt as though the mischief was already done, so went to bed and tried to sleep, before six they arose, ate a hasty breakfast and Michael drove Richard and Norwood into town, leaving the carriage they tried to make their way down Summer St., by jumping from one large granite boulder to another, although the stones were so hot that it was almost dangerous to attempt to pass over them. After going some distance, fortunately, they saw one of their porters who, having been over the ground, could tell them where Federal St *was the day before* – all traces of *any street* were gone – nothing but a mass of smoking ruins everywhere. After much difficulty they reached their own store – nothing standing but two row posts that had supported the fire-proof. Literally nothing there but heaps of ashes to represent the 250,000 lbs of wool they had left there the day before!!! Their iron-fire-proof safe lay in the midst of the ruins in the cellar with fires blazing all around it. The coal in the cellar seemed to be burning – the heat and smoke were so great that they did not try to find Norwood's store. There was not a wall standing in Federal St. Heart-sick they turned away. The porter told them that everything was in a blaze when he got there in the night and nothing had been saved. Richard walked out to Mr. Coburn's in 'Chester Square' (I think that's the name) and there he found most of their valuable account books and papers, deeds etc. Mr. Coburn had gone into town. Seeing the great light and when he found the fire seemed to be coming towards their store, with the assistance of two of their porters, he carried down the most valuable contents of the fire proof. The carriages from the bazaar opposite were being drawn out into the street for removal. He seized a buggy, put all the papers etc into it, and he and the two men dragged it to his house, nearly three miles distant. Twice they were stopped by ugly-looking people, but two loaded revolvers were potent and they secured their prize. It is a help to Richard to have his books, if anything can be called a help in his condition. He says he is 'irretrievably ruined' but I hope it will not prove so bad. (*Letter: Davis' mother to W. M. Davis, 14 November 1872; Philadelphia*)

In a letter to Shaler at about this time, Davis gives no clue to his private disappointment and dissatisfaction with Dr Gould but he clearly was immersed in his private entomological studies at Cordoba and full of plans to return home:

Since receiving the books you were kind enough to send me, I have many times wished to write and thank you again for them, and to give you, at the same time, something that might be worth reading to the Nat. Hist. Society, which I trust is progressing well – but the consultant application that we have all had to give to our 'Uranometry', has prevented my doing much for myself, in anything outside of astronomy. Now that my stay here is drawing to a close, I will tell you how much Packard's book has helped me – and I will also ask your advice in regard to plans for my route home. For the past year, I have devoted nearly all my Sundays to studying insects, with a view to learn something of their anatomy. I began with a large weevil, about an inch in length, that lives on the back of the 'Algarroba' and have made out most of the leg and head muscles, and have drawings of them. The wings were too difficult to begin with, so I have gone on with leg and head muscles in wingless forms of Copris and Carabus and find the former especially

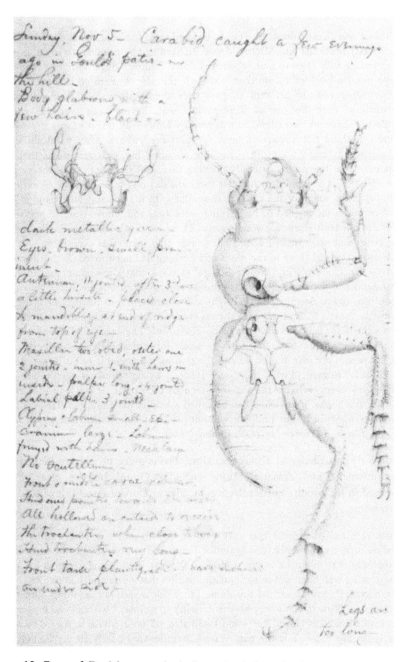

FIG. 12. Page of Davis' entomological notebook kept in Cordoba, 1871 (From *Entomological Notes, Cordoba*, 1871, p. 37) (Courtesy Museum of Comparative Zoology, Harvard University)

interesting. The complex growth parts being most remarkable. In the Carabus, I have followed the major nerves and its larger branches – In this, Packard has been of service in giving me starting points – the outline of insect-anatomy being given in the beginning of his 'Guide' from which I gained a general idea of what I was to look for. But in outdoor works, the 'Guide', and Harris's 'Insects impervious to Vegetation' (which I received from home) have been of great assistance. My collection will be very incomplete, because I can give very little time to hunting up any but the commoner forms – but of those, I intend to have plenty. Cicinde-lidae and Carabidae are rarer than might be expected here – of the first, I have only three or four specimens; various Searabaeids are common – especially the one I suppose to be Copris – black, with *very* large prothorax and front legs – of Buprestids there was an abundance last year, but few, I think, the year before – Elaterids are represented by the Pyrophorus – 'Tuco' it is called here – Several species of the flowers, of the flat-leaved cactus, and of the weevils, I have found but one or two, besides the large one, already mentioned, and another, smaller, with yellow strips, along the wing-covers, it may be *Otio and Lynchus*. Of the 'long-horns', I have some very beautiful species – and this summer shall get some more from the larvae that I have collected. I have perhaps one hundred specimens of butterflies (in paper). They were caught in 70–71 and will be very dry when you get them. They are so much more difficult to keep, that I have neglected them. Of bees, wasps and ants – I have quite a variety – and also a few species of Cynipids and Chalcid – one of the latter I have found as a parasite on the eggs of a larger green Maritas.

The two summers that I have passed here, I have lived in the City, so that a half-mile walk was necessary before getting to any hunting grounds. Now, by the kindness of Dr. Gould, all the assistants are together in a house on the observatory land, on the 'Altos' – and we have only to cross the ditch to be in the wilderness. So I hope to add many specimens in the coming hot months – and about the first of next March, and will send them to you.

I want very much to stop a month at Rio on the way home – to see the country a little and study some of the insects and it is about this that I would ask you – how can I best spend a month here? If Mr. Ben Mann is in Cambridge, he might be able to tell me where to go near Rio; and as Prof. Agassiz has been there I thought that you might have learned something of the place from him, so that you could give me directions about what and where to study – tho for so short a time as a month, it may be difficult to do this. I shall leave Cordoba the first of next March – and get to Rio about the 20th. A letter sent from the States later than Dec. 20 should be addressed to me, care of S. B. Hale & Co., 24 Reconquista, Buenos Aires – otherwise simply to me – Cordoba, Argentina Repub.

I have heard very little of Cambridge news since being here only that the addition to the museum and the memorial hall are about completed, and that many changes are being made in all parts of that University. I shall try to see it before a year is past.

Anything that you care to tell me of Rio will be very serviceable and I shall be extremely obliged for it. (*Letter: Davis to N. S. Shaler, 23 September 1872; Cordoba, Argentina*)

It is clear that his dissatisfaction with employment at Cordoba caused Davis to turn in earnest to his entomological studies, and he filled many field notebooks with detailed drawings and notes [Figs. 10, 11, 12 and 13]. After his return to Philadelphia he joined the American Entomological Society and remained a member for more than forty years (Calvert, 1934, p. 84).

Throughout his life Davis was not given to comments on his close associates though undoubtedly he had his opinions. The decision to leave Cordoba could indicate disappointment prompting him to think wistfully of happier times at Harvard. Nevertheless, as might be expected, his three years in the Argentine made a strong impression on him which lasted well into his later years. During his world tour, four or five years later, there are constant comparisons of scenery and plant life. The lack of remarks on his colleagues suggests that most of his off-duty time was spent alone in examination of the countryside, either on foot or by mule. After he had left Cordoba at the end of his duty he does not seem to have met any of his Argentine companions again except Mrs Rock and her two young daughters in Washington, D.C., in 1884 and forty years later one of these daughters long after she had married.

On the first part of his world tour in 1877, the dry Californian countryside and local Spanish names revived memories of South America which tell us several facts hitherto unrevealed of his life in Latin America. While stationed at Cordoba he was apparently entitled to short leaves during which he visited surrounding districts.

> The shrubs reminded me of Soconcho. You know the Rocks and I had our midsummer holiday of '71 and '72 at Soconcho, in the hills forty miles south of Cordoba, and the stunted growth there and here is very much alike. I wish I had been there with Dr. Hooker and Prof. Gray, for I am no botanist ... (*Tour; 2 October 1877; Yosemite Valley, California*)

Elsewhere in this world-tour correspondence there are descriptions of rides on horseback across the Pampas in search of flowers and insects. There is even a reference to his lodging with a Swede for part of his stay. All his Cordoban lodging places seem to have been with foreigners which was probably inevitable as South America was not then a sphere of United States' influence.

On the way south from Hong Kong to Singapore in 1878, the changing constellations brought back forcibly happy memories of observatory days in Argentina and at Cambridge, Massachusetts.

> After leaving Hong Kong latitude 22 degrees, we rapidly turned the earth around, depressing the North Star, poor thing, till he was nearly lost at Singapore, and raising the southern constellations, bringing up many old friends over the watery horizon, so you won't wonder at my having thought often of Cordoba and our party there in the house on the hill; of Dr. Gould and Mrs. Gould and the poor little girls, lost in the river. Pity I didn't see Thomas last summer. I have had out

F I G. 13. Drawing of a moth made by Davis in Cordoba, 1872 (From *Entomological Notes, Cordoba*, 1872, p. 42) (Courtesy Museum of Comparative Zoology, Harvard University)

the charts he sent me and made tracings of a few 'variables' and their surrounding sky and borrowed Taylor's glass to observe at faint charts in a dim light, picking out the stars up aloft, nearly breaking my neck, leaning back to see them, and ending with 'a b c 2 d = V f = g 2 h'.

I could almost see Rock and Thome and Hathaway by me on the roof, and hear our fierce disputes as to whether a star should be called 'six-nine' or 'seven-nought'. The Cross rises about ten o'clock. Tom is the only one on board who has not seen it before, but he knows it now. Nearly five years since I saw it, on the 'Patagonia' with those young Chilians, going to Europe to study medicine. How different the five years are from my imagination of them. I used to say on that voyage, 'Just let me get home once, and I will never go far away again', and here I am farther away than ever.

It is pleasant to see the southern stars again, they seem very natural, hardly changed at all. So I rejoice in looking at them in the warm evenings on deck, and pointing out my particular friends to Tom and Taylor, I don't think that the

> ... Earthly god fathers of heavenly lights
> That give a name to every fixed star,
> Have no more profit from their shining nights,
> Than they that walk and wot not what they are.

Being somewhat of a Godfather myself, I resent this slur, and could easily show if I choose, how untrue it is. For instance when I showed Le Puppis to the other fellows, it recalled a whole chain of pleasures, and if pleasures aren't profits what are they. The night I called down to Rock, rather excitedly perhaps, 'I've found another variable!' how I skirmished through all our note books to find old observations of the same star, so as to work out a period if possible before saying anything about it to Dr. Gould, but I couldn't do it. It brought up the older times of T Coronae, and the summer at Auburn after it, and Cambridge and the Observatory in the fall. 'No more profit' haven't they. If it wasn't the Bard who said so, I should say Bosh, but being a sort of an honorary member of the Medford Shakespeare Club, and not having been to Stratford, where the weather may be peculiar, I refrain. (*Letter: Davis to his parents; 12 February 1878*)

Davis took advantage of his journey home from Cordoba to combine astronomical sightings, foreign travel and a visit to his relations in England. Instead of going by sea direct to England he traversed the Pampas and then crossed the Andes into Chile, probably by mule. He took sextant observations at intervals along the route and determined longitude by telegraph signals to Cordoba. From Valparaiso he travelled on an English steamer to Liverpool touching at Montevideo, Rio de Janeiro, Bahia, Pernambuco and Lisbon.

Over sixty years later Davis recalls this journey home in a letter to his son:

We also had a letter from Bill written just before he landed; and he said he had been thru the engine room; a trick that his grandfather used to enjoy, in his youth. I recall once being much puzzled when on returning from the furnaces where coal was then shoveld in, one of the men leand down before me and made a chalk mark on the iron floor. But I caught on and gave him a shilling or so; that is what he meant.

The *SS Patagonia* on which I came from Valparaiso to Liverpool in 1873 had a compound or double expansion engine, which interested me greatly; just as a Corliss engine did that I saw at the Centennial Fair in Phila. in 1876 . . .

Dragonflies never caught my fancy. I didn't catch one down at Cordoba, Argentina. The chief memory I have of them dates back to July, 1863, on a walking trip thru middle Pennsylvania. A butterfly was flopping about clumsily; a dragon fly made a wide curving sweep at him; took him in his jaw; lit on a tree trunk, and muncht him up with apparent satisfaction. Never saw it done since . . .

The reference to Cordoba above means that while there I made a large collection of hymenops and coleops, which is now in the Harvard Museum, with Davis, Cordoba, on each pin. Thousands and thousands. The Museum has also some of my old note books with a lot of insects drawn rather nicely. I used to dissect them with a lance made by filing down a knitting needle. Learnd a lot about muscles and nerves. (*Letter: W. M. Davis to E. Mott Davis, 29 July 1933; Pasadena, California*)

The contradiction between Davis' retiring nature and the calm resolution with which he was prepared to undertake a trip half-way round the world and pass through areas which were hardly, if at all, developed is puzzling, but is a feature of his character which commands respect. In England he was met by members of the family. The summer was spent in Europe and included a walking trip through the Alps.

He then returned to Philadelphia where from late 1873 to the spring of 1876 (with a long break in the summer of 1875) he acted as book-keeper in the Barclay Coal Company of which his father was president. Although this meant that he could live at their home near Philadelphia, he soon realized that he had neither aptitude nor liking for this type of occupation. Brigham confirms this:

> Davis ... had no bent toward commercial matters and felt discontented at being thus separated from work in science. Such free time as he had was spent on a collection of hymenoptera, with special attention to the genera *Andrena* and *Halictus*, and he was in the way of describing a large number of new species, when his attention was suddenly turned in another direction. (*Brigham, 1909, p. 6*)

Another mention we have of this period is the following:

> A dinner party this evening is to give me the pleasant society of my former colleague, Wolff, with whom I stayed at Christmas time, of the famous firstbasemen of the Harvard nine in the late 60's. Robert Gould Shaw, brother of George who married your unknown cousin, Emily Mott and lived happily ever after, and Ernest Wright of Germantown by Philadelphia, with whose family I used to play in the interval between South America and Cambridge that is, between 1873 and 1876. (*Letter: Davis to his children, 19 March 1926; Valyermo, California*)

At this point of his life Davis stands silhouetted as a conscientious and earnest young man, somewhat alone, interested mostly in serious activities of the mind, especially scientific observation and analysis – but as yet uncertain which direction to follow. There is no hint at all of what was to come. His parents did not seem to press him to make any decision and he was allowed and possibly encouraged to follow his own inclinations. In this lack of purpose he was surprisingly immature and a person in a position of less financial security would have been obliged to make a choice long before.

It was not until Davis joined Professor Shaler's summer camp in 1875 that the pattern of his future became fixed. Whether this was chance, the result of deliberate choice by Davis, or contrived by Shaler, is not known. Once the choice had been made, however, the framework of the pattern was to be permanent and the remainder of his life merely served to fill in the details.

> You must hav mist the Alleghenies, great favorites of mine, if you left Pittsburgh to the south when crossing Pennsylvania but you took in another favorit, Cumberland Gap, where Virginia, Kentucky and Tennessee all have a corner together. It was there I attended Shaler's summer school of geology in 1875; and there it was that he invited me to return to Harvard as his assistant in field work. That was the beginning of my opportunity, and the Gap is therefore a place of prime importance in my story. (*Letter: W. M. Davis to E. Mott Davis, 2 November 1932; Pasadena, California*)

Fate or deliberation, the choice best suited his temperament and led directly to the highest pinnacle of success. Brigham describes this progression from unpaid assistant to his permanent appointment as an Instructor.

> This came about from his spending the summer vacation of 1875 with Professor Shaler, his former teacher at Harvard, in a geological summer school in Kentucky and Tennessee; at the close of the session he was offered the position of assistant in geology at Harvard, which he accepted the following spring. His first work in teaching was the conduct of field excursions, for which, as he himself avers, he was poorly trained. Teaching of that kind was then, indeed, little developed, and he had to work out its methods for himself. Good practice to this end was found in the summer of 1876, when he went as assistant in Shaler's second summer school across eastern Tennessee into the mountains of North Carolina.
>
> Several subordinate matters may here be mentioned, inasmuch as they made the beginnings of certain lines of work that were carried much further in later years. In the Spring of 1877, Shaler gave a 'normal course' in geology for teachers, and Davis had charge of the laboratory work, for which he made a series of plaster models showing the relation of structure to form, based chiefly on his work in the preceding summers; this was his first effort to interpret land forms. In the same season he visited the Hudson valley in south-eastern New York to study a belt of 'little mountains', which are only hills in altitude but which have a special value for instruction, because they possess typical Appalachian structure and form; features which were illustrated in one of Davis' earliest papers (The Little Mountains east of the Catskills, *Appalachia*, 1882) by a series of block diagrams, a graphic device that has been further developed and frequently and skilfully used in his later work. In 1877, when Shaler led a summer-school party across Massachusetts, Davis, again acting as assistant, first saw the Trias of the Connecticut valley, the subject of his later geological work for a considerable period. The year from September 1877, to September 1878, Davis was occupied with a journey round the world with his cousin T. M. Osborne, of Auburn, N.Y.; assurance having been given him that he might resume work at Harvard on his return. (*Brigham, 1909, p. 6*)

Davis began his third decade with high hopes, but the years between 1870 and 1876 must have been filled with frustration for him. The unhappy relationship with Dr Gould, followed by his lack of interest in the business world in which his father and brother were so successfully at home, must have been intensely frustrating to his creative intellect. In later years when he spoke of the period 1870–3 it was of the stars, plants and insects – he almost never mentioned his employment with his father. Nevertheless, as we have seen, his fortunes were now due for a decisive change.

CHAPTER FIVE

World Tour: The American West

The beginning of Davis' teaching career soon suffered an interruption, and it is a measure of the more leisurely academic atmosphere which prevailed a century ago that in 1877 he was able to take a year off to tour the world. Thomas Mott Osborne, of Auburn, New York, the grandson of a Martha Coffin and some ten years Davis' junior, was of a sickly disposition aggravated by attacks of chronic catarrh. He had been advised to take a world tour for his health before going to University, and his cousin W. M. Davis was chosen to accompany him. Osborne subsequently entered Harvard in 1880, became the Warden of 'Sing Sing' Prison and a noted prison reformer (Chamberlain, 1936). In the light of this future career, it is interesting that Davis referred to him on one occasion during the tour as 'Judge' Osborne (see pp. 104–5).

The tour is not an important part of Davis' geographical career; even the scientific observations in his letters are unremarkable and little if anything in advance of what one would expect from an intelligent and well-trained student. Writing in 1907 Davis remembers this early lack of professional perception.

> I am wondering whether you attempted to describe those cliffs at Boscastle and Clovelly according to the scheme – the only proper scheme, which you will find exposed in a certain textbook which you may have read. It has been too long the fashion with travellers like you to describe such cliffs anyhow instead of somehow. Just so when you got to the Alps at Chamonix, did you see that the *mer de glace* had its end in a very well defined hanging valley whose floor was a thousand feet or more above the main valley in which Chamonix lies? If you did see that, you got ahead of me, for I was there in 1868 and in 1878 and never recognized the real significance of that extraordinary feature until 1899, when I saw it in a photograph. (*Letter: Davis to J. K. Wright, 26 February 1907; Cambridge, Mass.*)

The tour letters are very detailed, deal mainly with travel impressions of things new and curious and give a fascinating insight into the intelligent, but rather opinionated and priggish young man which Davis had become. Scientific explanations appear, but geology has to share pride of place with botany and descriptions of insects. The many letters are valuable for their youthful extravagances about the countries visited and for the conclusions one can draw from them about Davis. In style they are neither really lively nor humorous and, as each topic is treated at length, after a few letters their instructional vein has much the same effect as an over-prolonged public lecture. The attempts at humour are either forced or verging on the pedantic, yet despite their stylistic dullness they have an interest apart from being written by Davis.

The tour started quietly enough with a cross-country rail trip to the west coast. Davis' father accompanied Davis as far as Auburn, New York, but thereafter the cousins were left to fend for themselves; Davis, as the elder, being very much in charge. Visits to friends and relatives along the route had been arranged and the family maintained contact with the two young travellers by sending letters to selected staging points. Other expedients to smooth the way were letters of introduction to persons of importance in the countries to be visited.

Davis' wife-to-be lived at Auburn and he writes approvingly of the impression she made on him.

> Then Nelly Warner, *she* is a walking miracle! Twenty-three (older than Florence) taught school five years! and looks like very sweet sixteen! She teaches at Miss Howard's in Springfield and is successoress in a measure, of May's teacher, Miss Lizzie Simmons, whom she excels in every way, I have no doubt. She was kind enough to take a little walk with me Monday afternoon, when I learned this, and a good deal more of charming interest.
>
> What a good thing to have that 'general susceptibility that preserves us the breed of bachelors'. (*Tour; 11 September 1877; Cleveland to Toledo*)

However his affections were not irrevocably committed and he writes of numerous other young ladies in even warmer terms. Throughout the tour both young men were readily attracted by a pretty face and certainly did not regard themselves as spoken for.

The travellers had carefully prepared themselves for the trip. In order to take full advantage they primed themselves with a detailed knowledge of the places they would stop at.

> Thomas reads 'Venetian Life', Howells – taking time by the fore-lock. I take short snatches at Whitney's 'Yosemite Guide Book' and map between. (*Tour; 11 September 1877; Cleveland to Toledo*)

The first part of the journey, by way of Cleveland, Toledo and Chicago, cannot have made much new impression because nothing is written on it. The prairies excite the first comments.

> For mowing machines and land-surveyors, the great prairies are doubtless fine, but they are no end monotonous. One look out of the window lasts for about twenty miles. Along side of the track, there was generally a muddy borrowpit, for a hundred miles west of the Mississippi. Then a good board fence, and beyond endless cornfields, choked with rag-weed. A sunny house and a few trees in the distance break the straight horizon.

The corn does *not* overtop the houses as it is drawn in railroad companies' hand bills, nor are the villages as neat as they are figured.

If Yankees have colonized this country, they have left their thrift with their rocky hills, but the southern influence may be strong in Missouri, and in Kansas, – we may find an improvement. No wonder the people are flat in so flat a country. Not a hill to suggest an aspiration, not a decent stream to suggest any sparkling emotions. The streams are dirty, muddy, and snaggy, everything that is horrid. (*Tour; 14 September 1877; Kansas City*)

The evening light softened this judgment.

After sunset the flat country takes a pleasanter look, the glare of universal sunshine is gone, the very low hills can then throw shadows, the streams don't seem so yellow and the snags fade out of sight; the landscape is almost attractive. (*Tour; 14 September 1877; Kansas City*)

Kansas City found little favour in any light.

We have just left the Union Depot, a busy place where many unknown roads assemble, where baggage smashes glory in endless trunks, whence three 'Pacific' roads start – the railroad men seem to mistake the plains for the ocean. Kansas City may be very active and enterprising, but it is not the place we care to linger in. I trust that Kersey Coates has made a very large fortune and buys everything money can, if he has to live there, on a dusty, inaccessible hill, with the muddy Missouri on one side and a horizon on the other. (*Tour; 15 September 1877; Kansas City to Pueblo*)

As Davis discovered, the very flatness of the prairies served to accentuate the abruptness of the Rocky Mountain slopes.

After so much flat country, the mountains have an added beauty, and our ride along the plain at their base this afternoon was enjoyable.

It is very strange to see how suddenly they rise from a sea of land to the east, almost precipitously, it seemed as we looked square at them. (*Tour; 15 September 1877; Kansas City to Pueblo*)

The stop at Colorado Springs was probably provoked as much by the presence of relatives as by its scenic attractions. Cast in a mould of eastern standards Davis is at first inclined to be critical of Western manners and environment.

The town seems spread out too much – it is too thin. The dwelling houses are all small and all of wood. The stores are in brick blocks and when we passed them Saturday afternoon, seemed very busy. A number of open, two-pony wagons were in the streets, giving them a lively look.

The people in the streets seemed of a much better class than I have generally seen in Western towns. (*Tour; 16 September 1877; Colorado Springs*)

This general condemnation was not pleasing to his Western relatives and whether on account of their complaints or through an improved understanding of the Western environment his views softened. The few days there were occupied by pony trips through the surrounding mountains and, in Davis' case, sketching. Argentina was the only subtropical country he had visited and Cordoba is frequently compared.

> We started with Tom P. at 8.30 this morning on ponies, had a good ride across the mesa, up into the canon; made sad attempts at sketching it (rather an impertinence it seemed to me, but valuable as practice and saving of description). Back again at 12.30. The country reminds me in many ways of that about Cordoba. There is the general aridity on the hills, washed-out gullies leading down from them to the more fertile valleys, irrigation canals making contour-lines around the slopes and green fields of grass where they are discharged. The water-canals along the streets in the town suggest South America too, and the Spanish names increase the resemblance. (*Tour; 16 September 1877; Colorado Springs*)

One of the climbing trips meant a very early rising.

> We mounted at 2.30. Tom has his ulster, in which he nearly disappeared. As my overcoat is thin, I made up by doubling underclothes and borrowing a blanket for an extra wrap [Fig. 14]. Thermometer 36 degrees! Air clear and still. Stars are brilliant. The moon had set behind a spur from the Peak, when we started, but Mars and Sirius and the Zodiacal Light showed us our path. We had no difficulty

FIG. 14. Drawing by Davis of himself ascending Pike's Peak, Colorado, during his world tour (Letter: Davis to Emma Maria Mellor, 21 September 1877; Georgetown, Colorado)

at all in following it, at first along a lateral moraine, then on the mountain slopes. We passed the timber about four and soon after saw the first streak of dawn.

By the time we were on the summit, it was light enough to see my map, and about fifteen minutes later, at 5.35, the sun made its appearance. Thermometer 26 degrees, sky perfectly clear, wind barely enough to turn the anemometer on the station. The minimum thermometer showed 22 degrees as the lowest for the night. This is about as usual, but the absence of wind was very favorable to us.

You know the blue shadows of the earth in the sky that rises in the east as the sun goes under the western horizon? We had the same effect reversed – the shadows sank in the west as the sun rose. Very soon after it came up, a transparent shadow of our Peak, very distinctly marked in color and form, appeared in the clear air above the western horizon, gradually sinking until it left the sky and fell only on the South Park, directly west of us. The effect was very peculiar.

(*Tour; 19 September 1877; Colorado Springs*)

Detail and beauty registered equally in his memory.

The average view is a ridge (apparently) on each side and rising into a few knobs on its summit lines, and rounded mountains before and behind us. The granite stands out in great broken masses at the top and rolls down the slopes of gravel in great boulders. The dark pines and balsams and firs contrast well with the rocks, but we saw more of them dead than alive, looking like gray toothpicks on a red background. The delicate green of the Aspen is already turning yellow preparatory to its autumnal red; it doesn't add so much to the view as the cone bearers do.

The most interesting point to me was the moraines that dam this valley and hold the lake. They are very distinct and give ample proof of glacier's being here some time ago. Moraines, you know, are the ridges of rubbish that glaciers push along at their front and sides. I have them mapped. (*Tour; 18 September 1877; Pike's Peak, Colorado*)

Mere description by itself he always said was valueless; it ought to be supported by scientific explanation. A striking view lost none of its qualities because you enumerated the geological strata.

I think little points of this kind add to the interest of the view, but *some* persons don't; they think that such an unfeeling, cold – in short – geological 'dissection' of a landscape destroys all its beauty, all its aesthetic charm, all its what do you call 'em. (*Tour; 21 September 1877; Georgetown, Colorado*)

While description probably does benefit from a scientific base it is extremely doubtful if beauty gains from analysis. Reducing the 'Pastoral Symphony' to mathematical formulae is hardly likely to explain why the emotions are affected; it is more likely to dull the emotional appreciation of such a masterpiece. Pursuit of scientific facts certainly decided what he looked at and avoided an aimless examination of the countryside. The railway trip to the mining area near Golden was an example. A large part of Davis' university course had been concentrated on mineralogy and it was not surprising that in several countries he made a special point of visiting mines.

Physiography was not yet a well-defined subject. His major contribution to geography was after all to achieve this.

> Saturday morning we had a fine walk up Griffith's Mountain, a hill east of the town. We followed a miners' trail up its slope for about two hours, and then seeing the top another hour above us and knowing we had to be down by noon to catch our train for Denver, and having a beautiful view where we were, we sat down and absorbed it, and tried hard to get it, in a shadowy way on our drawing paper, but no go. I can't find the lines of expression in a mountain covered with pines, or on a long sandy slope, and I run to shading and come out tenth best.
> (*Tour; 21 September 1877; Georgetown, Colorado*)

The same rail trip took them into canyon country and Davis' knowledge was obviously coloured by Powell's writings.

> In order to see the cañon better, we rode from Idaho, its upper end, early down to Golden, on the little engine, and had an excellent view of the hills. The walls are not by any means vertical, as cañons always suggest after seeing Powell's pictures, but they are steep. (*Tour; 21 September 1877; Georgetown, Colorado*)

The incidents of the journey also produced a store of personal reactions. Perhaps his youth made the expression of them seem sharper but it is interesting how many of these youthful attitudes remained in his middle and later years. There was a cold streak in Davis which had no time for human stupidity or weakness.

> The conductor thinks it is funny to say to us when we try to write, 'Boys remember me in your wills'. Some of the passengers stand on the back platform and look at the vanishing points of the tracks, through cinders and smoke. One old covey is reading 'Kismet' – it is absurd for him to read it, he is too old. What can he remember of the emotions of Youth? (*Tour; 11 September 1877; Cleveland to Toledo*)

His Quaker background predisposed him to expect a cosmic order. He certainly believed that there were reasons for all aspects of creation, that it was the duty of scientists to untangle these and explain the pattern to less perceptive citizens. Only the inventions of man could be illogical. You never find Davis taking the opposite position – criticizing nature and praising man's attempts to alter it. In architecture and other social values he was very conservative; his values were limited to the standards he had grown up with and which were accepted by his contemporaries.

> I suppose it is mean to contrast anything out here with a place as old and conservative and proper as Cambridge, but I was tempted to yesterday, when we passed a 'college', of which a sketch is given below. Anything more foreign to study I can't imagine. A ship at sea is very much preferable. When you see such a place it is easy to understand why western ideas and manners are verging on the crude. (*Tour; 11 September 1877; Cleveland to Toledo*)

This is all the more surprising when you consider how radical he was in his onslaught upon outdated geographical ideas.

Coarseness he disliked in any form, whether it was reflected in a low intelligence or a crudeness of behaviour.

> One of our passengers, with two ladies, sat at the end of the table opposite me. When we others were at our third piece of chicken, or thereabouts, this great Mogul was still unhelped. He cried out, 'Here! bring me some grub, damn it!' audible even around the corner. Neither he nor his ladies blushed, but I trust I did.
> (*Tour; 15 September 1877; Kansas City to Pueblo*)

> I called 'Good Morning, Mr. Black' which brought him to the door, a great big, unshaven fellow, in a suit of red flannel. He made a fire, by which we warmed ourselves, and was polite as far as he knew how, but he was mighty stupid. He had been on the Peak four days, 'only four days' he said, and hadn't yet learned any of the other mountains. He said that the observers who had been there before had let things run down, so that the station needed repairing now. Each one who came there, knowing he would not stay long, let the building and everything about go as it chose, only caring to be comfortable for the time of his stay there, and Black said he was not going to bother himself keeping the place nice, if the others hadn't! Very evidently he hadn't been troubling himself, for the rooms were very dirty and untidy, disgracefully so for a place of the kind, where a kind Government pays the bills, and where the observers have nothing to do except look at the instruments at certain hours, cook their meals and sleep.
>
> But Black was just the stupid kind of a fellow who would let things go to ruin because they weren't his, and because he would soon be sent somewhere else.
> (*Tour; 20 September 1877; Colorado Springs*)

While he criticized the defects in others he was not afraid, on occasion, to admit his own fallibility.

> After thinking this all over, I was somewhat disappointed to find, on looking at the map more carefully and taking bearings with my compass that I had been gazing at the wrong mountain! (*Tour; 19 September 1877; Colorado Springs*)

With such a sensitive nature and the influence of his grandmother still very strong, one could be excused the mistake of thinking his reading might be restricted. Several references show this was not so. Emerson, Thoreau and *Tom Brown's Schooldays* are all mentioned and Mark Twain seems a particularly favourite author. True his grandmother was unlikely to fault any of these on moral grounds but nevertheless it does reveal a temporary lowering of the intellectual temperature. In later years this attraction for lighter reading seems to have been lost.

Another miscellaneous item of information is the dating of the end of his interest in bees.

> The thistles recall a time long ago when I used to catch bees. The bees served a sort of *ad interim* to Astronomy and Geology. My bees are neatly pinned, care-

fully boxed, and I will sell them to the highest bidder. (*Tour; 15 September 1877; Kansas City to Pueblo*)

There is also mention of his activities on shipboard while sailing to Buenos Aires.

The effect was the same that I used to get of the sea, as seen from the deck of the 'Ella', or from her main-top or cross-trees. I think my ship-letter told how I could expand the horizon by climbing the mast. Our eastern horizon this morning was very much smoother than I have seen that of the sea on a stormy day. (*Tour; 19 September 1877; Colorado Springs*)

The matriarchal hand of grandmother Mott still kept in touch with members of the family, however distantly they might be scattered across the United States, and Davis found himself delivering presents and messages at various stops along the route.

I delivered the two packages that grandma sent out, they will probably call forth special acknowledgements. (*Tour; 15 September 1877; Kansas City to Pueblo*)

A revisiting of the Peaks recalled memories of his student trip and we are given one or two new facts.

So of course I looked carefully at the Sawatch Range this morning and thought I could make out the deep ravine we followed up Mt. Yale and the valley Sharples and I crossed to Mt. Harvard, and I walked that long pull up and down over again, and remembered how the darkness came on and Sharples' unhappy experience on the moraine, and our firing a pistol in hopes of hearing an answer from camp, and finally seeing the blaze of the camp-fire. You know we found our party all asleep, and not praying for us, though it was ten o'clock when we came in. (*Tour; 19 September 1877; Colorado Springs*)

We stayed overnight here in '69 and bathed in the hot springs. Brewer, Hoffmann, (Petler) Marvine, Gannett, Bridge, Sam Bowles and I, and the next day rode over the mountains to Golden, there was no railroad through the canon then, and the day after, galloped across the plains to Denver. (*Tour; 21 September 1877; Georgetown, Colorado*)

It was over two of the W.U. wires, made into a loop, closed at San Francisco and open at Cambridge, that the Coast Survey made the interesting experiments in 1869, to determine the velocity of signals used in longitude work. Don't you remember the good times our class used to have, going up to the cosy, to watch the work, and the suppers that Prof. Winlock had spread for the observers, to which he allowed us to go when through. (*Tour; 23 September 1877; Union Pacific R.R.*)

After Colorado Springs the journey was through mountains and the drier regions of the west, where the scenery was new to Davis and often surprised him.

We are nearing Sherman, the highest point on the railroad and where the proper thing to say is, 'Who would think we are over 8000 feet high', for the mountains in the north and south and small knobs of granite near us show we are among the Rocky Mountains; all the country about us is gently rolling and looks better grassed than the plains to the east below. The soil is gravelly, excellent for the bed of the railroad and smooth for the emigrant wagon trail that follows near the track. There are fresh wheel marks on it and I am told that it is still used by the unfortunates who want to go to California and who find it cheaper to pull slowly along over the plains, than to hurry across by rail. They camp by the water tanks, where they can have good water free, and meat can't be very dear, where great herds of cattle and flocks of sheep are grazing unwatched. (*Tour; 23 September 1877; Union Pacific R.R.*)

Lucky as they were to be on a world tour, their parents did not provide them with a lavish expenditure account to draw on: that or they were of naturally frugal habits. There are constant examples of deliberate economy at all points on the trip.

A man was sitting on the cowcatcher of the engine, with a gun on his lap and two rabbits by him. This suggests supper, therefore will I eat an apple. Why only an apple? Why, because we are hard up, for what do you think the fare from Denver to Salt Lake is, a thirty-six hour ride? It is $5, more for a sleeping car berth! So naturally, what I considered an elegant sufficiency until we could cash a check at Salt Lake, is running low. (*Tour; 23 September 1877; Union Pacific R.R.*)

Powell's study of the Colorado Canyon probably formed an important part of Davis' student reading and several references make it clear that he had read it with interest.

We have passed Green River, a breakfast station, where Major Powell started on his boat trip down the Colorado River. Nothing more desolate can be imagined than the surrounding country, dry, dusty, and barren, not a tree in sight for miles, nothing but the everlasting sage. Cattle can live only close along the river bottoms and hardly there unless the fields are irrigated. The hills are all of the mesa kind and all are fast decaying, mostly under air action, I should say, for everything seems too dry for much water to ever fall here. Occasionally a hard rock stands out on the hillside like a signal column, as Pilot's Butte, for example near Green River Station, but more often this is the landscape. (*Tour; 24 September 1877; Union Pacific R.R.*)

King is apparently another geological author he had read.

This lake has dried up; the glaciers are gone from these mountains, and Clarence King, in his Yale address last June, said it is all nonsense to think that the Colorado River cut its canon with its present amount of water. Such striking facts suggested to Prof. Whitney a rather slang title for an article he wrote for the *American Naturalist*, about a year ago. 'Are we drying up?' (*Tour; 26 September 1877; Salt Lake City*)

Salt Lake City was the next main stop. From the moment of arrival Davis seems to be antagonistic to every aspect of Mormon society. Perhaps he had been warned against the rival faith or perhaps polygamy jarred the fine mechanism of his Quaker belief.

> Through the valley we can see the Jordan for ten or twelve miles, not continuously, but only where it runs toward or from us; where it crosses our line of sight, its banks hide the water. The Plain along the river is chequered with fields. We can follow many of the irrigation streams and see where they overflow their banks for their good work. The cultivated fields are easily separated from the Sage Plain by their square shape and color, light yellow, where the harvest is done; green where lucern is growing for a late crop. Cordoba over again.
>
> The city is below us, all laid out in squares, doubtless a good and simple plan to begin on, but not productive of beauty. It is green and white, trees and houses. We recognised the Tabernacle, (it looks like the Roc's egg) the New Temple, Brigham's House, Co-op Store, Theatre, Bank and our Hotel. The railroad to Ogden the only diagonal line we find, and the Lake – It was a deep blue when we saw it from Camp Douglass in the morning. Now it has gone through all its colours and given us a double sunset; it has a very ill-defined, shallow, marshy shore nearest us and the Jordan hardly knows when it has ceased being a river. Between the sedgy islands the water is smooth, giving a clear reflection; out in the lake an unfelt breeze ripples it into blue. I was sorry we shall know it only from a distance. (*Tour; 26 September 1877; Salt Lake City*)

Davis' moral sensitivity varied widely. Hunting he disapproved of, as he did any form of cruelty.

> We scared three lots of antelopes, two of three and one of thirteen. The train was stopped for the baggage-master to shoot at the large herd. I was glad to see the bullets knock up the dust wide of the mark, but how they made the antelopes run! They flew over the ground, generally straight away, but sometimes doubling in a remarkable way. A man in this U.P. train has just told me that grayhounds can run them down – 'great sport'. (*Tour; 23 September 1877; Union Pacific R.R.*)

He never failed to report campaigns for women's suffrage. Whether he had any strong feelings on the inequality of the sexes is not certain but he knew that his parents would be interested, and particularly his grandmother. He was particularly hard on other religions.

> The Desevet National Bank and the 'Zion's Co-operative Mercantile Institution', the 'Coop Store' generally called, are among the best. The latter sells everything; over its front door is the Mormon legend 'Holiness to the Lord', under this, 'Z.C.M.I.' and between them an open eye. Probably the other eye (most things have two) is winking so hard you can't see it. I should like to draw that. (*Tour; 26 September 1877; Salt Lake City*)
>
> Perhaps it is very unfair and unkind, but what can we do but look on every man, woman and child here as a Mormon and rate them accordingly. An unhappy, plain middle-aged woman, we called No. 2, many successoresses being understood. All

the P.G.'s are the *last*, and all the women in black are Mrs. *Young*. Isn't it horrid? When one thinks of – however, you are Gentiles, it is unnecessary to contrast the two conditions. (*Tour; 26 September 1877; Salt Lake City*)

Pretentiousness or luxury were perhaps the strongest anathemas.

Perhaps it is all right for Brigham to have wanted a modern house with modern conveniences, but it is incongruous. He ought to have lived in the pagoda or a mosque or something. (*Tour; 26 September 1877; Salt Lake City*)

He expected very high standards of religious devotees, as he later did of his students, and not even the smallest semblance of human vanity was tolerated. Yet his approach to religion was not that of a zealot. He had no time for doctrine or ritual. Nor did he accept all that was written in the Old Testament. He took the modern view and derived his spiritual satisfaction from the pure essence of faith. He steadfastly believed in the wise benevolence of God and also that everyone, as his Quaker training had taught him, should find his own way to communion with God through moderate conduct and personal reflection. The same severe attitude towards moral failings made him set equally high standards in personal cleanliness and behaviour. Each new race he sees is scrutinized in relation to his own standards.

Indians were more numerous than we had seen them before. Dismal creatures, dressed in a half christian way, some painted, the men with blankets, toga fashion; most of the women bareheaded, some with papooses in the real picture-book style, swaddled in a sort of board cradle, and hung from the mother's forehead and shoulders. Miserable, dull stupid faces. Most thoroughly unromantic and unideal. A frequent gesture of hand, searching through long, dusty, matted black hair, suggests a forcible and appropriate adjective. Some of them rode on the platform of the baggage car from one station to another, free, I think. At the station, the women came back, squatted on the platform under our windows and looked 'beg'. The men stood around. (*Tour; 29 September 1877; Virginia City, Nevada*)

Begging seems to be a habit which he regarded as especially sinful. While he abhorred slavery and subject status, his attitude was still aristocratic. He had at this time little glimmering of the modern belief that all men are socially equal and was far from being a social reformer.

The newsboy on a Southern Pacific train, advertising in the generous, ridiculous way of the country, throws peanuts into the laps of the passengers, and even that little pleasantly, I know some of them feel like punching his head, but when he passed a sleeping Chinaman, he threw nuts into his face and grinned.

The case of the Central Pacific R.R. is well-known. If Chinamen could not have been employed at low wages, the road would not have been built. They were contracted for in China everything they were to have being stipulated, and were brought over by the thousand. At one time, 12000 of them were at work on the line. They are willing to live very moderately, and work for a small profit, so of course the labor-unions, who are regardless of the Cent, despise them. (*Tour; 17 October 1877; San Francisco*)

The social needs of the poor had not yet struck him as being a problem capable and worthy of solution. Writing as a much older man, on the occasion of his South African tour, he showed recognition of the problems of subject or underprivileged peoples but still without any firm notion of how to cure them.

From Salt Lake they pursued their trip west through the regions of Nevada and California. The disfigurement of the mountain scenery by the mine workings did not appeal to Davis.

> After a proper distance we left the river and mounted the sides of its ravine, with many a squirm, and started across the barren slopes to reach the steady up grade line we could see winding around hillsides far ahead of us. Barren fields were bad enough, but barren slopes seem worse; they were better before the miners stripped them of what little wood they used to have. Now the river hides itself from them in a deep canon; the clouds hurry their shadows away as if ordered to give the hot sand no respite from a glaring sun, and all is arid and 'orrid.
>
> We mounted a little above Gold Hill, turned a corner and there was Virginia City, as if caught in the act of sliding down hill. Garrets in one street look into sub-cellars on the next above. Fortunately we had seen Gold Hill, and an acute, abrupt surprise on beholding Virginia was spared us. (*Tour; 29 September 1877; Virginia City, Nevada*)

Deforestation is not recognized as a problem but perhaps this is unsurprising as it was not recognized by most of his countrymen until many years later.

> You have heard of the wholesale destruction of forests in the Sierras – we saw its effect. Almost a whole mountain front without a tree. A pity, but a necessary consequence of developing a mineral region. Most of the wood goes to Virginia for the engines and the mines. The railroad was as busy carrying it on Sunday as on the day before. The day isn't carefully 'observed' out here. (*Tour; 30 September 1877; Reno, Nevada*)

California might have made a more favourable impression had it not the misfortune to be suffering an exceptional drought.

> First we crossed the plain, which makes the bottom of the valley of California. On seeing it yesterday, I was very much surprised at its absolute flatness and complete dryness. I had expected dry fields, but not dreadfully parched and barren ones, and the plain appearance was news to me for I looked here, just as along the Arkansas, for something more like what is generally meant by 'valleys', and not dead level. There are not trees of any size about Merced. Yesterday, between Sacramento and the foot of the mountains, we saw plenty of the oaks of one sort and another, that grow in groups or singly as in a park, giving an effect which must be very pleasing in wetter weather; but this morning, except some planted trees, cottonwoods and eucalyptus, there were none in sight when we set out.
>
> Of course, it and the winter just past are unprecedented in dryness; '64 is mentioned as having been dry, but this year, people are sort of scared and don't know what will come of it. This is bad for us and will leave an unpleasant memory

of the state, for it is frightfully dry and dusty. We ought to have had a view of the mountains both last night and this morning, but they were quite invisible in the smokey air. (*Tour; 2 October 1877; Yosemite Valley, California*)

Merced was chosen as a convenient stopping place, being reached on 2 October 1877. The sailing of the steamer for Japan had been postponed for a few days and this allowed time for a few extra trips. The one they enjoyed most was to the Yosemite Valley. Preparations for this brought the first contact with some English travellers.

We brushed off our coats as well as we could on arriving, after the very dusty ride, and cooled our sandy faces in a tin-basin of cold water, and then joined the other travelers at table. They were all on their way out – four young Englishmen, a New Hampshire editor and a native lady who is very proud of having wintered in the Valley. There was a dead silence when Tom and I sat down, a striking illustration of the effect of American air on British manners. When it became painful, I ventured a mild question, which the Editor answered, and gradually a little talk was established. (*Tour; 2 October 1877; Yosemite Valley, California*)

The Yosemite trip was also the cause of some interesting geological reflections.

Its form is entirely unlike that of any water-cut valley, as that of the Merced, just below, where the sides are sloping at most 30 and 35 degrees and where loose debris covers a third or more of the slope. The steep walls of the valley have none of the rounded form given by water. Vertical faces of rock meet each other at sharp projecting or reentering angles.

The sides of the Valley are not at all grooved as glaciers would have left them – as the rocks in the high Sierras are – and there is no trace of moraines at the lower end, where the melting glaciers would have piled up their chips.

There is no trace of folded-in structure in the granite, and there is no similarity between the Yosemite and the down-folded valleys of the Alleghenies.

If the valley had been split open and pushed apart, the sides should agree in shape so as to fit if put together again, and they don't in the least.

But if you just suppose that a great slice of rock has dropped from its original position directly down into some abyss, then everything, most everything, can be explained. The dotted vertical lines in the section show the size of the slice 'most a mile thick'. With this supposition, precipices, with vertical lines or grooves in their faces, steep, solid rock slopes below precipices, steps showing several lines of breakage, and the very small amount of talus in the valley, all explain one another.

But how – why did the slice drop down? If the mountains here were in a state of compression, then the slice must have had this shape in order to be forced down. There are no such over-hanging cliffs as this requires, but a tension-strain will suit better, since many faces of rock are steep, but not vertical. (*Tour; 5 October 1877; Yosemite Valley, California*)

The careful examination of the separate features and contrasting explanations demonstrates a thoroughly analytical mind and affords a good example of how he would approach a lecture topic. The same perception of unusual details and spirit of inquiry, the hallmarks of a true scientist, are exhibited also in the following description of the chaparral, the characteristic plant-growth of those parts of California with a dry Mediterranean type of climate.

The animal intelligence of a native woodpecker momentarily excites his curiosity.

Of course you have heard of California 'Chaparral', a bush with small leaves, and sharp twigs set at right angles with the branches; – it makes dense impassable thickets on the Coast Range; here it grows less freely, but close enough. Sheep will eat its lower leaves; goats stand up and nibble from the higher stems. Chamiso, as Whitney has it in his Yosemite Guide Book, or Chamisol, as I hear it called, is another shrub, more worthless than the Chaparral; nothing will eat its fine narrow leaves; the bush looks like a worn out green feather duster. Manzanita, the last of the common bushes, has oval leaves that stand with their edges up and down on the dark red stem, a peculiar effect. These names, you see, show the early Spanish settlement of the state. (*Tour; 8 October 1877; Yosemite Valley, California*)

I haven't told you about the woodpeckers, have I? There is a black fellow with a red crown on his head, and white marks on his wings and back, that flies with a run and a jump; he was very common on the road in here, and somehow he has learned a very curious trick, or inherited a strange habit, or is provided with a marvellous instinct – as you like it – of drilling holes in the thick bark of the pines, and plugging up each one with an acorn. We saw some large trees peppered all over with these little store-rooms this way for fifty or sixty feet of large trunk, and many of the rooms stored. You may call the holes store-rooms or trunk-rooms, just a'cor'n' as you like it. In winter, when the fields are white, he takes them out for his delight, but he doesn't eat the acorns – no, only the grubs that have grown comfortably stout and white inside of them. Sometimes the jays or the squirrels get ahead of him and steal his provisions. How do you suppose the pert little peckers learned to do this? (*Tour; 5 October 1877; Yosemite Valley, California*)

The heights attracted the companions irresistibly and they made several scrambling explorations, some quite adventurous.

We saw nearly all of our trail as far as Glacier Point when we were on Eagle Point the other day, it looks quite Alpine, with its sharp zig-zags up green slopes, disappearing around a corner at the top; but we knew that the green was of bushes not of grass as in charming Switzerland. The absence of high pastures, of the rich grass that one sees in the Alps, is noticeable here, more so at this dry time of the year than in the spring, I suppose, but even when the snows are melting the grass is sparse, though enough to support cattle and sheep. (*Tour; 8 October 1877; Yosemite Valley, California*)

(On Half Dome) I reached the rope first and began walking up, foot over foot, hand over hand, astride the line, and went on for about eight hundred feet, to where I could walk the remaining hundred feet to the summit. By no means all at once. I stopped often, leaned on a ledge, or resting on a bolt, panting fit to blow up, going on again when my breath came back and my arms seemed rested. Looking back at Tom and Anderson, I gained some idea of where I was; they were coming up over the bulge in the Dome like spiders finding their way home along a web. No part of the base of the Dome could be seen; it was hidden underneath us. As I went on, the others were left out of sight, on the steeper slope, and I came to the end of the rope, laid down flat on my back and puffed like a house afire. It was the most novel and suigeneris experience of its kind that I have yet had, and extremely enjoyable. It took twenty minutes for me to get from the foot of the rope to the flag pole on the summit. (*Tour; 10 October 1877; Yosemite Valley, California*)

It will probably have been noted that Davis climbed on his own. On more than one occasion during the tour he showed an almost unreasonable determination to succeed alone. Whether this was all part of the need to prove himself exceptional it is difficult to be certain; consciousness of his smallness may have contributed. There was undoubtedly a strong competitive streak in him.

After I had questioned him on the trees, where he was all at sea, he asked about a string of fellows from Pennsylvania, and I was equally ignorant, but with better excuse I think. He didn't know anything about the Valley, had no guide and no book, so I taught him all the names I knew. We had something of a race up hill too, not avowedly a race, but neither one would stop and we made one pull from Union Point to Glacier, forty-five minutes, and most of it steep. Tom came along about fifteen minutes later. (*Tour; 8 October 1877; Yosemite Valley, California*)

They arrived in San Francisco on 17 October. From there they made one or two excursions, including a day spent inspecting a mercury well and mine. Their active disapproval of intemperance and absence of intellect inevitably produced the occasional difficulty.

The only 'Gentlemen's room' that we could find, was the bar room of the Station Hotel, where we sat for some time amid orders for drinks and loud and stupid laughing. I walked the platform awhile and the clock stood still – nothing to do. I made up to a good looking old man and said 'Boo!' to him, he answered pleasantly, walked into the ladies waiting room and went to sleep. (*Tour; 30 September 1877; Reno, Nevada*)

It took very little to discountenance Davis. He was very conscious of his personal convenience and reacted quickly against any move which seemed likely to jeopardize it. Also while very prone to emphasize the importance of his social position he could be conversely scornful of the status of others.

At Lathrop, on asking for tickets, we encountered the nuisance of coin currency. When we presented greenbacks, the clerk very rudely told us that the tickets had to be paid for in coin and acted like a fool generally. After we changed our money and paid him, I had the satisfaction of cooling him down . . . (*Tour; 30 September 1877; Reno, Nevada*)

There were moments when he became positively priggish.

At last the stage came, driven by George, a very popular colored driver, who has been on the line over ten years. For passengers he had three girls, a San Franciscan going home, a Yosemitan going with her and a Big Tree Stationer going back after a ball in the Valley the evening before. Tom and I took the back seat and listened to the unceasing talk of these belles, for so they made themselves out, and such talk! Stages, you know, are the best places in the world for striking up acquaintances with fellow-travellers, and we didn't want to be lacking in politeness or sociability, but until one o'clock we sat without saying boo! – except when the girls on the middle seat got out to walk. What could we do! Speaking for myself, when I prepared a nice little ice-breaking speech, one of the girls would probably call out 'Hi –' in imitation of a beau of the night before, 'come lets take a drink!' at which all three would laugh heartily, and George would kindly join in the chorus, adding that 'Ted' the beau, 'was pretty full last night'. At which my incipient stream of conversation froze solid and we remained snobbishly silent. As if to increase the coldness, Tom read me several letters from home and elsewhere, at which the contrast between what we had seen and what we saw was so violently shown us, that we might as well have been dumb. (*Tour; 11 October 1877; San Francisco*)

Yet the almost savage disdain with which he regarded some of his fellow-humans could quickly change to extreme consideration.

As the Grandma in the stage seemed tired of carrying the baby, I ventured to ask, if I might hold him to relieve her, and indeed I might, she was most tired to death. So if you had seen me anywhere along eight miles the other side of Snelling, you would have wondered why I was carrying a roll of blankets so carefully in my lap. The poor little boy had had the whooping-cough in San Francisco, and was coming out to his grandpa's ranch to pick up a bit. He fell fast asleep in my arms.
(*Tour; 2 October 1877; Yosemite Valley, California*)

The truth lay in an extreme sensitivity, which took forms which could be good or bad. From what we are told Davis was apparently acting as tutor to his cousin and not just as a companion. By temperament he was ideally suited to be a teacher. He had a tireless desire to improve others and was constantly on his guard lest he should express thoughts that lacked the maximum precision.

When I write winged I think again of Col. Adney and Cumberland Gap in '75. He sat on me for making it one syllable and an adjective. We were walking up hill at the time and he did it very neatly – and this calls to mind the system of punch-corrections that Tom and I have instituted. If one of us makes a mistake, and the other one corrects it, he may intensify the correction with a thump; but if the speaker corrects himself, he may thump his companion for negligence. (*Tour; 5 October 1877; Yosemite Valley, California*)

F I G. 15. Edward Morris Davis (From a Philadelphia newspaper, 27 September 1879)

He certainly expected everyone else to acquire these attributes, and was more than a little impertinent to the Bailey family who kept a lunch house in Yosemite and greeted them 'as if they knew all about our plans'.

> And in speaking of Mt. Dana which is roofed with slate, Mr. Bailey said it was all granite, his sister had been on it and had seen the granite in blocks. I may have been a little impertinent at this, for I wanted to know what they called granite, was it quartz, feldspar and mica, or would syenyte, gneiss, mica schist or diorite come under the name. When a fellow tells you wrong distances to places and wrong names for mountains among which he lives, you can't believe his knowledge of granite will be very exact. I don't see why a man who has *some* brains and much leisure, and knows how to read and write, doesn't learn something exact about his surroundings especially when they are so attractive as in this part of the country. (*Tour; 8 October 1877; Yosemite Valley, California*)

Davis' Western tour which ended at San Francisco revealed other facets of himself and the country. There are interesting sidelights in the ghost mining-towns and on the growing prosperity of some of the forty-niners who had taken up farming. In the exciting mountain setting, geology naturally held a high place in Davis' thoughts.

> Besides all these pushings and pullings, the rocks had a hard time of it long before they took their present shape, as the intrusive dikes on El Capitan and the veins on many of the cliffs show. I wish I had a month to work it all out. (*Tour; 6 October 1877; Yosemite Valley, California*)

Perhaps the most amazing omission in the letters is any mention, other than the one warm reference, to his future wife. Apparently she had not yet achieved the position of firm favourite, a rating held at this time by Miss Emma Mellor, the recipient of 'the bear letter' quoted in Chapter 8.

If we had known this morning that the Emma mine was at the head of Little Cottonwood Cañon, I think we would have risked the dullness of the snow-sheds and gone up to Alta. If it had been my Emma mine, I should have gone up on foot if necessary. (*Letter: Davis to Emma Maria Mellor, 21 September 1877; Georgetown, Colorado*)

We know almost nothing about Miss Mellor and can only conjecture that she might have been a young relative of the wife of Davis' elder brother. We are equally ignorant whether she discarded Davis or whether he found she did not match his intellectual requirements. While almost improvident with information on every other topic Davis is completely silent on affairs of the heart.

CHAPTER SIX

World Tour: Japan and China

Davis and Tom Osborne started their steamer journey across the Pacific from San Francisco to Yokohama on 23 October 1877. Steamer is perhaps a misleading description as ships were going through a transformation. While owners were equipping vessels with the latest means of propulsion, captains were still not prepared to rely on engine power alone and in most cases the traditional sails were retained and used. Like many travellers Davis was apprehensive of the effect of wave motion. The first day out of San Francisco he tells us he was seasick but after that initiation both he and his cousin were unaffected until much later during a stormy passage to Canton, and for a few hours off Aden.

The average daily itinerary was dull but undoubtedly productive.

> Our programme for the day is, reading in the morning, reading and writing in the afternoon, and writing and talking in the evening. We read on the hurricane desk and write in the smoking room, where I am now. It is hardly used by the other passengers, but the purser and Freight Clerk and Watchman sometimes talk too loud or too much for our comfort. (*Tour; 8 November 1877; 'S.S. Belgic'*)

They deliberately mixed very little with the other passengers and their remarks on their fellow travellers are distinctly adverse.

> Our passenger list is not an exciting one. There's the Missionary, who must have been called very loud to be awakened to his profession, and his wife, who deserves a better lot, judging by a few sights of her. The Lone Lady, just the kind we don't like, and yet a just heaven has put her place at table between the Doctor and me; she is still very sick and I have wicked hopes of continued rolling.
>
> Two Englishmen returning to their 'houses' in Hong Kong and Calcutta, pleasant enough, but very British. Mr. Salter, a New York Shanghai man, a big blower and generally undesireable; and a young fellow, as yet unnamed, going out to Yokohama to keep books for an American firm ... Not very attractive for twenty days close companionship. Indeed I think Tom and I have such good friends at home, that we are a little hypercritical now, and are apt to look down on fellows who so soon choose such miserable subjects for conversation as our shipmates do. But we are hardly acquainted and mustn't judge yet. (*Tour; 26 October 1877; 'S.S. Belgic'*)

The bigger Englishman is more knowing, but he also looks up when we look down and east when we look west. He is greatly impressed with his ability and position. He told us the other day that the British Quarterly Review was well worth reading, it had articles by all the leading men of England, and shortly afterward added that he had got an article on Indian famines into it. He tells us most gratuitously of his 'houses' in Calcutta that bought $3,000,000 of American Bonds when they were worth only 40%; of the £1,000,000 of Cashmir shawls they send home and to Paris every year; of his punka wallah, the unhappy Bengalee who has to fan him daily; of the other one, who stirs a nightly breeze. The latter, he says, sometimes falls asleep, on which the Nabob awakes, takes his riding-whip, which is conveniently placed by his bedside and gives the sleepy Zephyr a cut. It makes no difference whether all this is true or not, it shows just as well what sort of a fellow it is who talks it. He knows a good deal, but he is so insufferably arrogant that we don't care to learn from him.

The Unknown, going to clerk in Yokohama, is a mere cypher. He says nothing at all and can't be made to. Judging by his manners with the Chinese waiters – he snubs them unmercifully – I think he must have just come out of a restaurant.

The Lone Lady of whom we know nothing but that she is called Mrs. Price, is fairly good mannered, but not especially delicate in her perception. She and the Doctor keep up a stupid chaffing at meals, into which Tom and I are dragged.

(Tour; 4 November 1877; 'S.S. Belgic')

The only incident that made a strong impression was the discovery of some stowaways.

On the third or fourth day out, the search for stowaways was made. The Purser and Freight Agent, with a lot of white tickets, stood at the after hatchway. The ship's interpreter stood at the bottom of the steps leading out of it. Then a couple of officers and quartermasters went down the forward hatchway, leaving a guard at the top, and drove the flock aft. They came crowding up by the Purser, who had his tame elephant on hand, to let them out one at a time. Their yellow tickets were taken, the number read off and written by the Freight Agent on a white ticket, which was given in exchange for the yellow, and then they were passed along to the stern. A few came up who were sent forward instead, this meant they were working their passage. Some poor, miserable sick men staggered out, and looked round quite dazed, steadying themselves by the railing. A quartermaster stood ready to lead them off to where they might sit down on the soft deck. Occasionally an absent-minded fellow surrendered his ticket and without waiting for the other, walked off. This would always raise a laugh among those who had been through the mill and stood near by looking on, but he couldn't get far before a tame elephant would seize him with his trunk and lead him back. I saw nearly all of the four hundred come back on deck, and not more than twenty had any approach to good looks; the rest were positively, actively ugly, and some gave evidence that Lord Chesterfield was wrong when he said a man needn't be hideous, but perhaps he only meant Englishmen.

The result of the search below was the finding of the four stowaways, who as I said, were ironed and taken up on the bridge in disgrace, a great crowd of the honest ones looking up at them. They stood meekly together, and looked as if they expected to be thrown overboard next. The ship's interpreter was called up to translate the Captain's verdict. 'You stay there till you are ready to pay your fifty dollars!' The oldest of the four came down first he had the money in his pocket. They were all down before dark. The money had been found by themselves or by their friends. (*Tour; 8 November 1877; 'S.S. Belgic'*)

An interesting letter written on board to Emma Maria Mellor exhibits Davis in a self-consciously affectionate mood:

Here are some specimens on Inspiration, Emma dear, which I trust thy kind sympathy will appreciate. They are the 'originals' of the illustrations to a poem which I have written to my darling little Edna – and show what one goes throu' in writing on board ship. They remind me forcibly of the Cartoons of the Old Masters that one sees framed under glass, in the Louvre. As in those admirable studies there is an evident change in idea often shown between the first and the last of a series. In no. 1 and no. 2 [Fig. 16] the personality of the scribe differs strongly from that in the others. A premonition of the change is shown in the study of a head in the northwest corner of no. 1. The pigtail in nos. 3 to 7 completes the realization of the new idea. Another point is to be noticed. The artist had trouble in making his faces look alike – so he posed his figure in a way that only the back of the head wd. show, as soon as this difficulty was discovered. The prominence of Chinese characteristics in the sketches should also be re-marked – it is of course the result of seeing the Celestials on board ship. The novelty and force of this new experience causes a variation from the artist's ordinary methods. I trust thy tender interest in him will prompt a generous indulgence in thy criticism of his efforts.

Would thee like to hear the poem? – A still greater indulgence will be necessary in that case for I don't set up to be as much of a poet as of an artist, which would make my pretention in that direction exceeding small, wouldn't it? There is perhaps another reason why I shouldn't write thee the poem – for it would be violating a custom of great antiquity. It has long been usual, when one young man courts *two* (or more) young women, for him to conceal the duplex (or manifold) fact from them – and to a persuade each that she only is his 'object'. But I confess I don't feel constrained to observe this time-honoured deceit and besides, it would be rather late to begin now, wouldn't it, when I have abundantly confessed to each of you the power that the charms of the other have over me. Also, as both affairs are utterly hopeless, why should I not be candid! Again, as Edna showed no signs of jealousy, may I not hope thee will be as generously lenient and pardon thy 'Mr. Will'. There. I wish I hadn't said 'pardon', for, in bestowing on my dear Miss Edna a large share of the large remainder of my heart just as I did with thee before, I feel guilty of no crime. Why, it's just like Arithmetic – the more 'sums' you do, the better you can do them. The more charming little girls you like the better you can like them. Practise is everything, isn't it? Well – now for the poem – since the reasons against writing it come to naught ... If I get it crooked lay that to the pitching of the ship. This is our roughest day and one might as well write in a swing ... Interrupted ... Thomas, the coloured steward, comes in and says 'Well, I declare, here you are again, Mr. Davis – always at it'! ... I am writing in the smoking room, thee sees, and Thomas comes in to be sociable and talk provisions with the Purser. 'Always at work, Mr. Davis!' says he. 'Oh no, Thomas,

FIG. 16. Drawings by Davis illustrating the difficulties of letter writing during his crossing of the Pacific on his world tour (Letter: Davis to Emma Maria Mellor, 10 November 1877; S.S. *Belgic*, Pacific Ocean)

this isn't work,' I answer, 'this is play' . . . He comes along and sees the sketches . . . 'My eyes, jess look! . . . what *are* all those pictures, Mr. Davis . . . may I look at 'em?' Of course he may. After he is throu' I say 'Do you think *she* will be pleased Thomas, when she sees them?' 'Well, guess so,' he replies . . . (he makes lots of mistakes, doesn't he!) 'but that depends', he continues 'on how much she likes *you*. If she likes you very much, I dare say she'll be pleased – for that's the way with 'em . . . when they like a fellow, they're perfectly satisfied with anything he does!' . . . 'Well, old man, you are complimentary, now, aren't you . . . *anything* . . . hm –

I like that! . . . call my drawings *anything* . . . Catch me showing you any more of them' . . . Did thee ever here such impertinence, Emma! . . . I hope *thee* will show more discrimination and better judgment and taste, than this wicked old Cook does . . . To return to the poem . . .

Please remember, in reading it, that it was addressed to a young lady who has only just passed her seven and halfth birthday and its style and sentiment are adapted to that uneven age . . .

> Let the Band play! . . .
> 'Old man thought he'd better
> 'Write Edna a le . . .'

My jiminy! What a pitch *that* was, what a sea we shipped. Just hear the water swashing down to the lee-scuppers. See the Chinamen scampering away, getting under shelter – they hate to be wet – it takes all the starch out of their clothes, and out of them too, I should think, by the way they yell . . . To come back to our muttons . . . The poem, the poem!

> 'Old man thou . . .'

no need of wasting ink on that verse . . . thee can easily fill it out from above and then . . . (Remainder missing). (*Letter: Davis to Emma Maria Mellor, 10 November 1877; 'S.S. Belgic', Pacific Ocean*)

TOUR IN JAPAN

Yokohama was reached on 13 November. The people, their faces, size and clothing are the points that catch his immediate attention.

> The women and children grade into each other in looks and heights so that when you see one with a baby on her back, it is hard to say whether they are mother and daughter, or sisters (supposing that the baby is a girl). Their complexion is much brighter than the men's and their arrangements of hair more intricate. A majority of them are pretty, not like any pretty person at home, but according to quite another standard, of which photographs will give you specimens. The little girls and the babies they carry come near being of a size. They are wrapped in one shawl, and the one who has her feet on the ground jogs about with a sort of stutter between steps that lulls the other into quiet. (*Tour; 14 November 1877; Yokohama*)

Soon his interest changes to the countryside, for generally speaking Davis was more attentive to facts than persons.

> It is on a larger scale, much like what we saw yesterday (ages ago) in our morning walk from the hotel. Level-topped hills of clay and gravel, cut down in broad flat-bottomed valleys, with irregular branches coming in on all sides. Our narrow road would lead us to the head of a valley over a divide, generally deeply cut to lessen the ascent, down another valley a distance, up one of its feeders, and so on, about fourteen miles in all to this charming place. The fields have the right of way here, and the roads wind through them, so narrow in most parts that on meeting a horse, his leader halted him at a broad spot on seeing us coming so as to pass easily. Mostly smooth enough not to jolt, occasionally rough from dry mud or

small stones, sometimes interrupted by a still narrower plank bridge, crossing an irrigating stream, where the head 'k sha' would call out to the others to slow up as he carefully pushed his way between the cracks. The horses are small and shaggy, and have straw-shoes. Most of them were laden with rice-sheaves. We have seen now, a number of points in the cultivation of that universal grain; it grows in little tufts about six inches apart, to the height of two or three feet. Earlier in the year it is bright green, but is now yellow and gray. Where cut, the ground appears very wet or swampy, perhaps from the rain two days ago. Men, women and boys stand in this mud over their ankles, cutting the tuft with a small sickle, and laying them on the raised paths that run all through the fields a foot above their level. Then the sheaves are carried to a roadside or farmhouse, and the grain combed out by pulling the straw between long iron teeth, nailed on to a wooden frame, in which the old woman at work kneels. Winnowing is done by hand in a little cradle of bamboo-wicker work. The hulling I haven't noticed.

Although the valleys are naturally nearly flat, they have been terraced and graded flatter still, and every little patch of bottom is used. There is no farming, it is all gardening. (*Tour; 15 November 1877; Enoshima*)

Davis and his cousin were in Japan just under two months (fifty days according to their own account) and the country and especially the people charmed them. There is no single moment of criticism. The gracious behaviour and intelligence of the Japanese have often appealed to Americans and Davis and Osborne fell into the accepted pattern. As well as Yokohama, they stayed in Tokyo, and made two inland trips, a short one to Enoshima and a longer one of seven days to Nikko. On the latter they employed an English-speaking guide and found this greatly increased their understanding of native customs and history as well as reducing some of the prices. They also travelled overland when they went to catch their ship at Nagasaki, going via Odawara, Yumoto, Mishima, Numadzu, Shidzuoka, Fusiyada, Mitsuke, Hamamichi, Shinjo, Toyohashi, Okasaki and Nagoya. Shidzuoka was one of the stops and in the evening they sampled oriental theatre. Nagoya was another halt and Davis was full of praise for its castle, shops, theatre and the china factory at the neighbouring town of Seto. The impressions all express a vivid sense of delight. Moments of keen pleasure are experienced in India and Spain but are never so general nor so continuously sustained.

Language was a constant problem in Japan and even more so in China but Davis struggled manfully with the aid of his dictionary. Sometimes he depended more on the patience or persistence of the inhabitants.

'Ocha arimas?' says I, and the little waiting girl bobs and ... my eyes how she talks from that small provocation. I repeat 'Ocha arimas?' – have you tea – She goes out and soon comes back, first with an elegant little brazier, then with a charming tray holding two cups (no saucers) a teapot and kettle. Grand tableau representing 'Success'. Next we asked for rice and fish ... another flow of incomprehensibilities, ever so much longer than that. What could it mean! We repeated our second course. Girl goes out and appears again with a dish of fish, varied as to species. That's it, she had asked what kind of fish. Why somehow it had never occurred to me that there was more than one kind. She must think we have delicate appetites. We chose sole. Strange to say this didn't satisfy her, she went on talking. Tom suggested it meant 'How as to cooking?' – a lucky guess, for when I turned to *fry* and read out distinctly 'Abura de agera' (which I have since come to suspect means 'Oil in cook') she smiled and carried the fish away. I then wrote two lines and got to the middle of a good thought, when this pertinaciously obliging young woman came in again. At the end of a charming speech, I had to say, quite humbly 'Wakarimasen', i.e. 'nix fushtay'. But the persevering girl kept on and gently talking she approached me, put out her hand. Heaven, she was going to sound my stomach as to capacity ... did I blush? ... when she touches my watch-chain ... it was merely a question of time! (*Tour; 15 November 1877; Enoshima*)

Ultimately he gave in and, as we have said, they enlisted the help of a guide.

Temples and shops receive equal attention. The former excite descriptions:

With the steps are the gateways, of whose original use as rests for (sacred?) birds, Griffis makes mention. This is the general shape of many that we have seen, this particular one being from memory of a stone gate at Kamakura. The pillars and lower cross-pieces are monoliths. Notice the stone wedged to hold them together. Other materials are wood, in which the curved upper piece is often straight and bronze, as at Enoshima, on entering the town from the bridge-east hollow, with raised and sunken characters on it, a beautiful piece of work.

The gate at Katase was roofed as you may see in the plan, to protect its carving from the weather. Its panels were cut into open work representing various legends of which we have no explanation. The work was quite as good as much that is photographed to represent the art of medieval Europe. The Ni O (see Griffis, p. 380) are often enshrined on either side of the gate, but were absent at Katase. In the gate in front of the Dai Butsu they remain, although the temple covering the great bronze idol is no longer there; and near by where we saw the gilded idol of Kuanon, the Ni O are within the temple on either side of the door, perhaps because there is no room for a gate on the narrow terrace where the temple stands. The steps leading to it do not mount straight from the village below, but have to wind to one side to climb the bluff. These 'Two Kings' are often stuccoed with spit-ball prayers, little papers sold by the priests, chewed up into a pious consistency and thrown at the idols. If they stick, the prayer will be answered.

In the upper frame of the gate, dragons and birds are cut on the beams, uncolored except in the eyes and tongues, where little dabs of red or gilt add nothing to their beauty, but much to their fierceness. (*Tour; 16 November 1877; Enoshima*)

Shops particularly fascinated Davis because their open structure and the native mode of salesmanship were completely foreign to his American experience.

At one 'Banko' china and earthen ware store, we bought up all the pretty models, nearly eleven in all, and had half Tokio looking over our shoulders. You know the stores generally open their whole front onto the street. The customers sit on the floor, and all transactions are very public. The store keepers are very accommodating and return us a pleasant bow even when we buy nothing after looking all through their stock. At one place where we had bought something, we were shown through the whole establishment, a shop where all kind of lacquer ware is made. Instead of having a factory building, all the work is done in little rooms of ordinary houses, where the men sit on the floor together, and model and paint inside the thin paper-walls, like bees in a hive. (*Tour; 20 November 1877; Tokyo*)

Trees always interested Davis if only for botanical reasons. The Japanese pine impressed him for its beauty and its use as a recurring theme throughout Japanese culture.

The country about Tokio is a vast alluvial plain, with hardly a pebble, and yet the castle walls are unsparing in height and thickness and some of the stones are four or five feet cube, and about the shoguns' tombs there is much solid masonry. On top of the wall grow broad spreading pines, reaching far across the street; they have an irregular habit, very different from the characteristic tall pine at home. The same trees grow on the castle walls, with their branches hanging down over the moat, a charming subject, that no photograph that we can find does justice to. Pines are everywhere carved in panels, painted on lacquer, in the representative style, not a copy but an excellent suggestion. When with storks, they mean 'Long life to you'. (*Tour; 25 November 1877; Tokyo*)

The trip to Nikko took them beyond the limits of the popular tourist area. Even though Japan had not long been open to the visits of foreigners it had already adapted its main cities to meet the demands of foreign tourists. (Indeed, Davis' astute father had 'thot it might be a good idea for me to take some phone stuff to Japan and introduce it there; but when I tried to hear words, I couldnt. The machine was very primitiv then'. *Letter: W. M. Davis to E. Mott Davis, 19 February 1933; Pasadena.*) As is often the case, this adaptation was quickly taken to such lengths that the main cities soon began to lose their native character altogether. Davis many times remarks on the difference between the less popular towns and the cosmopolitan centres of Yokohama and Tokyo. The road to Nikko gave them an idea of what the real Japan was like. The plentiful supply of human labour is an obvious subject for comment.

As we rode into a town, a row of houses strung along the street, with backyards opening on to their farms, the coach boy, or 'betto' as we have learned to call him, would toot on his little horn, a warning to little children with babies on their backs, and to the little women at the tea houses. The betto is a great institution here and cuts a fine figure when he runs ahead of the nobby phaetons and dog-carts of the swell foreigners. What would coachmen do at home, if, besides harnessing and unharnessing and feeding and curreying, they had to run ahead of their bosses carriage to clear the track! The costume these little tigers wear is very effective; a loose, large-sleeved cape over a shirt, and as to the rest, simply tights and socks and sandals. All blue except the white figures on their backs; everything is blue in the clothing line here. (*Tour; 27 November 1877; Nikko*)

New dishes and strange tasting foods not unnaturally produced tart reflections.

When we stopped at Ozawa, ... something was passed in on a plate with the tea; it was neatly served in two-inch squares, with the thickness and color of underdone buckwheat cakes, and against Kimura's advice I tried it. Two bites were enough – indeed one ought to have been, but I had on at the moment a spirit of rigorous investigation ... when I began to go through the whole four inches. There are no English adjectives suited to its description, in my concise notes, taken on the spot, it is mentioned calmly as 'fried fat do', but that is tame. Its presence, vivid, living presence, accompanies me all yesterday, and its oily memory is still almost too much for a quiet and conscientious stomach. At another stop, pickled plums done up in salt, were tried, with better results, but what a shame to treat plums in such a manner.

Our dinner at Nakada clinched my conviction that Japanese food would not support a foreigner, even of my enduring and indifferent composition ... He of course ordered a regular Japanese lunch, as we said we wanted to see what the ordinary fare was. We sat on cushions on the matting and three pretty little 'no' lacquered trays were placed before us. Of course tea had been brought before; that was served as soon as we had taken off our shoes and walked in the room and squatted by the brazier. The rest of the lunch was a soup, in lacquered bowl, of marvellous flavor, stewed cuttlefish, pickled parsnips, and one or two other messes unmentionable, besides plenty of rice and fresh tea, also a pair of chop sticks for each. I tried everything with the decision resulting as above detailed, and used the sticks successfully on all but the soup, one of our spoons seemed more appropriate there. Imagine us on the floor, chopsticking those 'outlandishes' (portmanteau).

(*Tour; 27 November 1877; Nikko*)

The traditional tree-lined roads appealed to Davis.

Soon after dinner, we entered the double row of trees that made the glory of the old highways, and since then, for fifty miles, to this place, their line has scarcely been broken except at the villages, which are quite without shade. We shall get some photographs that will give you an excellent idea of these shaded roads, but they can't show everything; they can't show the red brown of the trunks, or green of the pine needles or the full effect of the warm sunlight streaming in between the trees. What an admirable idea of the needs of a good road the old daimios must have had to plant these trees, and wait for their shade till they grew to their present hundred feet or more of height. The pines are irregular, with crooked trunks and crossing bows, not at all like the yellow pine in California, the one I call typical there. (*Tour; 27 November 1877; Nikko*)

The poverty of some of the country persons they passed aroused his pity.

The hand-carts are seen best in the cities, where heavy loads are pulled and pushed by two or three men who grunt to keep step and cheerful, I suppose. They are of the lowest class, poor, wrinkled, hard-worked old fellows, some of them. What a life to get up to every day! This morning, in the beautiful tree avenue, we met a team composed of a family, father, mother and baby, hitched to a load of long bambooes, the mother carrying her baby besides pulling her share. I am glad we have few of such unhappy sights at home. (*Tour; 27 November 1877; Nikko*)

He admits to being baffled by the geology.

I don't make much of the Geology, the rock outcroppings are few and obscure. There seems to be something like a N.N.E., S.S.W. trend in some ridges, but not appalachianically distinct. Around the temples here, seen in this afternoon's walk, the walls are of volcanic rock, sometimes cellular, as lava often is. As the determination of such rocks is a matter of difficulty, I won't swear that it is trachyte.
(*Tour; 27 November 1877; Nikko*)

Overnight stops at local teahouses proved enjoyable in each case and confirmed their favourable impression of Japanese hosts. There was the inevitable embarrassment at the bath hour but we are already aware of Davis' extreme modesty.

Kimura left us to bathe and came back reporting that a warm spring opened directly into the bath room. We should find no better on the trip, better try it. Of course, let's do it. I went down along the little entry to the back of the house, found the two tanks steaming, very tempting indeed did these warm vapors look – but – the tanks were occupied. After a moments contemplation and deep thought, I concluded I was scared and backed out, and returned to our room – but nonsense! This timidity was absurd. I went back again and on the way met the young women of the house leaving the bath and walking to their rooms and clothes. Only a boy remained; he had one tank, I took the other, and stewed alone a minute, but that is evidently not the way that they do in Japan, for the boy soon came in and sat with me and we boiled together, conversing freely in Japanese on the war in Turkey, on Kimura's education, on our travels, and on many other things understood by only one of us, a profound secret to the other. It was very pleasant and sociable, like lenkerbad, and the rest of the girls of the house soon came in and made it more so: they took the other tank, because we boys filled the little one and about then – I – left – not precipitately, not hurriedly, but leisurely and moderately, just as if I had always been used to it, from the far distant days of County Line swims down to the present time. (*Tour; 1–3 January 1878; Steamer 'Saikio-Marn', Nagasaki to Shanghai*)

The mountains about Nikko tempted Davis to try a little climbing on his own.

In one of the views between the trees on Thursday, the mountains of the beautiful Nikko range showed with great distinctness and looked so temptingly near that I resolved on a walk for the next day, and made inquiry about paths and distances. The Nautaizan – Black Haired Mountain – so-called from the thick forest that reaches to the top, while other peaks of the same altitude are bare and snowy; this mountain proved too far and high to be ascended in a day. Its summit was fifteen miles from the village, and I reluctantly gave it up; but there were plenty of others nearer, so I took an early breakfast and a light lunch, chose my own path without saying nothing to nobody and had a long ramble and a delightful day's walk just where I chose to go. It was half past seven before I started; ice had formed on our out-door wash-stand, the thermometer was just above forty, and with an overcoat and a shawl wrapped around me, brisk walking made me very comfortable . . .

The Black Haired Mountain rose higher and higher. My mountain receded as I walked up its flank; its red rocky crest, which from Nikko seems joined directly to the spur I ascended, proved to be separated from it by an immense crater-like chasm opening into the side valley I had first followed. By going around two or three miles from a little peak that I struggled onto at half past twelve, the real summit could be reached, but I was too tired, there was too little time and too much cloud to try this; for the last hour I had been half the time in flying mists, and the path had become rocky and very steep so that I had to rest too often, and the ground was frozen hard and was snowy and slippery in places. I had to give up, without getting the view to the north-west that I wanted to see so much, so I sat down in a sheltered sunny corner, and ate my lunch, like an English Company, 'limited', three eggs, a chunk of bread, pepper, salt and snow. (*Tour; 3 December 1877; Tokyo*)

The presence of cows on the road (near Yumoto) puzzled Davis because he never saw any grazing in the fields.

It is a mystery to me where these animals come from, in all our out-door days we haven't seen a single one grazing. They must be brought up by hand, fed on rice-straw in the back-kitchens. Perhaps it is this mystery that makes their steaks so tender. Japanese beef is universally excellent, but until lately it was hardly ever used as food, being against all precedent (religion) and economy. Now it is better liked, we have seen butcher shops even in the interior. (*Tour; 1–3 January 1878; Steamer 'Saikio-Marn', Nagasaki to Shanghai*)

An absence of beggars commands his admiration.

Beggars used to be as common here as, perhaps, in Italy. Now, in all our stay we haven't seen ten; the change came with the revolution, the railroads and the telegraph (quite Henny Penny like) in obedience to an order of the new government. Those who can't support themselves are taken in charge, supplied with work, and fed and clothed; by working extra time, they can earn something, but they mustn't beg. Nor must any one give money to them; this is punished by a light fine. So notwithstanding the great poverty of the people, there is no begging.
(*Tour; 1–3 January 1878; Steamer 'Saikio-Marn', Nagasaki to Shanghai*)

Almost his last reflection on Japan is Fujiyama.

Our Wednesday morning ride, was the triumph of the Tokaido. The air was crisp and sparkling – 43 degrees at 8.00 o'clock – a few clouds floated over the mountains, sometimes lying against Fuji, but most of the time we had a clear view of him – wonderful – wonderful – most wonderful. We were on the plain close by the seashore, so we saw all of the 12000 feet of height of the Great Volcano, an advantage that one rarely has. At Chamounix you are over 3000 feet, at Zermatt about 6000 feet and at Colorado Springs nearly 8000 feet, so that Mont Blanc, Monte Rosa and Pike's Peak are heavily discounted, but here on the plain of Suruga, all of Fuji's rise of nearly two miles and a half, stand clear in the air, a continuous rise from base to summit, a nearly unbroken slope. (*Tour; 1–3 January 1878; Steamer 'Saikio-Marn', Nagasaki to Shanghai*)

After the thrill of its scenic grandeur standing supreme above all the surrounding land his attention shifted to its internal composition.

Where its slope ran against Actacyama, it seemed as if some deep ravines, running down from the summit of the latter had been filled up in their lower part by a lava flow from Fuji. I wish I could spend a few years studying this mountain, mapping and modelling it, and seeing whether Von Buch's or Lyell's theory would suit it best; but as I can't, I have advised Isawa to do it. (*Tour; 1–3 January 1878; Steamer 'Saikio-Marn', Nagasaki to Shanghai*)

His last two descriptions are perhaps befittingly of a china factory at Seto and a theatre at Nagoya.

We jogged the eight miles in about three hours and came to the little village in a narrow valley, between low gravel hills. The clay comes from layers between the gravel hills and is carefully chosen, washed and mixed – in this the secret lies – and turned, baked and painted and glazed, all in the plainest and simplest fashion. All hand work, with no labor-saving machinery, except of course the lathe, no labor-saving arrangement of buildings, everything irregular and confused on the hillsides, instead of even and exact, as one would expect to find the great head-centre of china-making for a large part of Japan. Nearly all the ware is blue, and most of it common, and we learned here, what we had suspected before and was confirmed afterwards, that the place of manufacture is by no means sure to be the best place to buy. (*Tour; 1–3 January 1878; Steamer 'Saikio-Marn', Nagasaki to Shanghai*)

I was surprised to see anything in Japan so dirty as this theatre, it was really a shock. Boys were constantly walking about with oranges, candy, cakes and tea to sell, for when a performance lasts all day, lunch becomes a necessity, and the élite in the boxes as well as the crowd in the pit were scattering orange skins and spilling their cups in a way that I grieve to write. Our shop-keeper treated us to oranges and we played natives very well.

The building has all its roof-beams and rafters showing, as in a barn, with flags and lanterns hanging from aloft, 'to add to the hilarity of the scene'. Between the acts our excellent Nagoyan took us onto the stage, by a side passage, back of the boxes on the left, which the actors use when going forward to make a front entrance, but there was little we couldn't see from the boxes. The curtain is pulled across the stage by a man behind it, there is very little scenery, no grooves for it, the slips are held up by props with iron points that are stuck into the floor each time they are used. A door may be stood up alone, without any wall, so that going around it would be much easier than following the imaginary necessity of sliding it open and going through, and the scene painting is poor, and yet this was the best theatre in Nagoya and the actors were the most brilliant stars from Tokio. The most peculiar part was the turn table in the middle of the stage, about twenty feet in diameter; it is turned half way around when a change of scene is made without drawing the curtain, in a way that delighted all the boy left in me. Back farther were dressing rooms where we saw some fine costumes, and property rooms and bath rooms, all showing the richness of the theatre; the accommodations for the actors were better than for the audience.

Actors are either all men or all women, but they must not act together in the same theatre. We saw two troops of men, who took women's parts excellently. The women are not allowed to speak in their plays, but do everything, even to moving the lips in dumb show, while a male chorus on the side of the stage does the talking. (*Tour; 1–3 January 1878; Steamer 'Saikio-Marn', Nagasaki to Shanghai*)

No country was left with more regret, and the far less congenial atmosphere of China only increased this. It was not merely novelty. Their liking for all things Japanese increased as they became better acquainted with the country. Somehow the Japanese shopkeepers and others they met struck the right note and despite the lack of linguistic communication managed to make them feel welcome. The pleasure of this experience had a useful side effect. Not only did it lead Davis to write rapturously of all he saw but it also evoked a greater number of personal reflections. We saw earlier his scathing comments on his shipboard companions. There is no doubt that he had a fine brain which was capable of retaining a great deal of knowledge but he never seemed to come off the boil. This inability to descend to ordinary levels made him a difficult person to speak to, particularly as vulgarity in speech or manner were equally offensive to him.

Having on an investigating spirit one morning lately, I ventured to address Mrs. Missionary – a most bold-faced proceeding – no introduction, no pleasant nod from her as a signal for beginning, nothing but cold cheek on my part. She is rather delicate looking, and has a weak little face, from which I argued something of refinement might come from her; a very wrong and inconsequent argument, as will appear. A Shanghai grammar lay on a chair beside her, and I ventured an interrogative comment on its difficulty; and affirmative, with a rising inflection, you know. There was nothing important or essential in the first subject of conversation, it might have been a Shanghai rooster just as well, in which case I should have made a parable on its legs, but fate wished me to begin with the Chinese language, so I did. Mrs Missionary calmly looked up and from her little mouth distinctly came these words, 'Yes, it must be a red-hot language to learn' – ! – ! Angels and ministers of Grace, I thought, defend us! Is this your delicacy and refinement, Mrs. Missionary? The contrast between what should have been and what was, was too dreadful to allow me to go on. I slunk away. As a converter of

Missionaries, I am a failure. I returned to Richthofen and abstract thought. (*Tour; 4 November 1877; 'S.S. Belgic'*)

Next and close by came Dai Butzu. You remember, doubtless, that this is the great bronze statue that Pumpelly found so fine, so wonderful, so completely 'expressive of the realization of rest and annihilization of care' and perhaps you haven't forgotten my wicked joke on it. Be surprised therefore when I join the line of its enthusiastic admirers! I shouldn't speak of it as Pumpelly does, but it is a magnificent work, and I am not yet cooled down from my disgust at seeing Mr. Dunn climb up on its knees, and grasp its hands, shouting 'How are you, old fellow!' Even Mrs. Hubbard's telling me to look at the contrast in the size of the giant and (confounded, miserable, desecrating, barbarian of a) pigmy, a (typical tourist of a horrid) six foot pigmy, could not make me attentive in this humiliation. I turned away and walked off. One of the ladies asked for a chrysanthemum growing near by, and was told by another she couldn't pick it, it grew on sacred ground, at which I, who have not yet learned the worthlessness of all these wonders impudently asked, 'Is anything sacred here?' and only Tom understood me. Isn't it a shame for foreigners to set such an example to the Japanese. (*Tour; 15 November 1877; Enoshima*)

Even Davis realized that his strictures were at times overdone yet his irritation was such that he was unable to contain it completely. Thus during the journey across the Pacific he admits.

If the generous sea air was more inciting to writing, I should revise the preceeding sheet, and let you have it in a milder form, but under the lazy circumstances I will let it go and be ashamed. It will serve to show to what conceited condition one can be brought when thrown among uncongenial companions. Tom and I only make fun of our fellow-passengers. (*Tour; 4 November 1877; 'S.S. Belgic'*)

In Japan he found for the first time a joy in shopping and retained his old joy in sketching although the results were, in his own words 'neat but weak'. However, we find no weakening in his aversion to formal functions nor to certain aspects of religions other than Quaker.

In the temple of Asakusa (asaza) in Tokio (where I am writing) the censer is under the main roof. The pious on approaching it, throw a copper into the grated box at its foot and put a pinch of incense on to the coals . . . it smells Catholic . . .

At Asakusa, an image of Birizura is rubbed smooth by rheumatic worshippers. If their toe is ailing or their nose is out of joint, they rub the corresponding part of the idol and then pass their hand over the afflicted part. One old fellow standing in front of Birizura was a little confused and took the left shoulder of the idol to correspond with his right. I hope the mistake won't kill him. (*Tour; 16 November 1877; Enoshima*)

TOUR IN CHINA

Davis and his cousin stayed a much shorter time in China than in Japan. They reached Shanghai on 4 January 1870, and left the country at the end of the month. There was only time to visit Hong Kong and Canton in addition to Shanghai. The severity of the weather came as a surprise and the very rough crossing from Japan and the more primitive facilities of the country combined to discourage any inclination to stay longer or visit other towns.

The weather and hotel together are not 'calculated' as R.G.W. would say, to inspire us with pleasant thoughts; one is cold, damp and cloudy, rainy and chilly even for the season; and the other is a compound of Portugese management and Chinese service, bare, barny, cheerless, with the table and beds, however, middling good and comfortable. There is a portico, overlooking the river and its boats, from which the city proper dimly mopes on the other side, excellent place to sit and write or sketch in summer weather, but the bare suggestion makes me shiver now. This frigid climate that we find where we always thought tigers tramped about tangled in the equator, is one of the most serious geographical reverses we have yet suffered. It began, you know, with the snow and ice, giving sleighing and skating, in Shanghai, where the thermometer went down around 16 degrees at night and barely reached 50 degrees in the day. Here there is no freezing, but with thermometer at 40–45 degrees morning and evening, and dampness equal to Boston's worst, I think we are warranted in placing Canton near the North Pole, in our *revised* atlas. (*Tour; 18 January 1878; Canton*)

Their appreciation of Chinese resources and customs was necessarily superficial and was carried through with none of the enthusiasm that was manifested in Japan. Shanghai provided little amusement and much of the time seems to have been spent resting or planning the next stages of the tour.

Compared to our great knowledge of Japan, we shall learn little of China. We are delighted we stayed in the Great Empire so long instead of coming earlier here to freeze and be hated by the Chinese. How different the two countries and people are. Knowing one, invert your knowledge and you have the other.

The best thing so far in Shanghai is the blazing coal fire in our grate, by which we sit and write. The buildings on the Bund are very fine, but they are simply importations from England. China doesn't attract us as Japan did, our dear Japan! As for its pongees and gauzes, that's another thing. (*Tour; 7 January 1878; Shanghai*)

There is a brief description of the town itself which chiefly illustrates the country's extreme backwardness.

Going up the Bund, past the English and French quarters, and through a strip of modified Chinese houses, where native merchants do a growing business, we came to the East Gate in the wall around the native town, and going through it were in uncorrupted China. Outside the wall is a muddy street, and a muddy, narrow moat, half frozen over when we saw it, the wall is fifteen or twenty feet high, with fantastic towers, hardly towers either, on the Northeast corner, the only one we saw; it was too cold to stop and sketch them. We are told that Canton has much the same, only better and more.

Where we entered there was a double wall, with a little space filled with shops

F I G. 17. Drawing made in China by Davis during his world tour. The characters denote the mountain Ch'eng Shan.

between; the passage through the walls is about eight feet wide and ten feet high (that is an extreme measure), it can be quickly closed by four gates, and through such ways as this must go in and out all the traffic and travel of the native town.

Inside the street was like a picture. We are sorry not to find photographs of such places, but they would be difficult to take. We walked some distance into the City, past fine silk stores with tempting goods, and little shops of all kinds; the houses nearly met over our heads; signs with complex letters hung from the eaves. 'Nothing' to see in Shanghai! We were half dazed with the strangeness of what we found. (*Tour; 13 January 1878; Steamer 'Geelong', Shanghai to Hong Kong*)

Davis seemed to realize, however, that a more thorough exploration could have provided as much interest as they found in Japan. At Canton they followed the normal tourist round and dutifully visited a flour mill, a fish garden, glass and jade factories, several temples and saw examples of silk weaving in progress. In Hong Kong they had less time and used this to climb the Peak and walk through the gardens. Much of what they saw was well worthy of attention but the choice of an opium den is peculiar.

Then going along a narrow little side street, we turned into a narrower passage-way, quite dark, that led into an opium room, a bare cold place, with a shelf around the sides on which the smokers lie. They are supplied with a mat, a bamboo pillow, and a smoking outfit, tray, lamp and pipe, and fifteen or twenty minutes enjoyment costs them ten or twelve cents. This was a common place, the 'respectable' people smoke at home, Mr. Ah Cum says, and some of the inveterates keep at it all the time, spending a fortune of three or four dollars daily, but of course this is rare. (*Tour; 18 January 1878; Canton*)

Of course, in Davis' era the dangers of taking drugs were not fully realized and the western nations still thought it right to exploit the commercial possibilities of this weakness.

To get to Calcutta, we can choose among the following steamers. Two going direct, via Singapore (not via Ceylon) to-morrow Jan. 15. They come from Calcutta, with cargos of opium, once a month, sailing as soon as possible after the auctions, and go back when they are ready. (*Tour; 14 January 1878; Hong Kong*)

Descriptions of the Chinese countryside are unfortunately few. The short trip between Hong Kong and Canton provided one.

For the first three or four hours it was misty and cloudy and the mountainous islands showed only in dim outline; about noon, there was better weather, and we had a good view of the Bogne Forts, rather gone to ruin now. Farther on, the banks of the river we entered were low, and only rice-fields and plaintain-hedges, and graveyards and villages attracted our attention. In the latter, the square towers of pawnbrokers' shops were most conspicuous; they are built high and strong as storehouses of many precious goods. (*Tour, 18 January 1878; Canton*)

The need for women to work was a fact that drew his attention to the average living conditions.

Susan and her family have no home but the boat. As on the ferry boat, the cooking is done under the cross-boards, where the old woman stands sculling, only Susan's boat is always clean; she is constantly scrubbing it. And there is generally a baby in the kitchen, who begins to learn its life's work very early; first, by nearly having its little head wobbled off, when strapped on its mother's back while she is rowing; later, when it can stand, by holding on to the scull rope and swinging back and forward as its grandma sculls. Under the covered stern, I sometimes would see tapers burning before a little shrine to the protecting Joss, and hanging over the water there was often a round, basket-work cage, with a pair of fowls in it. Generally only one woman sculls (though there are two in my sketch) and she looks over the houseroof and sees her way clear to steer. There is sometimes a man on these boats, a brother or a husband, but oftener he is away to work elsewhere, a porter or sedan-carrier, perhaps. I should think there were several thousand of the generic Susan on the river and how they all make a living is a mystery to me. As we come along the wharf to the ferry-landing, we pass twenty or thirty boats, each eager to take us to the end of Canton, if necessary; and so they are crowded up and down the shore. Susan or her little sister sometimes go up onto the street and waylay possible passengers there, but the possible passengers generally take the ferry. It always seems mean to pass them, but then think of the poor ferryman and his mother. I wish I could draw you a picture of Susan's

'Little Sister'. Every Susan seems to have one, all of much the same size and prettiness, though a few excel the others in the latter detail. Yes, you ought to have a sketch of her as she sits in front of us, with one bare foot braced against the house corner, pulling a splashy oar; her hair is braided and coiled behind, and 'banged' in front, per-these Canton boats are the original home of bangs – her eyes are narrow and black, her nose and mouth, small and cunnin', really she is very pretty for a little China girl. She has ear-rings, bracelets and anklets of imitation silver and imitation jade; perhaps I can find her photograph. (*Tour; 21 January 1878; Canton*)

Even the temples incur an unfavourable comparison.

Next we took a temple, but we haven't yet learned the arrangement of temples here, as we did at Japan, at Katase and Uyene. They are comparatively plain. The roofs make an attempt at graceful curves, but fail; just under them on the walls are some excellent carvings in which the figures are remarkably good; they are painted in imitation of life. The idols are poor, generally gilded, sometimes painted, surrounded with drums, tinsel and incense burners. Here no one need take off sandals on entering the temples, consequently it is dirty, like most other places in Canton; we don't feel like lying down on the floor to draw the ceiling as I did at Uyene. Praying is not a very delicate operation. The supplicant takes a little box holding a number of long sticks, like lamplighters, and kneeling before the image, shakes the box till a stick drops out. He notes the number and tries again, then he goes to a fortune teller and says, 'I want to do so and so on such a day and at such a temple I drew these numbers. Do you think they are favorable?' So Mr. Glover described the operation to us in Shanghai. Perhaps it is the weather or the dirt, or the crowding Chinamen and boys, perhaps sights are no longer novelties, but certainly we do not rave here as we did at Katase. (*Tour; 18 January 1878; Canton*)

The culture of goldfish apparently aroused no feelings either way.

You know that gold fish are great favorites here; they are raised in earthern tanks in the city and sold cheap enough when small, but $1.50 a piece, if five or six inches long and of good color. (*Tour; 18 January 1878; Canton*)

The custom of regarding small feet as beautiful evoked sympathy.

Ah Cum said he could recognize the Tarters by their faces; we could not at all. We saw only that they had no shops and that the women sometimes wore three pairs of ear-rings, and that their feet were simply ridiculous. The poor victims of this ugly fashion go tottering about, as if they 'walked on pins and needles, Horatio' (Parrish's Shakespeare, Edition 1864). Their shoes from heel to toe, are not over three inches long, but this is not a fair indication of the size of their feet, which are nearly bent double, with the instep broken by growing against its tight binding ... I believe I prefer small waists to small feet. (*Tour; 18 January 1878; Canton*)

The overall impression was of a low level of backwardness in every aspect of Chinese life, ranging from mechanical equipment such as the water clock, to the educational system with its examinations.

We followed the wall, eastward from the south-west gate, about half a mile, and then went down into the city to see the Examination 'Hall' and the Temple of Confuscius.

Examinations are held every three years generally, when thousands of students present themselves for degrees. For the higher ones, they must go to Peking, the lower can be obtained by satisfactory work in the cells here. In this sketch you look across the main, middle avenue, by the entrance gate, through which we passed, and found ourselves in a broad passage between endless rows of long, low sheds. I estimated the area covered as a quarter of a mile long by an eight wide, but like enough this is an exaggeration, (Why are such estimations always *over-measures?*) and excepting the gate, the passageway and a pair of temple-like buildings, this large space is filled with sheds. You can see how close they are built in the picture, and all are divided into cells about six feet by four, one alongside of the other. The walls are of brick, the floor and roof of tiles, and one side is open to the south. How hot it must be there in sunny summer and how cold in cloudy winter. Little grooves in the walls support a couple of shelves which serve for table and bed, and with this small supply of comfort, the students must be content for three days. 'The Middle Kingdom' gives a good description of the examination and of how the doubtful applicants try to cheat, even going so far as sending in substitutes who are poor and miserable enough to hire out their learning, and Andrew H. told us of small copies of the 'Classics' printed on very thin paper in a few pocket-volumes which are sometimes used as 'jokers'. The work to be done is writing essays in the most approved and classical style on very metaphysical subjects, about as dry and as generally useful as would be a *very* advanced and difficult Greek examination at home. The Chinese are not 'liberally' educated according to Huxley's standard. (*Tour; 20 January 1878; Canton*)

We learn that Davis had discussed the trip with his former professor, Pumpelly, who had a personal knowledge of China.

Mr. Pumpelly advised me to buy a complete set of all kinds of jade, and to learn the Chinese names for each and all sorts of things, but he did not accompany said advice with a letter of credit, so I don't follow it. (*Tour; 18 January 1878; Canton*)

Chinese geology seems to have been overlooked but the limited time was the probable cause of this as it had been in Japan. On the only occasion geology occupies his thoughts it is to concentrate his deductive abilities on the Yosemite Valley.

In answer to Prof. Lesley's question about the Yosemite Granite, forwarded by Henry ——

It is in no way stratified (did he say 'stratified'). In the mountains near by, and along the sides of the valley, it is very distinctly dome-shaped; and concentric exfoliation, along joint-planes parallel to the surface, is plainly shown. This is seen best in the 'Royal Arches' where it reaches some two hundred or three hundred feet below the top of the cliff – very likely more – I cannot remember now. I don't think this splitting can be due to expansion and contraction under the sun's heat and the night's cold, it reaches too deep. (A theory I heard Prof. Lesley quote at Dr. LeConte's one evening.) I should connect it with the formation of the domes, but for this I have no explanation. (The expansion and contraction theory only accounts for superficial (how deep?) splitting, parallel to some previously formed surface.)

As for the valley – ordinary water-erosion or glacial erosion are quite insufficient. Where they are known to have acted elsewhere, they have not made valleys at all like the Yosemite. They are quite insufficient causes. The bottom falling out seemed more plausible than any other explanation, and was better supported. It seems to me that such a double faulting would be accounted for by a tension, not due to upward pressure from below, but to contraction in the 'crust'. The wedge between the faults (the side cliffs) has fallen down, and if the fault-planes sloped as the cliffs do now, I think the first figure gives the best explanation. The sides of the valley at El Capitan and the Half (South) Dome slope in, instead of overhanging.

But all this is rather vague. Not the dome faulting, but the cause of it. (*Tour;* *20 January 1878; Canton*)

Perhaps the most significant part of this was his professor's request for information. If, as is probable, this was more than a courteous interest and the answers were intended to be relied on, it indicates that Davis was highly regarded as a junior lecturer.

These extracts from Davis' letters on his tour in China may well be concluded with two very readable half-humorous sketches which show his gifted power of observation. The way he coaxed the gatekeeper into allowing a portrait reveals ability, confidence and a laughable account of human reactions.

All this time the Gatekeeper looked on with much interest, and as his face was worthy of immortality, I told him I should like to draw his picture, that is, I pointed to him and began to scribble, and you should have seen his delight. The broadly grinning delight in his venerable countenance. 'Come on' he said, and led me into his shanty, gave me a table and chair and sat down on a bench opposite with the most fixed and staring expression he could find, and glared at me. I think the likeness is recognizable; the family certainly thought so and helped the old man to keep a pleasant and cheery, unconstrained face, by calling out, 'He's doing yer nose!' 'He's got yer left ear on!' 'There's yer mustach!' and so on. Once or twice this was too much for the old fellow, and as I looked down, he smiled openly, but you should have seen him 'recover' when I looked up. It was very jolly, the only unhappy occurrence being when I had to refuse, for your sakes, to give him his picture. He was very much disappointed, and I fear he did not join in the chorus of 'Can do' that followed my departure. (*Tour; 1 February 1878; Steamer* *'Mirzapore', Hong Kong to Singapore*)

Davis' description of Chinese beggars is particularly well done, being free from exaggeration and signs of personal irritation.

Beggars! Why I have told you nothing about them. They are superb: in numbers and perseverance they are good; in raggedness, unapproachable, finished, artistic, professional raggedness; and in misery, they are dreadful. Two or three sometimes beg in company and stand in a store front, groaning and howling their appeal in something like a song, or trying to attract benevolent notice by beating drum and clashing cymbals, once the respectable bottoms of brass pans.

If the storekeeper has given what he considers enough, for that day the calls in the morning may have been very numerous, he pays no attention to the noise nor cash to the beggars, but goes on with his writing and shouts out his bargains to his customers, for he mustn't drive the ragged rascals away.

They have a right to stand in his door as long as they like and make what noise they wish. If he likes his cash better than his quiet, he must endure the racket till the beggars are tired out. They manage this better in Japan. (*Tour; 1 February 1878; Steamer 'Mirzapore', Hong Kong to Singapore*)

CHAPTER SEVEN

World Tour: India, Egypt and Western Europe

After leaving Hong Kong by steamer at the end of January 1878, Davis and his cousin had little time ashore until they landed at Calcutta. They managed part of a day at Singapore and a short trip inland at Penang. They were luckier in Ceylon, where they spent four days at Galle and made several journeys inland and to Colombo. The stop at Madras was restricted to a walk round the town.

At this time work was proceeding to convert Colombo into a great modern port to supersede Galle which had a rockier and smaller anchorage. Davis noticed this and the well-developed appearance of Ceylon.

> Our road was excellent, smooth and well-made as all the roads in Ceylon are said to be; it follows near the shore all the way, sometimes nearly on the beach, then inland a little with palm groves between us and the rolling surf. There are many poor little villages of mud and thatch strung along it and numbers of poor people, but comfortably poor, for their wants must be very poor and easily supplied. Even in this winter time, keeping cool is more difficult than keeping warm, and to people who wear so little clothing, keeping dry must be of little more consequence than getting wet. Cocoanuts and plantains are common, so what more can they want, except occupations and ambitions. Their dress is enviably scanty. The cocoanut palm is very much commoner than anything else, it has followed us up here and is still in greater abundance than any other tree, it lines the shore and gives the character to harbor view, and 'Ceylon' will always suggest a grove of its tangled trunks and feathery tops. (*Tour; 17 February 1878; Kandy, Ceylon*)

Inland they were attracted by the elaborate cultivation of the fields and the luxuriance of the vegetation.

> For twelve or fifteen miles, the railroad rises with a grade of one in 45 – there I go again – along the steep slope of the mountain of foliated gneiss, the western part of an obscure anticlinal – goodness gracious! – rounding sharp points or tunneling through them, with the valley very much, indeed more than vertically below us – hundreds – millions of feet below us, with houses like hats and people like flies, seeming only a quarter of an inch long. When we didn't run through tunnels, we cut little grooved galleries in the rocky side of the mountain and hung on with our flanges, generally though, going round curves so sharp that half the wheels were in the air. And to all these glories add a sun that scorched our heads when we ventured to lean out the window a thermometer panting at 89 degrees, or let's say 90, or why not 100, just to keep it in round numbers; a road, branching in the

valley below us, one of its white smooth lines running south among the hills with outlines suited, in their jagged absurdity, to the occasion; the other rising with many curves and turns and zigzags to reach the pass we were aiming for, and then follow us to Kandy. Terraces around little hillocks beneath us, so even and exact that with a photograph of them, I should never have any trouble in teaching even obstreperous Sophomores what is meant by contour-lines. As we passed over these model terraces, the farmers were turning the water from little mountain-streams that cascaded down to them onto the flat steps, till they glistened in the sun ... Then going around a great concave curve, we looked back to them again, perfect in their regularity, with our gallery in the rock above them, half way up to the summit, a rounded dome with its coarse structure lines crossed into plaid by the vertical water stains, and not a photograph of it to be had. (*Tour; 27 February 1878; Steamer 'Poonah' off Madras*)

While he probably did not realize it, Davis at the age of twenty-eight still went on regarding the world as a large schoolroom. Some regard it as an experience to be enjoyed, others as an adventure, others as a game where you match your skill against the elements and the players taking part; others again believe it is a spiritual experience, a bewildering and prolonged attempt to resist the temptations of the devil: but for Davis it was an adult extension of his schooldays. The world and everything about it was a lesson to be learnt and understood. Personal pleasure was of secondary importance and the recording of learning was of first priority.

I have my usual wail to weep. Tom is on his tenth or fifteenth sheet of special Ceylon correspondence, and is probably only a few hours behindhand. I am nowhere. So I shall scribble through a personal narrative; then give you a sketch of Ceylon, historical, scientific and otherwise, and finally come on with my philosophical reflections, if there be time before getting to Calcutta. Taylor is with us, we are three, and all in one room just now amicable so far, and trusting in a quiet future. By the way, in a gush last evening, just after a silent spell, Taylor invited us to stop and see him in Birmingham on our way home, and we shall if there is time. (*Tour; 17 February 1878; Kandy, Ceylon*)

Taylor was a young Englishman who had come aboard at Hong Kong. Despite some preliminary scathing remarks on his low intelligence and general demeanour, the friendship seems genuine and Taylor accompanied the two cousins on a number of excursions until they reached Delhi. As he was the only acquaintance so indulged throughout all the tour it seems likely that he had been rather misjudged at first sight. Obviously Davis admired talent and industry in others as is seen equally in his attraction to the young assistant at the botanical gardens at Kandy.

So I excused him [the Director], and he introduced us to his assistant, Mr. D. Morris, and by Mr. D. Morris will I swear as long as there is any swear in me. He is the best fellow I have met on the trip, not excepting even any of our Japanese or Chinese friends. I believe he is as fine a man as any I know, and his acquaintance was the crowning glory of my trip in Ceylon ... Well he took us in charge, and walked and walked and talked and talked in a way that was perfectly entrancing to me, and how Tom and Taylor could tear themselves away to go to Kandy at half past ten, I can't imagine.

Morris had taken the regular classical course in Dublin University, studying for honors in Science, and coming off 'Senior Wrangler'. After that he studied with Dr., or Sir, Joseph Hooker at Kew, and at some time with Huxley and several other eminent London professors. He brought out his note-books to show me some special points he had worked out, and it was easy to see he had made thorough work of his studies. They were not only botanical, but they covered the whole field of Biology, after the modern, Huxleyan style, a most delightful and complete course. Don't think he recited all this to me in a conceited way, as showing how much he had done. It merely came out naturally as we compared our Universities and studies and methods. He was well posted on geology too, and gave me some interested hints about my 'obscure gneissoid anticlinal' rather confirming the little I had made of Ceylon rocks and structure. (*Tour; 27 February 1878; Steamer 'Poonah' off Madras*)

Despite these learned geological disquisitions and the reliance placed by a professor on his views, Davis openly admits that he could not interpret a foreign landscape.

We looked across the north and south valley before us onto the hills and low mountains beyond, and tried to make out their geological relations and failed, for one can't find Catskill geology everywhere. (*Tour; 27 February 1878; Steamer 'Poonah' off Madras*)

This we know to be true because both Brigham and Daly tell us how inexperienced the youthful Davis was. The same accusation is often thrown at him in later life but the reader must make up his own mind on that. Modesty was a garb for occasional wear. Confidence in himself was something he did not lack. Even inexperience did not cloud his vision; he knew how to get on and where he was going. A few years later Davis' ability to impress, not to say court, eminent men played a big part in obtaining for him a permanent appointment at Harvard.

'Do you see that man?' said he this morning, pointing to another old fellow with a big felt hat and spectacles and a double chin with a fringe of whiskers under it. 'Yes, I have marked him for my own', I answered, 'he is reading "Nature" and when he stops I shall speak to him.' Any one who reads 'Nature' I consider fair game. 'That man,' Old Bailey goes on, 'is Mr. Blanford, the head of the Meteorological Department of India and I'll ask Major Osborne (Not the Judge, not Thomas) to introduce you to him, if you like, I saw them talking together.' 'Do,' said I, 'I have no doubt that Blandy will be glad to know me'; so I suffered myself to be led up and presented, and now I am quite posted on many things I wanted to know, and when in Calcutta, am to be admitted to the Penetralia of the Museum, the Geol. Survey Office, The Surveyor General's Office, the Bengal Asiatic Society and lots of desirable places. It only takes a little cheek.

It is pleasing to watch Judge Osborne during these ceremonies; he looks on me with scorn, as an imposter and doubtless wonders what in the world Will means by such assurance. You see I am so kind and familiar and condescending with him that he hardly appreciates the importance of my office as 'Assistant in Geology'. I think he rather pities the victims of my impertinences. But then you know he doesn't understand 'Impersonality' and 'Impenetrability'. It will come in time, when he finds his doll is stuffed with sawdust. (*Tour; 1 March 1878; Steamer 'Poonah' off Madras*)

'Impenetrability' was a trick he played on unsuspecting persons whom he chose to tease by pretending not to understand something which was very simple.

The cousins' time in India was restricted to a month, far too short for so large a country. They certainly appreciated the break in the hills around Simla after an exhausting attempt to sample the various landscapes.

After the steady work since leaving Calcutta, Simla is a pleasant quiet rest, from which we have to go away to sight-seeing again only too soon. Understanding the beauties of Architecture and appreciating the beauties of nature are very different, and as far as I go, the latter is the easier, so we enjoy ourselves lazily here and shall stay just as long as we can. It is a strange place, almost as remarkable as Virginia City, and at present much harder to reach. The treasure sought here is health. (*Tour; 22 March 1878; Simla*)

Nevertheless they tried hard to conquer the distances. They covered the whole length of the Ganges and even penetrated as far west as Lahore and Amritsar. Only a small proportion of the sights could be taken in at any of the stopping places but they resolutely strode their way through the normal tourist itinerary. The East Indies had been sacrificed so that more attention could be paid to India and they valiantly suffered heat and dust to do it justice. At Benares they managed the ancient city, some shopping, a few selected ruins and a visit to a metal-wire workshop. At Lucknow it was the temple, observatory, a mosque and a glimpse of an Indian meeting; Agra provided the Taj Mahal, Akbar's tomb, the fort and a glance at inlay work; Delhi and the other cities produced similar diversions.

Their memories of India are a mixture of heat and dust, bright colours, the Taj Mahal, hair-raising views on the road to Simla and always the crowds of persevering, importunate beggars. With his strongly humanitarian views on subject status and Christian standards, Davis' attitude towards poverty is surprisingly harsh and unenlightened. Instead of seeing poverty as a state of degradation, he seemed to feel that a person's relegation to the poorest class was a matter of choice and the result of his idleness. Davis possessed the 'Yankee ethic' of the New Englander.

In these out of the way places, where there is not very constant travel, about all one can get at the bunglas is a bedstead, a little crockery, plenty of water and

> basins and wash tubs, and any number of attendants, needed by the highly
> differential state of service established in this lazy country. In leaving a bungalow,
> the water boy with his emblem, the water skin on his back; the sweep with a wisp
> of rushes for a broom; a stray waif who has brought you an unexpected orange;
> the cook, the various table-boys all appear at your carriage with many and low
> salaams, each hoping for an individual backsheesh, which as far as we are con-
> cerned they do *not* get; it is simply absurd the way these loafers use a small service
> as a pretext for unlimited begging. (*Tour; 20 March 1878; Delhi*)

This contrasts strongly with his criticism of colonial practices.

> And we don't think much of the system of living that we see so far. Everyone
> seems to have his private servant, who is to be scolded and sworn at from
> morning to night for most unintentional and invisible offences, and whose first
> duty is to salaam and cringe and get out of the way, and who sleeps on a mat on
> the tiles by his master's door. But this *so far* is only our hotel experience. We hope
> that gentlemen in India don't behave as everyone seems to here. It is almost
> intolerable at table. (*Tour; 5 March 1878; Calcutta*)

The shortcomings of their own institutions were perceived by some of the
resident Englishmen.

> We knew our fellow passengers pretty well by this time, and enjoyed talking with
> them. Mr. Blanford was entertaining on India generally, Major Osborne on the
> subject of his book, but he is a full blown pessimist and thinks we are all going to
> the bad and speaks sadly enough of the corruptions hid under the paternal
> despotism that had so pleasing an appearance in Ceylon, and of the listlessness and
> apathy of a people governed by a distant country of unsympathetic race and
> beliefs. (*Tour; 3 March 1878; Calcutta*)

While the main object of the tour was to revive his cousin's health and
build up a useful fund of knowledge before his entry to university, Davis did
not neglect his own advancement. He had only just started his career but he
was obviously keen to do well. His examination of local geology was necessarily slight because of the distances involved and the brief period available.
What he did in each country was to gain an idea of the general geology and
he usually obtained this by an approach to the principal geological organization. In Calcutta he consulted Blanford of the Geological Survey. Then to
improve his proficiency in field studies he would relate this knowledge to the
landscape of the provinces they passed through. At the same time as a very
junior departmental dog's-body and embarrassingly anxious to please, he
carried out various commissions requested by his professors and took or
bought photographs of features in which he knew they had a special interest.

> Bourne & Shepherd, *The* Indian photographers have their work shop here, with
> offices in Bombay and Calcutta. We went to look at their views yesterday and had
> eight large portfolios sent down here to inspect over today, and it is a treat, for
> they have made several excursions into the mountains north from here, and up

Cashmere way, and their snow pictures are superb. One of them shows the Ganges, the holy Ganges, as it starts on its pious flow from under a glacier ice-cave. And there are other glaciers, and moraines, and snow-fields, and peaks, valleys and villages; a splendid series on which I shall execute Professor Whitney's commission of getting 'everything that will illustrate the Physical Geography of the Himalayas', and I shan't neglect myself either. I shall be ruined with these photographs, for besides mountain views, we have all our cities and palaces and temples to illustrate, and several thousand copies of the Taj to get to give away to our friends who will be so glad to see us. But this kind of ruin is perfect larks.
(*Tour; 24 March 1878; Simla*)

Yet, although his appraisal of the landscape could not be detailed, some of his descriptions were surprisingly perceptive.

There were some common place pines occasionally, but all small, the large ones must have been cut out, only a few of the mountains are forest-clad. There was a striking peculiarity about the valleys; excepting one of the largest, none have any flat bottom-lands at the level of their stream; the water flows in a sharp V-shaped cut; but two or three hundred feet above it, there is nearly always a level terrace, where the village people find room for wheat fields. They take advantage of any gentle slope on the mountain side and cut it into steps; such little patches are green now with the fresh grain-grass, and they absorb nearly all the water from the ravine rivulets for their (?) irrigation (?).

But these small terraces are quite distinct from the first ones; they are smaller and not so flat and generally much higher than the streams and are evidently artificial, while the others are of the same height on either side of the valley, and are so massive that they are quite beyond artificial construction, and suggest an old period of deep valley-cutting, then a silting up with stones and sand to several hundred feet depth, and now a return of the early conditions of cutting, all of which was of course duly expounded between the jolts of the tonga. (*Tour; 24 March 1878; Simla*)

Much of their travelling followed the track of the Ganges across its flat alluvial plain and Davis gives a description of what he could see from the train window.

During the night, we passed little of interest, except some coal-fields of small extent, but great importance to the railway and to Calcutta, and some trap-rocks, and the limit of growth of the Cocoanut palms. It needs a sea-breeze, so we shan't see it again until near Bombay, where I shall look for its reappearance. In the early morning, for an hour or so of daylight, we had small hills cropping out of the plains, but we soon left them, and since then have seen nothing but a dead level surface, unending. Till noon, the Palmyra palm was its commonest tree. Then this half disappeared and clumps of trees of very home-like shape and growth were very common and the plain took the appearance of a western prairie in harvest time, the ground dry and dusty; the grain dry and yellow, stretching to the horizon and meeting there a white edged sky, hot to look at, but not oppressive, 84 or 86 at the highest and about 76 at night in the car. The ride is really very monotonous, but not at all uninteresting, and yet I dozed several times and even read books, instead of looking out the windows with all my eyes.

Patna was the largest city we passed, but there was nothing imposing in our view of its suburbs; crowded and scattered houses of low, mud walls and tile roofs, crooked, narrow dusty alleys, vacant lots with broken tiles and pots and unbroken dustheaps, one little dingy mosque with bulbous domes and stumpy minarets, enclosures of grey walls with castor-oil plant, a few palms and other trees over the houses, altogether a dull, dusty, poverty-stricken place, not the least attractive from where we saw it. The season is against us here as in California and we see dryness instead of verdure everywhere. Close to Patna was a stream channel almost dry. Thirsty oxen tramp it up around the green pools into rough mud, but this doesn't prevent its being used as a washing ground for the indefinite clothing of the natives. (*Tour; 7 March 1878; Benares*)

The main cities like Benares, Agra and Lucknow with their temples and palaces offered an entertaining contrast to the surrounding flatness. The scene on the banks of the Ganges where the ritual bathing takes place appears with especial clarity as he describes the coloured cloths brightening the monotony of the grey steps, and the profusion of wet and half-naked bodies.

So we felt a little disappointed from our first sight, but when we left the observatory and went out on to the river in a boat and saw the ghats all alive with people, it was highly satisfying. It wasn't beautiful; strange and wonderful are better adjectives, if any are necessary; it was so thoroughly different from anything else I ever saw, that our hour in the boat was too short. The Rajahs of different states build fine houses, or moderate palaces – I hardly know what to call them – and come to Benares for a few days every year to bath in the river and worship in this peculiarly holy city. These are strung along the bank almost to the exclusion of other houses, and being higher than the ordinary houses add a great deal to the effect. The lower part of their walls are plain, above they are ornamented with balconies and colonnades. Among them, or on the ghats below are a number of Hindoo shrines or small temples of the conical roofed kind, and a Nepaulese temple, with carved woodwork and at the lower part of the city is the Great Mosque, standing at the end of the princes' residences and famed for its tall minarets. All these are especially interesting, but are overshadowed by the ways of the people, which so absorbed me as we pulled down the stream that I have a very confused idea of the buildings.

Men and women lined the lower steps, coming down to the river, bathing, coming out again like drowned rats, with their clothing clinging about them. Many of the dresses are white, more dingy, some are of dull colours, and a few, of bright, but occasionally they are grouped so that the 'eastern sun' lit up a very pretty sight and there were always some red or blue or orange to relieve the grey monotony of the ghats and palaces. (*Tour; 9 March 1878; Benares*)

The Indian custom of burning the dead he found slightly revolting.

So now the river bank is quite given over to the bathers, except in a few small breaks, where building stone and fire wood are for sale, and at the Burning Ghat, where the peculiar rites interfere with the other forms of worship. At this Ghat there are no steps, the heaps of wood are on the dusty shore close to the water, and there is very little ceremony. We saw quite enough in a few minutes there. Our guide told us the fire was *sold* to the mourning family for from five to fifty rupees a light, or even more, if the deceased member was a person of high position. There seemed to be no priests, no services, nothing but lighting the wood, but our look was hasty. A corpse lay in the water wrapped in a white shroud; close by a heap of wood was being prepared for it. Next to this we saw the widow of another dead man light his funeral pile with a torch of dry reeds, she was dressed in white. Other piles were in various stages of burning from the full blaze to the dying embers, with unpleasant intermediates of charred feet standing out from the fagots. On seeing them unburnt, an attendant poked them into the middle of the fire. There was *no unpleasant smell* noticeable, although we were within a hundred feet from the place. (*Tour; 9 March 1878; Benares*)

The two travellers visited the Taj Mahal four times but, perhaps because of its exceptional renown, Davis seems disinclined to concede any extreme praise.

The difficulty in judging of the Taj is in casting aside the opinions of others which one can hardly avoid reading before getting to Agra. It has a considerable number of worshippers who call it 'the most beautiful building in the world', but I shall try to describe it independently of others. (*Tour; 1 April 1878; Agra*)

He was certainly impressed, though he emphasizes its size above all else. The approach and the garden setting are additional features which Davis feels account for its universal popularity.

There is no preface to the sight. As you step in to the doorway, the long straight alley is open ahead and the Taj is at the other end. The front arched-doorway and the dome in full view, the sides hidden by trees over which rise the minarets. From here there is always shade to walk in, and flowers by the paths. You can go up the middle of the garden with the great building in sight all the way, or you may turn aside under the trees, and not see it again till close by, where there is a little open lawn, with seats in the shade and there rest awhile to take it in. But if you are happily left alone, you can appreciate how much the garden adds to the beauty of the tomb. Outside of the walls is a desert of dust. Within, everything is green, and fresh and shady. Without the garden, the Taj wouldn't be half what it is.

What struck me most on entering the gate was the great size of the building, as we saw it at the other end of the garden; it is enormous. I may have read of its being 186 feet square and 115 feet high, and the point of the dome reaching 213 feet and the top of the metal pinnacle, 243, but I had not realized what the measures meant. An elaborate building about as large as 'Roadside' would have satisfied me, but instead it is about *twenty times* the size of that haven of rest, and comparatively plain. The Mosque and Tomb and gate at Kutab are much more decorated. By this plan, you can see what a strong element symmetry is in the arrangement. The court where the road enters, the garden, the platform and its buildings, are all symmetrical about a north and south line, the latter made so in a remarkable way by building the false Mosque to balance the True Mosque, alike in everyway, except that one is useless *as* a Mosque, because its niche is to the east instead of to the west. (*Tour; 1 April 1878; Agra*)

While he never commits himself to a final grading, Salisbury Cathedral was a favourite with him and is brought into use on other occasions when he wished to evaluate a particular architectural masterpiece.

And still, after all, when I think of Salisbury Cathedral, its close and its cloisters, the Taj must do its best to hold the preference. (*Tour; 14 March 1878; Agra*)

Simla, as well as providing a rest, offered a distant prospect of the Himalayas.

The mountains are so far away that none of them show a very striking outline, and the outliers and central peaks are all thrown together in one range, but they are fine enough to make us wish that we could take a fortnight's camping trip up to the Sutlej among them and see their glaciers coming down into the valleys. That is hidden from here by intermediate ridges, but I can only mention this to counteract the superlative delight that the sight of these Great Himalayas gives us. To think that we should come here and gaze on them! To think that we should be here only a year after Professor Whitney was telling us all about them, with maps and sections, but without photographs. Shan't I just bully him when I get back!
(*Tour; 24 March 1878; Simla*)

Shortage of time prevented any attempt to ascend the nearby summit and they had to be content with photographs and pony rides in the adjacent hills.

After the harsh magnificence of the Himalayas and the elevation of spirit and body they returned to the lower but livelier cities of the Ganges Plain. Here, in Lahore, Davis recognizes the primitive curiousness of an oriental market – as he says, perhaps the same as it appeared in the tenth century.

In some places, the carriage could not go; the streets were too narrow and crowded, so we got out and walked. I wish I could give you a faint idea of what we saw, but it is quite beyond me, especially as it is all now two weeks old, but it was completely and delightfully eastern. Not beautiful, but so strange, just like a dash into the Arabian Nights, turbans and tunics, men sitting cross-legged in the little shops, women sometimes wrapped up so they could hardly see to walk, narrow streets without side pavements – conveniences unknown in the east that we have seen, except where introduced by Europeans – houses of two or three stories, with balcony windows above. Everything so different from our western way. The shops are generally very small, and the stock on hand smaller. An ivory-carver, a metal worker, a cap maker, will sit in the alcove that serves him for a shop, raised above the street so as to be on a talking equality with his customers, cutting, hammering, sewing, all in full sight from the street. His stock consists of what he can make over what he sells. On several occasions we have bought out one or more articles.

The large shops, what we should call 'stores', are generally up a narrow, steep flight of stairs, in a room opening onto the street in front and an upper open courtyard behind. (*Tour; 1 April 1878; Agra*)

The working conditions of a shawl-weaving workshop in Amritsar were governed by a similar timelessness and, for once, Davis seems genuinely affected by the miserable lot of the native weavers.

One of the sights of Amritsar is the weaving and embroidery of shawls. Of course this suggests a factory-like building and rows of men working together, but you know they don't work in that way in the east. Just as it was in Japan and China, so we found it here. Dingy little rooms. Three or four miserable men working hard on delicate patterns and earning a few cents a day. First we saw the woven shawls, of course inferior to the hand-patched and embroidered. The looms are rickety-looking, the surroundings suspicious, and the men thin and hungry, but the shawl is fine. The work was very slow and laborious, no quick throwing of the shuttle back and forwards, something had to be lifted or spliced or untangled at every stop. There was nothing lively in it and when we were told that the poor fellows get only six cents a day, it seemed melancholy. I hope they don't tire themselves.
(*Tour; 1 April 1878; Agra*)

He was quick to notice that the crafts were localized and to suggest a reason for this.

There is a surprising localization of handicrafts. Each place has its speciality, so when you see a thing you like, get it at once for you will not find it elsewhere. This may be accounted for perhaps by the connection between crafts and castes. The father's business is inherited by the son, but it implies very little travelling about the country if that is the true explanation. (*Tour; 1 April 1878; Agra*)

Commercialization of religion always aggravated him. He expected believers to practise reverence. While his own religion taught him to suspect formal doctrine it also led him to expect natural piety to coexist only in association with a serious approach to worship. To him a casual attitude automatically indicated a lightness of faith. The Sikhs gained his respect for their pious humility and sincerity of belief.

This was the only place in India where we had to take off our shoes. You remember that this was always required in the holy places in cleanly Japan; ... Here, even in the mosques, *we* walk shod, although the Faithful always kick off their slippers on entering, but they do the same at any other house. The bhistis' at the railway stations left their shoes on the platform when we called them into the carriage. But the Sikhs are more particular and we respect them for it. (*Tour; 1 April 1878; Agra*)

Bombay was the last they saw of India. The short time they had to wait for the boat was spent at the library, some local caves, shopping and catching up on letter-writing. Now they were due to leave Davis was beginning to realize how much remained unseen. He promises himself to come again, though we know he never did.

On the way to Africa their seafaring confidence was temporarily disturbed by a bout of seasickness in a stormy approach to the Red Sea. To save time they disembarked at Suez and took the train to Cairo. Already they were beginning to anticipate their return.

I feel quite at home in Egypt, just as if we had been here often before, because I suppose, we are now so near home. (*Tour; 20 April 1878; Cairo*)

Egypt was as brief a visit as anywhere. Except for the railway trip from Suez via Izmailia and Zagazig to Cairo they saw little of the country, yet it was one of the more enjoyable parts of the world tour. There was a mixture of expeditions to the Nile, the palace and museum, mosques and the physical excitement of climbing the Pyramids. Davis' spirits invariably reacted to two kinds of stimulation, either a lively exchange of intelligent conversation or a test of physical exertion. You can sense when this happens because his letters become so much more amusing and enthusiastic. The rail journey north gives us a number of good descriptions which are sharper drawn than those of the mountain areas because, perhaps, the topography presents simplicity and greater contrasts.

Suez is hardly worth describing. It is on the flat sand at the head of its gulf, midway between two ranges of hills that on continuing further southeast bound this arm of the Red Sea. Its houses are grey, dusty, and dirty, tumbledown. Its people are so dirty that all their other character except their rapacity are hidden. That would shine through an iron mask. The boys carelessly say 'Buckcheesh' as a stranger passes, merely from force of habit. I wonder if anyone is ever foolish enough to give them anything. We are not.

Our last look showed the Red Sea to be a dark blue, a wonderfully clear, deep, dark blue. So were the Bitter Lakes, when we came to them, half way to Ismailia. The desert was yellow, glary, gravelly, sandy, sometimes rocky and hilly, and always barren. Along the edge of the fresh water canal, a few feet of dark green sedges made a well marked line, where they could be seen over the dredgings of sand on the banks. Away from this, only a few, scattered plants could live, miserable specimens, not so large as the sage bush on our western alkali plains. There were no trees at all. We saw a few vultures, several crows and some other birds, names unknown. In some places the sand was worn into many faint paths, as if by wandering cattle, but none were to be seen wandering. Perhaps donkeys had done it. We passed a few of them, browsing on the sedges. Most of the hills were of the flat bedded kind, do you know them? . . .

The valley of the Nile is surprisingly narrow here; not over ten miles of low level fields, which are covered by the high water. Then on either side rise yellow, sandy hills and the desert begins at once. There is very little debatable ground. The change from thriving wheat or corn to drifting sand comes in a hundred feet. Back of this line is a dry barren slope under the rocky hillface, then a more or less abrupt rise of one to six hundred feet and beyond this, the rolling desert, such as we saw about Suez. The highest bluffs over the river plain are on the eastern side, just south of Cairo. Close to the city are quarries from which much sandy lime stone has been taken across to the pyramids, as far as I can learn. I think there ought to be some other word than valley to name the Valley of the Nile, it is so different in form, though not perhaps in cause, from a mountain valley; it suggests too great a difference of height between the middle and sides, and too great a slope toward the stream. (*Tour; 21 April 1878; Cairo*)

The Pyramids interested Davis because of their educational importance. He had read up the subject carefully and is quick to explain to his relatives that some of the pyramids were constructed from the rock on which they were built, others from quarries in the hills a little behind Cairo and not from much more distant sources as is often thought. Despite this preoccupation with facts, he probably obtained his greatest thrill from climbing one of the smaller and lesser known pyramids. He completed this venture on his own by starting early in the morning and so avoiding the heat of midday and the main concourse of tourists. His motives were partly independence and partly a desire to reflect on the scene alone; economy was not a reason on this occasion.

The others seemed content with the assurances of the men that this was the best corner, so I was soon alone and before long, on looking up, I saw each set of guides with a victim, leading, lifting, pushing or hauling him up the steps in a way that made me blush for my country. Before starting up, I had another talk with my men, Abdul, Mohammed and Ali, telling them that I would give an 'extra' to them if they would be quiet, and speak only when spoken to, and considering their bad education they did remarkably well.

There is not the regularity in the courses of stones or the preservation that I had expected. The corners of the pyramids are pretty clean, swept off for convenience of climbing, but elsewhere they are half filled with dust and sand, making them uncertain footings, difficult, slippery, perhaps dangerous to mount. The highest steps are about four feet, most of these are in the lower courses, and I think everyone can be avoided if one wishes to have very easy work by taking time to look about, and walking along to one side or the other to where an intermediate stone or a break makes the rise less difficult. The Arabs, if allowed to have their way, grab your hands or arms, and go up on the rush, thinking that you are anxious to reach the top as quickly as possible. Then you have no time to look about, to examine the fossils that Pliny called something like lentils, to see how the soft stones are slowly flaking themselves to pieces, and above all to sit down quietly in the shade and look at the other pyramid close by, at the tombs on the platform, at all parts of the wonderful view, the unequalled view, not for beauty, but for thoughts. I took a particularly vicious delight in doing this, after seeing the surrender of the other fellows, so I was somewhat later than they were in getting to the top, but I am pretty sure from the time it took me, that I could easily make the ascent in eight minutes, including two breathing spells. Really, as a climb it is nothing at all, and how it has earned the reputation of being inaccessible without help from the fierce Bedouin, is most remarkable. (*Tour; 21 April 1878; Cairo*)

After careful preparation, including changing braces for a waist strap, as being easier to bend in, he also explored one of the tombs by himself, and greatly enjoyed the experience.

It was quiet and dark too before I lit the candle, the perfection of both. I thought of the great depth and weight of stone around me, the almost eternal safety of any treasure placed there, against everything but man; the ages it had lived through, the ages it would endure. There was the poor scarred coffer, silently bearing all the cruel blows from ruthless strangers. I could no more strike that venerable stone than – I know what – You may imagine – but I patted it and promised it safety as long as I was there. And those impudent names, painted, scratched, even cut on the speechless walls. What abominable impertinence! I hope they are all copied off in the Black Book and that the writers may sometime have a long, hot, orthodox chance to repent over their autographs. (*Tour; 26 April 1878; Suez*)

On their way by boat from Alexandria to Spain they stayed for five hours at Malta and slightly longer at Gibraltar, where they climbed Ape's Hill. The following description of Gibraltar is actually made from the boat.

Really its appearance varies according to the side from which you see it. Geologically it is quite wrong, absurd, impossible, it has no business at all to stick up as it does, with no visible companions, except across the sea; it is ridiculous. What a place for England to have, armed with hidden guns up on the hillsides under the rocks, that will carry balls over to Africa, and lower down, on natural or artificial parapets, with smaller artillery to pepper ships nearer at hand. In case of war, which by the way is not yet declared, I think Malta and Gibraltar at least are safe. The eastern side of the Rock needs no defence; it has not a flat shore, about a third of its length is bordered by a sliding talus of loose sand, the rest is sheer rock, standing in great vertical steps, quite inaccessible from below; a wonderful place, surprising in spite of all we had heard of its grandeur. We rounded the southern point, where the lighthouse, barracks and batteries occupy the lower ground, here not naturally so strong as on the two sides; then entered the harbor, and anchored opposite the town, a little way outside of the fleet of ships of war and traffic that are moored off the debatable land, the low ground between the English point and the Spanish Main.
    The west side of the rock is a great contrast to the east, a gentler slope of only about 35 degrees; a little flatter at the base where the town is built. The upper part all rough with outcropping edges of rock, tinged green and yellow with grass and flowers, a few patches of small trees back of the houses; slanting lines of paths zigzagging in long stretches to the signal posts or tower on the top; yellow and white houses, stepping one above the other from the heavy wall along the water's edge. The fort proper, enclosed by wall all around, except on the steep eastern face, occupies about half the point, the southern half. The rest is not quite so military, but is still well fortified along the shore. (*Tour; 7 May 1878; Gibraltar*)

Gibraltar's apparent geological anomaly is explained a little later.

Davis' years in the Argentine had given him a working knowledge of Spanish. This perhaps encouraged him to travel widely in Spain where they spent a full month and tried every form of transportation from railway train to mule and donkey. In all they made many stops including Ronda, Cordova, Málaga, Seville, Granada, Linares, Almadén, Madrid, Segovia, Ávila, Burgos and Zaragoza in a long traverse of the country from south to north. In the towns their attention was restricted to paintings, cathedrals, monasteries and other such interests. Davis did visit the lead mines at Almadén but it was the intervening journeys that were valuable as these gave the two young men a chance to assess the true character of Spanish life. They saw the villages, the means of agriculture and the peasantry in their natural state against the grim, dusty setting of mountains and plains. En route he was able to resolve the geological problem of Gibraltar:

> Gibraltar is no longer such a geological puzzle as it was before. Opposite Gaucin was a mountain, called the Sierra de Casares, after a pretty little town on its southern side, that is an inland repetition of the great fortress and about double its height, a great mass of hard limestone, breaking up through the surrounding sandstones and softer rocks. Today we have seen other cases somewhat similar, many of such evident interest that Tom says, when I come here with a geological camping party, count him in. Perhaps this isn't altogether on account of the rocks, but when you can see the exact plane of separation along which two huge masses have been pushed past each other; when the position of such a plane can be determined by the line it makes running over spurs and down ravines, a very rare occurrence; when the layers are bent and twisted and curled like – like everything, then even Thomas awakes to the charm of terrestrial dynamics. Doesn't it make even *Uncle* Thomas also perfectly wild to come and see it all? Really though, it is the best small patch of geology with variations that I have seen since last summer in the Catskills with the school. (*Tour; 9 May 1878; Ronda, Spain*)

The profusion of spring flowers and the geology filled his mind much of the time. Another reference to geology shows that his attention is already directed to valley formation.

> This little valley and the Nile and the Great Plain of Hindustan and Bengal on a still larger scale, are all alike in being sand and clay fillings in a hollow of greater depth formerly. The Charles and the Mystic are the same. I should like to find a river near the sea that is now cutting its way deeper, instead of silting up. (*Tour; 10 May 1878; Gobantes*)

There are some interesting references to chance acquaintances with lady travellers, including a charming Miss Scott, niece of Professor Prestwich, the Oxford geologist.

> You remember the *pleasant* English ladies at dinner table at Ronda, we have seen them again. They told us they expected to come back here, so as we came near

Gobantes yesterday afternoon, we say them on the platform and I jumped out and helped them into a carriage; they went first class, we were trying second as a streak of economy, but with such company in the other car, our resolutions vanished and at the next station, we asked permission to join them; they gave it willingly; we paid the difference of fares and had a very pleasant evening. Tom sat opposite Miss Ewarts, and I opposite Miss Scott, which suited me exactly, and Tom too, as he likes them equally. Miss Ewarts is extremely pleasant, but Miss Scott is perfectly delightful. She reminds me of Aunty Pat, is about of the same youth and even as pretty, and she blushes most charmingly. Her companion is the stronger-minded, the leader in the party, not at all pronounced in her manner, though both are beautifully English all over, but decided and quite able to manage a trip through Spain; while Miss Scott is perhaps not quite so timid, but as I said she blushes divinely. We talked about three hours and then slept three more. I heard Tom carrying Miss Ewarts all around Tokio, and over to Kioto, introducing her to Kimura, and afterwards discanting on Hindistani Architecture. Somehow, Miss Scott began to tell me of a trip she had made with her father and sister to the United States in 1859. They had met a Boston gentleman in England and he had invited them to visit him. 'Perhaps you may have known him, Dr. Charles Loring' – exclamations – 'and he had a stepson, whom I want to hear about, can you tell me?' 'You don't mean George Goddard!' – but she did, she really did, so a minute account of George Goddard was rendered her. 'And we visited a gentleman who owned an island.' At a venture I said, 'Mr. Forbes at Nanshon.' 'Yes that's the name.' By all that's strange! and she asked after his son who went into the army, of whom I could tell little besides his name, and one or two other things. Philadelphia was included in the trip 'and we had letters from Mr. Martineau to a Unitarian clergyman there' – 'Mr. Furners' It even was! He introduced them to some Quaker families and though she couldn't remember their names, I ventured to bet one of the houses was '338' only perhaps in 1859, there already wasn't any 338 any more, but the names couldn't be identified. Then we went over to England and fell to talking of walking and she told how she enjoyed walks with her uncle who is 'something of an antiquarian and a geologist. He is professor of Geology at Oxford now.' 'Is it Mr. Prestwich?', and it even was again! So I confessed everything, and if I only have time, Miss Scott assured me her uncle would be glad to pilot me over some interesting ground. Wasn't this all passing strange.

As we are without guide books, we have looked everywhere without finding any, these kind ladies lent us theirs for this morning. When we returned it at lunch, just before they started for Madrid, Miss Scott offered to buy us a book there, where they probably would be for sale, and send it by mail to us in Granada, or Seville. How many persons would do this! Not one in a million! So it was with hidden tears that we said farewell at noon, and with hopes of meeting again beyond the capital, at Segovia, Avila or Burgos. Indeed I am deeply, eternally inextricably in love with Miss ... but this is a secret. If for nothing more than to see what a delightful tour these ladies are having, it would do you good to see them, Nancy. They go where they like, carrying their sketch-books and paint-boxes, and drawing pictures that made us envious. (*Tour; 12 May 1878; Cordova*)

Davis saw southern Spain at a good time of the year and is often charmed by a panorama or particular feature.

> Occasionally a lone house, sat on the hillside, approached by slanting paths from several directions, but the little villages hold nearly all of the population, and we passed by or through or opposite to a number of them, all alike, red and white, compact, on a convenient shelf on the mountain side. No wheeled wagon can approach them, everything must be packed on donkeys or mules, and life there must be very primitive. Menali lay just below us; its clock in the church tower was trying to strike ten as we passed, but the quarters before and after seemed blended with the hour. This town also has its ruined little castle on a projecting point of rock, but it was too much below the horizon for a sketch. (*Tour; 13 May 1878; Cordova, from Gaucin to Ronda*)

The guide books had led him to expect poverty and he comments on the prosperous appearance of the villages and cultivated fields of Andalusia.

> These were very fertile districts and the crops indicate good farming. What we have seen between Cordova and Seville and Seville and Granada, refutes the Guidebook's statements about the backward condition of Spanish agriculture, but one can't judge well in a railway journey. The fields were clean and neat, and free from poppies. The olive orchards are regularly laid out and are kept tidy; the rows of trees made pretty lines of dull dark green over the farther hills. There were some vines too, which I was affably told yield excellent grapes. The towns are more thrifty looking than I had expected of Spain, close and compact, bearing the stamp of antiquity in their old castles, and of devotion in their great churches; of care in the freshly whitewashed walls and generally well swept streets. (*Tour; 19 May 1878; Granada*)

As he said himself, however, the prospect from a train can conceal much and some of his later journeys by donkey showed him the poor state of the peasant people and the meanness of their dwellings.

His description of the Moorish mosque at Cordova is quite moving but is followed by his normal antagonism towards other religions.

> It is certainly a wonderful sight, row after row of pillars, open arches all woven together, vistas direct and oblique, all across the mosque an uninterrupted nave, or diagonally among seemingly irregular columns to the walls of the cathedral in the center; and to add to the light and color from windows high and low, we had chants of music from the choir and organ, so beautiful, it was fit to make a fellow weep. On our long trip we have had no music but from piano, a few voices, and some indifferent military bands, sometimes interesting, but not moving, so these beautiful cathedral hymns were all the more touching. How well chosen the organ, and the chorus of men's and boy's voices are to give emotional religious effect. I can imagine only one thing that would increase their power. There were impressive processions with solemn chants, passing among the pillars, pauses for intoning holy prayers, bursts of music after quiet silences, when all but us unbelievers were kneeling and bowed down with reverence. We stood half hid by the columns,

or sat on benches in dark corners, taking as large a share of this descent of heaven on to earth as unhappy heretics may, and if a kind fate will sometime again take me on Sunday morning to the Cathedral in the Mosque at Cordova, I shall be very thankful. Confessionals are marked on the plan by a cross, not very many of them, and none occupied by so favourite a father as the one Camacoes has painted. If it was only not abused, if the father confessors were only the benevolent old men one likes to imagine instead of the old rascals they sometimes are, the confessional would be a beautiful part of the Church. Although the priests bow and kneel and observe many forms that we cannot, I think we feel more the sanctity of their cathedrals than they do. All the service is such a trite matter of fact, of every day's occurrence, they change so quickly from praying in their holy robes to joking in their ordinary clothes; the choir boys are such nasty, impudent little fellows. There is very little piety apparent in it all, and there is much levity that is very shocking to an outlawed heretic. (*Tour; 14 May 1878; Seville*)

His fastidiousness comes out well in his vigorous condemnation of a French traveller who had attached himself to an unwilling listener. For each article of blame we must assume that Davis acted otherwise, or at least thought he did.

At supper at the 'House of Guests', the Frenchman again favored me with an account of Etna, and a mention of the French Alpine Club, till I was about wild, and resolved to escape from him that night; so I hired a cart, drawn by two mules tandem, and driven by a stupid fellow who could hardly understand my elegant Castillian, and from nine o'clock till midnight, jolted to Almadenejos, dozing as we rambled slowly along the road, delighted to have left that nightmare behind me. You should have seen him! His grin and squint, his long slim fingers, the way he had of screwing up his face at the end of a sentence and bending his head to one side, and coming close up beside me till I was forced to put out my elbows to keep him off. You should have heard his constant talk, the same things over and over again, and indifferently in French, English or Spanish. His complete oblivion of time and place; he wore his hat at meals; when I left him, the railway dust of his morning's ride still begrimed his face and neck and hands, and as for his shirt and collars and cuffs, they must have ascended Etna with him. But I hope that they also are not members of the French Alpine Club. Do you wonder I fled! (*Tour; 27 May 1878; Toledo – Refers to Almadén*)

While Davis sometimes showed much talent in his sardonic descriptions of acquaintances he was unable to appreciate the humour of situations involving himself.

When we came here last night, to this comfortable 'Fonda of the 2nd Mayo, called of the Englishman', we were pleasantly welcomed by the landlady, and shown to our large front room by a brisk maid. After a short preparation we went down to a late dinner. A vegetable was served in due course, and having forgotten its Spanish name, I ventured in confidential whisper, to ask the waiter what it was called. 'They are called Beans, sir, STRING BEANS, Sir,' he shouted, really roared it through the Hall. He must be one of those misguided creatures who thinks all

foreigners are deaf. If I only get a solitary chance, I'll sit on him before I go.
(Tour; 4 June 1878; Avila)

No doubt for one who prided himself on his 'elegant Castillian' Spanish, it was rather aggravating to get such a disconcerting answer in English.

The cousins left Spain somewhat impressively by crossing the central Pyrenees in sight of the Maladetta. The journey took three days, by cart, by mule and parts on foot and they enjoyed it thoroughly. They reached France on 13 June and spent a little over two weeks there, mainly on the volcanic region of the Puy de Dôme and the chateau country along the Loire. Some of the enjoyment of the hill walking was lost through a number of rainy days but the fine ones that succeeded them repaid their perseverance. While in France the two cousins split up for a short time, Davis staying in the mountains and his cousin going to Paris.

For Davis, Paris and the Loire valley were not new; he had seen them twice before with his family. In the capital the cousins later rejoined company and their behaviour, perhaps fittingly, was almost libertine. They tried a circus and went to the theatre two or three times. In the other towns their pleasures were characteristically more circumspect. At Blois, Tours and Chartres they dutifully admired the cathedrals, at Fontevrault the abbey, and all along the Loire they inspected various famous chateaux. History was not allowed to dominate. The countryside about still caught their eye and previous acquaintance with France in no way weakened Davis' excitement. The Cere valley is picked out as exceptionally notable.

> ... the little valley that we followed from Aurillac up to its head is all simply a gem. It is narrow and deep as valleys ought to be, luxuriantly green, the fields divided by rows of poplars and ash trees, grass and wheat in excellent condition; cherry trees well fruited by the villages, and the villages the most picturesque possible, each with its old church, or ruined castle, or restored chateau. Note the place well, and some time, in the millenium, we poor Americans who have no such valleys will come here and walk slowly from village to village along the Cere. Tom and I were in high spirits that morning, and having a compartment to ourselves, we carried on like wild, rushing from one window to another so as to lose no little glimpse of rock or town or tower, and striking Anglo Saxon attitudes of admiration and delight, sketching at every station where anything could be seen worthy of our pencils, sketching even between stations with wonderful agitated result, sketching even from memory, which I once succeeded in carrying so far as to reproduce a chateau with a tower, two turrets, much roof, many windows, uneven doors, an outbuilding, a terrace and many trees. I dare say that it lacks the exactness that a Beaux Artist could reach, but considered geologically, it recalls the subject very nicely, and considered anatomically it suggests the *type* satisfactory. *(Tour; 18 June 1878; Mont Dore nr Clermont Ferrand)*

It is noticeable that Davis' geographical training aided his observation of the physical and cultural landscape.

FIG. 18. Photograph of Davis in his twenties taken in Paris
(Courtesy E. Mott Davis Jr., Austin, Texas)

The sides of the granite peak were too steep for easy walking, a slope of loose sand and gravel with projecting edges of rock, or else a matted growth of high grass under pines, larches and spruce near the top, and beeches and birch lower down. These trees are all small, and are arranged in very regular horizontal lines around the mountain side, almost as if they had been planted so, evidently the result of cattle grazing there years ago and treading the slope into paths on which of course all the young trees were killed. In the crater of Puy de Parion the same lines show very distinctly in the grass, real contour lines, expressing the shape very well indeed. On some other mountains, like Puy de Dôme, the land or pasturage on the slope has evidently been divided taking the centre of the top as a starting point for the fields at the base all radiate evenly, and the limits of growth of trees of different size, showing clearings at different times by the various owners, are almost geometrically exact. (*Tour; 22 June 1878; Angers*)

Perhaps the most appropriate memorial to their departure from France is Davis' description of lovers in Fontainebleau Forest. At twenty-eight, he was well behind the moral standards of his contemporaries, backward in fact of the morals of any age. His attitudes towards the opposite sex, and particularly his thoughts, are reminiscent of the novels of Jane Austen.

I dare say the young couples from Paris, so exclusively from Paris that the Bois de Boulogne is a wilderness, dare not trust themselves alone and unprotected in this unbroken solitude, where the deer still roams and the pheasant feeds in flocks. Therefore, at the entrance to the dangers of the region of the Rock which weeps, several neat old ladies clad in white caps and badges of authority, await the timid pairs, and lead them through the forest. I trust these guides are deaf and blind to all save the forest views and the cuckoo's plaintive sounds, otherwise they might see and hear what was not intended for more than one. *I* ought to have been blind and deaf once yesterday, but for my neglect of such precautions, I shall atone by being forever dumb. Nothing shall ever force me to disclose that I saw a mild young man put his arm around a milder young woman's waist in the full unshaded glare of a July sun. That shall ever remain a secret in my bosom. '*I* wouldn't have done it even in the dark.' (*Tour; 8 July 1878; Pontarlier, Jura Mountains*)

Switzerland was their last real stop. They managed almost a month in some of the main Alpine resorts – Chamonix (France), Zermatt, Eggischorn, Interlaken and Murren. It meant almost no time in England but this was one of several cuts they made to the original itinerary.

Thy letter, father, with Mrs. Bright's and Mildred Yamill's addresses was the one last night. I shall certainly try to get to Beckenham, but where is Coventry? Tom thinks it is up by Scotland. We shan't have time in England. $31 - 21 + 29 = 39$. Such is the equation that we are constantly solving. (*Tour; 21 July 1878; Martigny, Switzerland*)

Davis had been to Switzerland five years before.

Saturday evening we went to an extra concert at the Kursal in Interlaken where Anna and the others tried to escape me when I broke my leg on the Tschingel Tritt, at the end of that trip of Draper's and mine. (*Tour; 5 August 1878; Murren*)

They visited at least three of the same resorts, perhaps because of the convenience of previous knowledge of the hotels. The mishap of his earlier trip did not deter him and most of the time was passed climbing or mountain scrambling. They so enjoyed the scrambles around Mont Blanc that they stayed four days. The weather was poorer at Zermatt and hope of a clear prospect of the Matterhorn was abandoned. At Eggischorn they crossed glaciers with a guide and at Murren enjoyed the view of the Eiger and Jungfrau.

Switzerland made a happy impression on Davis. It had the advantage of fine scenery and glacial features which were particularly interesting to him because he had been taught them at Harvard by the principal originator of the glacial theory.

> I remember one of Professor Agassiz' lectures in which he described the mountains over his old home, where he used to study geology. He drew a figure on the blackboard, which I can recall distinctly and it is satisfactory to have compared my memory of it with the original, even though we had but a flying railway glance. But I am advocate of railway geology. On foot one sees more of course, but from the cars, with the aid of maps and sections of the country, one can learn more than all descriptions would teach. (*Tour; 10 July 1878; Martigny, Switzerland*)

The endless opportunities for walking trips were rounded off by comfortable modern hotels.

> It is remarkable to see how differently the Swiss treat themselves and other people, and the others get so much the best of it. An insignificant town called Plattin, half a mile above here, is about the dirtiest hole I ever had to walk through. Talk of Japan and China; we saw nothing there to compare with it. And yet, in this hotel, everything is spotless. Perhaps the keeper of it knows enough of what is expected of him to keep his house clean. I wish our hotel men at home were as wise. In New York state, there is a semi-decent town called *Moscow*. (Sometimes I should like to write an essay on the names of towns in New York.) The Hotel there is a disgrace to the Declaration of Independence and a blot on the Constitution. Possibly our noble sons of liberty think what is good enough for them is good enough for strangers, while the downtrodden republicans of Switzerland, descended from ancestors properly brought up under ducal and kingly sway, know very well how absurd it is for strangers to be slovenly merely because they are.
>
> I suppose the Englishman has done more than any one else to raise the hotel standard here. He grumbles well, and the landlords submit, and instead of swearing back at him as they would in freer countries they grovel and pander to his luxurious tastes. I like to follow in his emphatic tracks. (*Tour; 1 August 1878, Ried, in the Lotschenen Thal*)

At the time of their visit mountain tourism was only just beginning and Switzerland was the first country to become popular. As Davis pointed out the resorts were dominated by the English intruding their distinctive attitude of overlordship, even to the extent of establishing their own temporary places of worship.

> The two hotels take in all the strangers that come, and yet those practical worthy English are building a little chapel here so they can spend the Sunday quite to their mind. There is such a thing as carrying a thing too far. Why not use the

hotel parlors alternately, and give the money that the church costs to buy hand-kerchiefs for Japanese children. (*Tour; 5 August 1878; Murren*)

Davis, belonging to the American minority, was very ready to criticize the English.

We are now in the country of the Tourist. Switzerland is nothing but a summer boarding house, and *some* of its boarders, goodness gracious me, what people. There were two specimen sets in the cars to-day. First, some Americans, with what is known as the American voice, really most disgraceful beings. Tom and I talked Japanese and Spanish. Of the former, we say, 'I don't understand.' 'How much?' and count up to ten; it sounds well. In Spanish, I talk to Tom and he says, 'Como!' and 'Si, Senor.' Then we add 'Tum Hindustani bol-sakte?', so of course we are taken for distant foreigners. If Tom could only talk Spanish, we should have larks. The other party was an Englishman and two daughters. The fond father had been here before, and was now bringing his daughters to show them the beauties of the country. He explained everything thoroughly. There was an old French lady, or Swiss lady, in the car with us, on whom the Englisher looked with great suspicion. When he entered, she had politely said to him, 'Lay your bundles with mine if you wish,' and he not understanding a word, at once picked up his bags and carefully set them out of her reach. I daresay Uncle Tom and May would smile at my French, but I am glad that I am not like such Englishmen. (*Tour; 10 July 1878; Martigny, Switzerland*)

Switzerland had much to offer in spectacular geology and his tutelage under Agassiz had taught him what to expect and where to look for it.

We upset a good-sized rock table, just for the fun of seeing it fall, the pleasure of destruction. These tables are rocks large enough to shelter the ice under them and prevent its melting; so while the surrounding surface melts away, the rocks are left supported on a pedestal. The largest we saw were ten or fifteen feet in diameter and standing four or five feet above the average surface. There seems to be no limit to the size; the larger the rock, the longer its pedestal will endure. Finally the ice evaporates away, and they fall over, to begin their work again.

When a stone is ten inches or a foot in diameter, it is about neutral, but smaller than this, it absorbs and transmits heat to the ice, on account of its generally dark color, and so sinks into the surface instead of rising above it. The large rocks are too thick to be heated through in one day, and so serve as protection. Black sand melts the ice faster than anything else, and is generally found at the bottom of a little well six or eight inches deep. The shape of the well agrees with the outlines of the patches of sand, sometimes very irregular. (*Tour; 25 July 1878; Am Riffel, above Zermatt*)

The folded strata received mention simply because it was spectacular.

The wonder of this day's walk was the bending of strata in the rocks, almost incredible, but from our lunch station, we could see distinctly one stratum folded so as to appear *eight* times in a single vertical section, thus: from 1 to 8 was about 2000 feet! It was in the Ferden Rothhorn. After seeing this I think I can believe

anything. An hour or two later, we found some still more remarkable contortions.
(Tour; 3 August 1878; Interlaken)

England was over in a matter of days. A visit to Canterbury Cathedral, tea with Miss Scott and the boat home from Liverpool. No time for impressions nor even descriptions. Home was too close to make letters worth while. His last reflection was on the value of all the letters written.

What a stack of sheets you must have, and what fun it has been writing them, not quite time enough sometimes to do them up with care and completeness I wished. I am very glad we have kept at writing, for I am sure a great deal would be forgotten if we hadn't it down on paper. But I have no idea of making a book. I shan't have the time. Besides when I do write a book, I should rather have it something thorough and trustworthy, than merely a hit or miss set of opinions formed while rushing through a country. There are plenty of such books already, and there will be plenty or too many more soon. *(Tour; 15 August 1878; London)*

The reason for not putting them in a book is pure Davis. Our reasons for devoting so much space to them are twofold; firstly, because they greatly illuminate our understanding of the complex personality of the twenty-eight-year-old Davis, and, secondly, because this represents the first attempt to analyse the 900 or so typewritten pages which were subsequently copied from the letters.

CHAPTER EIGHT

Near Failure: First Marriage and University Apprenticeship

MARRIAGE

On his return to America in 1878, Davis resumed his duties in earnest as Instructor at Harvard. Just over a year later he was married. The bare details of the certificate are given below:

GROOM	BRIDE
Name. William M. Davis	Name. Ellen Bliss Warner
Color. W	Color. W
Age. 29	Age. 25
No. of Marriage. First	No. of Marriage. First
Residence. Cambridge	Residence. Springfield
Occupation. College Professor	Occupation. ——
Birthplace. Philadelphia, Pa.	Birthplace. Springfield
Father's Name. Edward	Father's Name. B. Frank
Mother's Name. Maria M.	Mother's Name. Charlotte E.

Place and Date of Marriage. Springfield – Nov. 25, 1879.
By whom Married. A. D. Mayo – Clergyman – Springfield.

Davis had known Ellen Warner for some time but the marriage path

> was interrupted by his invitation from Thomas M. Osborne, who was about to enter Harvard, on a year's trip around the world. There was a large family gathering at the time of departure and I believe that Miss Ellen Warner from Springfield, whom he later married, was among those present, since she was a close friend of an Osborne sister. On the return there was a similar gathering. I do not know whether that was 1877 to thereabouts, but they became engaged and were married Nov. 25, 1879. (*Interview: R. Mott Davis with R. J. Chorley, 7 September 1962*)

Characteristically his surviving correspondence gives not the smallest hint of all this. On the contrary, we have seen how he had been writing in affectionate terms to Emma Mellor while on the world tour.

> Going up Pike's Peak the other night, I was thinking what I should write to thee, Emma Maria – for it was high time to write – I had been anxiously waiting for an opportunity for about ten days – Don't think I meant a little joke when I said it was 'high time' altho' it was a very high time – I meant quite time.
> It was a good time to think – it was dark and cold – I was on a pony – all wrapped up in an overcoat and a white blanket – my hands under the blanket – the

reins loose – the pony picking his way for himself [Fig. 14]. Tom said the blanket was all he could see of me. It was as clear as it was cold – very both – and the stars were just beautiful – I hope, my dear, that thee will make as good friends with them as I have – and learn to recognise their many virtues. I thought out quite a long lecture on Astronomy, that I would have delivered if thou had been on the pillion – about Venus, Jupiter and Mars we had seen the evening before – how the old Shepherds called their 'wanderers' among the stars – how the old Greeks made complicated cycles and epicycles for them to wander on – how Mr. Brahe set them turning about the sun – how old Copernicus included the earth with them – how Kepler puzzled out their orbits – and how Newton at last showed why they did it all – It makes a very interesting series of discoveries, doesn't it – Venus had set before we had gone to bed the evening before – Jupiter went down while we were asleep and Mars dropped behind the distant Sawatch Range after we were far up the mountains – But there remained Sirius – The Zodiacal Light – which was very brilliant that night – The Milky Way and all the little stars. No lack of things to write Thee about – indeed, too many – but I think it would be more fun to *talk* about them – so, some nice sharp winter evening when I am home again we two – My Instruction and her professor – will take a nice little walk and gaze at the starry heavens and talk of them and other things.

It was very still – dreadfully still as we went up the mountain – even the nervous little leaves of the Quivering Aspen hung quiet on their twigs – no katydids – no mosquitoes – nothing but the crunching of the gravel under the horses' hoofs – the rattling of the bits – and a little squeak that I at last traced to one of my stirrups – when suddenly an enormous bear jumped out from behind a bush and rushed at my pony! He growled frightfully and opened his mouth so wide that we nearly fell down his throat before we could stop – 'Oh! Mr. Will, I don't believe it!' – Doesn't thee, Emma? Well, no more do I, there was no bear, I am glad to say – nothing at all of the kind. It was so quiet, tho, that I thought thee would enjoy the excitement of a bear-story just for a minute – Adventures don't happen so easily as that, thank fortune; – I dare say I shall go all the way round the world and come home as barren of good, dangerous, hairbreadth escapes as when I came home from [Cordoba?]. Only think of it, Emma – we didn't know each other then. (*Letter: Davis to Emma Maria Mellor, 21 September 1877; Georgetown, Colorado*)

Indeed one of the most exasperating features of his letters is the almost total absence of any reference to his future wife. Moreover, from the time he left college, his correspondence has only brief comments on members of the family or close friends, and becomes crowded with scientific arguments, detailed descriptions and the interminable narration of how Davis entertained his fellow-guests at the dinner table. It is difficult to decide whether enthusiasm for geography so absorbed his attention as to leave little time for sentimental expression; or whether his nature was such that he deliberately inhibited such confidences as being too private for insertion in ordinary letters. Or perhaps he began to suffer from a deficiency, commonly attributed to all scientists, of human emotion. Regard for his health, expres-

1879

FIG. 19. Drawing made by Davis of an unknown person (his father?) in 1879
(Courtesy R. Mott Davis)

sions of disappointment or disapproval, all occur at different times but
affectionate feelings towards particular persons occupy a relatively small
space in his letters until long after his retirement. As far as his surviving
correspondence is concerned the intimate linking of his life's fortunes with
another human being in marriage – normally a major emotional occasion for
most of us – makes little mark.

EARLY DIFFICULTIES AT HARVARD

As already noted, on his return to Harvard, Davis was appointed Instructor
in Geology. He continued his field work in connection with Shaler's lectures,
and was given sole charge of a course in Physical Geography and
Meteorology in 1878. For a university lecturer he was in no way equipped
either to teach or to specialize in the subjects allocated to him. He had not
received any formal training as a teacher and can have had only a very general
knowledge of the two subjects he was chosen to lecture on. Thus it is hardly
surprising that

His first efforts in this new work were not altogether successful or agreeable, for he was diffident and inexperienced as a lecturer, had no thorough instruction in the subjects he was given to teach and he was painfully aware of his shortcomings. In field geology his experience was larger; and in the interesting district of eastern Massachusetts, where many structural problems embracing stratified and intrusive rocks are associated with phenomena of glacial and marine erosion and deposition, he developed the method of placing second year students on their own responsibility, counselling them to use no term which they could not define, and to make no statement which they could not defend. (*Brigham, 1909, p. 8*)

In two letters written to Shaler just before he was to take up his appointment Davis recognizes his lack of qualification by his preparedness to perform virtually any duties which might be chosen for him.

My Dear Sir –

It is a pleasure to send a letter to you with a home date – We finished our travels a few days since by a quick run across the ocean on the 'Adriatic' of the White Star Line – and am now enjoying looking over the trophies we sent home from Japan and China – and explaining them to many kind friends who profess much interest. I shall be in Philadelphia in a day or two – and thence shall return to Boston about the 22. I fear I cannot get away from home much sooner. Will not the 22 be early eno' to arrive in Cambridge? Is there any extra work you wish me to do during the coming year? Anything beyond Quarries or models? If the Physical Geography still goes abegging, and you care to give it to me, I should be glad to have it – indeed anything will be welcome.

I have only heard that there is a summer school – and this indirectly – I hope it has been successful. Will you please send me a line to 205 Walnut Place, Philadelphia – saying when you will be in Cambridge and oblige. (*Letter: Davis to N. S. Shaler, 5 September 1878; Auburn, New York*)

My Dear Sir

I have your letter of the 30th from Yorks. Just before getting it I had written to you Newport, Ky. – and had spoken of the Physical Geography that you mentioned. I shall be glad enough to get the extra work, provided you think it is something I can do satisfactorily. If it should include an increase of pay, it would please me immensely.

Will it not do for me to arrive in Cambridge about the 23 or 24? – I have a very short time at home – but if it is necessary I can be there by the 20.

I am glad to hear of satisfactory work in the Summer School – and of the . . . down in Maine – it is rather strange that there has been no detailed work on them – I should like very much indeed to be down at Yale with you – but it will be quite impossible, on a/c of the short time between now and the beginning of work.

(*Letter: Davis to N. S. Shaler, dated 8 September 1878; Auburn, New York*)

It is fair to mention that even at the present time the lack of teaching experience is no bar to a talented student progressing to a lectureship. The reference to additional salary may indicate that Davis was now having to depend on his own resources with no supplement, or a reduced one, from his father's pocket.

These first years were probably hard for him and certainly marked by uncertainty. The following letter shows the terms of his appointment and Davis' own notes at the foot tell us the salary he received during the next three years.

> I beg to inform you that on the 30th of June 1879, you were appointed Instructor in Geology for three years from September 1, 1879, with salary at the rate of $1000 a year, and that the appointment was confirmed by the Board of Overseers at their meeting of July 9th 1879.
>
> (*Letter: President Charles D. Eliot to Davis, 21 July 1879; Cambridge, Mass.*)

To this letter Davis later added the following note:

1879–80 – $1000
1880–81 – $1200 ($200 extra for lectures in N.H.4)
1881–82 – $2000 (During Shaler's absence)

At this time he would be struggling to find his level as an Instructor. To do this he first had to master the subjects he was supposed to be teaching and then to devise some attractive means of imparting this knowledge. Nowadays a lecturer's task is made much easier. There are recognized training techniques; there is a greater range of media, some of which have a stronger appeal for the duller or less determined student. Above all there is the advantage of a vast source of written material, both American and foreign, and several well-tried schools of geographical thought. When Davis started teaching advanced educational methods were in their infancy and education itself had still to achieve the status of a public service. It was regarded even then more as a cultural adornment for the rich and a form of disciplinary activity for the poor. Much that was best in geomorphological work was basically descriptive in approach and related to individual areas. Consideration of the general pattern of the surface forms of the earth in a broadly methodical sense was largely untouched. As might be expected in a rapidly expanding new country, attention of earth scientists was mainly directed towards an assessment of the mineral potential and the best techniques for exploitation through geological mapping. The era was characterized academically by strong personalities rather than by method. When only descriptive analysis can be offered it is enthusiasm, individual charm, a vibrant personality, or a novel type of delivery that are likely to tell with students. Davis encountered all these difficulties in the early years and for a time chance of success seemed far beyond his reach.

This anxiety cannot have been eased by the additional responsibility of his first son, Richard Mott Davis, who was born on 4 December 1881. With the prospect of more children to come it was essential that his position in the academic world be made secure.

FIG. 20. Davis' first wife and his eldest son. Photograph taken in Boston during June 1883 (Courtesy R. Mott Davis)

The full reason for Davis' early academic difficulties is not clear. They may have been entirely due to University financial stringency. Possibly he disappointed Shaler, was not sufficiently outgoing and understanding with the students, or perhaps failed to meet the particular standards of the Harvard President, whose control over all staff was very close.

Eliot was not a great advocate of 'book learning', as such, and actually had experience of geological field work. During the summers of 1852–57 as a student, he assisted in geological field work in Eastern Canada, New England, New Jersey and Pennsylvania. (*James, 1930*)

Eliot kept the power to appoint and promote the teaching staff entirely in his own hands. He wrote in 1911; 'Each department, when making nominations, should be required to convince the president that their nominations are well grounded'.
(*Nielson, 1926, p. 225*)

It is certainly a fact that Davis was very dissatisfied with his inability to progress within the department and it certainly reached the stage where he considered his employment at Harvard to be in jeopardy. Writing in later years to a colleague who had found himself in similar circumstances, Davis, while advising him, recounts his own early experience.

Let me say in the first place that my experience is in favor of the squareness of university officers. I believe they strive to do the fair thing to all concerned, and I know that they have great difficulties and embarrassments. They are eager to advance men as fast as they can and as fast as the men ought to be advanced. Sometimes their ideas as to rate of advance do not agree with those of their professors; I know such was my case here some twenty years or more ago; I was extremely discontent, looked around for other places; yet wished to stay; was nearly upset on getting a letter from President Eliot, saying I had better go elsewhere, as there was little chance of my promotion, and so on. As it was, I set to work and built a house and camped on the ground here, to fight it out; and in the end it came along all right ... One of the difficulties in the case is that the President and his intimate councillors cannot and must not express all their opinions to all the professors. They have to keep quiet on many points. But I believe they are constantly striving for the best of their university; and that means they are doing what they think they ought to do, as closely as funds will allow ... There are few things so disappointing, as not being recognized to the measure of your own ideas; I assure you I have been thru that phase of growth; and perhaps I made a mistake, in not pulling up stakes about 1883 and going somewhere else; for example, Becker of the Survey offered me a place under him in field work in California. I am glad now on all sorts of accounts that I did not accept it. But at the time, conditions here were very irksome. (*Letter: Davis to Lawrence Martin, 14 May 1912; Cambridge, Mass.*)

The letter to which Davis refers was written by Eliot himself. The position is put most bluntly and its grim significance must have been upsetting. Davis, an ambitious young man, was suddenly told that, though competent as an Instructor, no more senior post was available for him.

The Corporation offer you a reappointment as instructor in geology at a salary of $1200 a year, and with work as follows:– To give the course called Nat. His. 1 and to assist Prof. Shaler in giving Nat. His. 4; and an advanced course in geological field work (as in '79 and '80) or to render an equivalent amount of assistance in the department of geology. The Corporation are quite aware that this position is not suitable for you as a permanency; but it is all that they are able to offer you now, with their present resources, and all that they expect to be able to afford for some time to come. In considering whether it is your interest to accept this offer temporarily I hope that you will look in the face the fact, that the chances of advancement for you are by no means good, although the Corporation have every reason to be satisfied with your work as teacher.

(*Letter: President Charles W. Eliot to Davis, 1 June 1882; Cambridge, Mass.*)

Administrative doubts must have continued because there was a further query in the following year

> President Eliot desires me to ask you what was done about the renewal of your appointment last fall. Your appointment in 1879 was for three years from September 1 and expired September 1 1882, and the President can find no record of the renewal.
>
> *(Letter: Edward Hale to Davis, 15 April 1883; Cambridge, Mass.)*

to which Davis made a very prompt reply:

> Dear Sir,
>
> In answer to your note of today, I quote the following from a note from Mr. Eliot, dated June 1, 1882.
>
> 'The Corporation offer you a reappointment as instructor in Geology at a salary of $1200 a year, and with work as follows' – etc.
>
> It was understood in the conversation that I had with Mr. Eliot on accepting this offer that no definite term of appot. should be mentioned, beyond what is implied by my retaining a place in the Faculty. *(Letter: Davis to Edward Hale, 16 April 1883; Cambridge, Mass.)*

THE BEGINNING OF THE CYCLE

When Davis joined the department he had no teaching training or experience, as we have seen, and was glad to be entrusted with any routine task. This situation soon altered, and when he had shed his early diffidence he must have looked forward to quick advancement. Whether his record justified such expectations is doubtful but the family tradition of distinction, as well as his personal ambition, prevented him from being reconciled for long to an inferior position. The letter from Eliot was a sharp awakening to the realities of his situation at the age of thirty-three and, though it is never pleasant to be squarely made aware of one's limitations, in this case it may have provided the necessary challenge which ultimately changed the whole course of Davis' career. However, with his obvious ability it is difficult to see prominence being denied him, but it might have been delayed or not have taken the same form. It is significant that all the rest of his life – for over fifty years – Davis kept a copy of Eliot's letter. Whether it was irony or pride which made him do this is not clear but the importance to Davis was obvious enough. It was an episode he never forgot. After 1882 Davis 'built his house, camped on the ground, and fought it out'.

Davis published nothing in 1879, which is perhaps hardly surprising as he was new to university life and also to marriage, and needed to adjust to both. It is often possible to rush off articles in order to impress one's professor but Davis was a conscientious and painstaking worker. His pride would have prevented him from submitting for publication anything that was in the least shoddy or inadequately verified. His first article appeared in 1880 and was on the mineral fillings of cavities. In the following year there was a joint publication with Professor Shaler on glaciers as well as a few remarks on the geology of Mount Desert in Maine. Thereafter the articles began to flow with an increasing regularity and in 1883 contained his first publication on climate. It will be noticed that these early writings, which are listed fully in an Appendix, contain an extensive classification of lake basins and an account of the Triassic trap rocks of the north-eastern United States.

At this stage in his career Davis appears in many ways to have been a comparatively unworldly person, rather immature in social outlook. He had a fine appreciation of nature but only the poorest understanding of his fellow-men. Buying success by conformity or good behaviour would not have occurred to him, and if it had was quite alien to his principles. Whatever the reasons for his lack of immediate academic success, it is clear that Eliot's letter acted as a powerful stimulus. Peattie even suggested that Davis was so mortified by this criticism that he immediately set about evolving a theory that would shake the geographic world.

> So he set about to get some idea with which to write. T. C. Chamberlin had published a three volume work on the geology of Wisconsin which was reported as being excellent reading. Davis read the volumes and noted that Chamberlin in discussing the valleys of the Driftless Area spoke of one as young and another as old. From his use of these terms of age Davis evolved the concept of the cycle of erosion and his job at Harvard was assured. (*Peattie, in Martin, 1950, p. 178*)

While it is clear that this suggestion is altogether too facile, it is certain that at about this time T. C. Chamberlin was using a simple anthropomorphic model in describing the fluvial landforms of the 'driftless area' of Wisconsin, although he never developed the idea in the sophisticated manner which Davis subsequently employed. Indeed, it is always tempting to seek out evolutionary overtones in textbooks which Davis would have studied at this time, such as that by Green (1876 and 1882) and, especially, by Huxley (1877). Chamberlin's *Geology of Wisconsin*, Volume 1 (1883), and his U.S. Geological Survey report with R. D. Salisbury on the 'Driftless Area of the Upper Mississippi Valley' (1885) contain identical topographic descriptions employing the terms 'meandering stage of degradation', 'indices of youth', 'topographic youth' and 'topographic old age' (Chamberlin and Salisbury, 1885, pp. 228-9). The full antecedents of Chamberlin's usage of these terms are not clear as there is also a reference back to his contribution to the *Annual Report of the Wisconsin Geological Survey* for 1878.

> Now falls are the indices of youth. They are expressions of rapid erosive activity, and that very activity hastens their disappearance. In obdurate formations they may have a greater endurance, but in soft strata their life is geologically short. In view of this fact, the presence and absence (of falls) above noted teach us that the degradation of the driftless region has passed beyond the time of youth which permits cascades and falls, while the adjacent drift region has renewed its youth,

through glaciation, in the formation of new valleys and new stream erosion ...)
(Chamberlin and Salisbury, 1885, p. 229)

The ultimate result [of the stages of degradational development] is a complete
reduction to a basal plain of erosion closely analogous to the original one from
which the evolution took its origin. It is old age declining again to the level of
childhood; but the level of age and the level of childhood are not the same.
(Chamberlin and Salisbury, 1885, p. 237)

Davis does not acknowledge the influence of Chamberlin, and Daly cites
another obvious source of inspiration:

On many occasions he told of his deeply-felt indebtedness to American geologists,
particularly Lesley, the staff of the Geological Survey of Pennsylvania, and
Powell, Gilbert, Dutton and Holmes of the great western surveys. It was while
reading their published writings that 'geography gained a new interest' for Davis.
That interest culminated in the development of his most famous idea, that of the
'cycle of erosion'. He visualized a structural unit in the terrestrial landscape and
then deduced the topographic results of erosion of this unit by rivers born on its
original surface or developed on the unit during the later, systematic evolution of
its river system. *(Daly, 1945, p. 272)*

This is equally credible, provided one allows the probability that there is
more than one source. As a student and lecturer Davis had necessarily
consulted the standard writings on his subject and inevitably what he wrote
would to a large extent be based on this earlier knowledge.

His rise to success was less than precipitate and certainly less calculated
than Peattie suggests. The threat of losing his job and the possibly implied
stigma of incompetence gave him a severe jolt, there is no doubt of that, but
it would be going too far to suggest that he then sat down and gave himself
the task of creating a revolutionary model of landform development, and
that when he got up again 'the cycle of erosion' had been born. Carl Sauer
supports Daly's view to some degree:

He told me how he had been put on notice at Harvard to find a position elsewhere,
how he had gone west to the Rockies in distress of mind, how he had read G. K.
Gilbert, and the erosion cycle took form in his mind. It was in the nature of
revelation, though he did not use the term, but I suspect a religious experience of
cosmic order. He was raised a Quaker and some of this stayed with him. *(Letter:
Carl Sauer to R. J. Chorley, 18 October 1961; Berkeley, California)*

We shall see in Chapter 10 how, in a commentary provided for Vera Rigdon,
Davis himself ascribes the birth of the cycle of erosion concept to his own
field observations in Montana in the summer of 1883. In his address to the
Harvard Alumni at the age of eighty Davis recalled these earlier days,
although with perhaps a tendency to over-dramatize the events of that
distant summer on the High Plains:

My second western experience, apart from a mere cross-country passage in 1877 on the way around the world with my cousin, T. M. Osborne, '84, came in 1883, when my former teacher, Raphael Pumpelly, then director of the Northern Transcontinental Survey which he had organized for the Northern Pacific railroad, invited me to undertake a special study of the geological formations below the coal horizon in the plains and mountains of central Montana. This experience came after I had begun teaching at Harvard and was a lifesaver for me; for not long before I had received a most discouraging letter from President Eliot, advising me to look elsewhere for promotion as there was little chance of it for me at Harvard. No wonder, for it is doubtful if Harvard ever appointed a less prepared teacher than I was, when, on Shaler's friendly recommendation, I was made instructor in Physical Geography in 1878. I 'read up' in approved textbooks and tried to keep ahead of the class; but my lectures had little life in them, and not until after the Montana summer of 1883, when the facts of that inspiring field led me to develop the scheme of the cycle of erosion – the natural history of rivers, as it might be called – had I any proper measure of success. In the memory of the hard jolt that Eliot's letter had given me I had a particular satisfaction sixteen years later, when, after having been advanced to an assistant professorship in 1885, and to a full professorship in 1889, I was appointed during absence on a sabbatical year in Europe to the Sturgis-Hooper professorship in 1899. Whenever reference is made, as it often should be, to Eliot's extraordinary sagacity in selecting men for appointment or promotion in the Harvard faculty, I am tempted to bring up my own experience as marking an exception; but I hesitate to do this publicly, for fear that my hearers might think that the exception was found not in his early discouragement of my untrained efforts but perhaps also in his last promotion of my status. (*Davis, 1930F, p. 396*)

Davis clearly had a heavy and not entirely congenial teaching load at Harvard. As well as giving demonstrations with models he was responsible for the lecture courses in physiography and meteorology. We are told that his first performances as a lecturer were disappointing. Although he knew the standard works from his student days he had so far evolved few ideas of his own, and inevitably his lectures may have lacked those essential personal ingredients of conviction, feeling and prejudice which come with experience, as well as that warmth and tolerance needed by the good teacher. Yet as he progressed with his published articles any uncertainty of purpose must soon have departed and a more individualistic view and enthusiasm became apparent. The youthful groping towards a method was beginning to prove successful and, as his ideas became more coherent and connected, the appeal of his lectures must have increased markedly. From the time of his first employment at Harvard he assisted with, and later organized, student fieldwork in the north-eastern United States. In the summer of 1883, as we have seen, he worked much farther afield in Montana under the direction of Pumpelly in conducting a geological survey of the route of the Northern Pacific Railroad, particularly in connection with the occurrence of coal deposits.

Davis saw that the Rocky mountains of Montana had had two periods of deformation; the present relief being due to moderate erosion following irregular warping and uplift of a previously deformed and much denuded mass, while the intermont basins were recognised as aggraded depressions. It was also perceived that the plains of nearly horizontal strata, east of the mountains, had been enormously denuded, as is proved by the volcanic mesas and dikes which now stand up in bold relief. Good examples of uplifted and somewhat dissected peneplains were seen and appreciated, though a special name was not then applied to such forms.
(*Brigham, 1909, p. 10*)

The years 1883–6 show a marked increase in published output. Ten articles appeared in 1883, fifteen in 1884, ten in 1885 and twenty-seven or more, his maximum annual turn-out ever, in 1886. Especially in this last year many of the publications were very slight, mere accounts of a page or two that added to knowledge but not to learning. The accumulative result, however, is impressive.

Although the interest of present scholars must first focus on the article on 'Gorges and waterfalls' (1884N), which showed Davis' embryonic interest in sequences of denudational forms (see Chapter 10), it was at the time merely one of a growing number of articles on a wide variety of topics. At this time meteorological themes strongly dominate in Davis' writings. The cycle theory was after all more a culmination or blending of ideas than a simple analysis of one set of facts. Where he succeeded, perhaps, was by gaining confidence in himself. Through these early articles he gradually found that he was beginning to transform his observations in the field into ideas which were both intelligible and at least comparable in originality with those of his colleagues. This growing ability was in turn recognized by his senior colleagues and success then became a matter of continued application. Davis' advice to a young lecturer was perhaps evoked by his own early experience.

I also remember clearly his farewell to me – then a young instructor in geology at the university; he tapped me lightly on the chest with his forefinger and said earnestly, 'Attend to production, young man, attend to production'. Certainly that is advice which he, himself, followed faithfully throughout his life. *Letter: T. S. Lovering to R. J. Chorley, 20 December 1962*)

It is doubtful whether the inception of the cycle theory determined his continuance at Harvard. After all he did not develop the theory at all fully until his article on 'The rivers and valleys of Pennsylvania' (1889D) and not definitively until 1899, long after his status at Harvard had become firmly established. However, by 1885 demonstration in published form of his scholarly competence must have transferred to his seniors the growing confidence that Davis had already found in himself. Also Davis was making powerful acquaintances, which undoubtedly did not impede his promotion to tenure!

THE GEOLOGICAL SURVEY

What made Davis' articles so remarkable was the explanatory method of presentation with which he was beginning to experiment. Where he excelled was in giving the details observed real meaning. He did this by suggesting hypotheses for the creation of each separate feature. Then he would weld together the separate explanations to demonstrate that they formed part of a much larger explanation. Finally he was able to show the reader how all these different features were clear indications of the processes which had shaped the present landscape and were continuing to fashion the forms of the future. In short his theories were simple to grasp because they represented a logical and believable sequence of events. Later, when he had established his general principles of analysis and had invented his own special terminology, the attraction of his work was to spring from the added comprehension and ready intelligibility that his explanatory methods supplied. The deductive approach provided an enhanced vision and stimulus that was missing from the old descriptive methods of teaching. At this stage colleagues must have been impressed by his evolving methods and their dedicated application.

In an interesting letter to his parents in early 1884 Davis shows how much his confidence had increased and how ardently he was courting influential professional contacts.

Last year's visit to Washington was very successful but this one is far more so. It has really been a great gain to me, not only in some rather good offers from Powell of the Geol. Survey, but in the extension of my acquaintance, and I go back home invigorated.

The half can't be told of my running about and seeing people; but several of the events – those that have puffed up my pride the most – are worth telling. In the first place, pretty nearly everyone I met seemed to know of me and referred to some of my papers – or to last summer's work. That is extremely satisfactory. Prof. Abbe, chief scientific man of the Signal Office, was sorry I couldn't stop longer – he wanted to arrange a few lectures for me to give on some meteorological subject. Well eno' to be asked, but I was glad to escape the critical, scientific audience that Washington might gather.

Major Powell, Director of the Geol. Survey, with whom I have just lunched at his office in the most comfortable, informal way – gave me a definite offer – $1800 a year, with growth to more – and my choice of two pieces of work. One to aid Gilbert (an excellent geologist) in studying the mountains from N.Y. to Alabama – the other, to work indoors, in preparing a geol. history of the United States. 'I have looked at your papers on Lakes, and other subjects,' he said – 'and would rather have you do this work than anyone else in the country, with perhaps two or three exceptions, who are on a par with you.' – But I told him I was attached to Cambridge, and determined to fight it out on that line, if it took all summer. So just now, in my last hour with him, I suggested an idea that Gilbert and I talked over Wednesday evening – namely, that I should attempt a smaller work first – a *subject-index* to American Geology – a classification, arranged under various heads, of all that has been done in this country. Powell said, 'send me a plan'. The work would be done chiefly by an assistant under my direction and supervision – in Cambridge – and it would be a pretty severe task, too. 'The Major', as all his assistants call him, was affable and cordial – the very opposite of Hazen, Chief Signal Officer, a stiff backed army man – whom I called on to see, and to give thanks to for aid in the Lowell lectures – He was reading *Science* (probably about the conduct of the Greeley relief party) when I went in – and I soon left him to it. The Hydrographic Office of the Navy was my main point of attack this time – and I accomplished more there than fondest hopes could have expected. Commander Bartlett is chief of the Office – and as soon as he finished showing a Board of Trade delegation around, he gave me a very polite attention, with introduction to the various rooms and officers in charge, and later, on finding my connection with *storms* in *Science*, he was especially polite and communicative – so much so that I was encouraged to bear witness against him – and give him a plain spoken piece of my mind about a growing rivalry between his office and the Signal Service on some work that both are undertaking. It really surprised me to hear myself talk, and to see the way he took it all. The main point was that co-ordination and agreement were better than indifference or opposition – that a definite under-standing should be reached – and arrangement made between him and Hazen as to the work each should do – that this should be done in harmony. I made him admit that a feeling of antagonism existed and then emphasised the necessity of working together by recalling the fate of the several geological surveys that hated one another so bitterly a few years ago, so that in the end it became a scandal and Congress had to interfere, and boil them all down into the present single and excellent organization – in which public money is not spent twice on one piece of work (The cars are whizzing – and the writing is pretty bad). Imagine me, preaching this way to the Chief of the Hydrog. Office of the U.S. Navy! It would have been impossible to talk in the same way to Hazen – but Bartlett didn't seem to mind it at all. Another thing he admitted, that showed the truth of my sermon, was that he had called on all sorts of officials to secure their co-operation in the publication of his monthly pilot charts – but had *not* been to see Hazen.

Lieut. Moore, in charge of Bartlett's Meteorological work, is an excellent fellow – not nearly so learned in theoretical matters as he should be – but everlastingly obliging. He assured me, on my leaving, that all their scientific publications would be sent to me – and anything they had would be lent me anytime it was wanted. Sample proofs of some fine meteorol. charts are to be sent to me as soon as finished – because, after looking over the copies in his office, I approved of his plan, and asked to have the chance of reviewing the charts for *Science.* Of course, much of all his talk may be with the special object of gaining floor for the new work undertaken by the Hydrogr. office in semi-opposition to the extension of the Signal Service on to the Ocean – but for all that, a good deal that he said was sincere – and acquaintance with him and Bartlett is among the chief gain of the trip. It gives me the feeling that if I should at any time walk into the south-eastern room on the ground floor of the immense Army & Navy building – warm welcome awaits me.

It is perhaps, rather surprising that out of all this, I go back more resolved than ever to hang on to Harvard. Half a dozen of the Geol. Survey men hoped I would come and join them – but I remain obdurate! After all, there is more freedom in the Cambridge work – and the social position there is vastly beyond what the survey would give in Wash. – for most of the men I know there are not only of moderate means, but with all their scientific attainments they haven't – not all of them – other things also important – and well – it's all snobbishness, I suppose – then Cambridge suits me better. Then, with the good prospect of a house now, why should we not stay – and blow dignity – and go to Washn. once a year and blow our trumpet – and come back 'inflated with pride'. (*Letter: Davis to his parents, 9 February 1884; in train from Baltimore to Philadelphia*)

Opportunities were now beginning to open up for Davis and he begins to exchange correspondence with the most prominent names in American geological society.

It seems desirable that we should have a conference of geologists who are interested in, and who have especially been studying, seismic phenomena. Would it be possible for you to come to Washington within the next ten days to meet Rockwood, Abbe, Paul, Dutton, and others at the rooms of the Geological Survey? I think, after looking over the ground pretty fully, that it will be possible to inaugurate systematic observations, and it seems that we ought first to give attention to the character of the observations most desirable, the instruments to be used, &c, &c. I hope that you will find time to come. Your travelling expenses will be paid.

I have written to Rockwood the same as to yourself and if you will confer with him and adjust the time to convenience of both, it would be well. (*Letter: J. W. Powell to Davis, 12 November 1884; Washington, D.C.*)

There is some mistake in regard to my opinion about the Adirondacks, for I have none. I have never seen them.

I should be very sorry to have you give up the matter of index, for it is a work of great importance; and in my judgment no one could supervise it better than you. I enclose my own tentative plan, hoping it will stimulate you to further thought and additional interest.

Sincerely yours
G. K. Gilbert.

P.S. I find that it is Powell who has an opinion about the Adirondacks. He and I have been discussing the index plan especially with reference to the logical classification of the subject; and as we are likely to agitate the matter for some time I will not longer delay my letter. In his judgment the classification should be strictly logical without reference to the convenient division of the card-list into packs of similiar size. (*Letter: G. K. Gilbert to Davis, 16 May 1884; Washington, D.C.*)

In the summer of 1884 Davis began field work for the U.S. Geological Survey in New England under T. C. Chamberlin, as the latter reports for the year 1884–5:

During the year Mr. W. M. Davis of Harvard University has become associated with the glacial division and has undertaken the special study of drumlins and certain selected subjects of monographic study. His other engagements have not permitted him to devote much time to the Survey, up to the present date, and his investigations can only be said to have begun. Besides observations in the vicinity of Boston, he has commenced the examination of the belt of drumlins that stretches from the vicinity of Brookfield and Spencer, Massachusetts, towards Pomfret, Connecticut, and has already demonstrated its lateral limits at several points. He found the drumlins of the belt about Spencer to stand on a base elevated 1,200 feet above the sea but to present all the pecularities of those seen at lower levels. (*Chamberlin, 1885, Administrative Report, pp. 36–7*)

Davis' personal report to Chamberlin at the end of the summer of 1885 shows the widening range of his work and his concern for 'theoretical explanations':

I have the honor to report the following statement of work accomplished during August.

In accordance with correspondence and instructions from you I visited Canajoharie, N.Y. to examine the gorges through which two streams there reach the Mohawk, and conclude that they are manifestly of very recent origin, even tho' the old drainage lines were not identified: the most significant point was that the divide between the upper waters of the two streams was distinctly lower than the rock-and-drift ridge at the points where the gorges were begun. I conclude therefore either that the gorges were begun as subglacial channels, or that they are survivals of streams on the surface of the melting and stagnant ice sheet.

During this work I had opportunity to examine the surrounding country, and made considerable changes in the map of the district prepared by Messrs. Hall and McGee.

The direction of the striae and drift ridges about Canajoharie and Palatine Bridge was from S 60°–80° E. There was no surface that gave suff. indication of the direction of motion decisively.

On the way to Wayne Co., N.Y. I stopped a few hours at Little Falls, there also determining considerable changes needed in the map by Messrs. Hall and McGee, finding an interesting deposit of coarse quartz conglomerate about two feet thick between the Archaean rocks and the Calciferous rocks, and examining the channel of the river with respect to its age. It seemed to me that the general adaption of the Utica shale country to the present or even to a lower level of drainage was a difficulty in the way of believing that the Little Falls channel at the Noses below Canajoharie, where altho' there is direct evidence of the passage of a large river (the Ontario overflow?) the valley on the Utica Shales seems too well opened to be post-Glacial.

I next proceeded to various points in Wayne and Cayuga Cos. to examine the parallel drift ridge – and here – in accordance with permission from you, expressed in Mr. Salisbury's note, paid the expenses of Mr. Whittle, a student who was with me, and by whose assistance the area examined was much increased . . .

In regard to the form and position of the hills I could arrive at no definite conclusions, altho' a great number of these were mapped. The excursion has added materially to the knowledge of fact, but has not yet greatly advanced theoretical explanations.

The map serves to show where much additional time could be spent to advantage, and I shall hope to pursue the study next summer.

During the coming winter, it will be a great advantage to me to have opportunity to discuss with you the more detailed report and maps that I am now preparing. (*Letter: Davis to T. C. Chamberlin, 1 September 1885; Cambridge, Mass.*)

THE BEGINNING OF THE FLOOD

Although, by modern academic standards, Davis began publication at a comparatively late age, his production both in volume and range soon rose to an impressive torrent which was to flow on with few recessions until after he died, more than half a century later. Within only the first five years of his output, until he became established at Harvard with some tenure as Assistant Professor of Physical Geography in 1885, many of the interests which preoccupied his mature years were already apparent. Structural geology, glacial geomorphology, meteorology, fluvial geomorphology and theoretical speculations all formed part of these early years.

Davis' first publication was a short paper (1880A) describing mineral fillings of amygdalid rocks and their use in inferring subsequent tilting of the strata. His description of the geology of Mt Desert, Maine (1881B), foreshadowed his developing interest in glacial action; and his analysis of the 'miniature Appalachian' structures just east of the Catskills (1882E), resulting from his summer fieldwork with Shaler in 1877, led him to draw extensive structural and topographic analogies with large mountain ranges, including the Rockies and the Alps. In 1882 (C and D) he began his extensive publications on the sedimentary and structural history of the Triassic rocks of the Connecticut Valley, as a result of fieldwork during the summer of that year; treating the history of their investigation and being particularly concerned with their monoclinal structure and the age and structural relations of the associated igneous 'trap' rocks. This was followed by a more extended treatment of the same subjects (1883L). A second student field trip to the Catskills in the Spring of 1882 led to three more publications (1883I, J and M) relating to the miniature 'Appalachian' folded belt of the Helderberg limestones between the Hudson River and the Catskills in New York State. These are thoroughly geological treatments of stratigraphic and structural relationships but, in the light of Davis' later work, it is significant that great stress was placed on the identification and interpretation of the nonconformity between the Lower and Upper Silurian rocks. In 1882 Davis followed up the glacial interests shown in his previous educational publication with Shaler (1881A) and in that on Mt Desert (1881B) by a long paper (1882B) concerned with glacial processes in an historical and classificatory framework, and later dealt specifically with the distribution and orientation of drumlins in parts of Massachusetts (1884F) and with their assumed sub-glacial origin which he compared with that of sand-banks in a broad river (1884O).

Davis' meteorological publications began in 1883 and in quantity soon swamped the rest of his work. He was at this time particularly concerned with the history and construction of charts showing ocean winds and currents (1883H; 1884D; 1884I), the deflective effect of the earth's rotation (1883B; 1885G) and the structure of airflow within depressions and tornadoes (1884A; 1884B; 1884H; 1884J; 1884K). More important than this, however, were the significant beginnings of his work on fluvial geomorphology and, particularly, in his groping towards the idea of the cycle of erosion. We shall see in Chapter 10 how his work on the Northern Pacific Railroad Survey in the summer of 1883 (1886T) apparently led him to his cyclical speculations, and how these were first manifested in his articles on gorges and waterfalls (1884N) and geographic classification (1885D) which were presented in the following year. The same theme also appeared in one of his articles on drumlins (1884O) where he uses such terms as 'geographic evolution', 'embryonic features', 'adolescent plateaux', 'precocious development', 'second childhood', 'cycle of development' and 'polygenetic mountains' in briefly proposing a two-fold classification of landforms based on structure and 'degree of development by erosion'.

In 1885 Davis was promoted and given renewable tenure for five years. While it is not clear whether a chance vacancy had appeared in the Faculty or whether his scholarly work and influential connections had been noted, Davis was at this time established on his lifetime's vocation. However, he never forgot his feelings on receiving President Eliot's letter of 1882, publicly referred to it at the age of eighty and a copy of it made by Davis was found among his personal effects long after his death. Although now firmly launched into the scholarly maturity which was to bring him world-wide acclaim, Davis' early years at Harvard had not been tranquil.

PART TWO

Maturity

FIG. 21. Davis (4th from left) on an unknown (European?) field trip
(Courtesy R. Mott Davis)

CHAPTER NINE

The Rise to Fame: University Promotion

METEOROLOGY

If Davis' promotion to Assistant Professor of Physical Geography in 1885 was the result of his developing ideas relating to the study of landforms, his publications during the years immediately following do not reflect this. Until the publication of his classic paper on 'The rivers and valleys of Pennsylvania' (1889D) his scholarly output was overwhelmingly meteorological, reflecting his association with the New England Meteorological Society which he helped to found in 1884 and of which he was Secretary until it was consolidated with the U.S. Weather Bureau in 1896, during which time the Society built up a corps of over three hundred local weather observers. This meteorological phase, which resulted in some eighty publications in the period 1884–94, culminated in the production of his *Elementary Meteorology* (1894D) but his research into the subject continued until 1901 (1901J), by which time it had become eclipsed by his physiographic studies. Referring to the sequence of his scientific interests:

> Davis used to tell students that in his work he 'came from the heavens, through the air down into the earth, and finally found himself back with his feet on the ground'. (*Rigdon, 1933, p. 61*)

Davis' work on meteorology covered a wide spectrum of topics. In 1885 and 1886 he organized a network of volunteer observers who reported on New England thunderstorms during the months June to August, and he reported on the spatial distribution of summer thunderstorms in belts, together with their individual paths and resulting precipitation (1886R and S). Davis, as usual, drew on the results of foreign studies of the occurrence and synoptic associations of thunderstorms (1886I), and later extended his own observations to the features of winter thunderstorms (1892G), which he regarded as different in character and origin from those of summer. He similarly organized a hundred volunteer observers throughout coastal Massachusetts and New Hampshire in 1887 to report on the features, inland extent, thickness and deflection of sea breezes (1889L).

Even in his meteorological phase Davis exploited the topographic possibilities of atmospheric studies, giving a thorough analysis of meteorological phenomena at high elevations (1886A and W), with particular reference to mountain and valley winds (1886W, pp. 332–6 and 1893U) and the foehn/ chinook phenomena (1886B; 1886W, pp. 336–48 and 1887I). Later he broadened this topographic theme (1896D) in relating climate to the general distribution of land, sea and relief and to possible polar displacements; giving sedimentary and topographic evidences of climatic changes and the possible topographic consequences of various glacial theories. His meteorological interest in climatic changes (1890E), exemplified by his rather favourable review of Croll's glacial theory (1894L and 1895M), clearly paved the way for his work on arid and glacial 'climatic geomorphology' during the first decade or so of the twentieth century.

Another focus of Davis' meteorological work was the pattern of airflow within weather systems. He had a long-continued interest in the relationships between local wind observations and large-scale pressure distributions

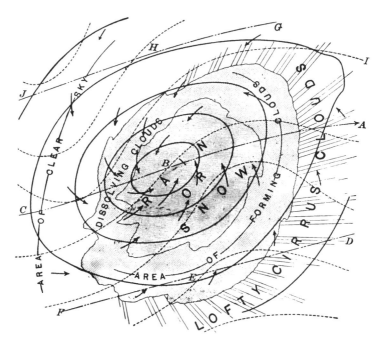

FIG. 22. An ideal diagram representing the distribution of the various weather elements around a well-developed centre of low pressure (From Davis, 1894D, Fig. 72)

(1893E and U), and in the reasons for the deflection of winds by the earth's rotation (1891K), together with the rotational influences over broader atmospheric circulations (1893H and I). His writings on tornadoes include a review of Ferrel's convective theory which contains the following significant statement, so characteristic of his own aspirations:

It is manifest enough that its deductions [i.e. of Ferrel's theory] as to the processes of tornado action run far beyond the facts of direct observation; but this is as it should be, for the success of a theory may be, in a measure, gauged by the light that it throws ahead on unseen phenomena. Although led by fact, the theory as a whole is highly deductive or imaginative; but it can fairly be said that the use of the imagination in framing the theory is thoroughly scientific; that is, it is guided by a full knowledge of associated facts, and of physical and mechanical laws pertinent to the discussion. (*Davis, 1890D, p. 462*)

Continuing his interest in the atmospheric circulation, he published on the oceanic wind systems of the Atlantic (1893E) and the Indian (1893J) oceans, and on the general relationship between winds and ocean currents (1897G).

Davis' work also embraced such topics as:

a. The heat budget of the earth (1895B); with particular emphasis on the effect of atmospheric water vapour (1887D and 1894B) and of the albedo of snow (1887C). It is noteworthy that his block diagram on the global receipt of insolation [Fig. 23] is still used in modern textbooks.

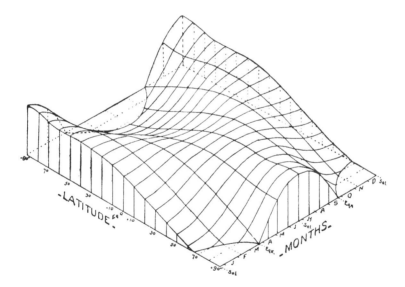

FIG. 23. Davis' influential diagram showing the theoretical variation of insolation over the globe by latitude and season (From Davis, 1894D, Fig. 2)

b. Rain-making; represented by a rather critical treatment of the proposal by Edward Powers, supported by Congress, to use explosives to try to produce artificial rain in Texas (1892D).

c. Classifications of meteorological phenomena; giving a four-fold genetic classification of winds based on the source of energy exciting the motion, the arrangement of contrasting temperatures, the period of occurrence, and their relation to larger disturbances (1888F). He also reviewed with approval Clayton's classification of clouds (1893V).

d. Forecasting; reporting on European work (1891C), and using the unpredicted cold wave of January 1886 (1886D) to call for more detailed observations (1893F), together with an increase of observers (1893G). The following letter from Gilbert reflects this interest:

> The agricultural Secretary bill with amendment transferring the weather bureau has passed the Senate. Powell is assured that the House conferus will accept the amendment, and it will be passed before the session ends. The chief is to be appointed by the President and confirmed by the Senate.
>
> Abbe has been at our office today and Powell has written Mendenhall asking permission to nominate him for the place at the proper time. If you are assured of his willingness there is no reason why the boom should not be started at once. Otherwise it is best not to waste our powder.
>
> The appointment cannot be immediate because the transfer is not made till July 1, so the only need we have is to get attention turned to the right man early. (*Letter: G. K. Gilbert to Davis, 27 February 1887; Washington, D.C.*)

Perhaps in the long run, Davis' main contribution was to the teaching of meteorology. He produced many works on the history of American meteorology (e.g. 1889I), particularly of the work of William Ferrel (1892C and U) but also singling out Coffin, Espy, Tracy, Loomis and Redfield. He made pleas for the more extensive use of climatology maps in schools, similar to

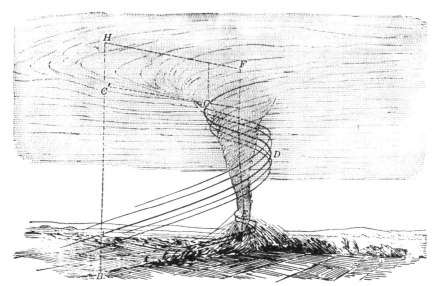

F I G. 24. A tornado funnel cloud, comparing the conditions of air spiralling inwards with that rising vertically (From Davis, 1894D, Fig. 104)

those produced in Europe by Hann (1888E) and for a more extensive university syllabus (1892F). In his *Elementary Meteorology* (1894D) he neatly and logically set forth the most up-to-date ideas on the science, supported by his own observations, imaginatively illustrated, and making very effective use of deductive methods. His following sequence of treatments was to set a pattern which is still influential – atmospheric composition, solar energy, temperature, pressure and circulation, winds, moisture and clouds, cyclonic storms, local storms, rainfall, weather observation and prediction, climate. Needless to say *Elementary Meteorology* soon replaced in the United States Loomis' physically-orientated and outdated *Meteorology*.

Brigham cleverly summarizes Davis' main achievements in meteorology:

The text books then available did not satisfy him, and he proceeded to develop a plan of teaching of his own. His efforts were early directed to the explanatory treatment of total phenomena, and toward their rational and systematic association, rather than to the instrumental description of items and to the statistical obliteration of non-periodic features. His wish was always to cultivate an intelligent appreciation of atmospheric movements and changes, rather than to drill routine weather observers. His plan of presentation resulted in the enchainment of a long series of meteorological processes in a rational sequence, as set forth first in his article on 'Meteorology in the Schools', published in the *American Meteorological Journal* 1891, to which he was a constant contributor, and translated in *Das Wetter* (1892); the same plan was followed in his text book of Meteorology (1894), a work which ranks high for clearness of statement and completeness of view. No other text book has yet [1909] taken its place in America or England. In all his meteorological work, Davis recognises the preponderating influence of Ferrel in Washington and of Hann at Vienna; it may be added that he has had much to do in making their results more widely known, although his own studies never advanced to the purely mathematical, physical or statistical sides of the science.

Davis was associated with Upton of Brown University, Niles of the Massachusetts Institute of Technology, and Rotch of the Blue Hill Observatory, in the organisation of the New England Meteorological Society in 1884, and in the management of its affairs. A set of observing stations was at once established, and with the aid of the National Weather Service and of the Harvard Astronomical Observatory, a monthly bulletin and an annual summary were issued, under the direction of Upton and Davis. Assisted by a grant from the National Academy of Sciences, and with the support of the National Weather Service and of some 300 volunteer observers, Davis made special studies of Thunderstorms in New England (*Proc. Amer. Acad. Boston,* 1886) and, with Ward, of sea-breezes on the Massachusetts coast (*Annals Harvard Coll. Observ.* 1890). He also wrote a general account of New England weather (*Annals Harv. Coll. Observ.* 1890), in which the cyclonic control of weather changes is given full recognition; and he prepared a classification of the winds according to their causes (*Amer. Met. Journ.* 1888). His election as corresponding member of the German Meteorological Society followed these studies and was the first formal recognition of his scientific work in

Europe. The New England Meteorological Society was dissolved in 1896, when its more routine work was taken over by the National Weather Bureau. After teaching meteorology at Harvard for some fifteen years, Davis gave over this subject to his former student and assistant, Ward, who has since then carried it well forward, with increasing attention to climatology. (*Brigham, 1909, pp. 16 and 18*)

When Davis handed over his meteorology course to his brilliant student Robert DeCourcy Ward in 1896, he maintained his interest in meteorology and climatology, but turned it more productively to geomorphic purposes. In his many contributions to glacial action, arid processes, climatic controls over coral reef development and the effects on landforms of climatic changes, much of his authority and expertise stemmed from his decade of work in meteorology.

WIDENING VISTAS

The decade which followed Davis' appointment as Assistant Professor at Harvard witnessed a marked widening of his interests and a strengthening of his scholarly authority, accompanied by his promotion in 1890 to Professor of Physical Geography. This development was, of course, most marked for us by his profound contributions to fluvial geomorphology which will form the subjects of Chapters 10 and 11. It was, however, also characterized by an increase of scholarly breadth, counterbalanced by a growing tendency to treat his wide range of problems with an eye to the study of fluvial landforms. It is during this decade that Davis became primarily a geomorphologist, as distinct from a geologist in a petrological and stratigraphical sense. All his life he collaborated with geological colleagues and from 1899 to 1912 he occupied the Sturgis-Hooper Chair in Geology at Harvard and in 1906 and 1911 presided over the Geological Society of America, but the amount of non-physiographic geology he accomplished was small. Nearly all his geological researches were ultimately directed to physiographic ends.

During this time Davis produced a number of important papers on fluvioglacial topics. He described the sub-glacial origin of certain eskers (1892L) drawing on field work in 1891 on a railroad cut near Auburndale, Massachusetts, analysing fluvio-glacial process in general and concluding from observations of sedimentary petrology that these deposits were produced sub-glacially, rather than as a supra-glacial sand plain. Drawing on his earlier Harvard summer fieldwork he described a fluvio-glacial gravel delta in the Hudson valley near Catskill, New York (1892J), overlain by postglacial clays. An examination of the deposits on the south side of Cape Cod (1893L) revealed wind-faceted pebbles and, together with other evidence, this led him to propose that these deposits (together with the similar ones of Long Island) were formed as confluent sub-aerial fan deltas or sand plains built out beyond the ice margins. Earlier (1886V) Davis had published a paper on the New Hampshire earthquake of 1886, inviting earthquake reports from members of the public.

In the summer of 1891 he conducted a party of geologists attending the International Geological Union in Washington to view the ancient shoreline features of Lake Bonneville, described by Gilbert in his monograph the previous year, together with its outlet at Red Rock Pass (1892N). Continuing this theme, Davis (1891L) disagreed with J. W. Spencer on the interpretation of the ancient overflow channel near Rome, N.Y., proposing that it drained an ice-dammed lake above sea level and not an arm of the sea. This disagreement led to a long-standing mutual dislike, as we shall see. In 1893 Davis took the opportunity of a trip to the Chicago World's Fair to examine the former outlet of Lake Michigan (1894F) into the Illinois Valley, and the 'geographical' importance of such channels as canal and railroad routes.

G. K. Gilbert was by far Davis' most influential associate at this time, and we know that he thought highly of the ideas of slope recession in 'The glacial origin of cliffs' (1889K). Gilbert must have been even more impressed with Davis' paper on the profile of badland divides (1892B), in which creep was invoked to explain an apparent anomaly in Gilbert's 'law of divides', for he employed similar reasoning in his important study 'The convexity of hill-tops' (Gilbert, 1909) seventeen years later. A letter as early as 1886 shows the close association of the two scholars:

> What I said about submerged embankments in the great lakes was ill considered, and I repented after the pages had been cast – not that it may not be true, but that I am not sure of it. I have told you some time of the evidence from fiords that an old southern shore of Lake Ontario is now submerged. There is similar evidence on the south shore of Lake Erie, at least near its west end, and I have heard of submerged cliffs in the northern part of Lake Michigan. I therefore jumped at the conclusion that the subaqueous sand ridges described by Whittlesey and others had been formed at the shore and afterward sunk beneath the water by a change in the system of levels. I have not seen them myself, and on that account should not have expressed dissent from published opinions concerning them. Russell saw them on his photographic trip to Lake Michigan, after my paper had been set up, and his observations led me to question the competency of my explanation.
>
> I saw the statement with reference to the sinking of the Rhone waters as they enter Lake Geneva, but have no cognate phenomenon in the range of my own observations. I should suppose that the density conditions in Lake Bonneville would be similar to those in Lake Geneva, at least so far as the ice-field streams from the Wasatch and Uintah mountains were concerned, and yet the Bonneville deltas have the form and structure which I have observed in the small deltas on the margins of pools, where the stream waters visibly flowed over the standing water. (*Letter: G. K. Gilbert to Davis, 18 March 1886; Washington, D.C.*)

Davis had by this time clearly attained major stature and his first public award, the H. H. Warner Medal for scientific discovery, was given to him in this year. Yet he still had some way to go. While his abilities were recognized his conclusions were not accredited with any protective coating of infallibility. Gilbert shows us this later when he questions some of Davis' principal deductions on the ground that insufficient attention had been paid to the details of the underlying strata – the kind of criticism which is levelled against Davis by many modern geomorphologists. Gilbert's comments also show that Davis was already extending his cycle theory to the problems of coastline formation and desert landscapes:

Yours of Jan. 7 is received, and so is the Miller pamphlet.

I have not studied the drainage in its relation to the Triassic, and so have no ideas to exchange with you in regard to your hypothesis at Harrisburg. For other reasons however I believe that the Trias was widely extended somewhere, dotted at numerous points from Nova Scotia to Carolina, having everywhere the same physical characters, having great thickness, and being associated at many or all points with systems of faults, it seems to me highly probable that its existing exposures are remnants of those points where subsequent dislocations have left the formations below base level.

I have read your physical geography of New England with the most lively interest and pleasure. It is practically novel in its method, and it seemed to me to be sound throughout, as it is certainly most aptly put. I have read also something about cliffs in the *American Geologist*, and the article brings up questions partly old and partly new in regard to which I am by no means satisfied. I do not understand *a priori* why a glacier should accentuate a cliff when the direction of ice motion is toward the face of the cliff, and yet I suspect that such accentuation really takes place. In New York the Niagara escarpment and the Water-line escarpment face the striae at about 60 degrees. The capping rock is in each case limestone, and south of the glaciated region such limestones do not in general make cliffs. On the other hand, the cliff at the Water-line escarpment has at its base an almost continuous ridge of drift with ground moraine habit, showing that over the outcrop of the shales the last work of the ice was deposition. I question too whether you have given the true explanation, or at least the full explanation, of the relative abundance of talus in various situations. In the plateau region of the West there are many cliffs wonderfully bare of talus. Without referring to my notes, my impression is that these are in all cases due to sand-stones overlying shales, and that where a volcanic rock is capping stone, talus is always abundant. The Cretaceous and Triassic shales at the bases of the sand-stone-capped cliffs are usually so well exposed that there is no difficulty in their study, and sometimes the bare area far exceeds that covered by talus.

Your statement that in a dry climate transportation is reduced to its lowest terms and the waste from the mountain slopes accumulates in maximum of volume, etc, I believe to be radically wrong. It appears to me that in a moist climate disintegration has the advantage over transportation, while in a dry climate transportation has the advantage over disintegration. The difference is doubtless chiefly in disintegration but it is nevertheless true that the scant precipitation of a dry climate includes storms of great violence during which rate of transportation is high. There are few places in our arid West where the water channels, usually dry, it is true, are not as large as the channels corresponding to equal catchment areas at the East. (*Letter: G. K. Gilbert to Davis, 25 January 1889; Washington, D.C.*)

Correspondence over the years shows that Davis placed great importance on Gilbert's views and this habit of consultation persisted until Gilbert's death.

> He was affectionately devoted to Grove Karl Gilbert, and his biographical memoir of Gilbert for the National Academy of Sciences is one of the finest analyses of method and mental process in scientific literature ... (*Bowman, 1934, p. 177*)

In 1889 Davis was invited to give a series of lectures at the Johns Hopkins University and to share in the summer course at the University of Wisconsin. There is also a record of an invitation from Chamberlin to lecture at the latter.

> It has been urged by the State Superintendent that it would be wise to organize our core of instructors for our next summer school as early as practicable, and send forth circulars inviting those who wish to attend to undertake preliminary studies in the lines they intend to select, so we can reap larger results from the work we undertake and do that work more satisfactorily, because the students will, in some measure, come fresh and alive to the subject. Now I do not need to say that we are very anxious to have you form one of our faculty, and it would be of great service to us to be able to advertise this in the proposed circular. I do not need to urge our wishes because you know what they are, and how sincere they are. I know that you did not give encouragement that you could do so when you left, but I write in the hope that the outlook may be more favourable now.
>
> Dr. Kimball has agreed to be with us another year. (*Letter: T. C. Chamberlin to Davis, 26 October 1889; Madison, Wisconsin*)

In five short years Davis had progressed from near-failure to scholarly prominence.

THE GEOLOGY OF THE NEW ENGLAND TRIASSIC

Davis' summer fieldwork for the U.S. Geological Survey in New England which began in 1884 combined with his Harvard field classes in the northeastern United States to provide a fruitful source for his researches for the next decade. More important than the glacial studies which have been previously referred to, were Davis' sustained studies of the Triassic basin of Connecticut and Massachusetts, as both his biographers emphasize.

> But the young instructor knew fully well that effective, authoritative teaching of geology, the principal subject of his first instructorship, demanded close personal touch with Nature. To get such experience he selected for field study in detail the Triassic formation of New England and New Jersey. On those regions he published fifteen preliminary papers (1882–1896), and a monographic summary of most of his results in 'The Triassic Formation of Connecticut' (1898). This gave the first full account of the Triassic volcanic history of the region, announced criteria for proving the extrusive character of some of the 'trap sheets' and the intrusive character of others. He also showed how the analysis of topographic forms could be used in explaining the underground, invisible structures of Connecticut and similarly faulted areas of the earth's crust. (*Daly, 1945, p. 269*)

Several summers of the decade 1880–90 were devoted to work on the Trias of Connecticut for the United States Geological Survey, under Pumpelly and Gilbert. A number of new points were determined: the occurrence of both intrusive and extrusive trap sheets; the identification of two fossiliferous shale belts; the presence of numerous oblique, curved, sub-parallel faults; monoclinal structure; and the passage of two chief cycles of erosion separated by a period of uplift, one cycle of Jura-Cretaceous date, essentially complete, another cycle of Tertiary date, as yet far advanced only in the weaker rocks. This determination of geological dates for the production of erosional forms was a novelty in New England, where previously no clear understanding of cycles of erosion had been acquired. The term, peneplain, was first formulated and used here in 1889. Many short articles relating to this field were followed by an extended final discussion with a geological map, which appeared as a part of the Eighteenth Annual Report of the U.S. Geological Survey, for the year 1896/97.

A characteristic feature of this discussion was the attention given to the methods of proof employed in geological problems, and to the evaluation and the probable correctness of the results; a problem in which Davis acknowledges his great debt to the admirable studies of Gilbert and Chamberlin. (*Brigham, 1909, pp. 10 and 12*)

This study of the Triassic depression of New England, and by implication that of New Jersey, resulted in some fifteen publications on the subject between 1882 and 1898, culminating in 'The Triassic Formation of Connecticut', forming part of the 18th Annual Report of the U.S. Geological Survey for 1897 (1898G). This research was mainly concerned with four themes: the character of the igneous rocks and their intrusive and extrusive relationships with the Triassic sandstones and conglomerates (1882C; 1889F; 1891F; 1896I; 1898G, pp. 40–82); the use of present topography to interpret underground structures, particularly faults (1889E; 1898G, pp. 87–137); the employment of sedimentary and structural evidence to interpret the conditions associated with the sedimentation, monoclinal warping and faulting of the basin (1883L; 1894H; 1898G, pp. 19–40 and 144–57); and the Cretaceous and subsequent erosional history of the region (1898G, pp. 157–84). Davis' thesis was that post-sedimentation compression of the basin produced fractures of the metamorphic basement, perhaps guided by steeply-dipping schistose structures, leading to the differential displacement of narrow north–south aligned blocks. This displacement was transmitted to the overlying Triassic sediments and the interbedded trap volcanics to produce the composite faulted monoclinal structure in which the beds dipping at 15°–30°E are interrupted by a series of north–south faults with variable upthrows of the down-dip blocks (1886P; 1889E; 1896I); terminated by faulting on the eastern boundary of the basin (1894H). These relationships he held to produce a repetition of surface outcrops, of which the volcanics are the most topographically significant (1887J; 1888G), and Davis (1891J) employed palaeontological evidence to demonstrate this repetition in the case of shale outcrops near Meriden, Connecticut. Even with this strictly geological work, however, Davis was engaging in topographic speculations, as when he considered the original forms and structural relations of the volcanoes fed by the intrusive igneous bodies (1891I). In the same characteristic manner he marshalled in his final report (1898G) a mass of evidence concerning the summit accordance of the trap ridges and the surrounding crystalline uplands, the wind gaps, and the anomalous courses of certain streams (notably the Connecticut River below Middletown), into a coherent theory of denudation chronology. The following extracts speak for themselves:

There are three stages in the geographical development of the Triassic valley lowland that deserve special attention. The first includes the form initiated by deformation and more or less modified by contemporaneous erosion. The second is the peneplain to which the initial form was reduced by long-continued denudation, this being inferred from the remnants of the peneplain still preserved in the Eastern and Western uplands. The third is the form of to-day, carved in the peneplain after a gentle slanting uplift to about its present altitude. Shortly before the existing form was assumed came the glacial invasion commonly recognized in New England, and associated therewith are certain changes of level, especially significant near the shore lines; but these are minor episodes compared to the long and almost complete cycle of erosion by which the general peneplain was made, or even to the partial cycle during which the valley lowland was worn down between the uplands.

For ease of description and explanation, the geological dates of these critical stages may be given at once. The topography produced by the monoclinal tilting and faulting may be roughly dated as falling at the initiation of Jurassic time; Jurassic being used here not to indicate a very definite epoch, but rather to imply a succession after Triassic deposition. Jurassic time included a cycle of denudation that sufficed to reduce the forms initiated by tilting and faulting to a peneplain; and as the latter is found to be associated with Cretaceous strata, it will be spoken of as the Cretaceous peneplain. Its preparation appears, however, to be chiefly the work of Jurassic time, and a record of this division of the geological scale on our Atlantic slope is more likely to be found in the work of obliteration here manifested than in the usual record of deposition. The uplift and the carving of the peneplain into the form of to-day were accomplished in Tertiary time; hence the Valley Lowland will be spoken of as the work of the Tertiary cycle of denudation.

(Davis, 1898G, p. 153)

A view of the Eastern and Western Uplands from any commanding summit of the trap ridges, such as West Peak, near Meriden, or Mount Tom, in Massachusetts, discloses the even sky line to which their rolling hills so regularly ascend. The same persistence of accordant altitudes over large areas may be seen from any of the eminences that surmount the general level of the uplands by a small amount, such as Great Hill, near the village of Cobalt, north of the Connecticut gorge in the Eastern Uplands. The upland has a gentle southward slant, so that it becomes a lowland near the coast line, and its further continuation would pass beneath sea

FIG. 25. North-westerly view of quarries in a trap ridge (composed of two faulted lava flows) about one mile north of Meriden, Connecticut, visited by Davis on Harvard field trips in the early 1890s (From Davis, 1896I, Fig. 3)

FIG. 26. Detail of the southern quarry in the Meriden trap ridge shown in Fig. 25, looking north-west; showing the lower lava flow (a), the upper flow (b), and (c) breccias of angular trap and sandstone traversing the quarry. Cat-Hole Ridge is in the background (From Davis, 1889F, Fig. 18)

level. The steep-sided narrow valleys of the uplands are sunk beneath the general summit level thus marked. Before the excavations of these valleys the uplands must have been much more continuous then now; they must, indeed, have constituted a gently rolling region of moderate relief – a peneplain. (*Davis, 1898G, p. 157*)

It is only in the even crest line of some of the trap ridges, such as Totoket Mountain, closely accordant in altitude with that of the neighbouring Eastern Upland, that the peneplain is recognizable within the Triassic area. Yet it can not be doubted that the even surface of denudation represented in the gently rolling uplands was once continuous across the region of the present Valley Lowland. The process of peneplanation is not local but far-reaching. If it finds an application in the Eastern and Western uplands, it surely must be applicable over the intermediate area, and the more so when it is remembered that the Triassic strata are, with the exception of the trap rocks, notably weaker than the general body of the crystalline schists. The Triassic belt must have been more than a peneplain; it must have been a true plain of denudation during the later phases of the Jurassic cycle of denudation, relieved only here and there by low swells of stony soil marking the outcrops of the trap sheets and dikes. To this ultimate form must the initial faulted blocks have been at last reduced, passing on the way through a whole series of intermediate forms. During so long a cycle of denudation, acting upon a mass of greatly varying hardness, many spontaneous adjustments of streams to structure must have taken place, and at the close of the cycle it is not to be conceived that any but the larger streams could have preserved the courses that they had possessed, either through antecedent or consequent origin, at its beginning. (*Davis, 1898G, p. 159*)

It is, moreover, entirely by such denudation, causing a retreat of the outcrop in the direction of the monoclinal dip, that the extrusive trap ridges are now so systematically offset from one another, and a similar process controls at least to a considerable extent the relative location of the intrusive ridges. The offsets of the ridges have therefore nothing to do with lateral faulting; this is excluded by the nearly vertical position of the slickensides occasionally revealed on the fault planes, as in the Meriden quarries. (*Davis, 1898G, pp. 159–60*)

During the long process of subaerial peneplanation there was excellent opportunity for the development of subsequent streams along the strike of the relatively weak strata and for the diversion of the upper courses of transverse streams by the growing subsequent streams. (*Davis, 1898G, p. 162*)

It should be for these reasons expected that on the essential completion of peneplanation all the area of the lower sandstones should have become tributary to a few large rivers; that all the area of the upper sandstones should likewise have a relatively unified drainage system; that only a few of the stronger transverse streams would cut across the trap ridges by which the upper and lower sandstones are divided; and that only such longitudinal consequent streams as ran along thoroughly shattered fault lines could still maintain their courses through the low notches between the trap outcrops. (*Davis, 1898G, p. 162*)

The unconformable superposition of the Cretaceous strata on the seaward border of the peneplain shows that after peneplanation there must have been a time of depression and partial submergence, the unsubmerged area on the north and northwest furnishing the waste now seen in the strata that were deposited on the submerged area to the southeast and south. By those who do not accept the subaerial origin of the peneplain another interpretation might be given here. Instead of arguing that Cretaceous deposition followed depression and submergence, it might be contended that the sea cut its way into the land and that the Cretaceous strata are the products of its destructive advance. (*Davis, 1898G, p. 164*)

As far inland as the Cretaceous cover reached the pre-Cretaceous streams of the peneplain would be more or less completely extinguished. When post-Cretaceous elevation raised the landward part of the Cretaceous peneplain into an upland and converted the adjacent sea floor into a sloping coastal plain, the streams from the upland would be extended across the plain, guided in their new courses by its slope toward the new shore line ... The usual sequence of processes normally obtaining on an oldland and a coastal plain elevated together from a former lower stand would then take place. As the streams incised their valleys through the coastal plain and discovered the buried part of the oldland beneath, they would flow in paths entirely independent of the oldland structure and indifferent to the stream courses that obtained in the later stages of peneplanation before submergence, and they would thus take on the characteristics of superposed streams. As the covering strata gradually wasted away the stripped belt of the oldland would be characterized by these superposed streams, except in so far as new adjustments to oldland structures might be gained, particularly by the smaller streams. The Connecticut below Middletown may thus be an example of a superposed course, and not of a course consequent upon initial warping ... The lower course of the Housatonic may have a similar origin. (*Davis, 1898G, p. 165*)

As soon as the rolling uplands are seen to be remnants of a peneplain either of subaerial or of marine denudation, the uplift of the region since peneplanation becomes evident. A peneplain must at the time of its completion stand close to the sea level, and its streams must have grades so flat that no further trenching of the surface is possible. A peneplain whose surface is now several hundred feet above sea level and is deeply dissected by numerous valleys is manifestly an uplifted peneplain; and such is the case here. Therefore, as has already been stated, the present position of the Triassic formation is not the result of a single initial uplift; it has been gained by two uplifts, and the dates of the uplifts were separated by a period long enough for the base-leveling of the forms initiated by the first before the second took place ...

The greater height of the uplands in the northern part of the State than near the coast, and the greater depth of the chief valleys in the interior than near their mouths, suggest that the peneplain, with the Cretaceous cover of indefinite width along its southern margin, was elevated more in the interior than near the coast, so that it slanted southward. (*Davis, 1898G, p. 166*)

As we shall see, Davis employed not only the denudational history of the Connecticut Triassic in his wider thesis on the denudation chronology of the Appalachians, but was even able to use the very formation of the Triassic basins themselves to assist him in the assumed reversal of the Appalachian drainage. All geology was to become grist to the Davisian physiographic mill!

CHAPTER TEN

The Cycle of Erosion

Although Davis constantly acknowledged his debt to such predecessors as Powell, Jukes, Dutton and Gilbert, in later life he came to refer to his first notion of the cycle of erosion, while working on the Northern Pacific Railroad Survey in Montana in 1883, as rather like the blinding flash of understanding experienced by a prophet in the wilderness. Indeed, depressed by President Eliot's letter he must have pictured himself something of an outcast. As already mentioned in Chapter 8, Davis never forgot his mood of this period and kept a copy of Eliot's letter for the rest of his life (see also Volume I, pp. 621–41). He took every opportunity of reminding his Harvard audiences of Eliot's decision to terminate his appointment, a decision which Davis' later eminence seemed to ridicule. However, at the age of thirty-three on the high plains of Montana with no substantial publications to his name and the threat of dismissal hanging over him, Davis was perhaps in a receptive mood for the idea which was to dominate and mould his academic life.

In the spring of 1883 I had a very discouraging letter from President Eliot, in which he said that my chance of promotion at Harvard was small and that I had better look elsewhere for it. By great good fortune my old teacher Raphael Pumpelly invited me to spend the summer of that year in Montana, where he was in charge of the Northern Transcontinental Survey for the Northern Pacific Railroad. I went out there with the idea that the Plains of eastern Montana were smooth because they had recently emerged from the sea in which their strata had been laid down. To my great surprise it was soon discovered that the Plains were fairly smooth because they had been almost completely worn down from a much greater original altitude; not down to sea level, but to so small an altitude that their rivers had gentle fall; also that since that their down-wearing their rivers had begun again to cut new valleys in consequence of some change, perhaps a small new uplift. The evidence for this conclusion was, first, the repeated occurrence of low outcrop scarps of faintly inclined resistant strata, all of which had been worn back from their original extension. Second, the occurrence of several dike walls from 100 to 500 feet in undulating height; they must have been enclosed when driven up for otherwise their lava would have spread over the Plains far and wide. Third, the presence of several lava-capt mesas, which evidently represented slight hollows in the land surface at the time of the outflow of their lava, and which survived today over the lower land worn down where not lava protected. Fourth, the Crazy Mountains, which stand 3000 or 4000 feet above the plains, because their nearly horizontal strata are knit together by a multitude of dikes.

F I G. 27. Davis (4th from left) on an unknown (European?) field trip
(Courtesy R. Mott Davis)

F I G. 28. Davis examining a volcano. (Courtesy Howard M. Turner, Boston)

It was thus forced upon me that the plains are *old*, not young; also that young valleys are now incised in the old plains because of revived erosion by their rivers. The scheme of the cycle of erosion, published in the following year, was a natural outcome of this summer's observations. The following years at Harvard I introduced the natural history of rivers into my course on physical geography, and thus enlivened what had been before a very dull topic. (*Comment by Davis, written for Rigdon, 1933, pp. 65–6*)

As it turned out, the direct result of Davis' thoughts on the landforms of Montana had to wait three years before they were published, obscurely buried in a massive government publication. Even then they seem to owe more to Powell's concept of baselevel than to any cyclic ideas:

On account of the absence of strata of effective hardness and of master-dikes there seems to be no system or regularity in the form of the (Highwood) mountains or in the arrangement of the valleys. The former culminate in ridges and rounded peaks, from which serrated spurs wander out to the lower country. The upper valleys are all steep-sided, V-shaped, and without benches; evidently the processes of deepening and widening are here going on together, and there is no evidence that these upper streams have ever reached a base-level of erosion. The lower valleys – for example, that of Highwood creek – show on the other hand a distinct flood-plain between well-marked benches or terraces that are quite independent of the rock structure, and thus prove a former relatively lower level of the mass, preserved long enough to allow the streams to cut down their channels close to the then base-level of drainage, and accomplish something of the widening of the valleys by lateral cutting; all this followed by a general uplift which restored the deepening action of the streams and enabled them to cut cañon-like clefts in their old flat bases; now succeeded in the outer valleys by a second period of lateral cutting. (*Davis, 1886T, p. 710*)

The plains from Fort Benton to Sun river consist of essentially horizontal Cretaceous strata, with gently-rolling surface. Excepting near the rivers and *coulées* the local relief is seldom more than 30 to 50 feet in a mile, and this only in undulations of great breadth. The drainage lines are of older and newer origin. The older are seen in the broadly-open, flat depressions extending with curvature of long radius for miles across the country, followed by insignificant streams, or sometimes sinking into faint basins for very shallow alkaline lakes or sloughs. All these we consider the work of a river system now extinct, a system that, when in activity, cut and carried away an unknown thickness of overlying strata and brought the original surface down near its base-level of drainage, thus producing the comparative smoothness of the existing plains. But later, and very probably at the time when the intermontane lakes were formed, the rivers of this old drainage system were turned from their flat courses by a tilting and elevation of the country, and then the second and still-existing system of drainage was instituted. The large rivers, such as the Missouri, have cut deep, but comparatively narrow and steep-sided trenches, often in soft rocks, to 400 feet below the old level; but the work thus far accomplished in this new altitude is a very small share of what can be done if the existing elevation above the level of discharge be maintained.

On the Missouri there is evidence of a pause in the assumption of the present altitude in the existence of a broad bench – which seems to be a temporary 'base-level' held during the change of altitude – 250 to 300 feet above the river; but on the Teton, where seen, about 25 miles from Fort Benton, this intermediate bench is not visible. Keeping pace with the deepening of these larger channels, the small lateral streams have cut ramifying ravines in the river banks, making a very rough, almost impassable country, from 1 to 3 miles wide, known as 'the breaks'; and still smaller streams, probably following new directions since the last tilting of the country, have cut distinct, narrow but shallow channels on the steeper slopes of the old eroded surfaces. In general the new drainage lines have channels proportional to the volume of their wet-weather streams, while the broad old channels are either dry or swampy, or are followed by streams of comparatively small size. Where they hold shallow lakes the basins may plausibly be ascribed to the obstructive action of new streams depositing their alluvial deltas in the old channels. The thickness of strata destroyed in carving out the present general surface of the plains can only be inferred by an extended study of the Cretaceous strata over a broad region; and this we have had no opportunity of making. But the Highwood and Crazy mountains probably give a minimum measure of 3,000 to 5,000 feet of erosion on the country around them, and the plains are therefore to be regarded as plains of denudation, cut down to an old base-level of erosion. (*Davis, 1886T, p. 710*)

When the erosion began the Cretaceous strata must have been at least as high as, and probably higher than, the present summits of the buttes and mesas; so we may infer a subsequent denudation of at least 1,000 feet over all the neighboring plain country. The contemporaneous erosion over many of the adjoining mountain ranges must have been certainly vastly in excess of this measure, the difference in the two amounts being most likely due to their different altitudes above the level of drainage discharge.

The relation of the several features of the plains now described is of importance. Their comparatively smooth surface is not a form of original deposition, but results from the denudation of a broad sheet of strata, as is shown by the attitude of the mesas and dikes. There is every reason to think that this denudation is the work of rain and rivers; but in order that these agents should produce a smooth surface they must perseveringly act until their valleys widen and consume the intervening hills and reduce the surface nearly to the base-level of erosion. It thus follows that the present elevated surface of the plains stood for a long period through Tertiary and early Quaternary time close to the level of drainage discharge. The same conclusion is suggested by the broad distribution of ice-rafted boulders of north-eastern origin. Subsequent to the general erosion and to the importation of the boulders, the plains gained their present elevation. We are, therefore, led by two independent lines of evidence toward the same series of events in the history of the plains. (*Davis, 1886T, p. 711*)

The vagueness of Davis' cyclic ideas at this time is underlined by his article on 'Gorges and waterfalls' (1884N), which incidentally includes the notion of comparatively sudden uplift and makes use of very evolutionary terminology.

Waterfalls are also found in countries quite free from glacial action; but such countries are generally high, and the cascade, although old, is not yet old enough to have worn its hard bench-rock back into a smooth, gentle slope. (*Davis, 1884N, p. 123*)

An old stream, having its volume, its load of silt and its slope so related to one another that its channel is eroded with extreme slowness, comes as near to a condition of stability as streams can. It has, after a long age of endeavor, at last adapted itself to its environment, and has very little tendency to variation. But any change in its condition of life sets it at work again, seeking another attitude of satisfied equilibrium. If, for example, the land across which an elderly, conservative stream flows, is elevated with comparative suddenness, the stream finds its point of discharge lowered, and thereupon sets to work, with all the strength gathered from its whole drainage basin, to cut the lower end of its valley down to the level of quiet water again; and the deep cutting will work its way backward up stream, requiring a readjustment to the new slope up to the head of every little creek and rivulet of its basin. For a time, during the change, the deepening of the channel will go on faster than the widening, and the old flat valley will be marked by a gorge-like trough along its median line. Thousands of valleys of this compound form may be seen in the streams of the Rocky Mountain region. Indeed it was in that region that this important relation of the cutting power of a stream to its 'base-level of erosion' was first well appreciated; and its illustration includes some of the finest work of our government geologists. (*Davis, 1884N, p. 131*)

Finally, if the overflow be turned aside (by glacial blocking), from the old valley way, and find its line of escape leading over steep sloping rocks, the gorges, cascades and waterfalls will mark its line and give character to the new channel. Even in soft, shaly rocks, the down-cutting in such cases is much faster than the lateral widening; and in the relatively early stages of development the gorges will be narrow and steep-walled. (*Davis, 1884N, p. 132*)

It was at the 33rd Meeting of the American Association for the Advancement of Science in 1884 that Davis gave the first broad idea of the cycle of erosion in his article on geographic classification. With hindsight we can draw attention to some remarkable features. The valley-form will vary with the rate of uplift; the stage of maturity is introduced and elaborated; the terminology is sadly inadequate and, for example, includes (in old age) 'the kernels of the longest enduring outliers'. However, the word 'cycle' is introduced as 'a new cycle of life'.

But of greater geographical importance than these early and constructional characteristics, are the later, destructional ones, determined by erosion, inasmuch as they comprehend the topographical form that we actually observe. It is this division of the subject that is worthy of most attention from the comparative geographer, and in assigning a region of horizontal structure to its proper place in this classification, the following sketch of the life-history of such a land mass may serve as a guide. Just as the surface of the deposit rises above its base-level of erosion – or, in the case of lake deposits, as the governing base-level sinks below the surface, a smooth, unbroken plain is revealed. Its drainage is imperfect, for the newly formed streams have not yet had time to establish their channels, and much of the rainfall stands on the ground until it evaporates. Faint depressions on the surface, the effect of slight irregularity in the original distribution of the deposit, hold the water back to form shallow marshy lakes. The smoothness of the surface and the shallow lakes are indeed truly infantile features, retained only during the earliest life of the plain, and soon lost in its further development. This develop-ment finds its opportunity in the gaining of an effective elevation above the base-level of the region and in the persistent down-cutting of the streams. Rivers establish their courses, the smooth plain is trenched across by their meandering channels, and all the lakes disappear. This is adolescence. The channels will be narrow and steep walled in regions of relatively rapid elevation, but broadly open in regions that have risen slowly, and I believe that rate of elevation is thus of greater importance than climatic conditions in giving the cañon form to a valley. Of course the channels must be shallow in plains of small altitude, while they may become deep in regions that have been raised to a great height; but great cañons can occur only in young plateaus, for cañons are marks of a precocious adolescence, and in this feature, the Colorado plateaus are truly American. Adolescence as thus defined includes that part of a plateau's early life in which the stream channels are narrow and well marked; but as the valleys increase in number and open widely so as to consume a good share (one-sixth to one-fifth) of the plateau mass measured above the existing base-level of discharge, then adolescence merges into maturity. Maturity may be said to last through the period of greatest diversity of form, or maximum topographic differentiation, until about three-quarters of the original mass are carried away; and through this period when the geographic names for this group of forms should be most applicable, there are none whatever that can be properly applied, nor is there any that includes this group of forms as a whole.

During maturity no vestige of plain surface remains (except in slow-rising, low plains that never acquire any marked relief), and the region is too broken to be named a plateau. To call it a hilly or mountainous country resting on a plateau would be an unappreciative, indeed a vicious, style of nomenclature. Indeed there is no suitable technical name yet proposed for such a surface, although it possesses a most peculiar and well-defined topography. Long lines of cliffs are the most pronounced features; their upper edge is of constant altitude, but their direction is exceedingly irregular, continually varying as they pass from reentrant to pro-jecting angles. When standing on the bench or platform of such a cliff, its extension can be traced for miles, contouring around the spurs that separate ravines and valleys. Every cliff marks the out-cropping of a more resistant stratum or group of strata than those that over and underlie it, and the natural variety in the sequence of harder and softer layers gives origin to the many varia-tions on a single type that are observed in the scenery of broken plateaus. The cliffs are separated by sloping surfaces, that in the younger plateaus are steep and largely covered with debris from the cliff above forming a talus almost down to the cliff below; but with the recession of the cliffs, the talus proper is restricted to a narrow space around their base, and the intervening bench land has a more gentle inclination.

Along the larger valleys, outliers are frequently isolated from the main plateau mass by the meeting of side streams behind them; these are often used as strongholds by savage tribes, and as we know in the west the cliffs also may afford protection to weaker races who seek shelter beneath them. The models of these peculiar dwellings, prepared by Messrs. Holmes and Jackson of our geological survey, illustrate admirable examples of the control exerted by topography on the conditions of human life. During the maturity of a plateau all the streams cut down nearly to their base-level and give a maximum of relief between the upper levels and valley bottoms. The stream system itself is very characteristic: the stream courses having been selected on nearly horizontal surfaces, and finding no lateral constraint in the attitude of the strata through which they cut, are necessarily meandering; they branch and subdivide in all directions, like the veins of a maple leaf, giving a consequent irregularity to the outcrop of hard strata on spurs and in ravines, and rendering a well-broken plateau extremely difficult to traverse. No small degree of skill and experience is required to extricate one's self from the labyrinth of almost impassable valleys in crossing such a country.

One of the latest phases of maturity reveals the higher levels of the plateau broken through by the ever increasing number of ravines and valleys, producing in some cases the peculiar pinnacled topography in which vertical lines are so numerous and so pronounced that it is at first difficult to associate them with a horizontal structure; but the pinnacles are simply remnants of the cliff faces that first appeared as bold steps on the valley slopes that were next seen wrapped around the gradually isolated outliers, and that now still retain their height, while their length has diminished as the outlier wasted away.

On leaving this stage, the plateau passes from maturity to old-age: the relief of the surface diminishes, for while the pinnacles dwindle away and the country loses in altitude, the valleys do not correspondingly gain in depth. The cliff faces are forced farther and farther back and the valleys increase greatly in width, the slopes become gentler and the whole aspect of the region is simpler and tamer than before. The kernels of the longest enduring outliers stand scattered about on the low plains into which the all-consuming valleys have grown and fitly take the form of great fortresses in their final struggle against the elements. At last even these remnants are carried away, reducing the once rugged country to almost as low, flat and featureless a surface as it was at birth; but its perfected system of meandering rivers separates it from new-born plains, and although now smooth when compared to the ruggedness of its maturity, still it rises gently between the streams in faint swells which distinguish it from the more nearly perfect level of unworn surfaces. It is now easy to traverse, but is still capable of misleading the traveller over its monotonous expanse, where every trough is closely like its neighbor, and all the intervening swells are too smooth and low to serve as landmarks or as points of view. This is simple old age, a second childhood in which infantile features are imitated and thus the decrepit surface must wait either until extinguished by submergence below the sea, or regenerated by elevation into a new cycle of life. (*Davis, 1885D, pp. 429–32*)

During the next quarter of a century Davis elaborated, refined and illustrated the cyclic concept, particularly in his 'The Rivers and valleys of Pennsylvania' (1889D) and 'The Geographical Cycle' (1899H). The former first contained the terms 'cycle in the life of a river', 'cycle of development' and 'cycle of erosion'. The publication of his *Geographical Essays* (1909C), three years before his retirement, brought a great deal of this work together and represents a statement of the 'classic' geographical cycle which stood as a guiding light for professed 'Davisian geomorphologists' for the next half century, despite significant modifications by Davis himself during the 1920s and early 1930s. These modifications are treated later in this volume, and here it is proper to let quotations from Davis outline the cyclic scheme which evolved during his many teaching years at Harvard University.

THE CYCLE OF EROSION: STRUCTURE, PROCESS AND STAGE

In the scheme of the ideal geographical cycle a complete sequence of land forms of one kind or another may be traced out. The cycle begins with crustal movements that place a given land mass in a certain attitude with respect to base-level. The surface forms thus produced are called initial. Destructive processes set to work upon the initial forms, carving a whole series of sequential forms, and finally reducing the surface to its ultimate form – a low plain of imperceptible relief. The sequential forms thus constitute a normal series by which the initial and the ultimate forms are connected. As a result, the sequential forms existing at any one moment are so largely dependent on the amount of work that has been done upon them that they are susceptible of systematic description in terms of the stage of the cycle which they have reached. Moreover, the correlation of all the separate forms appropriate to any one stage of the cycle is so intimate and systematic that any single form may be designated in an appropriate and consistent terminology, as a member of the group of related forms to which it belongs, and thus better than in any other way the features of the lands may be systematically and effectively described. (*Davis, 1905J, pp. 150–51*)

In the following article, which included treatment of the genetic classification of landforms, Davis writes on structure, process and time, but makes it clear that *time* represents the amount of change from the initial form or, in other words, its *stage*. However, he does not make it clear that time itself is directly important only where structure and process are absolutely constant.

All the varied forms of the lands are dependent upon – or, as the mathematician would say, are functions of – three variable quantities, which may be called structure, process, and time. In the beginning, when the forces of deformation and uplift determine the structure and attitude of a region, the form of its surface is in sympathy with its internal arrangement, and its height depends on the amount of uplift that it has suffered. If its rocks were unchangeable under the attack of

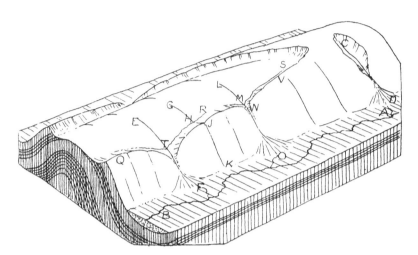

F I G. 29. The drainage of a denuded arch. A longitudinal consequent river (AB) is fed by lateral consequents (CD, LO, EF), which, in turn, are supplied by subsequent streams (TR, MS) developed on a weak stratum. TR has beheaded the lateral GK at H. The larger lateral consequents have formed alluvial fans (F, O, D) (From Davis, 1899H, Fig. 2)

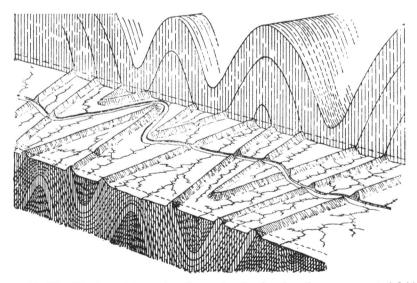

F I G. 30. The Allegheny Mountains, Pennsylvania, showing the reconstructed folds and (by the dashed line) the peneplain level from which the river was superimposed over the folds (From Davis, 1898E, Fig. 118)

external processes, its surface would remain unaltered until the forces of deformation and uplift acted again; and in this case structure would be alone in control of form. But no rocks are unchangeable; even the most resistant yield under the attack of the atmosphere, and their waste creeps and washes downhill as long as any hills remain; hence all forms, however high and however resistant, must be laid low, and thus destructive process gains rank equal to that of structure in determining the shape of a land-mass. Process cannot, however, complete its work instantly, and the amount of change from initial form is therefore a function of time. Time thus completes the trio of geographical controls, and is, of the three, the one of most frequent application and of most practical value in geographical description. (*Davis, 1899H, pp. 481–2*)

He never tires of advertising the all-embracing, deductive nature of his explanatory system of landform description, as the following extracts illustrate. The first is also notable for its treatment of drainage density and is so important that it is repeated in another context on p. 331.

Another advantage of the explanatory method is the ease with which a simple series of ideal forms, such as those of the plateau series above mentioned, may be expanded into a more elaborate series, so as to fit a much larger variety of natural cases. In the first place, a plateau mass of horizontal structure may be conceived as consisting of resistant and weak strata in any order; a group of resistant strata between two groups of weak strata, . . . ; or a group of weak strata between two groups of resistant strata, . . . ; or many alternations of resistant and weak strata, . . . ; and so on. The general plan of development being once understood, all the particular features that will be associated with each particular structure may be easily deduced. In the second place, a plateau of any special structure may, at the beginning of its cycle of erosion, stand high or low; the *relief* developed by its dissection will evidently be strong during the youth and maturity of the first case, . . . , but can never be strong in the second case, . . . Finally, the *texture* of dissection – that is, the spacing of the streams and their valleys, and the resultant breadth of the hills and spurs – may vary according to certain conditions, the chief of which are the perviousness of the plateau mass, or at least of its soil cover, and the climate under which its dissection takes place. Pervious strata allow much of the rainfall to enter the ground and thus diminish the number of small branch streams; as a result, the texture of dissection is coarse, and the hills are broad and full bodied, . . . If the strata are impervious and of fine-grained materials, and if the climate is so dry as to make vegetation scanty, then the texture of dissection during youth and early maturity will be very fine, and the surface will assume a 'bad-land' quality, with innumerable little hills and spurs between innumerable little valleys, . . . Thus the scheme of description really contains five elements: structure, process, stage, relief, and texture; all require consideration, but the first three suffice when giving the scheme a name.

. . . The few diagrams here offered may be supplemented by many others, illustrating earlier and later stages of development, in the case of coastal plains; diverse structures, varying from horizontal to inclined, faulted, folded, massive, and so on; varied combinations of resistant and weak strata, as well as different measures of relief, in the case of plateaus; differences in the scale of texture, from coarse to fine, and so on. It is manifest that the observer who is equipped, mentally or graphically, with a large number of type forms, correctly deduced and systematically arranged and developed, will be greatly aided when he comes to describe actual forms, and that the reader who has a corresponding equipment will be able to appreciate very closely the meaning of the observer. (*Davis, 1909G, pp. 307–8*)

The surface of the land may be regarded as composed of a number of individual forms, whose general character depends on the rock-structure which the processes of land-sculpture have worked upon, and whose more particular expression depends on the degree of advance in the degradation of the surface from its initial, constructional form to the smooth, low, baselevel plain to which it is finally reduced. Thus regarded, any geographic individual may be associated with certain others, to which it is related by similarity of structure, and the whole group of similar individuals, thus related, may be idealized in a type, which presents all the essential, but none of the accidental features of the group that it represents. The type is therefore an elastic conception, not limited in the way of size, nor in the number of its features, nor in any variable element; but always holding fast to those characteristics that distinguish it from the types of other groups. Moreover, in order that individuals of different age may be properly represented by a single type, every type must be conceived to vary systematically in passing through the cycle of changes that its individuals suffer from the time when the first attack is made upon them by the destructive forces of the weather, to the time when they are worn down to baselevel, the level of the standing water into which their drainage flows, below which land erosion cannot reduce them; ... (*Davis, 1889J, p. 367*)

Structure is a pertinent element of geographical study when, as nearly always, it influences form; no one would today attempt to describe the Weald without some reference to the resistant chalk layers that determine its rimming hills. Process is equally pertinent to our subject, for it has everywhere been influential in determining form to a greater or less degree, and it is everywhere in operation today. It is truly curious to find geographical text-books which accept the movement of winds, currents, and rivers as part of their responsibility, and yet which leave the weathering of the lands and the movement of land-waste entirely out of consideration. Time is certainly an important geographical element, for where the forces of uplift or deformation have lately (as the Earth views time) initiated a cycle of change, the destructive processes can have accomplished but little work, and the land-form is 'young'; where more time has elapsed the surface will have been more thoroughly carved, and the form thus becomes 'mature'; and where so much time has passed that the originally uplifted surface is worn down to a lowland of small relief, standing but little above sea-level, the form deserves to be called 'old'. A whole series of forms must be in this way evolved in the life-history of a single region, and all the forms of such a series, however unlike they may seem at first sight, should be associated under the element of time, as merely expressing the different stages of development of a single structure. The larva, the pupa, and the imago of an insect; or the acorn, the full-grown oak, and the fallen old trunk, are no more naturally associated as representing the different phases in the life-history of a single organic species, than are the young mountain block, the maturely carved mountain-peaks and valleys, and the old mountain peneplain, as representing the different stages in the life-history of a single geographic group. Like land-forms, the agencies that work upon them change their behaviour and their appearance with the passage of time. A young landform has young streams of torrential activity, while an old form would have old streams of deliberate or even of feeble current, as will be more fully set forth below.

... The sequence in the developmental changes of land-forms is, in its own way, as systematic as the sequence of changes found in the more evident development of organic forms. (*Davis, 1899H, pp. 484–5*)

The fully developed scheme of the cycle recognizes the passive mass of the earth's crust, raised here and there, and thus exposed to the destructive processes; the various destructive processes or agencies by which the passive crustal mass is systematically carved; and the waste or 'chips' that result from the carving processes. The waste is much less active in its creeping and washing movements than are rivers, glaciers, or winds in their flow, and yet the waste is much more active in its down-slope movements than is the passive mass on which it rests. The cloak of creeping rock waste that covers a graded hillside is as much deserving of systematic description as is the great rock mass of the hill as a whole, or the slender thread of the stream in the valley; and the suggestive correlations that result from giving a definite place in systematic physiography to the 'forms assumed by the waste of the land on the way to the sea' are sufficient warrant for this element of the scheme. (*Davis, 1905J, p. 160*)

Davis was at pains to identify what he considered to be the 'normal' process of erosion.

The destructive processes are of great variety – the chemical action of air and water, and the mechanical action of wind, heat, and cold, of rain and snow, rivers and glaciers, waves and currents. But as most of the land surface of the Earth is acted on chiefly by weather changes and running water, these will be treated as forming a *normal* group of destructive processes; while the wind of arid deserts and the ice of frigid deserts will be considered as climatic modifications of the norm ... (*Davis, 1899H, pp. 482–3*)

CYCLIC ASSUMPTIONS: INITIAL SURFACE; UPLIFT; BASELEVEL; CLIMATE

Once established, an original river advances through its long life, manifesting certain peculiarities of youth, maturity and old age, by which its successive stages of growth may be recognized without much difficulty. For the sake of simplicity, let us suppose the land mass, on which an original river has begun its work, stands perfectly still after its first elevation or deformation, and so remains until the river has completed its task of carrying away all the mass of rocks that rise above its baselevel. This lapse of time will be called a cycle in the life of a river. A complete cycle is a long measure of time in regions of great elevation or of hard rocks; but whether or not any river ever passed through a single cycle of life without interruption we need not now inquire. Our purpose is only to learn what changes it would experience if it did thus develop steadily from infancy to old age without disturbance. (*Davis, 1889D, pp. 21–2*) *

In the following quotations Davis shows clearly that he had considered rapid uplift, gradual uplift and slow uplift, but preferred the first as typical and the two others as special cases. Rapid uplift was more convenient, especially for the learner – not to mention for Davis himself!

The elementary presentation of the ideal cycle usually postulates a rapid uplift of a land mass, followed by a prolonged still stand. The land mass may have any structure, but the simplest is that of horizontal layers; the uplift may be of any kind and rate, but the simplest is one of uniform amount and rapid completion; hence plains and plateaus have an early place in a systematic classification of land forms; but all sorts of structures and all sorts of uplifts must be considered before the scheme is completely worked out. In my own treatment of the problem, the postulate of rapid uplift is largely a matter of convenience, in order to gain ready entrance to the consideration of sequential processes and of the successive stages of development – young, mature, and old – in terms of which it is afterwards so easy to describe typical examples of land forms. Instead of rapid uplift, gradual uplift may be postulated with equal fairness to the scheme, but with less satisfaction to the student who is then first learning it; for gradual uplift requires the consideration of erosion during uplift. (*Davis, 1905J, p. 153*)

... *it should not be implied* ... *that the forces of uplift or deformation act so rapidly that no destructive changes occur during their operation* ... even during uplift, the streams that gather in the troughs as soon as they are defined do some work ... The uplands also waste more or less during the period of disturbance, and hence no absolutely unchanged initial surface should be found ... we must always expect to find some greater or less advance in the sequence of developmental changes, even in the youngest known landforms. 'Initial' is therefore a term adapted to ideal rather than to actual cases, in treating which the term 'sequential' and its derivatives will be found more appropriate. (*Davis, 1899H, p. 487*)

A special case necessitating explanation by slow uplift may be easily imagined. If an even upland of resistant rocks be interrupted by broadly open valleys, whose gently sloping, evenly graded sides descend to the stream banks, leaving no room for flood plains, it would suggest slow uplift; the absence of flood plains would show that the streams have not yet ceased deepening their valleys, and the graded valley sides would show that the downward corrasion by the streams had not been so rapid that the relatively slow processes of slope grading could not keep pace with it. In such a case there would have been no early stage of dissection in which the streams were inclosed in narrow valleys with steep and rocky walls; the stage of youth would have been elided and that of maturity would have prevailed from

---

* Reprint page numbers.

the beginning but with constantly increasing relief as long as uplift continued. Examples of this kind must be rare; it is nearly always the case that a beginning of flood-plain development is made before the valley sides are completely graded to even, waste-covered slopes; and hence the usual supposition of rapid uplift – rapid, as the earth views time – is probably essentially correct. Moreover, it should not be forgotten that uplift must usually be much faster than the down-wear of general subaerial erosion, however nearly it may be equaled by the corrasion of large rivers. The original postulate of rapid uplift therefore requires only a moderate amount of modification to bring it into accord with most of the land forms that we have to consider.

The postulate of a still-standing land, unmoved until it is worn down to a plain, is like the postulate of rapid uplift, a matter of convenience for first presentation; but it is also something more. It is essential to the analysis of the complete scheme because only in the ideal case of a land mass that stands still after its uplift can one trace out the normal series of sequential events in which the real value of the cycle scheme consists, and thus learn the systematic correlation of forms that characterizes each stage of the cycle. It is only after the normal series has been analyzed that the peculiar combinations of forms which result from two or more cycles of erosion can be understood. The recognition of the systematic correlation of individual forms appropriate to any given stage of the cycle constitutes a marked advance over that earlier style of physical geography in which the various elements of form were described as if they had nothing to do with one another. *(Davis, 1905J, p. 154)*

It seems to me advisable to limit 'baselevel' to . . . an imaginary level surface, and to define it simply as the level base with respect to which normal subaerial erosion proceeds; to employ the term 'grade' for the balanced condition of a mature or old river; and to name the geographical surface that is developed near or very near to the close of a cycle, a 'peneplain', or 'plain of gradation'. A full understanding of the development of land forms can be gained only by tracing the progressive changes of a generalized example from the initial stage through the various sequential stages to the ultimate stage of an ideal geographical cycle. *(Davis, 1902A, p. 84)*

It suffices at first to recognize that in the ideal undisturbed cycle of normal erosion the baselevel must be more and more closely approached as time is extended.

This definition of 'baselevel' as a level base certainly has the advantage of being easily conceived. Once conceived in the study of the ideal cycle, it needs no modification so long as the relative attitude of land and sea remains fixed. If the land rise or fall with respect to the sea, the baselevel takes a new position within the land mass, and further progress of erosion is then continued with respect to the new limit.

As the study of the cycle advances it becomes desirable to speak of various local or temporary controls of erosion: a rock ledge or a lake on a river course, the central basin of a dry interior basin either above or below sea level, the surface of a lake in such a basin. Nothing can be simpler than to imagine a level surface passing through any one of these controls, and rising or sinking as the control

rises or sinks; and such a surface is naturally called a local or temporary baselevel.
(Davis, 1902A, p. 85)

The normal ideal cycle postulates no climatic change except such as accompanies
the decrease of surface temperatures and the increase of precipitation caused by
the initial (relatively) rapid uplift and the gradual rise of surface temperatures and
decrease of precipitation that accompanies the slow wearing down of the region
to a lowland plain. *(Davis, 1905J, p. 158)*

CHARACTERISTIC STAGES OF THE RIVER CYCLE

Youth (infancy and adolescence)

In its infancy, the river drains its basin imperfectly; for it is then embarrassed by
the original inequalities of the surface, and lakes collect in all the depressions. At
such time, the ratio of evaporation to rainfall is relatively large, and the ratio of
transported land waste to rainfall is small. The channels followed by the streams
that compose the river as a whole are narrow and shallow, and their number is
small compared to that which will be developed at a later stage. The divides by
which the side-streams are separated are poorly marked, and in level countries are
surfaces of considerable area and not lines at all ...

As the river becomes adolescent, its channels are deepened and all the larger
ones descend close to baselevel. If local contrasts of hardness allow a quick
deepening of the down-stream part of the channel, while the part next up-stream
resists erosion, a cascade or waterfall results; but like the lakes of earlier youth, it
is evanescent, and endures but a small part of the whole cycle of growth; but the
falls on the small headwater streams of a large river may last into its maturity, just
as there are young twigs on the branches of a large tree. With the deepening of the
channels, there comes an increase in the number of gulleys on the slopes of the
channel; the gulleys grow into ravines and these into side valleys, joining their
master streams at right angles. *(Davis, 1889D, pp. 22–3)*

Maturity

... [Maturity] is marked by an almost complete acquisition of every part of the
original constructional surface by erosion under the guidance of the streams, so
that every drop of rain that falls finds a way prepared to lead it to a stream and
then to the ocean, its goal. The lakes of initial imperfection have long since
disappeared; the waterfalls of adolescence have been worn back, unless on the still
young headwaters. With the increase of the number of side-streams, ramifying
into all parts of the drainage basin, there is a proportionate increase in the surface
of the valley slopes, and with this comes an increase in the rate of waste under
atmospheric forces; hence it is at maturity that the river receives and carries the
greatest load; indeed, the increase may be carried so far that the lower trunk-
stream, of gentle slope in its early maturity, is unable to carry the load brought to
it by the upper branches, and therefore resorts to the temporary expedient of
laying it aside in a flood-plain. *(Davis, 1889D, p. 23)*

It is only during maturity and for a time before and afterwards that the three
divisions of a river, commonly recognized, appear most distinctly; the torrent
portion being the still young head-water branches, growing by gnawing backwards at their sources; the valley portion proper, where longer time of work has

enabled the valley to obtain a greater depth and width; and the lower flood-plain portion, where the temporary deposition of the excess of load is made until the activity of middle life is past.

Maturity seems to be a proper term to apply to this long enduring stage; for as in organic forms, where the term first came into use, it here also signifies the highest development of all functions between a youth of endeavor towards better work and an old age of relinquishment of fullest powers. (*Davis, 1889D, p. 24*)

The extension of the graded condition over all parts of a river system introduces a thoroughness of organization in the processes of land sculpture that warrants the use of the term 'maturity', as the name of the stage of the cycle in which the organization of river systems is chiefly accomplished. The growth of organization goes with the development of grade. In every reach of a river in which the graded condition has been attained, the lowest point on the reach is always coincident with and dependent on a controlling baselevel (as above defined) either general or local, and river action at any point in the graded reach is then delicately correlated with that at every other point. River action in such a reach may justly be said to be organized, inasmuch as a change in form or action at any one point involves a change at every other point. Adjacent reaches, separated by a fall on an ungraded ledge or by an unfilled lake, are independently organized; a change in one does not necessarily call for a change in the other. But when all falls and rapids are worn down, and all lakes are filled up, and the entire river system is graded, as is characteristically the case in the late-mature stage of a cycle, the organization of the system is so complete that all its parts are correlated. A change at any one point then involves a change, of infinitesimal amount perhaps, all through the system. (*Davis, 1902A, pp. 92–3*)

There was a time not very long ago when a river with a maturely graded course could be regarded, even by mature geographers, as simply an existing thing without their feeling any concern about the long series of changes which must inevitably have run their course before maturity could be reached; but that time is not ours. No modern geographer who recognizes the evolution of the present from the past – and no modern geographer can fail to do that – no modern geographer can escape from imagining the young stages of a river as preceding the mature in order properly to appreciate the mature stages. Still more, if one speaks in concise terms of 'a meandering valley maturely incised in an uplifted peneplain of disordered structure', must the appreciative understanding of this descriptive phrase build up the conception of the present on a long succession of past events, and indeed not only on a succession of inorganic events, but of organic events as well; for ... there is a most intimate, systematic, and essential correlation between these inorganic changes and the organic changes that lead up to the present organic population of an uplifted peneplain of disordered structure, traversed by an incised meandering valley, and no proper account of the present populations can be given without an appreciation of their past (*Davis, 1912H, p. 114*)

Grade

The Davisian concept of 'grade' was taken direct from the great American geologists and from French civil engineers. Davis always had difficulty in persuading some of his German opponents that his 'graded' rivers could continue to lower their beds.

> ... the balance between erosion and deposition, attained by mature rivers, introduces one of the most important problems that is encountered in the discussion of the geographical cycle. The development of this balanced condition is brought about by changes in the capacity of a river to do work, and in the quantity of work that the river has to do. The changes continue until the two quantities, at first unequal, reach equality; and then the river may be said to be graded, or to have reached the condition of grade ... The idea of grade is not of almost axiomatic simplicity, like the idea of baselevel; its meaning must be gradually elaborated as it is approached. Moreover, a graded river does not maintain a constant slope; it changes its slope systematically with the progress of the cycle: ... but it may be noted in passing that the graded condition persists all through the old age as well as the maturity of an uninterrupted cycle. 'Grade', meaning a condition or balance, must not be confused with the same word used in another meaning, namely, the slope or declivity of the river when the graded condition is reached; for 'grade' meaning slope, varies in place and in time, while 'grade', meaning balance, always implies an equality of two quantities. In fine, grade is a condition of essential balance between corrasion and deposition, usually reached by rivers in the mature stage of their development, when their slopes have been duly worn down or built up with respect to the baselevel of their basin. (*Davis, 1902A, pp. 86–7*)

After a river system has attained a maturely graded condition, it will maintain a graded condition through all the rest of the undisturbed cycle; but it is important to recognize that the maintenance of grade, during the very slow changes in volume and load that accompany the advance of the cycle, involves an appropriate change of slope as well. Instead, therefore, of having to do with a fixed control of erosion, such as is found in the general baselevel of a region, we have here to do with a slowly, delicately and elaborately changing equilibrium of river action, accompanied by a corresponding change in river slope. For example, a large river in a mountain valley may reach grade in the early maturity of its region. It will then flow with a rushing current on a rapidly sloping bed of cobblestones; and it may stand hundreds or even thousands of feet above baselevel. In the old age of the region, the same river will flow with a sluggish current on a nearly level bed of sand and silt through a peneplain, only a few tens or scores of feet above baselevel.

This is a point that is not generally enough recognized. It is too often implied – in the absence of explicit statement to the contrary – that when a river is once balanced between erosion and deposition its slope thenceforward remains constant. The beginner would gather this understanding of the question from several of the definitions of 'baselevel' above quoted, but such is evidently not the case. When a stream is first graded, its channel is not level, and it has not reached the base of its erosive work. In virtue of the continual, though slow, variations of stream volume and load through the normal cycle, the balanced condition of any stream can be maintained only by an equally continuous, though small, change of river slope, whereby capacity to do work and work to be done shall always be kept equal. It might at first be thought that changes of this kind would be perceptible, and that there would be occasional departures of a river from the graded condition; but such is not the case, because the change in the value of any variable in a unit of time is only by a quantity of the second order, by a differential of its total value. Once graded, a river will never depart perceptibly from the graded condition as long as the normal advance of the cycle is undisturbed. The slope of a river must necessarily be steeper on the first attainment of grade in early maturity, when an abundant load is received from the steep valley sides and the active headwater, than in late old age, when the valley sides have been worn down almost level, and when even the headwater streams are weak and sluggish. Hence, just as a graded river has slopes of varying declivities in its different parts at any one time, so the slope at any one part of the river must vary at different times in the successive stages of the cycle. (*Davis, 1902A, pp. 95–6*)

Davis obviously felt that 'grade' demanded repeated explanations, as it depended so much on load, although he himself occasionally used the term when he meant 'gradient'. We are asked, for example, to accept that an adolescent stream may aggrade its valley trough while an old-age river may degrade its valley-floor.

Not only do graded streams vary in slope in different parts of a river system; the slopes may vary greatly in two neighboring river systems at the time of the general establishment of their grade. This may be illustrated by considering the unlike conditions obtaining in two rivers, alike in volume, but one flowing through an upland of resistant rocks, the other through a similar upland of weak rocks. The first would have to cut down a deep valley to a gentle slope before grade was reached; because its load would be slowly delivered from the resistant rocks of its valley walls, and high walls would have to be produced by deep valley cutting before a balance could be struck between the increasing load from the walls and headwaters and the decreasing capacity of the river. The second river could not cut so deep a valley, however weak the rocks of its bed, because it would have an abundant load supplied by the rapid wasting of its valley sides, even when they were of moderate height, and a strong slope down the valley would be required in order to maintain a velocity with which the graded stream could bear the abundant load away. Only as the whole upland is worn down in the later stages of the cycle could the second stream wear down its valley to a gentle slope; and then the valley would be still shallower than when grade was first attained (*Davis, 1902A, p. 94*)

The consequent streams proceed to entrench themselves in the slanting plain, and in a geologically brief period, while they are yet young, they will cut their valleys down so close to baselevel that they cannot for the time being cut them any deeper; that is, the streams will, of their own accord, reduce their valley lines to such a grade that their capacity to do work shall be just equal to the work they have to do. When this condition is reached, the streams may be described as having attained a 'profile of equilibrium'; or, more briefly, they may be said to be *graded*. It may be noted, in passing, that inasmuch as the work that a stream has to do is constantly varying, it must as constantly seek to assume new adjustments of grade. In the normal course of river events, undisturbed by outside interfer-

ence, the change in the work is so slow that the desired adjustment of capacity to work is continually maintained. It may be that during the adolescence of river life, the work to be done is on the increase, on account of the increasingly rapid delivery of land-waste from the slopes of the growing valley branches; and in this case, part of the increase of waste must be laid down in the valley trough so as to steepen the grade, and thus enable the stream to gain capacity to carry the rest. Such a stream may be said to aggrade its valley, adopting a good term suggested by Salisbury; and in this way certain flood-plains (but by no means all flood-plains) may have originated. Aggrading of the valley line may often characterize the adolescence of a river's life; but later on, through maturity and old age, the work to be done decreases, and degrading is begun again, this time not to be interrupted. The longer the river works, the fainter is the grade that it adopts.
(*Davis, 1895E, p. 130*)

The same theme is continued below; for example the reason why old age streams may degrade is explained. In addition, we are presented with Gilbert's concept that the slope adopted when grade is assumed varies inversely with the volume.

The altitude of any point on a well-matured valley floor must therefore depend on river-slope and distance from mouth. Distance from mouth may here be treated as a constant, although a fuller statement would consider its increase in consequence of delta-growth. River-slope cannot be less, as engineers know very well, than a certain minimum that is determined by volume and by quantity and texture of detritus or load. Volume may be temporarily taken as a constant, although it may easily be shown to suffer important changes during the progress of a normal cycle. Load is small at the beginning, and rapidly increases in quantity and coarseness during youth, when the region is entrenched by steep-sided valleys; it continues to increase in quantity, but probably not in coarseness, during early maturity, when ramifying valleys are growing by headward erosion, and are thus increasing the area of wasting slopes; but after full maturity, load continually decreases in quantity and in coarseness of texture; and during old age, the small load that is carried must be of very fine texture or else must go off in solution . . .

If a young consequent river be followed from end to end, it may be imagined as everywhere deepening its valley, unless at the very mouth. Valley-deepening will go on most rapidly at some point, probably nearer head than mouth. Above this point the river will find its slope increased; below, decreased. Let the part upstream from the point of most rapid deepening be called the headwaters; and the part down-stream, the lower course or trunk. In consequence of the changes thus systematically brought about, the lower course of the river will find its slope and velocity decreasing, and its load increasing; that is, its ability to do work is becoming less, while the work that it has to do is becoming greater. The original excess of ability over work will thus in time be corrected, and when an equality of these two quantities is brought about, the river is *graded*, this being a simple form of expression, suggested by Gilbert, to replace the more cumbersome phrases that are required by the use of 'profile of equilibrium' of French engineers. When the graded condition is reached, alteration of slope can take place only as volume and load change their relation; and changes of this kind are very slow.

In a land-mass of homogeneous texture, the graded condition of a river would be (in such cases as are above considered) first attained at the mouth, and would then advance retrogressively up-stream. When the trunk streams are graded, early maturity is reached; when the smaller headwaters and side streams are also graded, maturity is far advanced; and when even the wet-weather rills are graded, old age is attained. In a land-mass of heterogeneous texture, the rivers will be divided into sections by the belts of weaker and stronger rocks that they traverse; each section of weaker rocks will in due time be graded with reference to the section of harder rock next down-stream, and thus the river will come to consist of alternating quiet reaches and hurried falls or rapids. The less resistant of the harder rocks will be slowly worn down to grade with respect to the more resistant ones that are further down stream; thus the rapids will decrease in number, and only those on the very strongest rocks will long survive. Even these must vanish in time, and the graded condition will then be extended from mouth to head. The slope that is adopted when grade is assumed varies inversely with the volume; hence rivers retain steep headwaters long after their lower course is worn down almost level; but in old age, even the headwaters must have a gentle declivity and moderate velocity, free from all torrential features. The so-called 'normal river', with torrential headwaters and well-graded middle and lower course, is therefore simply a maturely developed river. A young river may normally have falls even in its lower course, and an old river must be free from rapid movement even near its head.

If an initial consequent stream is for any reason incompetent to carry away the load that is washed into it, it cannot degrade its channel, but must aggrade instead (to use an excellent term suggested by Salisbury). Such a river then lays down the coarser part of the offered load, thus forming a broadening flood-land, building up its valley floor, and steepening its slope until it gains sufficient velocity to do the required work. In this case the graded condition is reached by filling up the initial trough instead of by cutting it down. (*Davis, 1899H, p. 488–9*)

Reduction of relief

By 1899, in 'The geographical cycle', British readers are left in no doubt that the sequence of landform development applied to large areas.

The base-line, $\alpha\omega$ of [Fig. 31] represents the passage of time, while verticals above the base-line measure altitude above sea-level. At the epoch 1, let a region of whatever structure and form be uplifted, B representing the average altitude of its higher parts, and A that of its lower parts; thus AB measuring its average initial relief. The surface rocks are attacked by the weather. Rain falls on the weathered surface, and washes some of the loosened waste down the initial slopes to the trough-lines where two converging slopes meet; there the streams are formed, flowing in directions consequent upon the descent of the trough-lines. The machinery of the destructive processes is thus put in motion, and the destructive development of the region is begun. The larger rivers, whose channels

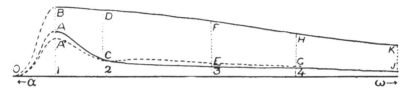

FIG. 31. Schematic diagram of the change of relief through the passage of the ideal cycle of erosion ($\alpha\omega$ indicates the passage of time). The surface is relatively rapidly uplifted to an average elevation B, with the main valleys at an elevation A. The development of relief is progressively shown by CD, EF, GH and IJ. The most rapid period of stream incision is 1–2, and that of the consumption of the uplands 3–4. The dashed line OA' indicates incision of young valleys during uplift, and its continuation after time 2 the possibility of valley aggradation. (From Davis, 1899H, Fig. 1)

initially had an altitude, A, quickly deepen their valleys, and at the epoch 2 have reduced their main channels to a moderate altitude, represented by C. The higher parts of the inter-stream uplands, acted on only by the weather without the concentration of water in streams, waste away much more slowly, and at epoch 2 are reduced in height only to D. The relief of the surface has thus been increased from AB to CD. The main rivers then deepen their channels very slowly for the rest of their life, as shown by the curve CEGJ; and the wasting of the uplands, much dissected by branch streams, comes to be more rapid than the deepening of the main valleys, as shown by comparing the curves DFHK and CEGJ. The period 3–4 is the time of the most rapid consumption of the uplands, and thus stands in strong contrast with the period 1–2, when there was the most rapid deepening of the main valleys. In the earlier period, the relief was rapidly increasing in value, as steep-sided valleys were cut beneath the initial troughs. Through the period 2–3 the maximum value of relief is reached, and the variety of form is greatly increased by the headward growth of side valleys. During the period 3–4 relief is decreasing faster than at any other time, and the slope of the valley sides is becoming much gentler than before; but these changes advance much more slowly than those of the first period. From epoch 4 onward the remaining relief is gradually reduced to smaller and smaller measures, and the slopes become fainter and fainter, so that some time after the latest stage of the diagram the region is only a rolling lowland, whatever may have been its original height. So slowly do the later changes advance, that the reduction of the reduced relief JK to half of its value might well require as much time as all that which has already elapsed; and from the gentle slopes that would then remain, the further removal of waste must indeed be exceedingly slow. The frequency of torrential floods and of landslides in young and in mature mountains, in contrast to the quiescence of the sluggish streams and the slow movement of the soil on lowlands of denudation, suffices to show that rate of denudation is a matter of strictly geographical as well as of geological interest.

It follows from this brief analysis that a geographical cycle may be subdivided into parts of unequal duration, each one of which will be characterized by the strength and variety of relief, and by the rate of change, as well as by the amount of change that has been accomplished since the initiation of the cycle. There will be a brief youth of rapidly increasing relief, a maturity of strongest relief and greatest variety of form, a transition period of most rapidly yet slowly decreasing relief, and an indefinitely long old age of faint relief, on which further changes are exceedingly slow. (*Davis, 1899H, pp. 485–7*)

Development of divides

Davis was well aware that having extended his sequential development scheme from rivers and drainage basins to vast areas he must pay special attention to the wasting and diminution of basin watersheds or divides.

There is no more beautiful process to be found in the systematic advance of a geographical cycle than the definition, subdivision, and rearrangement of the divides (water-partings) by which the major and minor drainage basins are separated. The forces of crustal upheaval and deformation act in a much broader way than the processes of land-sculpture; hence at the opening of a cycle one would expect to find a moderate number of large river-basins, somewhat indefinitely separated on the flat crests of broad swells or arches of land surface, or occasionally more sharply limited by the raised edge of faulted blocks. The action of the lateral consequent streams alone would, during youth and early maturity, sharpen all the vague initial divides into well-defined consequent divides, and the further action of insequent and subsequent streams would split up many consequent drainage slopes into subordinate drainage basins, separated by sub-divides either insequent or subsequent. Just as the subsequent valleys are eroded by their gnawing streams along weak structural belts, so the subsequent divides or ridges stand up where maintained by strong structural belts. However imperfect the division of drainage areas and the discharge of rainfall may have been in early youth, both are well developed by the time full maturity is reached. Indeed, the more prompt discharge of rainfall that may be expected to result from the development of an elaborate system of subdivides and of slopes from divides to streams should cause an increased percentage of run-off; and it is possible that the increase of river-volume thus brought about from youth to maturity may more or less fully counteract the tendency of increase in river load to cause aggradation. But, on the other hand, as soon as the uplands begin to lose height, the rainfall must decrease; for it is well known that the obstruction to wind-movement caused by highlands is an effective cause of precipitation. While it is a gross exaggeration to maintain that the quaternary Alpine glaciers caused their own destruction by reducing the height of the mountains on which their snows were gathered, it is perfectly logical to deduce a decrease of precipitation as an accompaniment of loss of height from the youth to the old age of a land-mass. Thus many factors must be considered before the life-history of a river can be fully analyzed.

The growth of subsequent streams and drainage areas must be at the expense of the original consequent streams and consequent drainage areas. All changes of this kind are promoted by the occurrence of inclined instead of horizontal rock-layers, and hence are of common occurrence in mountainous regions, but rare in strictly horizontal plains. The changes are also favoured by the occurrence of strong contrasts in the resistance of adjacent strata. In consequence of the migration of divides thus caused, many streams come to follow valleys that are worn down along belts of weak strata, while the divides come to occupy the ridges that stand up along the belts of stronger strata; in other words, the simple consequent drainage of youth is modified by the development of subsequent drainage lines, so as to bring about an *increasing adjustment of streams to structures*, than which nothing is more characteristic of the mature stage of the geographical cycle . . .

There is nothing more significant of the advance in geographical development than the changes thus brought about. (*Davis, 1899H, pp. 491–2*)

At first sight one would be inclined to think that the crest-line of a divide between adjacent river basins would merely waste lower and lower as it weathered away, without shifting laterally, and therefore without causing any change in the area of the adjacent drainage basins. It is probable, however, that this simple process is of very rare occurrence in nature. It is much more likely that the line of the divide will move more or less to one side or the other as it weathers away, on account of the unequal rate of wasting of its two slopes. The possible causes of unequal wasting are various. The declivity of the two slopes may differ, in which case the steep slope wastes faster than the other and the divide is very slowly pushed toward the flatter slope. The rocks underlying the two slopes may be of different resistance; then the weaker one will, as a rule, waste away the faster, and the divide will gradually migrate toward the more resistant rocks. Again, the agencies of erosion may be of different activities on the two slopes; one slope may have a greater rainfall than the other, or may suffer a greater number of alterations from freezing to melting. Although the last is generally a subordinate cause, it probably contributes in a small way to the solution of the problem as a whole.

The shifting of the divide as thus explained is generally accomplished by a slow migration. In some cases, however, when the divide is pushed to the very side of a stream whose basin it inclosed, then a little further change diverts all the upper drainage of this stream into the encroaching basin, and with this change the divide makes a sudden leap around the upper waters of the diverted river, after which the slow migration may be resumed. The movement of a divide may therefore be described as alternately creeping and leaping. (*Davis, 1896J, pp. 195–6*)

Slope development

Just as Davis widens his attention from rivers and river basins to vast areas, so he extends his comments on slopes from lower valley-sides to slopes in general.

When the migration of divides ceases in late maturity, and the valley floors of the adjusted streams are well graded, even far toward the headwaters, there is still to be completed another and perhaps even more remarkable sequence of systematic changes than any yet described: this is the development of graded waste slopes on the valley sides. (*Davis, 1899H, pp. 494–5*)

The transportation of the weathered material from its source to the stream in the valley bottom is the work of various slow-acting processes, such as the surface wash of rain, the action of ground water, changes of temperature, freezing and thawing, chemical disintegration and hydration, the growth of plant-roots, the activities of burrowing animals. All these cause the weathered rock waste to wash and creep slowly downhill, and in the motion thus ensuing there is much that is analogous to the flow of a river. Indeed, when considered in a very broad and general way, a river is seen to be a moving mixture of water and waste in variable proportions, but mostly water; while a creeping sheet of hillside waste is a moving mixture of waste and water in variable proportions, but mostly waste. Although the river and the hillside waste-sheet do not resemble each other at first sight, they are only the extreme members of a continuous series; and when this generalization is appreciated, one may fairly extend the 'river' all over its basin, and up to its very divides. Ordinarily treated, the river is like the veins of a leaf; broadly viewed, it is like the entire leaf. The verity of this comparison may be more fully accepted when the analogy, indeed, the homology, of waste-sheets and water-steams is set forth.

In the first place, a waste-sheet moves fastest at the surface and slowest at the bottom, like a water-stream. A graded waste-sheet may be defined in the very terms applicable to a graded water-stream; it is one in which the ability of the transporting forces to do work is equal to the work that they have to do. This is the condition that obtains on those evenly slanting, waste-covered mountain-sides which have been reduced to a slope that engineers call 'the angle of repose', because of the apparently stationary condition of the creeping waste, but that should be called, from the physiographic standpoint, 'the angle of first-developed grade'. The rocky cliffs and ledges that often surmount graded slopes are not yet graded; waste is removed from them faster than it is supplied by local weathering and by creeping from still higher slopes, and hence the cliffs and ledges are left almost bare; they correspond to falls and rapids in water-streams, where the current is so rapid that its cross-section is much reduced. A hollow on an initial slope will be filled to the angle of grade by waste from above; the waste will accumulate until it reaches the lowest point on the rim of the hollow, and then outflow of waste will balance inflow; and here is the evident homologue of a lake.

In the second place, it will be understood, from what has already been said, that rivers normally grade their valleys retrogressively from the mouth headwards, and that small side streams may not be graded till long after the trunk river is graded. So with waste-sheets; they normally begin to establish a graded condition at their base, and then extend it up the slope of the valley side whose waste they 'drain'. When rock-masses of various resistance are exposed on the valley side, each one of the weaker is graded with reference to the stronger one next downhill; and the less resistant of the stronger ones are graded with reference to the more resistant (or with reference to the base of the hill): this is perfectly comparable to the development of graded stretches and to the extinction of falls and rapids in rivers. Ledges remain ungraded on ridge-crests and on the convex front of hill spurs long after the graded condition is reached in the channels of wet-weather streams in the ravines between the spurs; this corresponds nicely with the slower attainment of grade in small side streams than in large trunk rivers. But as late maturity passes into old age, even the ledges on ridge-crests and spur-fronts disappear, all being concealed in a universal sheet of slowly creeping waste. From any point on such a surface a graded slope leads the waste down to the streams. At any point the agencies of removal are just able to cope with the waste that is there weathered *plus* that which comes from further uphill. This wonderful condition is reached in certain well-denuded mountains, now subdued from their mature vigour to the rounded profiles of incipient old age. When the full meaning of their graded form is apprehended, it constitutes one of the strongest possible arguments for the sculpture of the lands by the slow processes of weathering, long continued. To look upon a landscape of this kind without any recognition of the labour expended in producing it, or of the extraordinary adjustments of streams to structures, and of waste to weather, is like visiting Rome in the ignorant belief that the Romans of today have had no ancestors.

Just as graded rivers slowly degrade their courses after the period of maximum load is past, so graded waste-sheets adopt gentler and gentler slopes when the upper ledges are consumed and coarse waste is no longer plentifully shed to the valley sides below. A changing adjustment of a most delicate kind is here discovered. When the graded slopes are first developed, they are steep, and the waste that covers them is coarse and of moderate thickness; here the strong agencies of removal have all they can do to dispose of the plentiful supply of coarse waste from the strong ledges above, and the no less plentiful supply of waste that is

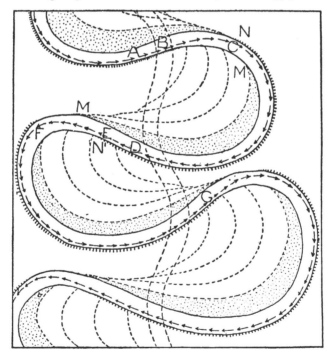

F I G. 32. The hypothetical development of a meandering river channel. The original slightly meandering course (central dashed channel) develops into the sinuous channel in which the line of arrows (A, B, C, D, E, F, G) marks the thread of maximum velocity. The dashed lines show the intervening stages of increasing sinuosity and downstream migration as a stretch of concave–bank erosion (NN′) is associated with a scroll of new floodplain (MM) (From Davis, 1902D, Fig. 7)

weathered from the weaker rocks beneath the thin cover of detritus. In a more advanced stage of the cycle, the graded slopes are moderate, and the waste that covers them is of finer texture and greater depth than before; here the weakened agencies of removal are favoured by the slower weathering of the rocks beneath the thickened waste cover, and by the greater refinement (reduction to finer texture) of the loose waste during its slow journey. In old age, when all the slopes are very gentle, the agencies of waste-removal must everywhere be weak, and their equality with the processes of waste-supply can be maintained only by the reduction of the latter to very low values. The waste-sheet then assumes a great thickness – even 50 or 100 feet – so that the progress of weathering is almost *nil*; at the same time, the surface waste is reduced to extremely fine texture, so that some of its particles may be moved even on faint slopes. Hence the occurrence of deep soils is an essential feature of old age, just as the occurrence of bare ledges is of youth. (*Davis, 1899H, pp. 495–7*)

The organization that at maturity characterized the water streams has come in old age to characterize the streams and sheets of waste all over the land surface. From the beginning to the end of this process, there is steady progress without break or interruption through the normal cycle. There is an essential unity of development through the whole of it. It is very desirable that this unity should be expressed in the terms employed in the description of land sculpture and land form; and that the balanced condition of water streams and waste streams alike should be expressed by such a term as grade ... (*Davis, 1902A, p. 100*)

Development of meanders

Davis recognized clearly the importance of river-meandering in widening valley-floors and cutting back the foot of the side-slopes of river valleys. But again there is a lack of technical terms; for example, under the section on migration of divides the words 'misfit' or 'underfit' and 'overfit' are not yet used. However, a relation between river-size and meander-size is suggested.

It has been thus far implied that rivers cut their channels vertically downward, but this is far from being the whole truth. Every turn in the course of a young consequent stream causes the stronger currents to press toward the outer bank, and each irregular, or, perhaps, subangular bend is thus rounded out to a comparatively smooth curve. The river therefore tends to depart from its irregular initial path ... towards a serpentine course, in which it swings to right and left over a broader belt than at first. As the river cuts downwards and outwards at the same time, the valley-slopes become unsymmetrical ... being steeper on the side toward which the current is urged by centrifugal force. The steeper valley side thus gains the form of a half-amphitheatre, into which the gentler sloping side enters as a spur of the opposite uplands. When the graded condition is attained by the stream, downward cutting practically ceases, but outward cutting continues; a normal flood-plain is then formed as the channel is withdrawn from the gently sloping side of the valley ... Flood-plains of this kind are easily distinguished in their early stages from those already mentioned (formed by aggrading the flat courses of incompetent young rivers or by aggrading the graded valleys of over-

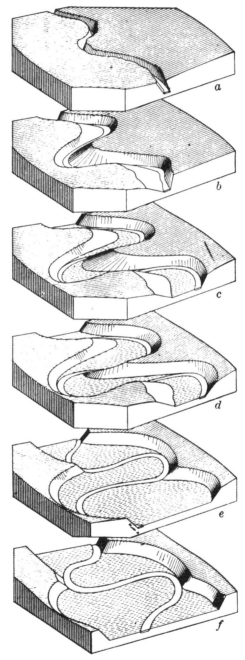

FIG. 33. A crooked stream widening its valley (b–f), after an assumed rapid incision
(a) (From Davis, 1898E, Fig. 152)

loaded rivers in early maturity); for these occur in detached lunate areas, first on one side, then on the other side of the stream, and always systematically placed at the foot of the gentler sloping spurs. But, as time passes, the river impinges on the up-stream side, and withdraws from the down-stream side of every spur, and thus the spurs are gradually consumed; they are first sharpened, so as better to deserve their name; they are next reduced to short cusps; then they are worn back to blunt salients; and finally, they are entirely consumed, and the river wanders freely on its open flood-plain, occasionally swinging against the valley side, now here, now there. By this time the curves of youth are changed into systematic meanders, of radius appropriate to river volume; and, for all the rest of an undisturbed life, the river persists in the habit of serpentine flow. The less the slope of the flood-plain becomes in advancing old age, the larger the arc of each meander, and hence the longer the course of the river from any point to its mouth. Increase of length from this cause must tend to diminish fall, and thus to render the river less competent than it was before; and the result of this tendency will be to retard the already slow process by which a gently sloping flood-plain is degraded so as to approach coincidence with a level surface; but it is not likely that old rivers often remain undisturbed long enough for the full realization of these theoretical conditions.

The migration of divides must now and then result in a sudden increase in the volume of one river and in a correspondingly sudden decrease of another. After such changes, accommodation to the changed volume must be made in the meanders of each river affected. The one that is increased will call for enlarged dimensions; it will usually adopt a gentler slope, thus terracing its flood-plain, and demand a greater freedom of swinging, thus widening its valley. The one that is decreased will have to be satisfied with smaller dimensions; it will wander aimlessly in relatively minute meanders on its flood-plain, and from increase of length, as well as from loss of volume, it will become incompetent to transport the load brought in by the side streams, and thus its flood-plain must be aggraded. There are beautiful examples known of both these peculiar conditions. (*Davis, 1899H, pp. 493–4*)

The most important process in the development of river meanders is the displacement of the line of fastest current by inertia from mid-channel toward the outside of every curve. As a result erosion tends to take place on the outside and deposition on the inside of the curve. This process is self-perpetuating. However slight the initial bends, they will be increased; and as the valley floor is broadened the curves will be developed into systematic meanders of increasing radius and breadth, ... The only conditions under which the river course will tend to straighten itself are: strong tilting in the direction of river flow, and downward erosion upon a weak stratum between two resistant strata in an inclined structure. Both these conditions are only temporary; for as grade is reached in either case and the valley floor is widened, the residual departures from a perfectly rectilinear course will be exaggerated again, and in advanced maturity the river must always be curved. The smaller the stream, the greater the effect of accidental causes, such as falling sods and trees, tributary deltas, etc., in forming new bends. The larger the river, the greater the effect of inertia in exaggerating pre-existent bends and in overcoming local or accidental irregularities. (*Davis, 1903K, p. 146*)

A river not only tends to increase its meanders; it also tends to push the whole meander system down the valley. This is because the line of fastest current, displaced toward the outside of each curve, enters the succeeding curve (or stretch between two curves) near the down-valley bank, which is therefore worn away, while the opposite up-valley bank is built out. . . . Cut-offs occur now and then, here and there, but the shortened course at a cut-off is not straight, and its faint curves are soon systematically exaggerated into meanders again . . .

Any river that we now see meandering in an open alluvial plain must have been meandering a long time. Its individual meander curves must have already advanced down the valley over considerable distances; and it is, I believe, chiefly for this reason that tributaries are taken in where the main stream bends toward them. (*Davis, 1903K, pp. 146–7*)

Their greatest dimensions [i.e. of meanders] are attained only where the curves are well organized, and such organization requires time for its accomplishment. A limit is set to the size of the curves, less by an equilibrium between current and bank than by the abandonment of the curves when short cuts and cut-offs are made. The river course is thereby made nearly straight again, after which a new series of curves is gradually established. (*Davis, 1902A, p. 92*)

Valley-side terraces are shown to indicate the downstream migration and valley-floor degradation of river meanders.

It thus becomes evident that in order to discover the number of times that a river has swung across its valley, making laterally sloping flood plains at lower and lower altitudes at every swing, we must not trust to the chance preservation of flood plain remnants in terraces here and there, but must seek a flight of terraces, systematically grouped on a long sloping ledge, which may preserve a lateral remnant of every flood plain that has been formed, . . . It is certainly a striking fact that the number of steps in a flight of valley terraces always reaches a maximum in just such situations. The preservation of numerous terraces of moderate height on long sloping ledges – however few such ledges there may be in the valley and however few terraces occur elsewhere on the valley sides – goes far towards excluding the theory of successive uplifts and pauses as a cause of terracing. It goes far also towards supporting the theory that the wandering river has been swinging from side to side across its valley, always degrading its channel but always acting as a graded river, during the whole period of terracing, whatever may have been the cause that determined the excavation of the valley drift. If uplift were the cause, the uplift must have been slow and relatively uniform. (*Davies, 1902A, p. 84*)

The flight of nine terraces in the Pochassic reëntrant, above described, constitutes the best series in the valley, as far as I have seen it. Closer study will probably increase the number of steps. The subequal height of the scarps suggests that these terraces record nearly every northward swing of the river in their locality. A similar number of swings has probably occurred elsewhere; the record of them is incomplete or wanting only because of the absence of defending ledges. It may therefore be concluded in general that it is only in localities well provided with ledges that one may expect to see preserved in terraces the lateral remnants of all

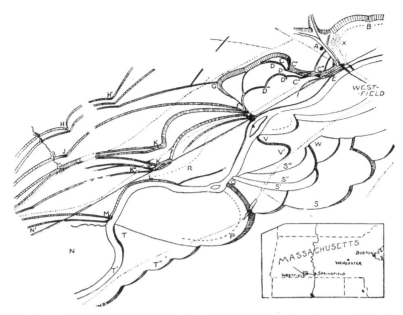

F I G. 34. Bird's eye perspective sketch of the terraces of the Westfield River, looking north east. A. Prospect Hill; F. Brown's Spur; K. Perry's Spur. C', C'', C''', D', D'', D''', K' and K'' are points where weak sandstone ledges defend terrace remnants. The terrace remnants at K', K'' and K''' are believed to correlate with S', S'', S''', respectively (From Davis, 1902D, Fig. 39)

the flood plains that were formed by the swinging river during the excavation of its valley; and that the maximum number of steps in a terrace flight is only the minimum number of lateral swings made by the river. All of these terraces testify to the graded condition of the degrading river at the time the terrace plains were made. The nine chance samples of river condition thus preserved may be fairly taken to show that the degrading river was meandering and swinging at grade during the whole period of terracing in this section of the valley. There is no indication that the individual terraces depend in any way whatever on individual uplifts. (*Davis, 1902A, p. 89*)

Old age

The penultimate stage in the cycle of erosion was represented by the peneplain but Davis could not conceive an ultimate stage. His term *pastplain* for an uplifted peneplain never caught on, and this part of the cycle concept was strongly attacked by German critics in the next three decades.

Maturity past, and the power of the river is on the decay. The relief of the land diminishes, for the streams no longer deepen their valleys although the hill tops are degraded; and with the general loss of elevation, there is a failure of rainfall to a certain extent; for it is well known that up to certain considerable altitudes rainfall increases with height. A hyetographic and a hypsometric map of a country for this reason show a marked correspondence. The slopes of the headwaters decrease and the valley sides widen so far that the land waste descends from them slower than before. Later, what with failure of rainfall and decrease of slope, there is perhaps a return to the early imperfection of drainage, and the number of side streams diminishes as branches fall from a dying tree. The flood-plains of maturity are carried down to the sea, and at last the river settles down to an old age of well-earned rest with gentle flow and light load, little work remaining to be done. The great task that the river entered upon is completed. (*Davis, 1889D, p. 24*)

Given sufficient time for the action of denuding forces on a mass of land standing fixed with reference to a constant base-level, and it must be worn down so low and so smooth, that it would fully deserve the name of plain. But it is very unusual for a mass of land to maintain a fixed position as long as is here assumed. Many instances might be quoted of regions which have stood still so long that their surface is almost reduced to its ultimate form; but the truly ultimate stage is seldom reached ... I have therefore elsewhere suggested [*Amer. Jour. Sci.*, Vol. 37, 1889, p. 430] that an old region, nearly base-levelled, should be called an almost-plain; that is a peneplain.

On the other hand, an old base-levelled region, either a peneplain or a truly ultimate plain, will, when thrown by elevation into a new cycle of development, depart by greater and greater degrees from its simple featureless form, as young narrow valleys are sunk beneath its surface by its revived streams. It therefore no longer fully deserves the name that was properly applicable before its elevation. It must not again be called a peneplain, for it is now not approaching and almost attaining a smooth surface, but is becoming rougher and rougher. It has passed beyond the stage of minimum relief, and this significant fact deserves implication, at least, in a name. I would therefore call such a region a pastplain. (*Davis, 1890K, pp. 8–9*)*

The peneplain is only one element in the theory of the geographical cycle. The systematic sequence in the development of land forms through the cycle is a much larger and more important principle than the penultimate development of a peneplain, considered alone; for the former includes the latter. One of the elements of the cycle is the development of the graded condition of streams of water during maturity, whereby an essential agreement is brought about between the ability of a stream to do work, and the work that it has to do. Another element, less generally recognized, is the development of the graded condition in the streams and sheets of rock waste or soil on sloping surfaces, where no running streams of water occur. By following out the ideal scheme thus suggested, it must result that just as the graded condition of water streams is normally propagated from the mouth towards the head, and in time reaches the source of all the branches, so the graded condition of soil-covered slopes is in time extended all over a land surface, from the valley floors to the divides. The supply of waste by the disintegration of the sub-soil rock is then everywhere essentially equal to its removal by all available agents of transportation. In a late stage of a cycle, when

* Reprint page numbers.

the surface slopes are small, agents of transportation are weak; hence the supply of waste must then be slow and the waste to be removed must be of fine texture. In order that the supply shall be slow, the waste comes to have a great depth, and the upper parts greatly protect the rock beneath from the attack of the weather. At the same time, transportation is facilitated by the refinement of the surface soil during its long exposure to the weather. Hence, under ordinary climatic conditions, normal peneplains must have deep local soils of fine texture at the surface, and grading into firm rock at a depth of 30, 50, or more feet. Moreover, it is only on a lowland surface of small slope that such a depth and arrangement of local soil can be normally produced.

In contrast to the deep soil of a peneplain, the steep sides of young valleys, whose graded waste sheets are not yet developed, must frequently reveal bare, rocky ledges. Only as the valleys widen and their side slopes become somewhat more gentle, will the ledges disappear; and even then the rock will be covered only by a relatively thin and coarse sheet of rapidly creeping waste. It therefore follows that the uplands of the Piedmont belt, with their deep soil, are of an essentially different cycle of development from the narrow valleys, with their bare ledges. The two elements of form remain mutually inconsistent, until reconciled by the postulate of an uplift of the region between their developments. But if this postulate is accepted, the plain is shown to have been a lowland of faint relief before the existing narrow valleys were cut in it. It is this double line of argument, based on deep soil and bare ledge, as well as undulating plain and narrow valley, that has convinced various observers of the verity of the peneplain in the Piedmont belt. (*Davis, 1899I, pp. 215–16*)

Davis discusses the difficulty of distinguishing between a peneplain (formed subaerially) and a platform or plain of marine abrasion that would be very flat and would be backed landward by a wave-cut cliff.

A subaerial baselevel plain is gradually completed by the action of ordinary forces on all parts of its surface. Reduction to baselevel is slowest along the divides on the harder rocks, and quickest along the streams on the softer rocks. The valley bottoms therefore approach and practically reach baselevel long before the interstream areas are reduced so low.

A submarine platform is essentially completed strip by strip, once for all, as far as it goes; its advance is rapid at first, very slow at last. Its landward margin is surmounted by a sinuous cliff or slope with a level base, facing the sea and separating an interior of greater or less relief from a smooth sea-bottom, unconformably veneered over with the deposits from the land. If the transgression of the sea over the land be aided by a depression of the land, many inequalities of the surface might be preserved beneath the unconformable cover of marine deposits. Such a surface, when again lifted and somewhat denuded, might be indistinguishable from one that had not been submerged. The occurrence of unconformable deposits on an even foundation cannot alone be taken as evidence that the foundation is a surface of marine denudation; it may be a subaerial baselevel plain depressed below sea-level and covered with sediments from an adjacent portion of the same that was not submerged.

The ultimate forms of the two kinds of plains are probably much alike, and it may be hopeless to seek to distinguish them after they have been elevated and roughened by subsequent erosion. But the penultimate forms of the two might be separated; one would be gently rolling, its residual inequalities being of the hill and valley type; the other would be smoother, and might be very smooth if its veneer be regarded as its surface, but it would have a definite margin, beyond which the penultimate subaerial plain would be found. The two forms are of course often associated. (*Davis, 1889J, p. 375*)

Residuals that survived the first cycle of erosion may dominate the general summit level of the second cycle and, when well defined, were called by Davis monadnocks.

It need not be supposed that the former cycle of denudation reduced the region to a dead level. Most of the peneplains that I have examined, even though now uplifted and somewhat advanced in the dissections of a second cycle, still possess residual elevations, rising somewhat above the general upland, and evidently to be regarded as unconsumed remnants of the denudation of the former cycle. From the need of reducing such matters as these to some simple form of statement for use in teaching, I have fallen into the habit of calling a residual mound of this character, a *monadnock*, taking the name from that of a fine conical mountain of south-western New Hampshire, which grandly overtops the dissected peneplain of New England. (*Davis, 1895E, pp. 140–1*)

I have as a rule given no particular attention to the composition of the monadnock rocks; indeed, it has generally seemed to me reasonable to infer their greater resistance on account of their form. But so far as attention has been given directly to this phase of the problem, the inference based on the peneplain theory is borne out by petrographic study ... In New England, the type Monadnock is, if my memory serves me, largely composed of an andalusite schist, which certainly has every appearance of being a resistant rock. Yet it must be freely admitted that, as far as I know, no artificial test has been made of its resistance as compared with that of many apparently resistant rocks around its base. (*Davis, 1899I, pp. 219–20*)

THE NATURE AND QUALITY OF THE DAVISIAN CYCLE

In the foregoing account we have allowed Davis to outline his scheme for the development of landforms through the simple geographical cycle, uncomplicated by gross structural variations, and uninterrupted by baselevel changes, and by climatic and volcanic accidents. These elaborations of the simple cycle will be treated in a subsequent chapter, as will extensions of the cyclic theory to semi-arid, glacial, shoreline and karstic environments. We will content ourselves here mainly with comment on the general methodological approach represented by the cycle concept.

A proper understanding of the specialized character of the cycle of erosion concept can be achieved only by considering it in the context of nineteenth-century thought in the natural sciences. At this time the writings of Herbert

Spencer and others were extending the concept of evolution from the biological into the physical, social and mental spheres such that it seemed to form a basic organizational framework for the whole world of experience. Although prominent Harvard philosophers of the time were resisting this extension (Leighly, 1955, p. 312), there is no doubt that the idea of organic evolution was one of the most important mainsprings of the cycle of erosion theory. In his first statement of the cycle notion Davis (1885D) termed it a *cycle of life* in which, as he later wrote, 'land forms, like organic forms, shall be studied in view of their evolution' (Davis, 1905J, p. 150), such that the cyclic concept has the 'capacity to set forth the reasonableness of land forms and to replace the arbitrary, empirical methods of description formerly in universal use, by a rational, explanatory method in accord with the evolutionary philosophy of the modern era' (Davis, 1922I, p. 594). The main problem arising from this association is due to the fact that in the later nineteenth century the highly attractive label of 'evolution' had practically become a synonym for any 'change', and often for 'history' in general. This identification has tended to obscure the special character of the concept of evolution, and in the same way the concept of the cycle of erosion has been identified with all types of change in landforms and with landform development in general. Thus it is only possible to grasp the special and restrictive characteristics of the cyclic idea by understanding some of the contemporary implications of the term 'evolution'. The late nineteenth-century view of evolution, particularly in its popular, non-biological connotation, implied an inevitable, continuous and irreversible process of change producing an orderly sequence of transformations, wherein earlier forms could be considered as stages in a sequence leading to later forms. 'Time' thus became, at least for many of those concerned with adapting the evolutionary notion to wider fields, almost synonymous with 'development' and 'change', such that it was viewed not merely as a temporal framework within which events occur but as *a process itself*. It was in this sense that Davis employed the concept of evolution as a basis for the cycle of erosion, and it is easy to see why what Fenneman (1936) termed 'non-cyclic erosion' seems just as inconsequential to the cyclic concept as the lack of sequential development of certain biological organisms through long time periods seemed to the theory of evolution.

In other ways, too, Davis' synthesis was typical of nineteenth-century scholarship in general and of geographical scholarship in particular. The emphasis upon historical sequences rather than functional associations, the reconnaissance and artistic basis of his fieldwork and the stress laid upon causal description are features of Davis' work which make it appear antique to the modern student of landforms, but which endeared it to his contemporaries trying to expand physical geography. Davis followed Ritter in his concept of the scope and nature of geography, wherein human activities were subordinated to, and largely based upon, the main features of the physical environment. There are many overtones of the 'landschaft' concept of geography implicit in Davis' reasoning, in that it is assumed that the landscape features contain within themselves the unambiguous evidence of their origin. While stressing the Victorian character of much of Davis' work it is only fair to note that he departed from the characteristic standards of much nineteenth-century work in the natural sciences in three important particulars; his lack of detailed field measurements, his unconcern with details of processes prompting change and the entirely qualitative nature of his methods.

This last characteristic leads us to a further feature of the geographical cycle which militates against its popularity with modern workers – its highly dialectical and semantic quality. Anyone at all familiar with the voluminous writings of Davis cannot but be struck with the essentially verbal logic which he employed, characterized by his concern over terminology. Much of this resulted, of course, from the theoretical basis or deductive nature of his work, but in its extreme form this preoccupation resulted in 'research by debate'. Again, to be fair to Davis, it must be recognized that he never intended his cyclic theory to be 'scientific' – at least in the sense that the term is currently employed.

The strongest and most compelling feature of the cycle of erosion concept is that it presents many of the features of a theoretical model. As distinct from classification, which merely involves the dissection and categorizing of information in some convenient manner, model-building requires the identification and association of some supposedly significant aspects of reality into a working system which seems to possess some special properties of intellectual stimulation. This stimulative quality, often resulting from the special juxtaposition of information which is the very foundation of the structure (one is almost tempted to say the 'artistic form') of the model, finds expression in an enlargement of what is thought of as 'reality' (i.e. involving the kind of scientific prediction which resulted from the construction of Newton's model). It was thus very characteristic of Davis' intellectual achievement that after he developed his cyclic model he was able to say that he could think of many more landforms than he could find examples of in the field! A moment's reflection on the magnitude of possible combinations of structure, process and stage in landforms shows exactly what he meant. It is therefore in the bringing together of certain aspects of the 'web of reality', stripped of other considerations, into a clear-cut theoretical model that much of the intellectual attractiveness and teaching strength of the cycle lies. Davis (1899H, p. 484) knit certain aspects of landforms together into a meaningful association both in space within a given landscape and in time throughout an assumed evolutionary history. In order to understand many of the special characteristics of the cyclic theory, and in particular to recognize both its strengths and limitations, it is profitable to consider it in the light of three of the properties common to all such models – their essentially theoretical character; the inherent need for the discarding or 'pruning' of much information; and the fact that no part of reality can be uniquely and completely built into any one model.

Davis himself recognized and defended the theoretical nature of his model, writing

> ... the scheme of the cycle is not meant to include any actual examples at all, because it is by intention a scheme of the imagination and not a matter for observation; yet it should be accompanied, tested, and corrected by a collection of actual examples that match just as many of its elements as possible. (*Davis, 1905J, p. 152*)

Rather than being surprised by such a statement, one should recognize this as a very characteristic state of mind for the model builder in a natural science, where the subject matter of even a small part of reality has usually to be accepted in uncontrollable mutual associations. The result is that there is an attendantly large 'elbow room' within which the researcher may select, organize and interpret his material, introducing a subjective bias into all work – good or bad.

The second model property is that much possible information relating to some part of reality has to be rejected in order that the rest (i.e. information and relationships which appear especially significant or interesting) may be presented in sharp outline. All models caricature reality by this pruning, but the most successful ones (e.g. that of Newton) are those wherein that which remains still retains some observable or testable significance as far as the 'real world' is concerned. Davis' cyclic model is heavily pruned, such that changes in the geometry of erosional landforms through time emerge as the central theme. When he excluded the possible effects of climatic change and of progressive movements of baselevel from his scheme, Davis was not (as he asserted, e.g. 1905J, p. 153) doing so to facilitate and simplify his *explanation* but to make the cyclic scheme *possible at all*! One can only imagine what would have remained of the cycle if Davis had permitted the possible effects of continuous movements of baselevel to have been superimposed upon those associated with the progressive subaerial degradation of the landmass. This is, in fact, just what Walther Penck attempted to do, and is the reason why his model is much more confused and unsatisfactory than the Davisian cycle.

The third property of models is the impossibility of building any part of reality uniquely and completely into any such concept. If discrimination and selection operate in terms of the building of information into a model structure then no single model can form a universally appropriate approximation to a segment of reality. One of Davis' most faithful supporters, Nevin Fenneman, almost inadvertently stated this:

> ... the cycle itself is not a physical process but a philosophical conception. It contemplates erosion in one of its aspects, that of changing form. But erosion does not always and everywhere present this aspect ... Cycles have parts and the parts make wholes, and the wholes may be counted like apples. Non-cyclic erosion can only be measured like cider. There is neither part nor whole, only much or little. (*Fenneman, 1936, p. 92*)

Another feature of the cycle of erosion concept, one which is basic to the whole reasoning underlying it, is the tacit assumption that the amount of energy available for the transformation of landforms is a simple and direct function of relief or of angle of slope. This unformulated, but nevertheless real, assumption on the part of Davis (and one which seems so logical in the abstract as to be unquestionably accepted as an axiom) is that rates of mass transfer by all agencies are greater on steeper slopes than on less steep ones. From this assumption many others are deduced – some apparently true and others, often not so apparently, untrue. The ideas, for example, that steep slopes are eroded faster than less steep ones and that stream velocity is solely dependent on bed slope, derive from this axiom, and lead inevitably to the conclusion that rates of change of landforms, as well as their geometrical magnitude, are direct functions of local relief. It follows, therefore, that the progressive changes of relief during the consumption of a landmass by erosion are held to be universally associated with a progressive landscape evolution wherein the geometry of individual landforms and the rates of their erosional change are both subject to sequential transformations through time. Considering individual valley-side slope elements and stream reaches, for example, this reasoning leads to the assumption that they are progressively transformed into lower and lower energy (i.e. gradient) forms as the general relief is reduced following late youth. The study of change is therefore the guiding purpose of the 'geographical cycle', wherein a sequence of changes leading to conditions of 'grade' in stream channels and slopes, is followed by a progressive, sequential and irreversible transformation of virtually all aspects of landforms as the potential energy (i.e. relief) of the system is dissipated. Many of the sequential changes were not treated in detail by Davis who also failed to embody a definition of equilibrium (grade) which was completely suitable to his temporal model. Characteristically, the timeless concept of grade was the one feature of Davis' synthesis which seemed least at home within the timebound cyclic framework. Many years were to pass before this apparent paradox was to be resolved by the application of systems theory by Mackin (1948). Davis, however, as we have seen, knew G. K. Gilbert's researches on river hydraulics and exploited the idea that river work depended partly on size and amount of load. This assumed relation of load to velocity or erosive power he used rather ruthlessly when it suited his arguments.

His cycle of erosion bears all the stamp of being a theoretical model of reality, dominated by certain implicit assumptions regarding the way in which the passage of time controls form and process, and supported by selected and 'correctly' interpreted field examples – often in the form of artistic interpretations of landscape features. It is only when one understands this basic quality of the cycle that one can appreciate Davis' difficulties in his attempts to subsume into his model the characteristics of grade borrowed from G. K. Gilbert's quite different equilibrium model, and in attempting to show that the German endogenetic/exogenetic model was simply a special case of his own model. Theoretical models in some ways are like delicately-constructed pieces of machinery, one can marvel at the beautiful and efficient manner in which they perform the tasks for which they were designed within their own distinctive terms of reference, but if we try to extend their range of operation by adding new parts from other machines or by arbitrarily linking them to other pieces of machinery they almost invariably disappoint us (Chorley, 1965).

Yet in these criticisms we are voicing mainly modern opinions. Most of Davis' contemporaries, except in Germany, saw only what a beautiful model he had devised. Moreover, Davis himself continued to advertise its high quality. In showing its defects we are expressing sentiments which did not become patent until the 1920s and certainly were relatively impotent until after Davis' death. By 1900 the geographical cycle or cycle of erosion had been launched and irreversibly entwined with the explanatory description of landforms. Its author enjoyed a further thirty years in which to improve and extend his early ideas. And Davis, as we will now show, was a man who never let slip any opportunity.

CHAPTER ELEVEN

The Davisian Analysis of 'Composite Topographies'

To Davis most landscapes were composite and needed more than observation to disentangle their complexities. He believed firmly at the same time in the use of deduction and in the complicated nature of denudational chronologies. It is readily apparent that, not only did he recognize the theoretical model (i.e. deductive) characteristics of the cycle (e.g. 1905J, p. 152), but also considered this attribute to be an advantage:

> The opinion prevails in many places that geology is chiefly an observational science. This is not correct. It is chiefly a speculative science, in that the great body of its statements go far beyond the field of observable fact into the field that can be reached only by means of speculative mental processes. The recognition of this truth carries with it two important consequences: first, that more systematic and thorough instruction in the non-observational side of geology should be given in the educational preparation of young geologists for expert work; second, that the published work of trained experts should make more explicit distinction between the inferred conclusions that they reach and the observed facts on which the conclusions are based. (*Davis, 1913M, pp. 686–7*)

Indeed, Davis' essay on the peneplain (1899I) shows that the real significance of this feature lies in its utility in interpreting the history of 'composite' or poly-cyclic landscapes:

> I have repeatedly insisted that it was only by recognizing the existence of a peneplain that uplift or deformation could be determined in certain cases; and that only in this way could certain stages of geological history be discovered, in the absence of what might be called orthodox geological evidence in the form of marine deposits. For example, it is by the remnants of an uplifted, inclined, and warped peneplain in the even crest lines of the Pennsylvania Appalachians that the post-Cretaceous uplift of the mountain belt has been determined: it was formerly supposed that the existing ridges were the unconsumed remnants of the ancient Appalachians, and by implication, that no uplift of the region had occurred since the mountains were crushed, folded, and upheaved. So in southern New England: there was no means of determining the date of uplift, as a result of which the existing valleys were eroded, until the peneplain of the uplands was recognized and dated ... Those who believe in the verity of peneplains will infer uplift, where they see a high-standing and dissected peneplain, as confidently as the geologists of the end of the eighteenth century inferred uplift when they found marine fossils in stratified rocks far above sea level. (*Davis, 1899I, pp. 212–13*)

Admitting the present to be exceptional in the lack of peneplains close to their baselevel of production, and thus postulating general disturbances by uplift and tilting in the recent past, I doubt if this condition is more exceptional than that which permitted the widespread deposition of the chalk of Europe upon its even foundation, or than that which determined the formation of the Coal Measures of Europe and North America. There does not seem to be any severe strain upon the reasonably elastic form of the doctrine of uniformitarianism in meeting the requirements of the peneplain theory. (*Davis, 1899I, p. 223*)

While it may be true that there are to-day no extensive peneplains still standing close to the sea level with respect to which they were denuded, the examples given in this and in the preceding section seem to me to prove that the earth contains many approximations to the peneplain condition, inasmuch as it preserves some excellent fossil peneplains; and that the stratigraphic as well as the physiographic method of investigation yields abundant and accordant evidence of their occurrence. (*Davis, 1899I, pp. 232–3*)

It would be as extraordinary to find no slanting peneplains as to find no inclined strata. Warped and faulted peneplains are no more unlikely products of coastal deformation than warped and faulted sedimentary formations. (*Davis, 1899I, p. 213*)

As to the time that has elapsed during the denudation or dissection of peneplains, there is apparently no way of measuring it but by the work done. Hence the question returns to the verity of the peneplains; whether much or little time is needed to produce them is a secondary matter. Above all, a preconception as to the insufficience of geological time should not in this day be urged . . . as a reason for not believing in the possibility of peneplanation. (*Davis, 1899I, p. 226*)

In terms of the practical interpretation of landscapes, most of which are composite, the cycle concept was used by Davis as a yardstick by which its complications and interruptions were assessed.

One of the first objections that might be raised against a terminology based on the sequence of changes through the ideal uninterrupted cycle, is that such a terminology can have little practical application on an Earth whose crust has the habit of rising and sinking frequently during the passage of geological time. To this it may be answered, that if the scheme of the geographical cycle were so rigid as to be incapable of accommodating itself to the actual condition of the Earth's crust, it would certainly have to be abandoned as a theoretical abstraction; but such is by no means the case. Having traced the normal sequence of events through an ideal cycle, our next duty is to consider the effects of any and all kinds of movements of the land-mass with respect to its baselevel. Such movements must be imagined as small or great, simple or complex, rare or frequent, gradual or rapid, early or late. Whatever their character, they will be called 'interruptions', because they determine a more or less complete break in processes previously in operation, by beginning a new series of processes with respect to the new baselevel. Whenever interruptions occur, the pre-existent conditions that they interrupt can be understood only after having analyzed them in accordance with the principles of the cycle, and herein lies one of the most practical applications of what at first seems remotely theoretical. A land-mass, uplifted to a greater altitude than it had before, is at once more intensely attacked by the denuding processes in the new cycle thus initiated; but the forms on which the new attack is made can only be understood by considering what had been accomplished in the preceding cycle previous to its interruption. It will be possible here to consider only one or two specific examples from among the multitude of interruptions that may be imagined.

Let it be supposed that a maturely dissected land-mass is evenly uplifted 500 feet above its former position. All the graded streams are hereby revived to new activities, and proceed to entrench their valley floors in order to develop graded courses with respect to the new baselevel. The larger streams first show the effect of the change; the smaller streams follow suit as rapidly as possible. Falls reappear for a time in the river-channels, and then are again worn away. Adjustments of streams to structures are carried further in the second effort of the new cycle than was possible in the single effort of the previous cycle. Graded hillsides are undercut; the waste washes and creeps down from them, leaving a long even slope of bare rock; the rocky slope is hacked into an uneven face by the weather, until at last a new graded slope is developed. Cliffs that had been extinguished on graded hillsides in the previous cycle are thus for a time brought to life again, like the falls in the rivers, only to disappear in the late maturity of the new cycle.

The combination of topographic features belonging to two cycles may be called 'composite topography' ... (*Davis, 1899H, pp. 499–500*)

The major legacy of Davis' work was to be the foundation which he laid for studies of what is now called 'denudation chronology'. During the first half of the twentieth century studies in the sequential development of erosional forms referred to changes in baselevel formed the mainstream of geomorphological work in Britain, the United States and France, under the influence of S. W. Wooldridge, D. W. Johnson and H. Baulig respectively, although there was a different emphasis on either side of the Atlantic regarding the eustatic or diastrophic nature of the baselevel changes (Chorley, 1963). However, the relationship of the concept of the cyclic development of landforms to denudation chronology is at once complex and ambiguous, such that the two are commonly confused; and it is important to realize that studies of denudation chronology, using the term in precisely the same way as it is currently employed, preceded or accompanied the formulation of the cyclical approach by Davis. We have already shown in Volume One of our *History of the Study of Landforms* that during the 1860s and 1870s Jukes and Ramsay proposed sequences of landscape development involving changes of baselevel; in the 1880s in the United States Joseph Le Conte (1880 and 1886) interpreted breaks in stream profiles as indicative of the discontinuous lowering of baselevel; and one year before the first really important statement of the cycle McGee (1888) developed an erosional chronology for that part of the Appalachians later to be made classic by Davis (1889D), Johnson

(1931) and many others. The notion of cyclic interruptions fitted so well into the interpretation of landforms directed pre-eminently towards an evaluation of baselevel changes that the two approaches merged, and provided an extension of geological reasoning, whereby the record of the last part of the last chapter of earth history was interpreted by means of physiographic, rather than stratigraphic, evidence.

ASSUMPTIONS AND METHODOLOGY

Davis' treatment of interruptions of the geographical cycle exhibits two basic assumptions:

(1) His belief that baselevel changes are best ascribed to regional changes in the level of the land (diastrophism), rather than to more general sea-level changes (eustatism).

(2) His reluctance to ascribe forms associated with those of the 'normal cycle' to climatic changes during the course of the cycle.

The strength of the first assumption is best displayed by the treatment by his pupil Douglas Johnson of Appalachian denudation chronology (1931), and by the attack by his disciple Emmanuel de Martonne (1929) on Henri Baulig's (1928) utilization of Eduard Suess' eustatic theory. In 1888 Volume 2 of Suess' monumental *Das Antlitz der Erde* appeared, suggesting that, apart from the orogenic belts, evidences of continental transgressions and regressions point to a remarkable synchroneity of swings of sea-level in widely-spaced areas of the world, indicating great continental stability (Chorley, 1963). In 1905(K) Davis attacked Suess' beliefs that local uplifts are restricted to mountain belts and are due to the tangential pressure producing an upward resultant force, whereas the 'elevation' of plateaus is due to the subsidence of surrounding areas. Drawing on his recently acquired observations of the Tian Shan Mountains, Davis gave evidence, based on the occurrence of high-level tilted peneplain remnants, of the reality of local vertical uplifts. However, as we shall see later, Davis' aversion to excessive eustatism was most clearly manifested in his support and elaboration of Darwin's subsidence theory of atolls.

Davis' reluctance to use climatic changes as mechanisms to explain features associated with those of the 'normal cycle' is best exemplified by his attitude to underfit streams (Dury, 1964). In an early article on the Osage River (1893Q) he suggested that valley meanders had been inherited from the floodplain meanders of an earlier cycle. In many subsequent papers he viewed the reduction of the size of river meanders (compared with that of valley meanders) as due to a reduction in stream discharge. This reduction was variously ascribed: primarily to piracy on homoclinal structure for some English rivers (1895E); to structurally-induced capture of the headwaters of the Meuse by the Seine and the Moselle (1896J); to the beheading of the Swabian rivers Schmeie and Lauchert, and of the consequent Cotswold rivers by the subsequent Severn and Warwickshire Avon (1899M); and to the progressive loss of discharge by underflow seepage into the floodplain alluvium (1913Q). True, he was worried by such features as the underfit character of the Warwickshire Avon itself (1899M) and thought that because a general decrease of stream volume seemed to be superposed on those changes due to river-capture, the chief reason for a decrease might be some climatic change of external and obscure origin (Beckinsale, 1970). Later (1909J) Davis also suggested that ice-dammed water may have discharged into the Cotswold valleys, but in his work on the cycle of erosion in the Alps (1923J) he again reverted to capture as the prime mechanism, with underflow and climatic change as additions.

To understand the treatment by Davis of composite or polycyclic landscapes it is necessary to have a clear view of his scientific method of work. This is treated by Douglas Johnson in several of his articles on 'Studies in scientific method' (1939–42). In general Davis did not favour the inductive method whereby there is a clear presentation of a series of facts leading directly to apparently sound conclusions. He preferred to set forth a tentative hypothesis (sometimes 'outrageous'), to deduce its logical consequences and then to confront these deduced consequences with the reality of observed facts.

> It is not easy to select from Davis' writings an essay that illustrates effectively the inductive method of presentation. Davis usually preferred either the deductive or the analytical method when presenting results of his researches. One essay in which the inductive method is employed to a degree unusual for him is that on 'The Seine, the Meuse, and the Moselle'. The author first presents an array of pertinent facts respecting the vigorously meandering Seine River on the west, the similarly vigorous Moselle on the east, and between them the sharply contrasted weak Meuse, a smaller stream staggering about in a valley characterised by a large meander pattern. The facts as presented point directly toward a conclusion (capture of Meuse drainage by tributaries of the Seine and Moselle) which the author ultimately states. But before completing the inductive passage from facts to conclusion he deliberately interrupts the treatment for a specific purpose. This purpose is to prepare his readers for the coming conclusion by demonstrating that stream captures do occur. The demonstration is partly deductive and partly inductive. Once demonstration of the reality of capture as a natural phenomenon is completed, Davis returns to his main theme, and completes the inductive presentation by finding in repeated captures of Meuse drainage by branches of the Seine and Moselle a satisfactory explanation of all the facts previously set forth. Additional observations in support of this major conclusion are then presented.
>
> (Johnson, 1939, p. 372)

> The essence of the deductive method, in presentation as in research, is to set forth a tentative hypothesis, to deduce the logical consequences of this hypothesis, and then to confront these deduced consequences with the facts actually observed.
>
> (Johnson, 1940, p. 59)

One of the best examples of deductive presentation in which deductions are pictorially presented is Davis's brief paper on 'The Sculpture of Mountains by Glaciers'. The discussion in some degree approaches the analytical method, for two contrasted hypotheses are considered: the first, that glaciers cannot erode; the second, that glaciers can erode. But the treatment lacks the use of induction in combination with deduction, as well as the extended checking and testing characteristic of the fully developed analytical presentation. The consequences of each of the two contrasted hypotheses are deduced, and then confronted with observed facts. A series of three extremely effective diagrams [see Fig. 60] present to the reader (a) the normal topography of stream-dissected mountains, which under the first hypothesis will not be profoundly changed by ice occupation if glaciers lack erosive power; (b) the same region during occupation of the valleys by glaciers which are assumed, under the second hypothesis, to be eroding their beds profoundly; and (c) the same region as it appears after the glaciers have melted away, leaving exposed to view the changes in topography which should be produced by glacial erosion under the second hypothesis. The topographic peculiarities of this third diagram, representing the logical deductions from the second hypothesis, are so perfectly matched by features well known to be abundantly developed in mountains formerly glaciated, that the demonstration is wonderfully convincing. The three diagrams have been frequently and widely reproduced; and it is probably not too much to say that this deductive presentation by Davis has done more to establish firmly the supremacy of the view that glaciers are effective eroding agents than has any other single contribution to the subject. (*Johnson, 1940, pp. 63–4*)

Davis, drawing on Chamberlin's method of multiple working hypotheses (1897), mainly utilized the *analytical method.*

There is a more complex method, employed in prosecuting research and in presenting the results of research, which makes conscious and balanced use of both induction and deduction ... It differs from the inductive and deductive methods not merely in the fact that both are deliberately employed in more or less equal measure, but in two other fundamental respects: it makes use of multiple working hypotheses in an effort to discover truth; and it consciously invokes the analytical powers of the mind in an effort to test the validity of every step in the procedure. (*Johnson, 1940, p. 156*)

Davis himself described it as follows:

The analytical method is characterized by the presentation, at least in outline, of the successive steps that have led the investigator from his original field of observation to the invention of various hypotheses, to the recognition of the most successful hypothesis, and if possible to its establishment as a verified theory ... This method is therefore most appropriate in the presentation of complicated problems which demand much theoretical supplement to observation, in the exposition of problems regarding which various unlike opinions have been held by different investigators, and before hearers who are fully able to appreciate rigorous scientific discussion. The essential feature of this method of presentation is that it should preserve the demonstrative quality of the investigation that it represents, and that it should proceed in such an order that the hearers may form a critical opinion as to the value of the conclusion reached at its end. Hence, just as in the usual presentation of a geometrical problem, so in an analytical presentation of a geographical problem, the conclusion or theorem to which the demonstration leads, is advisedly stated not only at the end, but also at the opening of the speaker's address, in order that the hearers may bear it in mind while observed facts, invented hypotheses, deduced consequences, and so on, are all set forth in proper sequence. Only when thus aided by being told the end at the beginning can hearers, who are not familiar with the problem under discussion, really form a competent and critical opinion as to the thoroughness with which it has been investigated . . .

In contrast with the inductive and other methods of presentation, the chief characteristic of the analytical method consists therefore in the candid completeness with which it reveals and discusses the various steps by which the investigator passes from the incomplete conception of his problem, based directly on observable facts, to the complete and comprehensive scheme which he has been led to believe is the true counterpart of the whole enchainment of facts, past and present, involved in this problem . . .

But there is another advantage possessed by analytical presentation, besides its comprehensiveness. It is well known that a speaker can best commend his work and himself to his hearers by a frank exposition of the reasons that have led him to certain conclusions rather than to others; and there is surely no way in which a clearer and more open exposition of the reasons for belief can be set forth than by presenting, at least in outline, the logical analytical method already described under the account of investigation. (*Davis, 1911P, pp. 228–9*)

It was always held that Davis' major work on Appalachian denudation chronology exemplified superbly the analytical method both as a research tool and as a method of presentation.

As an example of thoroughly analytical treatment of an exceedingly complex and difficult problem we may take Davis' classic essay on 'The Rivers and Valleys of Pennsylvania'. The task Davis set himself was to decipher the history responsible for many drainage anomalies encountered in the various structural belts of the Pennsylvania Appalachians. After describing the existing topography and drainage of the region, Davis outlines the geological structure and reviews the geological history responsible for the several structural provinces. Even this preliminary discussion becomes analytical in places, hypothetical interpretations of certain events being supported by deductions confronted with facts which seem to corroborate the interpretations offered. The ideal life history of a river system is then presented deductively, after which various possible types of adjustments to structure are deduced and illustrated by diagrams. In this part of the treatment Davis' great analytical powers are admirably demonstrated.

Then follows the application of principles developed in the first part of the essay to an elucidation of Pennsylvania drainage history. On the basis of certain postulates, which are first critically analyzed, Davis reconstructs the hypothetical drainage of Permian time. Next he deduces the adjustments which would normally take place in such a drainage system; compares the deduced consequences with observed facts; supplements or revises hypotheses to bring deduced consequences into more perfect accord with the facts; invents and tests several hypotheses to account for the reversal of a northwest-flowing to a southeast-flowing drainage, and to account for the course of certain rivers across rather than around the ends of folds of resistant rock; and finally reaches certain 'provisional conclusions' which for forty years were accepted as fundamental truths of Appalachian geomorphology.

If today there are those who doubt the validity of Davis' conclusions, they do not attribute such difference in view to any defect in the highly analytical character of Davis' study. On the contrary, they recognize that much that Davis deduced from his analysis remains an established part of our knowledge of Appalachian history; that Davis himself regarded his conclusions as purely tentative; that his analysis brought into strong relief certain facts which were distinctly out of harmony with that hypothesis of Appalachian drainage history which he favored; and that it was these discordant facts, emphasized by Davis with characteristic mental integrity, which ultimately led to the search for a new hypothesis. If there was any defect in Davis' use of the analytical method, it was his failure to push the invention of hypotheses far enough. He failed to think of one possibility (regional superposition from an overlapping coastal plain) which he later accepted as probably competent to explain all relevant facts. (*Johnson, 1940, pp. 261–2*)

Whether Davis' appraisal of his own methodology in relation to his researches into the denudational history of the Appalachians is completely objective is debatable. His severer critics have suggested that the approach to denudation chronology which he made so popular in the first few decades of the twentieth century involved, firstly, the selective identification of certain supposedly key geomorphic aspects involving the spatial and altitudinal features of the terrain together with drainage patterns, and, secondly, the uniting of these features in a postulated historical sequence involving recourse to as simple a cyclical explanation as was possible. This produced an 'explanation' which was apparently acceptable not only for the features concerned but for all the regional landforms, and was impressive for its 'clean', simple and incisive nature which owed more to the application of Occam's Razor than of Chamberlin's multiple working hypothesis. In this way Johnson's (1931) theory of stream sculpture on the Atlantic slope based on superimposition from a Cretaceous cover was more credible and compact than Davis' earlier rambling explanation. In the same way the widespread Pliocene onlap deduced by Wooldridge and Linton (1939) for South-East England performed a similar function with regard to the previous disconnected body of *ad hoc* local explanations for the evolution of the relief and drainage of that region. However, Davis' classic article, which we will now discuss, was written more than forty years before these more modern syntheses.

THE DENUDATION CHRONOLOGY OF THE APPALACHIANS

In his foreword to Johnson's *Stream Sculpture on the Atlantic Slope* (1931) Davis put into perspective his own important contribution to the understanding of the evolution of the landforms of the central Appalachians.

> The truth about the ridges and valleys of the Pennsylvania Appalachians appears, as it is traced farther and farther toward its entirety, to involve a more and more complicated succession of events. At the time of a walking trip in the summer of 1863, when I had my first sight of the water gaps by which the Susquehanna cuts through three even-crested ridges above Harrisburg, mention was still made with some respect of Rogers' daring comparison between the folds of the heavy Appalachian strata and the waves of a violently shaken carpet, and also of the belief so confidently expressed by Lesley that the great folds had been quickly worn down to their present forms, once for all, by the rush of a mighty oceanic flood. Twenty-five years later, when my acquaintance with the ridges and valleys was much more extended and intimate, the theory of the cycle or erosion had been only briefly outlined, the possibility that the Appalachians had suffered more than one such cycle was still hardly imagined, and the over-simple, three-fold genetic classification of rivers and valleys introduced by Powell was generally accepted, although it took no account of subsequents or obsequents. There was therefore something of the spirit of adventure in my suggestion in 1889 that the Appalachians had, after their folding, experienced a succession of erosional reductions to low relief, during which a progressive modification of an initially consequent drainage system had led to the gradual evolution of the adjusted drainage system which, with many long subsequent and short obsequent and resequent members, is seen in the rivers and valleys of Pennsylvania today.
>
> My error at that time of attributing a Cretaceous date to the peneplain represented by the ridge crests and my failure to recognize a later and lower peneplain were soon detected and corrected; but the general acceptance of – or at least the lack of successful arguments against – the rest of my conclusions during a period of forty years lulled me into a satisfied conviction that my scheme of Appalachian drainage development was essentially true. (*Davis, quoted in Johnson, 1931, pp. vii–viii*)

To understand and appreciate fully both the historical sophistication of Davis' contribution to the elucidation of Appalachian erosional history and the difficulties he confronted, it is helpful to review briefly contemporary and modern ideas on the palaeogeography of the eastern continental border of North America.

In 1859 Hall had been the first to identify a belt of accelerated Palaeozoic sedimentation in the northern Appalachians, and by 1873 Dana had introduced the term 'geosynclinal' for such a belt of long-continued subsidence in which there had been a steady accumulation of sediments. At the time when Davis was working on Pennsylvania and New Jersey, the recognition that both the prime sedimentary source for the rocks of the Appalachians and the dominant direction of thrust for the Appalachian folds

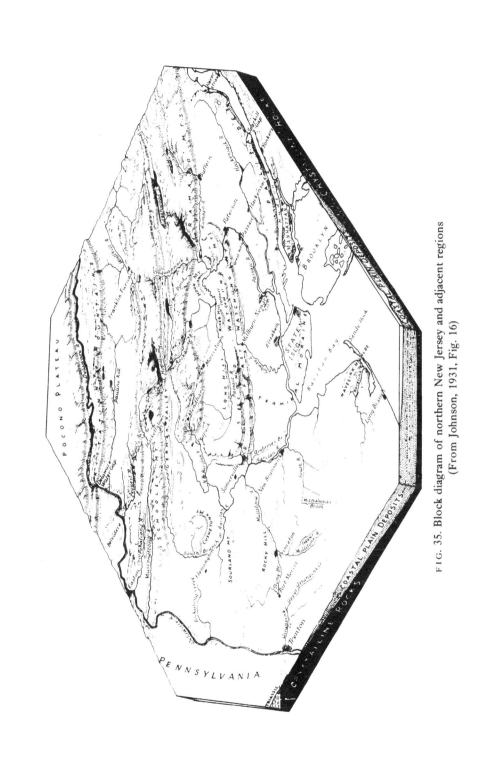

F I G. 35. Block diagram of northern New Jersey and adjacent regions
(From Johnson, 1931, Fig. 16)

was from the *east* had given rise to the view that from the earliest times until after the Appalachian orogeny (Permian) a substantial landmass covered at least the area of the present coastal plain. This belief was formalized by Dana in his theory of Archaean protaxes (1890), by Walcott (1891) and by Williams, who in 1897 named the particularly-active Devonian representative of this source area 'Appalachia'. The concept of such a large (now vanished) landmass which shed its sediments westward into the Appalachian geosyncline and acted as the more dynamic jaw of the vice in which the Appalachian mountains were compressed was developed by Charles Schuchert (1910). More recently this borderland theory was replaced by that of island arcs, and this, in turn, is today being modified in the light of continental drift theories. When Davis wrote about the evolution of Appalachian landforms and drainage 'Appalachia' hung like an albatross round his neck. It necessitated the postulation of an initial north-westward consequent drainage and also confused his ideas regarding the history of the rocks on which the sub-Cretaceous unconformity (i.e. Fall Zone peneplain) of the Coastal Plain was cut.

Davis' first publications (1882E; 1883M) relating to the Appalachian Mountains were concerned with a discussion of the miniature 'Appalachian' folds of the Helderberg Mountains on the north side of the Catskills. This area was used as a location for field trips by the Geology Department at Harvard for many years and R. E. Dodge is recorded as having attended a 3-stage trip in the summer of 1889 to the Genessee Valley with H. S. Williams and Shaler, then to the Catskills and the Black Mountains of southern Connecticut with Davis, finally ending up at Martha's Vineyard with Shaler (Griffin, 1952).

In 1883 Davis (1883D) commented on Löwl's (1882, p. 405) view that the Appalachian cross valleys could not be due to antecedence (Tietze, 1878, p. 600) but were due to a combination of lake overflow behind rising mountain barriers and, particularly, to headward erosion by backcutting fingertip tributaries. Davis gave qualified support to the antecedence theory in the Tennessee–Virginia–North Carolina area, and made the following comment on the region of Pennsylvania which he was later to make classic:

> Returning now to Virginia and Pennsylvania, we have to consider not only why the rivers there cross the mountains, but also why they flow to the south-east instead of to the north-west. Taking the last question first, we are forced to suppose that the north-westerly slope, which must have existed at least up to the end of the Carboniferous, was then or soon after reversed in the slow writhing of the surface. This is demanded by the lay of the land, and by the now small area of what must have been, in Paleozoic time, a large crystalline land-mass. The slope being changed early in the growth of the folds, or before their beginning, the streams tried to make their way to the eastward; and the Hudson, Delaware, Susquehanna, Potomac, and James are the descendants of those that succeeded. Their rectangular courses, alternately longitudinal and transverse, bear witness to

their defeats and victories. Lakes must have been numerous here once, though they are now all drained. It is known that rivers often chose cross-faults of small throw as points of attack in cutting their way through the growing ridges; and it is very probable that they made use of pre-existent valleys when they advanced over the old sinking land.

In considering the applicability of backward-cutting lateral streams to the production of our cross-valleys, we should test the past by the present, and examine such ridges as Kittatinny or Bald Eagle mountains in Pennsylvania, or Clinch mountain in Tennessee, rising between parallel longitudinal valleys, to see if they show embryonic cross-valleys in the more advanced stages of development. They do not. The continuity of their crest-line is most characteristic and remarkable: it very rarely departs from its line of almost uniform height. The exceptions are, first, the finished water-gaps, or transverse valleys, whose origin is in discussion; second, the occasional wind-gaps, or notches, which sometimes cut the ridge a third or half way to its base, and which are, we believe, always determined by small transverse faults; third, the less conspicuous serrations of small value. It is difficult to assign any reason why lateral streams should not now, as well as in former times, show us the later stages of breaking down the ridge on which they rise; and yet these almost-formed cross-valleys between adjoining longitudinal valleys are practically unknown in our Appalachian topography. The reason of their absence can hardly be, that there are now enough completed water-gaps for all practical purposes, and hence the lateral streams stop making any more; for this would imply a consciousness of the end that plays no part in geological operations, and we are therefore constrained to think that Löwl's explanation cannot apply to the Appalachians in any general way. (*Davis, 1883D, p. 356*)

It is interesting that Davis prefaced his most important contribution to Appalachian denudation chronology ('The rivers and valleys of Pennsylvania', 1889D) by a quote from Löwl.

A brief article that I wrote in comment on Löwl's first essay several years ago now seems to me insufficient in its method. It exaggerated the importance of antecedent streams; it took no sufficient account of the several cycles of erosion through which the region has certainly passed; and it neglected due consideration of the readjustment of initial immature stream courses during more advanced river-life. Since then, a few words in Löwl's essay have come to have more and more significance to me; he says that in mountain systems of very great age, the original arrangement of the longitudinal valleys often becomes entirely confused by means of their conquest by transverse erosion gaps. (*Davis, 1889D, p. 8*)

Being fully persuaded of the gradual and systematic evolution of topographic forms, it is now desired, in studying the rivers and valleys of Pennsylvania, to seek the causes of the location of the streams in their present courses; to go back if possible to the early date when central Pennsylvania was first raised above the sea and trace the development of the several river systems then implanted upon it from their ancient beginning to the present time. (*Davis, 1889D, pp. 2–3*)

Davis attempted to solve this problem by what Johnson (1931, p. xiii) later called 'a beautifully devised but highly complicated series of stream readjustments related to a succession of land movements'. In particular, he was at pains to explain the causes and interrelationships between the complicated drainage anomalies of the Appalachian rivers and the summit peneplain (Schooley) represented by the crests of the ridges in the Ridge and Valley Province and adjacent summits. His thesis was dominated by two assumptions, the questioning of which later led his pupil Johnson (1931) to produce a more acceptable denudation chronology for the region. These assumptions were:

1. That the present drainage developed directly from the original Permian rivers, consequent on the rising folds, and having a dominantly north-westerly direction away from Appalachia.
2. That the Schooley surface and the sub-Cretaceous unconformity of the Coastal Plain represent different parts of the same warped peneplain.

Davis met the problem of the assumed original north-westerly direction of the Pennsylvania drainage in two ways. Firstly, he assumed that the folded belt was so much and so rapidly raised that it, rather than Appalachia, became the divide for the initial consequent drainage. This, however, still left much north-westerly drainage to be reversed so, secondly, Davis ascribed this reversal to Triassic movements when the westward drainage had reduced the region to a very low slope which was reversed by the subsidence of Appalachia and the formation of the Triassic depressions – notably that of New Jersey. Davis reached this conclusion apparently as the result of continually testing rival hypotheses. For example, he contrasted the hypothesis of antecedence of north-westward-flowing rivers across the growing Appalachian folds with larger streams persisting to the present, although with flow reversal, with the hypothesis that the original master consequents were reversed and suffered heavy losses due to the extension of subsequent streams along the weak belts. Davis accepted the latter. The reversal of drainage could, in his view, be attributed either to earth movements associated with the Triassic faulting reversing the central segments of the rivers, followed by later adjustments due to capture, or to the progressive capture of the north-west-flowing by the south-east-flowing streams. Davis could find little evidence for the latter and therefore preferred the former explanation, but always recognized that this drainage reversal explanation was the least satisfactory part of his thesis.

Davis postulated that throughout the north-eastern United States there was evidence of the existence of a widespread peneplain formed at the beginning of the Cretaceous as the result of the Jurassic–Cretaceous cycle of erosion. This new cycle was initiated by the Triassic tilting which itself brought about the present dominantly south-easterly drainage pattern.

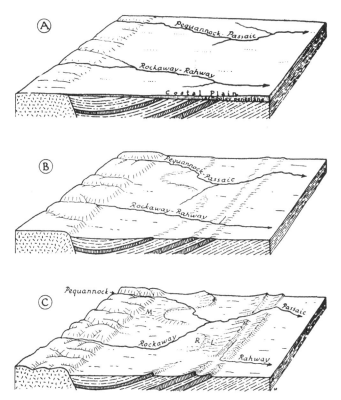

F I G. 36. Davis' idea of the superimposition and adjustment of the Watchung drainage, New Jersey (From Johnson, 1931, Fig. 18)

The reality of this peneplain was shown by the accordant ridge summits of the folded Appalachians (Davis, 1899D); and by the summit surface of Kittatiny Mountain (2000 feet) in western New Jersey, which warped down to the Schooley peneplain surface of the Reading Prong (1300 feet) and to the ridge crests of the New Jersey Watchung Ridges and the Palisades (Davis, 1889J and 1890K). The same surface, Davis believed (1896F), formed the summit peneplain of southern New England, and descended from about 2000 feet around the Green Mountains to sea-level at Long Island Sound. To the south-east this downwarped early Cretaceous (Schooley) peneplain was buried by overlapping Cretaceous rocks which today form the base of the Coastal Plain and of Long Island. It is particularly interesting, in the light of Johnson's more recent explanation, that Davis, on the basis of the structurally-anomalous courses of the lower Connecticut river and of the rivers in the vicinity of the Watchung Ridges of New Jersey, postulated a limited Cretaceous onlap over southern New England and the New Jersey Triassic rocks from which superimposition of drainage occurred as they were stripped back to their present out-crops.

Johnson summed up Davis' work as follows:

Davis' 'Rivers and Valleys of Pennsylvania' will ever rank as one of the most brilliant examples of close deductive reasoning to be found in physiographic literature. Fully to appreciate it one must not merely comprehend the utter vagueness of previous ideas on the erosional history of the Appalachians and on the behavior of rivers in general; he must also gain a clear mental picture of the detailed geological structure of eastern Pennsylvania, and then follow step by step the shifting images of changing forms developed at different levels upon a variety of structures. Who does this will have no doubt that the great value of the work lies in deducing the normal evolution of drainage in a region of folded mountains like the Appalachians. This value remains, whether or not one accepts the histories set forth as the most probable explanations of the present drainage features of Pennsylvania. (*Johnson, 1931, p. 56*)

It is impossible to describe Davis' methods of reasoning and exposition, his mental gymnastics, the subtle interplay of broad theory and detailed field and map observation, or his utilization of every possible piece of evidence which bears on the hypotheses which he is testing except by quoting passages from this memorable work. It is worth while at the outset, however, to point out a fundamental difficulty inherent in the method of 'historical explanation' used here by Davis, as distinct from the kind of 'functional explanation' often favoured by Gilbert. The use of the historical method requires the present to be explained by starting at some point in the past and then describing a sequence of relevant events culminating eventually in the existing situation. The problem is that, commonly, the further back one goes into the past the more speculative is one's understanding of the foundation upon which the subsequent events were built. As we shall see, many of Davis' Appalachian difficulties resulted from his initial postulates regarding the assumed Permian topography.

... it is difficult to believe that any streams, even if antecedent and more or less persistent for a time during the mountain growth, could preserve till now their pre-Appalachian courses through all the varying conditions presented by the alternations of hard and soft rocks through which they have had to cut, and at all the different altitudes above baselevel in which they have stood. A better means of deciding the question will be to admit provisionally the occurrence of a com-pletely original system of consequent drainage, located in perfect accord with the slopes of the growing mountains; to study out the changes of stream-courses that would result from later disturbances and from the mutual adjustments of the several members of such a system in the different cycles of its history; and finally to compare the courses thus deduced with those now seen. If there be no accord, either the method is wrong or the streams are not consequent but of some other origin, such as antecedent; if the accord between deduction and fact be well marked, varying only where no definite location can be given to the deduced streams, but agreeing where they can be located more precisely, then it seems to me that the best conclusion is distinctly in favor of the correctness of the deductions. For it is not likely, even if it be possible, that antecedent streams should have accidentally taken, before the mountains were formed, just such locations as would have resulted from the subsequent growth of the mountains and from the complex changes in the initial river courses due to later adjustments. I shall therefore follow the deductive method thus indicated and attempt to trace out the history of a completely original, consequent system of drainage accordant with the growth of the central mountain district. *(Davis, 1889D, pp. 220–1)*

Constructional Permian topography and consequent drainage. – A rough restoration of the early constructional topography is given in [Fig. 37] for the central part of the State [based on the reconstruction of folds], the closest shading being the area of the Trenton limestone, indicating the highest ground, or better, the places of greatest elevation, while the Carboniferous area is unshaded, indicating the early lowlands. The prevalence of northeast and southwest trends was then even more pronounced than now. Several of the stronger elements of form deserve names, for convenient reference. Thus we have the great Kittatinny or Cumberland highland, C,C, on the southeast, backed by the older mountains of Cambrian and Archean rocks, falling by the Kittatinny slope to the synclinal lowland troughs of the central district. In this lower ground lay the synclinal troughs of the eastern coal regions, and the more local Broad Top basin, BT, on the southwest, then better than now deserving the name of basins. Beyond the corrugated area that connected the coal basins rose the great Nittany highland, N, and its southwest extension in the Bedford range, with the less conspicuous Kishicoquilas highland, K, in the foreground. Beyond all stretched the great Alleghany lowland plains. The names thus suggested are compounded of the local names of today and the morphological names of Permian time.

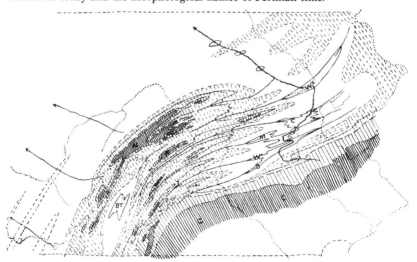

FIG. 37. Deductive reconstruction of the Permian topography and drainage of Pennsylvania (see text) (From Davis, 1889D, Fig. 21)

What would be the drainage of such a country? Deductively we are led to believe that it consisted of numerous streams as marked in full lines on the figure, following synclinal axes until some master streams led them across the intervening anticlinal ridges at the lowest points of their crests and away into the open country to the northwest. All the enclosed basins would hold lakes, overflowing at the lowest part of the rim. The general discharge of the whole system would be to the northwest. Here again we must resort to special names for the easy indication of these well-marked features of the ancient and now apparently lost drainage system. The master stream of the region is the great Anthracite river, carrying the overflow of the Anthracite lakes off to the northwest and there perhaps turning along one of the faintly marked synclines of the plateau and joining the original Ohio, which was thus confirmed in its previous location across the Carboniferous marshes. The synclinal streams that entered the Anthracite lakes from the southwest may be named, beginning on the south, the Swatara, S, [Fig. 37], the Wiconisco, Wo, the Tuscarora–Mahanoy, M, the Juniata–Catawissa, C, and the Wyoming, Wy. One of these, probably the fourth, led the overflow from the Broad Top lake into the Catawissa lake on the middle Anthracite river. The Nittany highland formed a strong divide between the central and northwestern rivers, and on its outer slope there must have been streams descending to the Alleghany lowlands; and some of these may be regarded as the lower courses of Carboniferous rivers, that once rose in the Archean mountains, now beheaded by the growth of mountain ranges across their middle. (*Davis, 1889D, pp. 222–4*)

Development and adjustment of the Permian drainage. – The problem is now before us. Can the normal sequence of changes in the regular course of river development, aided by the post-Permian deformations and elevations, evolve the existing rivers out of the ancient ones?

In order to note the degree of comparison that exists between the two, several of the larger rivers of today are dotted on the figure. The points of agreement are indeed few and small. Perhaps the most important ones are that the Broad Top region is drained by a stream, the Juniata, which for a short distance follows near the course predicted for it; and that the Nittany district, then a highland, is still a well-marked divide although now a lowland. But there is no Anthracite river, and the region of the ancient coal-basin lakes is now avoided by large streams; conversely, a great river – the Susquehanna – appears where no consequent river ran in Permian time, and the early synclinal streams frequently turn from the structural troughs to valleys located on the structural arches. (*Davis, 1889D, p. 225*)

Lateral water gaps near the apex of synclinal ridges. – One of the most frequent discrepancies between the hypothetical and actual streams is that the latter never follow the axis of a descending syncline along its whole length, as the original streams must have done, but depart for a time from the axis and then return to it, notching the ridge formed on any hard bed at the side instead of at the apex of its curve across the axis of the syncline. There is not a single case in the state of a stream cutting a gap at the apex of such a synclinal curve, but there are perhaps hundreds of cases where the streams notch the curve to one side of the apex. This, however, is precisely the arrangement attained by spontaneous adjustment from an initial axial course, as indicated in [Fig. 38]. The gaps may be located on small

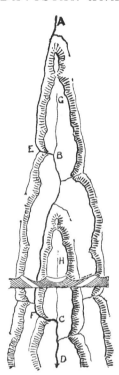

F I G. 38. Development of drainage on an exhumed syncline of alternating hard and soft strata (see text) (From Davis, 1889D, Fig. 13)

transverse faults, but as a rule they seem to have no such guidance. It is true that most of our streams now run out of and not into the synclinal basins, but a reason for this will be found later; for the present we look only at the location of the streams, not at their direction of flow. As far as this illustration goes, it gives evidence that the smaller streams at least possess certain peculiarities that could not be derived from persistence in a previous accidental location, but which would be necessarily derived from a process of adjustment following the original establishment of strictly consequent streams. Hence the hypothesis that these smaller streams were long ago consequent on the Permian folding receives confirmation; but this says nothing as to the origin of the larger rivers, which might at the same time be antecedent. (*Davis, 1889D, pp. 225–6*)

Departure of the Juniata from the Juniata–Catawissa syncline. – It may be next noted that the drainage of the Broad Top region does not follow a single syncline to the Anthracite region, as it should have in the initial stage of the consequent Permian drainage, but soon turns aside from the syncline in which it starts and runs across country to the Susquehanna. It is true that in its upper course the Juniata departs from the Broad Top region by one of the two synclines that were indicated as the probable line of discharge of the ancient Broad Top lake in our restoration of the constructional topography of the State; there does not appear to be any significant difference between the summit altitudes of the Tuscarora–Mahanoy and the Juniata–Catawissa synclinal axes and hence the choice must have been made for reasons that cannot be detected; or it may be that the syncline lying more to the northwest was raised last, and for this reason was taken as the line of overflow. The beginning of the river is therefore not discordant with the hypothesis of consequent drainage, but the southward departure from the Catawissa syncline at Lewistown remains to be explained. It seems to me that some reason for the departure may be found by likening it to the case already given in [Fig. 39]. The several synclines with which the Juniata is concerned have precisely the relative attitudes that are there discussed. The Juniata–Catawissa syncline has parallel sides for many miles about its middle, and hence must have long maintained the initial Juniata well above baselevel over all this distance; the progress of cutting down a channel through all the hard Carboniferous sandstones for so great a distance along the axis must have been exceedingly slow. But the synclines next south, the Tuscarora–Mahanoy and the Wiconisco, plunge to the northeast more rapidly, as the rapid divergence of their margins demonstrates, and must for this reason have carried the hard sandstones below baselevel in a shorter distance and on a steeper slope than in the Catawissa syncline. The further

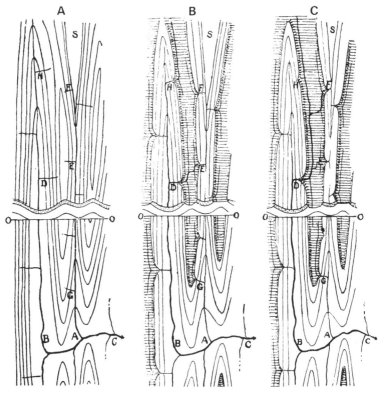

F I G. 39. Adjustment of drainage on a series of plunging anticlines and synclines (see text) (From Davis, 1889D, Figs. 16, 17 and 18)

southwestward extension of the Pocono sandstone ridges in the southern than in the northern syncline gives further illustration of this peculiarity of form. Lateral capture of the Juniata by a branch of the initial Tuscarora, and of the latter by a branch of the Wiconisco therefore seems possible, and the accordance of the facts with so highly specialized an arrangement is certainly again indicative of the correctness of the hypothesis of consequent drainage, and this time in a larger stream than before. At first sight, it appears that an easier lateral capture might have been made by some of the streams flowing from the outer slope of the Nittany highland; but this becomes improbable when it is perceived that the heavy Medina sandstone would here have to be worn through as well as the repeated arches of the Carboniferous beds in the many high folds of the Seven Mountains. Again, as far as present appearances go, we can give no sufficient reason to explain why possession of the headwaters of the Juniata was not gained by some subsequent stream of its own, such as G, [Fig. 39c], instead of by a side-stream of the river in the neighboring syncline; but it may be admitted, on the other hand, that as far as we can estimate the chances for conquest, there was nothing distinctly in favor of one or the other of the side-streams concerned; and as long as the problem is solved indifferently in favor of one or the other, we may accept the lead of the facts and say that some control not now apparent determined that the diversion should be, as drawn, through D and not through G. The detailed location of the Juniata in its middle course below Lewistown will be considered in a later section. (*Davis, 1889D, pp. 226–7*)

Reversal of larger rivers to southeast courses. – Our large rivers at present flow to the southeast, not to the northwest. It is difficult to find any precise date for this reversal of flow from the initial hypothetical direction, but it may be suggested that it occurred about the time of the Triassic depression of the Newark belt. We have been persuaded that much time elapsed between the Permian folding and the Newark deposition, even under the most liberal allowance for pre-Permian erosion in the Newark belt; hence when the depression began, the rivers must have had but moderate northwestward declivity. The depression and submergence of the broad Newark belt may at this time have broken the continuity of the streams that once flowed across it. The headwater streams from the ancient Archean country maintained their courses to the depression; the lower portions of the rivers may also have gone on as before; but the middle courses were perhaps turned from the central part of the state back to the Newark belt. No change of attitude gives so fitting a cause of the southeastward flow of our rivers as this. The only test that I have been able to devise for the suggestion is one that is derived from the relation that exists between the location of the Newark belt along the Atlantic slope and the course of the neighboring transverse rivers. In Pennsylvania, where the belt reaches somewhat beyond the northwestern margin of the crystalline rocks in South mountain, the streams are reversed, as above stated; but in the Carolinas where the Newark belt lies far to the east of the boundary between the Cambrian and crystalline rocks, the Tennessee streams persevere in what we suppose to have been their original direction of flow. This may be interpreted as meaning that in the latter region, the Newark depression was not felt distinctly enough, if at all, within the Alleghany belt to reverse the flow of the streams; while in the former region, it was nearer to these streams and determined a change in their courses. The original Anthracite river ran to the northwest, but its middle course was afterwards turned to the southeast.

I am free to allow that this has the appearance of heaping hypothesis on hypothesis; but in no other way does the analysis of the history of our streams seem possible, and the success of the experiment can be judged only after making it. At the same time, I am constrained to admit that this is to my own view the least satisfactory of the suggestions here presented. It may be correct, but there seems to be no sufficient exclusion of other possibilities. For example, it must not be overlooked that, if the Anthracite river ran southeast during Newark deposition, the formation of the Newark northwestward monocline by the Jurassic tilting would have had a tendency to turn the river back again to its northwest flow. But as the drainage of the region is still southeastward, I am tempted to think that the Jurassic tilting was not here strong enough to reverse the flow of so strong and mature a river as the Anthracite had by that time come to be; and that the elevation that accompanied the tilting was not so powerful in reversing the river to a northwest course as the previous depression of the Newark basin had been in turning it to the southeast. If the Anthracite did continue to flow to the southeast, it may be added that the down-cutting of its upper branches was greatly retarded by the decrease of slope in its lower course when the monocline was formed.

The only other method of reversing the original northwestward flow of the streams that I have imagined is by capture of their headwaters by Atlantic rivers. This seems to me less effective than the method just considered; but they are not mutually exclusive and the actual result may be the sum of the two processes. The outline of the idea is as follows. The long continued supply of sedimentary material from the Archean land on the southeast implies that it was as continually elevated. But there came a time when there is no record of further supply of material, and when we may therefore suppose the elevation was no longer maintained. From that time onward, the Archean range must have dwindled away, what with the encroachment of the Atlantic on its eastern shore and the general action of denuding forces on its surface. The Newark depression was an effective aid to the same end, as has been stated above, and for a moderate distance westward of the depressed belt, the former direction of the streams must certainly have been reversed; but the question remains whether this reversal extended as far as the Wyoming basin, and whether the subsequent formation of the Newark monocline did not undo the effect of the Newark depression. It is manifest that as far as our limited knowledge goes, it is impossible to estimate these matters quantitatively, and hence the importance of looking for additional processes that may supplement the effect of the Newark depression and counteract the effect of the Newark uplift in changing the course of the rivers. Let it be supposed for the moment that at the end of the Jurassic uplift by which the Newark monocline was formed, the divide between the Ohio and the Atlantic drainage lay about the middle of the Newark belt. There was a long gentle descent westward from this watershed and a shorter and hence steeper descent eastward. Under such conditions, the divide must have been pushed westward, and as long as the rocks were so exposed as to open areas of weak sediments on which capture by the Atlantic streams could go on with relative rapidity, the westward migration of the divide would be important. For this reason, it might be carried from the Newark belt as far as the present Alleghany front, beyond which further pushing would be slow, on account of the broad stretch of country there covered by hard horizontal beds.

The end of this is that, under any of the circumstances here detailed, there would be early in the Jurassic–Cretaceous cycle a distinct tendency to a westward migration of the Atlantic–Ohio divide; it is the consequences of this that have now to be examined. (*Davis, 1889D, pp. 229–32*)

Capture of the Anthracite headwaters by the growing Susquehanna. – Throughout the Perm–Triassic period of denudation, a great work was done in wearing down the original Alleghanies. Anticlines of hard sandstone were breached, and broad lowlands were opened on the softer rocks beneath. Little semblance of the early constructional topography remained when the period of Newark depression was brought to a close; and all the while the headwater streams of the region were gnawing at the divides, seeking to develop the most perfect arrangement of waterways. Several adjustments have taken place, and the larger streams have been reversed in the direction of their flow; but a more serious problem is found in the disappearance of the original master stream, the great Anthracite river, which must have at first led away the water from all the lateral synclinal streams. Being a large river, it could not have been easily diverted from its course, unless it was greatly retarded in cutting down its channel by the presence of many beds of hard rocks on its way. The following considerations may perhaps throw some light on this obscure point.

It may be assumed that the whole group of mountains formed by the Permian deformation had been reduced to a moderate relief when the Newark deposition was stopped by the Jurassic elevation. The harder ribs of rock doubtless remained as ridges projecting above the intervening lowlands, but the strength of relief that had been given by the constructional forces had been lost. The general distribu-

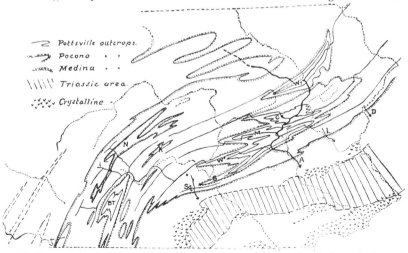

F I G. 40. General distribution of relief and drainage of Pennsylvania in early Jurassic time (see text) (From Davis, 1889D, Fig. 25)

tion of residual elevations then remaining unsubdued is indicated in [Fig. 40], in which the Crystalline, the Medina, and the two Carboniferous sandstone ridges are denoted by appropriate symbols. In restoring this phase of the surface form, when the country stood lower than now, I have reduced the anticlines from their present outlines and increased the synclines, the change of area being made greatest where the dips are least, and hence most apparent at the ends of the plunging anticlines and synclines. Some of the Medina anticlines of Perry and Juniata counties are not indicated because they were not then uncovered. The country between the residual ridges of Jurassic time was chiefly Cambrian limestone and Siluro–Devonian shales and soft sandstones. The moderate ridges developed on the Oriskany and Chemung sandstones are not represented. The drainage of this stage retained the original courses of the streams, except for the adjustments that have been described, but the great Anthracite river is drawn as if it had been controlled by the Newark depression and reversed in the direction of its flow, so that its former upper course on the Cambrian rocks was replaced by a superimposed Newark lower course. [Fig. 40] therefore represents the streams for the most part still following near their synclinal axes, although departing from them where they have to enter a synclinal cove-mountain ridge; the headwaters of the Juniata avoid the mass of hard sandstones discovered in the bottom of old Broad Top lake, and flow around them to the north, and then by a cross-country course to the Wiconisco synclinal, as already described in detail. Several streams come from the northeast, entering the Anthracite district after the fashion generalized in [Fig. 38]. Three of the many streams that were developed on the great Kittatinny slope are located, with their direction of flow reversed; these are marked Sq, L and D, and are intended to represent the ancestors of the existing Susquehanna, Lehigh and Delaware. We have now to examine the opportunities offered to these small streams to increase their drainage areas.

The Jurassic elevation, by which the Newark deposition was stopped, restored to activity all the streams that had in the previous cycle sought and found a course close to baselevel. They now all set to work again deepening their channels. But in this restoration of lost activity with reference to a new baselevel, there came the best possible chance for numerous re-arrangements of drainage areas by mutual adjustment into which we must inquire.

I have already illustrated what seems to me to be the type of the conditions involved at this time in [Fig. 41]. The master stream, A, traversing the synclines, corresponds to the reversed Anthracite river; the lowlands at the top are those that have been opened out on the Siluro–Devonian beds of the present Susquehanna middle course between the Pocono and the Medina ridges. The small stream, B, that is gaining drainage area in these lowlands, corresponds to the embryo of the present Susquehanna, Sq, [Fig. 40], this having been itself once a branch on the south side of the Swatara synclinal stream, [Fig. 37], from which it was first turned by the change of slope accompanying the Newark depression; but it is located a little farther west than the actual Susquehanna, so as to avoid the two synclinal cove mountains of Pocono sandstone that the Susquehanna now traverses, for reasons to be stated below . . . This stream had to cross only one bed of hard rock, the outer wall of Medina sandstone, between the broad inner lowlands of the relatively weak Siluro–Devonian rocks and the great valley

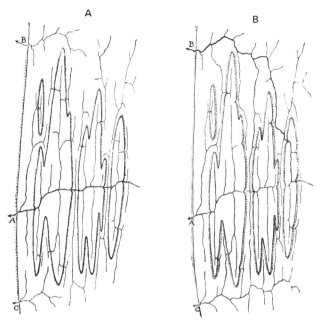

FIG. 41. Rearrangement of drainage on folds after rejuvenation (see text)
(From Davis, 1889D, Figs. 19 and 20)

lowlands on the still weaker Cambrian limestones. Step by step it must have pushed its headwater divide northward, and from time to time it would have thus captured a subsequent stream, that crossed the lowlands eastward, and entered a Carboniferous syncline by one of the lateral gaps already described. With every such capture, the power of the growing stream to capture others was increased. [Fig. 41A] represents a stage after the streams in the Swatara and Wiconisco synclines (the latter then having gained the Juniata) had been turned aside on their way to the Carboniferous basins. On the other hand, the Anthracite river, rising somewhere on the plains north of the Wyoming syncline and pursuing an irregular course from one coal basin to another, found an extremely difficult task in cutting down its channel across the numerous hard beds of the Carboniferous sandstones, so often repeated in the rolling folds of the coal fields. It is also important to remember that an aid to other conditions concerned in the diversion of the upper Anthracite is found in the decrease of slope that its lower course suffered in crossing the coal fields, if that area took any part in the deformation that produced the Newark monocline – whichever theory prove true in regard to the origin of the southeastward flow of the rivers – for loss of slope in the middle course, where the river had to cross many reefs of hard sandstone, would have been very effective in lengthening the time allowed for the diversion of the headwaters.

The question is, therefore, whether the retardation of down-cutting here experienced by the Anthracite was sufficient to allow the capture of its headwaters by the Susquehanna. There can be little doubt as to the correct quality of the process, but whether it was quantitatively sufficient is another matter. In the absence of any means of testing its sufficiency, may the result not be taken as the test? Is not the correspondence between deduction and fact close enough to prove the correctness of the deduction? (*Davis, 1889D, pp. 232–6*)

Superimposition of the Susquehanna on two synclinal ridges. – There is however one apparently venturesome postulate that may have been already noted as such by the reader; unless it can be reasonably accounted for and shown to be a natural result of the long sequence of changes here considered, it will seriously militate against the validity of the whole argument. The present course of the middle Susquehanna leads it through the apical curves of two Pocono synclinal ridges, which were disregarded in the statement given above. It was then assumed that the embryonic Susquehanna gained possession of the Siluro–Devonian lowland drainage by gnawing out a course to the west of these synclinal points; for it is not to be thought of that any conquest of the headwaters of the Anthracite river could have been made by the Susquehanna if it had had to gnaw out the existing four traverses of the Pocono sandstones before securing the drainage of the lowlands above them. The backward progress of the Susquehanna could not in that case have been nearly fast enough to reach the Anthracite before the latter had sunk its channel to a safe depth. It is therefore important to justify the assumption as to the more westerly location of the embryonic Susquehanna; and afterwards to explain how it should have since then been transferred to its present course. A short cut through all this round-about method is open to those who adopt in the beginning the theory that the Susquehanna was an antecedent river; but as I have said at the outset of this inquiry, it seems to me that such a method is not freer from assumption, even though shorter than the one here adopted; and it has the demerit of not considering all the curious details that follow the examination of consequent and adjusted courses.

The sufficient reason for the assumption that the embryonic Susquehanna lay farther west than the present one in the neighborhood of the Pocono synclinals is simply that – in the absence of any antecedent stream – it must have lain there. The whole explanation of the development of the Siluro–Devonian lowlands between the Pocono and Medina ridges depends simply on their being weathered out where the rocks are weak enough to waste faster than the enclosing harder ridges through which the streams escape. In this process, the streams exercise no control whatever over the direction in which their headwaters shall grow; they leave this entirely to the structure of the district that they drain. It thus appears that, under the postulate as to the initial location of the Susquehanna as one of the many streams descending the great slope of the Kittatinny (Cumberland) highland into the Swatara syncline, its course being reversed from northward to southward by the Newark depression, we are required to suppose that its headwater (northward) growth at the time of the Jurassic elevation must have been on the Siluro–Devonian beds, so as to avoid the harder rocks on either side. Many streams competed for the distinction of becoming the master, and that one gained its ambition whose initial location gave it the best subsequent opportunity. It remains then to consider the means by which the course of the conquering Susquehanna may have been subsequently changed from the lowlands on to the two Pocono synclines that it now traverses. Some departure from its early location may have been due to eastward planation in its advanced age, when it had large volume and gentle slope and was therefore swinging and cutting laterally in its lower course. This may have had a share in the result, but there is another process that seems to me more effective.

In the latter part of the Jura–Cretaceous cycle, the whole country hereabout suffered a moderate depression, by which the Atlantic transgressed many miles inland from its former shoreline, across the lowlands of erosion that had been developed on the litoral belt. Such a depression must have had a distinct effect on the lower courses of the larger rivers, which having already cut their channels down close to baselevel and opened their valleys wide on the softer rocks, were then 'estuaried', or at least so far checked as to build wide flood-plains over their lower stretches. Indeed, the flood-plains may have been begun at an earlier date, and have been confirmed and extended in the later time of depression. Is it possible that in the latest stage of this process, the almost baselevelled remnants of Blue mountain and the Pocono ridges could have been buried under the flood-plain in the neighborhood of the river?

If this be admitted, it is then natural for the river to depart from the line of its buried channel and cross the buried ridges on which it might settle down as a superimposed river in the next cycle of elevation. It is difficult to decide such general questions as these; and it may be difficult for the reader to gain much confidence in the efficacy of the processes suggested; but there are certain features in the side streams of the Susquehanna that lend some color of probability to the explanation as offered.

Admit, for the moment, that the aged Susquehanna, in the later part of the Jura–Cretaceous cycle, did change its channel somewhat by cutting to one side, or by planation, as it is called. Admit, also, that in the natural progress of its growth it had built a broad flood plain over the Siluro–Devonian lowlands, and that the depth of this deposit was increased by the formation of an estuarine delta upon it when the country sank at the time of the mid-Cretaceous transgression of the sea. It is manifest that one of the consequences of all this might be the peculiar course of the river that is to be explained, namely, its superimposition on the two Pocono synclinal ridges in the next cycle of its history, after the Tertiary elevation had given it opportunity to re-discover them. It remains to inquire what other consequences should follow from the same conditions, and from these to devise tests of the hypothesis (*Davis, 1889D, pp. 239–41*)

The theory of the location of the Susquehanna on the Pocono synclinal ridges therefore stands as follows. The general position of the river indicates that it has been located by some process of slow-self-adjusting development and that it is not a persistent antecedent river; and yet there is no reason to think that it could have been brought into its present special position by any process of shifting divides. The processes that have been suggested to account for its special location, as departing slightly from a location due to slow adjustments following an ancient consequent origin, call for the occurrence of certain additional peculiarities in the courses of its tributary streams, entirely unforseen and unnoticed until this point in the inquiry is reached; and on looking at the map to see if they occur, they are found with perfect distinctness. The hypothesis of superimposition may therefore be regarded as having advanced beyond the stage of mere suggestion and as having gained some degree of confirmation from the correlations that it detects and explains. It only remains to ask if these correlations might have originated in any other way, and if the answer to this is in the negative, the case may be looked upon as having a fair measure of evidence in its favor. The remaining consideration may be taken up at once as the first point to be examined in the Tertiary cycle of development. (*Davis, 1889D, pp. 242–3*)

Events of the Tertiary cycle. – The elevation given to the region by which Cretaceous baselevelling was terminated, and which I have called the early Tertiary elevation, offered opportunity for the streams to deepen their channels once more. In doing so, certain adjustments of moderate amount occurred, which will be soon examined. As time went on, much denudation was effected, but no wide-spread baselevelling was reached, for the Cretaceous crest lines of the hard sandstone ridges still exist. The Tertiary cycle was an incomplete one. At its close, lowlands had been opened only on the weaker rocks between the hard beds. Is it not possible that the flood-plaining of the Susquehanna and the down-stream deflection of its branches took place in the closing stages of this cycle, instead of at the end of the previous cycle? If so, the deflection might appear on the branches, but the main river would not be transferred to the Pocono ridges. This question may be safely answered in the negative; for the Tertiary lowland is by no means well enough baselevelled to permit such an event. The beds of intermediate resistance, the Oriskany and certain Chemung sandstones, had not been worn down to baselevel at the close of the Tertiary cycle; they had indeed lost much of the height that they possessed at the close of the previous cycle, but they had not been reduced as low as the softer beds on either side. They were only reduced to ridges of moderate and unequal height over the general plain of the Siluro–Devonian low country, without great strength of relief but quite strong enough to call for obedience from the streams along side of them. And yet near Selin's Grove, for example, in Snyder county, Penn's and Middle creeks depart most distinctly from the strike of the local rocks as they near the Susquehanna, and traverse certain well-marked ridges on their way to the main river. Such aberrant streams cannot be regarded as superimposed at the close of the incomplete Tertiary cycle; they cannot be explained by any process of spontaneous adjustment yet described, nor can they be regarded as vastly ancient streams of antecedent courses; I am therefore much tempted to consider them as of superimposed origin, inheriting their present courses from the flood-plain cover of the Susquehanna in the latest stage of the Jura–Cretaceous cycle. With this tentative conclusion in mind as to the final events of Jura–Cretaceous time, we may take up the more deliberate consideration of the work of the Tertiary cycle.

The chief work of the Tertiary cycle was merely the opening of the valley lowlands; little opportunity for river adjustment occurred except on a small scale. The most evident cases of adjustment have resulted in the change of water-gaps into wind-gaps, of which several examples can be given, the one best known being the Delaware wind-gap between the Lehigh and Delaware water-gaps in Blue mountain. (*Davis, 1889D, pp. 243–4*)

Davis was far from dogmatic in the belief in the veracity of all parts of his theory of Pennsylvania drainage and treated a number of so-called 'doubtful cases'. The apparent necessity for recourse to a very complex history for the region caused him to doubt the utility of such examples as teaching vehicles.

If this theory of the history of our rivers is correct, it follows that any one river as it now exists is of so complicated an origin that its development cannot become a matter of general study and must unhappily remain only a subject for special investigation for some time to come. It was my hope on beginning this essay to find some teachable sequence of facts that would serve to relieve the usual routine of statistical and descriptive geography, but this is not the result that has been attained. The history of the Susquehanna, the Juniata, or the Schuylkill, is too involved with complex changes, if not enshrouded in mystery, to become intelligible to any but advanced students; only the simplest cases of river development can be introduced into the narrow limits of ordinary instruction. The single course of an ancient stream is now broken into several independent parts; witness the disjointing and diversion of the original Juniata, which, as I have supposed, once extended from Broad Top lake to the Catawissa basin. Now the upper part of the stream, representing the early Broad Top outlet, is reduced to small volume in Aughwick creek; the continuation of the stream to Lewistown is first set to one side of its original axial location and is then diverted to another syncline; the beheaded portion now represented by Middle creek is diverted from its course to the Catawissa basin by the Susquehanna; perhaps the Catawissa of the present day represents the reversed course of the lower Juniata where it joined the Anthracite.

(*Davis, 1889D, p. 251*)

Davis completed his classic work on the denudation chronology of the northern Appalachians by a detailed analysis of the drainage of northern New Jersey (Davis, 1889J; Davis, 1890K). He used his knowledge of the Triassic stratigraphy and structure to deduce the conditions under which the Appalachian drainage was supposedly reversed, and explained much of the drainage of the region in terms of superimposition from a limited Cretaceous cover, followed by adjustments dictated by structure.

The Appalachian revolution, culminating in Permian time, produced mountain ranges strongly folded and presumably of great height for a certain period; but in the area of the Triassic belt their topography seems to have been reduced to a moderate relief before Triassic deposition began. During an ensuing epoch of mountain-building the great mass of sediments accumulated in the Triassic cycle suffered monoclinal tilting and faulting, with the result of a new mountainous topography of less elevation and much less structural distortion than that of Permian time. The Appalachian mountains are therefore not to be regarded as the residual relief of an elevation given once for all at the time of the great Permian folding. The original mountain ranges were worn down low before and during Triassic time, after which, in the Jurassic period, they were again uplifted and much eroded to a surface named by the authors the Schooley peneplain. Next the southeastern portion of this area was moderately depressed beneath the sea and covered by Cretaceous sediments. The ridges of the present time, showing remarkable evenness in their general height are parts of the old peneplain that have as yet withstood the erosion consequent on a third uplift, which as well as other oscillations of later date, seems to have been of moderate amount and gentle inequality.

Very interesting studies of changes in the course of streams are given in detail with numerous maps. The valleys in the Archaean highland portion of the peneplain are mainly coincident in position with those of an earlier geographic cycle, but in the Triassic area most of the streams were superimposed upon the present rocks from Cretaceous strata that formerly stretched across them, and in many instances they appear to have been much affected by adjustments required by their denudation of the soft Cretaceous formation and the uncovering of the hard trap ridges of the Watchung mountains. (*Davis, 1889J; Abst., 1890, pp. 195–6*)

Look now at the drainage of the crescentic Watchung mountains; the curved edges of two great warped lava-flows of the Triassic belt. The noteworthy feature of this district is that the small streams in the southern part of the crescent rise on the back slope of the inner mountain and cut gaps in both mountains in order to reach the outer part of the Central Plain. If these streams were descended directly or by revival from ancestors antecedent to or consequent upon the monoclinal tilting of the Triassic formation, they could not possibly, in the long time and deep denudation that the region has endured, have down to the present time maintained courses so little adjusted to the structure of their basins. In so long a time as has elapsed since the tilting of the Triassic formation, the divides would have taken their places on the crest of the trap ridges and not behind the crest on the back slope. They cannot be subsequent streams, for such could not have pushed their sources headwards through a hard trap ridge. Subsequent streams are developed in accordance with structural details, not in violation of them. Their courses must have been taken *not long ago*, else they must surely have lost their heads back of the second mountain; some piratical subsequent branch of a larger transverse stream, like the Passaic, would have beheaded them.

The only method now known by which these several doubly transverse streams could have been established in the not too distant past, is by superimposition from the Cretaceous cover that was laid upon the old Schooley peneplain. It has already been stated that when the Highlands and this region together had been nearly baseleveled, the coastal portion of the resulting peneplain was submerged and buried by an unconformable cover of waste derived from the non-submerged portion: hence when the whole area was lifted to something like its present height, a new system of consequent streams was born on the revealed sea bottom. Since then, time enough may have passed to allow the streams to sink their channels through the unconformable cover and strip it off, and thus superimpose themselves on the Triassic rocks below: we should therefore find them, in so far as they have not yet been re-adjusted, following inconsequent, discordant courses on the under formation. The existing overlap of the Cretaceous beds on the still buried Triassic portion of the old Schooley peneplain makes it evident that such an origin for the Watchung streams is possible; but it has not yet been

FIG. 42. Davis' drawing of a bird's-eye view of the topography and drainage of northern New Jersey
(From Davis, 1890K, Fig. 1)

independently proved that the Cretaceous cover ever reached so far inland as to cross the Watchung ridges.

Want of other explanation for the Watchung streams is not satisfactory evidence in favor of the explanation here suggested. There should be external evidence that the Triassic area has actually been submerged and buried after it was baselevelled to the Schooley peneplain and before it was uplifted to its present altitude; other streams as well as the ones thus far indicated, should bear signs of superimposition; and if adjustment of the superimposed courses has begun, it should be systematically carried farthest near the largest streams. I shall not here state more than in brief form, the sufficient evidence that can be quoted in favor of the first and second requisites. Suffice it to say that the overlap of the Cretaceous beds (which contain practically no Triassic fragments) on the bevelled Triassic strata at Amboy and elsewhere indicates submergence after baselevelling; and that the pebbles, sands and marls of the Cretaceous series point clearly to the Highlands as their source. The submergence must therefore have reached inland across the Triassic formation at least to the margin of the crystalline rocks. Some shore-line cutting must have been done at the margin of the Highlands during Cretaceous time, but the generally rolling surface of the old peneplain leads me to ascribe its origin chiefly to subaerial wasting. Moreover, the North Branch of the Raritan, between Mendham and Peapack [Fig. 42] and the Lockatong (L), a small branch of the Delaware on the West Hunterdon sandstone plateau, give striking indications of superimposition in the discordance of their courses with the weaker structural lines of their basins, so unlike the thoroughly adjusted course of the Musconetcong and its fellows, the Pohatcong, the Lopatcong, and others.

The third requisite of the proof of the inland extension of the Cretaceous, and the resulting superimposed origin of the Watchung streams may be stated in detail, as being more in the line of this essay: has the adjustment that accompanies superimposition systematically advanced farther near the large streams than near the small ones? The character of this adjustment should be first examined deductively. Given a series of streams of different volumes, flowing southeastward, in the direction of the present dip of the remnant of the Cretaceous cover, over the former inland extension of this superposed formation; how will these streams react on one another when they sink their channels into the underlying Triassic formation?

The conditions during the formation of the cover of Cretaceous beds are illustrated in [Fig. 43A], where the Triassic portion of the peneplain is submerged, and the shore-line of the transgressing ocean has reached the margin of the crystalline rocks. The waste from the crystallines is spread out as a series of gravels, sands and marls on the baselevelled Triassic area.

Then follows the elevation and tilting of the peneplain with the cover on its back; and with this regression of the sea, there is an equivalent gain of new land; a smooth gently sloping plain is revealed as the shore line retreats; streams run out across it from the crystalline area, or begin on its open surface, growing mouthward as the land rises. Three such streams, A, C, D, are shown in [Fig. 43B]; their opportunity for deep valley-cutting is indicated by the depth of the new baselevel, BL, below the general surface of the country. While these streams are deepening their channels in the Cretaceous cover, which is unshaded with marginal contour lines in the figures, their subsequent, autogenetic branches are irregularly disposed, because there is no lateral variation of structure to guide them; but after a time, the baselevelled surface of the buried Triassic beds is reached, as is shown by linear shading in the valley bottoms of [Figs. 43C–F]. The growth of the subsequent branches then developed, will be along the strike of the Triassic softer beds, that is, about square to the course of the three transverse streams under consideration. The most rapid growth will be found on the branches of the largest stream, A, because it will most quickly cut down its channel close to the baselevel of the time and thus provide steep sloping valleysides, from which the subsequent branches cut backwards most energetically. In due time the main streams discover the particularly resistant transverse lava sheets in the underlying formation; and then the subsequent branches of the largest transverse stream on the up-stream side of the obstructions, for example, F and G, [Fig. 43C], will have a great advantage over those of the smaller streams. The most rapidly growing subsequent branch, G, [Fig. 43D], of the largest transverse master stream, A, may grow headwards so fast as to push away the divide, X, which separates it from the head of the opposing subsequent branch, J, of the next adjacent smaller transverse stream, C, and thus finally to capture and divert the headwaters, H, of the smaller transverse stream to the larger one, as in [Fig. 43E]. The divide creeps while the two opposing subsequent branches are in contest; it leaps when the successful subsequent branch reaches the channel of the conquered stream. The first stream captured in this way must necessarily be the nearest to the large stream. The diversion of the considerable volume of headwaters, H, to the channel of the small subsequent branch, G, causes it to deepen its channel rapidly; the same effect is perceptible in H for a distance above its point of capture and diversion: the increased load of sediment thus given to G will be in great part dropped in a fan-delta where it enters the flat valley of the master stream, A, [Fig. 43E].

Gaining strength by conquest, other captures are made, faster for a time, but with decreasing slowness as the head of the diverting subsequent branch recedes from the original master: and at last, equilibrium may be gained when the headwater slope of the diverting branch is no greater than that of the opposing subsequent branch of the next uncaptured transverse stream. After the capture of a transverse stream has been effected in this way, the divide, Y, between its diverted upper portions, H, [Fig. 43E], and its beheaded lower portion, C, will be pushed down stream by the growth of an inverted stream, V. This goes on until equilibrium is attained and further shifting is prevented on reaching the hard transverse lava sheets, Z, [Fig. 43F]; here the divide is maturely established. In the case of a system of transverse streams, C, D, etc., [Fig. 43F], successively captured by the subsequent branch of a single master, the divides (Z, Y′), between the inverted (V, V′) and beheaded (C, D) portions of the captured streams will for a time present different stages of approach to establishment. The divide on the line of that one of the original streams, C, that is nearest to the master stream, A, may reach a final stable position, Z; while on the next stream further away from the master, the beheaded portion, D, may still retain a short piece above the gap in the upper lava sheet, not yet secured by the inverted stream, V′; and a third stream,

A

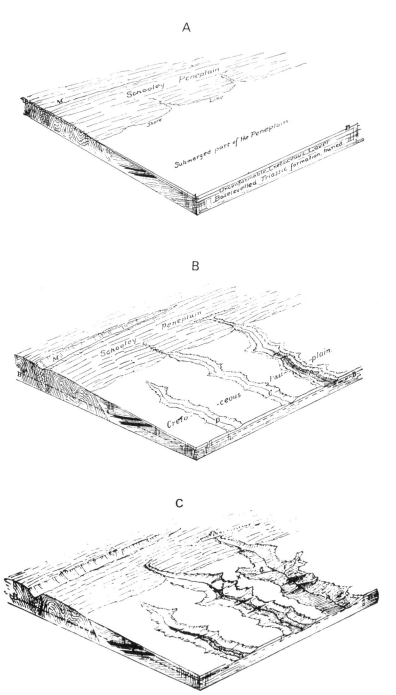

B

C

F IG. 43. The development of drainage in northern New Jersey following the

230

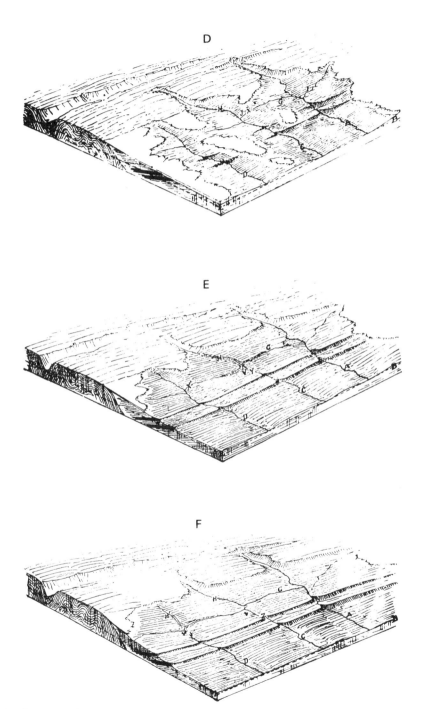

development of the Schooley peneplain (see text) (From Davis, 1890K, Figs. 2–7)

further away still from the master (not shown in figure [43F]) might remain uncaptured and independent.

It is by such tests as these that we may hope to recognize the occurrence of partial adjustment in the streams of the Watchung crescent as a result of their superimposition on the Triassic formation from its former Cretaceous cover. The greater the degree of complexity in the tests proposed, the more confidence we shall have in the theory when the tests successfully meet the facts. Hence the reason for deductively carrying out the theoretical conditions to their extremest consequences in order to increase the complexity of the tests that are to be confronted with the facts. This, as a matter of method, seems to me of great practical importance in any attempt to decipher the past progress of geographical development. (*Davis, 1890K, pp. 91–7*)

Throughout this important work on the Appalachian region Davis drew heavily on the newly published topographic maps of the region, his debt to which he acknowledged on many occasions. These maps formed a mainspring for his ideas on the evolution of relief and drainage, and it is more than coincidence that the first important denudation chronology of this region appeared at this time. In 1884 Massachusetts had begun, in cooperation with the U.S. Geological Survey, to map the state on a scale of 1 : 62,500 with contours at 20-foot intervals. The coverage was completed in 1888. New Jersey was similarly mapped by 1887, Rhode Island by 1885 and Connecticut by 1891. When Davis was working on the northern Appalachians he made full use of these newly-prepared topographic maps of New Jersey and of parts of eastern Pennsylvania at a scale of about an inch to a mile. Topographic surveying in the United States continued apace, although the rate of increased coverage dropped as scales and methods were refined. By 1904 some 31 per cent of the country had been mapped topographically, mainly at a scale of 1 : 125,000 or larger, and it is impossible to over-estimate the importance of the availability of these accurate, detailed maps in the development of geomorphological thought at the turn of the century.

In short, Davis' approach to denudation chronology was to select, usually from maps, certain pieces of morphological information which might be construed as evidence for cyclical evolutionary events – i.e. peneplanation, superimposition, capture and the like. He then used long periods of erosional history, usually comparatively lacking in dateable events, as temporal frameworks within which to suggest a sequence of events involving erosional cycles, parts of cycles and interruptions, the credibility of which depended not upon the demonstrable truth of any single piece of evidence but upon the skill with which he accumulated an apparent wealth of circumstantial evidence and wove it into a believable story. The credibility of Davis' theories lay largely in the elegance and simplicity of the sequence of events which he invoked, and it is significant that where his theories were later replaced it was not necessarily because better evidence had been subsequently accumulated but because a more simple and elegant story had been fabricated. Johnson's theory of Appalachian development, freed by the extended Cretaceous cover assumption from the need for the piecemeal reversal of the regional drainage, replaced that of Davis because of its more attractive simplicity and elegance rather than by its marshalling of significantly new facts. The best studies of denudation chronology have always been the products of minds fertile in narration, and often of those adept in extempore sermonizing and adroit in pastimes involving puzzles in which knots have to be untangled or pieces fitted into place. Davis, above all, possessed these facilities.

Although Davis subsequently studied aspects of the denudation chronology of many regions, including Central France, Brittany (1901K) and the Tian Shan (1904C), he never again equalled his work on the Appalachians in terms of weight of argument and subtlety of synthesis. Fenneman summed up exactly the impact of this work when he wrote:

> The simplicity and beauty of the conception of allotting all parts of an area to their respective cycles is alluring. So much so that we are prone to think in terms of diagrams, in which each higher level gives way visibly to a lower and younger surface, a newer peneplain which is constantly enlarging at the expense of the older and constantly losing by the spread of still newer and lower surfaces. The conception embodied in such diagrams is so simple, so illuminating, so useful, in many cases so true and it burst so suddenly upon the science, hitherto without it, explaining so many things, and introducing order where chance had reigned, that it can not be wondered at if its application was, for a time, made too broad.
>
> *(Fenneman, 1936, p. 90)*

CHAPTER TWELVE

Success Achieved

THE SUCCESSFUL MAN

In 1890 Davis was promoted to the Chair of Physical Geography, at the early age of forty achieving his main ambition as regards academic promotion, and in a few short years advancing from near failure to resounding achievement. In the process he had developed an eloquence, poise and fluency which inspired his academic audiences with his ideas and enthusiasm. His effect on Mark Jefferson exemplified this:

> When Jefferson enrolled at Harvard in 1896 he met a confident, well-read, much thinking, much published 46-year-old Davis who was finding a place for physical geography as a worthy discipline. Listening to Davis intently, seeing virtue in every idea, hungrily consuming Davis' every spoken and written word, travelling many a field mile with the man, and believing in the value of earth science, was a perceptive, intelligent and well-travelled Jefferson. Davis had an educated audience in the 33-year-old Jefferson: Jefferson had a teacher. Davis and Jefferson were drawn closely together in an enduring relationship which extended from 1896 to the time of Davis' death in 1934. (*Martin, 1968, p. 40*)

As we have seen, by this time Davis had acquired a high professional reputation, as is glimpsed in the following letter from G. K. Gilbert, who, far from questioning Davis' conclusions, limits himself to seeking an opinion.

> I have recently read your paper on the base levels of the Appalachian region. It is a fine generalization, and I think it will stand. When I was called east to take charge of the Appalachian division the part of the work I reserved for myself was the correlation of the coastal plain formations and unconformities with the base levels of the Appalachians; but I never got fairly at it, and so you and McGee have cut in ahead of me. As I do not believe in the establishment of scientific preserves, I have no complaints to make, and only a shade of regret that I am not in it; otherwise I am proud of the way the work is being done. I am glad you have discovered and recognized McGee's share, for he had so buried it under long sentences that I suspect no one would ever catch you at it if you ignored it.
>
> How do you interpret the rock terrace in the Highlands? From 150–175 feet above water there is a terrace cut into the rock on both sides so as to broaden the valley to a mile or more. At West Point there is locally a body of gravel on the terrace, doubtless added in the presence of the ice, but at most other points it is bare, and its configuration shows it to be entirely preglacial. It is somewhat dissected by the drainage, giving the impression that the deep part of the gorge

was cut through it after its formation. This seems in an obscure way to be the companion of the plain above the Highlands on which Poughkeepsie stands. It seems to me that the plain, as well as the trench through it occupied by tide-water, are pre-Tertiary. Have you seen the terrace? and if so how do you interpret it? (*Letter: G. K. Gilbert to Davis, 27 July 1891; Washington, D.C.*)

In 1884 Davis had bought a house, No. 13, Francis Avenue, in a district popular with members of the College.

Francis Avenue was in an academic neighborhood. The street started on Kirkland Street and ended at Norton's Woods, which was a large place of Professor Charles Eliot Norton's. On one side of the Davis house was the residence of Professor Hyatt, of the Massachusetts Institute of Technology. On the other side was Professor Davenport, a Professor of Botany. Farther up the street Warren A. Locke, the University organist lived. On the other side was John Graham Brooks, a writer on social questions. On Irving Street, the next street to the east, the two famous philosophers, William James and Josiah Royce lived. (*Letter: E. M. Turner to R. J. Chorley, 18 April 1962; Boston, Mass.*)

It remained the family home for the duration of his first marriage. Views appear to conflict on whether Davis was a regular or rare participant in Harvard society. In truth he was probably neither. To him society meant an opportunity for intelligent discussion; where he thought his intellectual interest was likely to be stimulated he would gladly fill the role of leading organizer. He is remembered as the founder of the Harvard Shop Club, which he started while still a young instructor. Here personal friends from the various faculties gave monthly talks on their own special studies and their papers were then discussed, sometimes light-heartedly, by the other members and their wives.

One of the social activities of the family was called The Shop Club. There were I believe twelve faculty members who with their wives met at the various houses in the evening. The host discussed his own subject, hence the name Shop, and approximately at 10 o'clock refreshments were served for the group. This activity as I recall took place every second week during the academic year and was the source of general entertainment and enlightenment for all concerned. (*Interview: R. Mott Davis with R. J. Chorley, 7 September 1962*)

We are fortunate in having a live extract from one of these meetings, in a diary of Mrs Wright, wife of the Professor of Greek.

Amusing Shop Club. At Davis'. He gave a new arrangement of the science of geography. His schedule made a sort of a box.
A.B. Natural phenomena.
C.D. Reactions on man.
The cross pieces in the telegraph wires and the uprights special divisions [see Fig. 160]. All the other men fell foul of him and warned him off their special provinces. Mr. Marks . . . for trenching on Biology. He was warned off religion by

Mr. Moore and Mr. Hall. Mr. Sheldon warned him off on general principles. John (my father) suggested a square instead of a round earth and the scrimmage became general. (*From the Diary of Mrs M. T. Wright, 5 February 1903; supplied by J. K. Wright*)

Davis' attention was not always concentrated on weighty or serious matters, and the lighter side of events often appealed to him. This humour broke out frequently, both in public and private, and, as would be expected, was exclusively witty and not rabelaisian. An example of verse shows this. Although by no means his best, it is typical.

Old and New

When the earth was still young and the sun's brighter rays warmed the land and
  the ocean,
And the waves of the sea danced with joy all day long in a rhythmical motion,
Then the great sea god, Neptune, rode forth in his chariot, drawn by sea-horses,
Surrounded by sea-nymphs and gambolling Tritons and all his sea-forces.
And three nymphs, so bewitching, so charming and fair, swimming gracefully after
The chariot, singing their songs of the ocean with peals of sweet laughter,
These nymphs are the Nereids, daughters of Nereus, unchangeable, olden,
And they lived in the youth of the world long ago, in the age called the golden,
But those times are long past, and their fancies must fade better matter of fact,
Their mythical lore is replaced now by learning precise and exact,
Mythology yields, and Zoology tells us in definite terms
That the Nereids no more are sea-nymphs, but chaetopod worms.
Nasty chaetopod annelids, armed with jaws, O alas and alack!
With protusible pharynx and tentacles slimy and gills on the back.
O misery, misery! Sad is the change from the fancies of old
To the facts and the figures of natural science, so dreary and cold.

(*Poem by Davis, 10 November 1883*)

However, in later years Davis did not himself like to be the butt of poetical humour:

But at a meeting of the Geological Society of America some years later, I saw him lose his temper. Davis was a short, portly, bald-headed man who wore chin whiskers. At a smoker, some of the more frolicsome of the geologists were improvising rhymes, taking off the idiosyncracies and peculiarities of the various men. Someone chimed in with the following couplet:

> D is for Davis with peneplaned crest
> His chin is prograded, and also his vest.

It should be explained that 'peneplaned' and 'prograded' were two of Davis' pet terms. Anyhow, he didn't appreciate the poetry. He spluttered. (*Gould, 1959, p. 133*)

Hugh Robert Mill, in his *Autobiography*, remembered the mature Davis as a more complex character:

William Morris Davis, Professor of Physical Geography at Harvard, was small, dark, alert and wiry. He was a hard man, with stern, logical mind, and he aroused

F I G. 44. Davis with a group of family and friends (*c.* 1900)
(Courtesy Harvard College Library)

great opposition by his dogmatic presentation of theory and his unusual and rather uncouth terminology. But he had depths of cryptic humour as he told impossible tales in a mirthless voice and with impassive face. (*Mill, 1951, p. 90*)

As a parent Davis seems to have been less successful than as a professor. He was clearly remembered by his eldest son when the latter was eighty but not, one feels, for reasons he would have chosen himself. Maybe it was because his sons were less studious than he or perhaps, as children often do, they reacted consciously or subconsciously against their father's seriousness.

My earliest memory (of my father) and the first word I can remember, I think, was 'don't' – 'don't interrupt'; when out walking, 'don't stop'; and, as the younger brothers grew up a little, 'don't tease'. He was always kindly and generous in his attitudes, and just in his decisions whether or not he would punish . . .

None of Davis' three sons had any academic or scientific learning – except the youngest who was interested in birds when young – but some of his talent has appeared in the second generation which contains one member of the Arctic Institute [the son of Richard Mott Davis] and two anthroplogists. (*Interview: R. Mott Davis with R. J. Chorley, 7 September 1962*)

More likely Davis made the same mistake as many other intellectual parents. In his anxiety for their welfare he sought to mould their natures by wordy exhortation and by what to him was subtly concealed moralization.

Another memory which marks a turning point in my personal mental attitude took place when my father and I were taking a Sunday walk, in the neighborhood, when I was post 10 or 11 years old. He was always a great story teller and had probably been relating some anecdote when all of a sudden he said, 'If you call a cow's tail a leg, how many legs has a cow?' Being a young liberal I naturally said five. 'No', said my father, 'calling a tail a leg does not make it a leg, the cow still has four legs.' Thus I became a conservative. I forgot in the early memories to mention my father's story-telling habit. When it was my bedtime he would come and lie down on the bed and tell me stories about imaginary boys. One villain in the piece was named Jacob who disobeyed his parents and ate green apples and became very sick. This was good propaganda since we were living at his parents home outside Philadelphia.

Later his stories were about a group of boys he called the Steady Boys. These boys always minded their parents and were helpful and played games in a friendly manner, much in the spirit of today's boy scouts. As the years passed his bedtime talks were on astronomy and at that time we lived in Cambridge at the foot of the hill where the Harvard Observatory was located. His entry into the Observatory, based on his previous relations there, made it possible for him to take me and point the small telescope at the moon, at Saturn, Jupiter, and various interesting situations among the constellations, and he gave me a book for children on the subject of astronomy, which I read at the time with interest. However, I have not pursued the subject since. (*Interview: R. Mott Davis with R. J. Chorley, 7 September 1962*)

Certainly he did not succeed with his eldest son, who candidly admitted that he had 'graduated in the pool room'; was poor at exams, had scraped an M.A. in accounting at Harvard; and, like his father, had in early life little idea of how to get along with people and little concept of human relations (Interview: R. Mott Davis with R. J. Chorley, 7 September 1962). Yet Davis was by no means a stern parent – he was not a Mr Barrett of Wimpole Street. He had a kind nature and took a strong interest in what his children did, but often failed unwittingly to strike a true spark of human kindness.

The Davis's lived opposite my family on Francis Avenue. My father was not alive, but Mrs Davis was a very great friend of my mother's. You have records, I assume, of the three sons, Mott, Nathaniel Burt, and Edward Mott. The youngest one of these, Edward, though somewhat younger, was a great friend of mine so that I saw a great deal of the whole family.

I remember Professor Davis as having a great sense of humor. He was always great fun to see when we met. His house was a three-story house; down stairs there was a living room, dining room and Professor Davis' study. I remember going in and seeing him there. He was sometimes playing Solitaire. I remember once he took exception to the comic valentines his boys sent to their friends, and he took the trouble to draw ones that they could send, having little poems and colored pictures. He was a very good draftsman. We boys were most interested in bird watching, and had a bird club to which we elected Professor Davis an honorary member. I cannot remember the circumstances, but I do remember that we got a very nice written acceptance of the honor we had conferred on him.

I do not know why, but he didn't like to have his boys playing billiards and pool in Harvard Square in the public billiard rooms, though as a matter of fact these were very respectable places, being mostly used by Harvard students. To keep the boys from this he built a pool room in a separate building in his backyard. I can't remember that we played in it very much, but it was more for the older boys than for my friend Edward. (*Letter: Edward M. Turner to R. J. Chorley, 18 April 1962; Boston, Mass.*)

His grandson's memory also recreates a picture of Davis' personality as very complex and many-sided:

There was nothing really warm about my grandfather, and although we enjoyed him very much, I suspect we were never really fond of him. In retrospect, it is likely that his reserve prevented him from being freely outgoing. The many things he did for us when we were small children, and the letters he wrote later, were done as if by the stern but indulgent patriarch. This is a little difficult to express.

Incidental anecdotes: he once wrote us from Tucson that he had just had a most enjoyable time talking to a high-school convocation about the time he had gone around Cape Horn in a square-rigger. My father's reaction was, 'I never knew he had gone around Cape Horn in a square-rigger!' I have the feeling that my father grew up in a relatively cool, reserved atmosphere. I know he said that whenever they (the boys) were sick, they were simply left alone in bed to entertain themselves; no one paid much attention to them...

A personality sketch of Davis ... Good Heavens, it would take a book. I remember him vividly, but have also heard a great deal about him from people like Kirk Bryan, under whom I studied physiography around 1940, and it is not always possible for me to disentangle my personal memories from what I have heard from others. In any case, it is certain that one of my grandfather's more evident traits was that he loved being the center of attention, the dominant figure in any group. I imagine this is one reason he liked so much to entertain children. In my early childhood we used to go in to Cambridge each Christmas for a big family party. Grandpa would make what he called 'spiderwebs' for us ... we would find hanging on the banister, clothespins with our names on them. The job was to take the clothespin and wind up the string it was tied to. This string would lead up the stairs, from room to room ... all over the house, until finally at the end would be a gift. I can imagine it must have been trying for Aunt Molly (the then wife) to have had all those strings winding all over the place! But of course we loved it. I recall winding my string and having to go all the way up into the attic; but I can no longer remember what the gift was ... Another thing he used to do was draw pictures for us, and I can remember how he used to switch the pencil from one hand to the other as he pleased. (I think I have been told that he was left-handed so taught himself to use his right hand.) These pictures (and I am speaking now of times later than the Christmas parties, on his rare trips east) were of special nature. One of the chief subjects was a dragon. It had the body of a long snake, the tail of a fish, two front legs with long talons on the feet, a head looking mostly like that of a cow but very fierce and breathing fire [see Fig. 94]. We used to copy, these things with loving care, and with a little practice I could probably draw one now. He also used to draw us elaborate mazes, occupying an entire $8\frac{1}{2} \times 11$ sheet; that is, we called them mazes; he always called them labyrinths. They were

extraordinary complex. My father used to do them for us too, and I do them for
my own sons.

   You may also have heard, perhaps from my brother, or from my uncle with
whom I understand you have also been in contact, of my grandfather's extraordinary facility at handicrafts. Give him some wood, wire, and cloth, and he would
make a working model of anything he pleased. He made my brother a derrick out
of two 1 × 1 pieces of wood; the upright was some 4 ft. high with a tack at the
bottom to stick into the floor and guy ropes to hold it erect; the boom was hinged
to it. The boom cord and the hook cord were wound on crank-drums made of
spools which were notched to receive spring-wire catches so that they acted as
ratchets. The whole thing was simple, ingenious, and worked beautifully. At the
end of his life he was working on a most elaborate model of Humpty Dumpty; it
was sent to me afterwards in case I wanted to finish it, but I never did. Apparently
he had done a skit at a party or some other occasion, in which he made a simple
half-egg with arm-holes; he put his head and shoulders in it and set the edge on a
'wall', and carried on a dialogue with a woman who played Alice. Whether this was
done for children or adults I don't know. In any case, it inspired him to do
something really elaborate, and the results were almost beyond belief. He was well
on the way towards making a Humpty Dumpty that operated entirely by pulling
strings. The egg part was about 3 ft. in diameter; the mechanism for making the
mouth smile was, as I recall, completed, as were the controls for making the
eyebrows (which were of cotton) rise and lower. One hand was complete, although
not covered with 'skin' each finger had three joints of wood, articulated to the
next with wire. A cord ran down the dorsal and ventral side of each, so that the
hand could be opened and closed. It was uncanny. (*Letter: E. Mott Davis Jr. to
R. J. Chorley, 13 January 1963; Austin, Texas*)

These kindly acts were, as we have noticed, too often marred by a semi-
deliberate – a truly Victorian – note of improvement which children are
quick to detect. Moreover it is quite obvious that Davis' duties at the
university filled much of his day and satisfied nearly all his interests. It was
natural that he should leave the management of the house and control of the
family to his wife.

   Generally he worked at the Museum at Harvard in the mornings, retired home for
   lunch and worked in his study at home for the rest of the day. He was available to
   his children for a time after supper. Davis collected stamps . . . [His wife] was a
   gentle person and Davis said that she softened his personality. (*Interview: R. Mott
   Davis with R. J. Chorley, 7 September 1962*)

However, with university studies occupying so much of his thoughts, he
must, without realizing it, have relied on his home as a place of retreat and
regeneration. His own childhood had probably made him dependent on a
home atmosphere and life to him was intolerable spiritually without a wife.
They had all-told three sons, Richard, Nathaniel and Edward, the last-named
being born in 1888 and, although Davis enjoyed the benefits of two more
marriages, he did not add to his family.

THE CYCLE MATURES

The 1890s witnessed the flowering, strengthening and ramification of the cyclic concept. During these years Davis' work and his attitude to its importance undergo a noticeable change. He ceases to be a persuader, a proposer of solutions to landform problems. He drops the role of the impartial researcher and shows clearly that in his mind the method of analysis and the solution to all problems lies in the trilogy of structure, process and stage. The cyclic sequence is the magic lamp which dispels darkness from all areas. It merely needs to be rubbed. And Davis rubbed it with a vigour and confidence reminiscent of the determination of his grandmother.

Moreover, his focusing distance seems to lengthen. Often, no longer content with areal studies, he advances on a broad front into generalizations. The geographical cycle or cycle of erosion was his first generalization but now the deductive habit spreads through most of his studies. Yet, as already noticed, at the same time he regards his ideas as representing an established school of thought which deserves constant defence and explanation. Criticisms of his cycle theory readily brought forth restatements of his views. In the following extract he admits that the cycle is an ideal concept used as an aid to illustration; yet he goes on to restate that this in no way denies the validity of the cyclic process. The insistence, towards the end of the statement, on the importance of intelligent guesswork cannot have been very reassuring to the professional purist.

> The geographer must generalize in order to bring the observable items within the reach of descriptive terms, and as soon as he generalizes, the use of idealized types is practically unavoidable. Such types have long been in current use, but they have been too few and too empirically defined for the best results. They need to be greatly increased in number, and at the same time they must be correlated with structure, process, and time; for only by following the path of nature's progress can we hope to store our minds with types that shall imitate nature's products. It may be fairly urged that the larger the store of types a geographer possesses, and the more careful and numerous the comparisons with nature by which the types have been rectified, the better progress can the geographer make in new fields of observation.
>
> But the geographer who adopts the explanatory methods in a whole-souled fashion will find himself called upon not only to imagine a large series of type forms; he must also call into exercise his deductive faculties and employ them to the fullest, if he would make the best progress in the newer phases of his subject, however purely inductive he had imagined it to be. In setting up a store of types there is need of deducing one type from another at every step; and it may be confidently urged that whoever hesitates to recognize this principle will fail in his efforts to describe through explanation. But as a matter of fact, geography has for some time been more deductive than geographers have supposed it to be; and the newer phase of the science is not characterized so much by introducing deduction for the first time, as by insisting on its whole-souled acceptance as an essential process in geographical research.

It is only by giving the fullest exercise to the faculties of imagination and deduction that the cycle of erosion becomes serviceable. Here the geographer who hesitates is lost . . .

Thus comparing the partial view of the landscape, as seen by the outer sight, with the complete view of the type as seen by his inner sight, [the geographer] determines, with great saving of time and effort, just where his next observations should be made in order to decide whether the ideal type he has provisionally selected fully agrees with the actual landscape before him. When the proper type is selected the observed landscape is concisely and effectively named in accordance with it; and description is thus greatly abbreviated.

Even the best surveys are necessarily sketched in great part; and the topographer must appreciate his subject before he can sketch it. He must have a clear insight into its expression; his outer eye must be supplemented by his inner eye . . . Let us therefore strive to complete a deductive geographical scheme . . . until it shall at last be ready to meet not only the actual variety of nature, but all the possible variety of nature. (*Davis, 1902E; quoted by Daly, 1945, pp. 274–5*)

Although there were early critics of the cyclic concept, most of them European (Tilley, 1968), criticism within the United States was surprisingly lacking. A notable exception to this was a paper by R. S. Tarr (1898) who, drawing on German work, attacked the peneplain concept, which Davis replied to in the following year (1899I). Tarr's points are effectively dealt with one by one:

1. That certain regions show no trace of peneplanation. – The concept does not apply to all uplands, some of which are monadnocks anyway.
2. The uplands of southern New England and northern New Jersey are not of uniform elevation. – This is the result of imperfect peneplanation followed by tilting during uplift.
3. The remains of certain peneplains are fragmentary. – What of it? Stratigraphic continuity is freely implied with much less area of outcrop.
4. Certain so-called peneplains are inclined. – 'It would be as extraordinary to find no slanting peneplains as to find no inclined strata' (Davis, 1899I, p. 213).
5. The asserted discordance between peneplain surfaces and rock structure is open to question. – The various resistant structures of the New Jersey Highlands *are* truncated.
6. Monadnock rocks are not more resistant than those surrounding them at lower elevations. – No tests have been made on this matter.
7. No peneplains are now found standing close to baselevel. – This is the natural result of recent disturbances of the earth's crust.
8. The crust will not stand still long enough for peneplanation to be completed. – The occurrence of upland plains shows that it must have done!

9. No part of the earth reveals even an approximation to a peneplain. – The unconformities in the geological record clearly point to peneplanation as a common feature of earth history.

Davis followed this paper with another defence of the peneplain concept in his 'Baselevel, grade and peneplain' (1902A) in which he was particularly concerned with terminology, and identified uplifted and dissected peneplains in the Scottish Highlands (1896H), as well as in central France and Brittany (1901K). Of all Davis' concepts this was the one which continued to excite the most scepticism:

> You are no doubt aware of the running feud between the elder Penck and Professor Davis. I was fortunate enough to take classes from each of them during the same term and to listen to some of their exchanges which were sometimes a little on the sharp side. For example, I remember Penck calling Davis 'Peneplain Davis' to the latter's face. (*Letter: H. F. Raup to R. J. Chorley, 5 January 1962; Kent, Ohio*)

Another very important and lasting feature of Davis' geomorphic work of this period was that on the development of river systems. Something of the philosophy underlying this was stated in his article on the 'Bearing of physiography on uniformitarianism' (1895I), which consisted of a plea for the study of the evolutionary aspects of physiography which he believed would enliven the subject (he uses the term 'geomorphology' on page 8) just as evolutionary studies had botany and zoology. All aspects of physiographic evolution, not least the migration of divides and river capture, were held to support the uniformitarian hypothesis. A significant feature of Davis' research at this time was his use of maps to interpret the evolutionary aspects of areas which he had not studied in the field; and his articles on the development of certain English rivers (1895E) and on the Seine, Meuse and

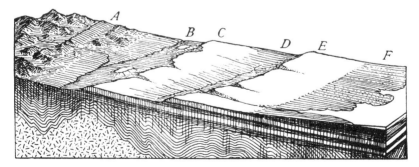

FIG. 45. Diagram of the ancient coastal plain of middle England between the Welsh oldland (A) and the young coastal lowland (F). The resistant ridges of the Cotswolds (C) and Chilterns (E) alternate with the lowlands of Worcester (B) and Oxford (D) (From Davis, 1898E, Fig. 87)

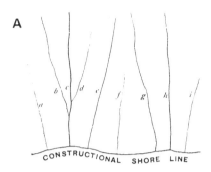

A

CONSTRUCTIONAL SHORE LINE

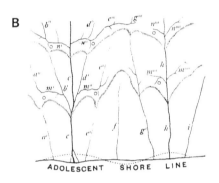

B

ADOLESCENT SHORE LINE

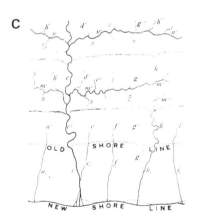

C

OLD SHORE LINE

NEW SHORE LINE

F I G. 46. The adjustment of rivers on a belted coastal plain
A. Consequent drainage
B. Later development after capture by subsequents, and the formation of obsequents
C. Final development after renewed uplift has resulted in the development of subsequents

(From Davis, 1895E, Figs. 1, 2 and 3)

Moselle (1896J) were prepared from the study of maps at scales of 1 : 126,720 and 1 : 80,000. Davis postulated that the rivers of central and south-east England are revived and mature successors of a well-adjusted consequent and subsequent drainage inherited from an earlier and far-advancing cycle of erosion (1895E, p. 128). In the first cycle the eastward-flowing consequents from the Welsh Uplands were captured by 'the patient progress of inorganic natural selection' of the extending subsequents, with their 'obsequent' tributaries (Davis coined this term on page 134), and finally a surface of faint relief was produced exhibiting considerable remants of maturely-adjusted drainage. Uplift instituted a second cycle, postulated Davis, in which this adjustment to structure, marked by extended subsequent drainage along the weak outcrops and more capture, was even more marked. Thus the main evidence for a two-cycle development lay in the bevelled escarpments of the Chalk and

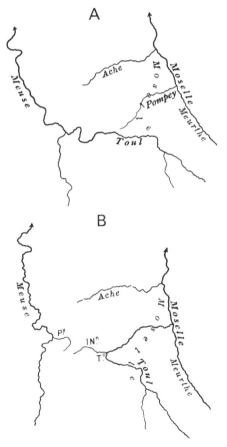

F I G. 47. Stages in the capture of the Toul by the Moselle
(From Davis, 1896J, Figs. 3 and 4)

Oolite (p. 138) and the proliferation of subsequent drainage (p. 139). His most important proposal, later adopted by many British authorities, was that a series of north-west- to south-east-flowing master rivers had been dismembered by the development of subsequent rivers like the lower Severn, Warwickshire Avon, Trent, Bedford Ouse and the Upper Thames. Similarly, Davis (1896J) identified the Seine, Meuse and Moselle as representing a second cycle of development. This view was supported in his mind not only by the incised meanders of these rivers and the supposed peneplain remnants but by the many evidences of capture by which the Seine and Moselle had grown at the expense of the Meuse. A series of captures were readily identified from maps, e.g. the Meuse tributary river Toul by a tributary of the Moselle [Fig. 47] and the Meuse tributary river Bar by a tributary of the Aisne (itself tributary to the Seine) [Fig. 48]. Davis pointed to the vigorous meanders of the augmented Seine and Moselle and contrasted them with the misfit river Meuse, saying that climatic change could not be involved because it would have affected all rivers equally, and suggesting vaguely that differ-

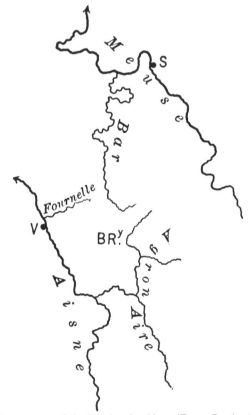

FIG. 48. The capture of the Bar by the Aisne (From Davis, 1896J, Fig. 5)

ential uplift had given increased efficiency to the Seine and Moselle. In 'The drainage of cuestas' (1899M) Davis reviewed his complete concept of the orderly development of drainage on uplifted coastal plains of homoclinal strata, including the development of cuestas by subsequent drainage extension, beheading and capture, and the migration of divides down-dip. He recapitulated his ideas regarding the French and English rivers, describing in detail the assumed misfit valleys of the Cotswolds (particularly the Evenlode) which, as we shall see in the next chapter, he had seen in the field for the first time in October 1898. Extending his thesis to the rivers of central Germany he was, however, mildly worried by the observation that, even above the assumed elbows of capture, many beheaded rivers appear to be misfit, pointing to 'some climatic change of external and obscure origin' (1899M, p. 93). Nevertheless, he concluded forthrightly:

> The systematic features repeatedly found in the drainage of cuestas point to a systematic order of development, and establish a well-defined class of geographical forms. Anyone who becomes persuaded of the correctness of the general scheme by which this class of forms is explained, will find therein a new instrument of research ... He will use his time to best advantage, by directing his steps at once to points where the most significant elements of form are expected, and if they are found, he will quickly and concisely describe them as members of a well-known class. (*Davis, 1899M, p. 93*)

A few years later (1903M) Davis was to describe another instance of 'the patient progress of inorganic natural selection' in the stream contest along the Blue Ridge which resulted in the westward migration of the divide and captures by the aggressive east-flowing rivers in North Carolina.

During the 1890s an interesting extension of the cycle of erosion concept was made as the result of Davis' interest in marine action. This included an important paper comparing marine and sub-aerial denudation (1896F), in which he reviewed the so-called marine planation (English) and peneplanation (American) schools of thought concerned with the production of features of low relief. Naturally he supported the more liberal American view which allowed marine planation but stated that 'My own studies lead me to believe that subaerial denudation has reduced various mountainous plateaulike uplifts to lowland peneplains' (1896F, p. 390). Adopting an *a posteriori* argument, from effect to cause, he reviewed the logical consequences of both types of planation (e.g. degree of drainage adjustment, form of the surface, etc.), tested these against known upland areas, and found it difficult to point to any secure upland plains of marine denudation. A more important outcome of his marine interests stemmed from an excursion with a Harvard Summer School to Cape Cod in July 1895 (1896G). In analysing the outline of Cape Cod he pictured it as a pro-glacial outwash plain modified by marine erosion at the southern end and by successive deposition at the northern end, the zones separated by a moving fulcrum. Davis took the opportunity to suggest a cycle of development of shore profiles (1896G, pp. 312–18) with young coasts, 'where the sea is able to do more work than it has to do', being worn back to a stage of adolescence, when the coast is straightened, when offshore bars have been pushed back to the shore and 'when there is an equality of work which the waves can do and the work to be done'. He drew the direct analogy between the profiles of graded rivers and those of post-adolescent graded shorelines, which continued throughout the rest of the marine cycle. The full importance of this embryonic shoreline cycle was only to be recognized as the result of researches by his students, notably by Gulliver (1899) and Johnson (1919).

FIELDWORK, ADMINISTRATION AND TEACHING

During his decade as Professor of Physical Geography Davis was untiring both in his own fieldwork and in conducting student field excursions, indeed, as we have seen, he made a practice of combining the two. For example, in the Autumn of 1895 he conducted a trip for New York Teachers' College to the Hudson River Highlands, where he remarked on the uplifted peneplain and noted the 150–200-foot bench in the Hudson Valley previously brought to his attention by Gilbert (letter dated 27 July 1891), which he ascribed to a stillstand during uplift (1895C). Barely a week later he took a party of Harvard students to the Connecticut Valley.

Davis was constantly travelling in connection with his vocation and undoubtedly without the promotion and increased salary he would have found travel much more difficult, as it seems that occasionally he had little spare money. Davis' father, though his fortune fluctuated more than once, died, leaving extensive and flourishing business interests. As Davis had always maintained excellent relations with his parents one would have expected some of this money to have come his way. While the details are uncertain, the following account affords some explanation why his inheritance was smaller than expected. It also demonstrates the high moral rectitude which Davis exercised in his work and in his private life.

> The death of his father in 1887 left a considerably involved estate of diverse holdings. The estate was put in order in the following fifteen years by the elder son Henry, upon whose death Henry's son Charles took over its management together with another cousin. It was decided that Charles should buy out the other two heirs Anna Hallowell and William Morris Davis, but serious differences arose within the family as to how much should be paid to them. The amount offered was $150,000, $75,000 apiece, but some family members felt that $100,000 apiece would be both fairer and feasible. A painful impasse resulted which was finally resolved by Davis who decided that, for the sake of family harmony, he would take $50,000 and let his sister's family have $100,000. (*Adapted from interview: R. Mott Davis by R. J. Chorley, 7 September 1962*)

It is known that Davis felt particularly badly regarding this family row, for he himself always placed family ties far above financial considerations. A grandson recalls that, despite his generosity to his sons, Davis always gave the impression of having 'to watch pennies fairly closely'. Certainly this incident can have done nothing to combat the general air of disillusion which, as we shall see, seemed to characterize his last years at Harvard prior to his premature retirement. It is interesting that the flamboyance of Edward Morris Davis, which found only fleeting manifestation in W. M. Davis, seems to have been transmitted through the elder brother Henry. The latter's son Charles, an early electrical engineer, business tycoon, candidate for the Republican Presidential nomination and honorary member of the Iroquois tribe, was the author of the proposal that a gigantic statue of Winston Churchill, cigar held aloft, should be erected on the shore of the English Channel after the Second World War!

Much of Davis' earlier geological research was on areas relatively near his home. He worked, as his many publications show, until the early 1890s on the Triassic formation of Connecticut and on contiguous or nearby regions. It was from the detailed surface observations made in these districts that he primarily evolved his cycle theory, and the mass of evidence uncovered in these studies went far to promote its acceptance. As we have already noticed, his last major contribution to the geology of this area was 'The Triassic formation of Connecticut' (1898G), published after the usual Survey delay.

After 1890 Davis extended his direct observations to more distant terrains, and it was during his visit to the Rockies with the International Geological Congress in 1891 that he first became interested in the origin of the Tertiary freshwater formations of the region. In the summer of 1897 he travelled west with Albrecht Penck and took the opportunity of examining the features of the sedimentary petrology of the freshwater deposits and to compare them with Penck's experience of the European molasse rocks. Concentrating particularly on the Arapahoe and Denver formations, he ascribed them to piedmont fluvial processes (1897I) which produced laterally-confluent alluvial fans on subsiding foundations (1900N). Davis could find little evidence for the proposed lacustrine origin of these deposits but suggested that the problem might be attacked by applying the deductive approach.

> The consequences of the theories of lacustrine and fluvial deposition should be confronted with the assembled and generalised facts that have been determined by observation of the western Tertiaries, in order to determine how the latter are best explained. (*Davis, 1900F, p. 600*)

On a visit to the classic Tertiary formation of Green River, Wyoming, in the summer of 1902 he concluded that even these lake beds were made up of strata deposited in many successive and fluctuating lakes of moderate area combined with the deposits of numerous aggrading streams (1903D). As Brigham concluded, his treatment was such:

> for the reason that the observational side of the problem had already been well set forth by various geologists, and therefore needed no repetition, while the deductive side had been almost entirely neglected. The fluviatile plains of California, northern Italy and northern India, were cited in evidence of the constructive efficiency of rivers, a process that had been too little considered. The paper certainly had a share in directing attention to a new interpretation, now very generally accepted. (*Brigham, 1909, p. 14*)

Brigham also commented on the three-weeks' trip to the Colorado Canyon which Davis organized in June 1900 by wagons and pack train from Flagstaff, Arizona, northwards to Milford, Utah. This huge fluvial chasm became almost his favourite lecturing topic in later years and incited a constant flow of lectures and publications.

> In 1900 Davis organized a party, including Gregory of Yale University, Dodge of Columbia University, Barrett of Chicago and Anderson of York, England, to visit the region of the Grand Canyon of the Colorado in northern Arizona. The excursion was extremely profitable, and led to the publication of two articles (*Amer. Journ. Sci.*, 1900; *Bull. Mus. Comp. Zool.*, 1901) in which several modifications of the geological history of the region, as interpreted by Dutton, were announced. The chief of these concern the date of faulting and the origin of certain valleys: the great escarpment known as 'Hurricane ledge', previously explained as immediately due to faulting, was shown to result from renewed erosion of a faulted and baseleveled mass; and several valleys, which had been explained as the work of antecedent rivers, were shown to be due to subsequent rivers. In reaching these results the deductive complement to the inductive side of the work was essential. A number of huge landslides, not previously noticed, were described and figured. The Grand Canyon was revisited in 1901, 1902, and 1904.
> (*Brigham, 1909, p. 14*)

Davis was particularly concerned with the erosional history of the canyons and the light shed on this by the two major unconformities, one truncating the basal schists and the other underlying the Palaeozoic rocks. Contrasting the planar features of the former (possibly marine) surface with the irregular (peneplain?) unconformity at the base of the Palaeozoics, Davis reviewed the previous literature, particularly the work of Clarence Dutton. The plateau surface, despite its structural accordance, was interpreted as a peneplain the dissection of which by the antecedent drainage during the present canyon cycle was initiated by early Tertiary uplift (1900K). The features of the canyon walls, particularly the Esplanade he thought to be mainly structural, rather than representing significant stillstands during uplift. Davis maintained his attitude of scepticism to climatic change by rejecting the suggestion that certain valleys (e.g. Toroweap) were cut in more humid times and interpreted the whole topography as having been formed under distinctly arid conditions characterized by occasional thunderstorm activity.

The fact that Davis had published an article on the canyon in 1892 (1892O) – extolling the efficiency of running water, discussing the geomorphic significance of unconformities and exhorting students to piece together the sequence of field evidence without being only concerned with absolute geological data – draws attention to a very significant feature of his technique: his habit of initially interpreting landforms from maps without having seen the terrain. By his cyclical method of analysis he was able to provide a composite explanation of all the associated landforms in so simple a way that it was readily comprehensible even to persons with little knowledge of geography and geology. He could do this solely from the exhaustive consultation of large-scale maps, a technique often derided as little better than charlatanism but which has been, and still is, highly praised by eminent geographers for the excellent results it can achieve. Albert de Lapparent, a famous French geologist, in an article entitled 'L'art de lire les Cartes Géographiques' (*Revue Scient.*, 1896) writes:

> Savez-vous, en effet, qui a eu le mérite de mettre ces ingénieux aperçus en lumière? C'est un savant américain, M. Morris Davis, le même qui, depuis sept ou huit ans, a donné aux Etats Unis une si vigoureuse impulsion aux études de ce genre. De l'autre côté de l'Atlantique, il examinait curieusement nos cartes de l'état major, s'efforçant d'y appliquer la pénétrante analyse qu'il avait employée avec tant de succès a l'étude des réseaux hydrographiques du New Jersey et de la Pensylvanie. L'allure si particulière de la Bar lui semblait tellement caractéristique qu'il n'avait pas hésité à en déduire, sur la seule inspection de la carte, la série des conséquences dont je vous ai entretenu. Venu en Europe en 1894, il s'est donné le plaisir de vérifier sur place le bien fondé de ses conclusions, tout heureux, lui Américain, d'apprendre à ses collègues de France ce qu'ils avaient ignoré jusqu'alors, faute d'une clef pour déchiffrer couramment les hiéroglyphes de la géographie. (*Quoted in Brigham, 1909, p. 28*)

Indeed, in his research and teaching maps and models played a very significant role. He extolled the value of large-scale maps as geographical illustrations (1896E), particularly maps on a scale of about 1 : 100,000 for physiographic work. Reviewing American and foreign maps, he described the map evidence relating to the geography of certain characteristic regions (e.g. the Scottish lowlands and Champagne), pointing to features both of physiographic history and human responses to the terrain. Davis showed how he had been able to interpret the relationships of the French rivers Bar and Aire from maps (1896E, pp. 503–5). In his use of models, Davis praised their value over maps of poor quality in the teaching of physiography by illustrating sequences of landforms, and called for each class of landforms to be accurately described and illustrated by models, as were botanical forms (1889H). In 1897 (1897F) he described the construction of a set of relief models during the winter of 1895–6 and how they could be used to give rational explanations for landforms, as well as helping to introduce notions of denudation chronology, physical controls over human occupancy of the earth, and even some history!

Davis' trip to the Grand Canyon followed closely upon his promotion to the Sturgis–Hooper Professorship in Geology at Harvard in 1899. This appointment, as with many senior posts, required more than mere professional ability. Harvard President Eliot perhaps sensed this potential weakness in Davis' character for he was certainly trying to anticipate a possible clash of personalities when he wrote:

> Dear Mr. Davis,
>
> You will enter on an inheritance of some troubles when you assume the Sturgis–Hooper Professorship. There is a strong interest on the part of Mr. Agassiz in the status and work of the Sturgis–Hooper Professor; and on the other hand there has been for years a great deal of friction between him and Professor Shaler, and the irritation is mutual. Mr. Agassiz has nominally resigned the Curatorship of the Museum; but of course his views about its management must have great weight as long as he lives. I doubt if we can discuss the situation to much profit by letter; but when you come home let me talk with you before you talk with Mr. Agassiz.
>
> I enclose a copy of a letter which I received today from him. It was written after reading your letter of May 10th to me from Munich.
>
> Mr. Agassiz no longer lives in Cambridge for any part of the year. He is now in Newport, and is said to wish to buy the Edward Austin house on Beacon Street, Boston. He is soon going to Australia for another expedition to the barrier reef.
>
> You will see that he wishes very much that you could immediately give up the elementary instruction, and I confess that I think that would be a prudent measure on your part. I am glad he thinks well of Daly.
>
> As to the double title of your professorship, I see no chance of bringing that about. Mr. Agassiz and the descendants of Mr. Samuel Hooper will surely object to any mingling of the title of his professorship with another, particularly as the whole salary will be paid from the Sturgis–Hooper Fund.
>
> I hope you will continue to enjoy your European experiences.
>
> Very truly yours, Charles W. Eliot
> *(Letter: Charles W. Eliot to Davis, 26 May 1899; Cambridge, Mass.)*

We have evidence that even Davis' faculty friends found him 'difficult' and his own son comments on this side of his nature. Many of Davis' difficulties came from his high principles which prevented him from compromising on anything. To him a fact was either true or incorrect; there could be no intermediate condition.

> Davis played 'solitaire' but was no good at bridge on account of his lack of any 'partner relationship' – the most unfortunate trait of his personality. In the name of 'truth' he would attack his scientific associates savagely and lost many friends on this account. He did not learn to get along with people and to organise until late in life. . . .

One feature of his [Davis'] character and attitude was, I believe, based primarily on his grandmother's saying 'Truth for authority, not authority for truth'. It is my personal opinion that in human relations tact outweighs truth, but I do not believe that this ever occurred to him unless it was in his later life. It is entirely possible that adherence to truth for authority may have occasioned my mother some hard times, but if this was so no evidence of it was ever demonstrated for the children. (*Interview: R. Mott Davis with R. J. Chorley, 7 September 1962*)

Whether knowledge can be as positive as Davis believed is doubtful but it seems certain that in everyday dealing with persons, expedient concessions are essential if substantial and timely progress is to be made. We have no further record of how successful Eliot's manoeuvres were but the absence of reports of friction with Agassiz may indicate some success. The letter shows that the professorship was granted to Davis while he was in Europe enjoying a sabbatical year. This journey and that of 1894 and the numerous travels abroad after 1900 are described in following chapters. There were, for example, many trips to Europe (twice as an exchange professor), two visits to Mexico and an international meeting of scientists which provided him with the chance of seeing parts of north, central and southern Africa. In a younger man confrontation with these new environments might have evoked a whole resurgence of new ideas. For Davis they produced no new images; in each fresh environment he merely saw his cycle confirmed; each successive sample of foreign relief proved a justification of what he had taught himself to expect from his American experience. Those must have been happy years for him. Light-hearted verse written in honour of a colleague's forty-fifth birthday seems to echo some of this contentment.

> Today the merry earth
> > Vibrates with very mirth
> > > Its equatorial girth.
>
> It crosses here the spot,
> > Marked with a nodal knot,
> > > On its perennial trot.
>
> Twist Crab and Capricorn
> > Where on a winter morn
> > > Sweet little Eph was born.
>
> Since then the earth has run
> > Its course around the sun
> > > Davidian times – less one.
>
> Now on the mundane stage
> > Eph plays the part of sage
> > > In hearty middle age.
>
> And makes appropriate wages
> > By reading ancient pages
> > > About the Middle Ages.

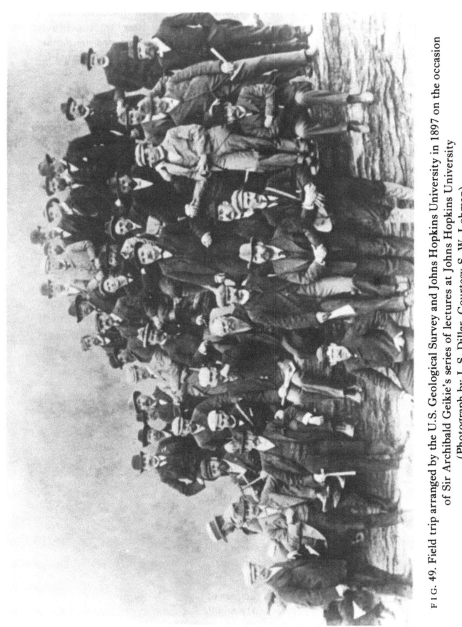

FIG. 49. Field trip arranged by the U.S. Geological Survey and Johns Hopkins University in 1897 on the occasion of Sir Archibald Geikie's series of lectures at Johns Hopkins University (Photograph by J. S. Diller. Courtesy S. W. Lohman)

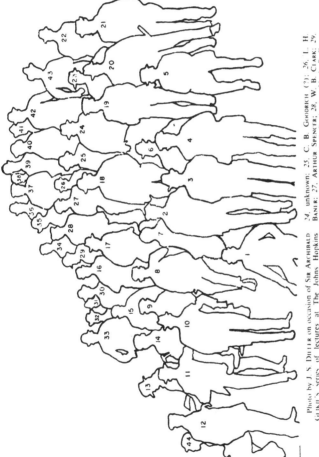

Photo by J. S. Diller on occasion of Sir Archibald Geikie's series of lectures at The Johns Hopkins University, Baltimore, Md. 1897. Taken on field trip arranged by U.S. Geological Survey and Geology Department of Johns Hopkins. 1, J. Stanley Brown; 2, J. A. Holmes; 3, C. W. Hayes; 4, I. C. Glenn; 5, H. S. Williams; 6, F. J. H. Merrill; 7, J. W. Powell; 8, C. D. Walcott; 9, unknown; 10, E. B. Mathews; 11, G. O. Smith; 12, G. F. Becker; 13, S. F. Emmons; 14, J. F. Kemp; 15, Bailey Willis; 16, H. Reid; 17, Robert Hill; 18, J. W. Spencer; 19, Thomas Watson (?); 20, Cleveland Abbe, Jr.; 21, C. R. Van Hise; 22, H. F. Reid; 23, G. W. Stose; 24, unknown; 25, C. B. Goodrich (?); 26, I. H. Banir; 27, Arthur Spencer; 28, W. B. Clark; 29, F. H. Knowlton; 30, F. D. Adams; 31, A. P. Coleman; 32, T. W. Stanton; 33, unknown; 34, Rufus M. Bagg; 35, W. J. McGee; 36, C. C. O'Harra; 17, Sir Archibald Geikie; 38, W. M. Davis; 39, H. B. Kummel; 40, G. B. Shattuck; 41, R. D. Salisbury; 42, A. C. Veatch; 43, L. M. Prindle; and 44, unknown. Identifications by Bailey Willis, N. H. Darton, A. C. Spencer, A. C. Lane, and T. W. Stanton, Nov.-Dec., 1947. Photo loaned by S. W. Lohman.

FIG. 50. Key for Fig. 49. Identifications by Bailey Willis, N. H. Darton, A. C. Spencer, A. C. Lane and T. W. Stanton made November–December 1947

Tis true, no toe fantastic,
    Likewise no feat gymnastic
        Maintain his joints elastic;

For simpletons athletic
    He's not apologetic
        But quite unsympathetic.

To rest on holidays
    He does not cathode rays,
        But chickens – says it pays.

May his hens forever lay;
    May his books forever pay;
        May his friends increase each day.

And when the earth comes back
    To this spot upon its track,
        No joy may Ephraim lack.

(Poem by Davis to Ephraim Emerton, dated 18 February, 1896)

Davis was now at the top of his profession and one of the leading geographers in the United States. Outside America his renown was already spreading fast. His early papers on the nature of rivers and the cycle had called forth a note by Emmanuel de Margerie on the 'Evolution of geographic forms' (*Nouvelles Géographiques*, 1892) which closes as follows:

Telle est, dans ses grands traits, la méthode d'analyse proposée par M. Davis; elle permet de classer les formes géographiques en leur attribuant à nous fournit un puissant moyen d'investigation pour reconstituer, dans chaque cas particulier, l'histoire des surfaces emergées – histoire sur laquelle l'ancienne géologie, laissant le sol au moment où il sort du sein des eaux, restait muette d'ordinaire. Enfin, et surtout, au point de vue philosophique, elle représente un progrès réel dans la compréhension de l'Univers, une substitution de plus a enregistrer du déterminisme rationnel à l'empirisme aveugle. (*Quoted in Brigham, 1909, p. 24*)

After 1899 many of Davis' articles appeared in foreign languages and he already had personal acquaintance with numerous geographers abroad. The next two letters from Emmanuel de Martonne indicate that the great French physical geographer was virtually following Davis' method of systematic classification, though whether consciously or otherwise is not immediately apparent.

I should like to show you the last chapter of the geomorphological part of my book on Physical Geography. It is something new – I think – but a little bold: a general classification of all types of countries according to the relief including Africa, Asia, and Australia.
    I was greatly helped in this work by the experience that I got for a few years of African geography, in reading many books of travels for studies of human

geography. One of the greatest difficulties was to find general terms for every type of country according to its morphological features and the genesis of these.

I regret now that this chapter was not finished when you asked for communications. I should have liked to bring this essay before the Geographical Congress. But it would have been very difficult to explain the whole thing in the time accorded for the communications – I used three lessons to expose that to my students.

Please excuse this awful long letter. It shall be likely the last one that you will receive before I have the pleasure of talking with you. (*Letter: E. de Martonne to Davis, undated; Laboratoire de Géographie, Université de Rennes, France*)

In my last letter I spoke of American Coldness in scientific works, and promised to explain my thoughts on this subject in another letter. I will not say that it is a peculiar quality (or defect) of American writers; but it seems to be really more common among American geographers and geologists than among French or German ones. Physical Geography and Geology are descriptive and historic sciences, the study of which involves a large amount of hypothesis and of suppositions, for we do not know all that we have to describe and very rarely are able to directly observe the physical process that we imagine to explain the evolution, only the result of which we can see. Do you not think that could be considered as a good reason for the fact that Physical Geography progressed more quickly in America than in Germany or in France, although it has been studied for a longer time by us?

Boldness in scientific works must proceed for the same facts as boldness in practical life. And – as far as I can judge – it is not so much in connection with the condition of natural environment – as it was very well emphasized by Ratzel in his book on the Anthropogeography of the U.S. of Am. as with the past social life.

We are old civilized peoples, enriched, but sometimes hampered, by experiences and the remembrance of the various and often bad effects of too much self-confidence.

We think a *quite new* idea cannot be absolutely exact; and indeed I do not know *one* example to contradict this belief in the whole history of sciences. Even the theories that seem to have been entirely new, and are considered as having brought the scientific evolution in quite new ways, proceeded in reality from other theories either too exaggerated or mixed with various errors. The germs of the most original ideas – *if they are true* – can always be found in other ideas. There is no '*generation spontanie*' either for living creatures or for ideas.

You said once to me you were a revolutionary in Geography. Let me say I do not think it is quite true. I was very struck by the way in which you began your reply to the attacks of Tarr against the peneplain, and showed that the germ of this idea could be found in other writers. (*Letter: E. de Martonne to Davis, 28 July 1904; Rennes, France*)

Unlike his ideas, the range of Davis' articles has not weakened except on climatic topics. There are still several detailed studies of particular regions within the United States and travel abroad has extended such studies to foreign areas. There are also articles where the treatment is broader, and these include the discussions on baselevel, peneplains, the erosion cycle and glacial phenomena. Interspersed are other articles and books on geographic teaching methods for the guidance of those presenting the subjects in schools. We will return later to all of these aspects.

Administration must have claimed some of Davis' time. His position as professor inevitably involved him in periodic committee meetings with senior colleagues of other faculties and in negotiations with the College President. How he fared in this type of situation may be glimpsed from lines of his poetry:

The Dutiful Chairman

All of the story of dutiful Ed'n majority Chairman;
Ed'n the Chairman, chosen in Smith's sabbatical year-off;
Call the suff'ring committee every Wednesday to hear a new project:
Boldly replying when President Eliot suddenly asked him
Why he had voted in this way or that on somebody's motion;
Calmly serene while Goodwin tried with irrelevant urging
Still to have Greek, man's soul saver, counted at more than its value;
Unmoved when our too stout James Mills, Faculty spondee,
Recklessly urged us to grant boys almost absolute freedom;
Scared a bit possibly when for a few days Jagemand Grandgent
Led off some of us under the spell of insidious Plan B;
Steady again when Plan B fell by the vote of its best friend.

Although vague and oblique they tell us what he thought of Greek, and incidentally reveal that a problem of student discipline existed then. The never enviable task of grading students is wryly described in further lines of typical Davisian humour.

Marks at Harvard and Remarks in the Shop Club

Return your marks in grades from A to E;
Five even grades: – So votes the Facultee
Regardless of the vote, we ease our minds
By giving A's of several different kinds.
We grant the heavy grind of minus B;
We cheer the dullard with a Radcliffe C;
Drive to despair the late-repenting cuss
By giving him an almost-pass E plus;
And e'en the Corporation sins in some degree
By giving Briggs, our Briggs, an LL.D.

Like many great men Davis was fortunate enough to possess a physical capacity that matched his mental exertions. In spite of the time involved in extra-mural activities, he did not allow himself to withdraw from active lecturing. Instead he made the two possible by restricting his attentions to the senior groups.

Two masterly, advanced courses in physical geography, one on the United States and the other on Europe, claimed the unfading admiration of those who listened. Illustrated with a host of large-scale topographic maps of United States and European countries, these lectures showed the solid worth of Davis' philosophy, though in scholarly fashion he gave full weight to the opinions and methods of other investigators on the two continents. Probably because of the difficulty of adequately reproducing the maps around which the discussion centered, the material of these unique lectures was never published. To spread his gospel Davis relied chiefly on what he used to call 'the rapid-fire gun', propagandizing with hundreds of papers a number of which were written in French and German, and printed in Europe. (*Daly, 1945, p. 277*)

Between approximately 1890 and 1900 most of his academic life was concerned with delivering lectures to undergraduates. After 1900 his work was more to graduate students with less time for lectures and more time for research and writing. (*Interview: R. Mott Davis with R. J. Chorley, 7 September 1962*)

THE CONTENT AND TEACHING OF PHYSICAL GEOGRAPHY

The 1890s witnessed a flowering of Davis' concern to define the subject matter and treatment of 'physical geography', and to establish the manner in which it could be taught at all levels. In defining the discipline his three major problems were, firstly, to distinguish physical geography from geology; secondly, to demonstrate that, nevertheless, physical geography was still a science in the true sense and, thirdly, to relate physical geography to developments in human geography.

Davis defined physical geography as the physical, rational, explanatory study of those existing features of the earth that enter into the relation between the earth and life in general, and man in particular (1902B, p. 18); as contrasted with physical geology which is concerned with the processes whereby changes in the earth have been brought about with the progress of time. Defining geology as being composed of a series of past geographies, he followed Mackinder's dictum that: 'geology is the study of the past in the light of the present, while geography is the study of the present in the light of the past' (1902B, p. 18). Many scholars felt that Davis went too far in divorcing physical geography from geology, as the following letter from R. S. Tarr shows:

I wish to write, in a most friendly way, a protest against your notice of my First Book in a recent number of *Science*. I have the feeling which I have always had, since I first talked with you about the matter, that you do not make sufficient allowance for the possibility of a difference of view, which may perhaps be correct. There is a good pedagogical reason for introducing astronomical facts into a physical geography which cover the general field. It is an introduction in science, and when the earth is being considered it seems not out of place, to say the least, to give some idea what the earth is. Since you first criticised this point some years ago, I have talked with many teachers about it, and find most of them in favor of the plan which I have followed.

As for the question of process and form, I feel even more strongly that I am right. I first felt the need of a preliminary knowledge of process when I attended your lecture in 1882, and later when I knew people who took your work without any geology. It has become even more impressed upon me as a teacher, and beginning with next year I make my course in geology a *required* course for *all* who would take my physiography. I have been firm on it because I feel certain that without some knowledge of geology most college students cannot grasp the points of physiography as fully as they should. I have come to this view by watching the difference in result between those who have taken geology and those who have not. Form is the result of process, and without appreciating the cause I find that student's cannot readily understand the result. This is even more true of immature students. You have very positively stated that this is a wrong view, and the weight of your authority will have effect. *I* believe on the basis of my experiences that you are entirely wrong. Should I not be allowed to follow my conviction, after full thought, without being convicted of a failure to clearly recognise 'what is essential and what is unessential in a physical geography'? The conviction comes without apparently giving me the least opportunity to hold a view of my own. You say that beginners 'cannot really appreciate' my geological chapter; but in reality, as I know from experience in High School classes, they come very much nearer appreciating this chapter than they do that of rivers even though that is shorn of some of its most difficult features.

As a result of my experience I have come in the main to place very little dependence on reviews, and now take no notice of them usually. I used to feel somewhat dismayed when I found myself hammered somewhat unmercifully; but I have now come to believe that I may be not so far in the wrong as might be supposed. I am going to go on, for I have the encouragement of the pupils and teachers, and I believe I am doing good and can do still more. I believe that a book that can get into the hands of 25,000 people in less than four months from the date of publication, notwithstanding the presence of another book on the same subject in the hands of 10,000 more, deserves more praise than that 'In more respects it presents a good view of the subject' while practically all the rest is blame.

I know that you are very busy and I do not write this with any idea of calling upon your time for a reply. I do wish to let you know that I cannot accept the review as a merited criticism. You will, I am sure, make allowances for the author's sensitiveness, which I have not entirely overcome, and accept that as an excuse for this letter, which is intended in the most friendly manner. (*Letter: R. S. Tarr to Davis, 10 December 1897; Ithaca, New York*)

While the first part of the letter is a genuine explanation of Tarr's point of view, his anger at Davis' criticism causes him to abandon his professional reserve and degenerate into boastfulness towards the end. Yet his argument on the desirability of geological knowledge seems most valid in hindsight.

It is natural that Davis rationalized physical geography in a scientific sense by concentrating on physiography and on its deductive treatment within the framework of the cycle. He quotes Guyot: 'To describe without rising to the causes or descending to the consequences is no more science than merely or simply to relate a fact of which one has been a witness' (1900L, p. 392). Pointing out that, often, geological teaching does not measure up to this requirement, he proposed that research should consist of these steps: observation and description, classification and generalization, invention and theory building (1897K). He noted that the 'outer eye' must be supplemented by an 'inner eye':

> The more complete the mental scheme by which an ideal system of topography forms is rationally explained, the more clearly can the physical eye perceive the actual features of the land surface ... (*Davis, 1894E, p. 68*)

> Further illustration of the growing recognition of form as the chief object of the physiographic study of the lands is seen in the use of the term, 'geomorphology', by some American writers; but more important than the term is the principle which underlies it. This is the acceptance of theorizing as an essential part of investigation in geography, just as in other sciences. All explanation involves theorizing. When theory is taken piecemeal and applied only to elementary problems, such as the origin of deltas, it does not excite unfavorable comment among geographers. But when the explanation of more complicated factures is attempted, and when a comprehensive scheme of classification and treatment, in which theorizing is fully and frankly recognized, is evolved for all land forms, then the conservatives recoil, as if so bold a proposition would set them adrift on the dangerous sea of unrestrained imagination. They forget that the harbor of explanation can only be reached by crossing the seas of theory. They are willing to cruise, like the early navigators, the empirical explorers, only close along shore; not venturing to trust themselves out of sight of the land of existing fact; but they have not learned to embark upon the open ocean of investigation, trusting to the compass of logical deduction and the rudder of critical judgment to lead them to the desired haven of understanding of facts of the past. (*Davis, 1900O, p. 168*)

> ... geography ... [should] include not only a descriptive and statistical account of the present surface of the earth, but also a systematic classification of the features of the earth's surface, viewed as the result of certain processes, acting for various periods, at different ages, on divers structures. (*Davis, 1889C, p. 11*)

> The phases of (topographic) growth are as distinct as inorganic forms ... We do not believe that an oak grows from an acorn from seeing the full growth accomplished while waiting for the evidence of the fact, but because partly by analogy with plants of quicker development, partly by the sight of oaks of different ages, we are convinced of a change that we seldom wait to see. It is the same with geographic forms. (*Davis, 1889C, p. 16*)

As far as the relationship between physical and human geography was concerned, his views are distinctly deterministic by modern standards, stemming from those of Ritter:

> ... geography as a mature subject is capable of a higher development than it has yet reached. In this connection it will be well to review briefly the three stages of development recognizable in the progress of our venerable subject. Until within

about a hundred years the content of geography consisted of a body of uncor-
related facts concerning the earth and its inhabitants. The facts were described
empirically and as a rule very imperfectly. Their location was noted, but their
correlations were overlooked; it had not indeed been clearly made out that
correlations existed. This blindly inductive first stage was followed by a second
stage which was opened by Ritter's exposition of the relationship between the
earth and its inhabitants; ... such relationships as were noted had to be explained
on the old doctrine of teleology – the adaptation of the earth to man – instead of
on the modern principle of evolution, – the adaptation of all the earth's inhabitants to the earth. It is this principle which characterizes the third stage of
progress and along with it goes a principle of almost equal importance; namely,
that all the items which enter into the relation between the earth and its inhabitants must be explained as well as described, because explanation aids so power-
fully in observing and appreciating the facts of nature ...

Geography has today entered well upon its third stage of progress. The 'causal
notion' is generally admitted to be essential in the study ... Thus understood,
geography involves the knowledge of two great classes of facts; first, all those
facts of inorganic environment which enter into relationship with the earth's
inhabitants; second, all those responses by which the inhabitants, from the lowest
to the highest, have adjusted themselves to their environment. The first of these
classes has long been studied as physical geography, although this name has
been used as a cover for many irrelevant topics. In recent years there has been a
tendency to compress the name into the single word 'physiography'.

The second of the two classes of facts has not yet reached the point of being
named, but perhaps it may come to be called ontography. Ecology, to which
increasing attention is given by biologists is closely related to what I here call
ontography, yet there is a distinction between the two, in that ecology is concerned largely with the individual organism, while ontography is intended to
include all pertinent facts in structure, physiography, individual and species.

Neither physiography nor ontography alone is geography proper, for geography
involves the relation in which the elements of its two components stand to each
other. Each of the components must be well developed before geography can be
taken up as a mature study. (*Davis, 1902C and 1909C, pp. 32–4*)

He pointed out that the study of plants and man should be introduced into
physical geography after that of the physical background, on which the
former depend (1900L, p. 393). The following passage on the analysis of
human settlement gives very much the flavour of his approach:

In geographical problems, the location of settlements and the growth of cities at
the head of navigation on rivers are repeatedly noted. By putting together all
relevant facts it comes to be believed, in explanation of this habit of man, that
settlements have been made at such localities because of certain advantages that
they have over others; that cities have there grown up because of the persistence
of these advantages. Yet this belief is based only on an explanation of extreme
probability, for we cannot possibly secure direct evidence of the reasons that led
one individual after another to live in these cities instead of going elsewhere.

When we see the scanty population of a cliffed seacoast, it is explained as a result of the difficulty of maintaining maritime activities on so inhospitable a shore. The sufficiency of the demonstration in this elementary problem will depend largely on the fullness of knowledge in the mind of teacher and student regarding the facts of geographical environment, and regarding the habits of mankind. One who is poorly equipped with this preparatory knowledge may contend that the explanations offered are insufficient, and that other explanations may be equally probable; but no hesitation will be felt by another who has walked along the crest of a sea cliff, and appreciated the consequences of its form, and who has at the same time learned from observation of many races in many regions how incessantly the habits of men and their manner of living are determined by the advantages that they bring; to him the explanation is sufficient. (*Davis, 1897K, pp. 437–8*)

Davis was untiring in his lecturing and writing on the teaching of physical geography at all levels, bringing to bear a professionalism and an ethical quality of the highest order.

The first excursion that I commonly take with a class leads us to an old quarry in Somerville, near the Agassiz Museum in Cambridge, where a large dike, some forty feet wide, cuts across the beds of the Somerville slates. There may be ten or twenty students in the party, and it should be remembered that they have had a preliminary course in elementary geology, in which nearly all the terms that we have occasion to use have been defined: they are also provided with hammers, compasses, clinometers, note-books, and outline maps of the district. On entering the quarry, I select two fragments of rock: one exhibits a fine, granular texture, with bands of alternating color, and is shown as the type of a bedded, stratified, or aqueous rock; the other is of crystalline texture, without arrangement in layers, and represents the groups of massive, crystalline, or igneous rocks; and without further explanation than this, the students are asked to search out the area occupied by each rock, the line of contact and the phenomena exhibited along it, and to determine the relations of the two and the sequence of events in their history. Emphasis is given to the importance of personal work, and I take pains to say how much more valuable is the ability to determine the facts than the facts themselves ... I therefore throw them at once on their difficulties; my own endeavor being rather to suggest observations and give encouragement than to answer questions. The questions are to be answered by the rocks. Even the best students are almost helpless at first, – so little has their general education taught them of independent, original observation even of a simple kind ... A student will sometimes come up to me, after a very insufficient search for facts on the ground, and, presenting a piece of the dike, say, with the idea that he is doing his full duty, 'Isn't this melaphyr?' Now, as a matter of fact, he is, in this particular case, quite right; at least, so I am assured by competent lithologists, and later on he should be told so, but not at first. He has perhaps heard the name, melaphyr, associated with dark-colored, fine-grained rocks, and makes a lucky venture in using it; but as the object of the work is to train his observation, not mine, I throw the burden of proof upon him by asking in return, 'Why do you think so?' 'It looks like it.' 'What is melaphyr?' This may sound, as I now read it, very much like snubbing a

F I G. 51. The Harvard University geological laboratory (c. 1900)

(From T. A. Jaggar, *My Experiments with Volcanoes*; Honolulu, 1956, Fig. 1)

praiseworthy inquiry; but see the results. Nine times out of ten the students say, 'Oh, melaphyr? Melaphyr is – I don't remember'; and this clears away the false knowledge that places names uppermost, and brings us down to a solid foundation for good work. No words must be used that cannot be defined: no suggestions must be made that cannot be justified by actual observation. The question may now be returned, 'If you cannot say that this is melaphyr, what can you say about it?' 'Well it is a dike.' 'Why do you think so?' 'It looks like one.' 'What is a dike?' 'A dike is a mass of igneous rock filling a fracture in the country-rock.' 'Does this igneous rock fill such a fracture?' Again it appears how much easier it is to make assertions than to defend them; it is very seldom that a student will on his first endeavor suggest and apply the tests that furnish him with answers to such questions as these. He must make a second and third attack before he really gets possession of all the facts about the dike – its width, dip, and direction: the detail of its contact phenomena; its texture and its joints; but by persevering attention all these facts can be discovered from his own seeing, and then a good lesson is learned. (*Davis, 1887K, pp. 812–14*)

One of my advanced students once came to me in high spirits, saying: 'I want to tell you what X and I have done. We have sent cards to all the fellows in Professor Blank's course, announcing a seminar for the night before the examination. We are sure to make a good thing out of it.' Certainly there was no lack of candor and no sense of impropriety here. I inquired: 'Has Professor Blank asked for your assistance in his teaching?' 'Oh, no.' 'Do you think that his lectures need to be supplemented by your seminar?' 'No, no, we are only – we are just going to' – Our conversation went somewhat further, finally bringing me the question, 'Do you think we ought not to hold a seminar?' and to this I replied that each of us must determine such matters for himself. The end of it was that a second set of cards was sent out, cancelling the announcement made on the first set. The candid complacency of the student-tutor had been overturned by the awakening of his own conscience, which never before had consciously inquired about the ethics of seminars ... Illegitimate tutoring may be described in terms opposite to those that define legitimate tutoring. It is sought for by students who are perfectly competent to do good work in the courses that they have elected, but who prefer to neglect their work so that they may do something else. It is not undertaken for intellectual profit, but merely under the constraint of an examination. It is not carried on systematically through a reasonable time, but is postponed to the latest possible day and hour. It is essentially a device by which the negligent fellow seeks to avoid doing even the moderate amount of work that is reasonably expected of him. (*Davis, 1903L, pp. 364–5*)

Whether he is designing the detailed lay-out and equipping of a practical classroom (1898C), or drawing analogies between the blowing of dust and the melting of snow in the streets of Cambridge, on the one hand, and the Sahara and the Laurentian shield drainage, on the other (1900J and 1901C), all is pursued with missionary zeal. His constant exhortation is that the teaching of physical geography must be explanatory and evolutionary in character and backed up with imaginative fieldwork (1900M, 1902B, 1903J).

His partisan and single-minded involvement was keynoted by his employment of his grandmother's expression: 'There is no other door by which one can really enter the domain of knowledge, where the motto is written: "Truth for authority, not authority for truth"' (Davis, 1900M, p. 78). He even advised Gilbert when the latter was thinking of quitting his irksome administrative post with the Geological Survey and taking a teaching post at Cornell:

My Dear Davis,
Your letter about Cornell versus Washington gives me great consolation in that it states the case very much as I see it. I have never consciously cared for the 'renown that comes of remaining in large affairs', and am disposed to think I have not really cared for it as I am perfectly conscious of caring for some other kinds of renown. At any rate I took my present office reluctantly and as an accommodation and my chief reluctance at the thought of leaving it is the difficulty to which you allude, of filling the hole. It is that difficulty which makes it impracticable to return to Geologist. If I could name an entirely satisfactory successor – who could be prevailed on – the Major would let me swap back.
If C. should take me in and give me a chance to develop a strong department I suppose I might attract some men who would otherwise get to you, but I imagine the chief result would be an increase of the output of geologists. Of late years there has been a fair demand for them and it ought to increase, but there is unquestionable danger of overstocking – especially if Chamberlin goes to Ithaca and opens the big training school he talks about. Perhaps if I go to C. I would better limit my ambition to liberalizing the education of future preachers, doctors and engineers. (*Letter: G. K. Gilbert to Davis, 5 June 1892; Washington, D.C.*)

Davis' belief that his own ideas of improvement could be carried through were undoubtedly encouraged by his service on two important educational committees. Certainly it would have taken more than a committee to convince him that he was wrong!

In 1892 the National Educational Association appointed a Committee, now historically known as the Committee of Ten, to consider secondary school studies. Among the various sub-committees appointed by this body was one which should consider geography, and of this committee Davis was an active member. The committee's report everywhere bears evidence of his acumen and of his progressive views. It may be said to mark the beginning of a new era in geography instruction in America leading the way to new methods and marking a complete revolution in the spirit and aims of the subject. The Committee of Ten in receiving the report on geography recognized its revolutionary character, for geography in the report means something entirely different from the term geography as generally used in school programs ... More distinctly than any other Conference they recognized that they were presenting an ideal course which could not be carried into effect everywhere or immediately ... There can be no doubt that the study would be interesting, informing and developing, or that it would be difficult and in every sense substantial.

In 1899 another committee was appointed by the National Educational Association to consider College entrance requirements. This led to the appointment of a sub-committee on geography, the members of which independently nominated by various organizations included besides Davis, Messrs. Rice, Cobb, Dodge, Snyder and Brigham, all of whom had studied with Davis. The report then prepared has been effective in directing and stimulating the progress of rational geography in the United States. (*Brigham, 1909, p. 36*)

Davis exhibited no little ambiguity in his approach to physical geography in that, although he propounded the merits of studies of the earth in relation to man and into the relation between human occupation and the physical features of the earth (1897L, p. 496), his own research was virtually exclusively on geomorphology. Leighly (1955, pp. 309–10) suggested that Davis exhibited a marked change of approach from that of advocating the study of the earth 'in its own terms and for its own sake' as proposed by the Committee of Ten in 1893, to his support for a man-oriented physical geography in the later part of the decade. This suggestion has been questioned by Hartshorne (Personal communication). As was shown by the previous letter from Ralph S. Tarr, Davis in his review of Tarr's textbook (1897M, p. 835) was very critical of studies which did not make 'a clear recognition of what is essential and what is unessential in physical geography'; but the cause and effect relationship between environment and man, shorn of its supposed astronomical and geological irrelevancies, which was preached by Davis was poorly supported by his own example, except in the most naively deterministic sense. Salisbury of Chicago, who disliked Davis personally, and Tarr advocated as well as practised a physical geography much more in line with modern definitions of earth science (Platt, 1957).

The culmination of Davis' work on physical geography during this period was provided by the publication of his book of this title (1898E), a carefully-structured and lavishly-illustrated book. Although the first chapter was devoted to the relation of man to the earth, exemplified by very deterministic considerations of African dwarfs and Greenland eskimos, and each chapter included some passing reference to the physical controls over human activity, the treatment was essentially that of what would be termed today 'earth science'. The following order of chapters, which was to be so influential in structuring similar works for the next half century, tells its own story: the earth as a globe; the atmosphere; the ocean; the lands; plains and plateaus; mountains; volcanoes; rivers and valleys; the waste of the land; climatic controls of landforms; shore lines. *Physical Geography* served to underline Davis' primitive determinism in geographical matters and, indeed, two years later he stated clearly that he did not consider the study of plants, animals, man or 'areal geography' as parts of physical geography (1900L, p. 403). We will have occasion to return to his geographical views in a later chapter.

CHAPTER THIRTEEN

Davis Abroad: The Cyclic Crusader

Ruskin's dictum that 'the eye of knowledge is as a telescope to the common eye whereby more is apprehended than is immediately seen' did not apply to Davis' journeys to and from Cordoba nor to his world tour, at least in a landform sense. He developed the cycle theme in New England when struggling for promotion and academic recognition at Harvard. Once established he carried it as a telescope to all parts of the world.

For those who hate puns it is unfortunate that the cycle-imbued Davis made his first extensive foreign expedition by bicycle. It was a vigorous choice for a man in middle age but he always had a latent capacity for refusing to be deterred by exceptional exertion. Needless to say he made it without his family, presumably because two of their sons were too young to keep up with him. This 1894 trip to Western Europe, though made in unconventional manner, produced several articles which were important mainly because they gave French and English geographers a first opportunity to test the values of his techniques within their own stamping grounds. Davis, of course, had already studied many of the areas at home by means of topographical maps and reports. He was especially concerned with the development of river patterns within drainage basins and more particularly with the phenomenon of river capture. To this end, he applied methods used in his studies of the Appalachians to the river network of the scarplands or cuestas of the English lowlands and of northern France. In 'The development of certain English rivers' (1895E), he suggested that the Severn system had developed at the expense of the Thames by beheading the headstreams of the latter. Similarly the Meuse was postulated to have suffered some dismemberment to the advantage of the Seine and the Moselle. Time has not dealt kindly with Davis' simple generalizations on the development of the Severn–Thames network, which happens to be very tricky chronologically, but the capture of the French rivers still adorns modern textbooks. He also studied the Swabian Alps in Württemberg. All the articles that ensued from this European tour, including a short account of 'The peneplain of the Scotch Highlands' (1896), were welcomed by geographers. The journey through England was recorded by Davis in his own brand of verse which, although not of high poetic merit, reveals strikingly his unremitting attention to landform studies.

Sing, Muse, the Sage who walks Pavonia's boards
Conning, with thoughtful brow, his notes and means –
List to his theme – no nobler earth affords –
The waxing Severn and the waning Thames!

What are to him King, President, or Pope,
Your Bostons, Londons, Romes, Jerusalems?
Nature's developments his higher scope,
The waxing Severn and the waning Thames!

Philology a trivial thing he deems,
Prefixes, bases, radicals and stems,
He finds a broader text in Britain's streams,
In waxing Severn and in waning Thames!

No astrologic puzzles rack his brain,
No pedigrees of Japhets, Hams or Shems,
He speaks of obsequent and peneplain,
Of waxing Severn and a waning Thames.

Davis' next foreign excursion was not until 1897, when he crossed Canada as a guest of the British Association. Other members of this party were Albrecht Penck, Keltie and Mill. Though there are many references in literature to the animosity that existed later between Davis and Penck, the letters written at this time indicate a warm friendship. The first letter also indicates that Penck seems to have accepted, with reservations, the value of Davis' cyclic concept.

Dear Colleague,

I received your letter of 21 March, together with the many stamps for my son Walther, just as I was making preparations for my Easter journey. [*Transl. Note:* Easter 1897 was on 18 April.] It made my parting from my family the easier to see how much Walther enjoyed receiving the stamps. He was very happy about the large increase in his American collection, especially about the older stamps which your sons were kind enough to cede to him. He has made great efforts in order to bring together some old Austrian stamps in order to show some tangible evidence of his gratitude. I enclose them herewith. I hope they will please your sons as much as Walther enjoyed his – that is their purpose. You were kind enough to ask what other stamps Walther would like: he would rather like some Newspaper Stamps; but I don't know if they are as frequently found in America as over here – I enclose a selection.

I took your letter on my journey. I hoped to answer it in the evenings. But I was always so tired at night that there could be no question of answering it. The purpose of my journey was to determine the recent folds of the alpine foreland (Alpenvorland). I had always considered the country between the Alps and Upper Danube to be an originally inclined plane and, therefore, that its rivers would be consequent rivers (Abdachungsflüsse) (cf. the region between Iller and Lech). It now appears that there exist quite recent folds. The country shows flat 'undulations'. It is no longer thus

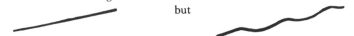

but

but the course of the rivers has not been affected by the repeated anticlinal and synclinal formations (Sattel – und Muldenbildung) – at least as far as I can make out at present. However, the determination of the folds is a rather lengthy task and I have by no means solved the problem yet.

Dr. Gulliver accompanied me during the last week. I have learned to appreciate his qualities, both as a man and as a scientist. I enjoy his presence since this enables us to hold valuable discussions during which he can inform me more fully about your views than can be gathered from a printed paper; for no matter how clearly you may write, the spoken word is more informative than the printed. Conversely, I hope to point out to Dr. Gulliver some of the phenomena which have decisively affected my views. The fact that you stress the developmental cycle far more than I do is obviously due to the circumstance that your studies were carried out in a very old, stable country, while here the cycles never develop in an orderly fashion, because new disturbances affect the formation of valleys. In your case the question arises as to the 'age' of the valleys within the cycles, whereas I here deal with the geological age of the valleys which I can trace back to the Cretaceous system. It thus often happens that a valley is 'old' in your sense, but young (i.e. post-Tertiary) in mine. We must avoid such terminological collisions, which is why the translation of some of your terms (e.g. youth) into German meets with considerable difficulties.

One can only expect the most beneficial results for the development of our science if a worker trained in one discipline engages in a different one. We two work in similar directions, so to speak; it is therefore all the more to be welcomed that in Dr. Gulliver I have met an independently thinking observer who looks at our local problems through American eyes. Of course, this needs time, for the circumstances are complicated and the literature is abundant. I, therefore, fear that this summer will not be sufficient merely to acquaint Dr. Gulliver with the problem in hand. I would greatly welcome it if he could spend another summer on alpine studies. For this reason, I strongly share his wish to extend his fellowship for another year. I put this quite spontaneously, after hearing from him the circumstances under which he travels. He has continued to influence my views.

I am still undecided as to what to do in the summer. I am greatly tempted by the Geological Congress with its excursions into the Urals and Caucasus, but the fact that 200 participants have registered deters me. I am therefore seriously considering whether I should accept the invitation of the British Association and go to Toronto. I have long since wished to visit America, visiting first the USA and then Canada. Unfortunately, I am unable to make any detailed preparations for either of these journeys since I am, as usual, overwhelmed with meetings, etc. In any case I shall not travel to Canada without at least visiting New England, when we shall meet!

I remain with my best wishes

(Letter: A. Penck to Davis, 22 May (?) 1897; Vienna)

At his port of departure, Penck expresses sincere gratitude to Davis for making his American visit so interesting.

My dear Friend,

Such a steamship affords not only the occasion of English exercises, she makes also thinking, and I pass in review the 10 weeks of my visit to America. With increasing vivacity I am aware of how much you have done in this time for me, how much you've influenced my plans and therefore also the results. I feel the necessity of writing you a last greeting from North America ... I arrived last night at Quebec. I slept the night on board and made this morning a very pleasant trip to the old city which looks rather fossil ...

I repeat all the hearty wishes I already expressed with a thankful feeling and remain with best greetings to Mrs. Davis and your boys.

Yours sincerely,

(*Letter: A. Penck to Davis, 17 October 1897; Steamship 'State of California', in the St Lawrence River*)

Seven years later Penck is still writing to Davis in a far warmer tone than would be expected from a mere expression of good manners. They were indeed firm friends.

My dear Davis,

Your letter of 21 May which I received today has brought back all my longing for the Far West, almost painfully so, for it cannot be fulfilled. As I wrote you in the spring I have been overworked and was unable, during the winter, to actively pursue my work on the Alps during the Ice Age. Instead, I had to travel at Easter, primarily to alleviate my restlessness. You will have gathered from the greetings I sent you from Faido [Switzerland] in April that I used the occasion to obtain some scientific benefit. In Faido I easily discovered all of Garwood's errors. He made his observations at the very point where glacial over-deepening [glaziale Uebertiefung] is discontinued [aussetzt] at a glacial threshold [Riegel] by running water.

Three weeks ago I returned to Vienna, but I am disappointed in my hopes that my work would now advance more rapidly. In all that time I have written one proof sheet [gathering; Druckbogen] on the Alps during the Ice Age! Before leaving for America I must write another seven! I hope that with better training I shall be able to quicken the pace, but this much is clear: it is impossible for me to leave Vienna in July and I must content myself with travelling directly to the Congress in August, hoping that I may perhaps be able to see a bit of the West by going to St. Louis. This hope is a very poor substitute for what I shall miss, namely Davis and Gilbert. I am really desolate, but am powerless to alter the circumstances.

The more heartfelt are therefore my wishes which accompany you, the more heartfelt the wish for a happy reunion in Washington and, perhaps, later in St. Louis and farther west and, finally, in Boston. Mott will be very welcome in our house. I am writing to him in Munich that the Penck family will be at home during the whole of June and that in July my people will be in Welschmofen [? illeg.] bei Rosen. Many kind greetings from family to family from

Your deeply sad

(A.) Penck.

P.S. You will have received Part VI of the Alps during the Ice Age through Professor Brückner. I herewith enclose a work – finally achieved – on Alpine maps and reliefs.

We hear nothing at this end about the excursion by the Congress to the West. Prof. Oberhummer [? illeg.] complains that he has received no answer.

(Letter: A. Penck to Davis, 2 June 1904; Vienna)

Another acquaintance which began on the Canadian trip of 1897 and which Davis maintained for many years was that of John Scott Keltie (later knighted), the Inspector of Geographical Education, and later Secretary, of the Royal Geographical Society, as the next letter shows. It also illustrates the growing importance of geography in British universities and schools.

My dear Davis,

I am writing to you in strict confidence to say that there will probably be a vacancy in the Readership in Geography at Cambridge University, either in October next, or October twelve-month. As you may know things have not been going satisfactorily there, as they have been at Oxford. We want a new man to direct the whole thing, as well as to lecture in certain departments. It is difficult to get a suitable man here; there has been a sudden demand for Lecturers and Professors in Geography in this country. There is not only Cambridge, but Sheffield, Edinburgh, Aberystwyth, and moreover, teachers are required also, and we are not prepared to supply the demand.

I write to you, as I say in confidence, to ask whether firstly, you would consider that Ellsworth Huntington would be a suitable man for Cambridge, and secondly if so, whether you think he would be disposed to come to England for a term at least, say for five years. I believe that if we recommended him to the Cambridge authorities, they would be disposed to accept him. The salary is very low, only £200 a year, but if a thoroughly good man were found there is quite a possibility that in addition he might be made a Fellow of one of the Colleges with free rooms, and an additional £250 a year. This latter, of course, I could not absolutely promise.

If you thought that Huntington would not be a suitable man, then you need not say anything about it, but judging from what I have seen of him, and from the papers that he has given terms, and especially from his most interesting book on the people of Asia, it appears to me that he has a good grasp of the subject both on the physical and human side, and whether he can lecture or not, he certainly can write. If you think he is a suitable man, would you mind sounding him cautiously whether he would be disposed to come forward as a Candidate for Cambridge when the vacancy is announced? According to the rules of the University, the vacancy must be advertised and Candidates invited, but I feel pretty sure that if Huntington offered himself backed by you and by us, he would get the post. Pray do not mention the matter to anybody except Huntington, and ask him to keep it quiet, we don't want it to get abroad at all as yet.

(Letter: J. S. Keltie to Davis, 20 December 1907; London)

FURTHER EUROPEAN TRIPS, 1898-1901

When Davis was appointed Sturgis–Hooper Professor of Geology at Harvard in 1899 he was, as we have already noticed, enjoying sabbatical leave in Europe. On this occasion his family went with him and there is a slight hint of mockery in his son's recital of the places visited. Probably the family did not always share their father's enthusiasm for geological exploration – not while on holiday at least.

> His sabbatical year I have mentioned, 1898–1899, with the family along, covered England, Scotland, Paris, southern France, three months at Cannes on the Riviera, then Italy for practically a month, where we stopped at Rome, Naples, and Verona on the Adriatic, where he wanted to see some particular item, but it rained for 2 or 3 days and thwarted his efforts; followed by Milan, Venice, the Austrian Tyrol, and eventually Munich. He and his wife made a side trip to Vienna, and after Munich, Switzerland, and then Paris, London, and home in the middle of September 1899. (*Interview: R. Mott Davis with R. J. Chorley, 7 September 1962*)

Davis should perhaps be credited with some consideration for the differing tastes of his wife and family but a suspicion remains that what he enjoyed he expected his family to enjoy also. On this visit many of the same places were re-examined and earlier impressions confirmed or amplified. Other areas were seen for the first time. Throughout attention was primarily concentrated on glacial erosion. The excursions made

> stretched from Brindisi in Italy to Bergen in Norway, with special studies in England, Auvergne, the Karst district north-east of the Adriatic, and the Alps. In connection with excursions near Oxford, the previous work on English rivers was revised and the Drainage of Cuestas was written (*Proc. Geol. Assoc.*, London, 1899): this paper, dated at Cannes, affords a good example of the author's clear description of generalised forms, followed by an ample citation of local examples of the types under consideration, taken from the Swabian Alps in Württemberg, visited in 1894, and from the Cotswold Hills of England, visited in 1894, and 1899. The excursion in Auvergne afforded much material for the article on 'The Peneplain' which like the proceeding was also dated at Cannes. The visit to the Karst district was made with Penck and a party of students from the University of Vienna; the observations then made were recorded in an article on 'An Excursion in Bosnia, Hercegovina and Dalmatia' (*Bull. Geogr. Soc. Phila.*, 1901), giving descriptions of graded slopes and peneplains of bare limestone, of sinks, basins and springs, of gorges and travertine cascades and of young coastal features due to submergence . . .
>
> The most profitable result of this year abroad was the new evidence obtained regarding the efficiency of glaciers in the sculpture of mountain valleys. Interest in glacial erosion had received a notable impulse at this period from the studies of Gannett, Tarr, Gilbert and others in the United States, and of Penck, de Martonne and Blandford in Europe: but it was essentially independent of the essays by these writers and on the basis of his own observations that Davis came

to recognise a much larger measure of glacial erosion than had seemed to him justified when discussing the same subject eighteen years before. His new views were not based on any physical considerations as to the movement of glaciers or on the capacity of ice to erode, but purely on a comparison of forms found in glaciated districts with forms found where only normal erosive processes have acted. Among the localities that were visited in this connection were the valley of the Rhue in Auvergne, during a winter excursion with R. L. Barrett of Chicago; the valley of the Ticino, the basin of Lake Lugano and other parts of the Alps; and the Hardanger Fiord in Norway, in company with Reusch of Christiania. The results of these studies are given in 'Glacial Erosion in the Valley of the Ticino' (*Appalachia*, 1900) and 'Glacial Erosion in France, Switzerland and Norway' (*Proc. Boston, Soc. Nat. Hist.*, 1900). (*Brigham, 1909, pp. 40–4, passim*)

Many years later, Davis confirmed that glacial processes were the main focus of his studies at that time.

It is perfectly true that according to my original scheme, no such wide young valley heads were mentioned. I first came to know them in central France in the winter of 1898–99. I am sorry that, on returning home the following autumn I was so plunged into teaching that the only thing I wrote in detail was Glacial erosion; the open young valleys remained in my note books. (*Letter: Davis to A. Penck, 3 April 1921; Cambridge, Mass.*)

Davis, however, did produce a number of non-glacial publications as the result of his year in Europe. One on the fault scarp on the north-east side of the Lepini Mountains (1900E) gives an excellent example both of his eternal vigilance and of his increasing use of analogy in research and description. He first observed the scarp on a train journey from Rome to Naples and returned later to take two walking excursions along its base. He immediately compared the dissected fault scarp and its buttressing spurs with the west face of the Wasatch Range, and the convex-up rock fans at their base with features described in the Sonoran desert by W J McGee. Davis postulated that canyon cutting first produced alluvial fans and that renewed uplift of the block, evidenced by a prominent recent fault scarp near its base cutting across the fans (which Davis first recognized as not due to structure), led to the stripping of the upper fan alluvium and the exposure of the rock fans.

His excursion to Bosnia, Hercegovina and Dalmatia (1901G) took place between 19 May and 6 June 1899 with Albrecht Penck and a party of his students. Davis' visit to Vienna was planned when he crossed Canada with Penck on the British Association field trip in the summer of 1897, returning via Grand Coulée and the Black Hills. Travelling by train, car, horseback, steamer and foot during a very rainy spell, and sometimes being on the move for twenty-one hours a day, the party made their way from Vienna to Sarajevo, Mostar, Gacko, Ragusa, Fiume and back to Vienna. This was Davis' first opportunity to study first-class karst features at close quarters and in his description of the trip he is noticeably reticent about theorizing as to the origins of the landforms. The party examined numerous dolines, together with the glacial features of Mt. Orjen. While supporting the general view that the bareness of the limestone slopes was probably the result of deforestation, Davis suggested that the bare karrenfeld might possibly be the result of

F I G. 52. Davis on an unknown (European?) field trip
(Courtesy R. Mott Davis)

present processes (1901G, p. 33). The origins of the poljes north of Trebinje clearly puzzled him but he related their flat floors to local baselevel control. Among his many other observations were that many of the gorges were due to uplift, that there is an elevated peneplain behind Sebenico cutting across limestones dipping at 30° and surmounted by monadnocks, and that the Dalmatian shoreline was 'young'.

By chance we have recently discovered an important letter which throws much further light on this European trip of 1898–9. It shows clearly how far Davis planned ahead, how strongly he was influenced by German geomorphologists and how strong at the time was his own influence on Albrecht Penck, their leading exponent of landforms.

My dear Davis,

I received your kind letter of Aug. 15 just as I was starting my first mountaineering expedition with Walther. Your letter accompanied me up hill and down dale, from Tyrol to Switzerland, Württemberg, Bavaria and Baden but everywhere I seemed to lack the leisure to reply to you in the same detail as you wrote to me. We have been home now for almost a week, recovering from my tours, and I am planning my work for the coming winter: I can thus easily find the time for a chat with you. I therefore bid you welcome in Europe once again, just as I did when I called to you from the Berliner Hütte.

I have been on the move, almost without interruption, since the beginning of September and have looked at my old haunts [*Transl. Note*: literally 'Untersuchungsgebiete' = research areas or regions] with American eyes. This has shown me a great many new things, and the whole problem of Alpine lake formation suddenly became clear to me. I started from the Alpine forelands where the fluvio-glacial gravel [Schotter] lies on a peneplain [Rumpfebene] and I wondered whether I would not find its morphological equivalent in the mountains. I re-examined a few gradeplanes [Thalsimse, terraces] in the mountains at the place where I was holidaying and found that the gradeplanes cannot be of Tertiary origin but must be younger. I correlated the gradeplanes with the peneplain, found that they both matched, and that what I had previously assumed to be due to glacial crustal movements are not necessarily to be interpreted as such. The result is given in the following sketch:

[missing]

[Erosionsleistung in der Glazialperiode = erosion during glacial period]

You will see from this sketch that I assume a very important erosion during the glacial period which does not obey the laws of valley-formation by running water and can only be explained as being due to ice. In addition, I studied the peakforms of the high mountains and, here too, I recognised in the cirques [Karen, corries, cwms] works of glacial erosion. The result is that the Alps owe their present shape mainly to the glacial period, that the valleys are certainly deepened and 'overdeepened' in the direction of the ice flow, and alpine lakes are merely overdeepened valley floors.

You may easily imagine how much I ran about in order to follow these points in detail: I climbed to 3500 m and investigated the entire Rhine valley from the Via Mala above Chur until below the Rhine-falls at Schaffhausen. This getting-about did me a lot of good, for in July and August I had to do a lot of writing. I had to do some work on a small morphology which I wrote in 1894 [1891? illeg.]. Here too, my American journey, and especially my acquaintance with you proved to be most useful and I adopted some of your terminology, viz.:

Initial forms	= Urformen
Consequent forms	= Folgeformen
Subsequent forms	= Unterfolgeformen

I have noticed that by introducing these three concepts, and also by introducing the concepts of youth, maturity and age of the forms the description gains immensely in lucidity, in particular, some chapters in valley-formation become comprehensible as soon as consequent and subsequent forms are strictly differentiated. But I have not gone any further, I did not distinguish between obsequent and insequent [?? illeg.], these are special cases. I am having reprints made of this work, for which I shall write an introduction and then distribute them.

Now, how about you? I am anxious to know your further travel plans and would be pleased to hear that they include Austria. Now, *my* plans for you are as follows: you will come to Vienna during May, and in June you will participate in a students' expedition to Bosnia and Hercegovina. There you will see some quite wonderful karsts which you will not find anywhere else; you will also travel along the Adriatic coast with its long drawn-out islands, etc.; finally we shall reach the Orient. During this excursion, which is planned to last 14 days, you will visit the parallel Hungarian valleys as well as our friends Richter, Loczy and Audure [? illeg.], after that you and your family will accompany my family to Kirchbichl in the Tyrol. We have just spent a very happy summer there and although there are many places in the Alps which can boast of grander surroundings, Kirchbichl is so situated that one can reach the most beautiful parts of the eastern Alps within a very short time and it also has the advantage of allowing a free and easy as well as a cheap stay. (The bed one night 16 cents) [English in the original, *Transl.*] From Kirchbichl you may show Alpine lakes and glaciers to your sons and from there you can also reach Berlin, for the Geographical Congress, in 13 hours. While you are travelling about it is best for your wife to stay in Vienna, but for your sons I once again recommend Prof. Miesch in Schaffhausen. There they will live en famille, learn German and French and there will be no compulsion exercised on them, particularly not in school, which is sometimes the case both in Germany and here. Again the best wishes from family to family. Au revoir!

Yrs.

Penck

P.S. I cannot send off this letter without mentioning Walther's joy about the wonderful exhibition stamps which you sent him. Only this evening he looked at them again, full of pride.

(Letter: A. Penck to Davis, 22 October 1895; Vienna)

In 1900 Davis made another excursion to England. In the next year he planned a further tour to Britain and the following letter from Penck shows that they arranged to meet. Strangely we also get from Penck an all too rare a glimpse of Mrs Davis.

My dear Davis,

Your long letter of the 22nd August gave me an immense pleasure. 13 pages written in an excellent German, full of news, full of ideas. I got it yesterday and read it on an alpine meadow close to the hotels where I stay with my people since three weeks, in face of the Cimone della Pala climbed yesterday by my son (3186 m), situated in nearby vergin (sic) forests. Alas that this pleasant spot must be left in a few days . . . in a fortnight we shall (illeg., return?), and in the mean time I must go to Salzburg [*Transl. Note*: the letter now continues in German] in order to deliver the speech at the unveiling of the Richter memorial. This is to be on Sunday the 15th. I shall then go to Berlin and then, via Amsterdam, to London in order to attend the celebrations of the London Geological Society. I wish I could join like you, one of the Society's excursions. But, as you can see, my time is fully occupied and I have so much to do that I even feel guilty about my visit to England. After the festivities I shall join the excursion to Oxford and, after that, go for a few days to Cornwall to observe the effect of wave action (Brandungswirkungen). Then, in mid-October, to Stuttgart where I have to deliver a lecture, and so back to Berlin where more work awaits me. It would please me greatly if after the London Meeting we could be together in Cornwall (beginning of October) and we could profit from your presence in Europe by discussing many of the points you mentioned in your letter. Some of these I have touched upon already in my lecture to the Academy which – together with much else – I sent when I left Berlin. I hope you have safely received the ribbon of the great star ('grosses Kreuzband': a decoration?].

Your letter brought back beautiful memories of Wianno, and when I now hear the roar of the waterfall I think of the surf on your beach. In my mind's eye I see Mrs. Davis, the careful housewife, I hear her talk about Pfleiderer and see her playing bridge. Let us also nurture our beautiful memories of New England, so that this may contribute more to our friendship. I look forward to our reunion: but first of all, welcome to Europe.

My addresses will be: Up to Sept. 15 incl., Hôtel Wolf Dietrich, Salzburg, then, until 20.9, Berlin, W.15, Knesebeckstrasse 48.

With best wishes,
Your devoted   (A.) Penck.

(*Letter: A. Penck to Davis, 8 September 1901; Paneveggio, Southern Tirol*)

We cannot let these European excursions pass without drawing attention, no doubt quite unnecessarily, to the fact that Davis was consciously in thought and flesh crusading for the cycle. It was at the turn of the century, before the Seventh International Geographic Congress at Berlin, that 'The geographical cycle' as such made its last solo appearance (1901I). This magnificent model was a late-nineteenth century design which in the twentieth century was to become seriously outmoded and had to be retrimmed to a new academic environment. Its modifications also coincided with the widening of Davis' fieldwork.

WIDENING HORIZONS: THE TURKESTAN EXPEDITION

So far Davis' travels had been confined to better known, and more civilized parts of the world. His first study of a more remote region came in 1903 when he renewed his earlier association with Raphael Pumpelly and acted as physiographer to Pumpelly's Carnegie Expedition to Turkestan. Ellsworth Huntington also went as an assistant. The journey was too hurried for detailed local studies but it provided some profitable general impressions and the raw material for several articles. The chief physiographical results are given in 'A journey across Turkestan' (1905G), which is embodied in the main volume published by the Carnegie Institute as *Explorations in Turkestan* edited by R. Pumpelly (1905). The expedition in its mobility and celerity was like a commando wartime raid. The period actually spent in Turkestan was too short for more than the most superficial examination. The party left New York on 18 April 1903, and did not start their exploration of the Tian Shan Mountains until 27 June. They completed the survey by 22 July, and were back in Boston on 28 August. Davis gives the following itinerary:

On April 17, 1903, accompanied by Mr. Ellsworth Huntington, who had been appointed research assistant by the Carnegie Institute of Washington, I left Boston; sailed from New York, April 18; landed at Cherbourg, April 24; spent April 25 in Paris, April 28 in Vienna, and May 1 to 3 in Constantinople; crossed the Black Sea to Batoum, May 4 to 8; and went thence by rail to Tiflis, May 10, and to Baku on the Caspian, where we arrived May 12. We crossed the Caspian on the night of May 22, and started from Krasnovodsk on the Central Asiatic Railway on the afternoon of May 24. After making short stops at Jebel, May 25, Kizil Arvat, May 26, and Bakharden, May 27, to examine the piedmont border of the great plains of Turkestan, we delayed at Askhabad, May 27 to June 9, long enough to make a five-day excursion, May 30 to June 4, into the Kopet Dagh, the mountain range along the Russo-Persian frontier. Leaving Askhabad by train the evening of June 9, we stopped at Merv, June 11 to 14, and Samarkand, June 16, and on June 17 reached Tashkent, where we remained three days. On June 20, accompanied by Mr. Huntington and Mr. Brovtzine, interpreter, I went by rail to Andizhan, where we stopped from June 21 to June 27, to outfit for an excursion across the western ranges of the Tian Shan Mountains to Lake Issik Kul. We set out from Andizhan, June 27; spent two days, July 8 and 9, at Lake Son Kul; reached Issik Kul on July 14; made a short trip into the mountains on its southwestern side; and then moved along the northern shore to the Russian settlement of Sazanovka. Here, on July 22, Mr. Huntington turned southward to begin his excursion to Kashgar, with the object of continuing over a large district of the high ranges the study of old moraines and terraces that we had begun together on the road to Issik Kul; while I turned northward with Mr. Brovtzine and began my homeward journey. Vyernyi was reached July 26; we went in tarantass to Semipalatinsk, August 2; by boat down the Irtysh to Omsk, August 7; by train to St. Petersburg, August 15, where Mr. Brovtzine resided; I continued by train to Ostend and London, August 17; and by steamer from Liverpool to Boston, August 20 to 28. (*Davis, 1905G, p. 23*)

In a brilliant piece of reconnaissance writing Davis (1905G) sets out the physiographic panorama of central Asia against the background literature. He describes the abandoned shorelines of the Caspian Sea near Baku, Krasnovodsk and Jebel, with cobble spits 600 feet above the present water level; and the piedmont plains of south Turkestan which, like the High Plains of the United States, Davis thought to be probably more fluviatile and less lacustrine than was hitherto believed. Davis took a five-day excursion south of Askhabad to the folded limestone ridges of the Kopet Dagh Range, climbed a 9300-feet summit, and was particularly struck by the sequence of rejuvenation terraces in the shale synclinal valleys, the highest extending on to the limestone anticlinal ridges. On the outward journey Davis had visited Penck in Vienna who had suggested that he should look into the problem of whether the loess was still accumulating or was fossil. The source of the loess was recognized around Samarkand to be the aggrading fluviatile plains and valleys, and Davis noted that the supply of loess ceased when stream trenching or vegetation growth began. By far the most productive part of the trip consisted of the crossing of the Tian Shan Mountains from south to north. The Bural-bas-tan Range Davis interpreted as a former peneplain on crystalline rocks uplifted to 13,000 feet and still at a very rare youthful stage of dissection (1904C). This uplift was accomplished without lateral folding and compression and Davis later took issue with Suess' view that unfolded plateaus are formed by surrounding subsidence (1905K). Davis believed that the Tian Shan was direct proof of rapid vertical uplift of a peneplain, such as Gilbert had previously proposed for the Basin Ranges and as he himself had for the southern Colorado Plateau. It did not worry Davis that no convenient mechanism was available for this uplift, believing that we cannot argue from a knowledge of forces producing uplift for we are largely ignorant of them (Davis, 1905K, p. 271). Noting the glacial features of the Tian Shan, relatively dating the moraines by their degree of fluvial dissection, and confirming the efficiency of cirque backcutting and peak sharpening, Davis moved on to the terraces of the Kurgart and Narin Valleys. This was the point where Davis was to leave Huntington to physiographic researches and there is no doubt that he was influenced in his interpretation of those terraces by the ideas of the latter. It was noted that, unlike those of New England, none of the terraces is bed rock-protected and it was tentatively proposed that they were due to climatic changes successively reducing the stream activity.

The following extract describes part of the crossing of the Tian Shan mountains.

On July 7 we forded the Narin, and turned northward along a trail up a side valley, camping in the mountains with a party of Khirgiz, who were driving their flocks to summer pasture by Lake Son Kul. The next day, July 8, we went on with the Kirghiz, crossing the Dongus-tan range at about 10,000 feet elevation, reaching the lake about noon, and camping above 9,300 feet in one of the summer villages on its southern border, after an afternoon ride to a small glaciated valley ... On July 9 we forded the outlet of the lake at its southeastern corner, went northward along the plains at its eastern side, visited two glaciated valleys of the Kok-Tal range in the afternoon and then had our view of the flat-topped Bural-bas-tan range to the southeast; we camped in another summer village for the night. On July 10 we crossed the Kum-ashu pass in the Kok-tal range and descended northward to the Tuluk valley, camping again in a Kirghiz village. Having seen during the descent a large moraine in the Chulai range, north of the valley, we

FIG. 53. Davis in Turkestan (Courtesy R. Mott Davis)

went up to it . . . on the morning of July 11, and in the afternoon followed down the Tuluk-su . . . (*Davis, 1905G, pp. 68–9*)

The trip was not spent in primitive conditions. In the larger towns the travellers were hospitably entertained by the Russian authorities and in the field they could always rely on the generosity of the Kirghiz tribesmen.

The leaders of the villages seemed to be men of energy and ability. They always received us with courteous attention and gave us of their best. In the midsummer season of our visit, the rude, mud-house villages in the valleys were almost deserted. We saw many of the houses open and empty; only a few men remained there to look after the irrigation of the wheat and grass fields. The rest of the population, with all their possessions, were found in the high valleys. The habit of life was that of seasonal migrants rather than strictly nomadic. The chief men were certainly well-to-do, and seemed to want for nothing. One of these, Kuve Gen Shigai-el . . . had been a judge among his people . . . He invited us to lunch in his yurt Akh Tash (White Stone) on the Son Kul plain. The yurt was one of the finest we had seen, with a hundred sticks supporting the clean felts of the roof.

(*Davis, 1905G, pp. 66–7*)

The party followed the main wagon roads and only once did snow conditions prevent their passage, and then on an attempted short cut across a mountain range. On 27 July they called on General Yonof, governor of the province of Semiryetshensk, and

> on July 28 started a ride of 1,000 versts northward across the steppes in a tarentass, or springless post wagon. We . . . reached Semipalatinsk, on the Irtysh, in the afternoon of August 2. The guest rooms on the post stations on the road were, with very few exceptions, clean and neatly furnished . . . At Semipalatinsk we waited two days for a boat to go down the river, starting in the early morning of August 5, and reaching Omsk on August 7. The fast express on the Siberian railway carried us westward from Omsk at midnight August 8! (*Davis, 1905G, p. 70*)

Davis did not revisit Turkestan but he was directly responsible for persuading Robert L. Barrett, one of his wealthy students, to finance the Barrett–Ellsworth Huntington expedition to inner Asia (1905–6) which resulted among other things in *The Pulse of Asia*.

THE MEXICAN JOURNEYS, 1904 AND 1906

In the summer of 1904 Davis attended the Geographical Congress at Washington, D.C., and played a considerable part in its organization. He also spoke at a World's Congress at the St Louis exhibition and took part in a congress excursion to the Colorado Canyon and to Mexico. The members of the Mexican trip included Vidal de la Blache, Emmanuel de Martonne and Albrecht Penck. The journey was rapid but the dissection of the eastern main escarpment of the Mexican plateau was observed briefly.

In 1906 Davis, together with his son Edward, revisited Mexico for the Tenth International Geological Congress held there and writes in one of his letters that he had 'a fine time'. On this occasion he studied both major escarpments of the great Mexican plateau as well as the volcanoes of Toluca and Jorullo in the western scarp, the Cordillera Neovolcanica. In proposing a toast in Spanish at a banquet given in the Municipal Palace in Mexico City, Davis revealed some interesting sidelights on his attitude both to his profession and to his fellow-men:

> Geologists, because of our necessity to do our studies in the field, are accustomed to wearing simple clothes and we know little of wearing official uniforms or the clothes of etiquette. Not for us the silken sombrero but rather the resistant felt hat or the light panama which protects us from the rigours of sun and rain. To come to the city necessitates clean shoes but usually ours remain dusty or caked with mud. It is not often that our work associates us with human activities but rather with the silent rocks. As a result we sometimes begin to think that we are apart from other men, that we alone follow a path a little distant from our fellows. In these conditions of isolation there is something sad, against which we must fight and, even though our interest in our studies is so great, this would not be enough without the enthusiasm and sympathy our fellow men pay us.

... Although isolated we are proud of the progress which geology has contributed towards the sum total of human knowledge.

That which astronomy has done for the extension of space, geology has done for the extension of time. If the face of our mother earth smiles, as it does in this so marvellously beautiful country, it is not the smile of youth but of a benign immortality. That which biology has done for the structure of animals and plants, geology has done for their predecessors and for their development. In this way we learn that death is not the punishment of sin but is, like birth, an essential part of the natural system of this planet. The discoveries of geology have not served solely to enrich the mining fraternity but have added new sources of knowledge to the total of modern philosophy.

Of all this we are proud, but at the same time we must be humble, humble at knowing our smallness in the trifle of time and space we occupy, although proud in understanding the knowledge and ability, with which the good Lord has endowed us, to sound out the immensity of time and space.

THE SOUTH AFRICAN JOURNEY, 1905

Whereas Davis wrote little relating to his journeys in Mexico, his excursion to South Africa which involved much boat travel, called forth an excess of descriptions. These are, however, particularly interesting because they extend to a wide range of topics including the problems of colonialism, white supremacy and of other current affairs. True comments on current problems also occur in his world tour letters but on the African excursion they appear much more frequently and reveal a keener, abler, professional observer and, inevitably, the descriptions of landscape are more exact in terminology. At times Davis becomes a trifle tedious, as when in the midst of an otherwise entertaining narrative he stops to explain how a word should be pronounced, but he shows a good memory for facts and his prose is eminently readable. Present-day events in South Africa verify many of his judgments on racial problems and biographically these sidelights on social and political themes have a secondary value as they reveal aspects of Davis that normally are denied us.

We can safely let him explain the purpose of the excursion and the qualifications necessary for joining it.

The visit to South Africa this summer was, however, a much more exceptional affair than a mere transatlantic jaunt; indeed, it was the largest scientific excursion ever carried successfully through a long-distance itinerary. Three vessels of the Union-Castle line brought detachments of the party from England, the voyages varying from seventeen to twenty-three days of fair weather. The oversea membership was of four classes. There was the 'official party', of about one hundred and seventy, including the general and sectional officers of the Association and a number of representative scientists selected by the Council from the eleven sections, from Astronomy to Education; for all these there were liberal reductions of steamship fares, free railway transportation over long distances, and much private hospitality, as well as a generous subvention towards expenses from funds that had been placed at the disposal of the Association by the colonial Governments in order to ensure a strong attendance. Then there were foreign guests to the number of sixteen from Europe and America, invited by the Council from nominations made by the sectional committee. These *fortunati* were practically on a par with the official party where they were not given even greater facilities and privileges; among them may be named Engler, botanist, and Luschan, ethnologist, of Berlin; Penck, geographer, of Vienna; Backlund, astronomer, of Pulkova; and Cordier, Orientalist, of Paris; and among the five Americans, Scott, palaeontologist, of Princeton; Brown, mathematician, of Haverford; Carhart, physicist, of Michigan, and Campbell, botanist, of Stanford. Third came the non-official party, for whom the concessions as to travelling expenses were less generous, but who, as far as the meetings and receptions were concerned, stood on an equality with every one else. Fourth were the wives, sons, daughters, and relatives of members, who were in sufficient number to give the excursion the appearance of a large family party. To all these should be added the colonial members, who, however, as it seemed to me, usually stood aside to let the procession pass. (*Davis, 'The Nation', 16 November 1905*)

The tour was done in style and probably as much emphasis was placed on social activities as on the scientific programme.

As to the voyage itself, seventeen days sufficed to cover the distance of 6000 miles. The weather was fair, the sea was quiet, except on the last day, the vessel large, roomy and steady. A morning at Funchal on the south coast of Madeira gave us a glimpse of an earthly paradise: a few of us had a closer sight of a little Garden of Eden, for a generous Madeiran fellow-passenger invited us to call at his summer residence up on the south-facing mountain slope, 2000 feet above sea level; it was reached by a cog railway, and the descent was made in toboggans in which all visitors are expected to coast down the paved lanes at a lively speed, guided by expert runners. We had a passing sight of Teneriffe at sunrise one morning, and of Cape Verde on the African coast. The several wind belts were duly crossed with their appropriate temperatures and humidities, necessitating a wide range of clothing with the changing latitude. Our highest temperatures were from 80 degrees to 84 degrees, at about 15 degrees north; on crossing the equator the temperature soon fell to 70 degrees or lower to the surprise of many passengers. This due to the cool West African current that there comes up from the south. The southern stars rose higher and higher on successive evenings. Scorpio climbed toward the zenith and beneath it came Alpha Centauri, our nearest stellar neighbour in the whole sky; with Beta of the same constellation we were led to the Southern Cross, a fine star group which would be more enjoyed if it were not for its overgrown reputation which is met before the Cross itself comes in sight. To my regret I found that thirty odd years of absence from the southern skies had greatly weakened the close acquaintance that I made with them in 1870–73, as an assistant to Dr. B. A. Gould in the National Argentine Observatory at Cordoba, South America, where many a night was spent on the Ozoten, studying the brightness of the stars.

F IG. 54. Rogers' geological field party for the South Africa British Association Meeting in 1905. Davis is on the extreme right (Photograph by Dr Tempest Anderson)

There was much activity on board the *Saxon*. Games and dances were planned by a committee of the less elderly. Good sport for players and lookers-on was found in a cricket match – Ladies vs. Gentlemen – in a space about twenty feet broad, under a roof that kept down the high balls, and within a net along the rail to hold in the long drives. Competitive games were played in great variety. There were several evening dances, with a good band playing in a glare of extra electric lights on a smooth deck, the dancers in full evening dress, even to white kids on the gallant men. The final affair was a brilliant fancy dress ball. Indeed in the matter of dress there was not only a variation with latitude, but also a strong diurnal range for the men, from nocturnal pyjamas, in which the more hardened travellers lounged on the open deck till the late breakfast, then neglige flannels over noon, to black togs that everyone put on for late dinner. This wasn't so bad for the cooler southern latitudes, but the togs were something of a nuisance in the few days of rather sweltering air that we had in the neighbourhood of the heat equator. (*Davis, 'Boston Evening Transcript', 16 September 1905*)

Clearly Davis was more interested in the evening lectures and impromptu discussions than in the general merriment. The following extracts represent parts of lectures which he remembered because they particularly interested him.

Colonel Bruce, recently chief of a commission to study 'sleeping sickness' in Central Africa, described ... the spread of this fatal disease a few years ago into Uganda, on the headwaters of the Nile, where it has killed the natives by hundreds of thousands, laying waste large districts previously populous. Sleeping sickness has long been known as of occasional occurrence on the Guinea coast, but so long as the natives of that region were usually murdered when they crossed over the divide to the Nile basin, the disease did not spread into the far interior. Of late years, however, under the so-called 'Pax britannica', many slaves liberated in the infected districts were brought over to Uganda, and this apparently beneficent action was followed, apparently as effect follows cause, by a violent outbreak of sleeping sickness in the new territory, with the most direful results. The studies carried on by Bruce demonstrated the cause of the disease to be a minute flaria, which is carried from the blood of the diseased to that of the healthy by a biting fly, similar to the Tsetse fly. Further spread of the disease is gravely apprehended, as every victim dies and as no means of immunization are yet known . . .

Ferrar, the young geologist of the recent English Antarctic expedition, gave an excellent account of the voyage and of the two years within the southern polar circle. Seals and penguins were the chief companions of the explorers in that desolate land. The penguins were described as excellent swimmers under water, where they used their degraded flipper-like wings, while their feet were held out motionless astern. The speed of the heavy birds is so great that on rising near a cake of floating ice, they shoot several feet into the air and land on the ice on their feet, standing upright. Their wings are reduced to hard and featherless arms, useless for flight, and of little protection to the head while the bird sleeps; yet the habit of tucking the head under the wing at night is still preserved.

This recalls a story told me by Mr. Belton, for many years the chief cartographer of Stanford's well-known geographical establishment in London, and as such the collaborator with many British explorers in the preparation of their books. He drew up Livingstone's maps for publication from the very rough drafts sent home by the famous missionary; the drafts were often made on scraps of paper, stationery being poorly supplied in central Africa; on one occasion two maps were drawn on a piece of newspaper, one superimposed on the other, while the north lines turned 90 degrees apart. Deciphering such a record was difficult, yet the maps as finally published do not appear to be very faulty when tested by later surveys.

Professor Haddon, ethnologist of Cambridge University, and a previous acquaintance of mine at the hospitable table of Graham Bell in Washington, told the little party how he became a crocodile on one of the Australasian islands; that is, how he was adopted into the tribe of which the crocodile is the totem . . . It was during one of his Australasian voyages that this versatile investigator became interested in the cats-cradles made by the natives, and since then he has collected a large number of patterns from many parts of the world; one of the most ingenious was learned from the Navajo Indians at the St. Louis Exposition last summer. Hardly a day passed on shipboard without a group of devotees learning new tricks from this pastmaster of the art; and some enthusiasts memorized twenty patterns. Haddon's stories of the 'Ickneald way' an old pre-Roman road in East Anglia, and of the pre-Christian 'holy wells' in Ireland, now bearing the names of Catholic saints, but still resorted to by the peasantry to pray for good crops or luck in fishing, have often served me for quotation since I heard them in English Cambridge five years ago. (*Davis, 'Boston Evening Transcript', 16 September 1905*)

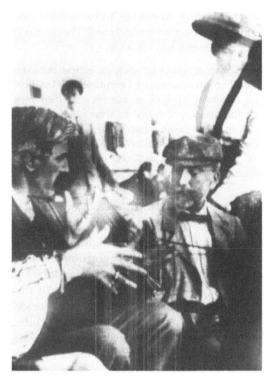

F I G. 55. Davis taking part in Professor Haddon's 'cat's cradle' class on the voyage out to South Africa in 1905 (Photograph by J. T. Bottomley)

We have already noticed that Davis had a gift for constructing working models and know from the above account that he was fascinated by cat's-cradles [Fig. 55]. One of his 'unexpected triumphs' on this excursion was to meet a native boy who taught him a new cat's-cradle which he

> had the pleasure of adding to the large collection made by Professor Haddon . . . It afforded me great pride to see 'Ambra', an unusually complicated pattern, included in the ceremonies of the childlike – but not childish cult, whose devotees were afterwards known on the steamer as 'Haddon's string band'. (*Davis, 'Harvard Co-operative Society Magazine', 1905*)

There seems no doubt that Davis was fascinated with all kinds of puzzles.

> He was a most interesting person outside of the classroom as well. He was a good conversationalist and could draw on a long lifetime of experience. You may not know that he was also a rabid fan of riddles and puzzles. He would sometimes work for hours on a 'simple' coin or match trick. He would never consent to be shown a solution. (*Letter: Arthur D. Howard to R. J. Chorley, 13 February 1963; Stanford, California*)

The cat's-cradle and lecture discourse seem to have been Davis' main escape from the lighter social activities on board. We get the impression that he leapt ashore like a cat from a hot tin roof.

An hour after coming ashore I was off on a geological excursion with Penck of Vienna, Coleman of Toronto, Lomas of Liverpool and Du Toit of Cape Town, walking along the tide-swept ledges of ancient stratified rocks and intrusive granites, on which the heavy surf rolled in from the broad South Atlantic. At noon I lunched with my host in his country house in the suburb of Seapoint, and now in the evening this writing is finished in a comfortable sitting room near a cheerful wood fire. Tomorrow the sectional meetings begin, and there will be little leisure until we embark for the homeward voyage at Beira, a month hence.
(*Davis, 'Boston Evening Transcript', 16 September 1905*)

The problem which most provoked his interest and to which he constantly returns, under one heading or another, was the general economic future of the country. The Boer War had just been fought. Among the confusion of feelings generated by it Davis picked out the position of the Boers in relation to the British, their conquerors, calculated the agricultural and mining prospects and above all else considered the future status of the native within this multi-racial, polyglot society. Looking far ahead he sensed that the lowly position of the native presented a very grave problem. He offered no solution though he believed that a continuing policy of white supremacy would only satisfy one section of the society and would do much to aggravate the other unfavoured section. As a foreigner he was not emotionally affected by what had happened before and his analysis is clearer for that. Some of his conclusions express his own opinions; others are a careful evaluation of what he had learned from speaking with residents and government officials.

From Rogers's party in the Karroo, three of us, Penck of Vienna, Coleman of Toronto, and myself, made a hurried run northeastward, and joined another excursion led by Anderson, geologist of Natal, and Molongraaff, formerly geologist of the Transvaal. This was in the Brijhoid district, transferred from the Transvaal to Natal at the close of the Boer war, and here we saw something of rural Boer life, as well as of geological problems. The few Boers whom we met impressed us as men of strong character and good intelligence; and we gained the impression here and elsewhere that, while they were not altogether happy in the present situation, yet they accepted it in good faith, and worked with good will toward the future. They did not appear to harbor bad feeling against the British as such, although there were abundant signs of political antagonism between the scattered farmers and the concentrated mine owners; but there were not words enough to express the contempt felt for the National Scouts, or Boers enlisted under the British flag in the latter part of the war. We saw something of the Zulus on this occasion, looked into a kraal (village of stick and straw huts) and heard much from the colonists as to the necessity of maintaining white supremacy – a subject on which men of a rougher nature expressed themselves violently, but on which those of finer fibre spoke with much feeling of responsibility, and with perception of the difficulty of the problems ahead and of the need of well-tempered justice in attempting to solve them. Nearly every one, except the missionaries themselves, described the Christianized natives as less honest than the unconverted, and we thus gained the impression that the missionaries had gone too far and too fast. The natives taught in Jesuit schools were less criticised . . .

Farming and irrigation in the veld were fruitful subjects for discussion. None of the party felt ready to advise the young British farmer of small capital to emigrate to the inner colonies of South Africa, the difficulties were too great: if he went, he should begin by learning to do as the Boers do, and make a living, at least as well as they can, in the dry, open country, far from supplies. Irrigation offers difficulties because of the individual independence of the Boers, as well as because of the rarity of perennial streams. One of our party declaimed against the gold-seeking population, saying that as they were so generally antagonistic to the farmers they were, on the whole, an injury to the country; they plundered it instead of developing it. Another, an Australian by birth, took the opposite view, and instanced Australia and California as countries to which large populations had been attracted by the gold fields, but in which the gold fever had now given place to steady agricultural and industrial work. The fact is that, were human nature otherwise, self-sacrificing immigrants might populate the veld for the sake of expected prosperity for children and grandchildren; but with human nature as it is, some visible temptation, some immediate reward must be offered to bring a new population into an open wilderness: and the gold of the Rand certainly serves this purpose, whatever harm it may otherwise cause.

THE 'YOUNGER SONS' FAILURE

The administration of the Crown colonies by 'younger sons' as the phrase is, mostly university men sent out from the Colonial Office in London, was found to cause much dissatisfaction among the people in the Transvaal; and this feeling was echoed by one of our party, a university man himself and a competent judge of the problem, who said that from what he had heard on this journey he felt that the system of university training at home should be reformed: for instead of developing the expected executive and administrative capacity, it seems to have produced an inept awkwardness in the treatment of South African affairs. The management of the native races, everywhere in great majority, is a problem that was much discussed, but with no very helpful solution. It is so serious a matter that, simply on the ground of avoiding a country of mixed races, some other colony than South Africa might well be chosen by the intending emigrants. It is agreed down there that the white race is to direct and dominate the black that the blacks are to labor and obey. Here is a curious contrast with conditions in Australia, where the blacks, few in number, are practically ruled out from all chance of labor; witness the difficulty about the contract for carrying the mails home from Australia. No steamship lines with other than white crews were allowed to compete for the contract. The P. & O. and the Oriental lines both had Lascar stokers, and for a time the mails were delayed. The P. & O. had the contract for outward mails and would not change its ways; the Oriental line gave in after a time, discharged the Lascars and got the homeward mails . . .

On the high Veld as well as in the Karroo, barrenness and emptiness were the prevailing impressions; a gray-brown country, no trees, the stream beds nearly all dry at the season of our visit, and very few houses except at the stations, and very few cattle: sheep were more abundant, though 'Cape wool' no longer has the place in the London market that it held formerly. We saw the country near the end of the dry season, when everything was at its worst, dry and dull. We were told that with the coming of the rains, the aspect of the plains changes: dead looking plant-stems are refreshed into life in a few hours: a green tint soon appears over the plains, and flowers are abundant for a while, quickly developing seeds before they dry away again. The Boers, who have been called unprogressive, really deserve great credit for developing a manner of living in this country: and one of our party who most fully recognized this was an English captain of yeomanry who had for two years fought against them. The Veld is divided into large estates, with a house and a plantation of Australian trees at a point where water can be had, and plenty of unfenced grazing ground on all sides. The land has not been plundered by overstocking, as has been too often the fashion in our arid West, especially with the sheep men: it has been lived upon in a frugal manner as a permanent home: the methods of occupation are well adapted to the dry climate, which prevents agriculture, and to the abundance of untutored natives who serve as herdsmen. It was well said that if English farmers propose to colonize there, they should first learn how from the Boers, and not attempt to introduce the ways of a wetter country. The few Boers whom we met were far better than the reputation given them during the war . . .

A feature of Natal is the large number of turban-wearing (East) Indians profit-ably engaged in the smaller industries. They make good house servants, they are largely employed by the railways, many of them are market gardeners on small holdings, and they are said to be displacing Europeans in the way of small shop-keeping. The point of all this is, that they work more steadily and intelligently than the Kaffirs, and live on a lower scale than the whites. Many of them have come from India as five-year contract laborers, but the Indian Government insists that at the end of this time they may settle permanently in South Africa, and be free to take up any work that they like. They may become citizens of the colonies, and acquire the right to vote; but suffrage for Africans and Indians is so restricted in Natal that practically none of them exercise it. In Cape Colony a large number of non-Europeans are qualified voters, but their representation in the Assembly is so restricted that it can never reach a majority. In both of these self-governing colonies, white domination is an undisputed principle by the race now in power.

(*Davis, 'The Nation', 16 November 1905*)

Davis' generous view of Boer character and his optimistic belief that the more liberal representatives within their ranks would ensure fair treatment for the blacks has not been supported by the present constitutional frame-work of the South African state. The repression and inequality which he hoped would disappear has instead been strengthened and granted the force of law.

The great question in Johannesburg today is the employment of Chinese labor. This question arose after the war, when the Kaffirs – the common name for all kinds of native laborers – elated with war-time wages, were slow to return to the lower terms that the mine owners thought enough; and when, moreover, the available natives did not seem numerous enough to supply the labor demanded in the mines as fast as work was resumed with the establishment of peace. In order promptly to secure sufficient cheaper labor, a bill was passed at the instance of the mine owners by the Legislative Council on February 10, 1904, signed by the King, and put into effect on May 19, allowing the importation of Chinese laborers for periods of three years to work in the mines alone. The bill contains numerous specifications and restrictions as to manner of recruiting in China, transportation to, reception in, and ultimate return from Durban, working days and holidays, wages, manner of payment, space in compounds, right of appeal to magistrates, permits for brief absence from the compounds, conditions under which families may be sent for, and under which any Chinaman may resign at any time and return home at his own expense, and particularly as to the kind of work he must not do in the Transvaal. If he so desires, re-enlistment for a second three-year term is permitted, but this contingency has not yet been reached. A third enlistment is not allowed; the coolie must then be returned to China if he has not been taken back at the end of his first term. The mines and miners are always open to governmental inspection. Nearly 50,000 Chinamen have been brought over under these conditions.

This is not slavery, as it is excitedly called in some of the English papers, nor anything like it. It is not so bad as signing on many a sailing vessel, where there is no inspection possible and where the rough tyranny of captain and mates cannot be escaped till the end of the voyage. It is not out of comparison with enlistment as a private soldier, except that there is no bright uniform. It is vastly better than the condition of many Italian laborers under railway contractors in the United States, where the work is hard, and where the manner of living is often dirty and wretched in the extreme. The compounds, in which the Kaffirs as well as the Chinese are kept, were regarded by all who saw them as large, well arranged and relatively clean. The Chinese compound that I visited had a large washroom, where the men could have a tub on coming up from work; a good kitchen where Chinese food was prepared by Chinese cooks; a smaller kitchen where the men could prepare their own food in their own way if they chose; plenty of hot water to be drawn, steaming, from a tap into individual teapots whenever wanted; abundant clear space among the buildings; and barracks clean to the eye and without offensive odor. We saw hundreds of the men coming up from the mine and wandering about the compound; good humor was their prevalent expression, and only a few looked sulky. There was abundant chance of the escape of any stray Chinaman without a permit; . . . a number of them have run away from work, and of these the few who have become marauders, stealing and murdering among Kaffirs, and Indians, have caused, naturally enough, a great commotion. They are the black sheep of an industrious flock.

This is not slavery; but whether the importation of thousands of Chinamen on contract labor will prove to be a wise economic act is a very different and very difficult question . . . The device of importation appears, however, to be more expensive than was expected, and less satisfactory as a remedy for real or imaginary evils, and in this respect Chinese labor has some likeness to the present British regime, which has proved more costly to the mine owners, who were essentially responsible for the Boer war, than the less elaborate government that it replaced. There is already a strong feeling in favor of a change from a crown colony, administered from London, to a self-governing colony, and of a replacement of 'younger sons' sent out from England, by resident Afrikanders or Colonials as officials. (*Davis, 'The Nation', 16 November 1905*)

It is interesting to see how accurate some of Davis' conclusions have proved. The following opinion is also strikingly modern in significance:

An official told us candidly enough that, inasmuch as transportation charges were very heavy, and as the export of products (apart from metals) would therefore be commercially unsuccessful, the policy of the company was to develop the needs of the natives as to clothing, household utensils, and so on; the native then would be led to work more steadily than he does now, in order to get the money with which to satisfy his growing needs, and industry and trade would thereby prosper. As a commercial policy, this is commendable; but it was well pointed out by one of our party that, as needs are cultivated and satisfied, intelligence will increase; and with increasing intelligence there will be growing ambitions; and as intelligent ambitions are aroused, the blacks, numerically in great majority, may not be content to remain subordinate to the whites, as they now are. (*Davis, 'The Nation', 16 November 1905*)

On the other hand, Davis' gloomy deterministic forecast for Johannesburg has proved wrong, and the city has continued to grow long after its original reason for growth has gone.

Johannesburg itself was immensely interesting from its extraordinary artificialness, a metropolis in a wilderness, with a bustling population of about 80,000 Europeans and as many natives – not counting the Kaffirs and Chinese in the mining compounds, with many large buildings and fine shops, tramway service, and residential suburbs, all set down in a high and dry country, nearly 6,000 feet above the sea, bare of trees, scant of water, empty of people for miles and miles together – a country apparently fit for little more than cattle-raising, and now almost deprived of cattle by rinderpest and other diseases, unsuited for wheat farming because of the rust that comes with the summer rains, and without high mountains to supply perennial streams for irrigation; a city where the working population is recruited from India and China, as well as from the African colonies, and where many of the European and American population would seem to be held only by high salaries or high wages, ready to escape homeward as soon as possible – for who would make a permanent residence at this centre of dust in a region of dreariness, if the possibility of living in a more verdant and versatile country were open to him? (and yet this is, I fear, only the stranger's view); a city where the interest of every one centres in the mines, from which the monthly output is £1,700,000 and yet from which the profits to shareholders are said on the average to be small, where 'Kaffirs Rising' as a headline in the morning papers suggests to the uninitiated outsider the need of getting a gun and joining the militia to suppress a native insurrection, but shows to the residents that their mining shares are advancing in price (and every one has some stock in the mines); a city so civilized that it would soon go naked and starve if supplies of all sorts were not received from thousands of miles over seas and lands, and yet so dependent on the mines that it will remain only as strange ruins when its population dwindles away after the Banket is worked out some thirty or fifty years hence, unless some unsuspected source of wealth is then discovered. (*Davis, 'The Nation', 16 November 1905*)

Among the serious memories of the tour Davis introduced an occasional lighter note. Always his humour is rather parched and pedantic but its appearance helps to restore one's faith in his human qualities. However, over the incident of the lost spectacles, our sympathies are on the side of the unfortunate Englishman:

When the train stopped for water, it was amusing to see the party emerge with butterfly nets, collecting bottles, hammers and plant boxes, scatter among the trees, and return unwillingly when the whistle sounded . . .

VICTORIA FALLS OF THE ZAMBESI

The falls and the new railroad bridge across the gorge were the special sights here. The former I shall attempt to describe elsewhere, in their physiographic aspects; the great arch of the latter is a fine engineering feat, the strength of whose single span is far beyond the comprehension of the natives. One of the chiefs who had watched the construction of the bridge with much curiosity predicted that it

FIG. 56. The British Association field party leaving Livingstone Island on the Zambesi in 1905 (Photograph by Mrs Ada Cleland)

would not be strong enough when finished to support a man's weight; when an engine and train crossed it, he said, 'The finger of God holds it up.'

We walked down a zig-zag path to the river in the gorge below the falls; boats rowed by natives took us out to Livingstone Island, just above the falls [Fig. 56]; one of the boats was paddled by a dozen men standing in the bow; a fine sight their black arms and backs made, with paddles all lifted, and I snapped a picture of them from the stern of the boat first requesting a lady passenger to lower her parasol; an unimportant incident, truly, except as it affords the following illustration of a certain brusque frankness that is sometimes met with among our English cousins. On returning from the island, I had as companion an eminent professor who, on hearing my satisfaction at having caught a picture of the black paddlers, exclaimed, 'Oh, it was you who did that, was it! Well, I damned you!' 'What for?' 'Why, when that parasol was lowered, it knocked my glasses off into the river, and I shall not be able to see anything till we get back to Bulawayo.' (*Davis, Harvard Co-operative Society Magazine, 1905*)

Darwin's presidential address was given in two parts, one at Cape Town, the other at Johannesburg; it treated molecular and stellar physics in a style that must have conveyed a somewhat severe impression of popular science to the large audiences by which the halls were crowded on both occasions, but this was alleviated by the lighter vein of his frequent informal remarks – as, for example, at Pietermaritzburg, when he repeated the story of a dispute reported by a lady's maid: 'The butler says as 'ow we're all descended from Darwin, but the cook says we hain't, and they can't agree.' (*Davis, 'The Nation', 16 November 1905*)

Davis' fondness for word-painting, in the style of the great Western explorers Powell and Dutton, often shows itself in his writings on South Africa. He was frequently moved by the beauties of nature and tried hard to pass on his own emotional reaction. At times, as in the following extract, he is truly successful:

Table Mountain was so good as to exhibit itself in varied expressions; sharp and clear with westerly wind for most of our five days, once half hidden in rain, once with its 'cloth' of cloud hanging down over the town. The cloud wisps at the ravelled edge of the overhanging cloth could be seen dissolving away as they were swept down in the cataract of a south-east gale; and damp, as the wind must have been in its cloudy upper portion over the flat mountain top, it was agreeably dry when it had descended to sea level. The southeast wind which forms the 'table cloth' cloud blows in so violently from the southern ocean and across False Bay on the farther side of Table Mountain range, as to sweep sand up from the occasional beaches and clothe the windward mountain slope with patches of dunes. (*Davis, 'Harvard Co-operative Society Magazine', 1905*)

The next extract shows that his early botanical knowledge has not been forgotten even though we know that concentration on his career had put an end as early as 1880 to his studies in botany and entomology.

The botanists told of the extraordinary richness of the South African flora; for example, of four hundred and twenty known species of heather in the world, four hundred are found in the southwest coast belt of Cape Colony, and three hundred are found on Cape Town peninsula, an area about equal to that of the Isle of Wight. The dry interior afforded many examples of plant adjustment to aridity: small leaves, covered with down or felted hairs, to reduce evaporation; quick growth of annual plants, which reach maturity in the short wet season and remain as seeds for the rest of the year, perennials which shrivel as if dead in the dry months, and become bright, and green after a few hours of rain; a double set of roots, the tap root going down deep for ground water, and a film of surface roots, as if to gather moisture from dew. Very curious are the plants (mostly Proteas) in which pollination is effected by birds instead of insects, not by birds of the humming type but by perching birds, and as if to meet them the flowers are turned backwards from the end of a twig toward the branch. While the gums (Eucalyptus) and wattles (Acacias) have been beneficially introduced in great number from Australia, the prickly pear (Opuntia) from America, has become a pest, even causing the abandonment of certain farms. (*Davis, 'Harvard Co-operative Society Magazine', 1905*)

The scientific results of Davis' trip to South Africa were published in two papers, 'Observations on South Africa' (1906E) and 'The mountains of southernmost Africa' (1906C). As we have seen, between the middle of August and the middle of September 1905 Davis travelled from Cape Town, to the Karroo, Natal, Johannesburg, Kimberley, Bulawayo, Victoria Falls, Salisbury and Beira. His scholarly observations were, in the main, limited to three topics:

1. The Cape Colony Ranges. Drawing constant analogies with the Appalachians, Davis described the water gaps, stream capture and showed that the present drainage divide must have been the result of a long period of stream competition. Although he found that evidence for more than one cycle of erosion was less clear than in the Appalachians, he proposed that the present drainage was developed from that super-imposed from an uplifted earlier surface, remnants of which exist on the ridge tops. In one of the major valleys, that of the Buffels River, he thought that the trenching of the terrace plain was more likely due to uplift than to climatic change.

2. The Dwyka tillite. Davis described the features of the sedimentary petrology in minute detail, comparing them with those of the New England tills. He suggested that evidence of both glacial and non-glacial epochs was present and postulated that the deposit was laid down by the action of a broad and continuous ice sheet. Considering the possible causes of glaciation, he found difficulty in those related to changes of the present climatic regime, of land and sea, of ocean currents and in a general refrigeration. In the light of the present status of the continental drift theory, it is interesting that on the subject of 'shifting of the poles' he concluded that further evidence might force 'this daring,

gratuitous, and discredited hypothesis to be taken seriously into account' (1906E, p. 420).

3. The Veld. Visiting this region with Albrecht Penck, Davis proposed that this was an uplifted peneplain of some relief acted upon by sheet-floods. The flat areas of higher elevation he interpreted as those involved in the central differential upwarping of the surface of the former low-lying peneplain and not, as Passarge had suggested for similar regions of southern Africa, as desert surfaces produced at high elevation independent of grand baselevel. The occurrence of marginal gorges and raised marine terraces were evoked by Davis in support of his thesis. Following his usual pattern of rejecting the possible effects of climatic changes during the present cycle of erosion, Davis proposed that occasional thunderstorm activity was the cause of recent trenching of the valley alluvium of the Veld.

A later result of this trip to South Africa was an analysis of the 'geographical factors' in its development (1911O). After a rather deterministic treatment of the influence of distance, remoteness and climate upon the human geography, Davis discussed the industrial and racial diversity of the country, concluding 'let us hope that the minority may act so justly as never to tempt the majority to violent revolution' (1911O, p. 146). Davis' geographical determinism was further underlined by an address which he gave at the St Louis World's Congress of Science and Arts in 1904 (1904I). In the course of this lecture Davis called for an increase of employment of deductive imagination and theory building in geography and geology; stated that races, history and religion are 'seen as if in a mirror held to nature'; and proposed that the inorganic features of the earth may be studied either for their own sake or 'with continuous attention to the controls that they exert over the inhabitants' (1904I, p. 683). The following quotes are very informative as to Davis' geographical views:

> The evolution of the earth and the evolution of organic forms are doctrines that have reinforced each other; the full meaning of both is gained only when one is seen to furnish the inorganic environment, and the other to exemplify the organic response. (*Davis, 1904I, p. 675*)

> The races of mankind ... are obviously determined by the larger features of the land. (*Davis, 1904I, p. 675*)

> So wonderful is the organization of these land and water forms in physiographic maturity and old age, so perfect is their systematic interdependence, that one must grudge the monopoly of the term 'organism' for plants and animals, to the exclusion of well-organized forms of land and water. By good fortune, 'evolution' is a term of broader meaning; we must share its use with the biologists; and we are glad to replace the violent revolutions of our predecessors with the quiet processes that evolution suggests. (*Davis, 1904I, p. 679*)

When regarded objectively, the geography of today is nothing more nor less than a thin section at the top of geology, cut across the grain of time; and all the other thin sections are so much more like the geography of today than they are like anything else, that to call them by another name – except perhaps 'paleogeography' – would be adding confusion to the earth's past history instead of bringing order out of it. (*Davis, 1904I, p. 681*)

For even if man's will sets him high above other forms of life, it must not be forgotten that his will often leads him along physiographic lines, and that he possesses many structures and habits entirely independent of his will, and similar to the structures and habits of lower animals as examples of ontographic responses. Even human houses and roads are only different in degree from the houses and roads made by animals of many kinds. (*Davis, 1904I, p. 684*)

The homeward journey from Beira was hindered by a delay in the Suez Canal and so, when the ship reached Marseilles on Tuesday, 17 October, Davis with many of the passengers left it and hurried across France instead of going around Iberia to Southampton. He found time, however, for a day in the Cévennes (1905M) and to recognize an uplifted peneplain there. A week later Davis sailed from Liverpool:

> ... and that week was as full of entertaining variety as any week on the trip. Tuesday evening was spent with a geological companion at Tarascon, just for the fun of it, and there in a spirit of bold adventure we ventured across the bridge over the hurrying Rhone to Beaucaire. Running up the Nîmes–Clermont line that night, we had a grand walk on Wednesday in clear and cold weather, from La Bastide de St. Laurent along a good road that rose over the upland and then ran down in slanting contours around the hills and spurs of the Cévennes. There in the deep valley of the Borno we came on St. Laurent les Bains, a village that is fitted in the most charming fashion to its picturesque surrounding ...
>
> Thursday we were on the train again northward, with a view of the Puy de Dôme on the way to Paris, where Friday was agreeably spent on making and receiving calls.
>
> Saturday morning was given to London, where through the kindness of friends I was put up at a comfortable club. It was the centenary of Nelson's victory, and Trafalgar Square was so crowded that a detour had to be made on the way to Charing Cross station to take a train for a week-end visit down in Kent. Returning on Sunday I found time for an afternoon call in Chelsea and for an evening supper far out 'N.W.' On Monday, there was a morning errand in connection with an apparently British but really American enterprise that is sheltered in the eminently respectable office of *The Times*, a luncheon with geographical friends at the club, and an afternoon run to Liverpool, where I lectured in the evening on the Colorado Canyon before the local geological society. On Tuesday, there was a geological excursion in the morning and a delightful luncheon at noon with new friends, before the *Saxonia* carried me off from the landing stage in the mid-afternoon. (*Davis, 'Harvard Co-operative Society Magazine', 1905*)

CHAPTER FOURTEEN

Ramifications of the Geographical Cycle

COMPLICATIONS OF THE GEOGRAPHICAL CYCLE

By 1899 Davis had carried the ideal geographical cycle about as far as any simple qualitative model could be carried. He first applied it to rivers, then to drainage basins and so to whole regions or vast physical landscapes. It was a visual method of analysis meant, unfortunately, to obviate the need for quantitative analysis.

The few years around the turn of the century witnessed a striking ramification of the cyclic notion. As we have seen, Davis' tentative suggestion of a shoreline cycle (1896G) had been developed by his student F. P. Gulliver (1899) and its complications were later outlined by Davis (1905J, pp. 159–60). Despite the Master's own reticence regarding the development of karst topography, other hands (Richter, 1907; Sawicki, 1909; Cvijić, 1909) were soon to turn the cyclic concept to good account in the classic limestone region of Yugoslavia. It was significant that the two papers presented at the British Association meeting in South Africa dealt with applications of the cycle of erosion to arid (1905L; 1906H) and glacial (1906J) environments. It is most noticeable that after 1900 Davis published no more major articles on the ideal geographical cycle as such, and when his regional knowledge broadened with travels in Turkestan, South Africa and Europe he takes more account of smaller landforms. The development of river-patterns, of river-terraces, of other incised valley forms, of river floodplain meanders, of glacial landforms and so on, are all incorporated into a cyclic framework. The cycle is used as a tool of description or classification; its transmutations are often assumed to be both cause and effect. But the astonishing thing always is how readily Davis now modifies and discards the simple cycle when it suits his arguments.

In 1899 he wrote,

> The sequence of forms developed through the cycle is not an abstraction that one leaves at home when he goes abroad; it is literally a *vade mecum* of the most serviceable kind. During my visits in Europe the scheme and the terminology of the cycle have been of the greatest assistance in my studies. Application of both scheme and terminology is found equally well in the minute and infantile coastal plains that border certain stretches of the Scotch shore-line in consequence of the slight post-glacial elevation of the land, and in the broad and aged central plateau of France, where the young valleys of to-day result from the uplift of the region

and the revival of its rivers after they had sub-maturely dissected a pre-existent peneplain. The adjustments of streams to structures brought about by the interaction of the waxing Severn and the waning Thames prove to be even more striking than when I first noticed them in 1894. (*Davis, 1899H, p. 503*)

Yet when the reader investigates the purity of the cyclic scheme that Davis is applying to these and dozens of other examples he is often staggered by its flexibility. Thus to take the earlier classic papers (1895E; 1899M) on the contest between the Severn and Thames, whereby the former acquired a large drainage that once went to the western branches of the Thames' system. The development of cuestas and of a trellis drainage originated, according to Davis, when the old mountains in the west of Britain after being submerged and covered with new sediments were affected by 'gradual and intermittent elevation' which was greatest in the west so that the new sediments were exposed as a gently sloping plain. Here the reader will notice two peculiarities. For some unknown reason the uplift is 'gradual and intermittent'; and rather inexplicably but most conveniently, the west rises more than the east. To explain why the present subsequent or lateral streams such as the Severn had developed such wide vales and why the summits of the Cotswolds appeared to be tablelands or 'upper plains', Davis postulated that

> the land has been at least once worn down to a lowland of faint relief (peneplain), and afterwards broadly uplifted, thus opening a second cycle of denudation, and reviving the rivers to new activities; and in the second cycle of denudation, the adjustment of streams to structures has been carried to a higher degree of perfection than it could have reached in the first cycle (*Davis, 1895E, p. 127*)

He then adds with a disarming impartiality

> it must not be assumed that the once nearly level surface of the peneplain was lifted up with perfect equality of elevation. Not only was there probably a greater elevation in the west than in the east; there may have been also widespread or local warpings; ... [in addition] the effects of other movements, as well as of glacial episodes, must be carefully examined when the subject is minutely studied, instead of being broadly sketched. (*Davis, 1895E, p. 141*)

Thus before 1900 the ideal cycle is seen to be readily modified when applied to certain landscapes. Glacial episodes are already considered important. In addition, by 1905 the study of desert climates had also advanced to the stage when the postulation of a geographical cycle in an arid climate was deemed necessary. These numerous variations of the cyclic theme inevitably evoked a masterly performance, a sort of magnificent *apologia*, from Davis.

It seems fitting that an important general statement of complications of the geographical cycle (1905J) should have been presented in 1904 at the Eighth International Geographical Congress at Washington, because it is

Davis' answer to objections against the final scheme of the cycle in its simplest form that had been read before the previous Congress at Berlin in 1899. This reply opens with the demand that landforms, like organic forms, should be studied in view of their evolution, that the processes of evolution should be considered in the main orderly and productive of a systematically-related sequence of surface forms. The strong objections that the cycle is too deductive and too rigid are countered as follows:

> ... the scheme of the cycle is not meant to include any actual examples at all, because it is by intention a scheme of the imagination and not a matter of observation; yet it should be accompanied, tested, and corrected by a collection of actual examples that match just as many of its elements as possible. (*Davis, 1905J, p. 152*)

> It has been urged that the scheme of the cycle is so rigid and arbitrary that it cannot be of service in describing the manifold phenomena of nature. This criticism is a result of regarding the ideal cycle alone, without going on to the modifications by which it is easily adapted to natural conditions. (*Davis, 1905J, p. 153*)

Davis then proceeds to consider the modifications that can be easily made to the ideal scheme so that it meets the complicated examples found in nature. The simple presentation of the ideal cycle postulates a rapid uplift of a landmass followed by a prolonged stillstand, but almost any variation can be conceived.

> Instead of rapid uplift, gradual uplift may be postulated with equal fairness to the scheme, but with less satisfaction to the student who is then first learning it; for gradual uplift requires the consideration of erosion during uplift ... A special case necessitating explanation by slow uplift may be easily imagined. (*Davis, 1905J, pp. 153–4*)

The simple postulate of a still-standing land that is worn down to a plain is said to be convenient for the learner and essential to the analysis of the complete scheme because it allows the normal series of sequential events to be traced and the systematic correlation of forms that characterizes each stage to be recognized.

> It is only after the normal series has been analyzed that the peculiar combinations of forms which result from two or more cycles can be understood ... Only after the norm has been established can the effects of various movements – uplift, depressions, warping, breaking – be duly considered. (*Davis, 1905J, pp. 154–5*)

Davis now proceeds to make it quite clear that *interruptions*, or movements of the landmass, may occur at any stage in the progress of the cycle and that physical landscapes may be 'revived' and 'rejuvenated' by relative uplift, or 'drowned' and so on by relative depression. The movements may be of a simple vertical nature or tilted or block faulted or folded. Because they vary in kind and degree and may effect various stages in a cycle at any time, the variety of possible landform effects becomes tremendous.

> Certain it is that when various kinds and degrees of interruption at various stages in a cycle have been considered, the variety of possible combinations becomes so great that there is no difficulty whatever in matching the variety of nature. The difficulty is indeed reversed; there are not enough kinds of observed facts on the small earth in the momentary present to match the long list of deduced elements of the scheme. A notable example of the deficiency of observations may be noted in connection with belted coastal plains. A number of examples are known in which the upland belt or cuesta is separated from the oldland by a continuous inner lowland, with appropriate drainage by longitudinal subsequent streams, diverted consequent headwaters, short obsequents running down the infacing slope of the cuesta and beheaded consequents on the outlooking slope. The elastic scheme of the cycle easily matches these facts of observation, but there are no known examples of belted plains in earlier stages to match the several deduced phases of cuesta development which are familiarly included in the scheme of the cycle. Rigidity and deficient variety can therefore hardly be regarded as defects of the elaborated scheme of the geographic cycles. (*Davis, 1905J, p. 156*)

Thus the elaborated scheme of the geographical cycle had ample variety provided that the rigid ideal conception of a single cycle is learnt in conjunction with combinations of interrupted cycles. But Davis goes on to argue that it is not true that the ideal cycle is only an abstraction. Although many landscapes show repeated earth movements or interruptions that result in a succession of partial cycles, and although baselevel peneplains are rare today:

> ... cycles of erosion have in some cases reached at least the penultimate stage, without significant interruption, for the explanation of certain uplands as now uplifted and partly dissected peneplains is supported by a large array of strong evidence ... the facts of observation on partly dissected uplands find no explanation save that which carries them all uninterruptedly through the stages of short youth and longer maturity far into very long old age. Thus in these cases at least there is full warrant for the original postulate of a stillstanding land mass. (*Davis, 1905J, p. 157*)

The interruptions discussed above are, however, not the only complications of the geographical cycle. Volcanic eruptions or 'accidents' may occur locally or regionally at any stage of a cycle. More important and widespread are special agencies and accidental climatic changes. The normal cycle tacitly implied that land sculpture was effected by the rain and rivers associated with temperate climates.

> ... the greater part of the land surface has been carved by these agencies, which may therefore he called the prevailing or normal agencies, but it is important to consider the peculiar work of other special agencies, namely ice and wind ... It is only recently that the conception of a whole cycle of glacial erosion has been discussed, and a whole cycle of wind erosion is as yet a relatively neglected

consideration; yet it cannot be doubted that both of these special ideal cycles deserve deliberate analyses, for until such analysis is made the next step, and one of more frequent application, cannot be safely taken, namely, the combination in a single cycle, uninterrupted by land movements, of a succession of normal and special agencies. (*Davis, 1905J, pp. 157–8*)

We are thus led to the important complication of climatic changes as distinct from the normal changes of temperature and precipitation that are associated with the uplift or lowering of landmasses. Climatic changes that are independent of the ideal cycle, as, for example, from non-glacial to glacial or from humid to arid conditions, may occur at any stage of a cycle and are therefore noted by the 'semi-technical term accidents'.

Davis goes on to discuss the special nature of the cycle of shorelines and the possibility of systematically describing the cloak of creeping rock waste that covers a graded hillside. He ends by defending the terminology of the cycle and attempts to disarm all future criticism by the following amazing claim.

It thus appears that the scheme of the simple ideal cycle may be gradually and systematically modified until its deductions cover all manner of structures, agencies, waste forms, interruptions and accidents. When thus conceived it is a powerful instrument of research, an invaluable equipment for the explorer. It is not arbitrary or rigid, but elastic and adaptable. It is a compendium of all the pertinent results of previous investigations. (*Davis, 1905J, pp. 160–1*)

It was to prove the versatility of the cyclic scheme, and not to disprove the reality of the ideal cycle, that Davis in the first decade of the twentieth century wrote so extensively on the landforms of deserts and of glaciated mountains.

THE GEOGRAPHICAL CYCLE IN AN ARID CLIMATE

Davis set forth this theme in one long article (1905L) and in a summary of it (1906H). Its conclusions were based largely on the findings of J. Walther in *Das Gesetz der Wüstenbildung* (Berlin, 1900), of S. Passarge in *Die Kalahari* (Berlin, 1904) (Davis, 1905E) and on his own observations in the arid regions of the western United States and western Asia. But Davis also knew well the writings of W. Bornhardt on East Africa (1900) and of Albrecht Penck on the influence of climate on the land surface which we intend to discuss in Volume Three of this history. Thus there was already a fairly imposing literature on desert landforms.

Despite the statement that 'no special conditions need be postulated as to the initiation of the arid cycle' (Davis, 1905L, p. 382), in fact severe constraints of high relief and enclosed basins were assumed, and this assumption proved vital in predetermining the sequence of erosional and depositional events which followed it.

Let consideration be given to an uplifted region of large extent over which an arid climate prevails. Antecedent rivers, persisting from a previous cycle against the deformations by which the new cycle is introduced, must be rare, because such rivers should be large, and large rivers are unusual in an arid region. Consequent drainage must prevail. The initial slopes in each basin will lead the wash of local rains toward the central depression, whose lowest point serves as the local base-level for the district. There will be as many independent centripetal systems as there are basins of initial deformation; for no basin can contain an overflowing lake, whose outlet would connect two centripetal systems: the centripetal streams will not always follow the whole length of the centripetal slopes; most of the streams of each basin system will wither away after descending from the less arid highlands to the more arid depressions. Each basin system will therefore consist of many separate streams, which may occasionally, in time of flood or in the cooler season of diminished evaporation, unite in an intermittent trunk river, and even form a shallow lake in the basin bed, but which will ordinarily exist independently as disconnected headwater branches. (*Davis, 1905L, pp. 382–3*)

Davis' ideas of original desert structural forms were undoubtedly influenced both by his study of Gilbert's early work on the Basin Range region and by the events of 1901–4. In the first of these years Spurr (1901) had published an attack on the dominantly block faulting Basin Range theory of Gilbert and of his supporters Powell, King and Dutton. Spurr had identified, quite correctly as it transpired, thrusts in the mountain blocks of Nevada and California which had no topographic expression and went on to explain the terrain as having developed from the erosion of broad folds, without much subsequent boundary block faulting. In 1902 and 1904 Davis visited the Basin Ranges of Utah and Nevada accompanied, firstly, by his Harvard students Huntington and Goldthwait, and, secondly, by the French geomorphologist Allorge. These trips resulted in two important publications (1903G and 1905B) in which, by the use of a high degree of deduction, Davis defended the faulting theory of Gilbert and presented massive physiographic evidence for the recognition of faults marginal to the present ranges while, at the same time, showing that erosion had in some places destroyed the topographical evidence of marginal faults. Davis was to abandon this structural theme for another twenty years but its immediate effect was to ensure that the arid cycle was to have much more constraining structural assumptions than had the geographical cycle.

In the youthful stage of the arid cycle:

. . . the relief is ordinarily and rapidly increased by the incision of consequent valleys by the trunk rivers that flow to the sea. In the early stage of the arid cycle the relief is slowly diminished by the removal of waste from the highlands, and its deposition on the lower gentler slopes and on the basin beds of all the separate centripetal drainage systems. Thus all the local baselevels rise. The areas of removal are in time dissected by valleys of normal origin; if the climate is very arid, the uplands and slopes of these areas are either swept bare, or left thinly

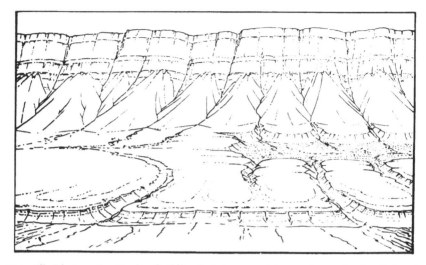

F I G. 57. Diagram by Davis of cliffs of shale capped by sandstone in the Colorado
Plateau, with basal pediments dissected by streams

veneered with angular stony waste from which the fine particles are carried away
almost as soon as they are weathered; if a less arid climate prevails on the uplands
and highlands, the plants that they support will cause the retention of a larger
proportion of finer waste on the slopes. The areas of deposition are, on the other
hand, given a nearly level central floor of fine waste, with the varied phenomena of
shallow lakes, playas, and salinas, surrounded with graded slopes of coarser waste.
(*Davis, 1905L, p. 383*)

Streams, floods, and lakes are the chief agencies in giving form to the aggraded
basin floors, as well as to the dissected basin margins in the early stages of the
cycle; but the winds also are of importance: they do a certain share of erosion by
sand-blast action; they do a more important work of transportation by sweeping
the granular waste from exposed uplands and depositing it in more sheltered
depressions, and by raising the finer dust high in the air and carrying it far and
wide before it is allowed to settle. Wind action is, moreover, peculiar in not being
guided by the slopes or restrained by the divides which control streams and
stream systems. (*Davis, 1905L, p. 384*)

Maturity is reached when erosion and deposition begin to nullify the
original structural controls over interior drainage.

Continued erosion of the highlands and divides, and continued deposition in the
basins, may here and there produce a slope from a higher basin floor across a
reduced part of its initial rim to a lower basin floor. Headward erosion by the
consequent or subsequent streams of the lower basin will favor this change, which
might then be described as a capture of the higher drainage area. Aggradation
of the higher basin is equally important, and a change thus effected might be
described as an invasion of the lower basin by waste from the higher one; this

corresponds in a belated way to the overflow of a lake in a normal cycle. There may still be no persistent stream connecting the two basins, but whenever rain falls on the slope that crosses the original divide, the wash will carry waste from the higher to the lower basin. Thus the drainage systems of two adjacent basins coalesce, and with this a beginning is made of the confluence and integration of drainage lines which, when more fully developed, characterize maturity. The intermittent drainage that is established across the former divide may have for a time a rather strong fall; as this is graded down to an even slope, an impulse of revival and deeper erosion makes its way, wave-like, across the floor of the higher basin and up all its centripetal slopes. The previously aggraded floor will thus for a time be dissected with a bad-land expression and then smoothed at a lower level; the bordering waste slopes will be trenched and degraded. At the same time, the lower basin floor will be more actively aggraded. . . .

The higher local baselevels are thus, by a process of slow, inorganic natural selection, replaced by a smaller and smaller number of lower and lower baselevels; and with all this go a headward extension of graded piedmont slopes, a deeper dissection of the highlands, and a better development of their subsequent and adjusted drainage. (*Davis, 1905L, pp. 386–7*)

As the processes thus far described continue through geological periods, the initial relief will be extinguished even under the slow processes of desert erosion, and there will appear instead large, rock-floored plains sloping toward large waste-floored plains; the plains will be interrupted only where parts of the initial highlands and masses of unusually resistant rocks here and there survive as isolated residual mountains. At the same time, deposits of loess may be expected to accumulate in increasing thickness on the neighboring less arid regions. The altitude at which the desert plain will stand is evidently independent of the general baselevel – or sea-level – and dependent only on the original form and altitude of the region, and on the amount of dust that it has lost through wind transportation.

The most perfect maturity would be reached when the drainage of all the arid region becomes integrated with respect to a single aggraded basin-baselevel, so that the slopes should lead from all parts of the surface to a single area for the deposition of the waste. The lowest basin area which thus comes to have a monopoly of deposition may receive so heavy a body of waste that some of its ridges may be nearly or quite buried. Strong relief might still remain in certain peripheral districts, but large plains areas would by this time necessarily have been developed. In so far as the plains are rock-floored, they would truncate the rocks without regard to their structure. (*Davis, 1905L, pp. 388–9*)

The beginning of old age in the arid cycle is distinctly indeterminate.

During the advance of drainage integration the exportation of wind-borne waste is continued. At the same time, the tendency of wind-action to form hollows wherever the rocks weather most rapidly to a dusty texture would be favored by the general decrease of surface slopes, and by the decrease of rainfall and of stream-action resulting from the general wearing down of the highlands. Thus it may well happen that wind-blown hollows should be produced here and there, through the mature and later stages of the cycle, and that they should even during early maturity interfere, to a greater or less degree, with the development of the integrated drainage, described above. In any case, it may be expected that wind-blown hollows would in late maturity seriously interfere with the maintenance of an integrated drainage system. Thus it appears that, along with the processes which tend toward the mature integration of drainage there are other processes which tend toward a later disintegration, and that the latter gain efficiency as the former begin to weaken. (*Davis, 1905L, p. 390*)

As the drainage becomes more and more disintegrated, and the surface of the plain is slowly lowered, rock masses that most effectually resist dry weathering will remain as monadnocks – Inselberge, as Bornhardt and Passarge call them in South Africa. At the same time, the waste will be washed away from the gathering grounds of maturity and scattered in the shallow hollows that are formed here and there by the winds as old age approaches. (*Davis, 1905L, p. 392*)

The surface ever wearing down, the waste ever washed irregularly about by the variable disintegration of the drainage system and continually exported by the winds, a nearly level rock-floor, nowhere heavily covered with waste, and every-where slowly lowering at the rate of sand and dust exportation, is developed over a larger and larger area; and such is the condition of quasi-equilibrium for old age. At last, as the waste is more completely exported, the desert plain may be reduced to a lower level than that of the deepest initial basin; and then a rock-floor, thinly veneered with waste, unrelated to normal baselevel, will prevail throughout – except where monadnocks still survive. This is the generalization that we owe to Passarge; it seems to me secondary in value only to Powell's generalization concerning the general baselevel of erosion. (*Davis, 1905L, p. 393*)

In the development of his arid cycle Davis drew heavily on Passarge's idea of 'levelling without baselevelling', which he termed 'Passarge's law' (1905E), although accepting its general application with reservations:

The deductive method by which most of the preceding paragraphs are charac-terized may be regarded by some readers as reaching too far into the field of untestable speculation. It is true that the examples of observed forms, by which the deduced forms of every stage should be matched, are as yet not described in sufficient number; but this may be because desert regions have not yet been sufficiently explored with the principles herein set forth – particularly Passarge's law – in mind. On the other hand, the examples of desert plains in South Africa, described by Passarge as plains of the Bechuana (Betschuana) type, suffice to show that the stage of widespread desert-leveling has actually been reached in that region, and thus justify all the earlier stages; for, however many land movements may have interrupted the regular progress of preceding cycles, the occurrence of widespread rock plains proves that at least the present cycle of arid erosion has been long continued without disturbance. . . .

If Passarge's views be now accepted, it follows that no truncated uplands should, without further inquiry, be treated as having been eroded when their region has a lower stand with respect to baselevel; the possibility of their having been formed during an earlier arid climate as desert plains, without regard to the general baselevel of the ocean, must be considered and excluded before baseleveling and uplift can be taken as proved. (*Davis, 1905L, pp. 394–5*)

Passarge holds the opinion that the plains of the *Inselberglandschaft* are smoother than any peneplain can be; for he describes the desert plains as true plains, not as gently undulating surfaces. He states that water is not competent to produce such plains; its power of erosion works chiefly downward, and only by exception laterally; and he concludes that, although long-continued normal erosion may produce a peneplain – that is, a low, undulating hilly surface – it nevertheless cannot produce a surface like that of the plains in the *Inselberglandschaft*. But, however difficult it may be to wait, in imagination, through the ages required to wear a low hilly region down to less and less relief by the weakened processes of weather and water erosion in the latest stages of the normal cycle, there are certainly some truncated uplands, ordinarily taken to be uplifted peneplains, whose interstream uplands are astonishingly even, and whose surface must have been, before dissection, very nearly plain over large areas; hence it does not seem to me altogether certain that a greater and less a degree of flatness can be taken to distinguish the two classes of plains. (*Davis, 1905L, p. 397*)

Davis concludes with speculations regarding interruptions of the arid cycle by uplift, depression, tilting or warping; and by modifications due to climatic change. Although he does allow for a lack of initial interior basins, his treatment of this possibility is cursory in the extreme:

The absence of inclosing mountains around a continental arid region would permit the development of escaping drainage systems, so that when mature integration was reached it might be developed with respect to normal baselevel, instead of with respect to a local interior baselevel; the Sonoran district of Mexico, as described by McGee, seems to offer examples of this kind. (*Davis, 1905L, p. 402*)

However, it is clear that without initially block-faulted terrain the sequence of forms characteristic of the arid cycle, together with the passage from youth to maturity, rests on much less secure foundations than does that of the 'normal cycle'. One of Davis' major contributions to the study of arid geomorphology at this time was to emphasize the role of fluvial erosion and deposition in deserts, as distinct from the wind action which Walther relied so heavily upon.

THE CYCLE OF GLACIAL DENUDATION

From as early as 1882 Davis (1882B) took a keen interest in mountain glaciation. This was partly due to his love of mountains, for, without being in any way a mountaineer, he liked the physical exertion of climbing. However, from a scientific point of view Davis (1900I, p. 137) wrote: 'Mountain tops are indeed worthy objects of a climber's ambition, but if one wishes to get at the bottom facts let him examine the valleys.' It is not surprising that he chose Europe as his glacial laboratory because of his frequent visits to the continent and because the contemporary geologists who knew most about glacial landforms lived in Europe. True the Swiss, Jean Louis Rodolphe Agassiz (1807–73), who was the prime founder of the glacial theory, had visited Boston in 1846 and lectured at the Lowell Institute and two years later migrated from Neuchâtel to New England where he became professor of zoology at Harvard, and where the youthful Davis greatly enjoyed his lectures on glaciated landscapes. When Agassiz died here in 1873 a boulder from the moraine of the Lauteraar was placed beside his grave but, in fact, during his American life he worked mainly on zoology and was a biologist rather than a geologist. In the meanwhile, in Europe, as described in Volume One of our *History of the Study of Landforms*, a fierce controversy had raged over the occurrence and nature of ice ages. Great progress was made in the knowledge of landforms sculptured by ice and European geologists were soon far ahead of their American counterparts in geomorphological concepts on the action of glaciers and ice sheets. Thus Davis became a mouthpiece in the United States for European ideas on glaciation. Needless to say when the chance arose later he also spread American ideas eastward and lost no time in adding his own indelible stamp: the deductive cycle concept and a series of simple, illuminating sketches which bear the mark of immortality.

Some of his writings on glacial landforms are largely elementary fieldwork and need hardly concern us. All told he wrote twelve main articles on glacial topics but re-publication in various journals, abstractions and the calving off of topics into separate articles brought the number of individual items up to at least twenty-one. There were also the chapters on this theme in his general textbooks, beginning with his first *Physical Geography* which he compiled with the assistance of William Henry Snyder in 1898(E). This included under 'Climatic control of landforms' a succinct well-illustrated account of present ice sheets and glaciers and of the work of ancient glaciers and ice sheets. It must have reached a wide circle of readers as we know that

> my two books, *Physical Geography* and *Elementary Physical Geography*, sold last year to the number of 40,000 or more . . . it enables me to leave Mrs. Davis and the boys in a comfortable summer house near the shore – with a good boat for sailing – and sparerooms for friends to occupy – that is what I call 'Applied Geography'.
> (*Letter: Davis to J. S. Keltie, 13 May 1903; Bakow, Russia*)

Davis' later more detailed articles on glacial features in the American West will be discussed as fine products of more advanced age; of his earlier writings on glaciation those of 1900(H) and 1906(J) were the most influential, partly because both were reproduced in Davis' *Geographical Essays*, edited by Douglas Wilson Johnson in 1909 and after several republishings still enjoy a wide popularity.

The long article on 'Glacial erosion in France, Switzerland and Norway'

(1900H), opens with a confession by Davis that he had formerly been far too conservative in accepting the true erosive power of moving ice, and it is clear that the three-fold purpose of the paper is to illustrate the reality and efficiency of erosion by valley glaciers; to point to the special topographic features resulting from this action distinguishing them from fluvial forms; and to assemble these forms into a cycle of glacial denudation. Most of Davis' heavily-deductive examples were drawn from the incised meandering valley of the Rhue in the Massif Central of France which he visited in January 1899, and the Ticino valley which he examined during the wet May of the same year while based at Lugano. In the former area Davis recognized that the truncation of interlocking fluvial spurs and their reduction to rugged knobs, or roches moutonnées, was due to a considerable amount of erosion by a valley glacier.

> The more or less complete obliteration of the spurs was the result of the effort of the ice stream to prepare for itself a smooth-sided trough of slight curvature.
> *(Davis, 1900H, p. 275)*

> The ice action sufficed to rasp away the greater part of the weathered material and to grind down somewhat the underlying rock, often giving the knobs a rounded profile, but it did not nearly suffice to reduce the rocky surface to an even grade.
> *(Davis, 1900H, p. 277)*

The Ticino Valley had previously been visited by Davis during the family European trip of 1868.

> Thirty years is a long enough time for one to learn something new even about valleys, and on my second visit it was fairly startling to find that the lateral valleys opened on the walls of the main valley of the Ticino five hundred feet or more above its floor, and that the side streams cascaded down the steep main-valley walls in which they have worn nothing more than narrow clefts of small depth.
> *(Davis, 1900H, p. 278)*

He complains of the relatively little attention paid to these valleys compared with the great attention given to the possible scouring-out of deep lake basins by glaciers. But gives (1900H, pp. 310–21) a long review of previous writings on hanging valleys, including the observations of W J McGee, E. Richter, A. de Lapparent, James Geikie, Albrecht Penck and many others. He later gave another review of subsequent work (1907D). What is more, he had written to his 'esteemed friend' G. K. Gilbert of Washington, telling him that all the lateral valleys seemed to be 'hung up' above the floors of the trunk valleys. Gilbert answered from Sitka, Alaska, where he was working with the Harriman Expedition and described the impressive lateral valleys of the Alaskan fiords. He suggested that such discordant laterals should be called 'hanging valleys' and Davis adopted the term. Both fully agreed that hanging valleys 'presented unanswerable testimony for strong glacial erosion' [see also Chapter 15].

F I G. 58. The Val d'Osogna, a hanging lateral valley of the Ticino
(From Davis, 1900H, Fig. 3)

Drawing analogies with the Rhue valley, Davis describes the characteristic features of strongly glaciated valleys as trough-shaped, with broad floors and cliffed, comparatively straight-walled sides. Lateral valleys enter the main trough near the summits of the cliffs. The broadness of the trough and the hanging nature of the lateral valleys cannot be ascribed, as some authors thought, to river action.

> The bottom troughs of the larger Alpine valleys were deepened and widened by ice action. This belief is permitted by the abundant signs of glacial erosion on the spurless basal (trough-side) cliffs, and required by the persistent association of over-deepened bottom troughs and discordant hanging lateral valleys with regions of strong glaciation. (*Davis, 1900H, p. 284*)

This trough, Davis is careful to point out, often lies within a much wider valley with gently sloping sides; it is an inner trough. Hanging valleys resulted not only from the widening of pre-glacial fluvial valleys, but more from their deepening. The glaciation of the tributary valleys may have obeyed Playfair's Law in the accordance of the glacier surfaces, but the valley floors were discordant due to less erosion by the smaller tributary glaciers. Davis wrote another article (1900I) on the Ticino in much the same vein postulating glacial deepening of the inner valleys within previous mature fluvial valleys.

Davis never felt easy in his mind on the question of excessive overdeepening causing the excavation of large lake basins where a mountain glacier met a lowland plain or flat piedmont. The idea had been put forward by A. C. Ramsay, the British geologist, as long ago as 1862 and had then aroused tremendous controversy. It had since been steadily gaining adherents under the lead of Albrecht Penck but Davis believed, with Lyell, Heim and others, that the large lakes at the junction of mountain and plain in northern Italy 'resulted from what has been called valley-warping'. In 1899 he re-examined lakes Maggiore, Lugano and Como with that hypothesis in mind and failed to find evidence of warping but was impressed with the signs of severe glacial erosion. However, he noticed quite accurately that some of the bigger Italian lakes were deepened by a distal enclosure of moraines and adds, probably inaccurately, that

> it seems especially out of proportion to suppose that the maximum erosion by a glacier takes place near its end, as has been done by some authors, on account of the prevalent occurrence of lakes in this situation. (*Davis, 1900H, p. 303*)

A very unsatisfactory attempt is made to explain the occurrence of small rock basins 'that are so often found in the floor of cliff-walled corries'. A change of climate, presumably to a warmer phase, is assumed to cause the trunk glacier to disappear while many of its blunt head-branches remain in the corries.

> Each little glacier thus isolated will repeat the conditions of erosion inferred for the trunk glacier; and if this style of glaciation linger long enough, rock basins

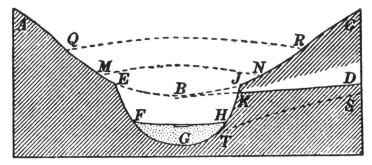

FIG. 59. Features of a strongly-glaciated Alpine valley. The steep-walled minor trough (EFHJ) is buried by gravels (FGH). The floors of the hanging lateral valleys (DK) are graded to the lowest point (B) of the upper valley (AEJC). The uppermost limit of glaciation (QR) seems to be less associated with valley-side steepness than is a lower limit (MN) (From Davis, 1900H, Fig. 4)

may very generally characterize the floors of the corries when the ice finally melts away! (*Davis, 1900H, p. 305*)

There is always an emphasis by Davis on the need to know what the physical landscape was like before the ice occupied it, and as a result of a ten-day walking excursion in Norway he hints at the different combinations possible, such as young, mature and old river valleys, modified by young or mature glaciation.

> The fiords of Norway result from the sub-mature and intense glaciation of a river-valley system whose stage of pre-glacial development is not yet well determined ... The variation of form between the main fiords and their branches gives some indication that the glacial work was accomplished in several successive epochs, with the inter-glacial epochs of normal river work between. (*Davis, 1900H, pp. 307 and 310*)

This quotation hints at sub-maturity and the assumption of a cycle of glacial erosion. In fact Davis precedes this with a clever but rather contrived comparison of river valleys and glacier valleys which are shown to have more resemblances than differences when due allowance is made for their individual peculiarities. Davis uses these points of resemblance to explore the possibility that mountain glaciers, during their ideal life history, might develop an orderly succession of features – a cycle of glacial denudation (1900H, pp. 294–300).

This cycle is not, Davis explains, a development of small glaciers (young) into large glaciers (mature) and thence to the melting (old age) of the ice mass. It relates to the life history of a glacier under a constant glacial climate, from the beginning to the end of a cycle of denudation. Thus young glaciers will be those newly formed in courses consequent upon the slopes of a newly-uplifted land surface; mature glaciers will be those which have eroded their valleys to grade and so have dissected the uplifted surface; and old glaciers will be those which cover the whole lowland to which the upland has been reduced or which are disappearing slowly in the milder climate of the lowered relief.

A critic may perhaps agree with Davis thus far even when a 'constant glacial climate' is specified because the cycle depends on uplift and the wearing down of an upland to a lowland. However, to give practical examples of the working of this scheme is difficult unless the lowlands have a glacial climate. To postulate a milder climate because of the lowered relief would demand a relatively vast reduction of the relief in any area where the lowlands were not under glacial conditions.

Davis, however, did not allow difficulties of this sort to deter him when he had set his mind upon a cycle. He begins

> Imagine an initial land surface raised to a height of several thousand feet, with a moderate variety of relief due to deformation. Let the snowline stand at a height

of two hundred feet. As elevation progresses, snow accumulates on all the upland and highland surfaces. Glaciers are developed in every basin and trough; they creep slowly forward to lower ground, where they enter a milder climate (or the sea) and gradually melt away. (*Davis, 1900H, p. 295*)

Perhaps we should here interpolate that the unkind critic could only assume that as the snowline is at 200 feet and as glaciers normally descend far below the snowline, the glaciers must either enter the sea or spread out as ice sheets near sea-level.

In the youthful stage, Davis states that each young glacier will proceed to cut down its consequent valley until it has graded its bed.

Each young glacier will then proceed to cut down its consequent valley at a rate dependent on various factors, such as depth and velocity of ice stream, character of rock bed, quantity of ice-dragged waste, and so on: and the eroded channel in the bottom of the valley will in time be given a depth and width that will better suit the needs of ice discharge than did the initial basin or trough of the uplifted surface. The upper slopes of the glacial stream will thus be steepened, while its lower course will be given a gentler descent. Owing to the diminution of the glacier toward its lower end, the channel occupied by it will diminish in depth and breadth downwards from the point of maximum volume; this being analogous to the decrease in the size of the channel of a withering river below the point of its maximum volume. A time will come when all the energy of the glacier on its gentler slope will be fully taxed in moving forward the waste that has been brought down from the steeper slopes; then the glacier becomes only a transporting agent, not an eroding agent, in its lower course. This condition will be first reached near the lower end, and slowly propagated headwards. Every part of the glacier in which the balance between ability to do work and work to be done is thus struck may be said to be 'graded'; and in all such parts, the surface of the glacier will have a smoothly descending slope. Maturity will be reached when, as in the analogous case of a river, the nice adjustment between ability and work is extended to all parts of a glacial system. In the process of developing this adjustment, a large trunk glacier might entrench the main valley more rapidly than one of the smaller branches could entrench its side valley; then for a time the branch would join the trunk in an ice-rapid of many séracs. But when the trunk glacier had deepened its valley so far that further deepening became slow, the branch glacier would have opportunity to erode its side valley to an appropriate depth, and thus to develop an accordant junction of trunk and branch ice surfaces, although the channels of the larger and the smaller streams might still be of very unequal depth, and the channel *beds* might stand at discordant levels. (*Davis, 1900H, pp. 295–6*)

Water streams subdivide toward the headwaters into a great number of very fine rills, each of which may retrogressively cut its own ravine in a steep surface, not choked by waste. But the branches of a glacial drainage system are much more clumsy, and the channels that they cut back into the upland or mountain mass are round-headed or amphitheatre-like; but the beds of the branching glaciers cannot be cut as deep as the bed of the large glacial channel into which they flow: thus corries, perched on the sidewalls of large valleys, may be produced in increasing number and strength as glacial maturity approaches, and in decreasing strength and number as maturity passes into old age. As maturity approaches, the glacial system will include not only those branches that are consequent upon the initial form, but certain others which have come into existence by the headward erosion of their névé reservoirs following the guidance of weak structures; thus a maturely developed glacial drainage system may have its subsequent as well as its consequent branches. (*Davis, 1900H, p. 298*)

As usual, old age is not clearly defined by Davis, except for the following:

As the general denudation of the region progresses, the snow fall must be decreased and the glacial system must shrink somewhat, leaving a greater area of lowland surface to ordinary river drainage. When the upland surface is so far destroyed that even the hill-tops stand below the 200-foot contour, the snow fields will be represented only by the winter snow sheet, and the glaciers will have disappeared, having normal agencies to complete the work of denudation that they have so well begun.

If a snow line at sea-level be assumed, glaciation would persist even after the land had been worn to a submarine plain of denudation at an undetermined depth beneath sea-level. The South Polar regions offer a suitable field for the occurrence of such a surface. (*Davis, 1900H, p. 300*)

If the initial form offer broad uplands, separated by deep valleys, snow fields of the Norwegian type may have possession of the uplands during the youth of the glacial cycle; but when maturity is reached, the uplands will be dissected, and the original confluent snow field will be resolved into a number of head reservoirs, separated by ridges. On the other hand, as the later stages of the cycle are approached, the barriers between adjacent reservoirs will be worn away, and they will tend to become confluent, here and there broken only by Nunataker. If the snow line lay low enough, a completely confluent ice and snow shield would cover the lowland of glacial denudation when old age had been reached. If the glacial conditions of Greenland preceded as long as they have followed the glacial period over the rest of the North Atlantic region, who can say how far the ice of the Greenland shield has modified the forms on which its work began! (*Davis, 1900H, p. 300*)

The reader may find a disconcerting vagueness or unreality in some of Davis' ideas of a glacial cycle of denudation. But the concept itself is neither easy to express nor to justify because Ice Age climates are generally considered to be vast latitudinal or regional phenomena largely imposed externally. Tectonic uplift may be involved in the formation of broader 'glacial climates' but it is generally thought of as only one of several factors. Davis imposes uplift as the basis of the glacial cycle; whereas modern meteorologists impose glacial climatic accidents upon any pre-existing relief. Yet Davis wisely insists that the student of glacial landforms must know the pre-existing relief before attempting to assess the effect of glacial denudation. Typically he also insists on the powerful deductive value of this cycle:

As a practical instance of the value of the glacial cycle, we may consider the aid given toward the solution of certain problems by the careful reconstruction – or at least the conscious attempt at reconstruction – of the form of the land surface on which the pleistocene glaciers began their work, and by the legitimate deduction of the characteristics of maturity in the cycle of glacial erosion. Beyond the mature stage, we may seldom have occasion to go, as there do not seem to be actual examples of more advanced glacial work. The initial form on which pleistocene glacial action began is in no case known to be that implied in the opening paragraphs of the section on the Glacial Cycle; namely, a land mass freshly uplifted from beneath the sea and not previously carved by the streams of an ordinary or normal cycle of erosion. (*Davis, 1900H, p. 309*)

Indeed, Davis was always at pains to emphasize the heavily deductive nature of his glacial studies. This is well exemplified by the question and answer technique he employed when writing on the glaciation of the Sawatch Range in Colorado (1904D), which he visited in July 1904 on the way to Utah – the first time since he was in the region with Whitney's party in 1869. However, the best exposition of his ideas on glacial erosion is contained in his 'The sculpture of mountains by glaciers' (1906J), one of the most adored of Davis' writings and in so many ways the quintessence of mature Davisiana. In method, style and illustration it shows the master at his best.

It opens with a brief summary of the fine comparative studies of mountain glaciation done recently in the Alps, western America, the Carpathians, New Zealand and Skye. These have shown *inter alia* that overdeepening of glacial valleys can occur far upstream from their terminal lakes and that 'the retrogressive glacial erosion of cirques (or corries) carries with it the sapping and sharpening of the culminating ridges and peaks'. The progress has been such that

the sculpture of mountains by glaciers has been given that degree of extreme probability that we may fairly call demonstration. It should not, however, be overlooked that certain investigators still remain unconvinced, notably Heim and Kilian in the Alps, Bonney and Garwood in England, and Spencer and Fairchild in the United States. (*Davis, 1906J, pp. 76–7*)

Davis then sets out to convince these objectors by the application of strict deduction:

Now if glaciers have no erosive power, then maturely dissected mountains that have been glaciated should present no features significantly unlike those above described for mature non-glacial mountains. But if glaciers have strong erosive power, special and significant features should be found in mountains where glaciers have had time enough to do their work. The most notable features of this kind that one would expect to find may be stated as follows: A large part of the cross-section of a glaciated valley would be included in the trough-like channel that was scoured out and occupied by the heavy, sluggish glacier; the bed of such a trough would have rock steps and rock basins similar to those in the bed of a river channel, but appropriately of much greater size; the sides or walls of such a trough would be comparatively even and parallel, like the sides or banks of a river channel; the troughs of small side glaciers would necessarily be of much less depth than the troughs of large trunk glaciers, and hence the bed of a side trough would hang hundreds of feet over the bed of the trunk trough; the valley sides above the trough would, in mountains of mature sculpture, be less steep than the trough walls themselves; the heads of the glacial beds would be broad-floored cirques, because the heads of glaciers are broad and leaf-like instead of being divided minutely like the headwater streams of rivers; the summits and ridges between cirques which head near each other would be sharpened into peaks and arêtes by atmospheric weathering, induced by the retrogressive glacial erosion of the ice in the cirques, and by the glacial widening of the troughs. (*Davis, 1906J, p. 78*)

There is, of course, nothing new in this exposition but it was illustrated by superb, simple diagrams, which are idealistic expressions of the 'deduced consequences' of the hypothesis that glaciers can erode [Fig. 60].

Moreover the figures ambitiously attempt to exhibit the changes that a given mountain mass would suffer from a pre-glacial time of normal sculpture, through a pronounced glacial period, to a post-glacial time in which the work of the glacial

A

B

C

FIG. 60. Stages in the erosion of a mountain mass

A. Not affected by glacial erosion

B. Strongly affected by eroding glaciers which still occupy its valleys

C. Shortly after the glaciers have melted from its valleys

(From Davis, 1906J, Figs. 1, 2 and 3)

period has as yet been but little affected by the return of normal conditions; in this respect they are necessarily only ideal examples. (*Davis, 1906J, p. 79*)

Davis, with typical mature self-deprecation, calls these diagrams 'hard and crude' but they became the Mona Lisas of the glaciological world and entranced all beholders. Few noticed that their simple charm carried an enigmatic smile. But we will return to this later. The arguments used by Davis to explain the diagrams may be judged from the following headings: consequences of theories confronted with facts; insufficiency of arguments against glacial erosion; insufficiency of the explanation of special forms without glacial erosion. The final heading, 'A fallacy leading to a quandary', refers to those persons who postulate normal erosive causes for features that can only rationally be imputed to glacial action.

The sculpture of mountains by glaciers is indeed now proved by so many facts, widely and yet systematically distributed, that it savors of extreme conservatism any longer to deny the efficacy of glacial erosion. (*Davis, 1906J, p. 89*)

This limpid prose with its logical aura and convincing diagrams must be treasured as pure Davisiana. It is a masterpiece of method and of advertisement by a master propagandist. But the scientific cynic could unkindly notice that it contained no shred of anything that was new or was not already accepted by the great majority of geologists. However, it reached the masses who, appreciating simplicity and veracity, made 'The sculpture of mountains by glaciers' a comfortable part of their fundamental tenets on landforms.

But what an enigma was concealed behind the charming curves of Fig. 60 showing the mountain mass shortly after the glaciers have melted from its valleys! How does this agree with the cycle of glacial denudation which ends with the destruction of the upland almost to sea-level? Has Davis, to achieve reality, now abandoned the idea of a 'constant glacial climate' and introduced, without in any way saying so, the idea of periods of glacial climate interspersed with warmer interglacial phases? These short-term climatic changes were, as we have already noticed, superimposed on normal climates and on normal relief by external factors and not by the sudden elevation and denudation of landmasses. Davis himself recognizes these short glacial phases where, for example, he writes

the glacier system of the second epoch would find the valley troughs so well adapted to its needs that there would be relatively small necessity of modifying them. The amount of sculpture effected in a first glacial epoch may therefore be reasonably estimated as of much greater volume than in a second glacial epoch. (*Davis, 1906J, p. 82*)

Such climatic oscillations may perhaps be included as a complication of the elaborate cycle of glacial denudation but in that case where would the geomorphologist look for the ideal cycle except perhaps in the coldest polar regions? It then becomes a question of time and rate of lowering, and Davis was none too keen on these mathematical conundrums.

An equally difficult problem facing the ideal cycle of glacial erosion is inherent in the lack of detailed consideration given by Davis to the profound structural effects on such erosional features. Similar glacial action on massively-jointed, resistant rocks can produce very different topographic effects from that on weak, poorly-jointed rocks. Davis' cycle applied very much to the former case. On his way to Norway in 1899 Davis visited the English Lake District but, despite his study of a model of the region on show at Keswick, he failed to detect the considerable differences exhibited by the equally-glaciated Borrowdale Volcanics and the Skiddaw Slates. There is a strong built-in lithological assumption in the cycle of glacial erosion, just as there was a tectonic one in the arid cycle.

CHAPTER FIFTEEN

Davis in Germany: The Cyclic Defender

One of the most taxing events in Davis' long career was his tenure of a visiting professorship in the University of Berlin during 1908–9. Here the chief difficulty was the language and the problem gives us the opportunity of discussing his linguistic ability in general. He was competent at composing in German, as may be judged from the following extract from one of Albrecht Penck's letters.

Your long letter of the 22nd August gave me an immense pleasure. 13 pages written in an excellent German, full of news, full of ideas. I got it yesterday and read it on an alpine meadow close to the hotels where I stay with my people since three weeks, in face of the Cimone della Pala climbed yesterday by my son (3186 m), situated in nearby vergin (*sic*) forests ...' (*Letter: A. Penck to Davis, 8 September 1901; Paneveggio, S. Tyrol*) (see pp. 277–8)

He was also competent at general conversation in German but was not sufficiently fluent to lecture impromptu, answer unexpected questions and convey complicated sequences of ideas. Lawrence Martin's assessment of Davis' linguistic ability, given below, needs heightening for we know that Davis gave many lectures in German and in French although they were carefully prepared beforehand.

Was Davis effectual with foreigners? Did he fully comprehend their lectures and publications and their field conversations and expositions? Indeed his linguistic ability was unusual for an American geographer. He lectured at the University of Berlin in English, at the Sorbonne of the University of Paris in English, but he had an adequate speaking knowledge of German and French, as was demonstrated during the transcontinental excursion in 1912. He undoubtedly acquired Spanish during his three years in Argentina. I have heard that he had a reading and speaking knowledge of Italian. Although he sometimes conversed in German, Davis wrote his 'Erklarende Beschreibung der Landformen' in English, having it translated for him by Alfred Ruhl; so also with his 'Grundzuge der Physiographie', in which Davis collaborated with Gustav Braun, and also his 'Balze per Faglia nei Monte Lepini' which was put into Italian by Fr. M. Pasanisi. The 1930 edition of Davis's 'Elementary Physical Geography', published in Japanese and taken from the 1902 or the 1926 edition of that work, was obviously put into Japanese by someone else. Naturally, then, Davis read geomorphological, geological, and geographical papers and books in European languages rather than having to depend upon translations or the summaries in English-language reviews. No great number of Davis's American contemporaries possessed these gifts. (*Martin, 1950, pp. 174–5*)

Davis was thorough in everything he undertook, and acquiring a competent degree of fluency was a drudgery that would be accepted when he agreed to the appointment. A description of his method of learning conversational German can easily be believed.

He taught himself (German) on trolley-car rides from Boston to Cambridge, speaking the words out loud because the trolley was so noisy that no-one could hear him. (*Letter: W. M. Davis II to R. J. Chorley, 30 January 1962; Bass River, Mass.*)

His own relation of how he learned Italian in 1907 and 1908 bears out this assiduity.

For my own part, I never ceased to rejoice over the spare hours that had been given, during the previous winter, to reading stories by Fogazzaro and other Italian writers, and to copying choice sentences in my notebooks and reading them over and over half-aloud, so as to gain a little facility in saying as well as in understanding Italian phrases. It happened fortunately that on our steamer between Boston and Naples there was an Italian who was willing to give me daily lessons during the voyage. What with this preparation and with the extraordinary facility of Italians in understanding foreigners, it became possible to get along very fairly in that pleasing language, and occasionally even to hold geographical discussions with the Italian members of our party. (*Davis, 'Boston Evening Transcript', 1909*)

That is not to suggest that all difficulties and mistakes were avoided but even accepting these the achievement is highly creditable.

The courtesy of the student audience, during the dry reading of written lectures and the addition of impromptu variations, was indeed most remarkable. On one occasion only do I recall seeing a smile excited by the innumerable mistakes that must have been made. As a rule the hearers generously overlooked all errors, and attended diligently to the subject in hand. As acquaintance increased and constraint lessened, one of the English-speaking young Germans in the front row used to come to my aid with a needed word, when I hoisted signals of distress.

But on the other hand, the students must have been many a time more or less mystified as to what I was driving at in German; and thus must have been especially the case with two Russians and two Roumanians, or at least with three of those four, who were struggling as I was with a language foreign to all of us. It is not so much the essence of the language that is a trouble, as the unnecessary complications of the grammar.

And so it is with the whimsical distribution of genders, and the arbitrary declensions of adjectives, to say nothing of the peculiar prepositions that are selected to join verbs to their indirect objects. (*Davis, 'Boston Evening Transcript', 1909*)

While resigning himself to the need to speak German, Davis was experienced enough to realize that he might save himself unnecessary labour if he were able to gauge how much he need know. So he sought guidance from a colleague who had accepted a similar appointment the year before. The advice given seems to have been very helpful and was closely followed.

> With regard to the Berlin affair I am sorry on your account that you cannot wait until the summer, for I am sure you will find it somewhat strenuous doing Professor Penck's work. I agree with you that the only thing to do is simply to make the best of it. In spite of my somewhat greater time for preparation, I too found myself in rather a large contract, and I found that the only way in which I could get through was to write all my lectures in English, and have them translated by someone else. It took me too much time to write them in German, and I could not trust to the inspiration of the moment for the whole lecture, although I was perfectly well able to fill in details in 'pigeon' German. Perhaps you will find this the best way to do in your case also. If so, I venture to suggest that you will do well to get the English manuscript entirely written before the opening of the semester – for I found it extremely bad for my German to have to spend a couple of hours a day in writing English. If you have the English entirely out of the way, you can give the whole mind to the German, which you will find will need thorough revision before being read, for no translator is able to get exactly the meaning in every sentence. English stenographers are not very easy to get and rather expensive in Berlin. It would almost pay you to import a student for August and September to take down your lectures in English and to travel around with you. Of course you might not need him after you got to Berlin, but if I were doing it again, I confess I should feel like having both an English and a German stenographer at my elbow most of the time, I don't know what I should have done if I had not spent a month in dictating before I left for Germany. (*Letter: T. W. Richards to Davis, 8 June 1908; Harvard Chemical Laboratory, Cambridge, Mass.*)

Davis' own account supports the belief that the advice was taken.

> On leaving Cologne we went to the charming University town of Marburg, on the Lahn, for the month of October, and here began in earnest the preparation of lectures for Berlin. To appreciate the stress of work during this month a personal explanation must be made. Professor Penck my colleague and adviser in Berlin, had written to me during the summer, making most urgent request that I should give at least one course of lectures at Berlin in German. This was entirely contrary to my expectation, but Penck insisted that the worst German would be better understood by most students of geography than the best English, and so there seemed nothing for it but to make a try. Partly to make a test of this possibility I had accepted the invitation, which also came through Penck, to speak on the Colorado Canyon in German before the Naturalists at Cologne, and although that ordeal was survived, the reception of the lecture was by no means so reassuring as to enable me to look forward to the effort of the winter without solicitude. Indeed it must be confessed that until almost the end of the semester the necessity of getting those German lectures ready beforehand was a very heavy responsibility.

Marburg was an admirable place for quiet study, because there were sympathetic friends to welcome me in the university and an excellent translator available for my English manuscript. (*Davis, 'Boston Evening Transcript', 1909*)

His friend also offered counsel on the niceties of official and social visits, subjects on which the very orthodox Davis had probably asked for enlightenment.

Professor Paszkosi (a minor official) will help you about many details, unless the new Minister has displaced him. He is the Hofmarshall for Austauschprofessoren, and will probably help you also about all sorts of things. You know, of course, that in Germany the visitor makes the first call. In Berlin you can call at any hour of the twenty-four, although I believe calling between one and four in the morning is not usual! The usual times are between twelve o'clock noon and six. Meal times in Berlin are in a state of anarchy. Some dine at twelve, some at one, some at two, some at four, some at six, some at seven, and some at eight-thirty, and you may have a supper at almost any time after that and before dinner the next day. The other meals are shifted around to suit. As a general rule Government people and bankers dine at four; society at eight-thirty; suburbanites at six-thirty or seven, but there are exceptions to all rules. As you know, the invitations always tell you what to put on, even as to the kind of necktie. You probably know that Ueberrock means not overcoat, but Prince Albert. You will find this capacious garment the most overworked part of your wardrobe. Fortunately in Berlin one does not have to pay dinner calls. (*Letter: T. W. Richards to Davis, 8 June 1908; Harvard Chemical Laboratory, Cambridge, Mass.*)

The arrangements for the professorial exchange were made mainly during 1907 when Albrecht Penck and Davis were visiting Britain. The latter spent some time examining glacial features in North Wales and the English Lake District and appears from the following letter to have been lecturing at Cambridge. Penck regrets that they cannot join in a visit to Lord Avebury, the famous physiographer and naturalist who wrote many popular scientific books.

A very thorough study of all circumstances gave yesterday the result that I cannot go with you to Lord Avebury. I must avoid to travel too much with Zoe so I shall stay with her at Oxford with the Townsends (117 Banbury Road) until Thursday morning, then we start over Bristol to Plymouth where we spend the Friday the 7th (Chubb's Hotel) and from where we shall go to Falmouth (Green Bank). Let me know your train so that we can join en route on Saturday. This day and Sunday the 6th and Monday the 7th are reserved for you.

Give my best compliments to Lord Avebury to whom I wrote excusing that I cannot come to him this week but that I can do so the 12th and 13th.

For a moment I was very sorry that we are separated in this country; that you will belong to Cambridge while I shall become an Oxonian. But further reflexion showed me that it is better so. Now we shall continue our morphologic races with the spirit of the members of both Universities. (*Letter: A. Penck to Davis, 28 September 1907; London*)

When Davis originally accepted the Berlin appointment, he planned to spend the summer of 1908 travelling through Italy and France, allowing time for attendance at the Oxford Summer School and a flying visit to Dublin. Afterwards he meant to return to Harvard for the winter and give himself adequate leisure in which to assemble his German notes. The following year he would travel to Berlin in time for the summer term of 1909. Unfortunately all these plans were upset because two days before he left for Europe he learned that the lectures were scheduled for the preceding winter term. This meant there was no time to return to Harvard and enforced some curtailment of his European travel itinerary. It also meant that the entire preparation of his German lecture notes would have to be completed abroad. Obviously he would have preferred to have done this preparation at Harvard, where all the necessary material was to hand and where, from long association, he was more conversant with the choice of translation facilities available. Nevertheless, quite undeterred, he quickly made the necessary rearrangements and still completed a very full programme.

Davis' new plans entailed a very circuitous approach to his main objective, which happens to be a characteristic of his journeying since his earliest travels. He loved planning routes and making arrangements to be in widely separated places at very exact times. He did not travel, he rushed; he could not walk, he strode. On this occasion he was accompanied by his wife. Their sons are not mentioned but they would now be adult and we know that by 1911 they had left home.

> It is now a long time since similar fortune came to me, my three boys are now grown men, and are all away at work; one, a chartered accountant in New York, one in woollen yarn mill and office work in Boston; he is happily married, living in Brookline, the third son is a farmer, with a class-mate, they have an apple farm in Shirley, Mass., a beautiful place of 169 acres, and they are finding out by experience the meaning of weather changes, pests and eternal vigilance. I hope that this experience of mine is a foreshadowing of a similar one for you, years hence.
> (*Letter: Davis to R. W. Goldthwait, 10 June 1911; Cambridge, Mass.*)

The sea passage was by direct route between Boston and Naples. We are not regaled with a description of the shipboard activities, probably because the account was not written until some time after Davis' return from Berlin. One episode certainly stood out. The reader may by now have a mild allergy to Davis' poetry but this is by far his best humorous poem.

The True History Of The Norwegian Barque

Late in the night o the twenty nine
Of Hapril, nineteen ha eight,
Hon the Rheu-ma-tic o the Black Tar Line
There befell what I now rela ate.

In 39 nor and west 34
With the wind ha blowing free ee,
From the nor nor west by no alf nor
Hand a little northerlee e.

On the bridge hayloft the Captain paced
An hattended to is bi iz,
At is bridge below was the Doctor placed
An the dummy and was i is,
When the lookout cried, 'Hi vow I've spied
Two pints on the starboard bow, Sir,
A queer sort o spot, but I can't say what,
Hi can't make it hout no ow, Sir.'

Said the Captain, 'Hi may try and try,
But I can't see in the da ark,
And I can't 'ear well for the ollow of the swell
But hit sounds to me like a ba ark,
Like a bark or a yelp, or a cry for elp,
Like a girl when something shocked er,
Hi can't make out what it's hall about,
But it seems to be a case for the Doctor.'

Then the Doctor chap e touched is cap,
An e says, 'What d'ye want o me, Sir'
Says the Captain, 'Ark, hout there in the dark
There's a bark huppon the sea, Sir
An I can't tell at all if its pug or spanial
Or a terrier, fox or scotch, Sir.'
Then the Doctor leered, 'Taint no bark you eared
But the tick of a lost dog-watch, Sir.'

Hat the break o day, hit appeared that they
Ad both made a bit of a blunder,
Twas a bark marine not a bark cayneen,
An the Captain said, 'By Thunder!
Hi'll bet my at that a rig like that
Haint often in this region.
By the cut of er jib and the curve of er rib
I fancy she's Norwegian.'

Then our Captain bold, e eaved a sigh,
An the Nor-wegian eaved to oo.
Hi couldn't eave to, an no more could you,
For this is what e ad to do oo,
Leggotoggallanalyards on the foremastanthemain
An clue up the mainsel an the fore,
Cast off main braces by the starboard chain
An aul em in on the por'.

Said the Captain, 'Bless my heyes, distress
Is spelled by the flags they're aflying,
We must lower a boat an set it afloat,
For perhaps there is someone ha dying.'
Then four of the crew an a hofficer too
Steps hup an says they are ready.
An the Doctor, says e, 'Keep you heyes upon me
Hill take em some food and some meddy.'

When the Doctor stepped aboard that bark
There wasn't a soul in sight.
He went down below and there in a row
Was the crew in a pitiful plight.
An one of em said, 'Hi berry near dead.'
And one, 'Berry, berry sick me.'
'Ich berry, berry sad.' 'Me berry, berry bad.'
Twas the worst sight the Doct ever see.

When back came the Doctor e looked very shocked
For says e 'Hi beats dis-entery.
Fore they reach the orizon they're all dead with pizen
They're struck with the plague beri-beri.
Hit's a fearful disease from the Japanese,
An there's no way known for to cure it,
Hit works on the brain an drives em hinsane
They all heither die or hendure it.'

Meanwhile, on the bark they were aving a lark
An a making hexceedingly merry,
Hall the crew, well and trig, were dancing a jig,
They ad shook off the plague, beri-beri.
The provisions we sent with Christian intent
They were eating like Pagans hecstatic
The wine they did quaff with a cheer and a laugh
For the Doctor on board the Rheumatic.

(Poem by Davis presented at a Ship's Concert on board the 'S.S. Romanic'; 6 May 1908; sung by Mrs A. Hall Aarons)

Davis apparently shared the same opinion of its quality.

Cate was a passenger on the steamer in which your mother and I crost * to Naples in 1908, spring; and he was my orchestra, or pianist, with 'cello obligato, when a Vaudeville songstress sang my verses on the Norwegian Bark; one of my very best efforts, ever. I put a chromatic scale in the accompaniment, and Cate stumbled over it; but he did the rest all right. (*Letter: Davis to his children, 15 December 1927; Stanford, California*)

The facts on which it is based did not end so happily.

* Note the phonetic spelling experimented with by Davis in his later years.

One incident of the voyage from Boston to Naples, April 25 to May 8, 1908, deserves some mention. The day before we touched at the Azores, a sailing vessel came in sight from the south, flying signals of distress, which read 'Medical assistance wanted'. She proved to be a Norwegian bark, bound from Valparaiso to England; when we neared her her lower sails were clued up, and the sails of her mainmast were braced around so as to be taken aback – if that is the right way of saying it – and thus her headway was checked while we lowered a boat and sent the doctor over for a visit. It was a pretty sight to see that boat ride over the crest of the waves and disappear in the troughs. When the bark was reached it appeared to be no easy task for the doctor to board her. There was a great array of field glasses along the side of our steamer, which went circling round the bark at low speed; so we saw that the doctor disappeared below for a time, then came on deck again, embarked with some difficulty in the dancing boat, and brought back the news that the Norwegian captain was ill with beri-beri; that no one else on board understood navigation sufficiently to determine the vessel's position, so that she had lost her reckoning and was astray on the ocean, and that she had been delayed by head winds, and her provisions were running low. The doctor had urged the captain to come on board the steamer and thus reach a hospital on the Azores the next day, where he could be well cared for, and to leave his bark in charge of one of our younger officers who would take it to destination; but this offer was refused. However, a boat was sent over to us, which our captain loaded down with provisions; we gave them as well their position and a course to steer; and so parted company. (*Davis, 'Boston Evening Transcript', 1909*)

The excursion through Italy and France was a strictly professional enterprise and cannot have been much entertainment for his wife. With customary fanaticism Davis campaigned hard until what had begun as a private trip became in reality an unofficial concourse of geographers.

On the occasion of a sabbatical year spent abroad in 1898–9, a first acquaintance was made with a number of interesting districts in northern Italy, but they were then seen so briefly that, when opportunity came in the spring of 1908 to visit Italy again, it was planned to examine more at leisure several of the places previously seen in a rapid journey; and as this plan took shape it occurred to me that it would be more profitable to visit the selected localities in the company of other geographers, so as to discuss our problems together on the ground; all the more so because the physical features of the classical peninsula have not as yet received adequate treatment in view of the modern advances of the study of land forms. With this object I prepared a circular letter and sent it, in March 1908, to my correspondents at home and abroad giving an outline of the trip across Northern Italy and into South-eastern France, with a brief account of the proposed method of study; and invited them or such of their advanced students as they could recommend, to join me for June and the first half of July. (*Davis, 'Boston Evening Transcript', 1909*)

The invitations could not always be accepted as the following letter shows.

Dear Colleague,

Many thanks for kindly sending me your plans for the journey to Italy. I am afraid none of my students will be able to join your interesting expedition since it would take place in the middle of the semester. This is why I, too, cannot join you.

I hope to be in Geneva this summer, where I hope to be able to make your acquaintance.

(Letter: S. Passarge to Davis, 26 May 1908; Breslau)

But Davis' efforts resulted in a small international party which met for the first time at Ancona on the Adriatic on 1 June and disbanded at Le Puy-en-Velay in the volcanic district of central France on 18 July.

The party numbered some twenty persons in all, although not so many were together at any one time, with the exception of a few days at Grenoble where one of the professors who then joined us brought a large party of his own students who were not counted in my list. The universities, of Paris, Lyon, Grenoble, Freiburg, Genoa, Marburg, Cincinnatti, North Carolina and Michigan, Williams College and the Lyceum of Oran in Algiers, were represented by one or more of their teachers, while Berlin, Vienna, Lille, Bern, Sydney (Australia) and Harvard Universities and two American normal schools furnished students or recent graduates. As a result we had members of most varied training all united in the common object of trying to learn how best to describe the forms of the lands.

(Davis, 'Boston Evening Transcript', 1909)

Despite Davis' explanation of the purpose of the tour a secondary motive seems to have been present. His theories had not really spread beyond the United States. Very few of his articles had appeared in foreign languages and discussion of his ideas had only just begun on the Continent where his concepts were far from established or even understood. Davis realized that the best and quickest way of advancing his cyclic themes was by giving practical demonstrations of the application of his method at selected geologic sites. In America a similar advertisement of the Davis technique was already taking place.

The method of 'structure, process and stage' has been subjected to several tests. At the Chicago meeting of the Association of American Geographers, in December, 1907, it was presented at a 'Round Table' conference, led by Davis, in which a number of members took part, some of them accepting the method as of practical value, others being unwilling to adopt it, but no one proposing any other generally applicable method of treatment, although request for such other methods was explicitly made. *(Brigham, 1909, p. 58)*

This conference at Chicago showed clearly that Davis' cyclic ideas were not completely accepted within his own country. Italy could have been a deliberate choice as a European testing ground because of the lack of detailed landform studies there.

This correlation of cause and effect constitutes a distinct advance in geographical description, for the reason that an element of land form is much more easily apprehended and remembered if it is presented in relation to its origin. It is certainly curious that in a land which has been known as long as Italy has been, there should be opportunity still for scientific conquest by invading barbarians from a distant continent; but as a matter of fact there is hardly a country in Europe – unless it is Spain – where geographical exploration may be more richly rewarded by discovery – discovery of explanations – than it may be in Italy. (*Davis, 'Boston Evening Transcript', 1909*)

The absence of previous work conveniently avoided any conflict with local opinion, and his conclusions, representing the first serious study of the areas,

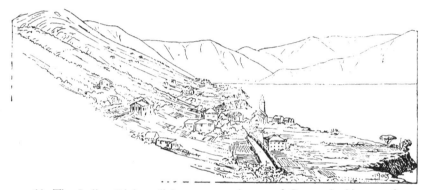

F I G. 61. The Italian Riviera di Levante south-east of Genoa, looking south-east
(From Davis, 1912J, Fig. 2)

would be of great interest to Italian geographers. Some idea of the striking analyses Davis was able to make can be judged from the description of the coastline south-east of Genoa.

From Florence we went down the Arno as far as Pisa, to see the flat delta of the river, over which a formerly insular mountain rises as much like an island today as it did before it was attached to the mainland; thence along the coast to Spezia, whence some of us drove over the mountain to the picturesque village of Sestri Levante on the Italian Riviera, half way to Genoa; the rest of the party going by rail through tunnel after tunnel. Sestri Levante is one of the most picturesque spots on that wonderful coast; it is built partly on the inner slope of a former island, partly on a sand-and-gravel reef by which the island is now attached to the mainland. From this point as a centre we made excursions by train to several stations along the coast, not omitting to ascend the mountainous peninsula of Portofino, and to study its marvellous view; nor to descend to the extraordinary little village of the same name in a small harbor at the southern corner of the peninsula. No district visited during the excursion afforded a better test of the method for the description of land forms with which we were experimenting, than this fine stretch of mountainous coast. One peculiar feature may be mentioned, because when once pointed out it may be easily recognized by even the hurried traveller who follows the coast by rail; and thus the greater enjoyment of what he sees during his lucid intervals may lessen the exasperation that he must feel in the darkness of the countless tunnels. After the sea had cut back the mountains to a late mature cliff of rather even front, interrupted by valleys that came out from the back country, the whole district was very gently tilted on an axis that ran through the Portofino peninsula, about at right angles to the general shore line; thereby the Genoa end of the coast was raised 150 or 200 meters, while the Spezia end was correspondingly depressed. As a result, the former wave-cut sea bottom or platform is now exposed as a rather flat bench (more or less dissected by new-cut valleys and worn back or narrowed by the new attack of the sea), which gradually increases in altitude from the peninsula to Genoa; while in the other direction the sea has risen on the sloping face of its former cliffs and steepened them, and has invaded or drowned the former valley mouths in which flat delta plains are being deposited. Nervi and its gardens, not far from Genoa, find plenty of room on the uplifted platform, which is there about half a mile wide, in spite of a considerable loss by sea erosion alone, so that it now terminates in a low and ragged cliff; here the railroad has no need to tunnel, though its track is often cut a little below the platform level, as well as bridged high over the new-cut valleys. Little Zoagli, on the other side of the axis of tilting, is crowded in the mouth of a small valley, between two cliffed mountain-spurs that now descend to the shore line with increasing abruptness where the sea is battering them back; here the railroad runs through long tunnels back of the cliffs, and comes out only now and then to breath as a ravine is crossed on a viaduct. Farther southeast a large valley broadly drowned by the tilting of this half of the district, is now as broadly floored up by delta filling; and here is Chiavari, the largest town of the district; largest because it has the most room to grow and because it stands in open communication with the inner valleys, where its streams head. Here the railroad lies on the plain, and here the traveller has the longest uninterrupted view during the ride between Genoa and Spezia. (*Davis, 'Boston Evening Transcript', 1909*)

Davis lost no chance of putting over his own ideas, and especially the cyclic concept, to his European audience and seemed to be pleased with the general results.

It is the search for easily explainable items to be used in reformed geography, that makes the re-exploration of old countries so entertaining; we thought and talked of all this at Bra . . .

A few words about the scientific results of this excursion. For some years past it has been a hobby of mine to try to develop a more systematic method for the description of land forms than is commonly employed in geographical writings. I had become fairly well satisfied with the method as it developed more and more fully, and was at last fearful of being over-confident as to its value. It was in large measure to gain an impartial test of its merit by using it in company with others of different training that the international excursion was planned; and this test was the more important because the systematic method of treating land forms was to be the chief subject of my lectures in Berlin in the following winter. It was, therefore, particularly gratifying to see that the members of the party, coming from so many different universities, after giving the proposed method a serious trial, felt favorably disposed towards it, while some of them definitely said they proposed to adopt it for their future work. This was felt to be great encouragement in preparation for the winter at Berlin. (*Davis, 'Boston Evening Transcript', 1909*)

Thus the trip from Ancona to the Massif Central was designed to test the method of explanatory description of landforms with an international group and, incidentally, to popularize the cycle concept in Europe. The results of this experiment were described in three very influential geographical papers (Davis, 1909G; 1910B and H).

A general favourable consideration was given to the method of structure, process and stage, during the excursion, but this must not be taken as counting altogether in its favour. A definite method naturally makes headway as against indefinite, unformulated methods; and more-over, as I was the leader and oldest member of the party, my views probably received a greater consideration than they would have gained if I had been a junior and a follower. Still, all allowances made, the excursion gave me great encouragement, and I resolved to persevere in carrying the development and the application of the method as far as possible; but always in the hopes of meeting other methods developed by my colleagues; and always with the promise, to myself at least, to make careful trial of other methods as far as I could learn them. (*Davis, 1910H, pp. 568–9*)

The essential principles here are, first, that the reader's mental picture cannot be well formed, unless the observer describes what he has seen in terms that are susceptible of definite interpretation; and, second, that the mental picture cannot be easily formed, unless the observer presents the results of his observations in a reasonable order.

Only after a definite description of the landscape has been presented, is it fitting to mention by name subordinate items, such as single villages and individual streams. It is altogether inappropriate to use unknown local names of villages and streams as a means of locating unknown structures and forms. This is a general principle that is too often overlooked. (*Davis, 1910H, p. 564*)

Whenever an observer attempts to tell what he has seen, so that a landscape or a region may be conceived by his readers, he must describe the observed forms in terms of certain similar forms previously known to him, and hopefully known also to those for whom he writes. It must always be in terms of something previously known that a verbal description is phased. Hence the most accurate verbal description will be made by that observer who is equipped with the largest variety of previously known type forms. It is important to consider how a young geographer is to obtain such an equipment. The ideally perfect method would be for him to travel about the world and see with his own eyes a great variety of actual forms, from which he might gradually develop a complete series of type forms. Then all other forms could afterwards be described in terms of these types.

But this method is manifestly impossible for general application. Some equipment of types may be secured by observation of actual forms; and this beginning may be significantly enlarged by the study of descriptions, pictures, models and maps of actual forms, as prepared by other observers.

The geographer who follows the empirical method stops here. The geographer who follows the explanatory method goes much farther. He extends and systematises the equipment, thus far gained, by deducing many related forms; and thus fills his mind with a series of more or less ideal forms. It will then be chiefly in terms of the ideal types, largely developed by deduction, familiarised by diagrams, and confirmed or corrected by experience, that his explanatory descriptions of actual landscapes will be phrased. But whether the geographer follows the empirical or the rational method, it will be only in proportion to the completeness with which his series of ideal forms provides him with counterparts of actual forms, that his descriptions of actual landscapes can be true to nature. Only in proportion to the compactness of the terminology in which the ideal forms are verbally expressed, can the observer's descriptions be tersely stated. Only in proportion to the correspondence existing between the ideal forms as conceived and named by the observer and by his reader, will the reader be able to apprehend the observer's meaning. (*Davis, 1910H, p. 575*)

Another advantage of the explanatory method is the ease with which a simple series of ideal forms, such as those of the plateau series above mentioned, may be expanded into a more elaborate series, so as to fit a much larger variety of natural cases. In the first place, a plateau mass of horizontal structure may be conceived as consisting of resistant and weak strata in any order; a group of resistant strata between two groups of weak strata, . . . or a group of weak strata between two groups of resistant strata, . . . or many alternations of resistant and weak strata, . . . and so on. The general plan of development being once understood, all the particular features that will be associated with each particular structure may be easily deduced. In the second place, a plateau of any special structure may, at the beginning of its cycle of erosion, stand high or low; the relief developed by its dissection will evidently be strong during the youth and maturity of the first case . . . but can never be strong in the second case . . . Finally, the texture of dissection – that is, the spacing of the streams and their valleys, and the resultant breadth of the hills and spurs – may vary according to certain conditions, the chief of which are the perviousness of the plateau mass, or at least of its soil cover, and the climate under which its dissection takes place. Pervious strata allow much of the rainfall to enter the ground and thus diminish the number of small branch streams; as a result, the texture of dissection is coarse, and the hills are broad and full bodied . . . If the strata are impervious and of fine-grained materials, and if the climate is so dry as to make vegetation scanty, then the texture of dissection during youth and early maturity will be very fine, and the surface will assume a 'bad-land' quality, with innumerable little hills and spurs between innumerable little valleys . . . Thus the scheme of description really contains five elements: structure, process, stage, relief, and texture; all require consideration, but the first three suffice when giving the scheme a name. (*Davis, 1909G, pp. 307–8*)

Pleading for the supplementary use of illustrations, particularly block diagrams, Davis gave a number of examples of the application of his method of the explanatory description of landforms, some taken from the international excursion of the summer of 1908. Here are some examples of his explanatory terminology:

1. The Adriatic Coast near Ancona: 'a coastal plain of imperfectly consolidated sands and clays, which has reached a stage of late mature dissection under the action of normal and marine erosional processes' (Davis, 1909G, p. 302).
2. The coastal plain of east Virginia and Maryland: 'a maturely dissected coastal plain, recently somewhat depressed. The new cycle thus introduced has not passed beyond early youth' (Davis, 1909G, p. 304).
3. The western Massif Central in the neighbourhood of the valley of the Lot:

This district ... may be described as a maturely dissected plateau of indurated horizontal strata, in which the valleys of the smaller consequent and insequent streams show the work of only one cycle, while the main valley of the Lot shows the work of two, in the first of which the late-mature meandering river had nearly consumed its valley-side spurs; and in the second, introduced by a moderate uplift, the river has now consumed about half of the previous flood-plain in reaching maturity again. The mental picture of the district, constructed from this description, must embody a great number of rounded hills of equable height, divided by many irregularly branching valleys; for in a plateau of horizontal strata, most of the drainage must be insequent, without systematic control, and hence following no definite direction. The hillside slopes, smoothly covered with creeping waste, must descend with gracefully curved profiles to the narrow valley-floors of the well-graded smaller streams. But the main valley of the Lot must be conceived as broadly opened and as showing a well-defined gravel-covered terrace ... below which the curving river has eroded a narrower valley, with well-developed and systematically placed flood-plain scrolls. (*Davis, 1909G, p. 305*)

The explanatory technique of description is further applied to the more complex terrains of North Wales and the Italian Riviera. We have no direct evidence as to the effect of this so-called objective experiment on those who accompanied Davis from Ancona but it is clear that, as far as Davis was personally concerned, the only experimental feature of the operation was in terms of his own ability to propagandize his previous ideas.

We have, however, yet to describe the journey. In approximately forty-five days they went by a circuitous route from Ancona on the Adriatic coast to the Puy de Dôme region of central France. After examining the coastal plain around Ancona, they went north to Faenza. From here they crossed the Apennines to Florence where they stayed a week. From Florence they descended the Arno valley to Pisa and then, as described in the extract already quoted, moved north along the coast to Spezia, the Riviera di Levante and

Genoa. At Savona they crossed the coastal mountains to Bra and so north-ward to Milan and Lake Como, where they stayed at Menaggio which served as the centre of excursions for one week. Here Davis was anxious to test and advertise his newly-acquired theories on glacial erosion that we have dis-cussed in detail in Chapter 14. We get no hint in his own account of the excursions that he himself had only recently been converted to accepting the formation of large lakes by glacial overdeepening. Nor have we yet been able to trace definitely the objectors to the glacial overdeepening theory he men-tions, although it seems certain that Davis would have sent invitations to the 'certain investigators' who still remained unconvinced in 1906, 'notably Heim and Kilian in the Alps, Bonney and Garwood in England' (1909C, p. 618). Anyway, at least two of those who did attend appear to have been impervious to the arguments of both Davis and Nature, whereas the English professors were not surprisingly bowled middle stump in their absence.

> We went up the valley above the lake (Como) as far as Chiavenna, where great glaciated knobs of rock are wonderfully displayed in the valley floor, down one of the lower arms of the lake to Lecco and the heavy glacial moraines beyond; and down the other arm towards Como, where we climbed one of the hills and had a commanding view of the lake waters below. The chief problem here in discussion was the work of the ancient glaciers, in the excavation of the lake basin. We all became convinced that only in this way could the basins be successfully explained, but there are others who deny this possibility. We had hoped that a representative of the other side might have been with us to present his views; but he was detained in England by college examinations, and we had ourselves to try to present his ideas as impartially as possible; evidently not with success, because we so easily bowled them over...
>
> A particularly interesting feature of this part of the excursion was the presence of two professors, one from a university in France and the other from one in Switzerland, whose opinion on this long-mooted question was far more conserva-tive than that which had been reached by a large majority of the party, and with whom we had the great advantage of discussing the problem of glacial action with the facts before us; apparently with the usual result of fortifying each party in his previous views. (*Davis, 'Boston Evening Transcript', 1909*)

From Lake Como the party went westward to Lake Maggiore where problems of overdeepening were equally evident. Hence, after further glacial excursions, they finally crossed into France by way of the Little St Bernard pass and stopped at Grenoble. From here they went through the Cévennes to the Le Puy district of the Massif Central, where the party broke up.

If the hectic traverse across Italy and south-eastern France had now ended for most of his companions, Davis' travels were merely about to begin. After Le Puy he allowed himself the luxury of a whole week's relaxation at Annecy, a lakeside resort in the Savoy Alps, before entering upon the bustle and exchanges of the Geographical Congress at Geneva. The Congress must have proved exhausting even for him judging from the fact that he did not take part in the excursions after it but spent the next ten days quietly at St Goar in the Rhine valley. Then, apparently refreshed, he rushed across Belgium in order to keep an engagement in August at the Oxford Summer School of Geography.

Here he revisited the valley of the Evenlode, an incised meandering tributary of the Thames, and was told by Professor Sollas and other Oxford geologists that glacial drift had been found on the nearby uplands. Davis saw immediately a possible bearing of this on his ideas of 'misfit' and 'underfit' rivers. He wrote up his conclusions in 'The valleys of the Cotswold hills' (1909J, pp. 150–2). Here he assumed that an ice sheet of the glacial period had once filled the lowland of the English Midlands and had overflowed the adjacent Cotswold hills. He imagined that at one stage ice-water streams had accumulated in lakes between the northward scarp-slope of the Cotswold cuesta and the southward slope of the ice. These lakes spilled over through the lowest depressions of the cuesta crest and greatly modified the pre-existing river valleys. As this temporary discharge was added to that of the existing streams, it followed that the curves of the valleys eroded would be much larger than the curves of the present streams.

> It is manifest that this discussion is based on the postulate that when the curves of a stream are too small for the curves of its valley, a diminution of stream volume is to be inferred; and this postulate is not accepted by all observers. Nevertheless, no other explanation of the disproportion of stream curves and valley curves can account for the valley curves and for the peculiar form of the trimmed spurs that project into such valleys. (*Davis, 1909J, p. 152*)

This brief article is especially interesting as it demonstrates how ruthlessly and quickly Davis seized on any ideas, often of others, that furthered his own concepts. Obviously most of the underfit streams in Cotswold valleys have never experienced glacial-like overspill such as affected the Evenlode but he was willing to generalize from it. Yet it is pleasing to be able to record that the presence of melt-water lakes in the area suggested by Davis was virtually confirmed by detailed geological evidence some forty years later. But satisfactory conclusions on the 'fit' of streams and on the relationship between the windings of a river and of the valley in which it flows have proved elusive. They remain among those numerous Davisian oversimplifications which are so exasperatingly naïve and yet so tantalizingly attractive (Beckinsale, 1970).

From Oxford Davis went on an excursion to the Snowdon district of North Wales, one of the classic fields of mountain glaciation. This also led later to a publication. Called 'Glacial erosion in North Wales' (1909E, pp. 281–350), it added little if anything to existing knowledge and much, because of its great length, to the already extensive literature on the locality.

F I G. 62. Davis (fourth from the right) and a field party at Wyndcliffe Hill, Gloucestershire, near Chepstow, in 1911 (Courtesy Dr G. Martin, Southern Connecticut State College)

We have discussed in Chapter 14 Davis' full conversion to the concept of glacial erosion and his determination to convert others both to its efficacy and to the possibility of applying cyclic terminology to it. This 1909 article, read before the London Geological Society on 24 March, was intended for beginners and is indeed a remarkable exposition, illustrated with thirty-three figures, of the fundamental facts of mountain glaciation.

Davis had visited Snowdonia in September 1907 going straight there with Joseph Lomas, F.G.S., who met the *Saxonia* when it berthed at Liverpool. Later, in October, he returned to the area previous to the presentation of a general account of it to the Liverpool Geological Society, just before leaving for Boston. In August 1908 he revisited the district with members of the Oxford Summer School of Geography and a few weeks later presented a summary of his work there to the British Association at Dublin. The 1909 paper was based largely on these short spells of fieldwork, on the excellent official 1-inch and 6-inch maps and on the famous Memoir of A. C. Ramsay (*The Geology of North Wales*, 1866; 2nd edition 1881). But, as the following letter shows, Davis had also taken the precaution of consulting G. K. Gilbert on certain tricky points.

My Dear Davis,

I have read Bonney and Garwood, and put into my Harriman script a brief statement of their theories for the hanging valley of the Alps. I do not discuss the theories (because I am not acquainted with the Alps), but merely remark, substantially, that the papers, while chiefly defensive of well-established views, yet had been disturbed by Davis' innovations, indicate an interest in glacial physiography and its interpretation, and may be but the beginning of a discussion like that inaugurated by the publication of Ramsay's theory for the glacial origin of rock basins.

(1) The greater part of Bonney's paper seems to me not at all to the point. (2) If glaciers were as feeble eroding agents as he believes, so feeble that they could prevent the deepening of tributaries while trunk valleys were not only deepened but broadening, then they would not be competent to do the work ordinarily ascribed to them of transforming V gorges into U troughs. (3) The assumption that tributary glaciers terminated accurately at the ends of hanging valleys in all the districts of hanging valleys, and for the long period necessary to accomplish the hypothetic stream work, is an incredible amount of coincidence and constancy.

Garwood's second hypothesis of long continued lacustrine level is offered without a particle of evidence. It implies the preexistence of the lake, and therefore of the lake basin, and thus fails to explain the phenomenon of hanging valleys, because the existence of the trunk valley is part of the phenomenon. It of course demands the subsequent appearance of a glacier to scour out the deltas and other deposits complementary to the erosion of hanging valleys.

Garwood's hypothesis for the retreat of cirques is based on a false statement of fact. When a glacial valley contains only a small body of snow or ice, that body does not lie upon the flat bottom, as he states, but does lie at the angle between the flat bottom and the heading cliff. It thus covers the tract in which it is essential that there be erosion in order that the valley lengthen by the retreat of the cirque.

Both authors seem to me have the mental attitude of the lawyer or the preacher. They are committed to certain conclusions, and seek arguments in support thereof.

All these things it is impolitic to say in print; but I trust they will afford you the consolation you crave.

(Letter: G. K. Gilbert to Davis. Undated)

In the London lecture of 1909 Davis describes at length Ramsay's ideas on plains of marine denudation and suggests that the main 'plain' was early or mid-Tertiary in age, and not 'older than the New Red Sandstone' as suggested by Ramsay. Because of the absence of a recognizable line of sea-cliffs or bluffs, Davis thinks it desirable that 'the upland of Wales may be treated as an uplifted and dissected plain of subaerial erosion' (p. 289). He adds:

A later generation of geologists seems generally to have given up marine erosion in favour of 'subaerial' erosion – called normal erosion in this essay – as the agency by which the great ancient mountains of Scotland and Wales were for the most part worn down: marine erosion being nowadays appealed to – if appealed to at all – only in order to give the last touches in the work of planation. Sir Archibald Geikie's essay 'On Modern Denudation' (1868, *Trans. Geol. Soc.*, Glasgow, 3, 153–90) appears to have played an important part in causing this

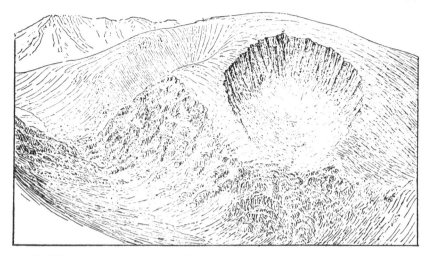

F I G. 63. Diagram of a cwm in a subdued mountain; based on a sketch of Cwm Du in Mynydd Mawr, looking southwards (From Davis, 1909E, Fig. 3)

change of opinion in Great Britain; nevertheless, explicit reference of the sufficiency of subaerial erosion to reduce a mountain-mass to a peneplain without supplement by marine erosion is seldom found in British geological literature; and still more rarely does one find there a deliberate and thoroughgoing analysis of the consequences of this simple idea. (*Davis, 1909E, p. 287*)

Davis proceeds to make good this deficiency in the literature by establishing to his own satisfaction, the pre-glacial features of the Snowdon massif.

At the time just before the Tertiary cycle of erosion was interrupted by uplift, the then low-standing group of Welsh monadnocks presumably had well-subdued crests and ridges, only here and there ornamented by outcropping ledges; the ridges must have been separated by fully mature valleys drained by perfectly graded, normally branching streams. Some time after the late Tertiary uplift, just before glaciers were formed upon the higher-standing monadnocks, the valleys previously eroded among them had presumably been somewhat deepened; but, if one may reason from the case of Dartmoor, the general aspect of the Welsh monadnocks must still have been that of subdued mountains, with dome-like summits and rounded spurs, drained by prevailingly graded streams of accordant levels at their junctions. Some of the larger streams of Wales may have been impelled to incise narrow gorges rather sharply in their former valley-floors, and thus in a very small way the valleys of the smaller side-streams may have come to hang over the narrow gorges cut down by the large streams. But, in the Snowdon area, there are no large trunk-rivers; all the streams are branching headwaters, and the disparity of volume among them is not sufficient to have produced striking cases of discordant junctions. The features of pre-Glacial Snowdon, as sketched in the opening paragraph of this essay, are thus justified. It would be against all reasonable analogy to believe that the sharp ridges, the steep cliffs, the valley-head

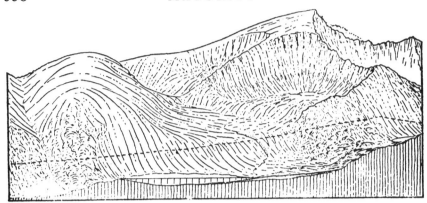

F I G. 64. Diagram of the present features of the Snowdon mass, based on a sketch from the flank of Mynydd Mawr, looking eastwards. This is 'a type-diagram ... in which the abnormal features, previously shown separately, are compactly generalized and presented in their natural association' (Davis, 1909E, p. 307) (From Davis, 1909E, Fig. 7)

> cwms, and the valley-floor steps and basins of today could have existed at that
> time. (*Davis, 1909E, pp. 292–3*)

In his discussion of pre-glacial terrain, Davis deals fleetingly with the idea of texture of dissection, and makes the following interesting comment of a quantitative nature all too rare in his writings.

> In a given stage of a cycle of erosion, landforms may vary as to the texture of their
> dissection, or number of stream lines crossed in a given distance. A stream-line is
> here understood to be a sloping line to which the converging slopes on each side
> contribute drainage. Such lines are very numerous if the dissected mass is of an
> impervious structure yielding a fine-grained waste on a barren surface; for then
> every little rill will carve its minute valley, as in the Bad Lands of Nebraska, where
> 1000 or 2000 stream-lines may be counted in a mile. Stream-lines are, on the other
> hand, relatively rare, if the dissected mass is covered with a coarse and pervious
> sheet of creeping waste overgrown with vegetation; for here most of the rain
> percolates beneath the surface, instead of running off in surface-rills. The subdued
> mountains of North Carolina offer good illustration of a thoroughly dissected
> region of relatively coarse texture; here many a mile of surface has not more than
> ten or twenty stream-lines. Pre-Glacial Wales should be pictured as of the latter
> coarse-textured kind of dissection; and this picture is warranted, because the
> Welsh moels (smooth, dome-like mountains) today still have large sweeping con-
> tours, seldom indented by stream-lines. (*Davis, 1909E, p. 295*)

Having deduced and depicted the pre-glacial landscape, Davis goes on to describe in detail the 'abnormal landforms' that may be attributed to glacia-tion. It is best described in his own words:

> An excursion around Snowdon ... in September 1907, led me to the conclusion
> that a large-featured, round-shouldered, full-bodied mountain of pre-Glacial time
> had been converted by erosion during the Glacial Period – and chiefly by glacial

erosion – into the sharp-featured, hollow-chested, narrow-spurred mountain of today. Or, to phrase it in somewhat more technical style, that a body of ancient slates, felsites, and volcanic ashes, greatly deformed in Palaeozoic time and greatly worn down in successive cycles of Palaeozoic, Mesozoic and Tertiary erosion, was reduced before the Glacial Period began to subdued mountain-form with dome-like central summit, large rounded spurs, and smooth waste-covered slopes, and with mature valleys . . . Also that this full-bodied mass was transformed during the Glacial Period, chiefly by the glacial excavation of valley-head cwms and by the glacial widening and deepening of the valleys themselves, into a sharp central peak, which gives forth acutely serrated ridges between wide amphitheatres: the serrated ridges changing into broad-spreading spurs as they are followed upwards; the wide amphitheatres, backed by high rocky cliffs, opening by great rock-steps to irregularly deepened trough-like valleys, with oversteepened, undissected sides, sometimes smooth, sometimes of a peculiarly roughened slope; the smaller lateral valleys hanging in a strikingly discordant fashion over the floors of the larger valleys; and the streams, far from following graded courses in steady flow, frequently halting in lakes or hastening in rapids and cascades. (*Davis, 1909E, pp. 281–2*)

The abnormal features of Snowdon and its neighbours appear to be in some close way associated with the glaciation. The cwm-floors, the valley side-slopes, the valley floor-steps are scored and striated. Similar abnormal features occur in other glaciated districts, all the world over, varying in intensity rather than in kind; but they are practically unknown in non-glaciated mountains. So persistent an association has naturally led many observers to look upon the abnormal forms as the result of glaciation: but it has been urged by other observers that the hanging attitude of lateral valleys over their main valleys, which so strikingly characterizes glaciated mountains, may be produced under normal conditions. The explanation thus offered is in essence as follows: If a normally sculptured mountainous area, drained by mature streams with accordant junctions, be uplifted, and especially if the uplift be so disposed as to increase the slope of the chief drainage-lines, then the main rivers will be impelled quickly to deepen the main valleys, and thus the lateral valleys will be left hanging over the newly deepened floors of the main valleys. There can be no question that hanging valleys of this kind may be normally produced in the early stages of a new cycle introduced by regional uplift, particularly where the main river is much larger than its branches, and where the uplift is rapid. But such hanging valleys can endure only so long as the main river has a narrow gorge-like valley; the lateral valleys must be worn down to accordant junctions by the time the main valley-floor has become open by the lateral erosion of its graded river and by the wasting away of its walls. Hence the suggestion, that lateral valleys may remain hanging over well-open main valleys which have been deepened as a result of land-tilting, is not acceptable . . .

It is, however, not so much the occurrence of any one abnormal feature, such as hanging lateral valleys, in glaciated mountains as it is the systematic combination of various abnormal features that is so strongly persuasive of the dependence of such forms in some way or other on glacial action. There is, indeed, to-day a general agreement among geologists as to the association of abnormal features, like cwms and hanging valleys with glaciation. (*Davis, 1909E, pp. 303–4*)

Davis continues this rather coy treatment by theoretically deducing the consequences of both the glacial protection and erosion theories.

Certain special consequences of the theory of glaciers as protective agencies need mention, particularly as applied to pre-Glacially subdued mountains.

(a) Protective glaciers occupying the upper parts of mature valleys in a group of monadnocks cannot produce cliffs by under-cutting the valley-head slopes; the glaciers can only preserve the moderate slopes of the pre-Glacial valley-heads without significant change. The mountain-tops, in so far as they rise above the protective névé-and-ice cover, must be reduced to very dull shape during the successive episodes of glacial growth.

(b) Even if a glacier be protective, the water-stream that is given forth by the ice would cut back a narrow slit in the steep face of the rock-step down which it cascades below the glacier-end. The streams that are today rushing down such rock-steps in glaciated mountains are actively at work trenching their courses in cleft-like gorges, but they have as yet made very little progress; had the streams been there during all the long interval of time required, not only for the deepening, but also for the widening of the lower valleys, one would expect that the stream-gorges would have attained a great length and depth.

(c) The number of rock-steps should correspond in each of the several radiating valleys of a mountain-group like Snowdon; for each step is only the local result of a well-maintained climatic episode which must have been of essentially uniform value all around the mountain-group. Furthermore, the height and spacing of the corresponding steps should be systematically related, for each set of steps is the result of a single climatic change and of a single uplift.

(d) It may be noted that the necessary postulate of increasing uplift with increasing glaciation runs counter to the evidence which in many regions associates increasing glaciation with depression.

(e) Each enlarged portion of a valley down-stream from an ice-covered rock-step should be of normal pattern; the several side-valleys should be well opened if the main valley is well opened; they should join the main valley at accordant levels; and the spurs that slope into the main valley between the side-valleys should be of normally-subdued slope, if the main valley floors are broad. But the normal development of accordant junctions for a group of side-valleys may be prevented by a highly specialized arrangement of side-glaciers, which shall conspire to hold their ends at the side-valley mouths, . . . Hence, in all such cases the floors of the group of hanging side-valleys must be the preserved parts of a previously eroded mature valley-system, and as such must, irrespective of the size of the side-valleys, hang at an altitude appropriate to a corresponding part of the main valley-floor preserved above the next upstream rock-step. Furthermore, (f) there can be no more groups of hanging side-valleys than there are rock-steps in the main valley floor; and all the members of a group of hanging side-valleys must occur between two successive rock-steps in the main valley-floor. On the other hand, (g) large or small lateral valleys not occupied to the mouth with side-glaciers, must enter the main valley at accordant level with it, and yet they may be in close neighbourhood with hanging lateral valleys.

It should also be noted that (h) the long-maintained halt of a group of side-glaciers at their valley-mouths, as required for the production of hanging side-valleys, is a much more specialized requirement of this theory than is an equally long-maintained halt of the main glacier at some unspecified point in its larger valley where a rock-step is to be made. The halt of the main glacier requires only a long-maintained uniformity of glacial climate, and while so long a halt is in itself certainly remarkable, it is not complicated by having to stand in a definite relation to any non-climatic element. The long halt of a group of side-glaciers at their valley-mouths, however, requires not only the same long-continued uniformity of climate, but the maintenance of the climate at just such a condition as shall cause a number of independent side-glaciers of unlike dimensions to end in a definite relation to a line of an altogether different quality, namely the line of the main valley to which the valleys of the side-glaciers are tributary. When it is noted that the drainage-basins of neighbouring hanging lateral valleys vary greatly in area and in altitude, it appears extremely improbable that any glacial climate could cause the independent glaciers of such neighbouring lateral valleys to end in a line of almost uniform altitude, on which the mouths of a set of hanging valleys must, according to this theory, stand. Furthermore (i), the climate of the Glacial Period, as a whole must, under this theory, be peculiar in advancing through several long maintained pauses to its climax, and then in moderating either with correspond-ing pauses or without significant pause to its close. (j) When the ice finally disappears, the valley-heads will be revealed, essentially unchanged from their pre-Glacial form. But certainly the most peculiar feature of the theory of glacial protection is (k) its fundamental postulate that glaciers do not erode, in view of the erosive action easily observed, even though slow, under the sides and ends of mountain-glaciers all the world over. (*Davis, 1909E, pp. 311–13*)

As in the theory of glaciers as protective agencies, there are certain special features in the theory of glaciers as destructive agents which need special men-tion. One of the most significant points to consider is the varying duration of glacial action.

(a) If the Glacial Period were short, the ice might disappear before the cwms were enlarged enough to destroy much of the pre-Glacial mountain-form, and before the glacial channels were well excavated beneath the pre-Glacial valleys; then the sides and floors of the channels would be left ragged and uneven, as a result of the more rapid work of erosion on the weaker parts and the less rapid work on the more resistant parts; and, so long as the young glaciers were actively deepening their channels, the channel-sides would be steep and cliff-like, with a U-shaped cross-profile.

(b) But, if the Glacial Period were long, the channels might be worn to such a depth that further deepening would be slow; then the floor and sides would be smoothed and the channels would be widened so as to lose their U-shaped cross-profile and gain that of a round-bottom V.

(c) Wherever a considerable mass of weaker or more jointed rocks occurred in the graded pre-Glacial valley-floor, young ice-streams would deepen their channels more rapidly than on harder or less jointed rocks next up-stream; thus rock-steps would be produced, and once produced they would 'retreat' up-stream by the removal of rock-blocks from their face. Similarly (d), the channels would be

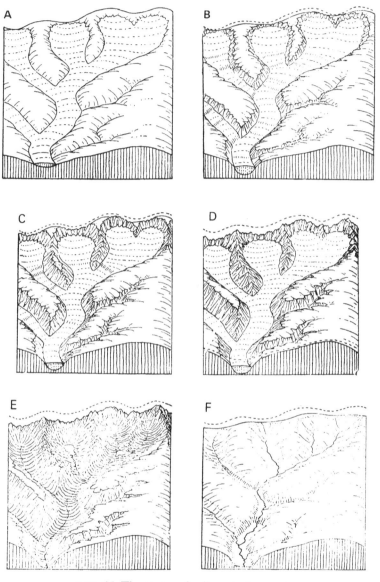

FIG. 65. The stages of valley–glacier erosion

A. A large eroding glacier filling the valleys of a subdued mountain

B. The same glacier at a later, sub-mature stage

C & D. The same glacier at progressively more mature stages

E. The landforms revealed by the disappearance of the glacier at stage D

F. The same landscape as it would appear had the glacier been protective, instead of erosive

(From Davis, 1909E, Figs. 14, 17, 18, 19, 20 and 21)

eroded to a greater depth in belts of weak rocks than in belts of hard rocks next down-stream, and thus rock-basins would be formed in the channel-beds.

If (e) the Glacial Period were very long, some of the cwms excavated in the valley-heads might grow so large as to reduce and consume the spurs between them, and thus destroy all traces of pre-Glacial form, leaving only sharpened peaks and ridges between the cwms, as in [Fig. 65D]; then (f) the channels would be so well perfected that their sides and floors would be nearly or quite smooth, and few rock-steps would remain in their course, as in [Fig. 65E].

But (g) rock-steps would still necessarily occur at the junction of small and large glacial channels, however long the Glacial Period endured; for it has already been shown that a discontinuity of channel-bed must persist at the junction of a small tributary with a larger trunk stream, whether the stream be of ice or of water, however maturely the channel beds are eroded to an even slope elsewhere.

(h) The hanging lateral valleys would, however, not be systematically arranged in sets related to rock-steps farther up-stream in the main valley; they would be irregularly placed as to altitude, for the depth of the floor of a hanging valley below the upper limit of glaciation would ordinarily be related to the depth of the main valley directly as the volumes of the lateral and main glaciers. Hence neighbouring lateral hanging valleys might open at different heights above the main valley-floor; and all the lateral valleys of a well-glaciated main valley should mouth or 'hang' at one height or another above the main valley-floor. (i) Non-hanging junctions should be expected only where two valleys occupied by glaciers of equal size come together; and hence each valley might be less deep than the united valley farther down-stream. (j) The valley-floor rock-steps should not correspond in number, in altitude or in spacing, along neighbouring valley-systems.

(k) After the final disappearance of the ice from its channels, water-streams will cascade in full view on the ungraded rock-steps in the channel-beds and on the hanging valley-mouths. (l) The hollows of the channel-beds, occupied by lakes, will be seen to indicate only the excess of erosion in one part of the glacial channel over the erosion in the part next down-stream. (m) Similarly, the ledges and knobs of rock that may rise here and there in the less matured parts of the glacial channels will not be regarded as surviving elements of pre-Glacial form and hence as witnesses to the inefficiency of glacial erosion, but remnants of much larger masses not completely worn down to a smooth channel form. (*Davis, 1909E, pp. 318–19*)

Going through the motions of confronting the two sets of deductions with the facts of observation in the Snowdon district, Davis continues to tilt at the windmill of glacial protection, and naturally concludes that the land-forms are fully commensurate with glacial erosion. Even Davis, however, has to admit the preconceptions which underlay this apparent exercise in objective deduction and testing:

Despite the efforts that I have made to present the alternative theories and their consequences impartially, it may be only too apparent that, in my own mind, the theory that glaciers are protective is essentially unsuccessful, and the theory that glaciers are destructive is, on the whole, eminently successful in leading to consequences that accord with observed facts. This opinion was not reached until after I saw in 1899 that the lateral valleys of the Alps prevailingly hang over the wide open main valleys, and that the piedmont Alpine lakes cannot be reasonably explained by the warping of normal pre-Glacial valleys. Since then, my own observations in Norway, the Tian Shan, and the Rocky Mountains have only served to confirm the opinion gained in the Alps; and at the same time similar evidence has been presented by observers from various glaciated mountains as far separated as Alaska (where Gilbert's studies are most convincing) and New Zealand (where the discussion by Andrews is no less unanswerable). But, however fully an author may be convinced of the success of one theory or another, it is always appropriate for him to revise his opinions in the presence of other theories; and that revision I have here tried to make, in view of the acceptance given to the theory of the protective action of glaciers by some geologists whose experience in the glaciated mountains of Europe is far greater than mine. In such a revision, the method here adopted seems logically correct and safer than any other as a means of solving the problem in hand, or any similar problem in which effort is made to determine the nature of invisible facts – invisible because of minuteness, or of remoteness in time or space, or of any other quality. It might, indeed, have been more satisfactory to have the consequences of the theory of the protective action of glaciers set forth by one of those who is convinced of its truth; but, as I have not been able to find from such a source a thorough attempt to carry the theory to its consequences and to confront the consequences with the facts, I have had to make that attempt in a more or less prejudiced manner myself. (*Davis, 1909E, p. 320*)

The arguments used by Davis in this paper are similar to, but rather more advanced and elaborated than, those employed in his classic paper on 'The sculpture of mountains by glaciers' (1906J), although he continues rather explicitly to play down the effects of structure (1909E, p. 285).

The discussion that followed the presentation of the paper was lively and professional. M. M. Allorge was charmed by the lucidity of the author's description and by his series of skilful diagrams on the blackboard, conveying so vivid a sense of three dimensions. But why did the author use the term 'cycle' which suggested a return to a primitive state of things, when in geomorphogeny the final peneplain was not entirely comparable with the original structural surface? Would not 'evolutive stages of erosion' be more accurate? Decidedly not, replied Davis, because in one sense cycle meant a long period of time, and in the other sense it was very appropriate for plains and plateaus and hardly less so for mountains eroded from uplifted peneplains as they started and ended with essentially the same form.

Professor Garwood, the leading glacial protectionist, congratulated Davis warmly and said that he thoroughly agreed with him about the assemblage of features characteristic of glaciated countries;

but, while never doubting that sea was an erosive agent, he [Garwood] had found it difficult to account for certain features in the Alps by simple differential erosion, and had been driven to consider the possibility of differential erosion as between ice and water. Was it not possible that ice, being relatively a protective agent, might have retarded the erosion of those portions of the surface on which it still remained after it had retreated from other parts of the district? (*Davis, 1909E, p. 346*)

Davis, however, demolished Garwood's examples of glacial protection by showing that they could be better explained on the glacial destruction theory. To other criticisms, Davis wisely admitted that cwm-head cliffs posed serious difficulties under any theory yet put forward.

All present admitted that the lecture was a masterly exposition of fundamental observations, as lucid as it was persuasive. The president, Professor William Johnson Sollas, who sometime earlier at Oxford had announced his guest as 'Professor Peneplain Davis', summed up the London meeting with his customary acuteness.

> He congratulated the Author on the successful manner in which he had rejuvenated a mildly-matured subject. His essay was an extremely consistent piece of deductive reasoning . . .

After North Wales, Davis went straight to Dublin where he gave a brief account of his latest glacial investigations and also contributed a rendering of his famous Colorado Canyon lecture. A week there, and a brief stop in London preceded his attendance at the German Naturalists' meeting at

FIG. 66. Davis (right) with a field party on a valley-side terrace near Bethesda, North Wales (1911) (Courtesy Dr G. Martin, Southern Connecticut State College)

Cologne where the Canyon lecture was again delivered, this time in German. He had in the previous June pointedly suggested to Albrecht Penck that the theme of the meeting should be the Grand Canyon and Penck, as usual, readily agreed with his wishes. After Cologne, Davis spent October at Marburg slowly piecing together the material for his winter lectures. Altogether a very exhausting experience for any man.

One wonders what part his wife played in all this and how far she supported his labours. There is just no evidence and one is left to make comparisons between Davis at this time and in the years immediately following his wife's death. Despite the equable assurance of his writings there are many indications that his energy had a nervous source and his equanimity or concentration could readily be disconcerted by the unexpected and unorthodox. Sharp noises certainly disturbed him.

> Apart from the meeting of the Naturalists, the feature of Cologne that most impressed us was not its cathedral, not the narrow little streets of the old Roman town surrounded by the wide thoroughfares of the modern city, but the quietness of its many railroad trains, and the almost noiseless passage of its electric cars. We stayed for nearly a week in a hotel fronting one of the squares through which several lines of street cars were continually passing even to a late hour in the evening, and just beyond which was the main railway station. The absence of engine whistling was in strange contrast to the prolonged and piercing shrieks that one must hear near every station in Italy, while the almost inaudible movements of the street cars was in most mortifying contrast with the disturbing clang of our electrics. There was nothing of the volleying down grades, of the slamming over poorly joined rails, of the screeching around curves, of the needless existence of the alarm bell, that torments all of us hereabouts, except those whose ears and nerves have become callous. It was a regret to me that I had no time to make inquiry as to the means of securing so much silence, but the subject is one that deserves inquiry from our city fathers. Surely it cannot be said that noise is an essential feature of our present convenient system of electric transportation.
>
> (*Davis, 'Boston Evening Transcript', 1909*)

Davis' susceptibility to strain or disquiet needed the constant attention of an understanding wife. His freedom from illness and the absence in his letters of any serious complaint provide the only real, though circumstantial, evidence of her very great assistance.

The real strain of the appointment in Berlin began with the formal address before the university at the beginning of November 1908. This was attended by a representative of the German Royal Family, the American Ambassador and senior administrative officials, in addition to numerous professors and students. The speech gave a general description of American geography, and, as advised by his friend, was limited to twenty minutes.

The students' course provided by Davis was a very full one. We know its contents from a letter from Albrecht Penck, who obviously had not yet been told at the time of writing that Davis' visit had been put forward to the winter term.

My dear Davis,

Yesterday evening I returned from an excursion with my students lasting several days, and on arrival found your kind post card of the 9th inst. I am pleased to learn from it that you will be at the German Naturalists Meeting in Cologne. In accordance with your suggestion my theme will be: 'The Colorado Grand Canyon'.

Together with your letter I received another from Geheimrat Schmidt where he tells me of his pleasure that President Eliot has granted your request for leave for the coming summer semester. He is especially pleased to have you with him during a whole summer semester here in Berlin. Your lectures will be announced in the lecture list, just as you wished:

Systematic Geography, Tue. & Wed., 4–5, practical exercises, Thur. 10–12.
Geography of the USA (in English), Friday, 4–6.
Geographical Colloquium, Tue. 6–8.

My students are very pleased about your forthcoming visit but they would rather you held your exercises on another day. This is inevitable here: this University is so large that certain collisions (in the time-table) are inevitable. One can only try to reduce them to a minimum, and this is achieved most easily by asking the students which hours are most convenient for them: it generally turns out that one's own suggestions are then adopted by the majority.

My excursion during Whitsun to Thuringia was entirely successful. I was quite surprised by the beauty of the river terraces. They show without any doubt that the mountains rose already during the Great Ice Age, by at least 100 m.

With best wishes,

Your devoted

(A.) Penck.

(Letter: A. Penck to Davis, 14 June 1908; Berlin)

The lectures on systematic geography and the practical exercises were on the systematic description of land forms. The lectures on the United States also dealt with physical geography but were given in English and were at first attended by a considerable number of students who apparently came to gain some practice in a foreign language. But, as Davis says,

these hearers soon disappeared and left me with a steady-coming audience of young geographers. All the rest of the work was conducted in German. In the colloquium the students and the younger teachers presented essays, which were then discussed by other members. On adjournment it was the custom of a number of us to go around the corner to a neighboring restaurant and there have supper, which was generally prolonged until half-past nine or ten in the evening. (*Davis, 'Boston Evening Transcript', 1909*)

Undoubtedly the German language presented more problems than the academic material, particularly as one-half of the course was devoted to his favourite method of systematic classification. An obvious consciousness of language deficiency must always have been present, and, no doubt, a partial means of evading this handicap, namely the use of blackboard sketches, must have proved a great help because he was exceptionally good at it.

> He had the remarkable ability to draw with both hands upon the blackboard while lecturing and when he drew a section through the Grand Canyon he would then tilt the blackboard to restore the attitude of Precambrian to horizontal. (*Letter: E. B. Knopf to R. J. Chorley, 27 January 1962; Stanford, California*)

Where a diagram was particularly complicated he often used to trace the outline by means of faint marks on the blackboard, some time before the lecture was due to be delivered.

> He was without question the most systematic lecturer I have ever heard. His outlines were perfectly prepared, perfectly timed in presentation and almost spectacular in results. To see him enter a classroom, proceed to the blackboard, and draw an involved physiographic diagram ambidexterously was astonishing for the first time. Closer inspection showed that a system of light dots in blue chalk had been placed on the blackboard before the class met, as a guide for his drawing. Nevertheless, it was an extraordinary experience. (*Letter: H. F. Raup to R. J. Chorley, 5 January 1962; Kent, Ohio*)

As well as economizing on his German, the use of sketches was a very vivid way of explaining geomorphic processes and forms. In Germany it had the added interest of being a comparatively novel technique.

> A graphic device known to many geographers in this country [United States] as a 'block diagram', by means of which the relation is immediately shown between underground structure on the side of the block and topographic form on its upper surface, was something of a novelty to my German hearers; and all the more so when several blocks were placed in perspective sequence, so as to exhibit successive stages in the development of a given structure. Blackboard pictures often served the further purpose of graphic lecture notes, and were a great aid in the German lectures when sufficient confidence was gained to turn from the manuscript and throw in extemporaneous remarks in explanation of certain details.
> (*Davis, 'Boston Evening Transcript', 1909*)

Despite the comprehensiveness of his courses and the extra strain of working in a foreign language, the inroad on his mental resources was insignificant. He still found abundant energy with which to participate in discussion groups, rather like his own Shop Club at Harvard, though taken perhaps more seriously. He also found occasion to compare American and German teaching methods within universities and as usual his comments are strikingly blunt.

> For example, with regard to university methods, a topic on which I had many conversations, there was a very distinct satisfaction with existing conditions on

the part of most professors, even to the point of being rather indifferent as to any other possible methods; and the excellence of German scholarship was instanced as good reason for this satisfaction. But my own opinion on this point came to be that German scholarship is high quite as much because of intense competition, as because of peculiarly excellent methods of instruction. The competition has two causes; one is the pressure for any reasonably good chance to make a living, a pressure that is vastly greater in Germany than it is with us, for it must be remembered that Germany has no half-occupied Western country, where opportunity for capable young men is still abundant. The other cause of intense competition is the secure and highly honored position of a professor in a German university. He is an official for life of the Government for all of the universities in which he may have office are governmental institutions, and the attainment of the professorial grade is therefore much more strictly limited as well as more permanent than it is with us, where indeed it is employed in so many feeble institutions as to risk bringing it into discredit. Moreover, the German university professor will pretty surely become a Geheimrat, if he lives long enough, and although many persons affect to regard this widely distributed official title as only a mark of distinguished mediocrity, no one, so far as I could learn, ever refuses it when it is offered. (*Davis, 'Boston Evening Transcript', 1909*)

These various academic happenings provided no entertainment for his wife but there were also numerous social functions which she may have enjoyed. Davis writes as if he did.

During the progress of the lectures and the abundant work required in the preparation of those given in German, we had the usual large amount of distraction in the way of social calls and invitations. It will be remembered that the German custom is for the stranger to make the first visit: it therefore lies entirely in the hands of the visiting professor to determine whether he will attempt a large acquaintance in Berlin or not. We intended to take a rather moderate course in this respect, and indeed had no time to spare for any other but before the winter was over we found ourselves in a social whirl where one engagement followed the other with a rapidity that would have been altogether delightful were it not that they had to alternate with the rather engrossing duties at the university. As it was, the engagements were delightful enough, but we were sometimes almost too tired to enjoy them. Too much cannot be said of the courtesy and cordiality of our German friends. With hardly an exception everyone upon whom we called invited us sooner or later to dinner and we thus had an admirable opportunity to see the Berlin University circle at home.

It would be in-appropriate to make personal mention of the many delightful evenings that we thus spent, but it may perhaps be permissible to describe the hospitality of the rector of the university for the current year, Professor Kahl of the faculty of law. It was incumbent upon him as rector to invite groups of university professors and other officials to a series of Saturday evening dinners, and the Roosevelt professor and I were included in an early one of these, a very handsome affair, during which we made acquaintance with a number of our new colleagues. On a later occasion the two visiting professors and their wives were invited to a mixed party, during which the rector took occasion to propose the health of all the ladies present, but particularly of certain ladies; namely two who had come with their husbands from other universities on call from the University of Berlin; one who with her husband had remained faithful to Berlin when he refused a call to another university and two others who had accompanied their husbands as visiting professors from America.

A UNIQUE FORM OF ENTERTAINMENT

The third occasion when Professor Kahl was our host was one of the most graceful entertainments that can be imagined. It was given in a large building originally used for an exhibition where several halls in succession served for the different hours of the evening about two hundred guests being present. We gathered at 6.30 in a large reception room, where tea was served. About an hour later we passed to a hall where platform and seats were arranged for a 'family concert', in which the rector, his daughter and some of their friends took part on various instruments, the gathering then adjourned to a larger hall, where dinner was served at many small tables and where our host and his son-in-law as well as another professor of law made three as jovial speeches as could be heard anywhere.

F I G. 67. Davis and unknown persons (Courtesy R. Mott Davis)

After dinner there was dancing and the entertainment was closed by a reader who rendered some poems and stories in a way that was greatly enjoyed by the guests. Throughout the whole evening there was manifest a delightful spirit of familiar acquaintance and comradeship that made the occasion enjoyable to all of us.

(*Davis, 'Boston Evening Transcript', 1909*)

Mrs Davis was denied the satisfaction of being presented at the German court and probably regretted declining this honour more than Davis himself affects to have done.

There appear to be certain traditions already established by previous visiting professors, as a result of which it has been repeatedly necessary for me to explain that I was not in the procession of Americans presented by our Ambassador and Mrs Hill at court, because on the evening when that goal of many ambitions was reached, I was lecturing in Leipzig before the Geographical Society of that city. If it be added that my evening there was much more enjoyable than it would have been in Berlin, the reader may exclaim, 'There is no accounting for tastes!' And yet it does not seem altogether surprising that one who is somewhat accustomed to public speaking should prefer an evening where he had it all his own way, rather than an evening where everything was cut and dried for him by order of someone else. Besides at Leipzig a visiting professor may see Geheimrat Partsch, the genial geographer of the Saxon University, and that alone was worth the journey for Partsch is one among a thousand. Moreover he not only brought a great audience together for me but arranged a fine dinner to follow the lecture, when Hermann Credner, the veteran geologist of Saxony and Hans Neyer, the mountain climber of tropical Africa and South America, were among the guests, and it must be remembered that Germans understand thoroughly well the fine art of enjoying themselves around the table. (*Davis, 'Boston Evening Transcript', 1909*)

The explanation Davis gives is reasonable but at the back of it one can hear the echoes of relief at the need to make a choice. Also it is hard to believe that he could not have missed the Leipzig meeting altogether had he wanted. We know his nervous aversion to extreme formality and he may have seized on his earlier commitment as an excellent opportunity for evading the tension and stiffness that such a function entailed. It is not likely he considered his wife in this. She appears throughout as an indispensable but impassive presence. Nothing in Davis' writings tells us what she was like. She must have been content to subdue her own personality and to rebuild her way of living to become a support for her husband's career and interests. Not surprisingly, the best brief descriptions of her come from Albrecht Penck and Mark Jefferson, both men of rare insight. They thought her charming and capable.

His first wife was very charming. I never saw a more delightful home circle than theirs with their ... sons all I think in their teens when I knew them at Cambridge in 1897–98. (*Mark Jefferson to R. M. Harper, 9 April 1935*)

She sounds, and must have been, a most dependable person with an intellectual stature of her own. Many of her characteristics must have merged with those of Davis but she undoubtedly had other qualities that counteracted his personal defects. As we have shown, their eldest son suggested that she had a heavy burden to bear but admits that this was never evident.

The remaining impressions of Davis' Berlin visit fall outside his normal interests but are perhaps important for this reason. At the time of Davis' appointment the Germans were beginning to thrust forward as commercial and colonial rivals of the British. Their newly-found unity had provided a rich soil from which was growing a proud and rapidly expanding industrial nation. Britain, as the most important established world power, feared this expansion by the Germans and there were several international incidents where their interests clashed. Yet while this scramble for concessions was bringing bitterness to the political scene, the family link between the royal houses preserved a semblance of friendship. The conversations which Davis had with his colleagues bring out these opposite views.

One is often asked on returning from an experience abroad, 'What do the Germans think upon such and such a thing?' My experience was that Germans, like Americans, have all sorts of opinions on all sorts of subjects. One man told me *apropos* of a question that is just now exciting much remark among our British cousins, that war with England was absurd, although of course, active commercial competition was to continue, another said that it was out of the question for any member of his family to be friendly with England. One evening a very competent speaker rather laughed at me for treating the excitement of last November too seriously, and soon afterwards a Berlin resident of wide acquaintance said that Germany had not been so much moved since the war of 1870.

(*Davis, 'Boston Evening Transcript', 1909*)

Today, unfortunately, we know which views prevailed. Davis also throws light on his compatriots abroad, a theme on which he had been eloquent at least since the start of his world tour thirty years earlier. Abroad he remains proud of being American and of the republican nature of his country's laws, but is severely critical of those who assume the benefits of American citizenship as a matter of professional convenience. In the remaining part of the extract he returns to the problems of inferior status, for he was genuinely displeased with any form of subservience.

A peculiar feature of these American gatherings was that some of the most pronouncedly patriotic of us talked our language with a distinct German accent. Such was the case when a number of American citizens filled the windows of the embassy to watch the passage of King Edward the Seventh with his host, Emperor William the Second. It was a fine sight, from which my attention was unhappily somewhat distracted by the broken English of some of my neighbors. That sort of thing is familiar and welcome enough over here; for where shall we find more devoted Americans than among the Germans who have left their fatherland to make our country their motherland. But one could not avoid raising the question on hearing broken English from American citizens resident in Berlin; if they want to live there, perhaps with the title 'dentist' under their name on a professional shingle, one could not help asking why they become citizens of a republic instead of remaining subjects in an empire. And speaking of 'subject' I recall feeling a peculiar shock one day when one of my students introduced himself as 'A Russian subject'. There seemed to be something almost unmanly in thus openly proclaiming himself a 'subject' of anything. As a matter of preference much is really to be said in favour of 'American citizen' when it comes to announcing one's standing in the world; and perhaps it is along this line that we may account for the presence in Berlin of Americans old enough to vote, but still uncertain about the sound of th. (*Davis, 'Boston Evening Transcript', 1909*)

However, the interpretation which has just been given could be wrong. Davis' early submergence in Quakerism would equally have encouraged standards of rectitude and bred a reaction against artificial positions of authority. Quakers do not allow themselves ministers because they believe that all are equal and no individual can be credited as having a better understanding of the word of God.

During his appointment at Berlin, Davis paid a brief visit to Greifswald on the Baltic Coast, where the 'Colorado Canyon' was one of the lectures given. Of more benefit to his wife was the fortnight spent on the Italian Riviera. This was purely a holiday and for once was free of geologic excursions. Presumably the couple enjoyed the more usual pursuits of the transient tourist. At the end of the term there was a walking excursion in the area of Göttingen, during which Davis' pride was perhaps a little hurt by a mocking but not ill-meant allusion to his age.

Before we left Berlin the students had held a Kneipe one evening in celebration of our winter's work together, and we had some capital fun in the way of verses and songs. In the course of which they introduced accounts of young mature and old valleys, burlesquing the geographical terminology that I had employed, and referred to me as a 'late mature' geographer. The evening we came down from the Meissner I had this little joke still in my mind, and when we reached the valley after sunset, with four kilometers of flat road before us I struck out at a good pace with a young American who had joined us from the University of Göttingen, and soon left nearly all of the party in the rear, so that except for two active youths who slipped into the hotel by a side door my young companion and I were the first to reach our goal that evening. After this the students by common consent changed my grade from 'late mature' to 'early mature' and I felt happily revenged. (*Davis, 'Boston Evening Transcript', 1909*)

He left Berlin immediately his appointment ended, with a feeling of relief as, not unnaturally, he had found the strain of constantly thinking in another language burdensome. The couple spent a few days in the Netherlands and

Belgium and a few more in England where several lectures in London, including that already discussed on glacial sculpture in North Wales (24 March), were compensated by a short rest in the countryside. Finally they embarked at Liverpool and arrived back at Boston after almost a year's absence.

CHAPTER SIXTEEN

International Acclaim:
Professor Peneplain Davis

COLORADO EXPEDITION 1910

The years 1908–12 mark the hey-day of Davis' work as an educator, scientist and international figure. Surely no man did more for his own cause than Davis. He wrote the play, chose the scenery, collected the audience and acted most of the parts. During this period he conducted a most sustained and elaborate campaign to advertise his theories. The European trip and visiting professorship at the University of Berlin during the years 1908 and early 1909 have already been described. They formed a mighty prelude to still greater efforts in the form of a much longer geographical excursion in Europe, a visiting professorship in Paris and an international trans-continental expedition in the United States. There were, of course, other events that might in lesser men have been highlights but in Davis become incidental and liable to be forgotten. Among these was a field course in advanced physiography in the Rocky Mountains of Colorado in the summer of 1910. It was sponsored by the Harvard Summer School and was attended by Mark Jefferson, a former student of Davis, and two other graduate American geographers. Jefferson made ample notes, sketches and photographs and, if we may quote from the fine book recently written on him by G. J. Martin,

> listened intently to Davis' exposition for more than three weeks in July. Following the departure of Davis, Jefferson spent a further few days with Gregory in the region before composing a 31 page paper entitled 'The Rocky Mountains of Colorado', which paper was later submitted to Davis for the latter's approval. The paper commenced with a statement of purpose:

> It was proposed to exemplify the method of description by structure, process and stage by applying it to a number of points in the Rocky Mountains of Colorado visited this summer. The general method has been to note salient items in the topography from the point of view of origin, with these to frame a theory of the genesis of the land forms of today, and then, turning the attention away from the present details, to deduce the logical consequences of the theory adopted, and finally to confront these deductions with the facts of nature. The last two items amount to prediction from theory of forms not observed and then looking for them.

> For a thorough account of the physiography of the party's mountain study one may read 'The Colorado Front Range' by W. M. Davis published in 1911. Probably this sixty-two page statement accounts for the reason why Jefferson never

355

published his own writing in whole or in part. More important than an account of the itinerary was the confrontation of Davis by a Jefferson thinking, teaching, and writing a variety of human geography. The letters Jefferson sent to his wife illustrate very clearly his great regard for Davis' powers of observation and interpretation and reveal his almost fiercely held belief that human geography had a greater appeal than physiography and would one day win for itself total respect from geographers. Excerpts from the correspondence provide:

> The rocks were splendid and the weather all right. W. M. Davis showed us many points I should not have seen . . .

> Davis made us see Pene-plains and cycles that I should never have seen without him . . .

> You would laugh to see our strenuous endeavors to draw and suit our leader . . .

> Davis is perpetually hammering at us to describe, to make a theory of origin, to deduce all imaginable consequences of this theory and then to compare these consequences with the facts; also to draw and I think we are making progress.

> To finish the work that he urges, I have to spend one week in field work and one in writing up report which I will then send to him. We have to see pene-plains everywhere! After that Gregory and I who are to work together, will spend a week or two out here seeing some of the more human interest.

(Martin, 1968, pp. 135–6; including quotations from letters written by Mark Jefferson to his wife, Theodora, 10 July 1910–28 July 1910)

It is obviously impossible in a few lines to do full justice to Davis' long report as 'The Colorado Front Range; a study in physiographic presentation' (1911I) is one of his most polished and brilliantly illustrated essays. It was dedicated to Davis' Harvard friend Archibald R. Marvine (1848–76) for the following reason. In the summer of 1869, Professor J. D. Whitney visited the Rocky Mountains of Colorado with a small party that included four of his students in the Mining School at Harvard – namely Archibald R. Marvine, Henry Gannett, Joseph H. Bridge and Davis. The main object of the expedition was to determine the height of the highest ranges that they could reach. They climbed and named Mt Harvard (14,375 feet) and Mt Yale (14,187 feet) in the southern half of the Sawatch Range. Davis paid nine more visits to the area before the summer trip of 1910 when with Professor Mark Jefferson of Ypsilanti, Michigan, Professor W. M. Gregory of Cleveland, Ohio and Professor E. W. Shuler of Fort Worth, Texas, he conducted an advanced course of fieldwork in physiography there. Davis spent more than two weeks with the party who stayed on about three weeks longer to complete the examination of selected problems. Fairly good governmental maps were now available and N. M. Fenneman, S. H. Ball and others had written official bulletins or professional papers on parts of the area. Consequently Davis' party directed their efforts almost exclusively to the preparation of 'systematic explanatory descriptions of the mountain landscape' (p. 23).

We will describe in more detail later (Chapter 17) his long discussion of 'the six chief methods of geographical presentation'. Briefly, the scheme adopted was that presented in the Round Table discussion at the Chicago meeting of American geographers in December 1907 under the proposed terminology of structure, process and stage. By this method a condensed, technical, explanatory, regional description of the Front Range of the Rocky Mountains in central Colorado was produced.

This terse verbal description, with its deliberate avoidance of geological matters as such, is later illustrated by line drawings so as to form the combined technical presentation and condensed diagram of the Front Range that is quoted below.

The Front Range of the Rocky Mountains in Central Colorado is a highland of disordered crystalline rocks, for the most part resistant schists and granites, as shown in the foreground block and section of [Fig. 68]. The originally greater rock mass long ago suffered more or less complete planation, depression and burial under a heavy series of strata; the compound mass thus formed being divided by a pronounced monoclinal displacement along a north–south belt, as in the faint-lined background block of [Fig. 68], into a lower eastern area – the Plains area – and a higher western area – the Mountains area. In the cycle of erosion thus introduced, both areas advanced to old age, as in the second, darker-lined block; the weaker strata of the Plains area at the east (right) end of the block being presumably worn down to very faint relief, the more resistant crystalline rocks uncovered in the Mountains area being less perfectly worn down and appearing as a peneplain, here and there surmounted by craggy or subdued monadnocks, five hundred to twenty-five hundred feet in relief, irregularly placed, singly or in groups. The whole region was then broadly up-arched into a highland attitude, as in the slightly shaded third block of the diagram; the broad crest of the arched peneplain forming the present crest of the Front Range at altitudes of from eleven to twelve thousand feet, while the higher monadnocks that chance to rise on or near the crest reach altitudes of fourteen thousand feet or more; the eastern slope of the arch descending gradually (about one hundred and sixty feet in a mile) for some twenty miles, to altitudes of seven or eight thousand feet at the mountain border. During the cycle of erosion thus introduced and still current in the mountainous highlands, the revived east-flowing streams and their wide-spaced, usually insequent branches have, as in the largest or foreground block of [Fig. 68], eroded young or early mature valleys from five hundred to one thousand feet in depth which sub-maturely dissect the highlands, giving them a coarse-textured form and a relief of medium measure; while the upper valleys and valley heads among the loftier monadnocks along the range crest have recently been submaturely or maturely glaciated, one glacier being represented on the further side of the foreground block, and some evacuated cirques and overdeepened troughs being shown on the nearer side. In the same cycle the weaker strata of the Plains have already reached advanced old age, being thus worn down some five hundred or one thousand feet lower than the mountain border; and since then have entered upon two later episodes of erosion in which broad, shallow valleys have been excavated

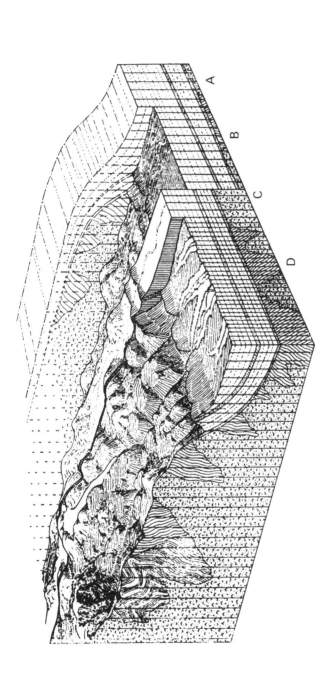

FIG. 68. Condensed diagram of the Front Range in central Colorado

A. Disordered crystalline rocks peneplained, buried under strata thickening to the east (i.e. right) and then subjected to monoclinal displacement along a north–south axis

B. Structure A reduced to an imperfect peneplain with a relief of up to 2500 feet

C. The surface B uparched to form the present crest of the Front Range at 11–12,000 feet, surmounted by monadnocks

D. The present terrain produced by the subaerial and glacial erosion of C

(From Davis, 1911I, Fig. 1)

beneath the still broader undissected interfluves, as shown in the right foreground. The work of erosion in the mountains, contemporaneous with these later episodes, seems to have caused only a narrow deepening of the chief valleys near the mountain border. (*Davis, 1911I, pp. 35–6*)

After some description of the regional subdivisions of the Front Range, Davis makes it clear that

its dominant feature is the elevated peneplain or highland, which gently ascends from the mountain border to the broad mountain crest; a highland surmounted here and there by craggy or subdued monadnocks, irregularly placed, singly or in groups; a highland submaturely dissected by submature or mature valleys. The highland must therefore be imagined as an elevated rolling surface between the monadnocks that rise above it and the valleys that sink below it; for even where typically developed a peneplain must not be conceived as a plain ...

The reasons for excluding marine action and for regarding the highland as an uplifted peneplain of normal erosion are, first, that its undissected form, as far as that can be reconstructed from its present undissected surface, must frequently have had a gently rolling surface, such as an old surface of normal subaerial degradation should have and not a smooth surface such as a plain of marine abrasion ought to have; second, that no line of ancient sea-cliffs is found, separating a smoothly abraded surface from a neighboring area of hilly land, such as the theory of marine abrasion either tacitly or explicitly presupposes; and third, that none of the numerous monadnocks exhibit sea cliffs on their exposed side, as they should do if the surrounding surface had been planed off by sea waves. (*Davis, 1911I, p. 40*)

The dissection of the peneplain is described as submature, even though some of its valleys are mature, because significant areas between the larger valleys still remain undissected; yet, in the neighbourhood of these valleys, many small ravine heads encroach upon the highland surface.

The Front Range clearly belongs in that lately recognized class of mountains in which the deformation that produced their original disordered structure has little or nothing to do with their present form.

The present form and altitude of the range is therefore not due to the monoclinal displacement of the compound mass, still less to the remotely ancient deformation of the crystalline basement, but to the broad and simple up-arching of a much later date. (*Davis, 1911I, p. 43*)

Davis proceeds to expound that numerous mountainous areas bear evidence of two or more cycles of erosion. The earlier cycle was initiated by the deformation which disordered the masses structurally; the later cycle was introduced by relatively orderly uplift.

It is therefore full time that conscious attention should be directed to a well planned arrangement of the ideal types, which shall be the mental counterparts of these many actual examples; and to the completion of a well developed, systematic series of ideal types, various intermediate members of which are developed deductively in view of the general principles established by analysis regarding two-cycle mountains. The most significant matters to indicate are first, as always, the general structure of the mass; then the stage reached in the earlier cycle before renewed elevation; next the character and amount of elevation; and finally the stage reached in the later cycle after elevation. An elementary graphic treatment of this problem is attempted in my 'Practical Exercises in Physical Geography'.

(Davis, 1911I, p. 47)

We describe later in this chapter how Davis returns repeatedly to this theme when revisiting in 1911 some of the mountain blocks of Western Europe and eventually decides to call the ideal type a 'morvan' or 'mendip'. In the article on the Front Range of Colorado he emphasizes the nature of the monadnocks (upstanding residual mountains) which he considers owe their survival and eminence to their greater resistivity to erosion. Difference of rock resistance is also assumed to explain why the valleys here, although usually 'submaturely dissected', vary from very young to late mature, widely opened forms. Glacial sculpture of the higher mountains and valleys is discussed and illustrated at great length because 'there is still some difference of opinion among geographers and geologists' upon it. But Davis wisely admits that 'my previous essays have said all that I have to say, for the present, upon the analytical aspect of the subject' (1911I, p. 54). However, certain statements appear to be of considerable importance to his biographers.

Thus he accepts the existence of at least two glacial epochs separated by an interglacial epoch of considerable duration and, to our surprise, throws in as a sidelight

It may be noted in passing that the Rocky Mountain region affords no sufficient evidence in support of the view that glacial episodes were brought on by an elevation of the region, and interglacial episodes by a depression. *(Davis, 1911I, p. 55)*

What then has happened to Davis' ideal cycle of mountain glaciation? He now makes use of the headwall retreat of cirques or cwms, which eats into the mountain mass until the sharp peaks and steep-sided arêtes themselves disintegrate into a rough-surfaced plain. But can the reader find in this a true resemblance to his former 'ideal case' in spite of the following explanation?

Cirques, like any other forms, ought to be treated with intentional regard for the stage of development that they have reached; a point that has lately been emphasized by W. H. Hobbs (The Cycle of Mountain Glaciation, *Geog. Journ.*, 1910, 146–163, 268–284). It may be noted that the term 'glacial cycle' as used by Hobbs, refers to what is here treated as a climatic accident or episode, introduced and closed by a change of climate due to external causes. The phrase glacial cycle, is here reserved for the ideal case of so long a continuance of glaciation under fixed climatic conditions – except for changes of climate with change of altitude due to degradation – that glacial erosion would be carried to its completion, truncating all the higher mountains at the snow line, and thereby causing snowfall to be replaced by rainfall, and glacial erosion by normal erosion, in a manner shown in [Fig. 69] . . . (*Davis, 1911I, p. 56*)

FIG. 69. Successive stages in the symmetrical development of cirques in a domed mountain (From Davis, 1911I, Fig. 3)

Towards the end of this article Davis returns to the theme of the mountain border.

What will be the appearance of the border of a mountain mass, in which the fundamental crystalline mass has suffered planation before a heavy series of weaker sedimentary strata was deposited upon it, and in which the compound mass was then divided by a pronounced north–south monoclinal displacement into a higher western area (the Mountains area) and a lower eastern area (the Plains area); then worn down to a peneplain which crossed from the crystalline rocks west of the monocline to the stratified rocks east of it; then uplifted long enough ago for the weaker rocks to be again peneplained while the harder rocks are only submaturely dissected?

Such a mountain border must be as rectilinear as the monocline that it follows, and almost as steep as the dip of the monocline; it must consist of a series of triangular or trapezoidal facets all standing in line, but separated by the notches of many superposed consequent valleys and ravines; somewhat the worse for wear near the top, somewhat protected by overlapping members of the stratified series near the base; its height not at all dependent on the much greater height of the monocline, but on the amount of the last uplift and on the resultant depth to which the weaker strata of the plains are now removed. Such a mountain border is in fact a fragment of a huge inorganic fossil, an ancient surface of planation, long preserved by being buried under later deposits, after the manner of organic fossils; then uplifted, attacked by erosion, its upper part destroyed, its middle part laid bare, its lower part still remaining buried. (*Davis, 1911I, pp. 64–5*)

There then follows an illuminating discussion on the problem of intersecting peneplains.

Physiographers have not yet given a special name to the combination of several peneplains, as seen in the highlands and the front of the Front Range, and in the

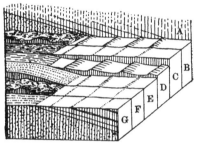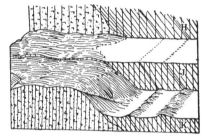

FIG. 70. Intersecting peneplains

Left. A. A tilted unconformity with resistant rocks below and less resistant above

B & C. Stages of erosion to a peneplain

D. Uplift of C and erosion to late maturity involving the partial stripping of the unconformity

E. Almost complete peneplanation of the weaker strata and extensive stripping of the unconformity

F. The resistant rocks reduced to late mature dissection

G. Peneplanation

Right. A variation of the above, with a steep and irregular unconformity

(From Davis, 1911I, Figs 8 and 9)

piedmont plains, although one phase or another of such a combination is a common occurrence, and a name for it would frequently be useful. One of my students has suggested that the combination should be spoken of as that of 'intersecting peneplains'. Their varying relations are shown in [Fig. 70, left]. The essential element of structure is a compound mass, consisting of resistant rocks below, and of less resistant rocks above, separated by a surface of unconformity, the double mass being tilted, as in section A. The chief variable elements of structure are the contrast of resistance between the lower and upper rocks, and the smoothness and inclination of the surface of unconformity. Contrasted resistance and a smooth surface of unconformity inclined about 30° are represented in this figure. The essential elements of later erosion are an approximate or perfect planation of the tilted compound mass, as in blocks B and C; the variable elements of later erosion depend in part on the nature and amount of the elevation by which the worn-down mass, B or C, is introduced into a new cycle of erosion, and on the stage reached in this cycle. After simple uplift of moderate measure, block D shows a late mature stage of erosion in the weaker strata and a partial stripping of the surface of unconformity; block E shows a new peneplain worn on the weaker strata, and the complete stripping of the surface of unconformity, while the hard-rock highland remains little changed; block F shows an aged plain on the weak strata and a late mature dissection of the hard-rock highland; block G repeats block C. [Fig. 70, right] is a variation of [Fig. 70, left], showing an uneven surface of unconformity, a steeper tilting of the compound mass, and a greater uplift after truncation. The border of the highland as here developed is uneven, chiefly because the stripped surface of unconformity is uneven. It is evidently possible to conceive a large number of variants on these simple examples. (*Davis, 1911I, pp. 69–70*)

THE PILGRIMAGE IN EUROPE OF 1911

The European excursion which began in southern Ireland on 1 August and disbanded in northern Italy on 5 October 1911, was called a pilgrimage by Davis because it visited many localities made famous in the history of the study of landforms by the writings of earlier geologists. Again this was a highly professional occasion and, although the party seldom exceeded ten at any one time, thirty-two persons participated in all, representing fourteen countries if England, Scotland, Wales and Ireland are counted separately:

H. O. Beckit, School of Geography, Oxford University, England.
Léon Boutry, University of Clermont-Ferrand, France.
Maurice Brienne, University of Lille, France. (student)
Abel Briquet, Douai, France.
G. G. Chisholm, University of Edinburgh, Scotland.
G. A. J. Cole, Royal College of Science, Dublin, Ireland.
J. Cvijić, University of Belgrade, Servia.
James Cossar, Training College, Glasgow, Scotland.
W. M. Davis, Harvard University, Cambridge, Mass.
Pierre Denis, University of Paris, France.
Albert Demangeon, University of Lille, France.
Lucien Gallois, University of Paris, France.
Gilbert Garde, University of Clermont-Ferrand, France.
Ph. Glangeaud, University of Clermont-Ferrand, France.
Walter Hanns, University of Leipzig, Germany. (student)
Amund Helland, University of Christiania, Norway.
Mark Jefferson, State Normal College, Ypsilanti, Mich.
O. T. Jones, University College, Aberystwyth, Wales.
Olinto Marinelli, Institute of Higher Studies, Florence, Italy.
J. E. Marr, Cambridge University, Cambridge, England.
Fritz Nussbaum, University of Bern, Switzerland.
Hans Praesent, University of Greifswald, Germany.
Giuseppe Ricchieri, Scientific and Literary Academy, Milan, Italy.

Alfred Rühl, University of Marburg, Germany.
Ludomir v. Sawicki, University of Cracow, Galicia.
Hans Spethmann, University of Berlin, Germany.
Aubrey Strahan, Geological Survey of Great Britain, London.
Antoine Vacher, University of Rennes, France.
Harry Waldbaur, University of Leipzig, Germany. (student)
W. B. Wright, Geological Survey, Dublin, Ireland.
Fritz Wyss, University of Bern, Switzerland. (student)
Naomasa Yamasaki, University of Tokio, Japan.

During the visit to each region a local expert, usually the local professor of geology, pointed out the most significant physiographic features and very often the participants would attempt a brief description of the area on Davisian lines. Davis' propagandist interest in such enterprises is only too clear and, while his enthusiasm for a common descriptive method did not achieve unanimous acceptance, it certainly persuaded many of his colleagues. The following extracts are from his own account of the pilgrimage.

There was abundant discussion directed to the origin of the land forms in the districts visited, and to the best means of describing them; but, as is usually the case when geographers are gathered together, many diverse opinions and methods were developed. Even the object and scope of our work were differently interpreted by the various members. Nevertheless there was a general acceptance of explanatory in preference to empirical methods for geographical descriptions; and frequent employment was made of the scheme of 'structure, process and stage' for the description of land forms, though with greatly varying proficiency and completeness. There was an interesting diversity of opinion in such problems as the origin of certain truncated uplands which we visited, some members ascribing them to marine denudation, others to subaerial degradation; hence this problem is discussed with some fullness on a later page. Frequent dissent was expressed from my principle, that geological formation-names, indicative only of geological dates in past time, ought to be excluded from purely geographical descriptions, since geography has to do only with the present. Some of the liveliest discussions arose when comparisons were made of one-page descriptions of a limited area that we had examined together; the order of the statement and the emphasis given to different elements varied greatly, in spite of the fact that we had all seen the same things. All this shows how far we are still from having standardized our geographical methods. (*Davis, 1912G, pp. 74–5*)

The naïvete of the final sentences seems unbelievable! But perhaps Davis meant that they had *looked* at the same features.

The first five days of the excursion were spent in southern Ireland under the guidance of G. A. J. Cole, Professor of Geology at Dublin and Director of the Geological Survey of Ireland. Among the notable landforms seen were the gap of Dunloe (a good example of a glacial overflow channel), the great sea cliffs at the west end of Valentia island, and

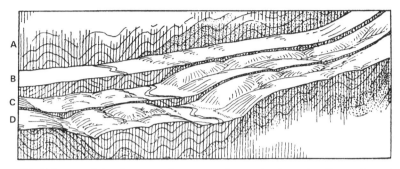

FIG. 71. Scheme of development for mountains, uplands and valleys of southern
Ireland, looking north-west

A. The generalized structure with sandstone to the north and limestone to the south

B. The area reduced to a peneplain with subdued hills on the anticlinal axes

C. Uplift initiates a second cycle of erosion during which a broad valley has been
excavated along the synclinal belt of limestone and the sandstone uplands submaturely dissected

D. Renewed uplift initiates a third-cycle main valley, flanked by terraces

(From Davis, 1912G, Fig. 2)

FIG. 72. A scheme of development of highlands, uplands and coast-line in west-
central Wales, near Aberystwyth, looking north-east

E. An extensive lowland with occasional residuals, produced by normal erosion or
marine abrasion in the first cycle

F. Uplift introduces a second cycle leading to fluvial dissection and marine abrasion

G. Renewed uplift introduces a third cycle and the withdrawal of the sea

H. The third cycle of erosion continues with dissection of two inland surfaces and
encroachment by a plane of marine abrasion

(From Davis, 1912G, Fig. 4)

the elbow of the Blackwater, where Jukes, just fifty years before, first gave correct explanation to the relation of transverse and longitudinal, that is of consequent and subsequent, streams and valleys in tilted structures of varying resistance. This explanation was for a number of years overlooked by observers in the United States, to the serious detriment of their work. (*Davis, 1912G, pp. 75–6*)

There follows a concise description of south-western Ireland and a composite schematic diagram of its physiographic development.

The next rendezvous was at Bethesda in North Wales where Dr J. E. Marr of Cambridge University led a greatly augmented party around the glacial features of Snowdonia [see Fig. 66].

On both sides of the Snowdon mass we saw fine examples of glacially over-deepened troughs, with hanging lateral valleys opening in the side walls and with rock steps and rock basins scoured in the floor. The lowering of preglacial passes by glacial cross-currents was noted at several points; the result in facilitating transportation through the mountains is here of evident economic importance on a small scale, as it is on a much larger scale along the line of the Canadian Pacific Railway across the Rocky Mountains and the Selkirks. (*Davis, 1912G, pp. 78–9*)

Two half days in central Wales under the guidance of Professor O. T. Jones brought to their notice the remarkable example of stream capture on the Rheidol near Devil's Bridge about twelve miles inland from Aberystwyth, and

the region made classic by Ramsay's famous essay in which he brought forward his theory of the origin of certain plains by the marine denudation of mountains. It was here a pleasure to find that our guide had made a new attack on this old problem, and had recognized two uplifted and dissected plains, separated by an escarpment, where Ramsay had seen only one; and further that he had good reasons for explaining at least the lower uplands as the result of marine denudation, a process that has been too generally neglected by British students since their recognition of the great value of subaerial degradation. (*Davis, 1912G, p. 79*)

They left Wales for south-western England by way of the incised meanders of the River Wye, also made famous by Ramsay's early work. The crossing of Devon and Cornwall terminated at the flat-topped upland near Land's End 'which proved fruitful as a subject for explanatory description'.

Some regarded it as a plain of marine denudation; but the absence of cliffs on the exposed sides of several low monadnocks convinced the present writer that the upland was chiefly the result of subaerial normal erosion; although the occurrence of gravels containing marine fossils, reported by British observers, shows that the plain was submerged for a time after it had been worn down and before it was uplifted and dissected. (*Davis, 1912G, p. 80*)

Davis then goes on to elaborate the difference, in his opinion, between the appearance of plains of marine denudation and of subaerial degradation, the prime criterion being the presence or absence of cliffs or scarps on the flanks of the uplands facing the plains.

These two processes must go on together, one on the seaward side and one on the landward side of a shifting shoreline; and the area of sea-cut plain must, so long as the land mass stands still, gain on the subaerial area. Furthermore, in order that the sea-cut plain may gain a width of ten or twenty miles, the retreat of the sea cliffs at its landward border must be more rapid, probably very much more rapid, than the degradation of the land area, as shown in [Fig. 73]. When both

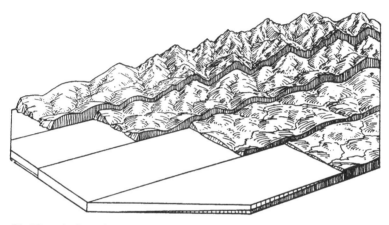

FIG. 73. The relation of normal degradation to marine abrasion, showing the situation in which the rate of horizontal encroachment by the sea is 5 or 10 times the general rate of downwearing of the land (From Davis, 1912G, Fig. 5)

processes are far advanced, the line by which their areas are separated must be of relatively simple curvature, must be marked by low cliffs, and some of these cliffs must have been cut back of the low summit-arch of vanishing hills; consequently such cliffs may be described as of 'decreasing height'. A simple line of low cliffs, some of which are cliffs of decreasing height, is therefore an essential feature of the boundary between a so-far finished plain of marine denudation and a nearly finished plain of subaerial degradation, or peneplain. Hence this inferred line of cliffs may prove to be a valuable criterion in uplifted examples where the presence of massive rocks prevents discrimination by means of adjusted or unadjusted drainage lines. (*Davis, 1912G, p. 81*)

The flat-topped uplands about Land's End were considered by Davis to be the work of two attacks of the waves;

first a rather well advanced attack, during or after which the cliffs gained a graded slope; then a lately renewed and vigorous attack, by which the base of the graded cliffs is sharply undercut in ragged and rocky forms, the platform in front of them not being yet widened enough to hold a beach, except in the larger coves. (*Davis, 1912G, p. 83*)

To explain the existence of this ragged, rocky coastline within a mile or so of an inner, older line of graded cliffs evoked a multitude of theories based on marine erosion, coupled perhaps with faulting, and so on. Davis suggests what he calls a 'round-about explanation' which we here give in part to show how elaborate his systematic descriptive method of structure, process, stage, had now become.

The deformed crystalline mass, of which Lands End and the greater part of Cornwall consists, was worn down to moderate relief, then depressed and buried under a heavy sedimentary cover; the compound mass was next unevenly uplifted and advanced to old age in the cycle of erosion thus introduced, so that the peneplain then formed traversed both the deformed older crystalline mass and the inclined strata of the sedimentary cover. The region was then rather evenly uplifted and the weaker covering strata were reduced to a new lowland or peneplain, while the resistant crystallines were only submaturely or maturely dissected; and in this condition the whole region was slightly depressed, so that the lowland of the new peneplain was submerged and the lower valleys in the dissected crystallines were drowned; thus the sea came to have its new shore line on the slope which separates the crystalline uplands from the drowned lowlands, this slope being nothing more than a part of the ancient worn-down land surface that had lately been laid bare by the stripping of the covering strata from it between the levels of the two peneplains. The irregularity of the slope, and hence also of the initial shore line that contoured in consequent fashion around it, would depend partly on the inequalities of the ancient land surface, partly on the unevenness of its uplift after burial; and a moderate recession of such a shore line would produce the existing irregular shore line with its immature cliffs.

Several facts may be cited in favor of this somewhat elaborate explanation. First, the eastern or landward border of the Cornwall–Devonshire upland is determined by an irregular slope where the crystalline rocks, long ago worn down to an uneven land surface, depressed, buried, and uplifted, have been, after the peneplanation which produced the even surface of the present uplands, again uplifted and laid bare around their border by the lower peneplanation of the weaker covering strata; that is, the uplands are here bordered by precisely such a definite but irregular slope as is postulated, in the explanation under discussion, for the initial north, west, and south borders of the uplands. No shore line follows the eastern border, because the depression after the newer peneplanation was not sufficient to submerge that part of the bordering lowland. Second, the valleys by which the uplands are submaturely or maturely dissected are now all drowned in their lower courses, and thus transformed into branching bays of most typical form, as may be seen near Falmouth and Plymouth. (*Davis, 1912G, pp. 83–4*)

The technical reader, having spent perhaps decades on some regional landform problem, may be astonished that Davis in a day or two could have the temerity to be so elaborately comprehensive. The non-technical reader may boggle at the virtuosity of the maestro who having progressed far beyond the simple music maker of sequential cycles now conducts a full-scale orchestra of structure, process, stage. Prospero in *The Tempest* never used the natural elements so ruthlessly as Davis. Some may say that under Davisian magic geomorphology has become 'full of noises, sounds and sweet airs that give delight and hurt not' but the cynic may see in it merely an exercise in the application of elementary concepts, a form of simple mathematics in which to achieve a known product the numbers are recklessly manipulated as negatives or positives. Before returning later to this theme, we must in all fairness to Davis quote his magnificent generalizations on the physiography of south-western England as a whole.

> After this brief inspection of the region it seemed to us all that its leading physiographical features could be concisely described as the products of two cycles of erosion on a disordered mass of crystalline rocks. The more even uplands of to-day, dissected by steep-sided valleys, are parts of the uplifted peneplain to which all but the most resistant rocks were reduced when the first cycle was interrupted by the uplift which introduced the second; while the large or small highland areas of broadly arched forms and coarse texture consist of the most resistant rocks, which therefore survived above the peneplain of the uplands as isolated or grouped monadnocks. On the other hand the districts of rolling hills with convex summits and with significant inequality of height, separated by well opened, late mature valleys, are areas of moderately resistant rocks, which were presumably worn to a smooth plain at the close of the first cycle, and are now already advanced to a late mature stage of subdued forms in the recent cycle. If to this simple scheme, we add the conception of a bordering mass of weaker strata, now worn down to a new peneplain and recently submerged, as stated above, the chief features of the coast as well as of the interior may be understood. Some British observers have recognized levels of erosion or abrasion at several different altitudes, on which they base a more complicated scheme of morphogeny; but as far as our journey gave us a view of the region, the leading features are all referable to two cycles followed by a recent episode of slight submergence. (*Davis, 1912G, p. 85*)

On leaving Devon, the party followed the English south coast route to Weymouth where they took steamer to Jersey. After 'a profitable day walking over the well-smoothed peneplain and along the fine cliffs of that tidy island', they crossed to St Malo in Brittany. Here, under the expert guidance of Professor Antoine Vacher, Davis inspected a physical landscape closely similar to that of Cornwall and Jersey.

> Cliffs of decreasing height were rare or absent. Hence, here again we found indications of a moderate measure of cliff retreat from an initial shore line produced by the transgression of the sea across a submerged lowland, until it stopped on the distinctly sloping border of an upland of more resistant rocks. Thus we felt for the third time the need of a more concise name for the total 'thing' of which Brittany as well as Cornwall and Jersey was supposed to be the present phase . . . Brittany, like Devonshire–Cornwall may be described chiefly in terms of two cycles of erosion: the first of which had reached an adianced stage when it (a subaerial peneplain) was interrupted by a broad elevation which

introduced the second; while the second reached a submature or mature stage when it suffered the slight and recent episode of depression, whereby the distal parts of the valleys were converted into little bays. (*Davis, 1912G, pp. 88–9*)

The next stop after Brittany was in the Limousin of central France where Professor Albert Demangeon pointed out the features he had described in an article published a year earlier. Here, prodded no doubt by Davis, some members recognized more cycles than did others but all agreed that the description of the observed landforms ought to be in terms of the elaborated cycles of erosion.

The physiographic pilgrims then entered the Auvergne under the enthusiastic leadership of Ph. Glangeaud, Professor of Geology at Clermont-Ferrand. The volcanic and glacial landforms were replete with geographic interest and some attempt was made to correlate the beginning of volcanic activity with the cycles of erosion in the adjacent Limousin.

The next halt was made in the Morvan, the north-eastern extension of the Massif Central of France. Here the general features and structure, well defined on east and west by fault scarps, so closely resembled those of Devon–Cornwall, Brittany and the highlands of the Massif Central, that Davis proposed to use the name *morvan* for the ideal type. Hitherto the party had called it provisionally 'thing', 'thingmebob' and 'skiou', the latter being coined in exasperation by Davis himself.

We quote the whole description as being typical of the complicated or the elaborate cycle and of the confidence with which Davis applies it to every problem. He continues to add new terms to his collection of definitions, but it must be noticed that these two additions relate to whole regions of such a complex nature that a wise scientist would have expected disagreement rather than agreement upon their long development. The example of the morvan is of exceptional interest because Davis cites it in his arguments with the Pencks in whose view it had induced the sort of false premise that characterized the use of the cyclic theory. We will, however, deal with the Penckian controversy in detail in later chapters. For the moment, the reader will find enough to stretch his imagination in Davis' description of the morvan.

The original Morvan may be described as a morvan in which the first wearing down of the crystalline foundation was advanced to senile planation, probably in part at least by the waves of an ancient sea; in which the covering strata are the marls and limestones of the Paris basin; in which the tilting of the compound mass was very slight, perhaps a few degrees only, to the north-west; in which the first truncation after tilting was well advanced even on the crystallines sometimes locally reaching planation, but more generally not having gone beyond peneplanation, and occasionally failing to obliterate subdued monadnocks, one or two hundred meters in relief; in which the next following uplift was accomplished in two phases, the first being followed by a pause long enough for the erosion of mature valleys in the crystalline area, and the second permitting the erosion of narrower young valleys of less depth in the distal parts of the mature valleys; in which the gently inclined covering strata on the north are so little eroded that the more resistant members still stand up, even crested and about as high as the Morvan highland, yet so much eroded that the weaker basal members are broadly excavated so as to produce a wide subsequent depression, enclosed by the inface of the first cuesta on one side and by the long stripped slope of the most ancient planation of the crystalline rocks on the other. On the east and west the highland is limited by fault-line scarps, as above stated; and on the south by a broad

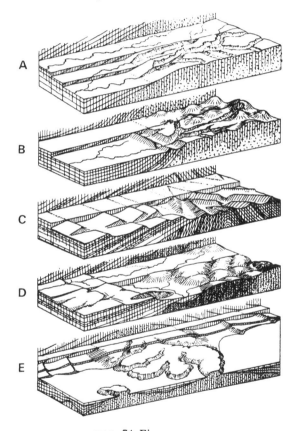

FIG. 74. Five morvans

A. A simple case of a low-angle plane unconformity and of double uplift giving a two-cycle landscape on the uplands

B. With a steep unconformity, glaciation and more monadnocks on the uplands, and one resistant stratum giving a monoclinal ridge on the lowlands

C. With more varied lithologies and structures on both the uplands and lowlands

D. With an excavated irregularity of the unconformity (i.e. a Mendip) interrupting the lowlands

E. A partly sea-girt morvan with a wholly sea-girt mendip nearby

(From Davis, 1912G, Fig. 9)

depression excavated on an obliquely transverse belt of down-folded or faulted weaker strata. The irregular rectangle thus defined measures about fifty kilometers on a side.

OTHER MORVANS – The uplands of Devonshire–Cornwall might now in their turn be described, following the explanation given for them above, as a morvan of irregular outline, measuring seventy miles north and south by sixty miles east and west, with a tapering south-westward extension about fifty miles in length; normally limited on the east by a distinct but irregular slope which descends to the newer peneplain that has been worn down on the weaker covering strata; and supposed formerly to have been limited in the same way on the north, west and south; but on those sides now cliffed by the sea in consequence of a recent depression by which the presumable lowlands thereabouts have been submerged: hence this example is on three sides a sea-rimmed morvan.

The evident advantage of this style of description lies in the abundant meaning that is packed into the technical term, morvan, for if the reader already knows what the term means, a long general introductory description is cut out and the attention is at once turned to the special features which characterize the particular example under consideration. The term morvan may thus come to have in geography a value of the same kind, perhaps of the same degree, that is possessed by such a term as logarithm in mathematics. No mathematician would to-day delay his statement by presenting in paraphrase the somewhat elaborate meaning that is so conveniently packed into the single word logarithm; and so perhaps no geographer, fifty years hence, will embarrass himself and his readers by the long circumlocution that is required if morvan or some convenient term is not employed, when the thing that is meant by the term is to be considered.

If this scheme of description is followed further, one might say that the Front Range of the Rocky Mountains in Colorado is a long morvan, of which the crystalline mass was long ago worn down to a remarkably smooth surface, and buried under a heavy series of covering strata; that the compound mass was then bent in an east-facing monoclinal flexure of great north–south extent, with a maximum dip of some thirty degrees, but that the flexure was sometimes ripped into a fault; that the truncation of the flexed mass produced a fairly good peneplain over large parts of the crystalline area, but left good-sized monadnocks standing singly or in groups; that after the later uplift the covering strata, already smoothly worn down a first time in the previous cycle, were again, because of the broad uplifting of the region, worn down to a plain of vast extent, except for a few monoclinal ridges near the mountain border; while the uplifted or up-arched peneplain of the crystalline rocks was only submaturely dissected, and in its higher valleys strongly glaciated, as has already been set forth without the use of the term morvan, in an earlier article in these Annals.

MORVANS AND MENDIPS – Whether the island of Jersey should be described as a sea-rimmed morvan is doubtful; for so long as the inferred surrounding strata are invisible, it can not be determined whether the crystalline rocks of the island gained the height that they once possessed above the level of the upland peneplain by deformation after burial, as is implied in typical morvans, or whether the former height of the crystallines above the level of the upland peneplain was simply due to the mountainous relief of the fundamental surface before and during the deposition of the covering strata. A standard example of the latter relation in England is found in the Mendip hills, which rise through the covering strata east of Devonshire without deforming their moderate eastward dip; an equally good American example is seen in Baraboo Ridge south of the crystalline highlands of Wisconsin. The small size of the Channel Islands may be taken to indicate that they are more probably sea-girt mendips than sea-girt morvans. (*Davis, 1912G, pp. 93–5*)

The excursion, which had previously been largely concerned with what are variously called today Hercynian or Armorican or Variscan upland blocks, now led over younger terrains. In the valley of the Armançon, north-west of the Morvan,

The occurrence here of several successive cuestas so close together that they overlap suggests the need of selected adjectives for the description of cuestas in terms of the distance that separates them. Thus we may speak of overlapping cuestas, as in the present case; of close-set cuestas, where the back slope of one, as it descends into the floor of a longitudinal valley, just reaches the inface of the next; and open-spaced cuestas, where a broad lowland is developed between the back slope of one and the inface of the next. The controlling factors for these three cases are chiefly the thickness of the several strata in the cuesta-making series, the dip of the series, the depth of the dissection, and the stage of erosion. (*Davis, 1912G, p. 96*)

Moving south to the Jura mountains, they were met by Dr Fritz Nussbaum and Mr Wyss, one of his students. Here they spent several days

in trying to determine whether these mountains bear the marks of two cycles of erosion, separated by a broad uplift with little deformation, as is believed by Brückner and Machatschek; and if so, what stages of erosion were reached in each cycle. The occurrence of several reduced or truncated anticlines of resistant limestone led us to accept the two cycle theory, and to regard the first cycle as having reached a far advanced stage over large areas, although strong residual reliefs seem to have survived where the strongest limestone anticlines, even if much reduced in mass, rose well above the level of truncation elsewhere detected. The second cycle has already produced old forms in the synclines of weak strata, but the valleys that cut through the strong limestone anticlines are still narrow and young, with abundant rock outcrops and frequent rapids in their streams. This conclusion was reached with some regret, for the Jura Mountains, considered as a one cycle range, had a high pedagogical value. Now that they must be displaced from the elementary position in which they have been long classed, where shall we find a one-cycle mountain range of folded structure with which to replace them? (*Davis, 1912G, pp. 96–7*)

The party crossed the Alps from Lucerne to Lake Maggiore by way of the Brünig, Grimsel, Furka and St Gotthard passes. Two days were spent walking from Muringen up the Aar valley and over the Grimsel pass to the source of the Rhône in its glacier snout. The severe glaciation was not new to them but the nature of glacial troughs was far from decided.

We looked with especial care at the form of the overdeepened trough for a score of miles or more, to determine to what extent its excavation had been preceded by the normal erosion of a sharp trench in the floor of the mature preglacial valley, as suggested by de Martonne; and we were obliged to reject such trenching as an essential precursor to deep glacial erosion of main valleys below side valleys, not merely because we found no signs of it, but because the occurrence of two-storied hanging valleys, of which we saw several examples, is essentially inconsistent with it. By two-story hanging valleys, we mean lofty branch valleys which hang over middle-height side valleys, which in turn hang above overdeepened main valleys; all these valleys having the well-defined trough form that is indicative of strong glacial erosion. The occurrence of two-story hanging valleys demands strong erosive power on the part of the mid-height side glacier, independent of any gorge that may be supposed to have been trenched in the floor of the main valley in immediately preglacial time; and if we are obliged to attribute strong erosive power to medium-sized side glaciers, we must attribute still greater erosive power

F I G. 75. Front view of the Guspistal, looking east, showing the connecting gorge
(From Davis, 1912G, Fig. 12)

to the main glacier, which therefore can have excavated its great overdeepened trough about as easily without as with the aid of a preparatory gorge of normal erosion.

It is perhaps true that, as de Martonne urges, the preparatory erosion of a normal gorge in the floor of a mature valley would facilitate the excavation of an overdeepened trough by glacial erosion; but even if so, it does not follow that this favorable condition was presented. (*Davis, 1912G, pp. 97–8*)

They were joined at the head of Lake Maggiore by Professor Olinto Marinelli of Florence and Guiseppi Ricchieri of Milan who demonstrated further glacial features, such as morainic amphitheatres around the ends of lakes, before the party disbanded at Lugano. In fact the excursion was scheduled to end at the International Geographical Congress at Rome but on 27 September the King of Italy had sent the excursionists a message stating that the Congress had been postponed – presumably because war had broken out between Italy and Turkey. We know, however, from the copious diary kept by Mark Jefferson (see Martin, 1968, pp. 137–8 and 356), the only other American present, that he and Davis went on from Lugano to Milan to see something of the North Italian plain. They then returned to Paris whence Jefferson returned via Brittany to the United States while Davis remained in the metropolis as a visiting professor.

PROFESSORSHIP AT THE SORBONNE, PARIS, OCTOBER 1911–APRIL 1912

We have suggested in Volume One of our *History of the Study of Landforms* that Davis' early ideas of the cycle of erosion made little impact on the thought of contemporary French physical geographers. There is no doubt, however, that his later more sophisticated or more elaborate cyclic concepts appealed strongly to French physiographers. Davis' philosophic and sometimes literary treatment of the subject soon found a close kinship with the French school of geographers. There was almost, they admitted, a Gallic flavour to his style. The following remarks by Professor Vidal de la Blache when introducing Davis to the faculties at the Sorbonne on 17 November 1911 express this clearly:

The Dean, detained elsewhere, has confided to me the great honor of presenting to this assembly Mr. William Morris Davis, professor at the illustrious University of Harvard, our colleague and fellow-worker at the Sorbonne –

He has been for a long time present among us by his thoughts and his writings and we have learned to know in him one of the masters who have contributed most to the strengthening of the rational element in Geography – and to the transforming of geography from a study which is limited to description, into a science which describes by explaining.

Familiar with the chief sciences concerned with our planet from Astronomy to meteorology and geology – without mentioning entomology – it is upon geography that for thirty years, Mr. Davis has concentrated his efforts and chiefly on the problem of the forms of the land.

He has attempted to show their origin, their associations, their evolution, and it would be well to say here, if time permitted, with what rigor of method he had co-ordinated scattered ideas, explained them by examples taken from his journeys, illustrated them by his drawings, and codified them in a terminology as expressive as it is ingenious – which puts the finger upon the correlation of forms and their successive share in the development of the surface.

It is no exaggeration to say that, after America, where his teaching had had results so fruitful, it is in France that the ideas of Mr. Davis have had the most hearty reception. He himself has found in our country many beautiful examples which support his explanations – I am glad to recall the interesting monograph that he published in 1895 on the relations, or rather the conflicts of three of our rivers, the Seine, the Meuse and the Moselle – following an hypothesis suggested by the study of our maps, confirmed by observation on the ground. (*Speech by P. Vidal de la Blache, introducing Davis; 17 November 1911*)

Davis came to the Sorbonne as an exchange professor under an arrangement between the French minister of public education and Harvard University. He was the first American exchange professor under this scheme to give a regular series of courses as distinct from seminars or conferences and was also the first to be attached to both the Faculty of Arts and Faculty of Science at the Sorbonne. He was, in fact, to act as a liaison or bridgehead between the two. The programme was similar to that already tried out on German students. It was given in French and consisted, as *Siècle* (3 November 1911) and *Le Figaro* (14 November 1911) tell us, of

Systematic Study of Landforms
Thursday at $9\frac{1}{4}$ hours a.m.
Friday 2 p.m.
— 3 p.m. Practical exercises on landforms (not obligatory)
All for the Faculty of Letters, 3rd stage students.

Saturday at $9\frac{1}{4}$ hours a.m. Discussion, alternating on all 15 days with practical exercises given on Friday. For 2 hours: to be held in the Faculty of Science, laboratory of physical geography, for second stage students.

Thursday at $10\frac{1}{2}$ hours a.m. in the Faculty of Letters, personal discussion with students on all kinds of geographical questions that interest them.

The inaugural lecture was delivered on Friday, 17 November 1911, to both faculties, and the editors of the *Annales de Géographie* (Davis, 1912A) were pleased to reproduce it. Called 'L'Esprit explicatif dans la géographie moderne', it is Davis at his best – and worst! He adores being an adventurer, an overthrower of old idols, a David with a sling facing a host of Goliaths. He sets up a vast semi-imaginary opposition in order to test the strength of his prowess; he does not ask for help nor expect anyone to listen to his cause!

It seems a pity to translate his simple French but it is the idea, not the tongue that matters.

> Instead of trying to present the principles of physical geography in a popular and attractive fashion in order to attract and please a wide audience, I want to express those principles in a scientific way, as a logical discipline, and so to contribute to the systematic and serious education that best suits the students of the Sorbonne.
> *(Davis, 1912A, p. 2)*

Davis admits that these methods have never attracted a crowd of students to his laboratory but they have stimulated several young men who have become professors of geography at various universities in the United States. He then recounts his geographical pilgrimage in Western Europe earlier in 1911 and tells of the various methods of landscape description that were displayed during it. Which was the best of these methods? The answer, young geographers of the future, is in your hands. There was at present a lack of modern books – 'je parle maintenant plutôt de la morphologie terrestre que de la géographie entière' – with modern (Davisian) descriptions of the physical landscape. The study of landforms lacked exact classification until its facts were better organized. Mental application is as necessary as physical exploration.

> On disait autrefois aux jeunes géographes: 'Allez voir.' On dit aujourd'hui: 'Voyez et pensez.' *(Davis, 1912A, p. 13)*

He does not, he goes on, belittle exploration and travel. He has done his share, from the Tian-Shan to the Transvaal, from Mt Harvard to the Argentine Andes, but his great aim is to establish a more exact, more scientific method of describing landforms. The principle of explanatory description alone supplies this method.

> We understand from the start that geography is concerned with the present appearance of the earth, and this in all respects is inherited from former landscapes; consequently, we should not treat its present appearance in an empirical way, as if it were unrelated to geological periods, but should always deal with it in an explanatory fashion, recognising at each step that the present is only the development of the past. *(Davis, 1912A, p. 14)*

Davis was well received by the students and by the Paris press. At the annual banquet of the American Club he admitted that after many visits he admired the metropolis more than ever. His only regret was the loss of its former quiet; the present streets were far too noisy, and the wailing of sirens on the Seine was unwelcome. Who, said the gracious reporters, would not agree with him? They described him thus in their exquisite journalese, under the title 'University Personalities', accompanied by a striking photograph:

> William Morris Davis is famous, in America, Europe, and Japan, for his great strides in synthesizing the results of local geographical researches, in combining

into a coherent doctrine its disorganized conclusions, in formulating laws which apply to all landforms, in devising a technical vocabulary, and in providing progressive working methods. His *Physical Geography* is an excellent exposition of the general achievements to date; his *Geographical Essays* are models of original investigations. He says he has founded the science of geography; he ought to admit that he has done much to organize it also. He has been at Harvard for thirty years and his white moustache and pointed beard tell clearly that he is no longer young. But his upright stance, erect head and nimble step bespeak a youthful spirit, a keen mind and a burning energy. And in his cold, severe countenance, shine two clear eyes whose piercing glances seem to follow the contours of things and to penetrate right to their core, at the same time building up in their vision practical concepts, shapes of the future and forms of the present.

Detailed in his investigations, consistent in his reasoning, W. M. Davis shuns brilliant but risky intuitions and retardative generalizations. His powerful intelligence is applied to the problems of landforms, particularly to the mystery of their origin. He knows how to select observed facts and to express their interrelationships without in any way abandoning his passion for truth to fanciful suppositions which imperil learning. (*'Siècle'*, *7 December 1911*)

It is perhaps surprising that this interlude in his beloved Paris was abnormally unproductive in a scholarly sense. However, a delayed result of a field trip taken in France with Dr Otto Lehmann of Vienna during the spring of 1912 appeared in the publication of 'Meandering valleys and underfit rivers' (Davis, 1913Q). We have already referred briefly in Chapter 11 to the development of Davis' views on this topic, and the idea of Lehmann's that mature rivers might spontaneously and progressively lose discharge by percolation into the valley-fill deposits caused Davis to treat the subject in some detail. Calling on Thomson's account of the centrifugal force of flowing water, Davis described the development of meanders, together with the simultaneous occurrence of their enlargement and downstream migration. This produces enlarging amphitheatres and eroding spurs, resulting in a wide mature floodplain in which the river freely meanders with curves related in size to its discharge. Rejuvenation may lead to incised meanders. It is in incised valleys that the best examples of underfit rivers are found, wandering 'aimlessly in small curves in the regularly meandering larger curves of the valley floor, instead of systematically following the base of the amphitheater walls and of the undercut spur slopes . . .' (Davis, 1913Q, p. 16). After treating various possible causes of the underfit condition (occasional high discharges, river capture, and climatic change), Davis sympathetically considers Lehmann's idea of percolation loss. Concluding that this is a perfectly logical result of the aggradation of a mature valley floor, Davis uses this as an example of how the logical consequences of his theoretical model of the cycle of erosion had not been pushed sufficiently far by the processes of deduction. This is yet another example of Davis' ability to fasten upon the theories of others and turn them to good account.

Much of Davis' attention was occupied during his stay in Paris by the organization of the European end of his planned American transcontinental excursion. As usual, he was looking several steps ahead and had at this time great plans under way at home. He did not spend much time in Europe after the end of his teaching duties:

> My work in Paris closes today. Tomorrow I am off with a party for an excursion of a week: then short visits to Universities at Dijon, Lyons and Grenoble, and after that two weeks, if you please, of well-earned rest in Italy. About April 13, 14, 15 I shall be in London on the way home and sail from Liverpool on 16th. (*Letter: Davis to J. S. Keltie, 9 March 1912; Paris*)

THE AMERICAN TRANSCONTINENTAL EXCURSION OF 1912

The idea of organizing an international geographical excursion across the United States arose in Davis' mind as a result of his European excursion in the summer of 1908 (Davis, 1915H). In December 1909 he mentioned the idea to the Association of American Geographers but no one commented favourably upon it. With typical perseverance Davis pursued the matter privately by circular letter and for nearly a year the replies received were mainly sceptical or playful. However, early in 1911, a correspondent told him of a man in New York who might take the matter up, and eventually this man proved a generous patron. The excursion had to be made in the name of the American Geographical Society, of which the patron was a member, and had to be announced as 'The Transcontinental Excursion of 1912' in celebration of the sixtieth anniversary of the actual foundation of the Society and of the completion and occupancy of its grand new building. Letters of invitation were sent out in June 1911 to the leading geographical societies cf the chief European countries, asking each to nominate a certain number of participants, some as guests and some as paying members. From July 1911 to April 1912, Davis conducted a massive correspondence concerning the Transcontinental Excursion from Europe, cajoling and appealing to patriotic motives, which proved a decided advantage in securing European members. Thus by his return home in the spring of 1912, the list of foreign members was well advanced. He spent from April to August working incessantly on the final details of the journey, and found, as he had surmised, that the printed list of European members proved a conjuring rod in the way of securing assistance. Individuals, commercial firms and clubs, chambers of commerce, government departments and other bodies vied with each other in supplying hospitality and facilities.

> It was a gratifying surprise to the American members, as well as a matter of astonishment to the Europeans, to see how we were welcomed and cared for, day after day, during our long progress. The supply of automobiles sufficient to carry our entire party across distances of 60 and 80 miles at various points in newly developed western districts was an astounding revelation of American enterprise

... The Pullman Company arranged to supply us with cars of the special pattern that we needed, and all participants will remember how much enjoyment we had from the two observation parlors, one in the middle car, one in the rear car; the same company moreover agreed to serve without extra charge a light afternoon tea. Our members will recall with satisfaction the safeguard of good health provided in our free supply of bottled Poland Spring water for table use, and ... the reduction of discomfort in a long journey, much of it across a dry country, by the abundant provision of Budweiser beer ...

The last days before setting out on a long journey are usually crowded with the settlement of many details by the director. The final day of preparation in New York was a tax on his patience and strength, although it was entertainingly broken by an excellent lunch, contributed by one of his former students at Harvard, in the lofty rooms of the Whitehall Club overlooking New York Harbor, and delightfully closed by a jovial dinner at the Harvard Club. (*Davis, 1915H, pp. 5–7*)

The excursion which Davis had so carefully planned lasted fifty-seven days, covered more than 13,000 miles and comprised forty-three European geographers representing fourteen countries and approximately seventy American geographers most of whom came and went at their convenience. Davis appointed Mark Jefferson the First Marshal of the Excursion and R. Dodge and I. Bowman second and third marshals, respectively. But Jefferson reported that his duties proved minimal, amounting to nominal leadership of the Excursion for one day in Yellowstone National Park when Davis was resting (Martin, 1968, p. 139).

Davis, as usual, thought up, planned, arranged the finances and conducted the excursion in his typical immaculate and tireless manner. Obviously at the age of sixty-two some strain and, at times of stress, a slight irritation were inevitable. The following song (to be sung to *Funicoli, funicola*) was written by him during the excursion:

Spare the weary, worried old Director

O Spare the weary worried old Director,
    and let him go, and let him go.
Have care of what you say, he may object or
    tell you no, no; tell you no, no.
Dont try him, asking easy teasy questions
    he doesnt know, he doesnt know;
As why the morning sun the wooly west shuns,
    'Twas always so, 'twas always so.

    Spare him, spare him, dont you hear him moan;
    Spare him, spare him, let him sit alone;
The only rest he has from morn to night is when you hear –
'Dinner now is ready in the dining-ining keer'.

How high is yonder peak? You haven't got to
    know every fact, know every fact.
The DI-rector begs leave to ask you not to
    be so exact, be so exact.

What matters it how high a peak today is:
    it can't remain, it can't remain;
It flattens down and then, as one might say, is
    a peneplain, a peneplain.

    Spare him, spare him, he doesn't know the height,
    Spare him, spare him, he tries to do what's right.
He has an hour of peace each day when no one guys him, for
Dinner then is ready in the dining-ining cor.

He cannot tell the age of that formation,
    You've got him there, you've got him there
For useless geologic correlation
    He doesn't care, he doesn't care.
What boots it if you call a rock Cretaceous,
    or Pliocene, or Pliocene?
He hoots at erudition so fallacious;
    What does it mean? What does it mean?

    Spare him, spare him, he's no geologist,
    Spare him, spare him, his head has no such twist.

(*Song by Davis, written during the Transcontinental Excursion, 1912*)

What is certain is that Davis never spared himself nor lost an opportunity of putting over his own views on the physical landscapes he knew so well. His great friend Jefferson recorded it all.

> . . . meeting with Professor Davis now in Lounge – Later. A very lively discussion it was that ended in Dr. Niermeyer telling Davis that he was no geographer but a geomorphologist. . . . Davis gave a talk last night on the physical features of the (Yellowstone) Park and then another on 'method of geographic description'. As usual he spoke only of physical geography. (*Letter: Mark Jefferson to his wife, Theodora, 5 September 1912; quoted in Martin, 1968, p. 140*)

E. de Martonne, the eminent French physical geographer, told A. Cailleux an amusing story of this journey.

> At one spot in the Appalachians, Davis, directing an excursion came to a magnificent viewpoint. The whole party rushed forward to see the landscape. But Davis made them turn their backs on it and then explained to them what the landscape ought to be according to his theories. That done, he allowed them to turn and face the view and demonstrated that in fact the landscape happened to conform exactly

with his theoretical construction of it. (*Letter: A. Cailleux to R. P. Beckinsale, 10 April 1966; Paris*)

The outward journey took a northern route *via* the Great Lakes and the Prairies to Oregon. The return was from San Francisco to the deserts and canyons of the south-west and so to the eastern areas near Metropolitan Washington. The longest stay in any one locality was the six days at Yellowstone National Park and considering that landforms had to share the time available with civic receptions and industrial and economic aspects, such as the Sears, Roebuck Company and bonanza farming, it was marvellous that so much physical geography was studied. The whole affair was, of course, typically Davisian – a miracle of comprehensiveness embodying the maximum distance in the minimum time.

No international geological or geographical congress, I believe, has displayed a country so thoroughly or authoritatively. Nor will Davis' achievement a third of a century ago soon be equalled or surpassed.' (*Martin, 1950, p. 174*)

We could hardly do better than take a few sample descriptions from the history of the excursion written by Albert Perry Brigham (1915) for the *Memorial Volume of the Transcontinental Excursion of 1912*, the first book published by the American Geographical Society (New York, 1915).

Early European arrivals were shown round parts of New England and Pennsylvania. On the evening of 21 August, forty-three Europeans and about a dozen Americans dined at the Harvard University Club in New York City.

The Transcontinental Excursion began at 8:30 o'clock on the morning of Thursday, August 22, at the Grand Central Terminal. The special train was made up of two standard Pullman cars, two Pullman observation cars, a dining car and a baggage car. The 'Circassia' and the 'Wildmere', the 'Huelma' and the 'Oronso', will always be pleasant names to members of the excursion, and to see one of these cars in future travel would be somewhat like a glimpse of a former home. A buffet car of the New York Central Lines accompanied the train to Chicago.

Throughout the two months there was usually attached to the train a private car carrying some official of the railroad traversed at the time. Several of these officials were in charge of the land holdings and of the industrial operations which are now so largely promoted by our great railroad systems in the West and South. They are accomplished men, and they gave to the party many valuable lectures and conversations throwing light upon the development of their respective regions . . .

For several months at the Society's house in New York, Mr. W. L. G. Joerg had been making preparations for the daily instruction of the excursionists. Maps and books were collected and so classified as to be made accessible at each stage of the journey, and throughout the time Mr. Joerg and Mr. F. E. Williams gave unremitting attention to this work . . .

It would not be easy to define in a single sentence the object and work of the excursion. The main aim, of course, was that every man might get as much first-hand knowledge as possible about the United States. This was attained in many ways. Observations from the car windows were continuous during the hours of daylight. The occasional exception was justified, when, for example, absorbing debate arose over some geographic problem, or when some member dropped into brief slumber through sheer fatigue of body, or when all were responding to the call for afternoon refreshment. Long after nightfall there was straining of eyes from the observation platforms to catch the features of the landscape. One of the joyous car-window experiences was in the early morning when the train toiled up the last curves of the Atlantic slope among granite outcrops, crossed the Continental Divide and brought many of the members for the first time within the domain of the Pacific Ocean. Special stops, sometimes two or three in a single day, gave opportunity for more deliberate observation of the facts of the physical geography. Thus, in Fishkill, a little more than an hour out of New York, the party alighted from the train and ascended by the cable car to the summit of the highlands, where one of the American members interpreted the topography of the mountains and the industrial and commercial interests of the Hudson River lowlands as they spread out northward towards Albany. At Little Falls a brief stop was made. The Dolgeville railroad was ascended to the top of the cliffs, the topography and history of the Mohawk Valley were briefly described, and a representative of the state engineer's office added an account of the Barge Canal. At Syracuse the party was taken by automobiles southward from the city to see on the hills the abandoned river channels and fossil Niagaras of the closing stages of the glacial time.

An entire day was given to the Niagara Falls and gorge, and a special car, stopping at will, gave opportunity for many expert lectures, for taking photographs and for discussions . . .

A strenuous day of fruitful observation was spent in Wisconsin. After a welcome to the University of Madison an excursion across the terminal moraine into the driftless area was followed by a climb over the Baraboo Ridge. Here, as at Camp Douglas later in the day, after struggling to the summit of a rugged sandstone outlier, it was evident that the plains of southern Wisconsin are not without relief. Under the station shed at La Crosse in the same evening three state geologists and an editor gave lectures of such interest that one distinguished foreign member, already garbed for the night, found it imperative to attend . . .

The topography around San Francisco was seen in journeys across the bay and in trips by motor cars to the Pacific side and along the rift valley, scarred by the earthquake movements of 1906 . . .

In crossing the desert areas of northeastern Arizona, a half day stop was made at Adamana in order that wagons might convey the party to the petrified forest lying a few miles to the south. The custodians of the forest were not required to have both eyes open on this occasion, and the geographers were permitted to help themselves somewhat freely from these most ancient wood piles of the Southwest. In the afternoon a wagon trip was made to Meteor Crater, and supper was served upon its rim by the host of the occasion . . .

The excursion gave an ideal opportunity for non-instrumental study of climate and of weather successions. Several members of the Weather Bureau were present during longer or shorter periods, or met the party at various points. Professor

Ward of Harvard University was present throughout, and presented in the Monthly Weather Review for December, 1912, 'Two Climatic Cross Sections of the United States', a paper packed with interesting observations on the relation of the atmosphere to human life and industry . . .

In addition to daily studies of the physical features from the train and by special excursions, a great deal of attention was given to the phases of economic and industrial development, for it is recognized by all true geographers, and it is especially emphasized by the geographers of Europe, that the science does not come to its full fruition until it has taken in, not only the lands, but the interests and relations of those who live upon them. Every one knows that a wide field for such study is open to one who crosses our continent. At the first stop made, standing at the summit of Mt. Beacon, the chief engineer of the project described to the party the huge siphon and twenty-five mile tunnel which are to conduct the Catskill waters to New York City. From Buffalo the party visited the Lackawanna steel plant and for an hour or more were transported up and down among the various buildings and furnaces upon flat cars provided by the company. At Niagara one afternoon was devoted to the power house and the various industries, and the whole of the following day given to the falls and the gorge, with many stops and brief lectures from experts by the way. In Chicago the party inspected in squads, according to their choice, the Stock Yards, the business methods of the Sears, Roebuck Company and the map-making plant of the Rand-McNally Company . . .

A few links in the excursion were made by trips over the water. At Toledo the party was entertained on the upper floor of a sky-scraper, then taken across the foot of Lake Erie and up the Detroit River by boat . . .

On the Mississippi River the party spent a happy day sailing from Memphis one hundred miles down the stream, watching the sand bars, snag boats, the means taken to protect the banks, the bordering forests, and, it must be said, – the lone steamer or two on waters that might carry the commerce of an empire. An old-fashioned landing was made in the twilight, head on to the shore, but it was not exactly old-fashioned to clamber up thirty feet of sand and find at the top a brilliantly lighted train of palace cars with dinner served.

There was much of a social and educational sort. Perfection of arrangement was shown almost every day, when promptly on the scheduled moment the train pulled into the station and with equal promptness a local committee stood upon the platform, and motor cars or electric cars awaiting the party stood in the street . . .

At the dinner in St. Paul, Governor Eberhart of Minnesota evinced his good wit by saying that he had 'never expected to see so many people who knew so much about the earth and owned so little of it', while Archbishop Ireland on the same occasion made a speech which, if a little long, was every word interesting and inspiring. It was not a little interesting to the French professors who sat at his side that he was quite their equal in the finished use of their native tongue. It is hardly to be believed that one hundred representative business men of any eastern town would come out by train at six o'clock in the morning a distance of fifty miles to meet a delegation of scholars, but this was done by the men of Fargo, in North Dakota, and the skidding of the automobiles in the wet gumbo outside of Fargo gave one a lasting remembrance of the quality of the black prairie soil . . .

One of the pleasant memories of the Pacific Coast is the hours of inspection and entertainment on the campus of the new University of Washington in a glorious suburb of Seattle. Indeed if any European came to the coast cities expecting to find things a little crude, he was obliged to alter his conceptions, for he saw paved streets, splendid buildings and innumerable blossoms and greenery that never fail during the twelve months. In Seattle there was keen appreciation of the magnificent utilization of hill and vale, of grading on a vast scale, of the combination of water and forest in the view, and of extraordinary activity and growth. The same impressions were made at Tacoma, by its aggressive business, its miles of attractive homes, its stadium and its park of native forest . . .

The amount of labor performed on the train and during the constantly recurring field excursions was prodigious. The convenience of the whole party limited the time which the specialist could give even to those things which most keenly interested him. The geologist could not deliberately pick his specimens and draw his sections, and the botanist could only make a dash at the flora before the fatal concerted call came to enter the stage or board the train . . .

Early in the morning of September 16th the train stopped at the station of Medford, Oregon. Through the kindness of the Medford Commercial Club thirty automobiles were in waiting, enough to carry the party, and two or three 'trouble cars' in addition. Soon began the eighty mile drive to Crater Lake . . .

Nearly four days were spent in the Great Basin. An afternoon was occupied in crossing the plains of Nevada and in a short excursion to one of the Lahontan beaches. The following morning Great Salt Lake was crossed by the great viaduct of the Southern Pacific Railway, which was remarked upon as being the one human structure by means of which the curvature of the earth's surface may be seen. It was here that the splendid beaches of Lake Bonneville began to come into view. A series of excursions took the party to the most interesting physiographic features of the region. One afternoon was given to an automobile trip to the mouth of Little Cottonwood Canyon and a climb over the faulted moraine at the base of the Wasatch, a range whose slopes revealed all possible delicacy and brilliancy of autumn color. There were excursions to Bingham, where the learned visitors at last succeeded in proving their innocence to the striking miners; to Garfield, for the Bonneville shores, and to Saltair, where some of the heavyweights of Europe demonstrated that even they could not sink in brines so dense as the lake affords. On the way eastward an afternoon excursion was made from Provo to see again the faulted moraines at the foot of the Wasatch . . .

Two days were spent at the [Grand] Canyon, occupied in descents to the river, in excursions to points of outlook along the brink, and in hours of vision and wonder, pondering the physical history, reveling in changes of light and color, and in trying to describe the indescribable – all in all, a place of which it is better to say only – go and see! (*Brigham, 1915, pp. 10–25 passim*)

The general results of the excursion were commensurate with the great efforts that Davis and his compatriots put into the venture. German, Austrian, Russian, Swede, Frenchman and Briton vied with each other in writing volumes of notes, forerunners of the crop of excellent papers and books on America that soon began to appear in the geographical journals of Europe.

Within the next two years, nearly fifty articles on the excursion had been published in European journals but only a few of these were concerned solely with physical geography. Henri Baulig wrote on the lava plateau of central Washington and the Grand Coulée (*Ann. de Géog.*, 1913, 149–59); E. de Margerie on the Crater Lake, Oregon and Meteor Crater, Arizona (*Ann. de Géog.*, 1913, 172–84); E-A. Martel on warm springs at the Roosevelt Dam, Mammoth Cave and on alluvial fans (*Comptes Rendus de l'Acad. des Sci. Paris*, December 1912; July 1914; *Spelunca*, December 1913); and E. de Martonne on a morphological study of Yellowstone Park (*Ann. de Géog.*, 1913, 134–48). Perhaps, however, the Dutch professor, K. Oestreich, hit the nail hardest on the head with his title 'In the classic land of geomorphology' (In het klassicke land der geomorphologie, *Tijdschr. Kon. Nederl. Aardrijksk. Genoot.*, Vol. 31, 1914, 176–87). The tide had indeed turned since the British led the way with Hutton, Playfair, Lyell, Jukes and Ramsay. Now the great American Western geomorphologists, Powell, Gilbert and Dutton, had found a worthy successor in Davis who had spread their fame and his own throughout the geological world.

The transcontinental excursion of 1912 had other indirect effects some of which we will refer to later. Many of these were beneficial to the Davisian cause. Eight years later he wrote to the patron who had backed the finances of the excursion, and with hints of a similar trip to South America, describes the unforeseen benefits of such a tour.

It has been on my mind, since seeing Bowman, Johnson and Martin, in the last few months, after their return from service on the staff of the Peace Commission in Paris, to write you in recognition of the unexpectedly useful results brought forth by your endowment of the Transcontinental Excursion of 1912.

You may remember that when I first met you – on the third floor of 15 W. 81, in April ? 1911 – Mr. Chandler Robbins then introducing me to you – you asked, with your customary directness: – 'What is the good of it?' that is, of the proposed excursion; and I replied in effect that while it had no special commercial value, it would be useful in giving European geographers a good first-hand know-ledge of much of the United States, and also in giving American geographers a helpful acquaintance with European geographers.

The last element of value has come true to a much greater degree than I could have anticipated; for tho' various persons talked of a war to be provoked by Germany, that sort of thing was out of my line and out of my mind. It seemed also to be out of the mind of the very agreeable party gathered on our delightful special train; for we got along famously together. No one asked to have his 'location' or berth in his Pullman changed on account of a disagreeable neighbor; no one asked to have his seat shifted at table (That I think was something of a triumph of my scheme of arranging table-mates, which need not however be explained here).

As a result of the acquaintance – I might say friendships – then formed, our geographers found their work in Paris and elsewhere abroad greatly facilitated, as I hope they have opportunity of telling you directly. Bowman on arriving in Paris, did not have to seek introductions to the geographers there; they welcomed him at once as already favorably known, and opened everything to him. Johnson, in 1918, had access to the 'geographical notices', prepared by Gallois and others, at the Geographical Service on the French Army, when our Army officers could not get them; not even the officers of our Military Intelligence Division! His later service in Paris, 1919, was facilitated in every way by his colleagues there, who have a high opinion of him and of Bowman also, as I know from their letters direct to me.

Lawrence Martin, Major, attached at first to Gen. Bliss, later charged with securing maps, etc. in Berlin and Vienna for use at Paris, was welcomed in both those enemy capitals by his geographical friends of 1912. He said it was invaluable in his work to have known those geographers before reaching their European homes. It was a saving of time and of work. Mark Jefferson, who had to prepare special maps for the Commission in Paris, was so confined to his office that he had little occasion to meet the European geographers; so his testimony does not go so far.

But I hope it will be a gratification to you, as it surely is to me, to see how practically useful the abstractly scientific Transcontinental Excursion has proved to be. Such things can hardly be evaluated in dollars; but if measured by the effectiveness in making the American Geographical Society pleasantly known to leading European geographers, and later in providing a warm welcome for the several American geographers during their stay in Paris last year, its value will I trust be satisfactory to you. (*Letter: Davis to Archer M. Huntington, 21 February 1920*)

CHAPTER SEVENTEEN

Master of Method

Davis was a great educator and probably deserves more praise as a pedagogue than as a peneplain professor. Ironically his outstanding advocacy of the need for explanatory geography did more than any other factor to ensure the eventual eclipse of his geomorphology. It seems to us that he has never been given half the credit he deserves for his stimulus to geographical education in the United States. This, however, is understandable as his methodological writings are voluminous and many of them are today not readily available. Moreover, method was his forte; it enters into so much of his work that anyone who attempts to summarize his ideas on it is faced with a colossal task. He propaganded, advised, wrote schemes and books, sat on committees and founded societies for the spread of geographical knowledge. At the same time he participated in university studies at home and abroad and, as we have seen, did far more than his share of fieldwork.

Twelve of his earlier methodological articles are reproduced in *Geographical Essays* (1909C, pp. 3–298). They deal mainly with the content and methods of geographical teaching in various grades of educational establishments. No doubt in the educational theory of the time, particularly in Europe, they held little new but in their insistence on proceeding from the familiar to the unfamiliar, on the use of visual aids, on the correlation or classification of facts and on causal explanations they were truly modern. It is, however, certain that a reader who keeps solely to *Geographical Essays* has missed Davis' finest writings on method. Between 1909 and 1916, irrespective of enlarged editions of *Practical Exercises in Geography* and of numerous articles on the uses of topographic maps and on descriptions of landforms, he published at least ten important essays largely or wholly concerned with methods of geographical presentation. Of these the following are of outstanding importance:

1909G   The Systematic Description of Land Forms.
1910B & H Experiments in Geographical Description.
1911I   The Colorado Front Range; A Study in Physiographic Presentation.
1911P   The Disciplinary Value of Geography.
1912A   L'Ésprit Explicatif dans la Géographie Moderne.
1912H   Relation of Geography to Geology.
1915K   The Principles of Geographical Description.

In two brief chapters we intend to try to give the essence of Davis' writings on educational matters and to assess the claim that he was 'an inspired teacher of teachers', a true 'master of method'. Our remarks largely concern landforms but they apply equally to meteorology and climatology and less so to most other branches of geography in which he showed some interest but lacked time for detailed exposition. In discussing the description of landforms he always in his later articles makes repeated use of his European excursions of 1908, 1909 and 1911, which we have already discussed. This is especially so when he is addressing European societies; American audiences were also presented with the Colorado Front Range.

METHODS OF RESEARCH

Map work

A fundamental feature of Davis' landform analysis was the study of topographic maps. Many of his ideas and theories, for example on the Seine, Meuse and Moselle, and on the Thames–Severn river-pattern and the drainage of

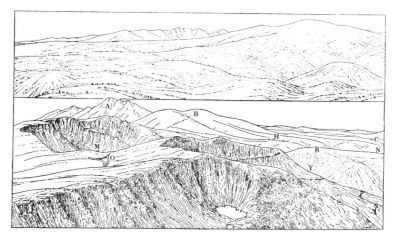

FIG. 76. Scenery in Colorado

Above. Bald Mountain (approx. 12,000 feet) (looking west), a smoothly graded, subdued monadnock, slightly dissected by normal valley heads, three of which in the right foreground descend steeply into the young submature valley of Lefthand Creek, eroded during the current cycle. On the left (i.e. south-west) is the dissected highland of Chittenden Mountain (10,500–11,000 feet). In the distance are the sharpened summits of Mr Arapahoe (13,520 feet), with its east-facing cirque

Below. View north along the Continental Divide crest of the Front Range near Corona station (O). The mass is dissected by east-facing cirques (M, J, S, Y) and surmounted by the non-glaciated Bald Mt. (B) and the glaciated Arapahoe Mt. (A). From the former, Middle Boulder trough (HN) discharges eastwards between Bryan Mt (R) and Caribou Flats (C)

(From Davis, 1911I, Fig. B)

FIG. 77. Scenery in Colorado
Top. Grays (14,341 feet: left) and Torrey's Peaks (14,336 feet: right), looking west
Middle. James Peak (13,282 feet) seen looking south from the crest of the Front
Range near Corona Station. Cirque heads of Mammoth Gulch (east background), Jim
Creek (west background) and South Boulder Creek (foreground) dissect the mass
Bottom. View eastward from the top of the cirque wall down the maturely-
overdeepened and widened trough of Middle Boulder Creek
(From Davis, 1911I, Fig. E)

cuestas were first completed indoors and later perfected in the field. Where
possible he used maps on various scales, with small-scale maps for regional
studies and large-scale maps for detailed local problems. Since by 1900 such
maps existed for large areas of Europe, at least the general outlines of their
relief and drainage were conveniently and accurately summarized in a very
convenient form for the use of physiographers. However, the detailed contour map with accurate contours at close intervals was largely a product of
the post-1920 world. It is important to emphasize that Davis worked mainly
with maps on which salient and basal heights were given and the intervening
relief sketched in by means of hachures. Maps were so essential a feature of

his own research and of his teaching that we do not hesitate to quote at length from his opinions of their adequacy and availability in the 1890s.

In 1893, when pressing for more and better geography in grammar and primary schools in the United States, he writes:

> Consider the case of the Empire State. In the first place, there is no respectable map of its area! There is no map to which the teacher can turn for a clear picture of its beautifully varied features. There is not only no map; there is not even a good written description of its surface forms . . . in the light of modern geographical science . . . The same is more or less true of nearly all the other states.
> (*Davis, 1893M and 1909C, p. 127*)

In 1895 he gives as one reason for the need of geography in the university, the lack of training in landforms of the topographers or surveyors.

> There is one phase of this subject of map making that is not generally understood. It is too commonly supposed that a map is made entirely by measurement; that it is necessarily accurate because the topographer carefully measured the forms of the land before him when the map was made. Now, while it occasionally happens that an elaborately measured survey has been made – as, for example, of the area of Central Park in New York City . . . – it is practically never the case in this country that a minutely elaborate method is employed on government surveys of large areas. The surveyor fixes by accurate measurement a certain number of points, but between these points he must sketch the intervening space. In sketching, he has to reduce the area of the ground before him to a much smaller area on his map sheet; and he must therefore omit many details and represent only the more important features . . . If the topographer actually measured every line that he draws, he might be as unintelligent but as faithful as a camera; but he *must* generalize and to do this he must have an intelligent understanding of this subject.
> (*Davis, 1895H and 1909C, p. 150*)

> If time were allowed them to run out all their contours by actual measurement, an exact map might be produced, but neither time nor money can be devoted to so slow and expensive a method. (*Davis, 1894E, p. 68*)

In his review of 'Physical geography in the university' (1894E) he discussed the topographical maps available. He selected maps with regard to their geographical features. In the United States the charts of the coastal survey offer admirable illustrations of littoral forms, including the delta of the Mississippi, 'a geographical gem'.

> The maps of the Mississippi River Commission offer remarkable illustrations of a large river on its alluvial plain. Its meanders, its cut-offs, and its ox-bow lakes are shown to perfection . . . But it is the topographical sheets of the United States Geological Survey that afford the greatest variety of illustrative material for this country; and it is not too much to say that the facts they present create a revolution in the student's knowledge of his home geography. We may well wish that they were more accurate, but, with all their imperfections, they present a great body of new information . . . These maps are simply indispensable. They

call forth much interest from the class. At first hardly translatable into words, their meaning grows plainer and plainer, until at the close of the course they are as suggestive as they were uncommunicative at the beginning.

Not less valuable and far more accurate than our own topographical sheets are those of various foreign topographical surveys. Unfortunately, the relief in most of these is expressed by hachures, altitudes being given only for occasional points, or by widely separated contour lines; but the general expression of the surface is admirably rendered in many of the surveys. The older maps are generally too heavily burdened with hachures, but the more modern surveys are very artistically executed. It has been my practice for several years past to select certain groups of sheets from the sets of foreign topographical maps in our college library and order extra copies of these groups, mount them on cloth and rollers, and thus prepare them for the most convenient use in the laboratory ... From the Army Staff map of France (1 : 80,000) there is a group of sheets showing the level plain of the Landes ... From the Ordnance Survey of Great Britain (1 : 63,360) one set of sheets includes the central Highlands of Scotland, with the Great Glen and Glenroy; two other sets include the fiords and islands of the southwestern and the northwestern coasts. These three sets agree in showing an old peneplain of denudation, once elevated and maturely dissected, but now somewhat depressed, with cliffs nipped on its land heads and deltas land in its bay heads ... A group of sheets for southwestern Ireland exhibits bold mountain ranges running directly into the sea, forming a strongly serrated coast. The English sheets are of older date and are not of particularly good expression, and for this reason I have not yet ordered any of them, although the ragged escarpment of the chalk and of the oolite treading northeast on either side of Oxford should be represented, and the Weald offers excellent illustration of well-adjusted consequent and subsequent rivers on an unroofed dome of Cretaceous strata.

The map of the German Empire (1 : 100,000) supplies many examples of striking features. The plateau of the middle Rhine has already been mentioned as a subject for lantern slides ... From Norway (1 : 100,000) the district of the Christiana fiord is already received in ten sheets of most delicate execution ... from Russia (1 : 400,000), the lakes of Finland and of the lower Danube; from Austria, a portion of the flood plain of the Danube and a strip of the fiorded coast of the northern Adriatic. This is only the beginning of what I hope the collection may be in a few years.

I cannot speak too highly of the educative quality of these grouped sheets ... No verbal descriptions from the teacher suffice to replace the portrayal of geographical relief on good maps. (*Davis, 1894E, pp. 91–5*)

Davis frequently had already made up his mind about the key problems of certain landscapes from map study before ever visiting the areas. He was in all senses a superb interpreter of topographical maps.

Fieldwork

For Davis fieldwork involved visiting a district about which he had already studied available maps and available information. Thus he would first read up any geological and physical descriptions that could be procured, either of an official or amateur nature. On Snowdonia, for example, he studied in advance, and on the terrain, the official geological maps, the governmental topographical maps and the fine geological *memoirs* written by A. C. Ramsay. We know of only one occasion when he failed to do this.

> I am working hard on my Turkestan report, which has, unfortunately, been postponed to this late date. One of our students who can read French, German and Russian, although not very fluently, is helping me in looking up previous descriptions of the region. What I am learning is exactly what I ought to have learned before going out there last summer. (*Letter: Davis to Ellsworth Huntington, 17 February 1904; Cambridge, Mass.*)

Also, whenever possible, Davis visited the locality with the leading geographical or geological experts on it. Thus, as we have already noticed in previous chapters, he visited the upper Coln near Cheltenham in England in company with the local geologist, S. S. Buckman, whose views he accepted and publicized. Or again in France and Italy he invited, and usually received, the guidance and opinion of the local experts on landforms. On the Colorado Front Range, upon which he wrote so lucidly, he knew well the pioneer 'peneplain' ideas of Archibald Marvine. Davis willingly acknowledges this assistance, of which, in some instances, he was the first to put into print or at least to place before a wide public.

As a fieldworker he was remarkably indefatigable and loved to exert himself. Mark Jefferson in his diary often remarked on Davis' 'stocky frame' and 'barrel thick legs' that sustained their leader long after most of the students had had more than enough. Davis' general policy was not 'Go and see' but 'Go, see and think'. In fact, however, the thinking was usually directed along Davisian methodological lines as must have been obvious in our accounts of his European and American transcontinental excursions. We will summarize his methods of description later but suggest here that even an uncritical reader would find it hard not to believe that Davis told his research students what to look for rather than what to look at; and that he always tried to convince the local experts of the error of their views if they did not conform to some sequential pattern. Needless to say the Davisian scheme as expressed by the master was so elastic that failure to comply with it simply meant lack of imagination!

But to us Davis' fieldwork methods comprise four and not three imperatives. They were: Go; see; think; and draw. If at every spot that he sat down and sketched the view there were erected a notice stating 'W. M. Davis drew here', with the date, he would be the best recorded international traveller in history. Often these sketches were developed and altered later but to those who knew the viewpoints they reflect the high quality of his artistic talent (when judged on the average attainments of geologists and geographers) and the slight emphasis on proportions favourable to Davisian themes.

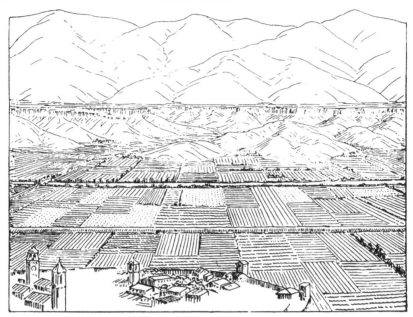

F I G. 78. The valley of the Arno near Figline, Italy, looking north-east (From Davis, 1912J, Fig. 1)

In later years, after 1908 or 1909, photography was also used but it was intended to supply details, especially of cultural and organic features, which the simplified physiographic diagrams deliberately avoided.

The obvious lack in Davis' methods of fieldwork was the instrumental measurement of terrain and of rates of processes. His awareness of the inaccuracy of many maps and indeed his undoubted ability to use scientific instruments would have led us to suppose that he would have undertaken many instrumental geomorphic surveys. But probably he judged the results would not have been commensurate with the time consumed. His hurried programme, covering vast distances and widely-separated localities, and his innate wish to be a human spirit-level encouraged him to avoid the tedium of detailed landform measurements.

There was an obvious lack also in Davis' laboratory work. He considered experimentation by physical models to be of little use in his day largely because laboratory techniques were as yet inadequate.

Experimentation, Davis thought, is of little avail in such matters. One can, indeed, reproduce, or rather imitate, some simple phenomena at their natural scale; but this is hardly more than making observation and measurement easier. When it comes to experimenting at a reduced scale, all the 'dimensions' of the phenomenon, lengths, volumes, densities, viscosities, velocities, must be adjusted in such a way as to maintain their natural ratios: this cannot be done, in general, for all the variables involved. At any rate, experiments are very different from the much more simple and abstract procedures of physical laboratories, and seldom lead to definite conclusions. (*Baulig, 1950, p. 195*)

Today both Davis and Baulig would at many national hydrological laboratories soon be shown to be grievously wrong, although their general contention on the difficulty of reproducing complex natural processes would be loudly applauded. The fact is that Davis did not like the detailed study of processes but was content to observe the forms they created. We have already shown in Volume One of this *History* that long before his day there were excellent laboratories in Europe where great progress had been made in the knowledge of river work solely from physical models. But the Americans either ignored or were ignorant of these great findings from hydrological experiments.

WRITTEN METHODS OF TEXTUAL PRESENTATION

Davis once he had entered the academic world at the age of twenty-six took the standpoint that anything he wrote was worth printing; that anything supporting his cause was printable. Some of his methodological articles – for example 'Experiments in geographical description' – were reprinted in almost the same form in two or three different journals. What is more, many of them contained sections of ideas or factual matter already printed several times previously in almost the same words.

Style and scale

His style has been variously described as lucid or limpid, marked by skilful and facile expression and enriched by a wide range of analogies. We have, however, shown enough of it for the reader to judge for himself. He often exhorted would-be authors to be brief, as printers were already overburdened with matter. But a study of his own articles shows how little he considered this applied to himself. Many of his articles are of great length; they may be well composed and comprehensive unities but no word is spared, not a jot removed for the sake of brevity and when seen against their predecessors and successors they become appallingly repetitive. Perhaps in an international spread this repetition may be forgiven but the matter would have been more forceful and more attractive if the examples used had in total been more numerous and if those given had been less often repeated.

He loved euphony – who does not? – and often let it take control, as in our quotation above where 'land heads' and 'bay heads' presumably mean 'promontories' and 'embayments'. Strange to say, whereas his technical terms often irritated his contemporaries, none seemed to dislike his constant imputation of physiological properties to inanimate things. For instance, H. J. Mackinder and other English geographers were explosive over 'peneplain' but they never objected to Davisian rivers being 'contented' or 'mating'.

Inevitably he exerted a strong influence on the style of his students, as Ellsworth Huntington affirms:

> One of the things where I can feel your influence most strongly is literary style. Your insistence on brevity and clearness, and on a brief but comprehensive statement of the purposes of an article or chapter at the beginning, keep coming to mind. (*Letter: Ellsworth Huntington to Davis, 5 February 1925; New Haven Connecticut*)

Davis wisely considered that the style, scale and content of verbal geographical description were interconnected and varied also with the audience aimed at.

> Hence the importance of giving conscious practice to the preparation of verbal description of a given district or region on different scales; one might be ten lines long; another might fill a page; a third, a chapter; a fourth, a volume. A geographer who proposes to make himself proficient in his science ought to practise himself as thoroughly in writing descriptions on different verbal scales as in drawing maps on different graphic scales . . .
>
> The style to be adopted should be first determined according to whether the description shall be technical, for trained geographers; or popular, for intelligent, mature, non-technical readers; or elementary, for young beginners . . . In view of the style and scale as thus determined, the critical selection of certain items to be included and of others to be excluded may come next; and with this should go the careful determination of the order in which the included items shall be presented . . . It is chiefly the generalised treatment of dominant or of recurrent elements that deserve verbal statement, with subordinate place for the most significant exceptional features. (*Davis, 1910H, p. 577*)

The value of technical terms

The quality of Davis' descriptions of landforms leans heavily on the use of technical terms, in which respect we can hardly do better than quote from his last great analysis in 'The principles of geographical description' (1915K).

> The value of explanatory technical terms, such as delta, volcano, cuesta, monadnock, moraine, ria, and so on, in geographical description does not lie only in the avoidance of cumbersome paraphrases, but also in the avoidance of geological distraction – provided that the introduction of the terms has been preceded by the explanation of the forms that they name, warranted by competent analysis and familiarized by careful systematization. Of course, a series of terms such as consequent, subsequent, obsequent, insequent, and resequent will seem difficult to anyone – a geologist, for example – who without interest or patience enough to study the things that the terms conveniently name tries to make a hurried etymological attack on physiographic problems; but to those who learn such terms as names for things, just as a mineralogist learns orthoclase, plagioclase, periclase, and loxoclase, or as a botanist learns glossopteris, dryopteris, goniopteris, actiniopteris, arthropteris, and hecistopteris, they will present no serious difficulty.

So long as physiographic descriptions contain elaborate explanations and cumbersome paraphrases, in which the action of process on structure in past time is the leading idea, it is difficult for the reader to hold his attention upon the geographical present. But if all the considerations which involve past time have been previously analyzed and systematized, and if the resulting concepts of land forms, thus made clear and familiar, have been graphically represented and technically named, the mere use of a name will suffice to bring forward the intended concept; and thus distraction by geological considerations will be reduced to its lowest terms. In illustration of this statement I take the liberty of referring to two of my articles,* in which the use of the term morvan serves in large measure to exclude geological distraction from a geographical description, where such distraction would otherwise be an embarrassment. It is perhaps wise to state at the same time that a recent writer† has condemned the term morvan, on the ground that different morvans are very unlike; but I may also state that this writer is a historical geographer who has made little use of modernized physiography; that he has not made practical trial of the value of morvan by using it and its long, paraphrased equivalent in rival descriptions; and that although he rejects the new term, morvan, he accepts the old term, volcano, in spite of the greatly diverse forms that it names. My own belief is that he rejects morvan because it is new and unfamiliar. In spite of its novelty morvan will probably be adopted by those who are thoroughly familiar with the well-defined group of features thus named and with the explanation of their origin, so that they can use the term with an easy mind; and who at the same time have so frequent need of thinking, writing, or talking about the grouped features, that they must use either a term or a paraphrase. Those who have no such occasion to use the term, or who feel uncertain as to the validity of its explanatory content, will probably condemn it on abstract grounds, without ever making a practical trial to ascertain its value.

It is however desirable that technical terms should be as simple as possible; it does not appear necessary to go so far as Falconer, who says: 'A moraine landscape much dissected by stream and run-off is ... to be classified not as an exogenetic aggradation landscape, but preferably as an exogenetic degradation landscape'; and it is surprising that so pronounced a Germanist as Passarge, who, although he uses Vulkan, Lakkolith, Kame, Drumlin, Delta, and Atoll, objects to other foreign terms, such as Cuesta, Bolson, and Playa in German books, should employ such complex exotic phrases as 'Monodynamische und polydynamische Einzelformen', and 'Das Studium physiologisch-morphologischer Karten', in which only und, einzel, and das are 'deutsch'. (*Davis, 1915K, pp. 86–7*)

We shall return again to the subject of Davisian terminology in our epilogue (Chapter 30) where we attempt to summarize its total contribution to geomorphological thought.

Davis' landform descriptions were remarkably devoid of mathematical expressions and statistics. He did occasionally give suggestions on the

* Davis, 1912G, pp. 89–92, and 1912H, pp. 112–18.

† Ricchieri, G.: 'Sui Compiti attuali della Geografia come Scienza', *Riv. geogr. ital.*, XXI, 1914.

amount of uplift, depth of erosional incision, and the gradient of slopes but generally he shunned attempts at mathematical precision. He seems to have particularly disliked mathematical equations.

> Davis, who had studied mathematics, never made use of it in his geomorphological work; not a single formula is to be found in his some three hundred papers on land forms. This is all the more significant as it strikingly contrasts with the general tendency of the nineteenth century: orography originated in Europe as orometry; Albrecht Penck's *Morphologie der Erdoberfläche* (1894) is full of equations ... (*Baulig, 1950B, p. 195*)

The use of diagrammatic aids

Davis, as we have already shown, skilfully combined visual and verbal descriptions. We hope that we have included enough of his drawings to demonstrate his artistic ability. All diagrams appealed to him but he particularly commended in systematic presentation block diagrams arranged in series to show successive stages in the erosion of a single structure.

> Block diagrams are more immediately understood than maps are; they are vastly superior to mere profiles, which of all graphic devices are of least value to the geographer; for he is concerned with surfaces, not with lines; yet if profiles are wanted, they are found along the side of block diagrams, in their proper position with respect to the adjoining surface ... Block diagrams can be drawn from any desired point of view, so as to show the features represented in the best possible relation to each other. (*Davis, 1910H, pp. 583–4*)

For Davis, however, block diagrams were more than mere physical models, they were theoretical models in a very real sense.

> Just as block diagrams aid in giving graphic illustration to the members of series of deduced type forms, as has already been mentioned, so they aid in the understanding, the description of actual regions, because they serve so immediately to present the generalised type forms with which the observer compares the actual forms ...
>
> Diagrams of this kind are not and are not meant to be mere pictures of observed landscapes, for they must always be simplified by the judicious omission of much unessential detail, and greatly compressed by the omission of many repetitions of similar elements. They may indeed be rather fanciful, in being designs rather than copies of nature ... (*Davis, 1910H, pp. 582–3*)

> It need hardly be pointed out, after what has been said above, that the block diagrams here recommended are not intended to represent actual forms; for they avowedly represent ideal, imaginary forms: nevertheless, the mental concepts thus presented by the writer and acquired by the reader, together with the names by which they can readily be brought to mind, supply the best means of describing the actual forms of which they are the counterparts, and therein lies their value. They are practical helps in showing the results of geographical exploration when it comes to preparing geographical descriptions. (*Davis, 1915K, p. 86*)

Although he considered sketches and condensed diagrams worth all the labour of designing and drawing them, he recognized that they may have certain disadvantages. They often had to be too specific, too definite; for example, in showing the overlap of strata upon an ancient planation surface when the details of the overlap had not yet been determined; 'But on the whole the advantages of a diagram far outweigh the disadvantages' (1911I, p. 36).

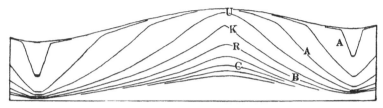

FIG. 79. Successive profiles between two developing river valleys (From Davis, 1912J, Fig. 25)

We interpolate diagrammatic aids here between our accounts of his technical terms and of his approach to description because he depended heavily on diagrams to support his written work. So much of what he wrote was a combination of a *condensed technical description* and a *condensed diagram*.

NATURE OF APPROACH TO DESCRIPTION

In Davis' copious writings on methods of geographical description and on the systematic description of landforms he often includes geography as a whole, that is organic as well as inorganic features. In all his writings the physical landscape dominates, but it is too often forgotten that he was both a physiographer and a geographer. As we wish here to emphasize his contributions to landform study, on which, in fact, he deliberately concentrated his efforts, we will first demonstrate the width of his geographical learnings. In 1902(C), when discussing the progress of geography in schools, he considered that the main need in the future was for the introduction of explanatory or causal analysis – 'on the modern principle of evolution' – of the adaptation of all the earth's inhabitants to the earth. The inorganic environment he called physiography and the organic 'ontography' which should include 'all pertinent facts in [organic] structure, physiology, individual and species'. The latter should be pursued even into forms of language and habits of thought. Geography involves the inter-relationships of physiography and ontography.

It is especially the factor of relationship of earth and inhabitants that characterizes geography as a subject apart from other sciences, and that gives an essential unity of content and discipline to all its varied parts. (*Davis, 1902C and 1909C, p. 36*)

Every category of physiographic elements should be accompanied by examples of the responses made to it by organic forms. It is not enough to take up the organic responses afterward; the habit must be formed of associating these responses with the study of the environing elements. (*Davis, 1902C and 1909C, p. 46*)

The most important principles established in physical geography during the nineteenth century are that the description of the earth's surface features must be accompanied by explanation and that the surface features must be correlated with their inhabitants. (*Davis, 1900O and 1909C, p. 70*)

We rather labour these points because a few years later Davis is forcing Jefferson and other research students to keep their eyes only on the physical landscape. In 1912 he is so often thrusting his physiography on visiting savants that he is accused publicly of not being a geographer but a physiographer, which he had to admit. Could it be that the great empire of physical geography which he had so laborously built up on scientific lines was now being submerged by a new human geography, by the poor, badly-taught 'ontography', the weakness of which he himself had deplored a decade earlier? But we repeat he was always convinced that a carefully arranged classification was essential in the study of both organic and inorganic features; as a trained geologist and meteorologist, he happened to pay more attention to the inanimate side.

We will try here to summarize the Davisian approach to physiographic and geographic description as it is of permanent interest. In the late nineteenth century Davis developed in the United States the idea of the geographical cycle or cycle of erosion as it was called later. As usual, he can be trusted to describe the progress of this theme himself:

only by beginning at the initial form can the systematic sequence of the changes wrought by destructive processes be fully traced and the existing form appreciated. This had often been done before in individual cases, but it now became a habit, an essential step in geomorphological study. Naturally enough, the terms of organic growth, young, mature, old, revived, and so on, came to be applied to stages in the development of inorganic forms; and thus gradually the idea of the systematic physiographic development of landforms has taken place. This idea is to-day the most serviceable and compact summation of all the work of the [nineteenth] century on the physical geography of the lands. (*Davis, 1900O and 1909C, p. 85*)

We have already described how Davis developed the simple cycle in type and in depth until it became a most complicated and elastic scheme. Inevitably, because the theoretical basal theme of 'development' or 'evolution' had become excessively complex and faced a real world of landscapes affected by frequent phases of relative uplift, denudation, climatic change and so on, he began to stress the complications. At the same time, as today, geographers and physiographers alike had not found any descriptive approach or method acceptable to all. Davis, therefore, proposed to expand and popularize the 'elaborated cycle'.

At a Chicago conference in 1907 he conducted a discussion centred chiefly on the possibility of developing and adopting a systematic method for the description of the lands, which he called 'structure, process, stage'. No doubt the title of the conference 'Uniformity of method in geographical investigation and instruction' was far too wide and, in any event, several members were not prepared to use Davis' method although they did not propose any alternative. Davis was not perturbed and, as we have seen, arranged an excursion in Europe in 1908 to test out his method on the participants. He also changed the title of the theme to 'Experiments in the systematic description of land forms'.

After his term of lecturing at the University of Berlin, when he exchanged professional duties with Professor Albrecht Penck, he returned home *via* England where on 18 March 1909 he lectured on the systematic description of landforms to the research department of the London Geographical Society. His avowed aim was to replace empirical description by thorough-going explanatory description based on the method of structure, process and stage, and to persuade English geographers to use the latter.

> Every feature of the land may be treated as the surface form of a certain *structural mass*, accumulated under certain past geological conditions, and placed by crustal movements, with more or less deformation, in a certain attitude with respect to baselevel, so that it comes to be acted upon by various external destructive *processes*, which have now carried forward their changes to a certain *stage* of development. (*Davis, 1909G, p. 301*)

He illustrates this with various examples which are described verbally and illustrated with block diagrams and photographs.

> The Adriatic border of Italy, near Ancona, consists of a coastal plain of imperfectly consolidated sands and clays, which has reached a stage of late mature dissection under the action of normal and marine erosional processes. The term 'coastal plain' implies that the strata of which the district consists were laid down as marine deposits, sloping gently away from the land mass from which they were derived; also that they were revealed by a broad uplift which did not significantly disturb their simple structure . . . The significant words 'late mature dissection' indicate at once that both of the erosional processes, normal and marine, are well advanced in the series of changes which, have as a goal, the complete destruction of the coastal plain. The main streams must in late maturity have opened wide-floored valleys; the side streams must branch elaborately, thus dissecting the original plain into a multitude of hills and spurs; the axes of the hills and spurs must trend in a general way toward the coast, because the streams which have carved the valleys between the hills must for the most part have had their courses determined, directly or indirectly, by the initial slope of the plain; the hills must have been weathered into smoothly arching crests and smoothly sloping sides; the

shore-line must have been cut back so as to truncate the hills in cliffs, all standing in line. Thus a general mental image of the district may be formed, in which all the features are systematically correlated . . . (*Davis, 1909G, p. 302*)

It happens, however, that the main valley-sides are moderately terraced inland and the shoreline now lies not along the base of the mature cliffs but along the seaward edge of a sandy strand-plain, 100 or 200 metres wide. Davis imputes the terraces to a recent, gentle slanting uplift of zero near the coastline to 10 or 15 metres at the inner border of the plain.

There is good reason for regarding the action of the streams in terracing or degrading the former valley-floors as the cause of the progradation of the strand-plain. The slanting uplift revived the mature streams, and caused them to degrade the valleys through which they were previously flowing contentedly; the degradation of the valley-floors increased the quantity of waste washed out from the river-mouths, and thus compelled the waves to abandon their former task of cliff cutting or retrograding, and for a time at least to take up the contrary task of distributing the river-waste along the shore, thereby building forward or prograding the strand-plain. The previous description may now be expanded and supplemented as follows: the coastal plain of imperfectly consolidated sand and clay strata had reached a stage of late maturity by normal and marine erosion, when a gentle slanting uplift occurred, which caused the extended consequent streams to terrace their main valley-floors, and the waves and currents to prograde the shoreline. (*Davis, 1909G, p. 303*)

He goes on to describe various localities and districts including a 'maturely dissected coastal plain, recently somewhat depressed'; the main valley of the Lot in the Massif Central of France; where the master stream is more mature than the side streams; and Snowdonia in North Wales, a late mature landscape recently submaturely glaciated.

The advantages of the explanatory method are extolled, especially to the reader who has already made a systematic study of an entire series of appropriate ideal forms, developed in regular order. Such a scheme can be expanded into 'a more elaborate device, so as to fit a much larger variety of natural cases'.

The general plan of development being once understood, all the particular features that will be associated with each particular structure may be easily deduced.
(*Davis, 1909G, p. 307*)

In describing Snowdonia, Davis makes use of 'a cycle of glacial erosion' but here the description involves a departure from the normal cycle unlike the interruptions caused by crustal movement.

The departure from normal conditions in the case of Snowdon does not appear to have involved any significant change in the altitude of the land, but only a temporary change in climate, where the snowfall was increased and glaciers were formed . . . The glacial accident was . . . a brief one; the normal cycle is now again in progress. (*Davis, 1909G, p. 312*)

Davis towards the end of this article makes the interesting suggestion – that rock masses should be distinguished from their waste products.

Land forms, treated as the surface of rock masses, may be described in terms of structure, process and stage . . . the description of land forms, thus considered, must always be associated with two other elements: one of these is the active process, which works on the land form and causes its change; the other is the land waste produced by the action of the destructive process on the passive structural mass. In a cycle of normal erosion, we have weather and streams as elements of the active process; and locally weathered soil, sheets of creeping soil, alluvial fans, flood-plains, deltas, and so on, as elements of the forms assumed by the waste of the land on the way to the sea. In a cycle of glacial erosion we have weather, snow-fields and glaciers as elements of the active process; and locally weathered super-glacial rock waste, moraines of various kinds, drumlins, eskers, kames, and so on, as the forms assumed by the waste of the land in association with glacial erosion. In a cycle of marine erosion we have weather, waves and currents acting on the coast; and beaches, sand-reefs, tidal marshes, and so on, as the associated waste forms. In a cycle of solvent erosion, such as prevails in limestone regions, traver-tine deposits gain an importance among waste forms, even to the point of building up barriers across valleys and causing lakes and waterfalls, such as occur in karst districts . . . In the cycle of arid erosion, sand dunes and loess deposits become of importance. In all these cases, the forms assumed by land waste vary systematically through the course of the cycle. In the normal cycle, for example, coarse rocky waste is usually associated with a young stage of the cycle, when valley-sides are steep; while fine waste prevails in the late stages, when the relief is faint and the slopes are gentle. A whole series of interesting considerations follow naturally upon the systematic description of the forms assumed by the waste of the land on the way to the sea, in association with the land forms on which the waste lies and with the processes acting upon it. (*Davis, 1909G, pp. 312–14*)

The criticisms by British geographers of this lecture are recorded in full. The genetic approach and great value of Davis' work on these lines were almost unanimously acclaimed. H. J. Mackinder thought Davis had rendered very valuable service to geography by his effort to introduce into it precise and consistent methods of description. But he and many others disliked some of the technical terms. A. J. Herbertson thought that systematic description and terminology demanded

something of a much more simple yet more comprehensive character than a series of descriptive terms. We might have recourse to letters and numbers, as the chemist does, and use say, numbers to indicate the different types of structure, letters to show the process of denudation, and another set of symbols to indicate the various stages in the evolution of the cycle of denudation. For instance, the late mature stage of volcanic plateau denuded by running water might be 2 *a* 8 . . .
(*Davis, 1909G, p. 322; see also 1910H, pp. 536–7*)

Some thought that the scheme was perfect in Davis' experienced hands but dangerous in the hands of learners, especially as geologists themselves were often uncertain on landform processes. As Dr H. R. Mill put it, 'His method appeals to me as one more for the master than the student, and I am afraid that his disciples will run away from him, and apply it in a way that will cause him anxiety at first and horror afterwards.' More important was the criticism that Davis had dealt only with landforms of the smallest order. Would not the theoretical simplicity of his scheme apply only to small areas? Would it not become confusing and difficult to follow amid the complexities of a large tract composed of rocks of very different durability?

Davis replied gracefully and logically to each point. The systematic description of a large area may be successful, if tried, particularly as knowledge of its detailed parts increases. The scheme, he admits, could not be applied to large areas consisting of rocks of very different durability if it is taken rigidly, but

> the method merely attempts to express the facts of the case systematically, be they simple or complicated . . . the intention underlying the whole effort is not to force the variety of nature into the narrow limits of a rigid and arbitrary scheme, but to adapt a reasonable and elastic scheme to the variety of nature. (*Davis, 1909G, p. 326*)

On 30 December 1909, Davis gave a modified version of the above lecture as his presidential address to the Association of American Geographers (1910B), and included European examples such as the dissected coastal plain near Ancona. The version, entitled 'Experiments in geographical description' was published with slight modifications, in three different journals, its final appearance being in the *Scottish Geographical Magazine* for 1910(H). This article shows how well Davis was acquainted with European scholarship, and how cleverly he plays the bagpipes to an Edinburgh gathering.

> In so far as the method has any novelty, it is to be found in the systematic treatment of well-known elements; and even in this respect it is not so novel as some have seemed to suppose. Its fundamental principles are to be found, for example, in the third edition of Sir Archibald Geikie's *Scenery of Scotland* (1901), where one may read: 'The problem of the origin of the scenery of any part of the earth's surface must obviously include a consideration of the following questions (1) the nature of the materials out of which the scenery has been produced; (2) the influence which subterranean movements have had on these materials, as, for instance, in their fracture, displacement, plication and metamorphism, and whether any evidence can be recovered as to the probable form which they assumed at the surface when they were first raised into land; (3) the nature and effect of the erosion which they have undergone since their upheaval; and (4) the geological periods within which the various processes have been at work, to the conjoint operation of which the origin of the scenery is to be ascribed'.

Here we have the very essence of which is implied under the terms 'structure, process, and stage' . . . Yet, obvious as these considerations are as regards the origin of scenery, it is seldom that they are completely and systematically employed by geographers in the description of scenery. (*Davis, 1910H, p. 568*)

He demonstrates clearly the weaknesses of the empirical method and the strength of the explanatory method.

All the types in an explanatory series, and particularly the deduced types, are learned in view of their origin by the action of some reasonable process on some specified structure through some limited period of time; and hence type-forms of this kind are necessarily considered in relation to their natural associates. The association may be regional, as in the case of the different parts of an ideal landscape produced by the imaginary action of process on structure to a given stage of development; or the association may be sequential, as in the case of a single element of form followed in imagination along its successive stages of erosional change, from the initial, through the sequential to the ultimate . . .

All the many members of an extended empirical series of ideal types must be learned arbitrarily and separately, for no mnemonic aid from explanation attaches to any of them. All the members of an extended explanatory series may be divided into groups, so that the groups themselves shall have certain highly suggestive general relationships, and so that the members of each group shall be treated as systematically interdependent and easily remembered. (*Davis, 1910H, pp. 576–7*)

Davis goes on to point out that neither the empirical nor the explanatory method was used with absolute purity but the empiricist introduces explanatory terms as it were by accident while the rationalist introduces them consciously and intentionally.

Under the stimulus of discussion and criticism Davis continued to develop these ideas on the method of geographical description and in 1911 published his detailed account of the Colorado Front Range that we have already discussed in Chapter 16. We here give the quintessence of its methodological, as distinct from geomorphic, matter.

An inspection of geographical articles in scientific journals discloses six chief methods of presentation, which may be called the narrative, the inductive, the analytic, the historic, the systematic, and the regional. In narrative presentation, the observer recounts the facts that he observed and the thoughts that they suggested, in the order in which they were encountered. In inductive presentation, the observed facts are arranged in some reasonable order . . . so that their leading characteristics may be generalized. In analytical presentation, the best theoretical explanation of the observed facts, as dependent on certain unobservable facts and processes of past occurrence, is selected from various proposed explanations, and set forth in such form that the hearers may judge of its sufficiency. In historic presentation, the gradual development of a problem is reviewed by summarizing the successive contributions made to its solution by various investigators. In systematic presentation, typical examples, as acquired by observation, induction, analysis, and history are arranged in a well considered order, so that they may be easily found when wanted. In regional presentation, the climax of geographical work is reached: here the attempt is made to describe, in their actual spatial relations, all the geographical features which occur together in a given district; these features having previously become well known by observation and induction, well understood by analysis, well appreciated by historic review, and well arranged in relation to similar features elsewhere by systematic classification. (*Davis, 1911I, pp. 23–4*)

Davis proceeds to suggest that these six methods could be variously combined and the presentation of all, except the analytical, could be treated either with the older-fashioned empirical motive or the newer-fashioned explanatory motive. Each could, or should, also be adapted as required to form (length) and to grade of geographical proficiency of the audience. Explanatory treatment is always better than empirical and depends heavily on analysis, which, after observation, is the mainstay of the explanatory treatment of geographical problems. In this respect, and particularly in the expansion of 'a systematic series of standardized types of land forms'

the mental process of deduction is of great service. After several related facts have been carefully observed and successfully explained by the analytical discussion of their mental counterparts, and thus shown to be particular instances of a general case, it is possible by deduction to form mental counterparts of new intermediate examples which shall complete an ideal series of standardized types. The types may then be named by nouns of generic value, and qualified by adjectives of specific value; yet, although thus standardized and named, the types are not made rigid. They are elastic and adaptable concepts, easily modified into endless subspecific varieties, and thus fitted to serve as helpful counterparts of the endlessly various facts of nature. Yet, elaborate as the mental equipment of a well prepared regional geographer thus becomes, it is easily understood by others, because all its parts are reasonably related. Herein lies the real value of the explanatory as contrasted with the empirical treatment of geographical problems. Explanatory descriptions . . . can be phrased in terms of a thoroughly elaborated equipment of standardized types . . . [which] can, by reason of well tested relationships, be so readily and accurately understood by those to whom the observer presents his results. This has proved by experience to be true of land forms, and I believe it to be equally true of all other parts of geography. (*Davis, 1911I, p. 30*)

However, Davis has yet to discuss the scheme whereby the regional presentation of landforms should be guided. He advocates the use of 'structure, process and stage'.

Under this scheme each element of the landscape is treated as the surface of a structural mass which has been carried forward from an initial form to some specified stage of development in the cycle of erosion by the action of some specified process or processes; the form thus genetically described being further qualified as to the strength of its relief and as to the texture of its dissection.
(*Davis, 1911I, p. 30*)

But, he says, he would like nothing better than to see abundant, conscious experiments in geographical description to test the quality of his own and other people's methods.

One of the clearest expositions of the method of explanatory description, which we have previewed in Chapter 15, was given by Davis in his article on the 'Relation of geography to geology' (1912H). Presenting his description of the Colorado Front Range almost as a religious text, he demonstrated 'the use of explanatory geological matter as a means to a geographical end':

> The Front Range of the Rocky Mountains in central Colorado, north-west of Denver, is a highland of disordered and generally resistant crystalline rocks, which show signs of having been long ago worn down from its initially greater mass to a surface of faint relief, slowly depressed and more or less broadly buried under a heavy cover of sedimentary strata. Then, as the result of a widespread uplift, a part of the compound mass west of a pronounced monoclinal displacement along a north–south line, came to stand above the rest, and thus the highland province of the mountains was marked off from that of the less uplifted plains on the east. The forms of the highland shows that the whole region advanced far through the cycle of erosion introduced by the monoclinal uplift, so that the resistant underlying crystalline rocks of the mountain area were stripped of their cover and worn down to a gently rolling peneplain, diversified by irregularly scattered monadnocks, rising singly or in groups, with a relief of from 500 to 2,500 feet, while the valleys of the highland show that a renewed uplift gave the whole region a greater altitude than before, with a gentle up-arching along a north–south axis in the mountain area 15 or 20 miles west of the monocline, whereby the peneplain, with its scattered monadnocks, gained the highland altitude of the present Front Range; the crest of the up-arching and the monadnocks that happened to lie near it defining in a general way the crest of the range, which here constitutes the continental divide. The weaker strata of the plains are now again worn down to small relief, but the harder crystalline rocks of the mountainous highland are only submaturely dissected by normal sub-mature or mature valleys, the higher parts of which have recently been strongly glaciated. Thus ends the description.
>
> (*Davies, 1912H, pp. 94–5*)

It must be apparent from the foregoing that the true geographical value of a condensed explanatory description can be reached only by expanding or translating each technical term or phrase into its full meaning with respect to the features of the existing landscape. The more successfully the translation is made the more fully will the reader's attention be brought forward from past conditions and processes to existing forms, and the more fully will the really geographical nature of this apparently geological description stand forth. (*Davis, 1912H, p. 100*)

Hence the strongest reason for advancing from the older-fashioned empirical treatment to the newer-fashioned explanatory treatment lies in the greater power of the newer treatment, in the power of deeper penetration on the part of the investigator into the real nature of the facts concerned, and, more particularly in relation to our present discussion, in the power of clearer and more intelligible

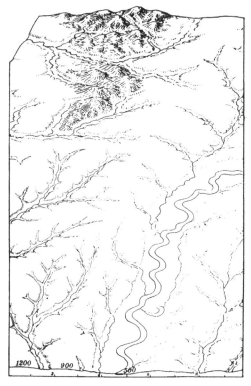

F I G. 80. A peneplain with monadnocks (From Davis, 1912J, Fig. 108)

presentation of the described landscape to the properly qualified reader. (*Davis, 1912H, p. 105*)

But the prime fact remains that explanatory concepts, deduced from general principles, are much more intimately and reasonably knowable than empirical concepts or even than facts of observation usually are, and in this quality of being intimately and reasonably knowable lies their highest value. It is as if one located them by sighting from many different points along the path of time, and thus fixed their position by the intersection of many converging lines of sight, while empirical concepts are located only by a single line of sight running in one direction from the viewpoint of the momentary present. (*Davis, 1912H, p. 106*)

Davis' last major summary of the principles of geographical description was presented to the Princeton meeting of the Association of American Geographers in 1913. Revised by censors during his absence in the Pacific in 1914, it was re-submitted for publication and again revised to incorporate the suggestions of censors in 1915. It was intended to bring together his various publications on methods of description since 1909, including 'Relation of geography to geology' (1912H, pp. 93–124) and 'Der Valdarno, eine Darstellungsstudie' (1914K, pp. 585–665).

In this long summary 'The principles of geographical description' (1915K, pp. 61–105) Davis again reviews at length the differences between empirical and explanatory methods. He admits the risk of error in the explanatory method but considers that a greater danger to it lies in the distraction from geography due to the possible introduction of complicated and irrelevant discussions on geological, biological and historical matters.

The descriptions of the Valdarno hardly concern us, although given first in the past and then in the present tense. However, the goal of geographers is said to be regional description.

> The description of existing landscapes, districts and regions of the earth's surface is the goal toward which other phases of geographical study, whether presented as personal narrative, historical review, analytical discussion, or systematic arrangement, all lead. Regional description is not systematic in the sense of describing things of a kind together, for it treats them in their unsystematic natural grouping. It is not analytical, in the sense of striving to find out the origin and meaning of existing facts, for it uses already discovered origins and meanings as an aid in setting forth the facts as they exist. It is not historical, either in the sense of tracing the progress of advancing knowledge regarding an area, or in the sense of following the discovery, settlement, and development of the area, though it may use the results of historical study in giving a better account of actual conditions. It is not narrative, for it seeks to present persistent and objective facts rather than temporary and subjective personal experiences. Regional geography is however synthetic in combining the helpful results of all other modes of presentation in a vivid description of a part of the earth's surface, so that all the geographical elements and activities there occurring, inorganic and organic, shall be appreciated in their true spacial relations. There are of course many other methods of geographical presentation than the five mentioned above, and these five and various others may be combined in any way that a writer desires; but it still remains true that pure regional geography is the final object of a geographer's efforts. (*Davis, 1915K, p. 62*)

It seems inevitable that, owing to the comprehensiveness of geography, the arrangement of a regional description will pose many problems.

> There can be little question that the least satisfactory feature of regional description lies in the necessity of presenting in separate, successive paragraphs or pages the many kinds of things that occur together in natural but unsystematic groupings. It is comparatively easy to describe together all the hills and valleys of a district in one section of an article, all the lakes, streams, and waterfalls in following sections, all the climatic and weather factors in a fourth and a fifth section, all the villages and railroads in sixth and seventh sections, and so on; it is much more difficult to present in a brief word-picture the situation of a single village in an open valley between well-watered, tree-covered hills, where the steady outflow of a lake, ice-covered in winter, supplies power at a waterfall for

factories in which the products of forest and field are prepared for shipment by a railroad that runs through farms along the valley of the lake-outlet to some other differently environed center of population; and so on through all the parts of the region concerned. Although all the geographical elements of a region occupy it simultaneously, they must be described in some reasonable sequence, for they cannot all be stated at once. A good sequence in essays by mature writers for mature readers treats, after briefly stating the location of the region considered, first, the land forms with their associated water forms; the ocean, in so far as it is included in the region under consideration, second; the climate, third; and then plants, animals, and men; but such an arrangement is rather for the convenience of the reader than for the restriction of the writer, who should certainly feel free to depart from this or from any other scheme if he thinks he can do so to advantage. (*Davis, 1915K, p. 99*)

Davis refers to physiographic analysis at many points in this fine essay,

The chief object of physiographic analysis is to provide a safe explanatory theory with respect to the origin of certain observed features, so that the imagined counterparts of the observed features and of many related features may be systematically deduced from the theory, and so that these deduced counterparts may be used, whenever needed, in describing the actual features which correspond to them: in other words, the chief object of physiographic analysis is to increase the number of terms in that part of a geographer's explanatory mental equipment which deals with physical features. When this important principle is recognized, it will be seen that both the methods and the results of physiographic analysis should be systematized as far as is reasonable, in order that they may more readily serve the important practical purpose of increasing the mental equipment in a well-ordered manner; for only after such systematization has been accomplished are the results of analysis ready for immediate use in regional description. This is equally true in all branches of geography . . . Let us see how this principle is carried out in the study of landforms . . .

As the result of many successful analyses, a certain number of geographers of the explanatory school attempt to state the result of analytical study of land forms according to some orderly plan of treatment. Thus Passarge of Hamburg has proposed a scheme under the title of 'Physiologische Morphologie', in which he elaborates a method of genetic classification for land forms; he presents a classification under Typus, Klasse, Ordnung, Familie, Gattung, and Spezialformen, with a large number of the last subdivision, such as Abgesunkene Schollen, Treppenbrüche, Kesselbrüche, Grabenbrüche; Schlammströme, Steinströme, Bergstürze; Grund-, End-, Seitenmoränen, etc.; and under 'ideale monodynamische Landschaftsformen', similarly subdivided, he includes various 'Oberflachenformen', such as symmetrische, asymmetrische Kettengebirge, Rostgebirge, Kettengebirgshochländer; Hügelländer, Landschwellen, wellige Ebenen in Tiefländern, Tiefland- und Hochlandbecken. More recently Falconer of Glasgow has presented a discussion of 'Land Forms and Landscapes' based on endogenetic and exogenetic processes, in which he classifies over 60 specific forms under various orders, families, genera, and subgenera. So far as I have seen, these schemes have not yet been put to use in the description of actual landscapes or regions, although such a test of their practical value is evidently desirable.

The scheme that I have come to prefer for my own use, after some years of experiment upon it, takes account of the attitude and internal structure of the land mass concerned, of the destructive processes that have worked to carve or erode it, and of the stage reached in the carving or erosion expressed in relation to the long series of stages that make up a completed cycle of erosion, from the form of initial uplift to the form of ultimate degradation. This scheme therefore bears a close resemblance to the scheme proposed by Hettner of Heidelberg, which also has to do with three prime factors: '1. Mit den Tatsachen des inneren Baus, 2. mit den Vorgängen der Umbildung, 3. mit den durch die Einwirkung dieser auf jene sich ergebenden Oberflächenformen und Bodenarten.' I have fallen into the habit of briefly stating these three prime factors as 'structure, process, and stage' ('stage' being preferable to 'time', earlier used), and find them essential in any thorough treatment of land forms. If two other factors are added, namely, relief of surface or local measure of vertical inequality, and texture of dissection or spacing of stream lines, the scheme thus completed becomes fivefold; but the three factors first named suffice as a handy name for it – except for those hurried students who think that they can themselves expand the name of a scheme into its full meaning; and for them even a five fold name would be insufficient. Any elaborate and adaptable scheme, such as this one, cannot be fully understood until it has been carefully studied and practically applied. It is not my purpose to explain the scheme of structure, process, and stage in these pages; that has been done elsewhere, most completely in *Die erklärende Beschreibung der Landformen* (Teubner, Leipzig, 1912), a volume embodying the lectures that I gave as visiting professor at the University of Berlin in 1908–9; my intention here is merely to set forth certain characteristics of the scheme.

The first and most important characteristic is, that by assigning the proper values to the variable elements of the scheme, they will combine so as to produce the counterpart of an actual land form: the determination of the proper values being the work of analysis. Thus the scheme enables those who use it to give intelligible account of physiographic features in terms of their origin. The next and hardly less important characteristic is, that, by playing reasonable variations on the assigned values of the several elements, the counterparts of a great variety of related landforms can be brought to mind. Thus the scheme enables those who use it greatly to extend their mental outfit, in so far as land forms are concerned.

(*Davis, 1915K, pp. 71–4*)

Davis then goes on to extol the elasticity of his scheme which will safely corporate all complications of natural occurrences. In it, every element of structure, every kind of process, and every phase of stage can be elaborated as far as need be. However, he realized fully that

The distraction caused by analytical explanations increases with their complication and their novelty; hence some means must be found of making them simple and familiar. This is best accomplished by a process that may be called systematization. It consists, first, in enlarging and generalizing the explanation, so that instead of being directed immediately to the facts in hand it shall cover all related facts; second, in deducing from the generalized explanation a full series of ideal or type examples; third, in arranging the ideal examples in systematic order from the initial to the ultimate stage, whereby they may all become familiarly understood; and fourth, in giving appropriate names to certain selected members of the type series, so that they may be easily called to mind. If drawings of the named members of the type series are added to supplement their verbal description, so much the better. When all this has been done, the mere name of any member of the series will recall it and the explanation that goes with it; the name can then be used in explanatory description with a minimum of distraction.

Let it be understood that the type forms thus established in a systematic series are not empirical imitations of actual land forms; they are imagined examples of rationally developed land forms, deduced from a general explanation, the validity of which has been previously established by a critical analysis; and herein lies their chief advantage. For if they were merely empirical examples, their inner structure and their past and future stages would be unknown; while as rationally developed forms they are transparent in space and time, they are known through and through, in the past and the future, and all the members of a series are helpfully related to one another. Moreover their elements are not arbitrarily combined, as must be the case in all purely empirical concepts, but reasonably and genetically associated. The separate elements will therefore be easily remembered under the name of the whole which suggests them. Although systematic methods are familiar in many branches of science, and although they are applied to geographical problems to a certain extent in many textbooks of geography, the application is too often empirical in its nature; or if explanatory, it is incomplete. I believe that great progress is yet to be made in the mature and disciplinary training of geographers by the more extended use of thoroughgoing systematic methods of the explanatory kind in all branches of geography. (*Davis, 1915K, pp. 77–8*)

Of course the most extensive recapitulation of the method of explanatory description was given in the German work *Die Erklärende Beschreibung der Landformen*, which resulted from Davis' Berlin lectures of 1908–9 and was published in 1912. As the chapter topics show, this extensive work represents the most complete synthesis of the geomorphic ideas which he held prior to his retirement:

The nature and scope of geography.
The cycle of erosion.
The frequent contradictions between theories and facts.
The deductive scheme of geography.
Landforms on simple structures.
Complex structures.
Volcanic forms.
The arid cycle.
The glacial cycle.
The marine cycle.

It was with the publication of this work that both Davis' reputation in Germany and opposition to his ideas there began to grow. The very title of the work was expressive of Davis' belief in the essential logic, reasonableness and truth of his vision of landform interpretation. Even his opponents must have envied the persuasive arrogance with which he argued his case. It was left to later generations to discover how merciless nature can be to her would-be classifiers.

CHAPTER EIGHTEEN

Propagandist and Pedagogue

Whether Davis was an advocate and teacher rather than a propagandist and pedagogue will never be decided. It may well be that he himself would have preferred the latter appellations.

DAVIS THE PROPAGANDIST

In writing for the cause of geography, both as a discipline and in its more specialized shape of landform description, Davis was indefatigable. He pressed his opinions on all kinds of readers and made full use of the local press. The repetition of his ideas in American, British, French, German, Italian and Japanese books or journals inevitably made him world-famous. He published sizeable articles in about ninety different journals and probably received summary notices in nearly treble that number.

He also constantly urged others to write for the same cause. These direct repercussions are, indeed, of great significance and, if added to the wider ripples of incidental comment and criticism, reach truly formidable dimensions. But typically he never asked others to do what he was unwilling to do himself. As early as 1886 (Y), he published, with N. S. Shaler and T. W. Harris, a series of twenty-five coloured geological models and twenty-five photographs of important geological objects, each accompanied with a letter-press description. Seven years later, in a more geographical vein, he wrote *Suggestions for Teaching Physical Geography: Based on the Physical Features of Southern New England* (Davis, 1893K). During 1896 (C, L and M) and 1897 (L) he published separately, with an explanatory text to each, as an aid to the study of geography in grammar and high schools, the State maps of Connecticut, New York, Rhode Island and Massachusetts. His practical exercises in geography prepared for various journals and booklets from 1900 onwards were extremely popular. In 1902, to assist geography in American schools, he advocated the elaboration of a series of fifty or more specific exercises for the laboratory study of physiography. He worked on this theme for years and eventually in 1908 (D) produced *Practical Exercises in Physical Geography* with a fifty-page atlas. In 1930 (J), when he was eighty years old, this book was published in Japanese.

A great deal of Davis' early writing was on behalf of geographical education but he rarely failed to advocate the importance of physiography. We have already described in Chapter 10 how he was a member of the Committee of Ten which recommended physiography as a subject for the high school curriculum. These students were succoured on Davisian themes. What could be otherwise when his own pamphlets, exercises and books provided the bulk of physiographic education in the United States until 1910 at least? Similarly his summer courses at Harvard for geography teachers were also concerned largely with systematic landform study.

In later years Davis wrote mainly to popularize his own explanatory system of landform analysis, although usually affirming that it could be applied equally well to other branches of geography. His main methods of propaganda continued to be publication, personal discussion, lecturing, and private correspondence. However, he also played a notable part in the actual foundation of clubs, societies and journals in the United States.

In 1902, Isaiah Bowman records joyously

> A new feature of the Department is the Harvard Geological Club, which meets every Friday evening at the house of some Professor. Informal papers, discussions, etc., are given after which refreshments are served. The first meeting was at the house of Prof. Davis. *He entertained.* The membership is about twenty or twenty-five, with some twelve or thirteen Professors and Instructors. I feel, deeply, the honor in becoming a member ... (*Letter: Isaiah Bowman to Mark Jefferson, 9 November 1902*)

A similar attempt to create popular participation in science was the Harvard Travellers' Club, which came into being in 1903 entirely through Davis' instigation.

> The club has done much to develop and maintain an interest in intelligent travel among young men in Boston and Cambridge, but it does not limit its interests to geography and does not claim to be a geographical society. (*Brigham, 1909, p. 50*)

A copy of an introductory speech made by Davis at one of the club's meetings is interesting because it typifies his style of after-dinner address. The guest was Sir Ernest Shackleton who just previously in 1908–9 had led an expedition that reached a point about ninety-seven miles from the south pole.

> In spite of careful research thru the anthology of the South Frigid Zone, only one verse has been discovered at all appropriate to this occasion; that is from the sympathetic hand of Oliver Herford who as you must remember, wrote the fine line:
>
> The pen-guin's mightier than the sword-fish
>
> But he goes on to say:
>
> He told this daily to the bored fish
>
> So after all it hardly suits my purpose; and my remarks must be unornamented.
> Allow me to remind you – I am now, Sir Ernest, addressing the middle aged

members and guests of the club – that the earth is a globe; and also to mention the fact that on a mere globe, there are no points naturally signalized as of interest supreme beyond all others; indeed a Baedeker devoted to the earth as a globe, a mere globe, would find nothing to star. In this respect Mother Earth reminds one of that exemplary lady of middle age, who was described by the Autocrat or the Professor at the Breakfast Table of so uniform a perfection as to offer no point of approach; and as being as unseizable, except in her totality, as a billiard ball. Just so with a mere globe. How fortunate then, for explorers, that the earth long ago formed the habit of turning round on its axis – what ever that is – and thus designating two points as pre-eminently interesting and two points only; which by very reason of their scantiness and remoteness make the bulging and accessible equator seem redundant and commonplace.

We welcome today an explorer who has been so near the South Pole that there is very little glory left for anyone who gets nearer. Therefore I beg leave to ask him if the earth is really, as far as he could make out, flattened at the poles. I used to think it was, because the assertion to that effect was so often repeated, apparently on good authority. But lately we have been hearing so much about pressure ridges near one pole and a 10,000 foot plateau around the other that I am beginning to think that 'flattened at the poles' is not a good description of the earth; and I am losing faith in the propriety of that old conundrum based on a supposed likeness between the earth and a defeated political candidate! At first thought you might think these two points would closely resemble each other; but in reality there is a vast difference between them; in this respect they remind me of the remark of that young girl about her sister Sal. She said; 'Me and Sister Sal arn't *no* more alike than if it twarnt us; she's just as different as I be, only the other way'. Just so, the North Pole is all water, while the South Pole is all land. Then there's another thing, we can all understand how it is an explorer sets to the north pole, just by following a meridian line, and climbing up over the curve of the earth to the very top, and just before he begins to go down again, there's the North Pole; but as to the *other pole*, I must confess to sympathy with a certain small boy. His father was explaining things to him with the aid of globe, and showed him the north pole up here, and the artificial horizon round the middle and the south pole down underneath, and told him that you, Sir Ernest had almost reached the South Pole; and the small boy, looked underneath and said: 'But papa; when he went down below the horizon, if he couldn't reach the pole, what *did* he hold on to?' I hope Sir Ernest you will make that clear in your lecture tonight, for it seems however to me that there was a mighty fine lot of holding on in that dash of yours. There are indeed two features of that extraordinary journey destined to stand out prominently in the annals of exploration; one was your wonderfully well balanced combination of impetuous daring and calm judgment; the daring that led you against fierce head winds, in benumbing cold and craving hunger, far across that icy desert, until you established, as has been well said, a *new record of human endurance*; and then, with an open prospect ahead of you, the goal almost in sight, your judgment turned you back; and not a day too soon, for tho you reached the coast in safety the difficulties of the return were terrible indeed. Brave leadership – wise leadership – all men admire and applaud – The other event was near the end of that return march, when one of the party, overcome, exhausted by illness, had to be left behind with a companion on the barren comfortless snow-field, while you, with the fourth member of the party, hurried forward by forced marches, reached the coast, signaled the waiting vessel, and then reenforced, without waiting for rest, returned at once, found the men on the snowfield and brought all hands out in safety. A gallant record this, after more than four months and over 1700 miles of the hardest kind of going; a brave rescue, a demanding dauntless courage and immense strength of body supported by true loyalty of heart. Body and will must be of the very finest temper to carry thru such a *tour de force*; Health must be yours already; your longest continued health I ask President Lowell to propose. (*After-dinner speech by Davis in honour of Sir Ernest Shackleton*)

The apparently provincial nature of these clubs must be seen against the fact that at the turn of the nineteenth century Harvard played a part never equalled since in the history of higher education in the United States. However, Davis' influence was destined to endure less through Harvard than through his creation, direct or indirect, of two professional bodies.

It was entirely through his efforts that the Association of American Geographers was founded in 1904, Davis told Ellsworth Huntington, who was then in the desert basins of Persia, of the project.

The preparations of the Eighth International Geographic Congress to be held in Washington next September, are taking a great deal of thought and time. I am a

F I G. 81. Davis on a Harvard field trip to the Catskills (*c.* 1910) (Courtesy Harvard College Library)

member of the Committee on Arrangements and Chairman of the Committee on Scientific Program, and in both capacities have much writing to do. Further, I am planning to organize at the time of the Congress an American Geographers' Club or Association, in which membership shall be limited to mature geographical experts. The plan is meeting with warm interest from just those whom I should wish to have take membership in the Club. (*Letter: Davis to Ellsworth Huntington, 17 February 1904; Cambridge, Mass.*)

Over twenty years later Huntington asked Davis:

I wonder whether you remember any more clearly than I do the day when you gathered some of us at your house, and suggested the formation of what later became the Association of American Geographers. I believe that the Association and the Harvard Travellers Club will last for generations as evidence of your inspiring leadership in American geography. (*Letter: Ellsworth Huntington to Davis, 5 February 1925; New Haven, Conn.*)

Previous to the creation of this geographical society, American geographers met professionally at meetings of the Geological Society or within the general concourse of the Association for the Advancement of Science or the general assemblies of the American Geographical Society at New York. The last named had been established in 1852 and was always a great aid to Davis. It was an august body of scholars of many disciplines, explorers and notable citizens. Davis wished for more professionalism, for a more purely professional geographical approach by academic geographers only. He considered that such an independent professional organization was more likely to raise geography into a discipline in its own right and more quickly rescue it from subservience, or an image of subservience, to geology and other sciences. The new organization was to be the vehicle for exchanging views, ventilating problems, publicizing the results of research and advertising the importance of geography to the government and education.

Within five years the American Association of Geographers was already beginning to be influential:

First let me thank you for your postal sending best wishes to the Association, meeting at Baltimore, also for copy of your address. My early departure for that place prevented me from getting it until my return.

Lest you might not hear promptly of our meeting I take this opportunity to say that it was very successful indeed. We had forty-three papers which averaged high. Brigham shows by the figures that we are getting each year a smaller proportion of pure physiography, or, in other words, of papers which *might* be read before the Geological Society, thereby showing more positively each year that there is not only a right but a need for our separate existence which fortunately no one any longer questions. We got a great deal of recognition this year from the G.S.A.

Penck gave a good evening lecture on 'Men, Climate and Soil'. It was not a startling event but was entirely creditable and called much attention to the A.A.G.

(*Letter: N. M. Fenneman to Davis in Berlin, 5 January 1909*)

We cannot resist interpolating here a few words on one of Davis' richest and most individualistic students, Robert LeMoyne Barrett who, having at his first sampling found Harvard 'too stuffy', quitted to live with Indians in the Rockies. He returned in 1894–8 and took his A.B. in science mainly under Davis. 'Gypsy' Barrett was an alfresco addict, a 'naturalist and globe trotter' who preferred Patagonia and mountain tops to stately mansions. He published a good deal but earned lasting fame by being in Davis' living room at Cambridge when the maestro suggested the formation of the American Geographers' Association. Barrett outlived all the other charter members and was prematurely reported dead several times. In 1912 the Harvard magazine announced his decease; in 1962 the *Professional Geographer* (Vol. 14, No. 3, p. 33) gave him an obituary notice lamenting his passing away in 'about 1960 aged about 89'. In fact Barrett happily survived to within two months of his ninety-eighth birthday and died in La Crescenta, California, on 5 March 1969 (Martin, 1972).

It is not surprising that Davis was also closely connected with the publication of the *Journal of Geography*. Its object was to popularize the subject by presenting new developments in a discursive form that would be understood by schoolteachers and even by lay readers. Technical articles are normally scattered throughout a large number of specialist publications with only a limited circulation and even the research worker finds it hard to locate them all. Then assimilation is no mean task. The average reader is not prepared to do either. He has the capacity to be interested but expects the bird to be caught, plucked and prepared for the table. This is what the *Journal* set out to do from 1902 onwards, with the help of at least fifteen sizeable articles by Davis.

On Davis' advice, the *Journal of Geography* (originally the *Journal of School Geography*) was founded by Dodge, one of his former students, now professor in Columbia University: the *Journal* has rendered large service to teachers. The publication of a series of 'Geographic Monographs' under the auspices of the National Geographic Society was undertaken in 1895 at Davis' suggestion, but was not carried out in accordance with his plans. He has been for some years associate editor of *Science*, the *American Journal of Science* and the *American Naturalist*. (*Brigham, 1909, p. 50*)

That he was more than associate editor may be judged from his contributions to the journals concerned. Between 1881 and 1932 he wrote about a hundred separate items for *Science* and at least twenty-five for the *American Journal of Science*.

That he took a vigorous and usually leading part in all meeetings he attended is almost equally true. He had a habit at gatherings of the A.A.G.

of rising from his seat after nearly every paper and saying 'If I had written that paper I would have put the matter this way etc. etc.'. Everyone recognized

Davis was an outstanding scientist who usually was correct and tolerated his ego. (*Eugene Van Cleef, in an address at Ann Arbor, Michigan, 13 August 1969*)

This professional consortium elected Davis their president on no less than three occasions. The Association proved an enormous success and became centred on the University of Chicago which gradually took over the Academic leadership of professional American geography on its eclipse at Harvard. The leading Chicago geologists and geographers happened by chance to be the chief opponents in the United States of Davis and his systems. For geography the A.A.G. was a constant stimulus; for Davis, it provided a theatre for reappearances as its prima donna and a vehicle for his finest articles. But, we repeat, he had created a society where his landform schemes were to face severe criticism and were, almost inevitably, destined to become a dwindling part of professional geography.

There is in this something ironic, a dilemma which Davis never really considered. To promote landform studies, he took geomorphology out of the clutches of geology and then within a few decades raised physiography to be the dominant branch of geographical teaching in the United States. It became there the logical, progressive branch of geography as a whole. So Davis, a geologist in his earlier years, turned against geology as such in his maturity. The man who once despised some innocent hotel-keeper for not differentiating between granite and other metamorphic or igneous rocks later spent a great deal of time showing the irrelevancy of most geological details to geographical studies. But he did not easily dispel his former allegiance.

Every geologist recognizes Prof. Davis as a geologist who has lent himself to the geographers for their benefit, and who looks at things essentially from a geological point of view . . . (*Mr Lamplugh: In Davis, 1909G, p. 322*)

His finest plea is in the 'Principles of Geographical Description' of 1915 (K), where among the examples are descriptions of the Valdarno both in the present and in the past tense.

EXCLUSION OF IRRELEVANT GEOLOGICAL MATTER. – The danger of distraction by the introduction of irrelevant complications into explanatory descriptions has already been mentioned. Such distraction is nowhere more common than in physiographic descriptions in terms of structure, process, and stage; for in statements concerning structure and process the author may be tempted to introduce an excess of geological matter, which gives the printed page truly an erudite appearance, but which too often contributes nothing to the better understanding of the described landscape. During the preparation of the description it may very likely be necessary to read many geological articles in order to learn what ought to be known about the structure of the district; but it is not necessary to publish the geological information thus gained, except in so far as it is directly helpful in geographical work. It is a sound principle of geographical description to omit all geological matter, however important it may be in some other relation, if it does not aid in picturing existing features. The use, for example, of the names of geological formations, such as Carboniferous, Triassic, or Tertiary, should be avoided, because these names tell only the date in the remote past when the strata concerned were laid down; they should be replaced by phrases descriptive of composition, attitude, and appearance, such as folded resistant blue limestones; inclined hard and soft red sandstones; or weak horizontal gray clays. An exception to this rule may be made if the geological name of a formation has entered so far into scientific usage as to have a geographical as well as a chronological meaning; but such exceptions must be rare, for it can seldom happen that the names proposed by geologists or paleontologists, to suit their needs, will serve a geographer's needs so well as names that he himself proposed. In descriptions of the Valdarno, presented above, the use of the term Pliocene in the first gives no aid in picturing the landscape; hence it is omitted from the second; but the phrase, 'series of imperfectly consolidated, horizontal gravels, sands, and silts', is retained, because the structure thus indicated is evidently helpful in understanding and therefore in conceiving the existing land forms. In so far as 'Pliocene' might implicitly aid a geographer by indicating the relatively short subsequent time that the deposits concerned have been exposed to erosion, that aid is much better given by the explicit statement 'maturely dissected series'. On the other hand, nothing is said in the present-tense description about deposition by torrents or in lakes, because all that a geographer needs to know of the deposits thus produced is implied by the announcement of beds of gravels, sands, and silts in a warped-valley, intermont basin. Whether the final sentence regarding fossils should be retained in the present-tense description is a matter of taste; but certain it is that the occurrence of fossil vertebrates has brought a good number of scientific pilgrims to the Valdarno, while the occurrence of fossilized plants in the form of lignite beds has sufficed to determine a small mining industry near the middle of the southwestern side of the basin; if the description were extended to include economic factors, such an industry would of course be mentioned.

It is only by attending scrupulously to details of this kind that a strong geographical discipline can be developed and maintained. In case an author is writing an article that is at once geographical and geological, there is naturally enough no reason for excluding geological matters; but in that case it is likely that geology, the better-developed science, will take the greater share of attention of both the writer and the reader, and geography, the less-developed science, will come in only as a poor relation, thankful to get whatever remnant of attention may be left over for it. (*Davis, 1915K, pp. 90–1*)

The wisdom of separating geography from certain aspects and details of geology can hardly be doubted. But in raising physiography to a dominant role in American geography, Davis constantly advanced the cause of geography as a whole and insisted that all its human branches would hinge on and benefit from the same explanatory treatment being advocated for landforms. Thus, almost unwittingly, he became a prime mover in the ultimate relative decline of physiography in geographical education.

DAVIS THE PEDAGOGUE

Davis was certainly not a 'born teacher' and few people can have made such rapid progress as he did in the art of teaching. By 1909 he was being hailed as a 'master' and a 'magician' in lecturing techniques. Yet the following letter to Ellsworth Huntington shows what an uncertain start Davis made and, in its content, demonstrates vividly that he had acquired some educational principles of lasting value.

Your letter of Oct. 28 was very welcome; I have been wishing to know how you are getting along down there, 'teaching the sons of Eli geography'.

Hard luck to have a disorderly section to begin with. But most of us have had that sort of thing at one time or another. Do not forget that every time they make you visibly angry or confused or disturbed, they have gained a victory. The great thing is to hide your feelings; grin at them in class; then the victory is yours. Also, do not attempt any reproofs before the class, never mind what they do; but fetch up the offender alone; go to his room, or call him to yours; or see him alone in the laboratory; then state the case frankly; show him how hard it makes your work; how little you can accomplish, and appeal to his sense of decency; ask if you cannot trust him in future to act on the basis of being a gentleman. If after such an appeal he continues to give trouble, he ought to be fired out of the class, double quick. I do not know what the Yale practice is in that respect; or whether you will have the support of the authorities in such action; that will of course depend largely on the confidence they feel in your judgment. Best find out what authority you have before trying to use it. But the chief thing is, to make the men see that they cannot faze you; that they MUST govern themselves; that if they do not they MUST leave the class. Warning is probably fair; tho I think it ridiculous that young men, presumably thinking themselves gentlemen, should expect to be warned before being summarily turned out for ungentlemanly behaviour. It only shows what an absurd tradition still obscures the real relation of teacher and student. It also shows what a lot of queer sticks there must have been, for generations, among the teachers; otherwise, why allow disorder; or why submit to it?

Thirty years ago I should not have written in this way. I had some very hard times then; disorder, no interest aroused, and all that sort of thing. My place here was very shaky; Shaler told me so directly once. But I stuck to it, and 'won out' in the end. This is for your encouragement, for I expect you to do the same thing; and in quicker order than I did. But the main thing is, not to let the boys gain victories; just grin at them and show that they cannot faze you. At the same time, you must work awfully hard, to make your lectures first-class; clear, sharp, interesting; never explaining things that the boys know beforehand; always explaining clearly things that they do not know; bringing in good illustrations; telling appropriate stories, adventures, items of all kinds; do not be too logical; yet avoid irrelevance; and above all things avoid reciting a piece; don't tell the same story in successive sections. I could fill pages with these truisms; but I dare say you will know them, unwritten. (*Letter: Davis to Ellsworth Huntington, 2 November 1907; Cambridge, Mass.*)

In another letter discussing Huntington's early teaching difficulties at Yale, Davis confirms his own uncertain start.

When I recall my own wretched beginning, with embarrassment, timidity, inability to talk before the class or to control the students, I am disposed to place little weight on similar difficulties in others . . . (*Letter: Davis to J. S. Keltie, 3 January 1908; Cambridge, Mass.*)

As a lecturer, Davis was always praiseworthily thorough in the preparation of his text, and if speaking in German or French had the matter and expression revised by natives. His delivery was clear and strong rather than rhetorical and he kept closely to his written text, but in debate and discussion he spoke easily, using the same phrases as occur throughout his writings. Consequently the same powers of persuasion and of sequential arrangement occur in each. He sized up audiences before he met them, an ability which says much for his general geographical knowledge and his careful investigation among friends and colleagues of the social qualities of the various audiences.

Occasionally he under-rated the technical knowledge of an audience, as was, for example, the case of his talk on the glaciation of Snowdonia to the London Geological Society.

Again I recall in his Presidential address before the Geological Society of America that he lectured the assembled geologists as a lot of incompetents or not much better. (*Letter: Richard S. Eustis to R. J. Chorley, 8 January 1962; Hancock, New Hampshire*)

But this misappraisal rarely happened on methodological topics in which the English-speaking world was sadly backward. In some lectures he made full use of the projecting lantern and had a large collection of slides which not unnaturally he always considered imperfect. He made constant and skilled use of the blackboard and sometimes with elaborate subjects prepared an outline for guidance before the lecture began. As Dr H. R. Mill said at the London Royal Geographical Society in March 1909: 'Professor Davis has a magical power of putting his theory before us. These diagrams of his are a real revelation in blackboard work'.

His practical tuition in the laboratory was exemplary and has rarely been equalled. He advocated strongly for special geographical rooms or laboratories at all educational levels. In high schools, as early as 1902, he suggested the setting aside of a special room for geographical laboratory work, and provided suitable instruction in a summer course for teachers at Harvard.

Among the elements most needed are wall maps . . . good pictures and maps of the actual examples by which type forms are illustrated, models of land forms, lantern slides in large variety, well-selected series of weather maps, plentiful large-scale topographical maps such as are published by our various governmental bureaus, and so on. (*Davis, 1902C and 1909C, pp. 57–8*)

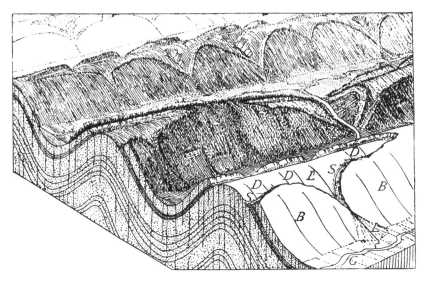

F I G. 82. The drainage of the Jura: young folded mountains. The ridges are composed
of both hard (B and D) and soft (S) strata. Drainage is by longitudinal rivers (G), fed
by short lateral streams (L), the latter occasionally cutting water gaps through the
folds (From Davis, 1898E, Fig. 105)

At universities, laboratory work was important because it gave an opportunity for deliberate and close observation of geographical facts. The exercises
were carefully graded and observation and description preceded the introduction of explanatory terms. The written results were subjected to thorough
criticism, at which Davis was a recognized master and soon acquired a vast experience. His unflagging industry, high standards and abhorrence of anything
slipshod were never concealed and he was rightly considered a severe critic
who spared no one who did not measure up to his standards.

Among the most marked characteristics of Davis' work are precision and
thoroughness. It is rather spontaneously, in the native habit of his mind, than by
any exercises of will, that these qualities characterize his attitude toward all
problems and all projects. What he finds thus in his own spirit and method he is
inclined to expect in others; hence he welcomes independence and the mastery of
facts and principles betokens a serious consideration ... His energetic action,
tempered with reserve and marked by unvarying poise, without haste, without
waste, suggests traits and temperament drawn from a Quaker ancestry. (*Brigham,
1909, pp. 62 and 64*)

Brigham knew Davis well and admired him greatly, but his judgment was
sound in most respects and he does reflect friendly contemporary opinion.

However, it is important to get other opinions. A personal letter from Douglas Johnson gives us a clearer picture of Davis' relations with his students. Johnson at least did not resent his tutor's criticisms, although time may have diluted somewhat their astringency.

If my conscious effort to apply rigorous analysis to shoreline problems is what brought me the award, and judging from de Martonne's review of the book [*Shore Processes and Shore-Line Development (1919)*] and other comments from French sources I judge this to be the case, then my indebtedness to you is all the more clear. I have always felt that no one of the teachers with whom it was my fortune to be associated did so much for me in the way of development of correct methods of investigation and exposition as did you. The one year of graduate work with you at Harvard, taken after I had received my Doctor's degree, was of inestimable value to me. The benefit of your criticisms were sometimes sharp, but they were always pertinent and searching; and I never submitted to you a piece of work but that you left it better than you found it. I am quite sure that I do not myself realize the full extent to which I am indebted to your training and your help; but I at least am fully conscious of the fact that the indebtedness is very great.

I would also like you to realize how much my association with you meant to me in a personal way. However vigorous your criticisms might be I always felt that in you I had a loyal friend who had a real, vital interest in my success. This in itself has been an inspiration at times when circumstances were profoundly depressing. That you had faith in me and in my abilities, I felt sure; and the assurance was worth more than I can ever tell you. Even in matters of personal criticism, when you called my attention to rough points which needed smoothing off, to peculiarities and mannerisms which affected unfavourably my usefulness, I always felt that you looked beyond these defects and believed you saw something worth while within me; and that your object was never to find fault for the sake of finding fault, but to give what was worth while in me a better chance to show itself. Some of these defects I think I have measurably improved. Others doubtless less so than you hoped, and certainly less than I have wished. But both in scientific achievement and in personal development I can not ignore the high debt I owe to you. And I have wanted you to realize the depth and sincerity of my appreciation.

I fear I have expressed it all most awkwardly; but perhaps you can read between the lines and discover what is in the mind and heart.

(*Letter: D. W. Johnson to Davis, 7 April 1921; Columbia University, New York*)

There is a fuller letter from Isaiah Bowman, whose praise of Davis is even more fervent and, judging from short side references, very genuine.

The announcement [of Davis' retirement] brought an unconscious review of my acquaintance with you and so this letter.

When I went to Harvard in 1903 I was loaded with enthusiasm but except in your work found in absolutely everything I took there scarcely more than one feeble sign of that something I had been led to believe hung around the intellectual home of Agassiz. One man made me redraw a Globigerina four times foolishly supposing that he was increasing my mental power. I was bitterly disappointed to find it was skill in drawing not in observation that he was after. Another drearily recited facts, likewise a third. I could not even attend the classes of a fourth. All of them had knowledge but alas! that is *not* always power. Yet these men were all very friendly; it is distinctly not their personal attitude that I am criticising. Shaler's striking personality appealed to me strongly and I count it the rarest good fortune to have been a student in his classes, though he was a source of inspiration rather than of mental power.

But in your teaching there was something which made each return to the classroom an intellectual joy, it was always certain that there would be something new and that even the old would be done in a new way. I envied your skill in original thinking and in the two years I was at Harvard I fairly well succeeded in getting close to your intellectual processes. I learned the value of real thinking and how to do it. My time has been so filled with teaching however that I am your poorest exemplar. Fortunately, my promotion this year to an Assistant Professorship of the 1st Grade carries with it much more time for research.

There are times when I feel so strongly dependent upon my training under you for results that I fear to become a replica, a 'we-too', and imitation is so cheap! Intellectual bondage is not a good thing. My palliative is a scrap. So once in a while I get impertinent and even 'sassy' to my mental creditors and refuse to pay the rent! I always assign large chunks of 'Davis' in my physiography! One day a senior came in, dropped a pile of pamphlets with a bang, and said bluntly 'I *hate* Davis' and to my astonished 'Why' replied, 'Because when you get through reading him there's nothing more to say!'. In a few cases it has turned near against you – men who cannot even say 'Yes, you are right' to an intellectual opponent.

Of your special kindnesses to me – and they are 'as the sands of the sea in number' – I wish to say nothing now: there will be a more appropriate occasion. I am however as keenly mindful of them as if I were in Cambridge now and it were 1903. You have created opportunities for me that have meant everything in my career. The original copy of your Physical Geography which I bought in 1901 is so filled with notes and invitations, etc. that I shall have to interleave it. It would do for an auto-biography. I want you to know that I appreciated those things, and that they are still among the warmest recollections of my two years at Harvard. They mean much more to a fellow who learned geometry principles from an old sea captain in a corner of a wretched hut and who did without an overcoat for one winter in order to buy apparatus for doing chemical experiments by himself than they do to a sleek lad from Andover or Hill.

Sometimes there seems to be something semi-religious in nature in my attitude toward your work. Part of it grows out of just plain personal regard for you and your results, part of it is related to the fact that your results, more nearly than those of any other man I know, seem to me to be real truth; the kind of truth that lasts after all the critics have had their fling – like the work of Huxley and Darwin and Newton.

I have chosen to say these things fully because I am not demonstrative, because I have never spoken like this before and it seems worth while once at least to uncork my vial of affection for the benefit of a man who has done so much for me and whom I expect to outlive. The form of expression is as cogent as McGee! But

I choose to say it so rather than in the conventional 'I enjoyed your work so much, don't you know'. (*Letter: Isaiah Bowman to Davis, 7 August 1912: New Haven, Conn.*)

Davis' concern for Harvard graduate students is shown by the following letter.

My dear Ellsworth,
  Yesterday must have been a severe strain on your share of the affectionate regard that has grown up between us. To me it was a horrid shock – and it is still a torment – It has made me feel as I never did before the difference between individual and departmental action. I had anticipated a great pleasure in this natural culmination of our work together – I cannot yet understand, much less accept the ground of the others who placed so much value on elementary matters (which to be sure have their proper place and time) as compared to proved capacity for large accomplishment in the world's work. I could hardly talk to you connectedly yesterday, for anger and grief – and it seemed pretty hard that I, who had advocated and defended your case, should have had the task of explaining how you had not satisfied others. Doubtless I made a mess of it.
  Well, you must live it all down, and to that let us now turn – I am thinking, as you probably are too, what Gregory may think when he comes to think of our action. To be sure it is nobodys business to tell him anything, but nevertheless the wretched thing may reach him – and I am wondering whether you had better not write him of it yourself. If you do, that will give me a chance to write also as I should like to do. Beyond this I do not wish to express myself at present. Indeed perhaps I have said too much already – but it is a relief to write you. I have been reading and writing Italian all day for an anesthetic.
  Go on with your work. I will shortly send you my notes on your climatic essay.
  (*Letter: Davis to Ellsworth Huntington, 26 May 1907; Cambridge, Mass.*)

The background to this letter has been explained by Geoffrey Martin.

Huntington failed his examination for the Ph.D. on May 25, 1907, before seven faculty members from the Division of Geology. The vote was four to two against Huntington, with one abstention. The examination committee included W. M. Davis, D. W. Johnson, R. DeC. Ward, R. T. Jackson, J. E. Wolff, J. B. Woodworth and Charles Palache. From careful work around the subject, I would venture that Woodworth supported Davis and Johnson abstained. Ward led the opposition. I do not believe that battle lines were drawn around persons, so much as beliefs. Ward was conservative in his approach to climate; Huntington loved the speedy dash into the unknown in quest of large vision . . . Ward's climate derived from statistics on paper. Huntington felt climate with his body . . . There is a curious sequel to this episode. On February 8, 1909, under the tutelage of H. E. Gregory, Huntington made formal application for the degree to Yale University . . . and a doctorate was conferred on him in June 1909. Isaiah Bowman wrote a lengthy dissertation to have the same degree conferred upon him at the same ceremony! (*Letter: G. J. Martin to R. J. Chorley, 5 July 1972; New Haven, Connecticut. See also Chapters 4 and 5 in Martin, forthcoming*)

Other memories of Davis are more critical. Carl Sauer knew him, 'though not too well', as a lecturer at Harvard and later in the west.

> Davis – many Welsh characteristics. Authoritarian with his students. Mellowed *greatly* after retirement. One of the most influential aspects of his work was his ability to draw block diagrams. Was always very active at meetings – discussing and organizing. (*Notes from interview with C. O. Sauer taken by R. J. Chorley at Berkeley, California, 30 August 1962*)

Sauer was not trained in the Davis tradition and some of his criticism is a decisive rejection of much that Davis taught. This is understandable as Salisbury continually admonished his students to do just this. Even then Sauer admits to Davis' impressive qualities as a lecturer and debator.

> I was a student of Salisbury at Chicago where Davis was not taken too seriously, though he was not held in disrespect. The two were of very different temperaments and physique and they were likely to meet in combat at the annual meetings, an event always looked forward to. Salisbury big and massive, direct and laconic of speech, never lecturing but the best master of Socratic discourse I have known. Davis almost petite, prim, nimble in movement and discourse and impressive in extemporaneous lecture and debate. I think we derived from Salisbury a distrust of systems, I'd like to say an open horizon; at least that is what he tried for. Davis trained disciples and he trained them well, but it seems to me that they did not leave the path the master had laid out.
>
> My impression of Davis while I was still in the eastern part of the country was that he was a sharp and sometimes merciless critic of work that departed from his formulations and that young men who were not his followers were turned away from the study of land forms. He had authority and dialectic skill and he would cut up a youngster who had neither. I felt sorry for them and I think physical geography in this country became poorer as Davis exercised his censure and approval.
>
> Salisbury came to see me repeatedly when I was doing my first job of field work. His comments were queries 'what do you make of this?', 'have you taken this into account?'. I was never in the field with Davis until after he moved to California when I took him out into areas new to him. At any stop he expounded the scene rather than ask questions, which, as I recall the occasion, he was inclined to brush aside.
>
> He was a visiting professor here in Geology at the same time that we had Penck in Geography. Davis remarked to me what a strange situation that was, the Geologist in Geography and the Geography over in Geology. Perhaps that is why I never warmed up to him for I am an historian at bottom and Davis was not. He didn't want you to use geologic time terms but only stages of cycles, not historical and not really recurrent series but recurrence. Nor did he like time scales. He built models as he thought the thing should be and thus commanded it to be.
> (*Letter: C. O. Sauer to R. J. Chorley, 18 October 1961: Berkeley, California*)

Sauer suggests that Davis' derisive treatment of any student, who might dare to question his ideas, prevented original thought and turned away candidates.

This is perhaps misleading. Preston James also worked under both Salisbury and Davis and though he supports much of Sauer's criticism he does make it clear that there was no real difference in their methodology. Salisbury's application of it might bear a different emphasis but basically the method was Davis'.

> Davis was a strict task-master – so much so that I suppose he actually turned a number of good people away from the field. Salisbury was also a task-master, but he handled his seminars so that one came away with a feeling of excitement. I always thought Davis was exciting too; but also I should confess that he never had a chance to take me to pieces directly, because I took my work at Harvard after he had left there. He came back for visits, and held a seminar one year, but he was then at Pasadena, California.
>
> On the other hand Davis was very fond of his students, and in the field he was absolutely marvelous. In 1933 I visited Davis at Pasadena, where he was then doing field work on marine terraces. He was wonderfully cordial, and we had a long talk about the status of geography. Later I sent him a paper I did on the surface features of Brazil, and he sent me valuable comments on it.
>
> Actually most of the people who attack Davis do so because they say he was over-simple, and inflexible in that simplicity. He was anything but over-simple or inflexible. If he could have gone into the field in Brazil I would bet that he himself would have come up with a new and more embracing theory. He recognized perfectly well that his structure, process and stage format was a simplification, and that actual landforms were always more complex. I think he would have revised some of his peneplains in the Appalachians, for he really had a keen mind and it was keen right up to his death.
>
> My own feeling about Davis is (1) he was a great intellect; and (2) it might be that he so dominated geography in America (and to a certain extent also in Europe) that the development of the field was retarded. No one dared to propose a new system for the treatment of landforms. Geographers were followers of Davis, or they were out of the field. Salisbury, who actually taught most of the geographers of the 1920's, taught Davis' system, as also did Atwood, and D. W. Johnson. There were no alternatives. This is probably not good for a field of study – to be so dominated by one mind, however great that mind might have been.
>
> (*Letter: Preston E. James to R. J. Chorley, 11 January 1962; Syracuse, N.Y.*)

It is interesting that in the second edition of Chamberlin and Salisbury's three-volume *Geology* (1909), some twelve pages were devoted to 'A cycle of erosion: its stages', yet on Salisbury's *Physiography* (1924) only 57 lines appeared on 'stages in the history of a valley', 'cycles of erosion' and 'peneplains'.

To experience the possible advantages of a different approach a student had to train within the German school

> In 1913, while I still was a graduate student, I took my bride to study a semester with Hettner at Heidelberg in Germany, financed by the profits of an incredible two years' field job in Northern Patagonia with Bailey Willis. During the two years in the field in Patagonia I had failed miserably at trying to apply Davis' ideas

on geomorphology. With Hettner at Heidelberg, and Philippson from Bonn on a two weeks' field trip in the Alps, I found out why I had failed, or at least part of the reason. Davis' ideas bore little relation to the facts of life, or more accurately, to landforms. While I was in Germany, Davis' ideas were sweeping over Germany as they had over USA like wildfire, but Hettner and Philippson refused to be engulfed in the conflagration. The result to me was that thereafter I looked at the Davis system with such a fishy eye that I never used it. Nevertheless, I always was and still am a great admirer of Davis for other of his ideas, not the least of which was heckling presenters of papers at AAG meetings. (*W. D. Jones, in Martin, 1950, p. 179*)

It is fair to point out, if the reader has not already noticed it, that Preston James only knew Davis in his later, mellower, years, and some of his comments must be hearsay. Most impressions are really a mixture of good and bad, with generally rather more to Davis' credit. No appreciation is an utter condemnation. Russell Smith portrays Davis' sharp side.

Davis' mind worked like a chisel cutting marble to make a statue. His sharpness and insight and preciseness were interesting to hear.

I recall an instance in one of the meetings of the Association when a member of the staff of the United States Geological Survey got up to make a little address, and he referred to something that had a 'typical XXX' . . . I don't remember what it was . . . some typical landscape that had a standard physiographic name. When the man was through, Davis got up and said that he noticed that the gentleman from Washington referred to a typical XXX, and Davis went on to state that he would like to know just what characterized a typical XXX.

The poor fellow from Washington blushed and stumbled and hadn't anything to say.

When the meeting was over, I spoke to Davis and told him I was interested to see him nail the skin of the Washington geologist on the barn door. 'Well,' said Davis, 'I wanted to *know*.' (*Letter: J. Russell Smith to R. J. Chorley, 5 January 1962; Swarthmore, Pa.*)

This acid approach was certainly a characteristic which periodically manifested itself; malice was not the motive. His mind was geared to such a fine calibration that even minor intellectual sloppiness would jar him where the average mind would not notice it. This fashion of precision in all he did was reflected in his use of simple explanations in ordinary language whenever possible. Davis' lectures were very easy to follow for that reason.

While he was near Columbia he was announced to give a lecture in the Geography Department of the School of Pure Science on the Geology of the Grand Canyon, the lecture open to the public.

I went. I was particularly anxious to see how Davis would handle his vocabulary on such an occasion. Believe it or not, any freshman entering one of your courses at Cambridge or anyone graduating from an American high school and with reasonably decent marks, would have had no difficulty following Davis completely. I think he did not use a single technical term they would not have understood. It was a marvel of clear, lucid, popular exposition by one expert in jargonese. (*Letter: J. Russell Smith to R. J. Chorley, 5 January 1962; Swarthmore, Pa.*)

Davis' behaviour towards his students and the low popularity of his classes seemed to have become a subject of campus gossip.

> I think I have heard men who had been with Davis say that they have seen grown men go out of his class in tears because he had handled them so roughly in their attempt to talk to him about physiography and geology.
>
> Now this is third hand information. I have been told by a man who did not go to him as a student that Davis had such small classes – almost none – that he was retired before his time by the university so that they could get somebody in that would teach classes. (*Letter: J. Russell Smith to R. J. Chorley, 5 January 1962; Swarthmore, Pa.*)

Other failings which provoke criticism were probably inherited family traits. The forthright expression of Lucretia Mott was famous, both within and outside the family circle, and apparently a similar directness was one of Davis' characteristics. While sometimes excusable, there is always a danger that forthrightness may be considered as deliberate rudeness.

> In brief, Davis was brusque. Thus I sent him a photograph I made in Death Valley of an 'hour glass valley', one of his brilliant incidental interpretations, and in reply received a postcard saying he had a better one he had himself taken. He apparently conducted much of his correspondence by postcard. On the other hand he sent me a postcard, unsolicited, commending me highly for preparing for publication (*Zeitschrift für Gletscherkunde*) post mortem, the account of Tarr's experiments on ice (with which I had been associated) both as to content and praiseworthy as a tribute to my chief. (*Letter: O. D. Von Engeln to R. J. Chorley, 8 January 1962; Ithaca, N.Y.*)

The recollections of another student of Davis confirm many of the points made by the writers quoted earlier.

> The best I can do for you is to give some general impressions, which I think would be shared by most of the men who were among his advanced students in the years prior to 1904.
>
> William Morris Davis was a great and skilful teacher, an exacting task master, and a most helpful critic. His only measure of effort was excellence and his standard of performance was perfection. So far as was humanly possible he imposed those tasks on himself, as well as expecting them from others.
>
> It was not always easy to face the fire of his searching comments and questions on one's place of special study, but the aftermath was generally a feeling of full gratitude for the lasting benefit to one's mental growth. (*Letter: W. S. Tower to R. J. Chorley, 8 April 1962*)

Tower's answers to particular questions contain nothing that marks Davis out as being unusual in the way he conducted his department. All agree that he demanded very high standards but nowhere is there any suggestion that he was eccentric.

Davis, so far as I know, did not assign, or choose, research topics for his students, but his veto power was skilfully used to obtain desirable results.

His supervision of treatment of research topics, as I saw it, was through close and detailed criticisms. Criticism from him was sometimes a trifle caustic, although correct.

On the question of being broad minded, where other's views were contrary to his, I think you will get both 'yes' and 'no' answers. My own would be 'no' in general. (*Letter: W. S. Tower to R. J. Chorley, 15 September 1962*)

Lawrence Martin knew Davis well and most of his memories are complimentary. Even so he cannot resist poking a little fun his old professor.

No college professor ever taught me so much or did me so much good as William Morris Davis did. He pulled me up by the roots, pruned me, fertilized me, and set me out again in the garden of geography. At Harvard in 1905–06 Davis gave only one course in the first semester, and Martin was the sole student in it. The course was called 'Advanced Physiography'. Davis had been to South Africa during the preceding summer, attending an international geological congress, and arrived in Cambridge late. I had been to Alaska studying glaciers and glaciation; and then, as suggested by Davis the previous spring, and, financed by a surplus from the Sturgis–Hooper fund, had stopped for a short period along the Front Range of the Rocky Mountains in Montana.

At Cambridge I studied all available maps and printed reports, wrote essays and summaries, took them to the room in the Geological Museum where Davis met his one-man class in Advanced Physiography twice weekly, and read them aloud to my professor, as directed. Sometimes he heard me through without comment. Sometimes he made me repeat phrases, or paragraphs, or arguments which he found incomplete, ambiguous, or illogical, and, being a master of logic and himself somewhat acquainted with the physiography of the northern Great Plains and Rocky Mountain Front Range in Montana, he was always right.

On one happy day he interrupted my discourse while I was reading a descriptive quotation from an author whose name I had not yet announced. 'What does the man mean?' said Davis. 'I cannot conceive' he continued, 'what the author had in mind when he wrote those words.' He dissected the quotation at some length. Poor Martin had been hanging on to his chair and holding his tongue all this time. Then I said politely: 'What did you have in mind when you wrote those words?' My quotation was from W.M.D. himself in a report on the Northern Transcontinental Survey under Pumpelly in 1877 [*sic*, 1886], some 28 years before, and my teacher had forgotten his own child. All he said, however, after a long awkward silence, uninterrupted by me of course, was 'Go on'.

Another remarkable morning in Advanced Physiography was the one when he arrived at the classroom happy and talkative, and said: 'You are a very lucky young man, Martin, as I hope you realize. Here is Harvard University providing you with a private tutor.' The Devil tempted me and I fell; 'Yes,' said Martin. 'Yes,

Professor Davis, Harvard University is even more generous since they pay me, as Edward Austin Fellow in Geology, a stipend to take the course.' That time Davis laughed heartily.

No university professor, either at Cornell or Harvard, I repeat, ever did more for me than William Morris Davis did in 1905–06. During my second semester as a graduate student at Cambridge, Davis gave his wonderful course on the Physiography of Europe, with lectures, laboratory work, and written reports. I was one of a substantial number of students in this course, and I received more information and less discipline; but I missed the ordeals of the first semester with its personal criticism and training. (*Martin, 1950, p. 176*)

From the account one receives an impression of stern gravity which, coupled with his very strong views, must have made Davis a favourite target for those offended by his strictures.

FIG. 83. Davis seated on the Continental Divide of the Colorado Front Range
(Courtesy Harvard College Library)

The recollections of Dr Winifred Goldring reveal many of the facets discussed above. In October 1908 she attended a geological conference of the New England colleges and went on an excursion led by Professor H. F. Cleland who,

stood on a high rock and gave a general picture of what had taken place geologically in the entire area. When he had finished, I heard Prof. Davis, who stood nearby, inquire very pleasantly, 'How do you know that, Prof. Cleland?' I don't recall Cleland's answer. (*Letter: Winifred Goldring to R. J. Chorley, undated, 1970; Slingerlands, New York*)

It appears that Cleland that evening wrote a furious letter to Davis who then apologized, denying that he wished to 'show up' Cleland but merely wanted a summary for the sake of those unfamiliar with the area.

In the spring of 1911 Dr Goldring had reason to suspect Davis of a certain amount of kindly understanding:

> His class was in session one day in spring when it was very warm; and he was lecturing. Suddenly he looked up and asked each one in turn, 'What was I just saying?' None of the men knew. I was the last one questioned and luckily had just 'come back' shortly before he questioned us. I knew what he had just been saying; but if he had gone much back of that I should have been lost. I felt sure he knew that, too, but spared me. He said nothing about lack of attention and just went on with his talk. (*Letter: Winifred Goldring to R. J. Chorley, undated, 1970; Slingerlands, New York*)

Over ten years later Dr Goldring was at a meeting of the Pick and Hammer Club at Washington at which Davis spoke. After the lecture he recognized her from a distance and came across the large hall to greet her and, being told that in the long interval since they last met she had become a palaeontologist, smiled, bowed and said, 'Geography's loss is Palaeontology's gain.'

We end deliberately with the remarks of Brigham and Daly because they both knew Davis intimately. Brigham wrote while Davis was still teaching but Daly's memoir did not appear till 1945. Even so they agree closely.

> He is an excellent traveller and makes nothing of hardship. His kindness is unheralded, and its source is often unknown to one who receives its benefits. His serenity and dignity usually mask from general view a humor and informality of manner, best known to a small circle of friends of many years in Cambridge. It may interest some of his scientific correspondents to be told, on the authority of one of his familiar associates, that his gift for versifying in a light vain is extraordinary. He has much enjoyment in music. (*Brigham, 1909, p. 64*)

> Davis had a wonderful capacity for continuous labor. Great physical endurance helps to explain his keen zest for life as well as his success in systematizing a world-embracing science. It took zeal and courage to attempt wholesale reform of the geography taught before his time; both qualities were confirmed as he saw his heresies become gradually accepted principles. His favorite tool was logic. Although at heart he was capable of deep emotion, he would rarely allow emotion to appear in his writings or in his college lectures. Partly for this reason the writings did not appeal to the general public, or the lectures to the rank and file of Harvard students. Davis was sometimes severely critical of student or colleague who, in order to lighten style of presentation, used simile, metaphor, or other figure of speech which could in the least obscure orderly expression of the thought. Rigorous with himself, he was rigorous with his students. He detested sloppiness and made disciplined thought and precision the outstanding aims of his courses in both college and graduate school. Yet he was sympathetic with honest endeavour

and spent much time and energy helping special students who through no fault of their own, had not been properly prepared for imaginative and logical attack on scientific problems. (*Daly, 1945, pp. 279–80*)

Whatever conclusion the reader may reach about Davis as a pedagogue, admiration can hardly be excluded from it.

PART THREE

Old Age

F I G. 84. Davis at about the age of sixty

CHAPTER NINETEEN

The Mysterious Resignation

In 1912 Davis resigned from the chair of geology at Harvard. His letters and actions neither explain nor prepare us for this act, the complex reasons for which will for ever remain his own secret. Yet it appears that years earlier he had mentioned the possibility of retirement to Isaiah Bowman.

> I read *Science* in large doses about once every two months and only the other day came across the news of your resignation. You had told me in January, 1908, of this possibility but since nothing came of it then I supposed you had reconsidered and would stay, to retire in the usual manner. (*Letter: I. Bowman to Davis, 7 August 1912; New Haven, Conn.*)

To resign suddenly at the early age of sixty-two at the zenith of his career seems to demand a powerful reason. Adverse critics may see in it a move to forestall enforced retirement but Davis never shunned an academic battle and would have gone down fighting. His son's recollection is more likely to include an element of truth.

> He retired from Harvard voluntarily because he said he would never be an old and incompetent teacher. (*Interview: R. Mott Davis with R. J. Chorley, 7 September 1962*)

Davis' tenure at Harvard included fourteen years as their leading geographer and thirteen years as professor of geology. For nearly three decades he had been a bridge builder between the two sciences and a notable prop of Harvard collegiate life. The closer the inquiry into the breaking of these links, the more mysterious and intriguing the action becomes. Were the motives concerned with his family, or finances, or with professional and collegiate difficulties or were they purely personal?

As far as we know both Davis and his wife enjoyed good health at the time of his resignation. Brigham, who knew them both well, wrote in 1909:

> It is pleasant to remark in concluding this sketch of a notable career, that Professor Davis is still in the prime of his years and his powers; it is therefore reasonable to expect that the science of geography will, for many years to come, be enriched by his labors. (*Brigham, 1909, p. 72*)

Davis' attitude to the company of children was not that of a preoccupied man, although he may have missed greatly the close companionship of his own boys all of whom had by now left home. He liked the inquisitiveness and

F I G. 85. Davis and his first wife, together with Mrs Davis' mother, Mrs Charlotte
Edward Warner, at Wianno, Cape Cod (Courtesy R. Mott Davis)

conversation of bright youngsters. Later in life he was fond of playing with
Sauer's children. At Harvard, his relationship with young J. K. Wright, no
doubt a scintillating child, is not that of a man facing problems about
resignation. Wright, shortly before his death, related how

> When about eleven or twelve I began to cultivate the acquaintance of Professor
> William Morris Davis, or he began to cultivate mine; at any rate he founded the
> Association of American Geographers soon after. Though he terrified his graduate
> students and frightened away all but the most courageous, he was very good to me
> during my boyhood and youth and I remember him with affection. My diary for
> 1904 mentions three calls by him on me while I was sick and five by him on me
> while I was well, all within less than three months. He gave me a drawing board
> and a copy of his 'Elementary Phisycal Geography' (as I spelled it), and he showed

me 'some fine map things.' which I presume were certain of his incomparable block diagrams. (*Wright, 1963, p. 2*)

The diary notes referred to are given below.

Jan 17, 1904 Sunday
No ice cream for dinner. After dinner I went to Mr. Davis' where I asked him about some maps, etc.

Jan 23, 1904 Sat. (I had been sick)
A.M. Stayed around. Mr. Davis came and showed me some fine map things. P.M. ... Mr. Davis brought me a drawing board.

Jan. 29, 1904.
Went to museum with Mr. Davis.

Feb. 19, 1904.
Went to museum with Mr. Davis and got the maps that I had ordered. Came home and displayed them; they are fine. Made some drawings from them.

April 3, 1904.
Went to Mr. Davis' about the C.I.C. dept. of travels.

March 2, 1905.
Went to Shop Club at Mr. Davis's. About glaciers. Very interesting.
(*J. K. Wright – 'Line a Day'*)

Similarly a letter written to Wright a few years later was full of good spirits and natural humour.

One of the things, however, that may be put to the credit of the superannuated family is a delightful week that Mrs. Davis and I spent lately in our summer house at Wianno, where we kept very comfortable indoors although the weather outside was extremely cold. There has been lots of good skating lately here too, and with that and an old man's gymnasium class in Dr. Sargent's gymnasium, the acting dean is keeping himself in pretty fine shape. Tell your pa that he surely must join the old men's class next year. It is the funniest thing you ever saw to watch the old fellows go through the antics that they imagine imitate the graceful movements of the teacher. The other day one Professor M. Warren was so highly absurd that his neighbor Archibald Howe just lay down on the floor and rolled over laughing. We do stunts at basket ball and that sort of thing too, and the result is that we are almost as tired laughing as exercising at the end of the hour.
(*Letter: Davis to J. K. Wright, 26 February 1907; Cambridge, Mass.*)

It may well be that a certain feeling of loneliness developed in subsequent years but we have no indication that Davis meant to migrate to fresh fields and pastures new. Perhaps the smallness of the family circle after 1910 helped no longer to distract from or offset other cumulative influences.

The possibility of the absence of the need for financial incentives has also been suggested as a contributory factor in Davis' resignation. He was able, it is said, to receive a Carnegie pension at this time. We hear little of the accountancy side but are sure that with private income, salary, royalties and lecturing fees Davis was comfortably off and had no serious drain on his income. In the twelve months ending 1 February 1912 Ginn & Co. paid him royalties totalling $1077 and on his retirement the University granted him a salary of $2475 from 1 September 1912 until he received a Carnegie pension. In financial matters he seems to have inherited the business acumen of his father and was excellent at achieving a credit balance. For example, after the great Transcontinental excursion of 1912 every participant was refunded $35. He did regard financial success as desirable and never depended solely on his professional salary.

> He was very much attached to his wife and sons and gave the boys constant thought. Spoke freely of them. Commented once that they wouldn't think of taking up work that led to such a profession as his. None of his colleagues, he said, had sons who thought a professor's life worth attempting. I gathered that they thought the material returns insufficient. (*Letter: Mark Jefferson to R. M. Harper, 9 April 1935*)

PROFESSIONAL EMINENCE

Davis could not have been displeased with his status in the world of geographical learning. His work had been widely recognized and honoured throughout the world.

> He was an honorary member of 15 geographical societies; a corresponding member of 5; a corresponding member of 4 geological societies; a foreign member of 4 academies of science; he was an elected member of the American Philosophical Society of Philadelphia ... the National Academy of Sciences in Washington, D.C., the Imperial Society of Natural History in Moscow, the New Zealand Institute, and the Geological Society of America. He was Acting President of the G.S.A. in 1906 and President in 1911. He founded the Association of American Geographers and was its President three times. He founded the Harvard Travellers Club and was its President from 1902 to 1911. He was an honorary life member of the American Meteorological Society. Davis deservedly received medals and decorations from many geographical, geological, and other learned bodies in the course of his life. (*Martin, 1950, pp. 176–7*)

In the United States, apart from isolated pockets of resistance, as at Chicago under R. D. Salisbury, the adherents of Davis' explanatory system of landform analysis ruled the field.

> It has been his privilege to train a larger share of the teachers of physiography and scientific geography in the United States; every worker in these fields is daily making use of the results of his labour. He is the leading American specialist in land forms, and led by a masterly knowledge of the methods of pure science he has become a leader in the wider field of geography. (*Brigham, 1909, p. 66*)

In Europe Davisian ideas had been widely accepted in Britain and France but had been received with scepticism in Germany. Brigham adequately describes the reasons for this, though allowance must be made for his bias.

Davis has exerted a great formative influence on geography, as it has been recently developed in England. The keenness of observation with which he detects the significant features of a landscape has struck everyone who has had the great privilege of joining him in a field excursion; his strong generalizing power enables him to combine details, which the ordinary observer might pass as insignificant, into essential parts of a great system. At first the British student felt considerable difficulty in getting at Davis' meaning through his new terms, introduced perhaps more profusely in his earlier than in his later work; but the appropriateness and convenience of these once unfamiliar expressions are now fully recognized. We cannot at once assign a reason for the unique position which he has won for himself as a geographer in the English-speaking lands east of the Atlantic, but perhaps the freshness and completeness of his system had most to do with it at first, and the courage with which he brought the most complicated phenomena in line with the simple expressions of theory confirmed his standing. The brilliant discussion of the rivers of the south of England was the first introduction of his methods to most English students; it produced a marked result on the teaching of physical geography which was then being undertaken for the first time in English Universities. His laboratory work at Harvard has also had influence on British geographic education. His tireless activity of mind and body make him simultaneously a teacher and a student; his occasions for giving instruction are bounded by no college rules; wherever he goes by land or sea, the stores of his accumulated and accumulating experience are at the service of every serious minded companion, he has thus influenced far more than ever attended his formal classes. Two features of Davis' work may be selected for special mention his theoretical reasoning and his systematic studies. Through his theoretical reasoning he has connected observations made in different parts of the world and thus created his cycle of erosion, which allows us to look backward and forward in the evolution of a landscape. Others had seen peneplains, but Davis, by inventing a name for them, brought the thing into the foreground and showed its working value. Some German geographers have not seemed to see the significance of Davis' systematic work; they regard mature for example only as a descriptive term without perceiving that it indicates a correlation of a number of features; this may be due to the comparatively rare occurrence of young valleys in Germany. Some have felt that certain parts of Davis' work was too theoretical; but we must recognize the necessity of theoretical work as essential to attaining a systematic point of view. It directs our eyes to facts which have a theoretical bearing; we shall arrive at good results when we observe with a good theory in our mind and with sharp eyes.

In France Davis is without doubt the best known of the American geographical school, and his influence is growing among French-speaking people. Among the chief reasons for this are naturally his studies dealing with French territory; but another powerful agent in disseminating the results of his work was their adoption by de Lapparent in his Leçons de géographie physique and their use as the cornerstones of the treatment of land forms as expanded in successive editions of this book. There is more than this, however, for there appears in the writings of Davis a personality particularly attractive to the French geographic public. His style is marked by skilful and facile expression. He has made the impression of having a unique ability to treat the subjects which he has undertaken. His originality is due in part to his deductive method, but is also due to an artistic taste quite his own, a vital and personal manner of presentation. The younger generation is impressed more and more with the ideas that Davis has put in circulation. Davis regards a peneplain not simply as a land form, but as a phase in the development of land forms; as soon as one appreciates this view, one gains a new conception of all land forms. Although he constantly makes use of past processes in explaining present forms, he separates the geological and geographical treatment of the land more clearly than is usually the case in Germany, by leaving to the former the intentional study of past events in their long succession and holding the latter strictly to the explanatory treatment of the existing surface. (*Brigham, 1909, pp. 68–72*)

We have a superabundance of private evidence showing the high esteem in which Davis was held by other scholars in the early twentieth century. G. K. Gilbert wrote frequently to him and replied promptly to his enquiries.

My dear Davis,
    Thanks for the proof sheets, which came to me even before the printed copy to Spurr's paper. I wish the review would have the same distribution as the paper, for it seems to me a complete antidote – at least so far as physiographers are concerned. Unfortunately some geologists have no appreciation of surface form and its evidence.
    And thank you for writing the review. Even without your felicity of presentation it would carry more weight from you than from me as I am prejudiced.
        (*Letter: G. K. Gilbert to Davis, 25 September 1901; Skull Valley*)

Paul Vidal de la Blache, professor of geography at the Sorbonne, went with Davis and others to Mexico in 1904. On the journey he writes to Henri Baulig, one of his protégées who had just landed in the United States.

I have just put in a word on your behalf with Mr. Davis who is arranging to return to Boston after being a useful and interesting companion to us. I have told him how serious and persevering you are and also about your educational training. I think Mr. Davis will be an interesting man for you to know; he is the most Europeanized of all the Americans I have met here. He has spoken to me about the excursions he is accustomed to make with his students, sometimes great distances; perhaps you would be interested in joining one . . . You would do well, I think, to present yourself on my behalf to Mr. Davis as soon as you hear of his return to Harvard. (*Letter: P. Vidal de la Blache to H. Baulig, 9 October 1904; Pearsall, Texas, en route*)

The remarkable outcome will be discussed more fully in Volume Three of our *History of the Study of Landforms*. Henri Baulig, then twenty-seven years old and trained mainly in historical methods, found in the Davisian school of geomorphology an interest that dominated the rest of his life. His first visit to the United States lasted eight years!

A letter from the Oxford School of Geography shows clearly that in Britain Davis was regarded as one of the foremost instructors in geography.

> You may recognize my name as that of a pupil of Mr. A. J. Herbertson, of Oxford, which was mentioned to you by him in connection with your proposed geographical investigation in the Alps and Apennines this summer. I need hardly say how highly I should appreciate the opportunity of working under you ... I am sure I am not alone in looking forward with delight to what will really be a unique opportunity for study. (*Letter: H. O. Beckit to Davis, 8 May 1908; Whitchurch, Salop.*)

We may conclude these eulogies with the remarks of the great Albrecht Penck, who in 1909 was in America acting as exchange for Davis.

> From Florida I went to Baltimore to be present at the meeting of the American Association and especially at the meeting of the Geological Society and the Society of American Geographers ... Among the geographers you were very much missed. I was very much surprised to see how much human geography now attracts your countrymen. And indeed some good papers were read, but there were others also. I had a fairly good review of the movement of American geographers without Davis; that is, of the body without the head. (*Letter: A. Penck to Davis, 18 January 1909; Columbia University, New York*)

All the above letters, we must point out, were written before many of Davis' finest articles and his best book had been published.

COLLEGIATE DIFFICULTIES

The academic empire that Davis had won had entailed wide travel and long absences from home. Almost inevitably as he widened his global influence, his hold on home affairs tended to weaken. He was like a king who had conquered a far-flung empire but had almost lost his throne.

It seems certain that he had few or no students for some of his classes and that most of those present were research graduates. This was so as early as 1908.

> Professor Davis regularly gives in Cambridge an advanced course on the physiography of Europe but most of his work consists of the direction of advanced research. (*'Harvard Crimson', 1908*)

> During several years before he was retired from Harvard in 1912, aged 62, he had no student. (*Letter: S. S. Visher to R. J. Chorley, 1 September 1962; Bloomington, Indiana*)

From Davis' own words it does seem as if his last years on the staff at Harvard were distinctly unsatisfactory to him.

The Department of Geology was thrown on its beam-ends by Shaler's death in 1906, and righted itself with difficulty. For six years Shaler's elementary course was conducted chiefly by Wolff and Woodworth. Davis had been appointed to the Sturgis–Hooper Chair in 1898, and this was followed by curiously opposed effects on the development of his work at Cambridge. The promotion relieved him from elementary teaching, but as the introductory course in Physiography was then for some twelve years given by a succession of younger and less trained men, his advanced courses were not well fed. The professorship gave him the freedom but not the funds for travel; yet from his fiftieth to his sixty-fourth year (1900–14) he travelled as never before, and thus continued to supplement deductive studies at home with inductive studies in many distant fields, including the western states, Great Britain, Turkestan, Mexico, South Africa, Central and Southern Europe and Alaska. In 1914 he crossed the Pacific on a study of coral reefs. His long absences from Cambridge worked to the disadvantage of Geography there. Davis resigned the professorship in 1912, to be succeeded in the chair by Reginald Aldworth Daly (A.M. 1893). (*Davis, 1930K, pp. 321–2*)

Davis was always rather defensive regarding his inability to attract large numbers of students, which must have been highlighted by the outstanding success of Shaler as a lecturer, and his last years at Harvard seemed to have been particularly disillusioning. Some twenty years later he was to write:

Davis's classes were never large, for his lectures were argumentative rather than descriptive, much emphasis being given to the reasons for accepting various theoretical explanations as well as to the explanations themselves. Moreover, he took seriously the duty of reporting upon the work of his students, and instead of depending chiefly on one or two announced examinations, he introduced various other tests, including laboratory exercises, although the employment of such exercises in physical geography was then unusual. (*Davis, 1930K, p. 315*)

Certainly a relative absence of students could be expected when the master was so often absent, as not all the graduates would wish to help him with his tours. But there is also the important fact that the applications of some of his students for a doctorate were rejected. These rejections, judging from the way the gossip about them survived, must have gained great publicity particularly as most of the unfortunates went on to achieve the degree and great fame elsewhere. Nothing was more likely to damn an institution in the eyes of research workers. The standard demanded may be, of course, (too?) high but the obvious corollary is that the tuition is inadequate. But surely previous writers on this matter have missed the very significant point that Davis was no longer king in his own castle and his subjects were no longer necessarily his disciples. As long as Davis' students wrote on land-form analysis he quite rightly could dominate a committee of assessors but once they wrote on wider geographical topics several professors of various other disciplines could assert their views. Davis' progeny included many who dealt with topics wider than pure geomorphology. Once a committee has acquired the habit of criticizing, the volume of criticism tends to increase, as academics are as a rule zealously competitive in the giving of advice. The unfortunate affair of Ellsworth Huntington, as recorded in Davis' letter quoted in Chapter 18, was obviously a great disappointment and defeat for Davis.

However, there is ample other evidence that Davis was not having things his own way at Harvard. He strongly disapproved of the way that Professor Jackson, a palaeontologist, was being treated and felt obliged to intercede on his behalf.

> I beg leave to submit the following statement for your consideration, because the feeling oppresses me that injustice would be done to Professor Jackson if his resignation were now insisted on; his unqualified five-year appointment having been made only two years ago.
>
> This feeling does not arise from any belief that it would be unjust to decide now that an assistant professor would not be promoted or re-appointed three years hence. What troubles me is the possibility that the opinion of the Corporation regarding his work may be so presented to him that he can have hardly any choice between resigning immediately and serving for the remainder of his term. Such an alternative would appear to be fully within the rights of a professor – indeed it does not appear that the suggestion of immediate resignation is based on any question of such rights. The ground for the suggestion seems to be the belief that it would be inexpedient for a professor, knowing he was not to be re-appointed, to continue working to the end of his term; inexpedient for the reason that his withdrawal at the end of his term would be tantamount to a confession that he could not get a re-appointment; while resignation during his term, without explanation, might be construed as purely voluntary.
>
> It is possible that this may be true in some cases; but I beg leave to state most earnestly that it does not seem to me fair to insist on this view of the case so emphatically as to leave a professor practically no opportunity to consult his own judgment as to what is best for his own interest.
>
> I therefore respectfully request the Corporation to consider whether it is not desirable that no further action be taken in this matter for the present; that Jackson be left to take advice from such friends as he may wish to consult, to come then to whatever conclusion (within the rights of his appointment) he individually deems best for his own interests, and to act without embarrassment on that conclusion. (*Letter: Davis to the President and Fellows of Harvard College, 25 June 1906; Cambridge, Mass.*)

Such action on his part must make it very doubtful whether he would have done this had his own position been in jeopardy. However, it may be that his intercession proved of so little effect that when his own turn was coming he preferred to resign rather than leave himself open to the same treatment. Mark Jefferson, in 1910 when on a field trip with Davis to the Colorado Front Range, tells his wife:

> He [Davis] seems disappointed at the prospects at Harvard. Jackson has retired,

there is no instruction in palaeontology, and he has 'no students', only two or three, he says, come to take his courses. We chaff him about the earth having been made on his plans but he takes it very well. As work this is interesting, but nothing like so much so as my human geography studies. Of course many of these are great earth features and of interest themselves for those who can get them . . . it is not practical to make studies of the human features much, for W. M. Davis will not be turned aside for them. That must come later. It looks as if he had not been successful. (*Letter: Mark Jefferson to his wife Theodora, 19 July 1910; quoted in Martin, 1968, p. 136*)

It seems also that Davis was not successful in negotiating his own successor, or at least in procuring a certain post for D. W. Johnson, his most loyal student. Whether Davis thought of Johnson as an ultimate successor is not clearly stated but the implication is there. He wrote to A. J. Herbertson about Johnson and received a cautious reply.

It is rather difficult to answer your questions about Prof. D. W. Johnson after so short an acquaintance, though if my impressions are of any use to you I am glad to give them freely as you ask.

He struck me as a carefree and capable student, trained in a good school, and not yet quite ready to walk alone. His lectures were clear and well arranged, but I wished he would throw aside his M.S. One lecture was dull; but when he spoke of his own work he became much more alive and interesting.

You do not say for what purpose you want the information, otherwise I might have been able to answer better. As a colleague under your direction I should expect him to be excellent. He would carry out your plans rather than supplement your work by introducing fresh ideas. He would be all the better for a year in Paris, Berlin and perhaps a little in London with Mackinder. For a professor in one of your colleges he should do well – probably better in the East than in the West, for he seemed to be the type of reserved American whom it takes time to know, though the more you know him the better you like him. I should have some hesitation about making him Head of a first or a new Department of Geography in a large University, until he had some more experience. Whether he would become the Geol. Chief or not under such circumstances, I do not know enough to judge. (*Letter: A. J. Herbertson to Davis, 2 April 1910; Oxford*)

Johnson had been a student of Davis' at Harvard since he joined the staff of the Geology Department at M.I.T. in 1903, and was appointed Assistant Professor of Geology at Harvard in 1907 (Bucher, 1947). He was a devout believer in Davisian ideas, making his own reputation largely by developing and modifying Davis' ideas on Appalachian development (1931) and the marine cycle (1919). Such dedicated acceptance of his methods may have given Davis confidence that under Johnson his school of thought would stand a better chance of perpetuation than if the position were entrusted to a more independent personality.

Johnson was in nowise a 'character' in the Jeffersonian sense, though he was a man of intellectual integrity and vigor. Of all Davis's disciples, he was the most profoundly influenced by the latter's thought. Davis had stopped teaching at Harvard just before I entered as an undergraduate and the interest in physiography he had aroused in me led me to study under Johnson and work for Johnson in other ways. Thus I got the essence, if not the quintessence, of many of Davis's most distinctive ideas. Johnson was a clear, systematic teacher, a demon for work, and an agreeable, conservative man but not a particularly venturesome or original thinker. Like that of Davis, his teaching was excellent discipline for those who don't rebel at being disciplined, but not too stimulating for the roving imaginative curiosity that science and scholarship need, fully as much as they need disciplined thought. (*Wright, 1963, p. 3*)

Johnson's own statement several years later supports the view that Davis had picked him out as a possible successor.

When a few years later, you called me to Harvard, to take up bit by bit portions of the work which you were relinquishing, I realized that another important milestone in my progress had been reached. (*Letter: D. W. Johnson to Davis, 4 February 1925; New York*)

Whatever may have been in Davis' mind there is no doubt that, at this time, opinions were divided on Johnson's ability.

It is my own private opinion that Professor Johnson's best friends would do well to advise his seeking a position elsewhere. That does not mean that he will fail to 'make good' here, in his two further years of probation (increase in number of students, and outside investigations, etc., may help him to keep his place here permanently), but it means that to me it appears to be a mistake for a man to insist on wedging himself in where experience of three years has shown that he has personal qualities which somehow or other do not seem to commend themselves to our students. That may be their fault, and not his. As for myself, I am going to be perfectly fair and open-minded about Professor Johnson's next two years. I do not think that, departmentally, he is any help, but rather a serious handicap. And the opinion of outside persons, whose opinion I value, is to that effect. I am interested in the Department, in the subject of physiography, and in the University. My own future here is unsettled, but at least I have done the best, as far as I have seen it, for my subject and for the Department, since I have been here. If I go, I shall feel that I go with a good record, of work as well done as I could do it, behind me. (*Letter: Robert DeCourcy Ward to Davis, 13 April 1910; Cambridge, Mass.*)

Johnson did not in fact succeed Davis in taking charge of Harvard physiography, and it is more than coincidence that he left Cambridge for Columbia in the same year that Davis retired. Indeed, as he was a junior member of the department and still on probation, his appointment would have been very surprising. Davis was probably looking several years ahead and at this stage only intended him to take charge of the physiographic section. As Johnson eventually became a very successful professor at Columbia University and did a great deal to extend the influence of Davisian ideas, Davis' judgment may not have been so misplaced as both Herbertson and Ward imply.

Shortly after Davis' resignation, Isaiah Bowman told Mark Jefferson of affairs that he had known confidentially since Christmas 1912:

> ... I must not close, however, without telling you a little about Harvard. You know, I suppose, that Atwood has been appointed a full professor there at $4000 ... The affair is involved in a dozen different ways with things that have no relation whatsoever to geography and make one sorry that important questions such as are necessarily related to a professorship of geography cannot be decided on their merits ... (*Letter: I. Bowman to Mark Jefferson; date unknown, but after 1912*)

The effect of this appointment on Davis may be partly judged from the following comments by S. S. Visher:

> My special teacher, R. D. Salisbury, had relatively little respect for Davis. I recall Salisbury's glee when he was employed to correct many of Davis' definitions in a new edition of *Webster's Dictionary*. Davis was 7 years older than Salisbury, and was harshly critical of him. Davis' successor at Harvard was Wm. Atwood, a special student of Salisbury's. (*Letter: S. S. Visher to R. J. Chorley, 1 September 1962; Bloomington, Indiana.*)

PERSONAL MOTIVES

It seems that up to about 1912 Davis' interest in landform analysis and in publishing articles on geomorphological topics did not lessen markedly. True some of the articles are repeats in German or Italian and his *Geographical Essays* (1909C) appeared solely through the labours of D. W. Johnson, but his later methodological writings are long and carefully composed. Fieldwork and excursions as well as the exchange professorships at Berlin and Paris took up much of his time and energy. These activities and private letters such as that to James Goldthwait printed below, contradict the idea of a waning interest in geography.

> Can you some day give me a few points on the plain, seaward from your St. Lawrence elevated beach? How far does it exhibit the typical features of a young marine coastal plain? Has it well defined consequent rivers? Have they opened flood plains across it, are their valley sides more or less frayed out by lateral insequent ravines? Do the valleys, near the inner margin of the plain disclose the floor on which the plains strata rest? Are normally expectable features there indicated? What sort of a shore line has the plain at its outer border? How far are human conditions influenced by physiographic conditions? Is there a marked contrast between the plain and its oldland? A brief statement on those points, at your convenience, would be a favor. (*Letter: Davis to J. W. Goldthwait, 10 June 1911; Cambridge, Mass.*)

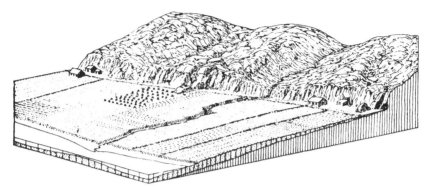

F I G. 86. An uplifted marine platform in western Scotland (From Davis, 1912J, Fig. 91)

The lessening in the number of publications must also be set against the gradual accumulation of *Die Erklärende Beschreibung der Landformen* which first appeared in 1912(J).

Yet, while Davis retained much of his earlier keenness and productivity, he probably experienced about this time a growing disappointment with the academic trends of many of his research students, quite apart from their Harvard difficulties. Isaiah Bowman, Mark Jefferson and Ellsworth Huntington are among those who went over to a wider geographical field. Partly under Davis' methodological and geographical propaganda, human aspects of geography made great advances in the United States in the early twentieth century. Albrecht Penck, as we have noticed, was in 1909 amazed at its progress. At Chicago, the hotbed of anti-Davisian geomorphology, Harlan H. Barrows was developing a successful branch of human ecology. But as so often happens in the development of doctrines, for Davis it was the friends within rather than the enemies without which caused his influence to wane. In many ways it was a tribute to and outcome of his own

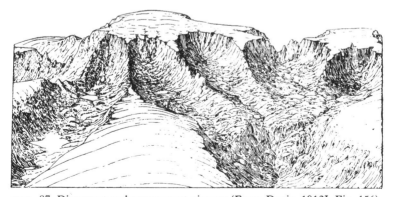

F I G. 87. Divergent and convergent cirques (From Davis, 1912J, Fig. 156)

achievements. Mark Jefferson when comparing human geography with Davisian physiography wrote prophetically:

> My people idea is a better idea and will tell better . . .
>
> I think he [Davis] has absolutely no interest in 'Human' geography, though he tries to speak of it with respect. He thinks I ought to write an elementary book with man first not merely because it would pay, but it would be he says, a good thing to do . . . We have learned a great deal from Professor Davis though not things in my own school line. (*Letters: M. Jefferson to his wife Theodora, 20 and 28 July, 1910; printed in Martin, 1968, p. 136*)

In our opinion, Russell Smith and Ellsworth Huntington size up the situation perfectly:

> You speak of the growth of geography away from the Davisian geomorphology. No one had more to do with the un-Davising of geography than did Davis himself. He went up and down the land between 1899 and 1903, delivering addresses before various scientific groups, laying out the point that geography was a relationship between the earth and the organisms that lived upon it. He is almost the father of the idea of relationship. The others had been geologists and interested in talking about the earth, describing the earth, telling how it got cut and chiseled and blown and washed into certain shapes.
>
> As a monument to that philosophy, go around to the library and look up Davis' physiography and the twin books by Salisbury. Put men through that and then put them to teaching geography and there isn't going to be much of the Davis philosophy in it. (*Letter: J. Russell Smith to G. J. Martin, summer 1962; Swarthmore, Pa.*)

> Professor Davis was a remarkable man, and had a great influence on geological and geographical thought in America.
>
> Curiously enough my own attitude toward Professor Davis' work has changed. In 1907, when I wrote *The Pulse of Asia*, I thought of him as the foremost of modern geographers. Today I feel that he never really became a geographer. He was always a geologist with a strong interest in the earth's surface. He was convinced that geography was the science of the relation of the earth's surface to man and his work, but that phase of it never enters into his own labors. One of the curious things about Professor Davis was that he inspired in his students a firm belief that the main part of geography was human. Yet when most of his students devoted themselves to anthropogeography he felt a sense of disappointment.
> (*Letter: E. Huntington to Kirk Bryan, 1 October, 1935*)

These divergences, when added to collegiate cross-currents, may well have nudged Davis towards resignation. However, there is yet another possible factor in the complex of influences. Davis may have considered *Die Erklärende Beschreibung der Landformen* his crowning work. It is certainly the most complete statement of his ideas on cycles of erosion and the explanatory method of landscape description.

This fine work really marks the end of the Davisian propaganda campaign.

Five years earlier Davis, in a letter to Huntington thanking him for a copy of *The Pulse of Asia*, had written:

> I have not yet been able to read enough of the book to answer your question as to your having carried theory too far but the whole impression is very favorable. As to your dedication and preface I fear that a good many German readers will not agree with you – but that is another question. (*Letter: Davis to E. Huntington, 14 November 1907; Cambridge, Mass.*)

Now, after some months in Berlin as exchange professor in 1909, Davis attacks the main stronghold that had been sceptical of his ideas. However, for once he was decidedly unlucky. The European academic world declined or stood still from 1914 to 1918 and the book had far less effect than it might otherwise have had. It was reprinted in 1924 but by then its chance had largely gone owing to the resurgence of topical economic, social and political geography.

Davis continued his methodological writings for a few years but with his resignation from the professorship of geology at Harvard in 1912 he soon begins to lead a new life physically and mentally. A year before his resignation Davis' disillusionment appeared in a reply to a religious questionnaire by the National Academy of Sciences, in which he wrote wryly: 'every man as he grows older finds how little right he has to call any one else to account for shortcomings'. Resignation, in contrast, seemed to revitalize him. He is freer, less rooted, less interested in method. However, whereas the resignation was by choice, the uprooting, as we will now show, was through unexpected adverse circumstances that seriously upset his main plan.

> Why after this excursion is over, I am going to work like thunder, writing books; that is what I have resigned for. The small number of advanced students I have recently had makes lecturing less useful than writing, for in books I can address a larger number. (*Letter: Davis to J. S. Keltie, 16 August 1912; Cambridge, Mass.*)

CHAPTER TWENTY

The Second Marriage

THE DEATH OF THE FIRST MRS DAVIS

On his resignation at the close of the year 1912, Davis had several publications in the pipeline. The two of these that especially concern us are 'Dana's confirmation of Darwin's theory of coral reefs' which in 1913 appeared in different-length articles in three journals (1913J, N and O); and 'Meandering valleys and underfit rivers' (1913Q), one of Davis' early topics. The first-named revealed his new interest in coral reefs and foreshadowed an important line of research which he intended to pursue during his retirement, as discussed in Chapter 25. The second article, which was treated in Chapter 16, represents the last of a series on fluvial processes with which, as we have seen, Davis had been occupied for many years. Thus Davis' retirement was marked by an academic hiatus, this was to be accentuated by events in his personal life.

We can only guess about Davis' plans for retirement. He probably looked forward to enjoying many happy years studying in the university library, as he had always done, taking a passing interest in departmental affairs, joining perhaps in an occasional expedition and keeping a watching brief on the general development of geography. Above all he was expecting to spend his evenings among the familiar comforts of his home. All prospect of this anticipated felicity was shattered by the sudden death of his wife on 29 April 1913 only a few months after his resignation. With his children grown up and living in homes of their own, this unexpected loss of his wife's companionship faced Davis with the need to make a radical change in his daily life, for which he seems to have been completely unprepared. His wife's death was a great shock, coming quickly and without warning.

> One is inclined to think of WMD as a decidedly crusty and in some respects 'hard' man, yet, after the first Mrs. D's death, he was terribly affected and revealed to my mother a very soft and sentimental side of his nature. He was very lonely and pathetic at the time until he married Miss Wyman. (*Letter: J. K. Wright to R. J. Chorley, 5 January 1962; Lyme, N.H.*)

When he had become resigned to managing without her presence, the hardest adjustment was the loss of a permanent home. Life had now become a choice between being looked after by an unfamiliar housekeeper or a succession of hotels and boarding houses. The security that he had grown to depend on

FIG. 88. Davis and his first wife at Wianno, Cape Cod (Courtesy R. Mott Davis)

and which was associated with the familiar surroundings of his home had suddenly vanished.

THE EXCURSION TO THE ROCKY MOUNTAINS IN MONTANA AND IDAHO

A logical solution for Davis to the loss of home life was a field excursion, and in the summer of 1913 he went to the Rocky Mountains at the expense of the Harvard Shaler Memorial Fund. Davis had first noticed the particular glacial features of the area from the train window between Missoula and Spokane on the 1912 Transcontinental Expedition. The field inspection of the area lasted for three weeks in August and many of the observations were from train windows, decks of steamboats and automobiles. On the basis of this rapid survey Davis published seven articles or abstracts. In our opinion, these show the master at his worst: full of self-vindications and of tremendous length due to a prodigious determination to leave out no word that could be included to persuade the presumed perverse intention of readers to misunderstand every sentence.

Davis visited the area of Montana where two north-draining trenches, the Rocky Mountain Trench–Flathead Basin and the Kootenai–Pend Oreille–Clark Fork Trench, had been invaded by Pleistocene ice tongues from Canada which had scoured the lower slopes of the faulted ranges (e.g. the Mission Range), the higher parts of which were at that time occupied by cirques and small valley glaciers. According to Davis' interpretation, the valley ice tongues had excavated lake basins and had probably dammed up proglacial lakes (e.g. Lake Missoula in the upper Clark Fork valley). Davis arrived at Paradise at the south-east end of the Clark Fork on 2 August, had three days of excursions and then moved to Spokane. The period 8–10 August was spent at Hayden Lake 'at the beautiful country seat of one of my students of thirty years ago' (Davis, 1920C, p. 84), during which he took car excursions to view the outwash gravels of the ancient Pend Oreille glacier. Starting back east on 13 August, he went from the Kootenai valley to the Flathead trough where he took a trip with Mr H. F. Smith, a Harvard student. Davis spent the next four days around Lake Flathead examining the west faces of the Mission and Swan Ranges, left for the east on 23 August, went by train to Duluth, by boat from Duluth to Detroit and reached Cambridge on 3 September.

In his paper on 'The Mission Range, Montana' (1916G) Davis described the features on the west-facing fault scarp of the range where there is almost 7000 feet of local relief. He recognized three altitudinal belts sloping towards the south:

1. The lower belt, existing only at the northern end of the range, consists of spurs imperfectly truncated by a valley ice tongue into knobs, hollows and ledges surmounted by lateral moraine embankments several hundred feet high trenched by the rivers flowing down the west face of the range. This morainic belt, marking the extremity of the advance of the valley ice tongue south from Canada along the Rocky Mountain trench, swings east into Flathead valley to enclose the southern end of Flathead Lake.
2. The middle belt, unaffected by either major ice tongues or local cirque and valley ice, consisting of a maturely-dissected fault face.
3. The high belt, particularly marked in the south, where this maturely-dissected topography had suffered local glaciation which is especially characterized by the southward displacement of the cirque heads by the retreat of the north-facing cirques. The extreme southern end of the range is, like the middle belt, unglaciated. Davis proposed that these local valley glaciers were initiated by climatic change but that a climatic amelioration occurred before the final stage of an uninterrupted cycle of glacial erosion was reached.

Cycles and Episodes of Glacial Erosion. Large as the McDonald cirque is, it does not represent the completion of a glacial attack upon a mountain mass; that demands a relatively rapid widening of the cirque floor and its slower lowering until the enclosing walls are consumed – the action of the weather on exposed surfaces here aiding the action of ice on covered surfaces – and the mountain mass is truncated; at the same time the thickness of the ice on the truncated surface should diminish by reason of lessening mountain height and consequently decreasing snowfall, until the thin and relatively inert glacial veneer almost or quite disappears, the glacial tongues descending from it shorten and vanish, and the truncated mass remains subject only to normal dissection by the retrogressive erosion of its flanks.

It was, I believe, Tyndall who first fancifully suggested that deglaciation might be the result of loss of height by glacial erosion; it is now generally agreed that deglaciation was the result of climatic change . . .

Thus two schemes of the life history of a glacier are suggested: one is the highly ideal scheme of a constant climate, during which an upraised mountain mass will, if at first high enough, be glaciated until it is worn so low that its snowfall is lessened and its glaciers disappear, . . ., this involves a complete 'cycle of glacial erosion' . . . The other scheme is the more expectable one of a variable climate, in which a mountain mass will be glaciated only as long as the snowfall is sufficient to form glaciers, as was the case with Pleistocene glaciation; glaciers were then extinguished long before their work was completed, and hence thus limited, the 'life history of a glacier' by Russell and the 'cycle of mountain glaciation' as presented by Hobbs include only a life history or cycle cut short by climatic change in its prime . . . In the Mission Range we evidently have to do only with an episode of glaciation due to climatic changes . . . (*Davis, 1916G, pp. 284–5*)

The account is illustrated with fourteen excellent line drawings and seven clear photographs taken by the United States Geological Survey. But many readers will be disappointed when Davis writes:

. . . I took a rapid trip 80 miles southward from Kalispell and return and made many notes and sketches of the mountain forms as seen from the lake and plains on the west at distances of one to five or more miles. The diagrams here presented are redrawn from my hurried outlines and represent the range as if it were seen from an elevated point of view several miles to the west. They are roughly generalized figures, suggestive of the kinds of forms there exhibited, rather than sketches of actual details; they undoubtedly exaggerate certain features . . . their uncompromising black lines cannot portray the softness of the graded slopes.
(*Davis, 1916G, pp. 268–9*)

Davis adds that this reconnaissance study was intended to stimulate someone else to write something better and also to show that, in spite of its incompleteness,

a systematic method of treating land forms is, sometimes, applicable in rapid work, where conservative geographers of the empirical school think it is inapplicable, their idea being that explanatory description must demand long and intensive study, and therefore cannot be based on brief inspection. (*Davis, 1916G, p. 288*)

However, a great many readers may not accept this self-vindication, which seems incompatible with Davis' earlier insistence on exactness.

In a subsequent paper, Davis (1917C) concentrated his attention on the twenty to twenty-four faint abandoned lake shorelines which had previously been described at the south-east upper end of Clark Fork, and named as marking the former levels of the ancient Lake Missoula. Davis proposed that this area had been occupied by a series of lakes up to 1500 feet deep dammed back by various still-stands of the Kootenai–Pend Oreille glacier tongue. He further propounded that, despite the tendency for ice flotation, deep lake basins were scoured out, supporting the view that fjords could be deeply scoured by ice debouching into the sea.

A much longer and more detailed account of this three-week field study was published as 'Features of glacial origin in Montana and Idaho' (Davis, 1920C). The long delay in publication was due to the European war and to the censors of the contents of the article, who altered some of Davis' basic ideas on the direction of the local ice-flow and, as he admits, greatly improved his findings. The article shows Davis' new preoccupation with the reform of spelling; for example, 'bard' for 'barred', 'imagin' for 'imagine', 'establisht' for 'established' and so on. It also shows some rare attempts at quantitative assessments; for example, the glacial overdeepening in Lake Pend Oreille is estimated at about 1000 feet at least. But, above all, it shows the worst side of Davis' explanatory description approach. The territory is described bit by bit; plain, lake, moraines, depressions, troughs, clefts, hanging lateral valleys, lakes, terrace, divide, transverse valley, glacial distributaries, outwash plain, deltas and so on. All told, sixty-two separate bits of topography are carefully named and described as separate items and the account includes a further eleven sub-sections of text. The general features of Glacier National Park; the description of the west face of the Mission Range; a long treatment of the glaciation of the Kootenai–Pend Oreille depression; catalogues of hanging valleys and outwash plains – all are treated in turgid detail. There are even misplaced and facile 'geographical descriptions' on such topics as the agriculture of the Flathead Basin and the location of the city of Spokane. Davis in some parts of this work shows a tentativeness and indecision which are both unusual for him and a measure of the general unsatisfactory nature of this piece of scholarship:

The Lake Missoula Shore Lines may Antedate the Valley-side Cliffs. The supposition that the valley-side cliffs of upper Clark Fork valley were eroded by a wholly submerged glacier is so inherently improbable and the mechanical difficulties that it involves as pointed out by my censor are so serious that notwithstanding my acceptance of this supposition in a brief earlier essay, an alternativ for it must now be examind; namely the possibility that the eroding glaciers of Clark Fork valley were not synchronous with the existence of the high-level stages of Lake Missoula.

... Let it now be supposed that while the Kootenai–Pend Oreille glacier made its greater advance and formd the heavy moraines at the southern end of Lake Pend Oreille, it servd, as already noted, as a high-level barrier across Upper Clark fork valley which was therefore filld with Lake Missoula. Let it be further supposed that the south-eastern distributary of the Kootenai–Pend Oreille glacier was shortend by flotation in the deep lake waters so that it invaded Clark Fork valley for a relativly short distance. Also, that any glaciers which in that epoch reacht the lake from the Cabinet mountains were likewise broken up by flotation shortly after they enterd the lake waters and before they reacht Clark fork valley. Evidently the lateral cliffs of Clark fork valley could not have been eroded under these conditions. Father east, in the Flathead basin, the Kootenai–Flathead glacier may have had thickness and weight enough not to be floated and broken up until it reacht the Mission moraine.

Now let the glacial conditions during the formation of the later moraines be considered. In that epoch the Kootenai–Pend Oreille glacier is thought to have ended at the Elmira divide between the Clark fork and the Kootenai drainage basins. When it was thus situated, it would not have held back any lake in Upper Clark fork valley; and the local glaciers from the Cabinet mountains might then, not being broken up by flotation, have advanced farther than before, even tho glaciation in general was less severe. One or more such glaciers might under these conditions have reached Clark Fork valley where they would proceed to scour and cliff the side slopes in adapting the form of the valley to their needs.

So far as a right-hand or down-valley distributary of a possible Cabinet-mountains glaciers in Plains basin is concernd, no special difficulty arises regarding its erosion of cliffs in the Woodin–Weeksville narrows, for it need not have encountered a lake there unless one was formed by another Cabinet-mountain glacier farther west. But if the Plains glacier could send out a right-hand distributary several miles in length, it must have formd a strong barrier across Clark fork valley in the Plains district, and thus have held up a good-sized lake in the Paradise district and farther up stream: and the left-hand distributary of the Plains glacier, by which it is here supposed that the cliffs of the Paradise district were scoured off, must have been more or less submerged in this lake. This would be particularly the case towards the up-stream extremity of the glacial distributary near Parma, where the valley floor is much below the top of the cliffs above Paradise, and hence much more below the barrier that the Plains glacier must have formd across the Clark fork valley. Hence while the suppositions here set forth diminish the difficulties that were encountered in the preceding section, they do not dispose of them altogether. Some sublacustrin glacial erosion still seems to be demanded.

As complete escape from the necessity of assuming at least a certain amount of sublacustrin glacial erosion is not provided by the above suppositions, it is worth while to inquire into the possibility of a valley glacier holding itself in a valley bottom beneath the waters of an ice-barrier lake. If the ice front in such a lake were broad and free, the opportunity for its flotation would be increased; but if only a narrow valley glacier is concernd, the manner in which it would wedge its way along the valley suggests that friction with the valley sides might seriously impede its flotation. The more the two valley sides were clift, the stronger would be the friction-hold on the ice between them. On the other hand, the presence of water above the ice would tend to counteract pressure from thicker ice in Plains valleys, and thus to retard if not to prevent the advance of the ice-distributary up stream. As to this point, all I can say is, that if no lake had been there, the distributary might have had a greater length than the 20 miles from Plains to Parma. An up-valley advance for such a distance would be singular enough, even if no lake were present to impede it, but it would not be so forbiddingly extraordinary as an up-stream advance for 100 miles, as I had previously supposed to be necessary when the ice supply was assumed to come from the Kootenai–Pend Oreille glacier. Regarding the fact of an up-stream advance of 20 miles, that is not more than was accomplished by many an Alpine valley glacier as it emerged from a piedmont lake basin to its terminal moraine. (*Davis, 1920C, pp. 144–6*)

This long article is illustrated by many line drawings with 'more or less fanciful' details. No attempt is made to present a regional synthesis or a reasoned thesis on the glaciation of the area as a whole. The production appears as a remarkable 'hogging of space' for it occupies nearly half of the annual journal, and readers who think highly of Davis are advised not to consult it. It demonstrates, even when allowance is made for the unexplored nature of the terrain, that Davis' methods needed more time for research and deduction than he suggests, and that they did not suit hurried regional surveys. Yet much of the disjointed superficial preparation of these essays on the Montana excursion of August 1913 must also be imputed to the shocked and upset state of Davis' mind on the death of his wife. The double loss of wife and of collegiate duties made his readjustment to society particularly difficult. The rest of his life was spent largely in creating a new role for himself.

A poem dated November 1913 shows that Davis visited relations in England * – perhaps an act of sympathy on their part – and that memories of home comforts were very much in the front of his mind.

> How happy he who leaving England new so far away
> Finds welcome in a house on Campden Hill
> Where gen'rous friends in England older let him come
> > and stay
> And lovely daughters call him Cousin Will
> How happy he for whom there's lots of porridge in
> > the dish
> And milk and sugar not to mention toast
> For whom there's always eggs & bacon when there isn't
> > fish
> And coffee, just the kind he likes the most

* He also gave his Grand Canyon lecture in Cambridge on 21 November, 1913.

And <u>who</u>, with <u>Lipsa</u> and with <u>Janet</u> by <u>his side</u>
Goes <u>forth</u> to seek a <u>Sunday</u> morning <u>walk</u>
Who after <u>taking</u> in a <u>motor bus</u> a penny <u>ride</u>
Alights to <u>stroll</u> along <u>within the Park</u>
How grateful he, who <u>sails</u> to England new from
                                        England old
Thru winds <u>and waves</u> so bleak <u>and rough</u> and chill
To feel himself <u>still</u> warmed <u>against</u> the <u>bitter</u>
                                        winter <u>cold</u>
By glowing memories of <u>Campden Hill.</u>

A footnote to this poem says that Lipsa (Elizabeth) Fletcher, (Ethel Parish Fletcher's stepdaughter) filled in the blanks (underlined) of the above verses written for her by W.M.D.

From a particularly abusive and rather wandering letter from J. W. Spencer we learn that his wife's death did not prevent Davis attending the Association dinners and functions. He may well occasionally have been short of patience and of tact, but whether he deserved the following castigation is doubtful.

It has come to the point where it is necessary to call you down upon the subject of your manners, or with regard to me, your need of good manners in my presence both privately, publicly and in my absence. Also in scientific matters you seem to be a bear with a sore foot.

You must remember your gruff rejoinder to my polite salutation to you at Princeton. This was doubtless meant to be offensive, but it was more amusing than provocative to see your ill-humor. I recall, how, at the Geological dinner at Washington you loudly announced, on seeing a cup still before me that 'when Spencer has finished' we will go on with the toasts. This was certainly discourteous and might have been taken as a public affront had it not been so pettyish and ludicrous, showing that you could not curb your impatience, under your overbearing disposition. You remind me of an American sea captain, of whom I was told in Guadeloupe, who during 5 years paid his sailors for only 18 months, by treating them so badly, using belaying pins etc, on them without cause, as to drive them to desert on reaching a port thereby losing their pay.

Furthermore, I understand that you do not hesitate to cold-shoulder one who speaks in a friendly way of another, who is non persona grata where this other is not obsequious, and has been compelled to distrust the soaring genie of speculative geology.

One cannot fail to have the conviction that your action arises from intense jealousy, which is really a tribute to the achievements of others, accomplished independently of and reaching a step outside of the official machine, with which merit may consist in speculating; and in revising in place of investigating, not even resisting a morsel not nailed down. Then you try to save yourself in abusing others and their labors, trusting that they will not be believed in opposition to your proscription. It is the story of every man who has accomplished anything.

Buckle, in his *History of Civilization*, has shown how official science leads to its own decay, by making the votaries servile and putting them in bondage to those who dispense the patronage.

You, yourself, have noted in regretful tone, the changes in geological research. Have you not put your interest in competition with duty, subverting independent research for official admiration, thus contributing to its decay and the dwindling of your own students. The decline of government science (in the bureau wherein there have been many of your students, one of whom said that probably the Germans failed to understand you, because they might not be familiar with your language) has been pointed out by President Van Hise. The official conditions, not conducive to making pre-eminent men are now somewhat ameliorated along with your retirement from active teaching, so as to permit of re-awakening individualism, which does not believe that $5 \times 1 \times$ official efficiency = 1, or $16 \times 1 \times$ official efficiency = 1. (*Letter: J. W. Spencer to Davis, 3 February 1914; Washington, D.C.*)

The letter must have worried Davis or have been something of a puzzle to him, for otherwise it is doubtful if he would have kept it. The concluding equation still baffles us! Davis' brief comment written along the top shows he was hurt by the reproach and was trying in his own mind to minimize its significance.

Receipt acknowledged 2/20/14. There must have been some hesitation in sending as it was not received until a day or two ago.

DAVIS' PACIFIC CORAL REEF EXPEDITION

There seems no doubt that Davis was still upset or unhappy in 1914 when Harvard University gave him a further grant from their Shaler Memorial Fund in order to inspect coral reefs in the Pacific. As a bridge-builder between geology and geography, it was natural that he should be attracted to the problem of coral reefs. In 1913 he published a review of Dana's recent contributions to the problem and almost certainly the grant was deliberate rather than coincidental. Apparently he asked Lawrence Martin, then a student, to go with him but Martin at the time was unable to accept. It may well be that the grant was augmented from other sources.

In 1914, a grant from the Shaler Memorial Fund of his university enabled him to visit many islands in the Fiji, New Hebrides, Cook, Loyalty and Society groups as well as Oahu, New Caledonia, and a long stretch of the Queensland coast inside the Great Barrier Reef of Australia. (*Daly, 1945, p. 278*)

However, the wife of a working colleague remembers that:

When Professor Davis had been retired for some time, his wife died. Shortly after this several of his ex-students arranged and financed a trip for him, to include an exploration of coral islands in the Pacific. (*Louie Hodge Lahee, 'Recollections of W. M. Davis'; provided for R. J. Chorley*)

Davis' first letter, written to his children when the trip was some way advanced, shows a restoration of his spirits. Several references indicate that

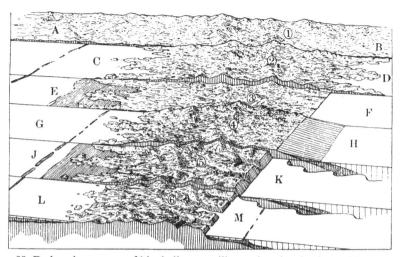

F I G. 89. Deduced sequence of block diagrams illustrating the physiographic development of New Caledonia. Unsymmetrical warping has depressed and embayed the south-west coast giving rise to a barrier reef and an absence of coastal cliffs (left: A, C, E, G, J, L). The north-west coast (right: B, D, F, H, K, M) has been raised and is fronted with sea-floor sediments and cliffs, having an absence of reefs (From Davis, 1928C, Fig. 117)

he had resumed his normal interests. On the S.S. *St Pierre* when skirting the coast of New Caledonia en route to the Loyalty Islands he notices:

> In the meantime drowned valleys were sailing past by the score; there was no counting them, but I sketched some of the best. The mountains are rather bare of trees, and some of them have no vegetation, but show all colors in their fancy soils. Their heights reach 3000, 4000, 5000 feet, so you can imagine something rather lofty and finely carved.

In the Loyalty Islands he describes the island of Maré, one of the four main raised reefs.

> Returning to Maré by night, I had a whole day there. The local délégué had been instructed by the governor to look out for me, which he did thoroly well. Off we went in his cart for a long drive, a saddle horse going ahead and another saddle stowed behind. On reaching a side path on the coral plateau, we mounted and went some three miles to a little hill, a few hundred feet across and thirty or more feet above the plain; Hagen had told me to see it, and the visit was well worth while, for it is the sole remnant of a great volcano, around which the reef had been formed. The volcano was pretty surely subsiding while the reef was growing up, but now the compound mass has been uplifted as a whole, and there you are, some 200 feet above the sea on a plain of lagoon limestone ten or fifteen miles across, and in the middle of it, this little volcanic summit. It was the easiest lofty mountain climb I ever made. In hammering off a specimen of the rock, a chip flew

up and cut my lip, so I am not so pretty now as usual. (*Letter: Davis to his children, 22 June 1914; Nouméa, New Caledonia*)

But to offset these there are sad passages which show he was still affected by the loss of his wife.

Good intentions instead of typewriting have played the mischief with my correspondence for a time past. Perhaps it is because I am so far away that writing seems useless, perhaps it is, if I must admit it, that I have been more lonesome than usual; in any case I have let the time slide away, just loafing on deck on my steamers in these remote parts, of which you don't know even the names ... I loafed a great deal, feeling disinclined to do anything else, except read a large French book on N.C. which Marshall of Dunedin has lent me for the voyage, and from which I have learned the French names of various kinds of reefs; but the author, poor fellow, rejects Darwin's theory on coral reefs and makes no application of the drowned valleys which abound all round the coast. His name is Bernard, and I knew him in Paris, where he was a professor at the Sorbonne. It would have been proper for me to use some of my abundant spare time in writing up previous notes, and I got as far as thinking of doing it several times; but it stopped just there: I wasn't in the mood, and when one is not in the mood for writing it is as well not to write. It would all have to be torn up. (*Letter: Davis to his children, 22 June 1914; Nouméa, New Caledonia*)

He seems unlucky too in his fellow-passengers or perhaps he was just not ready for company.

As for the other passengers, [on 'the fine old steamer, *Néra*' *en route* from Sydney to Nouméa] they did not attract me much; most of them, and not many at that, were half French, half English, and of the screaming kind. There was also an elderly Mrs. Manning; but when I heard her say to Mrs. White that she wanted to change her place at table, it seemed hardly worth while to explain coral islands to her. Among the more numerous second classers was a young fellow whom I had seen in the Sydney hotel and who turned out to be a Russian from the Caucasus, a mighty hunter by his own account, but not a mighty linguist, for his French is poor, his English worse, and his German nowhere. I took him at first for a travelling naturalist, judging by his many packages; finding him to be a hunter, I have not cultivated him much. The *Néra* is an ancient affair, once a great achievement I suppose, but now a back number. However, she brought us over without upsetting, and she afforded me some practice in French with our table steward, who tried his fragments of English on White, but spoke his own lingo with me: lucky for him he did, too, if he had tried English it would have cost some ducats at the end. The officers sat at one table together, and held themselves aloof; they did not care for our affairs. (*Letter: Davis to his children, 22 June 1914; Nouméa, New Caledonia*)

In other extracts Davis is more like his old self and seems to have enjoyed sporting his decoration.

June and July are to be devoted to French islands. Hence during these two months I am sporting the bright little red ribbon that indicates I am decorated. The effect is amusing. It was plainly visible when I went to the office of the French line, the Messageries Maritimes, in Sydney, to engage passage; still more so when I called on the French consul there – not decorated – to ask for information; he was expecting me, as word had been sent him of my arrival. That seemed rather impressive. (*Letter: Davis to his children, 22 June 1914; Nouméa, New Caledonia*)

In comparing British and French Colonies, the former receive something of a commendation.

He [i.e. Mr Hagen of Nouméa] telephoned to the governor's secretary, to know when I might call, and made an appointment for half past ten, and he gave me some excellent hints about the Loyalty Island trip. Next to the bank for some local money, and to leave instructions about letters. Then to the governor's residence, not a bad place, but so awful far below the residence I shared in Suva that it made me sadder than I was before. In fact Mousea is a depressingly forlorn sort of place, rather tumble down and ramshackle, whatever that is. The French are not in it with the English, when it comes to colonies. But the governor, I mean His Excellency, M. Brynet, was very cordial. He had been informed of my coming and wished to help me in every way possible. (*Letter: Davis to his children, 22 June 1914; Nouméa, New Caledonia*)

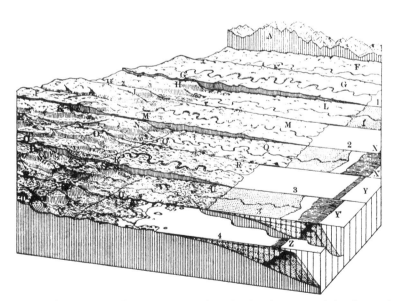

FIG. 90. Deduced block diagrams illustrating the development of the Queensland coast. A mountainous region (A) was reduced to a peneplain (FF′), which was uplifted and tilted (GG′). Then a lower peneplain (LL′) was formed along the coast. This uplifting, tilting and coastal peneplanation was repeated twice more associated with delta building and barrier reef development (From Davis, 1928C, Fig. 182)

A later letter to Lawrence Martin is freer of unhappy recollections and contains a humorous account of native dress but it also indicates that Davis is still finding it hard to concentrate on serious study.

My Dear Martin,

It was very kind of you to send me a letter – word from friends at home is very welcome on remote Pacific islands.

You may well wish you had come along, for I have had a wonderful trip and seen an extrordinary lot of fine things. Everything speaks for Darwin's theory of Subsidence – thus there are plenty of reefs that have been elevated since they were made. Travelling is very easy and usually comfortable eno. For example, I went all around the long island of New Caledonia in trading steamers and had a fine view of the coast as well as a lot of walks on shore. Finally I visited a selected part of the coast in a small sail boat for a week – and got some specially valuable points. The Loyalty islands were also seen – raised atolls! Now I am returning from a week in the New Hebrides – I bought most of a native's wearing apparel there, on the beach of the volcanic island. It consisted of a wild boar's tusk tied round his neck with a shoe string. When he took it off, I feared he would catch cold, there was so little left of him – chiefly a crescent of tortoise shell in one ear and a safety pin in the other. Such is life in the tropics. You would be amused at the way Governors and Resident Commissioners have been nice to me. In New Hebrides, the Resident's steam yacht was placed at my disposition for a very valuable bit of a trip. In Fiji I spent a luxurious month as guest of Government House – but the luxury was tempered by several absences on small sail boats, which pitched like . . . anything. A fair sized hurricane was included while on a steamer there.

In New Zealand, the professors of four Universities – Auckland, Wellington, Christchurch, and Dunedin, took good care of me. In Australia next month, I am to be guest of the British Association – and soon – Sept. 11 I sail from New Zealand for Tahiti – and may stay there a while – good chance to see some of the islands thereabouts. Date of return is therefore uncertain but probably November.

Good idea of yours to look ahead to Princeton, but I have no idea that Libbey thinks he is old eno' to retire. Let me know if I can help when the time comes.

I will try to write you something for the Journal – but my tries do not always succeed.

Give my regards to Whitbeck and other friends – not forgetting Austin Locke and Karl Young and their wives.

(*Letter: Davis to L. Martin, 28 July 1914; en route between New Caledonia and Australia*)

We shall return at length to describe Davis' work and travels in respect of *The Coral Reef Problem.*

THE SECOND MARRIAGE

Almost immediately after his return from the Pacific Davis married again. Details of the marriage certificate are given below from Volume 19, Folio 44, of the Record of Marriages in Cambridge, Massachusetts:

Number 1471. *Date and Place of Marriage:* December 12, 1914; Cambridge, Mass.

Groom: William Morris Davis	*Bride:* Mary Morrill Wyman
Residence: High Street	*Residence:* 31 Hawthorne Street
West Medford, Mass.	Cambridge, Mass.
Age: 64 years. *Color:* White	*Age:* 59 years. *Color:* White
Occupation: Teacher	*Occupation:* At Home
Place of Birth: Philadelphia,	*Place of Birth:* Cambridge, Mass.
Pennsylvania	
Father: Edward M. Davis	*Father:* Jefferies Wyman
Mother: Maria Mott	*Mother:* Adeline Wheelwright
Number of Marriage: Second	*Number of Marriage:* First
Widowed or Divorced: Widowed	*Widowed or Divorced:* Single
Name, Residence and Official Station	Samuel McChord Crowthers,
of Person by whom Married:	Minister,
	20 Oxford Street, Cambridge, Mass.

The lady was roughly Davis' own age and, from his son's account, comparatively well off.

> After his return in December 1914 he married Miss Molly Wyman, an elderly and absolutely ideally suited person for him. She had a comfortable home on Hawthorne Street in Cambridge.
>
> I have no idea of what the general circumstances were which led my father to marry Miss Molly Wyman, but I believe it was a most admirable choice because she was comfortably well-to-do, was sympathetic with his scientific activities, and he had no worries or domestic cares of any kind during the period of their marriage. She was a long time acquaintance of the family. Her father, Dr. Wyman, brought me into the world. She and her sister, or perhaps it was his sister, were members of what was called the Sewing Club, of which my mother was a member and so the casual acquaintance at least had carried on for many years. (*Interview: R. Mott Davis with R. J. Chorley, 7 September 1962*)
>
> The first Mrs D died about 1914, and soon after he married Miss Molly Wyman, a nice little old maid who had been a sweetheart of his youth and lived all alone in a big house right across the street from our house; WMD moved in and they were very happy for several years until her death. (*Letter: J. K. Wright to S. S. Visher, 29 March 1957; Lyme, New Hampshire*)

We happen to know from a written note at the end of Davis' typewritten account from Nouméa, New Caledonia (22 June 1914) that he was already very fond of Molly Wyman. To the typed copy sent to her he adds in ink:

> Dearest, you will understand better than the others who read this hurried letter, why I have been lonesome recently – I have done my best not to make too much of it – but it seems to me to stand out on all the pages – It is hard to write cheerfully – At every turn I ask myself, what will my dear Molly think and do and say – Will she send me an encouraging message? and if she does not, what shall I do – Go to the Solomon Islands for a year with Capt. Champion – *NO* – I will go to Cambridge and try to re-establish myself.

The happy outcome provided Davis with the wife and home he was longing for and it is significant that during the nine years of this marriage he made no further travels. So, except for his collegiate work, he largely restored his former way of life and for the next nine years happily slipped back into some of the old pattern.

For Davis the support and encouragement of a sympathetic wife was essential. With a wife beside him he felt more able to take his place within the family circle. The following poem reveals how much he valued presiding over such happy occasions – as well as his egocentricity!

OH arent we just the lucky ones to have an Uncle Will,
For in all entertainments he fills a double Bill.
He's just plain Billy Davis, or Prof. William D.
With a stack of golden medals as high as to your knee.
He tells children lovely stories – with a pencil in each hand,
He draws them awful dragons
And roars at their command –
He's mapped the Southern Heavens
Way down to Argentine
Globe trotted all the Continents
And Islands in between
He's dined with Fijee cannibals
And chiefs in Samarkand
And got degrees
From over seas

MY: Isn't he just grand!
Phi Beta Kappa sought him
And the audience sat up
While honors and distinctions
Pour down and fill his cup.
He paints the maple blossom
Makes hour glasses – or cranes –
And seeks new worlds to conquer
His castles are not Spain's!
In Institute or Layman's League
He's always in demand
Here's to his happy birthday
He surely 'beats the band'.

(*Poem by W. M. Davis, 12 February 1923*)

CHAPTER TWENTY-ONE

The First Rejuvenation

With remarriage Davis recaptured much of his former self-assurance and cheerfulness. He now continues to take a fatherly interest in the department's advanced students. Earlier writers have made many references to his generosity without giving examples; the following is an actual incident and gives some idea of the extent of his kindness.

> While Ellis W. Shuler was working on his doctoral degree at Harvard, the eminent geologist William Morris Davis became much interested in the young man. Just before Shuler's departure for Dallas in 1915 to begin teaching in the newly founded Southern Methodist University, Professor Davis called him into the book-lined study in the Morris home and said 'Here, Shuler, take all of these books. You'll need them in Texas'.
>
> Thus the geology collection in the University library owes its beginning to the interest of a famous geologist in his pupil. Included in this magnificent gift, which Dr. Shuler immediately made available to his students, were many items which a struggling university in its first year of classes could not have dreamed of owning. In addition to files of the U.S. Geological Survey publication, the first 12 volumes of the 'Bulletin' of the Geological Society of America and the complete run of the 'American Geologist', there were hundreds of monographs, studies and papers by geologists such as Lapparent, LeConte, Lyell, Dana, Jukes-Brown, Agassiz, Geikie, Marcou, Maury, and Viollet-le-duc.
>
> A stipulation in the gift made by Professor Davis prevented the legal transfer of these books to SMU, but that was only a technicality which never interfered with use of the volumes. The library catalogued and administered the collection even during the first few years when they were housed in Dr. Schuler's classroom. ('The Mustang Magazine', Southern Methodist University, Vol. VI, No. 5, March 1954, p. 12)

Similar generosity is shown by Davis' prompt response to a request from the daughter of a former Austrian colleague. In a note at the top of her letter he records that he had sent a $10 draft on 15 August and on 26 August had also sent $20 to U. Hess, the latter presumably being a kind afterthought of the plight of another German friend.

> Sir, You will kindly remember Edward Richter, Professor of Geography on the University of Graz, Austria (Stiria), of whom I am the daughter. In remembrance of my dear father, whom you visited in May 1899 at Graz, where I had also the honour to see you at our house, I take the liberty to ask you a favour.

You surely know very well the miserable conditions in Austria, therefore, I should ask you the great favour to send a dollar-packet to my sister, who lives at Vienna in great misery in respect of provisions.

With a dollar-packet you could help her a good deal.

I hope you will excuse my troubling you, and remain with infinite gratitude.

> Yours
>> devoted
>> Bertha Richter.

My sister's address:

Frau Dr. Mini Beil,

　Wien XIX Armbrusterstrasse 10.

　(*Letter: Bertha Richter to Davis, 26 July 1920; Kaslebrulh, Alto Adige, Italy*)

My Dear Miss Richter,

Your letter of July 26 reaches me today, in the country, and I have at once written to Boston to secure a food-draft in favour of your sister, Frau Beil, which will be sent to her in a few days.

You are quite right in thinking that my memory of your father makes me wish to help his daughters. I recall you both very well, at the time of my visit in Graz now over 20 years ago – so you are no longer young girls, but women grown. Your father's death was a personal loss to me but it sometimes seems that he is fortunate in not knowing the bitter unhappiness that has since then come to his country.

In reply to the compliment you have done me in writing for your sister, I do not wish to preach to you, more than to say that *I beg that both of you will do in future everything that lies in your power to avoid war.* Disputes are now settled reasonably, by legal means. (*Letter: Davis to Bertha Richter, 17 August 1920; Cambridge, Mass.*)

The lecture at the end is perhaps rather out of place but Davis was strongly against any resort to force and played an active part in peace campaigns.

The old habit of exchanging professional confidences also still continued. Apparently Gilbert was impressed by the importance of C. R. Keyes' study on wind erosion and wished to ensure that someone expert in the observation of such processes should be invited as a member of a field party organized by Keyes.

I am interested in what Keyes has been saying about eolian degradation – and I am quite skeptical. My skepticism depends partly on my acquaintance with our arid country and partly on a proneness of Keyes to see all things differently from other investigators – to put it mildly. His earlier papers on the subject were elusive because he omitted to mention localities, but a recent paper – or abstract – names several, and I see by the current number of Science that he is to lead an excursion for the study of desert erosion.

I hope there will be a competent physiographer in that party, and I am writing to suggest that in case you know one who hesitates whether to go you give him a push. It would be a fine thing for me to have someone whose judgment I respect take a look at Keyes' facts.

My motive in the matter arrives from the fact that I hope to write soon on Basin ranges, and I cannot either ignore Keyes, or accept him, or check him up by field examinations.

I've appreciated highly your coral reef studies. The 'home study' and field study combined make a fine example of the method of multiple hypothesis, and every such example fully set forth has educational value, and the problem itself was well worth while. (*Letter: G. K. Gilbert to Davis, 11 June 1915, Washington, D.C.*)

A later letter shows that Davis preferred to stay at home. Moreover he seems to have been hurt by Gilbert's failure to allude to his Powell Memoir (Davis, 1915F) and this despite a generous reference to his coral reef paper.

By all means send the mail to Johnson – and at the same time absolve me from the intentions of hinting that you yourself go with the desert erosion party. You have your own plans, and a plenty of irons in the fire. I merely tho't that you were likely to be in touch with men likely to join the excursion – and the same remark applies, of course, to Professor Johnson.

Yes, I read the Powell memoir and should have mentioned it but for the forgetfulness that is growing on me. I not only read and like it, but felt that the subject had great advantage from your fresh view-point.

If the biography was worth while it was worth while that you take the pains with it that you did – and I have the feeling that the life of Powell was worth recording, altho I am also conscious that I was too intimately associated to have unbiast judgment. (*Letter: G. K. Gilbert to Davis, 14 June 1915; Washington, D.C.*)

A further part of this correspondence is interesting because it shows that Davis made a habit of watching nominations to societies and was prone to canvas the merits of particular candidates.

I'll write Hagen about Lawson. Thank you for the suggestion. Not attending the meetings of the Academy, I do not hold in mind the list of geologists who are now members and so give little attention to nominations. (*Letter: G. K. Gilbert to Davis, 25 November 1915; Washington, D.C.*)

Though Davis had withdrawn from active participation in the teaching of geography, his name still counted for something abroad. A letter from England illustrates how his attempts to popularize physiography and raise its standing within the university syllabus had penetrated as far as the then largely undeveloped country of New Zealand.

I am sorry to bother you with so many letters; but a matter has cropped up, about which I should be glad to have your opinion.

The earth-sciences are considered of such slight importance in the syllabus of New Zealand. (I could not wish a better field for research; but the opportunities for teaching are very restricted and likely to be even more so.) I have been wondering some considerable time whether there might be a small opening for me in America; from which I might have the opportunity to advance to something better.

I should have reserved this matter until we met but for the fact that the chair of Geology at Otago University has become vacant owing to Marshall's resignation.

I have sent forward an application which may arrive there in time to be considered; but I seriously think of cabling to withdraw it.

Accepting the position, if I were appointed, would mean signing on for five years and this, I think, would be fatal to my chances elsewhere.

I am sorry to bother you with this; but I should very much like to know what you think of it.

Marshall has undertaken the headmastership of the Wangamin Collegiate School. (*Letter: C. A. Cotton to Davis, 16 November 1916; Rugby, England*)

This matter happened to be of great significance. Cotton became the chief disciple of Davis and from his insular fastness in the Southern Pacific poured a powerful stream of Davisian doctrines over the Northern hemisphere.

Applications to Davis for academic references as well as for advice continued to be frequent during this period of his second marriage. One of his

F I G. 91. Davis on a field trip in the western United States (Courtesy R. Mott Davis)

pet themes was the need to treat geography as an independent discipline in educational establishments. Quite how powerful his voice was within the realm of college administration will never be known but he must be credited with being the founder or co-founder of several existing university departments of geography. He continues to plead the case for geography with great skill and, in a letter to President Faunce of Brown University, makes the very apposite suggestion that geographical knowledge is an indispensable asset to those engaged in foreign trade.

You will recall that, for several years past, I have exprest the wish, in connection with recommendations concerning the Geological Department of your University, that Geography might be established on an equal footing with Geology. I now desire to present my feeling on that question more fully, and with the hope that such other members of the Visiting Committee as share the feeling in less or greater degree may indicate their position by supplementary statements.

Geology has a traditional position in our universities, because since about 1840 or 1850 its scientific and economic value has been imprest upon the American people by the work of state and national geological surveys. The science fully deserves the position that it has thus won.

Geography has no such position. It is still regarded by most persons in this country, even by most educated persons, as a school subject, not needing, indeed not capable of filling a place in our colleges. But the progress made by Geography in this country during the last 30 years, and in Europe during a longer period, suffices to show that this opinion is erroneous. Geography, properly presented, would be today of more general usefulness than Geology in a college education.

A general course on Geology with some field excursions, has a great enlarging value, in giving students some comprehension of the age of the earth, of the continuity of ordinary processes upon it all through its history; of the continuity and evolution of life; of the august antiquity of man. All colleges ought to offer courses of that general nature, and such courses ought to be taken by nearly all students, especially by students who do not intend to follow scientific careers.

But the special branches of petrography and paleontology, as well as courses on historical geology, altho now holding by traditional right an established place in many colleges, do not deserve that place as far as general education is concerned. They are special subjects, advanced extensions of certain parts of general geology; and as such, educational luxuries, because they are not called for by large numbers of students. True, they are essential for professional geologists, but even with the large demand that exists for geologists today it is not to be expected that any large fraction of college students will elect petrography and paleontology with a view of making geology their life work. The development of such courses, while geography remains undeveloped is, in my opinion, an educational mistake.

General courses on the Geography of the United States and of Europe, each occupying half a year or more, each open to freshmen, each a rounded entity, would not only have large educational value in general, but also larger application in other studies, such as engineering, history, and economics, which draw large numbers of students. It is indeed most unfortunate that most college students in this country have no college opportunity of gaining a better knowledge of their home country or of Europe than was given them by (mostly very inexperienced) teachers in their preparatory years – if indeed they studied geography at all after leaving elementary schools. The ignorance of important facts is commonly so great that it does not recognise itself. History is begun in school and continued in college, physics, chemistry, mathematics have similarly growing opportunities; so have the languages. Geography has not. This implies the absurd conclusion that young boys and girls can learn all that they need to know thru life about the geographical features of the world.

The above mentioned general courses on the Geography of the United States and of Europe should be untechnical; calling for no previous courses on Geography in college; with no special emphasis on scientific physiography, but with much attention to the general lay of the land, to climate factors, to natural resources, density of population, lines of movement, industries – all presented rationally, as a complex of interdependent facts. Further, these courses should be associated with other courses for more senior students, on physical geography, climatology, economic geography; and on other continents, especially at present on South America. The teachers of such courses should have some observational knowledge of the countries that they teach. They should be professional prepared geographers not necessarily geologists.

It should be noted that the course on physical geography may be about as broadening as the general course on geology, in the way of impressing the great age of the earth, the continuity of normal processes upon it, the conditioning of organic forms by their environment – but it would not include the evolutionary sequence of living forms. Like the courses on geology, those on geography should include some field exercises; the advanced courses should include field investigation.

Geographical courses have a distinct advantage over geological courses, in that the topics which they teach are much more easily visible to the general, non-professional traveller. In his later life he can recognise many of them from a train window; indeed, the very rapidity of a train has the advantage of giving a rapid summary of topographic features, which a prepared observer can easily appreciate – but which the unprepared observer fails to recognise. On the other hand, geological problems ordinarily require closer observation for their recognition than a passing traveller can give. The consequence of this is that geography will stay longer with a student than geology will; because geographical things are more readily seen than geological things – unless indeed the geological things are made to include physiographic features which belong equally to geography.

But more than this. Geography is destined to have an immensely larger economic use than geology in the coming half century and thus to extend more and more into the natural and commercial worlds. Geology will be called for in many lines of special exploitation. But geographical expertness, that is, a competent knowledge of various countries, will be called upon much oftener in the expansion of our commercial relations; an expansion which the long-visioned captain of industry is already preparing for.

Commercial expansion has thus far been largely accomplished by the try-and-try again method, without any scientific basis. Even the commercial agent who is sent to a foreign country with a view to establishing business relations there, ordinarily goes with next to no knowledge of its geography; and he is thereby heavily handicapped. A more intelligent method, in which expert advice and guidance is employed, must be developed, is, indeed, now in course of development of certain large corporations; which find it necessary to give instruction to their employees in the geography of the countries to which they are to be sent. But at present, this instruction is conducted at a double disadvantage; the teacher has no sufficient expertness, and the pupil has no sufficient preparation. There is a growing demand for better conditions on the part of exporting corporations. It is most unfortunate that so many of our colleges are not doing their share to remedy this backward condition.

For all those reasons, I am convinced that Geography should be given at the earliest possible date, a position at least as good as that which Geology has so well won for itself in Brown University. (*Letter: Davis to W. H. P. Faunce, 2 March 1920; Cambridge, Mass.*)

DAVIS' WRITINGS, 1915–23

During the nine years of his second marriage, Davis wrote – as distinct from published – about eighty-five major items, exclusive of repetitive reprints in various journals and scores of short reviews. To this period belong the half a dozen articles on the Mission Range, none of which approaches the high quality of most of his previous essays. The very fine discussion of 'The principles of geographical description' (1915K) although it first appeared in 1916 was essentially written during his first marriage – it is, as we have seen, mature Davis at his most majestic. It appears as an oasis in a desert, a gem among bric-à-brac. True some of the minor items of this dispirited period might have been claimed with pride by lesser men but Davis had reached an exalted state when undermined by widowerhood. During these bleaker, upset years he produced several articles on topographic maps and a few striking long reviews mainly of German works. His *Grundzüge der Physiogeographie* came out in a second edition in 1917(G), a truly unfortunate time for students in Germany. Among the best of his literary efforts were biographical memoirs of John Wesley Powell (1915F), John Peter Lesley (1915G), Grove Karl Gilbert (1918F) and Frederick Putnam Gulliver (1921E). These are competent, careful essays of considerable value to the historian of ideas on geomorphology, and we will return to them in a later chapter. Their personal tinge seemed to bring more out of the disturbed Davis than landscapes then could. However, a large part of Davis' writings from 1915 to 1923 consisted of thirty-six articles on coral reefs and on insular and coastal problems with particular relevance to the Pacific. Most of these are discussed together with *The Coral Reef Problem* in Chapter 25.

An increased amount of Davis' time during this period was taken up with reviewing, commenting and extending the work of others. This, for Davis, was a sure indication that he was passing through lean years as far as fluvial geomorphology was concerned and that all his intellectual vigour was being expended on the coral reef problem. Commenting on an article on Yünnan–Tibet drainage (1919C) he praises the fact that the article 'goes so much further into deduction than is usual in British geographical essays' (p. 413), in suggesting capture of headwaters by south-flowing 'second-order' tributaries fed by monsoon rains, but later castigates the author for not considering the possibility that the drainage might better be explained by postulating an earlier cycle of erosion during which many of the captures might have taken place – an explanation supported by the existence of incised meanders. Davis returned to the possible effect of topographic aspect on drainage development in his 'Deflection of streams by earth rotation' (1922A), a review of an article in which the author had suggested that the small-scale valley-side asymmetry of the south-flowing rivers of Long Island might be due to the action of the prevailing winds, rather than to the deflection of the streams. Davis reviewed similar suggestions for the asymmetry of the valleys in the Lannemezan plateau of south-western France, pointing out, however, that the steep valley sides are west-facing in Lannemezan and east-facing in Long Island, leaving the influence of aspect over slope processes very much in doubt. Continuing to propagate his ideas on the explanatory description of landforms, Davis (1922M) praised 'Dixey's physiography of Sierra Leone', although he felt that deduction had not always been carried far enough, concluding that 'the existing features are vividly presented and their explanation is subordinated to their description; therein lies the essence of well-conceived explanatory physiography' (p. 11).

During this period, perhaps because of his work on the biographies of G. K. Gilbert (1918F and 1927F) Davis interested himself once more in the problem of Basin Range Structures and faulting displacements. In a review of Suess' *The Face of the Earth*, which attempted 'to determine the general structure of the earth's crust and to infer therefrom the character of the larger movements by which the crust has been deformed' (1920J, p. 226), Davis again attacked Suess' denial of radial upheavals of fault blocks (as distinct from tangential ones) avowing that 'nothing better illustrates the ponderosity of Suess' mind than the tenacity with which he held to this thesis in spite of the enormous magnitude of the subsidences that it involves' (p. 232). However, Davis praised De Margerie's French translation of the work and suggested that the time was ripe to launch an international project, under De Margerie's direction, to carry out the revision and extension of Suess' work on the large geological problems. After rather unfavourably reviewing J. W. Gregory's subsiding keystone theory of African rift valley formation (Davis, 1920A), Davis at the Amherst Meeting of the Geological Society of America (1922H) returned to Gilbert's ideas on Basin Range structure. Describing Gilbert's later idea that block faulting might be the superficial result of stretching and underdrag on 'the other side of an over-thrust', he suggested a possible alternative mechanism involving upheaval, rather than horizontal extension:

> It may be imagined that spur-end facets, which have been regarded as parts of moderately inclined block-mountain fault-planes, are really landslide surfaces of much less declivity than that of true fault planes between the mountain blocks, and that great slabs of the steep-faced blocks slipped down these surfaces into the intermont depressions while the displacement was going on ... (*Davis, 1922H, p. 94*)

There are, however, some general undertones which must be considered. In 1917 and 1918 Davis became associated with American involvement in the European war. After this war he shows a distinct rejuvenation both of literary quality and propagandist energy; he largely regains his former composure and certainty, but there now appears also an explicitly religious note.

DAVIS AND THE FIRST WORLD WAR

The war which broke out in Europe in 1914 soon began to disrupt educational courses in the universities of the belligerent countries, as the following letter from Albrecht Penck shows:

My dear Davis,

Many thanks for your letter of 19 May and for your efforts to oppose the untrue reports about myself on the basis of the statements supplied by myself. I have decided to relate my experiences in Australia in a booklet, a copy of which I enclose. You will see from it what a very distressing situation I was in and what incredible lies were disseminated about me. But it is a fact with which we shall have to reckon increasingly: the English will ruthlessly twist the truth in order to gain certain objectives.

Your essay on Valdarno already appeared in 1914 in the *Zeitschrift für Erdkunde*, pp. 565 and 665, and I hear that you received reprints. But unfortunately, an entire mail shipment gets lost from time to time, when the English throw mailbags, coming from Germany and destined for America, overboard. I should be sorry if that had happened to a consignment addressed to you.

I am now busier than ever in Berlin, since all the young men are at the front: of six assistants only one remains; I must therefore perform a lot of work which I would not otherwise be required to do. But the great times in which we live also greatly increase productive capacity – it is astonishing what is being done here. In Berlin, for instance, the Friedericiastrasse railway station is being altered and considerably enlarged. Also an Underground line running north to south which supplements the line to the West running to the Nollendorfplatz. This is much more difficult here than in New York rock or London Clay. In the whole of Germany new towns are being built, growing to 60,000 inhabitants in 10 months: we have to house more than 1 million of Russian and French P.O.W's, and since we treat them well we build them decent houses. Recently, I saw such a town. It looks like a town in the Wild West – only it is well fenced-in and, therefore, the houses are rather close together.

My lectures are less well attended than before: only about one-third of my usual audience. During Whitsun I made an interesting trip to Thuringia with 12 participants. I came across some entirely different types of valleys. They are joined to anticlines (Sättel) and have arisen by the upper rock salt (Steinsalz) dissolving in the anticlinal region so that a cavity arose, which does not at all fit into the river network.

I have good news from Walther. His young wife stayed with us during the last weeks. Ilse also appeared to present her youngest child. This gave joy to the grandparents, both of whom send their heartiest greetings!

Your devoted

(A.) Penck

P.S. I recently had this letter returned with the remark: Retour – Pas de communication. They probably wanted to send it to Cambridge, England. Your separata from the *Zeitschrift der Gesellschaft für Erdkunde* have been sent off!

(*Letter: A. Penck to Davis, 24 June 1915; Berlin*)

Conditions were very different in the United States where the European war did not make a strong impact until 1917 when the German submarine attacks on trans-Atlantic shipping reached their peak. This coincides with the first mention of Davis' association with the preparations for war. He was too old to take any active part and as far as we know his sons did not enlist in any branch of the services. Despite the Quaker abhorrence of war and their strict rules against the participation, Davis followed his father's example. He had not officially belonged to the Quaker faith since his father's expulsion so the

FIG. 92. Diagram by Davis of a hanging tributary glacier

decision was hardly as convulsive in its effect on his life. Nevertheless, it may have meant some relaxation of his convictions unless, like his father, he felt that joining in a war for a right cause raised an exception. Davis, we know, eschewed war as a means of solving international problems. Yet where an attack had been made, the preservation of right and established principles became a greater responsibility. The part he played was very slight in any case. He sat as a member and chairman of a government committee on geography for the National Research Council and assisted in the one way he knew – the adaptation of geographical techniques to the needs of war. The committee's suggestions, though well intended, do not seem to have met with any enthusiasm from the Army authorities.

I suggested that the publication of the U.S. hand-books should be undertaken by the Geological Survey, and am glad the scheme went through. Ward thought it had been suggested before, but if so it did not impress the Committee and was not in the plans when I brought the matter up at Philadelphia. So if that suggestion was a benefaction, we will let it cancel any harm my letter to Keppel may have done.

As to the latter, I did not make the suggestion to Keppel merely as an individual, but as a member of the Committee. I stated first the Committee's decision at its Philadelphia meeting, reported the 'feeling of our Committee that fuller training by map reading and map interpretation for the prospective officers of our new armies' is desirable, said I thought it possible I could 'secure through our Committee the active co-operation of a body of volunteer instructors', and would be glad 'if in this way, or in any other way' my knowledge of maps could be made of service to my country.

Properly I should have communicated with Keppel through the Committee. You are *quite* right in your criticism. My failure to do so arose from three facts, which I state in explanation, not as excuses. First, I had plans for instruction in map reading for our soldiers under way before I knew the Committee would take the matter up, and should have taken the steps I did independently of the Committee had the Committee not considered it. In the second place, I wrote Keppel as a personal friend, believing his personal knowledge of me and my work would be more apt to secure a hearing for the plan than any formal recommendation from a geography committee. Finally, I was mainly interested in the result to be secured, and gave too little attention to the method of securing it.

I do not think the practical result will be in any sense harmful, and it may be helpful. I had another conference with Major Dorey, and am extremely doubtful if anything will come of the effort, whoever pushes it. He approves but thinks the army men will rely on army officers who can give the men some map work and drill them in other requirements at the same time (*Letter: D. W. Johnson to Davis, 26 April 1917; New York*)

Perhaps characteristically, Davis is mainly upset that Johnson should have acted on his own initiative and not under the authority of the committee. In all things Davis viewed correctness as almost more important than doing the job itself. The reader may remember how in Germany he took great care to ensure that he did not err in the use of titles.

The disagreement on approach did not end there. Davis took up the further issue that a committee should not stand on their dignity but be prepared to undertake any task, however trivial it might seem. Johnson took the opposite view; he thought that acceptance of obviously unimportant duties would lower the reputation of the committee.

Your letter of the 22nd duly received. I stand corrected as to the wording of the initial proposition regarding the National Research Council's Committee on Geography, but think your quotation from Chairman Hale's letter establishes the correctness of my general contention, namely that the Committee is supposed to concentrate its efforts for the time being on problems of national defense. Our object, of course, is precisely the same, namely to achieve the greatest possible good for the country; but we differ a little as to the best method of attack. I quite understand your point of view – that if the Committee's efforts to accomplish substantial results are not appreciated, we should do what you call 'baby work' in order to make sure that our existence is not wholly in vain. I have a very strong opposite opinion. It seems to me that our policy should be to aim at large ends compatible with the dignity of the National Committee. When these are turned down, then we should quietly bide our time and try again when a strategic opening appears. If the Committee devotes energy to the performance of baby work, we will, I believe, unfavourably affect our chances for doing men's work later on.

In my own case I have on several occasions offered my services for undertakings on a big scale, and have been as often turned down. In the meantime I have refused several possible openings where the opportunity for accomplishment of anything but trivial results seemed to me very small. I do not mean to imply that even the smallest task which really helps toward winning the war is any less dignified than the biggest one. I refer, of course, to tasks which seemed to me small because in this world crisis they would not contribute toward the end we all have at heart. I feel sure that in due time our Committee will have an opportunity to perform effective service, providing we use tact and patience and do not arouse unnecessary criticism by undertaking tasks which to most men might appear relatively unimportant under present conditions.

As I say, I have myself been disappointed, especially since the President's refusal to accept Roosevelt's services cut me off from special geographical work in his division – I have felt disappointed that my efforts to do something worth while had resulted in nothing tangible up to a recent date. I am now happy to say, however, that I am to go to Europe in two or three months on a special war mission which will probably take me to the French, Italian, and Balkan fronts and enable me to use my physiographic knowledge to some real advantage in the country's service. This you may regard as confidential for the present, until I am able to speak more freely about it.

In fact, I have rather an embarrassment of riches just at present, as some of the patriotic societies offered to send me to Europe under conditions approved by the War Department, for the purpose of making observations at the front and returning to participate in some of the speaking campaigns planned to arouse the patriotism of the country. The leaders of the American Defence Society have also asked me to take charge of a new publicity bureau which they expect to establish in Washington at the request of the Creel Committee, and with the express approval of President Wilson, with the object of assisting in the Government's campaign of education as to the causes of American participation in the war and the objects which we have in view. This work of national scope appeals to me very strongly and I am sorry that my plans for work in Europe will make it necessary to let such an opportunity for service pass. As my last lick in the American Rights League, I planned a series of advertisements in a campaign to educate the people to an understanding of the motives and objects of some of the pacifist organizations, particularly the People's Council; presented the matter to a Mr. Gregg last Sunday and he put down a check for $5000 as his contribution toward the work. I have the promise of several thousand more from two or three other parties. Think I can put ten or twelve thousand dollars back of the advertisements without much trouble. (*Letter: D. W. Johnson to Davis, 6 December 1917; New York*)

For once, Davis' views were perhaps the wiser. Johnson's attitude was probably coloured by satisfaction with his European appointment. The reference, towards the end, of the need to convince the American public of the reasons for joining the war is interesting in view of what has been said before.

Despite disagreements on policy, there can be little doubt that the course of Davis' geomorphic thinking was to be influenced by the leading political roles assumed by his two former students Isaiah Bowman and Douglas Johnson. Bowman, who had assumed the Directorship of the American Geographical Society in 1915, was appointed Chief Territorial Specialist to the American Delegation to the Paris Peace Conference (December 1918–May 1919) (Wrigley, 1951). Johnson, a major in the Intelligence Corps, also attended the Conference and during both the First and Second World Wars was violently anti-German, being the author of a number of bellicose political publications. Johnson was also a francophile and was in the habit of keeping a book of French literature on the breakfast table which he read daily to his blind wife (Lobeck, 1944; Bucher, 1947). Although Davis was very much his 'own man' in moral matters, there can be little doubt that the ideas of his students combined with his horror of war and political authoritarianism to give added zest to the geomorphological confrontation which he was to have with the Pencks.

Davis now busied himself with geographical literature for the benefit and enlightenment of the American forces in France. In 1917(F) he produced a booklet of 27 pages on *Excursions Around Aix-les-Bains* on behalf of the Y.M.C.A. National War Work Council. In the following year he published through the Harvard University Press a more substantial *Handbook of Northern France* (1918A). This is skilfully done, being precise, lucid and most clearly illustrated. The diagrams of the numerous scarps that encircle the

Paris basin have never been surpassed, even by the French – which is praise indeed. By the first of April 1919 Davis had made a net profit on this small book of $3127.

We hear more in Davis' correspondence about the National Research Council's Committee on Geography but many or most of his proposals seem to have been abortive. A year after the end of the War Johnson sends a long account of the important role played by geography at the peace conferences. Visher tells us that Davis eventually lost confidence in Douglas Johnson and in the later years of his life was highly critical of him but at this moment their relations are still very good.

> Your letter of September 30 was much appreciated particularly your commendation of my 'Shore-Line' book. I need not tell you that the superficial comments of those who scarcely trouble to look within the covers of such a volume do not particularly interest me; but words of approval from one who not only ranks first in the world in ability to express a competent judgment, but has also read and criticised freely the manuscript in preparation, are most highly prized. Nothing which anyone has written me about the book has given me so much satisfaction as the little paragraph in your letter.
>
> Pardon me for rambling on so indefinitely about personal matters. My only excuse is that you have always taken a personal interest in me and my work, and hence I am tempted to be more indiscreet with you than others. Mrs. Johnson joins me in sending cordial regards to you and Mrs. Davis. (*Letter: D. W. Johnson to Davis, 13 October 1919; New York*)

Johnson's impressions of the peace conference are of great interest and it is unfortunate they were not fuller.

> I must tell you personally, however, that the role played by geography at the Peace Conference was very much greater than I had hoped, or had reason to expect. At first the geographical and other experts did not get much direct contact with the Commissioners, and none with the President. As time went on, however, they demonstrated their value, and their role continually increased in importance. There was a time when some of the 'higher-ups' thought it quite possible to dispense with a large number of the experts, but in the end our difficulty was to force a reluctant consent to our return to University duties in America.
>
> I got really into the inside of things, both as a member of the Jugo-Slav and Roumanian territorial Commission, and the Central Territorial Commission (as one of the two American representatives on each of these commissions), and a member of the central steering committee which kept its finger on the pulse of all the work of the Conference and aided the Commissioners in pushing their work effectively. As the frontiers were in many cases entirely determined by the territorial commissions or by geographical sub-committees working under the direction of the territorial commissioners, I had an unexampled opportunity to make geographical factors count in the solution of territorial problems and in the delimitation of inter-national frontiers. I found my colleagues well disposed toward accepting geographical arguments on frontier matters, and it is some satisfaction to be able to point to a number of the new European frontiers and say

that they are where they are because geographical considerations which I urged upon the Commissioners were accepted and acted upon.

Our Commissioners came to have much respect for the experts' opinions, and practically every day I and one or more of our other experts were called before the Commissioners to discuss with them the problems which had come before the Supreme Council that day. We then sat in the Supreme Council, behind the President or Mr. Lansing, or whatever Commissioner was representing America, advising them during the progress of the debate. I may say confidentially that I was the principal adviser to the President on the difficult Adriatic question, and naturally saw a great deal of him in this connection and was also called into consultation by him on the Klagenfurt and other frontier questions. I must say for him that he has a very keen appreciation of the importance of the geographical factor in the problems of the European settlement, that he showed a surprising ability to grasp and use the material which we put at his disposal, and that he accepted and acted on our advice to a degree which was most flattering. The line which is known as 'the American Line' in the Adriatic tangle, is the line which I proposed to the President as the proper geographical frontier between Italy and Jugo-Slavia, and to which the President has steadfastly adhered throughout all the negotiations. The Italian papers commented freely on the extent to which President Wilson was relying upon his experts, and while some of them bitterly assailed us, and especially myself personally, others testified to the high quality of the work done by the American experts, even where their conclusions were adverse to Italy's claims. The London Times editorially paid a high tribute to the work of the American experts in connection with the Adriatic issue and advised the Italian government to rely on the American advisers for the real facts in the Fiume controversy. Take it altogether, I think you have grounds for gratification in the assurance that never before in the history of peace conferences has geography played so important a role in the solution of great world problems. (*Letter: D. W. Johnson to Davis, 13 October 1919; New York*)

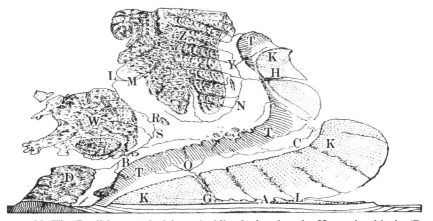

F I G. 93. The English coastal plain and oldland, showing the Hercynian blocks (D, W, Z) and the Jurassic (T) and Cretaceous (K) escarpments. The other letters relate to cities, rivers, gaps, etc. (From Davis, 1912J, Fig. 99)

It must be recorded that few, if any perhaps, of the boundaries suggested remain. This does not seem to say much for geographic method but in truth they probably suffered more from political illogicality and generations of racial tension. It was a little naïve, though possibly worth while, of President Wilson and his advisers to imagine that the European habit of changing frontiers by force (*faits accomplis!*) was likely to be replaced, with any permanent success, by a far more uncertain, if fairer, method.

Whereas Johnson tends to be gratified by the sudden political recognition of geography and perhaps even more impressed by his own vastly increased importance, Davis seems to have grasped the uniqueness of the moment and the opportunities offered to make valuable additions to academic knowledge.

> I doubt whether I shall attempt such a general account of the status of geography in the minds of the men I met at the Peace Conference as you suggest. I think you are right in saying such a summary would be enlightening, but my time is crowded with things which seem to be for the moment more necessary.
>
> My difficulty is to cut the work before me down to dimensions comparable to the time at my disposal. (*Letter: D. W. Johnson to Davis, 13 October 1919; New York*)

The suggestion could have produced a rare and very revealing study rather akin to the many official volumes on various aspects of the last war. It is a pity that Johnson was unattracted by the suggestion. The frequency with which Davis produced perceptive notions of this sort is one of the characteristics most praised by his students and a key to the rarity of his talent. As the American advisers at the Peace Conference began to return home much more became known of what went on and to what extent geography was taken into account.

In an earlier extract we have seen the relish with which Davis recounted to his sponsor the indirect benefits of the 1912 expedition. Though he had no intention of becoming the director, he was interested enough to press his sponsor to support a similar tour of South America.

> But there is another lesson that I am disposed to draw from it. Namely, that the possibility of a future greater excursion should be borne in mind, and its realization not too long delayed. The greater excursion that several of us already have had in mind is around South America, with trips inland from a number of ports on both coasts. And by peculiarly good fortune, you have in Bowman the very particular one man in North America most competent by training, by travelling and above all by native ability, to plan, organize and direct such a party. It will cost a little more than the excursion of 1912; a good deal more; but it will be worth its larger cost; for it may well be directed largely to commercial geography and to a triangular cementation of Europe, North America and South America.

An excursion of that sort will be much more widely known, over the world, than our excursion of 1912. It will have every opportunity, one may safely assume, from the S.A. countries. Also from the various N.A. commercial establishments interested in S.A., such as the National City Bank, and many others. If commercial geography be given the lead, as it well may be, the attention of many persons will be aroused, who never turned to look at the excursion of 1912. Indeed, if this Round South America Expedition be developed broadly, with such support as New York City can give it, it will be a resounding affair over the world, and the A.G.S. will step farther ahead than it is now.

One other reminiscence I must allow myself. Do you remember that some 12 years ago, after Bowman had asked for a fund from the A.G.S. for S.A. travel, and had been refused, I said to you: If you intend to use any money in that sort of work, I think you have made a mistake in rejecting Bowman's application. And on my request you allowed him to make a second application, which proved successful; and he got the needed money; but he got much more at the same time, he got the favourable opinion of your committee, and as a result, look at him now, your most successful director!

So I propose to regard his present position as – in small part – to my own credit, when things are finally added up. But at that time, altho you wrote me a time ago that you are not a geographer, your geographical credit will stand high, because you have had so large a share in making geographical things go. Do not be surprised, therefore, when I express the hope that you will have a hand, with others, in making the Round S.A. Expedition go in its good and due season.

(*Letter: Davis to A. M. Huntington, dated 21 February 1920; Cambridge, Mass.*)

The directness of his approach is instructive. There have been many allusions to his blunt and unconciliatory manner which it might have been hard to believe if it were not for known examples like this. Yet the arguments are well made and the point on commercial relations is a sound one. Though he was always primarily a geographer he kept up with allied subjects and current trends in politics and trade.

POST-WAR REJUVENATION

By 1919, after five years of retirement and remarriage, Davis had regained his old stride although his pace was now slower and more measured. His new interest in tectonic topics, such as faults and rift valleys, reveals his close connection with G. K. Gilbert who died in 1918 and whose biographical memoir was entrusted to Davis. These articles on tectonic themes have, however, little lasting value.

Also new, were certain religious themes which seem to be partly the aftermath of or reaction to the horrors of the war in Europe. But such an approach is often, as we have shown in Volume One of our *History*, subconsciously an integral part of the study by man of the natural environment.

Davis' religious beliefs had so far expressed themselves through his character – in a stern rectitude in all matters requiring a moral choice and an abhorrence of war. After his retirement he began to experiment with two lectures that were essentially religious in purpose. They were generally presented as public lectures and became almost as well known as his Colorado Canyon performance. The first of these was entitled 'The reasonableness of science' (Davis, 1922F), in which Davis examined whether it was possible to employ scientific means to discover the quality of goodness. He began with a fable regarding the interpretation of tidal oscillations to illustrate the fourfold nature of modern science

> The moral is that the observant hermit, the alert inventor, the thoughtful recluse, and the judicial onlooker represent not four different individuals, but only four different mental faculties in a single individual, the trained man of science, who uses his powers of observation to discover the facts of nature, his inventive ingenuity to propose various possible hypotheses for the explanation of the facts, his power of logical reflection to think out, or deduce, from each hypothesis, in accordance with previously acquired, pertinent knowledge, just what ought to happen if the hypothesis were true, and his impartial faculty of verification to decide which hypothesis, if any, is competent to explain the observed facts.
> (*Davis, 1922F, p. 196*)

Davis made great play of the importance of theorizing and of the exercise of 'rational credibility' in science.

> There is a popular prejudice against the use of the inventive faculty, ordinarily called theorizing. Theorizing alone, mere theorizing, is certainly of little value; but trained theorizing in proper association with trained observing is absolutely essential to scientific progress. The chief reason for this is that our observing senses are of limited power. We soon reach the conviction that many facts of nature elude direct observation, either because their medium is inherently transparent and intangible, or because their dimensions are submicroscopic, or because their time of occurrence lay in the irrecoverable past. And yet all of these unobservable phenomena are in their own way just as much a part of the natural world as observable phenomena are. If we wish really to understand the natural world, surely those of its phenomena which are not immediately detectable by our limited senses must be detected in some way or other; and the way usually employed is – theorizing. (*Davis, 1922F, pp. 197–8*)

> It is not to be denied that much credulity is called for in this daring search for the unobservable facts of the natural world. Science, however, is not alone in credulously building up an unseen world to complement the seen world. That has been done by non-science also for ages past. But the credulity involved in the two cases is unlike. In the latter the credulity is whimsical, fantastic, irresponsible, incoherent; in the former it is orderly, controlled, rational, coherent. (*Davis, 1922F, p. 199*)

> But what does a man mean when he says that he believes the scheme of the geographical cycle, with its imagined yet unseen changes of land forms and its inferred yet unobserved changes in the distribution of plants and animals. He

ought not to mean that the truth of the scheme has been absolutely proved, not only that it has been given a very high order of probability; for that is, as a rule, the nature of what is often called scientific demonstration. He ought to recognize also that many generalizations on which the argumentation of the scheme rests are likewise not absolutely proved: for example, the persistence of the present-day order of natural processes through hundreds of millions of years of past time; to say nothing of the unbroken continuity of time itself! Who can prove the truth of those generalizations in any absolute sense? Nevertheless one accepts their truth because he finds, after due inquiry, that they too appear to have a high order of probability. (*Davis, 1922F, pp. 201–2*)

In his definition of the 'natural history of goodness' Davis showed the inherently rational approach to ethics which was to lead him ultimately towards the Unitarian faith:

The natural history of goodness is therefore concerned with the concrete opinions and actions of ordinary men in commonplace, every-day life, and has nothing to do with the abstractions of metaphysics regarding absolute and eternal ideals. In that respect it might be compared with the natural history of mathematics, which would portray the efforts of early man in gradually and tentatively developing the multiplication table, but would have nothing to do with the metaphysical pre-existence and everlasting verity of 7 times 9 being 63. For in the same way the natural history of goodness would, if it could, describe the first recognition and the later modification of various ethical principles by certain peoples in certain places at certain times under certain conditions, but it would take no account of the metaphysical view that all ethical truths are eternal, as if they had existed by themselves somewhere in the interstellar spaces of the universe for untold ages awaiting recognition. (*Davis, 1922F, p. 204*)

The Israelites' view was, if we are to take their records literally, that their understanding of good and evil as well as their decrees for the reward of good and the punishment of evil, came to them by supernatural relevation; and this view was adopted by all Christendom in later centuries. The modern view, more and more widely adopted now, is that the Israelites derived their knowledge and their decrees concerning good and evil in a perfectly natural way from a perfectly natural source; namely, from their own perfectly natural experience, the same source from which all other human knowledge is gained. The decrees and commandments were not sudden acquisitions, but merely expressions of the morals and customs gradually developed among the people and formulated by this their leaders. (*Davis, 1922F, p. 206*)

Has science indeed anything to do with these religious matters? It has of course to do with the earth and the stars, with plants and animals, with steam and electricity; but by what right does science concern itself with questions of good and evil? It does so by the same right, precisely the same right that it studies the tides as governed by the moon and the sun, and the slow changes of earth's surface when lowlands are raised to highlands and when highlands are worn down to lowlands. For the observant and thoughtful study of mankind discovers many facts of opinion and facts of action concerning things that are regarded as good or bad; and all those facts, which together with their causes and consequences are included under the natural history of goodness are just as properly open to scientific inquiry, that is, to unprejudiced, reasonable inquiry, as any other facts in the world. Nevertheless, the feeling that science is a trespasser on such ground is often met; and also the allied feeling that science is too cold and hard to deal with such questions. (*Davis, 1922F, p. 210*)

[Some scientists] introduce their would-be rigorous methods of thought into daily intercourse with their neighbors, and are logical when they ought to be genial, argumentative when they ought to be sympathetic; in short, they are very tiresome fellows, and they do science a disservice. No wonder that a gentle-minded person would hesitate to trust scientists of that sort with decisions on delicate problems of right and wrong! But on the other hand, even the best of science is judged cold and hard with little reason by certain sentimental and emotional persons who, reclining on soft couches of prejudice and downy pillows of preference, are intellectually too indolent to face the problems of life fairly and squarely; too unreasonable to subject their opinions to candid scrutiny; too undisciplined to change their beliefs even in the light of compelling evidence. They know nothing of the calm and clear spirit of free inquiry; they are unwilling to follow free inquiry to an unwelcome conclusion; for example, they reject the philosophy of evolution because, as they fastidiously phrase it they do not like the idea of being descended from monkeys. I do not believe we need take their condemnation of science as being cold and hard any more seriously than their rejection of evolution because they do not like it. (*Davis, 1922F, pp. 211–12*)

Davis even concludes that the natural history of goodness can be taught by the case method.

In such instruction the nature of the subject should not be set forth so much in impersonal generalizations as by the 'case method', the same method which Louis Agassiz so successfully introduced into the study of zoology, which Langdell with equal success applied to the study of law, and which is now increasingly employed in the Harvard Graduate School of Business Administration as the best means of inculcating sound business principles. Indeed, the natural history of goodness lends itself remarkably well to this method of presentation; for its facts may be set forth in collections of concrete examples of various kinds of behavior, concerning which the pupils may make their own judgments and generalizations; and such collections of examples may be graded from elementary to advanced, so as to afford excellent material for individual exercises from early school years onward. (*Davis, 1922F, p. 213*)

While there may be an initial inclination to declaim his confidence in scientific method as naïve – after all, many centuries of research on the meaning of sin and moral values show the same contradictions that faced earliest man – nevertheless the present trends in delinquency and family welfare work illustrates that the need for some approach is undeniable. As a diviner of ideas, both within his own discipline and in other spheres, Davis was often far in advance of contemporary thinking.

FIG. 94. Dragon drawn by Davis for the children of his son Edward during a visit to Shirley, Massachusetts, in the early 1920s. Each half was drawn with a different hand (Courtesy W. M. Davis II)

Davis' second religious theme was 'The faith of reverent science' (1934C). As this was his final testament of faith, we publish it in full as an Appendix.

Of Davis' writings on landforms, other than coral reefs, at this time, we would select two as truly characteristic of his rejuvenation. The first, 'The young coasts of Annam and Northern Spain' (1919B) typifies an old method of instruction and propaganda; the second 'Peneplains and the geographical cycle' (1922I) is Davisian advocacy at its best.

In the article on coastlines Davis used a method he has followed in dozens of articles since 1910, treating the description of land forms. He takes a detailed description of a certain area by a well-known physiographer and, if in a foreign language, summarizes its content in English and recasts it as far as possible into a Davisian mould. Here his skill in languages is a great asset and enables him to present to his countrymen foreign literature they might otherwise miss. The first part of 'The young coasts of Annam and Northern Spain' is based solely on 'Plages soulevées dans le nord de l'Annam' by E. Chassigneux (*La Géographie*, Vol. 32, 1918, pp. 81–95). The gist is that the mountainous border of the Gulf of Tonkin has an embayed shoreline of sub-recent or recent submergence whereas the coast of Annam south of the great delta of Red River has two series of emerged longshore beaches, one at 3–4 m. the other at sea-level. Also, very strikingly coral reefs are absent. The second part of this article is based on J. Dantin Cereceda's 'Evolución morfológica de la Bahía de Santander' (*Trab. de Museo Nacl. de Cienc. Nat; Ser. Geol. No. 20*, Madrid, 1917). This account traces the physical history of the northern Spanish coast around Santander. Davis rather surprisingly admits the necessity of giving quantitative details of amounts of relative elevation and depression.

> As in all condensed explanatory descriptions, the qualitative statements regarding the Annam and Cantabrian coasts, given above, should be supplemented by detailed quantitative statements, in order to bring the local facts clearly before the reader: but the quantitative details are best apprehended by a mature reader if they are introduced by a concise qualitative summary. (*Davis, 1919B, p. 179*)

But his main concern is to demonstrate the difficulties of deciding on a general set of technical terms.

> The partly submerged coasts described in both of the articles above reviewed are stated by the authors to be in the stage of 'maturity'. It is gratifying to perceive in the use of this term that the systematic evolution of coasts, during which they pass through a well-defined succession of forms in orderly sequence, is recognized as of value in physiographic description; but it is questionable whether 'maturity' is the best term by which to designate the stage of evolution reached by the coasts in question. An embayed coast must, if it stands still through an entire cycle of marine abrasion, very soon reach the stage of slight headland cliffing and partial bay filling; it must somewhat later reach the stage of complete headland trunca- tion, when the steep cliffs, usually of increasing height, are cut back nearly or quite as far as the initial bay heads; and it will at a much later time reach the stage of far-retrograded, slanting cliffs of moderate height, when the slowly retreating shore line has receded well into the original interior highlands, which will then have lost a good share of their initial height by general degradation. A complete cycle of marine abrasion may thus be divided into early, advanced, and late, or young, mature, and old stages of evolution; and it is in this way that the marine cycle has been divided by those who have introduced organic terms for its description. In accordance with this scheme of nomenclature, the present stage of the Annam and Cantabrian coasts should be described as 'young'.
>
> (*Davis, 1919B, pp. 179–80*)

The discrepancy between describing the same coast as 'mature' and as 'young' is more than obvious. Davis concludes that either some standard cyclic terminology must be accepted by all, or aberrant authors must de- scribe their own evolutionary cycle in detail.

We now turn to Davis' 'Peneplains and the geographical cycle' (1922I) which was presented before the Geological Society of America in December 1921 and printed in their *Bulletin* in the following year. It is persuasive advocacy and an example of Davis' increasing mental flexibility. Whatever doubts the reader has of the logic and practicality of the arguments when dissected statement by statement, he must admit the concise cleverness of the whole. For example, geologists would loathe the idea of 'a large landmass of homogeneous structure', and geomorphologists would wince at the naïvety of the idea that peneplains 'in a region of varied structure will be first developed in areas of weak rocks, wherever they are situated' but, at the close of the argument, many of them despite their scepticism would admit that there might be something reasonable in so flexible a proposition. Perhaps, however, it is best to let the master speak for himself. We can only advise the unwary reader to notice the frequent subtle assumptions hidden away in qualifying adverbs and adjectives; for example, 'the *interior* divide where the initial upheaval was greatest', (Why should it not be coastal?); and '*weak* rocks' (for rocks that are eroded fastest: which seems a very different matter).

The essay opens with a description of the general features of a peneplain.

> If a landmass of whatever structure and large area be upheaved unequally to considerable altitudes in its interior area and if it then stand still for an indefinitely long period, it will be eventually degraded to a plain. In order to avoid the necessity of assuming so indefinitely long a period of still-stand, and in order at the same time to detain attention upon the gently undulating surface that such a region will have before it is worn down to a plain, the term peneplain was invented some thirty years ago. (*Davis, 1922I, p. 587*)

The definition of a peneplain has, Davis says, over the years become less vague and in his opinion could now be expressed as:

> A peneplain developed on a large landmass of homogeneous structure should be margined along its retrograding ocean shore by sealevel delta-free valley plains, alternating with low, wave-cut bluffs on faint inter-valley swells; and should very gradually ascend to greater altitude and greater relief at an interior divide, beyond which similar features would be repeated in reverse order to a farther ocean. The survival of large mountain-like hills along the interior divide where the initial upheaval was greatest is not inconsistent with the occurrence of a well developed peneplain – broad swells of gentle convexity between wide valley floors – over the greater part of the area between the divide and the ocean. Penck has proposed the term, mosore, for the residual hills that survive along the divide, not by reason of greater resistance, but by reason of representing a greater original mass to be consumed. From the mosores along the main divide, gradually dwindling trains of hills would follow the secondary divides. There is no break in the long sequence of slow changes by which the smaller hills of the secondary divides and the larger hills of the main divides are gradually reduced to so small a relief that they, too, may be regarded as part of a peneplain. The term peneplain should therefore be taken as especially applicable to certain advanced phases of land sculpture not sharply separated from the phases that precede and follow. (*Davis, 1922I, p. 588*)

Some attempt is now made to demonstrate the differences in general altitude that might occur, and so to suggest that an evenly-uplifted peneplain need not be flat. Moreover, rejuvenated valleys in a peneplain need not be evidence of uplift.

> If a peneplain have a breadth of 1,000 miles or more, or if a peneplain be developed in the interior of a large continent, its interior part – not merely the residual hills

of fairly strong relief, but also the gently undulating swells that rise but little over the broad valley floors – may have altitudes of from 1,000 to 3,000 feet or more above sealevel. Hence an evenly uplifted peneplain, now undergoing dissection, should not be expected to stand everywhere at the same altitude, and the present altitude of even its best developed parts should not be taken as necessarily giving a measure of its uplift, as if it had previously stood at sealevel. The uplift may have been several thousand feet less than the altitude. Cvijic has emphasized this point in his discussion of the physiography of the Balkan region. Similarly, if the coastal two-thirds of a broad peneplain be flexed or faulted down near or beneath sealevel, while the inner third remain at its former altitude, the river there will at once proceed to incise new valleys beneath their former valley floors; hence the mere occurrence of valleys incised in a peneplain should not be taken as evidence of uplift. Philippson has urged the importance of this interesting principle in an account of the Slate Mountains of the middle Rhine. (*Davis, 1922I, p. 588*)

Davis suggests that the rate of local formation, and presumably destruction of the peneplain stage depends on the resistance of the local rocks.

A peneplain formed in a region of varied structure will be first developed in areas of weak rocks, wherever they are situated; it will be later developed in areas of moderately resistant rocks; and by that time the weak-rock areas may be degraded to true plains. The most resistant rocks will long survive as knobs or ridges, now commonly known as monadnocks. (*Davis, 1922I, p. 589*)

He now adds a strong plea for the recognition of subsequent streams, a class proposed by J. B. Jukes in 1862 in his account of the drainage of Southern Ireland. For rivers, many American geologists and physiographers adapted J. W. Powell's trio of 1875 – consequent, antecedent and superimposed (later changed by McGee to superposed). But by failing to add a fourth or subsequent class Powell, Dutton and others had, according to Davis, made serious errors in the interpretation of the river development in the American West.

The spontaneous development of subsequent streams in regions of tilted, strong and weak strata is an essential process in the advance of the cycle of erosion. The addition of three other classes of streams – insequent, obsequent, and resequent – thus enlarging the original trio to a septet, is a refinement of much less importance than the recognition of the class of subsequents. (*Davis, 1922I, pp. 589–90*)

Davis emphasizes the uniqueness and significance of subsequent streams. The description is perhaps deliberately simple and disarmingly naïve. For example it deals only with stratified rocks uplifted evenly so that consequent rivers flow directly down the slope of the original surface rocks whereas J. B. Jukes, as we have shown in the first volume of our *History of the Study of Landforms* (pp. 391–401), had discussed consequent streams dependent on the slope of an uplifted plain of marine abrasion (planation). True the subsequent valleys would develop in much the same way as in Davis' scheme, but the belts of 'weak structure' were already widely exposed to subaerial erosion at the *start* of the 'cycle of erosion'. In fact Davis always seemed to turn a blind eye on the superimposition of consequent drainage by marine planation, just as he remained indifferent to eustatic (world-wide) changes of sea-level. Of course modern Davisians may argue that these are 'complications' of the cycle, but are they? For Davis himself no locality was too small to prove a principle! However, we expose the description to the reader's own judgment.

> The development of subsequent streams along belts of weak structure and the corresponding diminution of consequent streams is a characteristic feature of a well advanced cycle of erosion; and while subsequent streams are developing along belts of weak structure consequent divides will be largely replaced by subsequent divides on belts of resistant structure. But be it noted that the physiographic value of these two stream terms does not lie so much in the indications that they give of stream and valley origins as in the suggestions that they offer regarding the relation of the streams and valleys to their surroundings. A consequent stream or valley following the original slope of a body of inclined stratified rocks will usually have similar rocks and similar forms on both sides; but subsequent streams and valleys are usually characterized by different rocks and by different forms on the two sides; and, furthermore, in a region of slanting structure, where a master consequent is joined by one subsequent stream, it will usually be joined also by another subsequent stream coming from the opposite direction, both subsequents being developed on the same tranverse weak belt.
> (*Davis, 1922I, p. 589*)

Davis now turns to adverse criticisms of the 'geographical cycle'. Two of his European correspondents had lately abandoned the idea of the cycle of erosion because certain valley-forms they had found in the field did not agree with the system. First, in areas of relatively 'weak' rocks, the valleys were well-opened, having been widened as well as deepened while the region was uplifted. They had never been young in the sense of being narrow. Davis says that the nearest he had come to explaining 'this rather evident idea' was in 1905(J) when he suggested that if a region of resistant rocks be slowly uplifted its valleys will be widened as they are deepened and will be mature from the beginning. Also in 1912(J), he had stated that 'immediately mature' valleys were more likely to occur in weak than in hard rocks (*Die Erklärende Beschreibung der Landformen*, p. 147). He admits that 'young' valleys, in the sense of narrow, steep-sided valleys, must be rare in areas of weak rocks.

> In such cases the valleys are probably, Minerva-like, born mature. It is certainly a puzzling and regrettable omission not to have explicitly stated a matter as manifest as that; but what is still more puzzling is why the omission of a subordinate matter of this kind should be regarded as a reason for discarding the whole scheme of the cycle of erosion. The omission clearly gives reason for amending and improving the scheme; but, in view of the abundance of regions in which typical

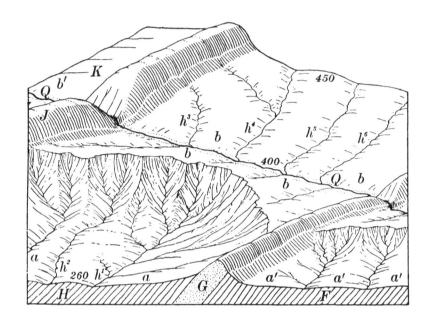

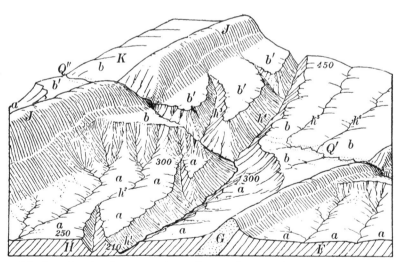

FIG. 95. Complex development of terrain associated with capture by subsequent drainage (From Davis, 1912J, Figs. 89 and 90)

'young' valleys are found in resistant rocks and in which a relatively rapid uplift is thereby proved, the scheme of the cycle seems to me still worth preserving, as well as worth improving as far as possible. (*Davis, 1922I, p. 590*)

With his usual ingenuity Davis proceeds to show that, in slowly uplifted areas of resistant rocks, the first-cut valleys of the larger streams may be born 'mature' while in weak-rock areas the first-cut valleys may be mature-born if the uplift is relatively rapid, and old-born of the uplift is slow. He goes so far as to imagine variations in the rate of uplift of weak rocks which would create old valleys, then cause incision to a state of maturity and eventually of youth. During an ensuing still-stand period the 'young' valleys would be converted through maturity to old age!

The second main geomorphological criticism by European correspondents was that some small headwater streams on 'an uplifted and partly dissected peneplain', occupy wide-open, shallow valleys even though the trunk rivers flow in gorges. Davis had noticed this in Central France in 1898–9 and explained it in his Berlin lectures of 1908–9 (*Die Erklärende Beschreibung der Landformen*, pp. 62 and 259). The obvious answer is that rejuvenation after uplift proceeds upstream and had not yet affected the little headwater branches which were 'mature-born' at the start of the new cycle of erosion. Therefore, Davis suggests, modify the scheme to include this principle but do not give up all the rest of the scheme because this modification is needed.

Climatological criticisms of the cycle of erosion were numerous and powerful, especially as the critics included Alfred Hettner of Heidelberg and Siegfried Passarge of Hamburg.

Passarge, for example, ascribes so great a value to past changes of climate that he is disinclined, to say the least, to regard any existing mature or old land form as the product of a one-climate erosion cycle; not merely that the climate would change during the progress of a cycle from cold and rainy on the initial highlands of an uplifted region to warmer and drier on the old worn-down lowlands; but that, by reason of the shifting of climatic zones, the tilting of the earth's axis, and other possible causes, the processes of 'normal' subaerial erosion in middle latitudes, for example, are not likely to endure through so long a period of time as an erosion cycle. He is therefore unwilling to look upon most existing land forms as the product of the prolonged action of their present-day erosional processes. (*Davis, 1922I, p. 592*)

Davis admits that the effects of changes of climate should be considered wherever necessary – as in glaciated regions and some sub-arid regions – but suggests that in most areas the Quaternary climatic changes were of relatively little importance, being mere transitory episodes in the immensely longer cycle of erosion. Thus in the Great Basin of Utah and Nevada the land-forms are essentially those of a long-continued sub-arid cycle of erosion in which a brief Quaternary humid phase has made little impression. His inquiries indicated that most American geographers and geologists agreed with him in this matter.

Many of the above objections to the general scheme of the cycle of erosion seemed, to Davis, to be founded chiefly on a misapprehension of the scheme as a rigid and complete concept. Whereas he had always considered it flexible and liable to correction, modification and extension as new facts and processes demanded. Even in the first statement of it (Davis, 1885D) he had suggested that the channels (valleys) will be narrow and steep-walled in rapidly rising regions and broadly open in slowly rising regions, and that rate of elevation was of greater importance than climate in giving canyon form to a valley. Possibly, Davis admits, he may have over-emphasized the elementary case of a rather rapid uplift followed by an indefinitely long period of standstill and so given less attention to its many complications;

> But even the brief mentions made of possible complications supplementary to the elementary case appear to me quite sufficient to show that they also inhere in the generalized concept of upheaval at various rates and in various manners, associated with degradation by various processes in various combinations (*Davis, 1922I, p. 593*)

Davis goes on to defend the cycles of normal erosion, of marine abrasion and of glacial erosion against the charge that they are based chiefly on deduction. All, he says, are composite results of seeing and thinking, and of thinking and seeing. However, in the arid cycle he had deduced a whole sequence of changes, many of which had not yet been observed in the field.

This scheme of the cycle of erosion is a means of describing the present landforms. But on this Davis does not bear paraphrasing: he must be allowed to express his own case.

> The essence and object of the scheme of the cycle does not lie in its terminology, but in its capacity to set forth the reasonableness of land forms and to replace the arbitrary, empirical methods of description formerly in universal use, by a rational, explanatory method in accord with the evolutionary philosophy of the modern era. All the older descriptions of land forms treated each form by and for itself. The idea that certain groups of forms may be arranged in a genetic sequence based upon structure, process and stage, and the further idea that the different form-elements of a given structural mass are at each stage of its physiographic evolution systematically related to one another, were not then recognized. The terminology by which these ideas are set forth is a subordinate matter, although it is of course desirable that some one reasonably consistent terminology should be generally adopted as a matter of convenience. But it is today inevitable that the fundamental ideas of the reasonableness of land forms and the systematic relations of their elements should be accepted and made use of in their description.
>
> The reasonableness is not exhibited only in the sequence of forms and the systematic relations of their elements during the uninterrupted progress of a

single cycle; it is also exhibited, although in a more complex manner, when one cycle is interrupted at any stage of its progress by a movement which introduces another cycle; or by a volcanic or climatic accident which for a time disturbs its ordinary progress. Great practical advantage follows from the recognition and utilization of these ideas, for they serve to impress, first, the helpful physiographic principle that every structural mass has, at any stage of its erosional development, a reasonable surface form; and, second, the equally helpful principle that if a structural mass is moved at any time, the form that it had gained when the movement took place must always be specified, as well as the new attitude into which it was placed by the movement and the changes it has suffered since the movement – always provided that the changes since the movement have not wholly obliterated the pre-movement forms. (*Davis, 1922I, pp. 594–5*)

Comments on advocacy of this quality are likely to appear anticlimatic but in fairness to Davis' critics we must point out that he is talking of landforms rather than of landscapes and of 'structural masses' rather than of continents. What is a 'land form', and what a 'structural mass'? When, for example, the Davisian valley-form can evolve from young to old and, as he admits, from old to young, and when every 'structural mass' will tend to be at its own stage of 'physiographic evolution', to what size of territory can the Davisian scheme be safely applied without confusion? Davis wisely puts the onus of testing this on others. In the meanwhile, as we will now show, German geographers worked steadily to provide an alternative to the Davisian system of landform description.

CHAPTER TWENTY-TWO

Growing German Opposition

During the first quarter of the twentieth century the Davisian scheme of landform description remained fairly secure in the United States except in isolated academic pockets as at Chicago University, the domain of R. D. Salisbury. Davis, as we have seen, was the chief interpreter of foreign ideas to American physiographers and made full use of his privileges as a contributory editor of the *Geographical Review* and as the founder of the Association of American Geographers, to whose *Annals* he and his students contributed extensively. His ideas also dominated landform studies in France and the British Isles. By a strange twist of fortune, he also now found a prolific disciple in New Zealand where C. A. Cotton soon acquired an international reputation. Davis, who as usual had had a finger in this pie (*see* p. 471), reviewed Cotton's first major work in 1923. It was as if Davis himself were speaking, writing and drawing. No wonder he welcomed the book. We give below brief extracts from this review of *Geomorphology of New Zealand. Part I – Systematic: An Introduction to the Study of Land-Forms.* The New Englander sings the praises of a New World melody echoed by a young New Zealander, who after the death of the master took over his mantle and wore it with great distinction for several decades:

> It has been remarked that, in a distant colony far removed from its mother nation, the element of remoteness tends to diminish conservative and to favor progressive legislation; a result to which emigrational selection contributes powerfully by keeping at home those of satisfied desires and the energy to satisfy them. The book above named, taken in contrast with certain books of the same object published in Europe, suggests that a similar consequence of remoteness may be found in science. The presence of preconceptions, preferences, and prejudices in an old country, where educational standards have been long established, may operate to retard the adoption of novel ideas and methods. Those grounds of objection to change have less force in a young country, where conventions are yet to be formed and where new ways of doing old things therefore may be freely judged on their merits and adopted if desired without encountering the opposition that is commonly aroused by the need of changing settled habits of thought.

> Cotton's book is the work of a young physiographic geologist who for the past fifteen years has had abundant observational experience in New Zealand, who has carried into his richly varied yet compact field a trained mind unusually competent in the rational study of land forms and a skilful hand exceptionally successful in reproducing landscapes in simple outlines, and who, when it comes to describing

what he has seen, has decided unequivocally in favor of a modern, genetic, explanatory method as against an old-fashioned and empirical method. Not only so: this decision does not remain an unapplied abstraction; it is given systematic application of the most practical and thoroughgoing kind in the book before us . . . All the chapters are copiously illustrated with ideal diagrams, landscape sketches, and half-tone views of local scenery, making 442 figures in all. The success with which type forms are matched with actual forms testifies to the remarkable physiographic wealth of New Zealand . . . The discriminating quality of Cotton's treatment may be indicated by a few extracts from his pages. The introduction states that the description of land forms must be explanatory in giving some account of their geological origin and yet that it must also be geographical in that it shall not distract from present form to a consideration of past geological processes. Brevity is secured by the use of an explanatory nomenclature. Concerning the relation of uplift by which a cycle of erosion is initiated, to erosion by which it is carried on, it is remarked that 'it simplifies the elementary study of landforms to regard this uplift as rapid. It is not to be regarded as ever sudden, or catastrophic, but it may take place so rapidly that the amount of erosion that goes on during uplift is small as compared with that which follows completion of the uplift. All uplifts are not so rapid as this, but the results produced by erosion will ultimately be very much the same whether the uplift is slow or rapid.' As a mark of the author's devotion to the truly geographical aspects of geomorphology it may be noted that, in strong contrast to the highly geological treatment of the subject by German geographers, Cotton excludes all geological time names even in the case of fossil peneplains in certain block mountains. (*Davis, 1923F, pp. 321–2*)

The reader may have noticed in the above review that Davis seizes on the chance to hurl a brick at German geographers, the majority of whom had always been antagonistic to his cyclic theme which they considered too deductive and inadequately concerned with underlying geological structures. The first major treatment of the cycle in German was by Martha Krug-Genthe in 1903. During the early years of Davis' exposition of the cycle their attitude had been negative or indifferent but as the method spread they could no longer ignore it. With justified caution they began to test it in the field and to compare its practical value with that of traditional methods. As we have seen in Davis' 'Peneplains and the geographical cycle' (1922I), the suspicions of some Europeans were deepened when they found in the field landform details which did not correspond with the features to be expected within the ideal cycle. This led in turn to the publication of many studies which conflicted with or even contradicted Davis' teaching. The Germans were beginning to see cracks and flaws in the cyclic scheme, whereas Davis himself regarded it as capable of expanding to absorb all possible occurrences.

This lack of enthusiasm in the German-speaking realms of Europe for cyclic descriptive ideas had long worried Davis and it was in Germany that he made his most strenuous propaganda campaign. We have already

F I G. 96. Davis on a field excursion (Courtesy Harvard College Library)

described (Chapter 15) his remarkable efforts in Berlin in 1908–9 and his intimate friendship with Albrecht Penck, the greatest German physiographer. Davis sent separately to Penck a poem specially written to be inserted in his copy of *Geographical Essays* (1909C).

> Well, well! I never tho't to find myself
> A standing in a row upon this shelf
>   Among so many learned books!
> I'd best shut up and never say a word
> Of English here, where what is mostly heard
>   Is German, judging by the looks.
>
> What joy, if once in every year or two,
> Someone would reach and set me down where you
>   Would come and meet me face to face,
> And read my inmost thoughts an hour or two
> Then would I mount to where I was before
>   And stand contented in my place.
>
> Fain would I turn about and make a bow
> When you come in the room, if I knew how.
>   My covers are too stiff. Alack,
> I'm bound to turn my face toward the wall,
> Lest, dizzy on my lofty shelf, I fall.
>   So won't you please excuse my back?
>   (*Poem: Davis to A. Penck, April 1910; Cambridge, Mass.*)

The friendship between Penck and Davis declined during, but survived, the war of 1914–18 when they steadily diverged in geomorphological views. No doubt there was a time when Penck gave some favourable attention to the cyclic scheme. Thus in his chapters on 'Die Erdoberfläche' in Scobel's *Geographisches Handbuch*, the 1895 edition has no mention of the cycle scheme, whereas the 1908 edition 'adopts it fully'. The latter phrase, however, is Davis' and perhaps 'outlines' would be more accurate. The divergence was not entirely on the German side, for Davis increasingly insisted on omitting from landform description all geology not strictly relevant to the present surface, and in his later years he heightened his insistence on the value of deduction. His great hope of converting German-speaking geographers to his system lay in *Die Erklärende Beschreibung der Landformen* but this elaborate work which contained his Berlin lectures had only appeared for two years when the European war broke out and its second edition (1924L) could hardly have been published at a more unfortunate time for central Europeans. Moreover, by now Albrecht Penck is only one of several influential German physiographers who have set their minds steadily against simple Davisian concepts. As already noticed, Davis in his subtle 'Peneplains and the geographical cycle' (1922I) discusses this opposition but refrains from mentioning its proponents.

Here we must give details of Davis' four chief antagonists in central Europe – Albrecht Penck, Walther Penck, Siegfried Passarge and Alfred Hettner. Needless to say each was a tower of strength in Europe and each led there an influential school of thought. Fortunately we are able to give an intimate and accurate account of what these eminent scholars thought of the cycle of erosion and what Davis thought of their own schemes.

SIEGFRIED PASSARGE'S OPPOSITION TO DAVIS

Davis acknowledged in 'Leveling without baseleveling' (1905E) that Passarge's early work on the deserts of southern Africa largely formed the basis for the Davisian arid cycle, in which cyclic ideas were applied to forms far beyond those already observed in the field.

> After extended observation on the desert plains of southern Africa, fully described in his book, *Die Kalahari* (Berlin, 1904), Passarge concludes that these plains are the result of leveling without baseleveling, through the combined action of wind and water erosion; and that such plains, nearly everywhere showing a rock surface independent of structure and interrupted only here and there by residual hills or mountains – which he calls by Bornhardt's term 'Inselberge' – may be produced over large areas at any altitude above baselevel. His article, Rumpfflächen und Inselberg (*Zeitschr. deut. geol. Gesellsch.*, LVI, 1904, Protokol., 193–209), in which this conclusion is announced, is well worthy of attention from American physiographers.
>
> The principle of leveling without baseleveling, or Passarge's law, as it may be called, in contrast to Powell's law of leveling by baseleveling, suggests that the

scheme of the normal cycle of erosion, so generally applicable in regions of ordinary or normal climate, should be systematically modified in such ways as will adapt it to the conditions of an abnormally dry or arid climate. This modification I have lately attempted in an article that will soon be published in the Journal of Geology; it is here presented in outline. (*Davis, 1905E, pp. 825–6*)

Every truncated upland that has been described as an uplifted and more or less dissected peneplain should now be reexamined with the object of learning whether it may not have originated as a desert plain at its present altitude above sealevel, and afterwards suffered dissection as a result of climatic change. (*Davis, 1905E, p. 827*)

In view of the possible change of interpretation now open for truncated uplands according to Passarge's law, it might be said by one who prefers to work on more purely inductive lines: 'Behold, here is another case in which deduction has led the investigator astray! He thought that he could deduce the sole conditions under which truncated uplands could be formed, and that these conditions necessitated uplift after degradation; now he finds a new series of conditions under which such uplands may be formed and all his previous conclusions are uncertain. Let us, therefore, beware of deductive or imaginative methods, and hold fast to the safer methods of observation and induction.' In reply to such a warning, one might say – besides pointing out that all problems which deal with unseen processes necessarily involve deduction and that the deductive side of the work should be conscious and systematic – that the fault in the method by which truncated uplands have heretofore been discussed lies not in the too free use of deductive methods, but in their too limited use. The mistake lies in our not having years ago set forth, by purely deductive methods, just such an analysis of the geographical cycle in an arid climate as has now been provoked by the discovery of rock-floored desert plains. (*Davis, 1905E, p. 828*)

Davis did not specifically defend himself against an early critical attack by Passarge ('Physiologische Morphologie'; *Mitteilungen, Geographisches Gesellschaft, Hamburg*, Vol. 26, 1912, pp. 133–337). The latter attacked the cyclic concept because it: (1) was unreal; (2) ignored climatic considerations, particularly those associated with climatic changes which have left vestigial remnants among the existing landforms; (3) attempted to describe terrain in terms of one set of processes, ignoring local causes which are mainly geological and broader genetic ones which are mainly climatic; and (4) led to quick, superficial and dangerous conclusions (Dickinson, 1969, pp. 138–40). Passarge's later main publications on the principles of landscape description were reviewed at some length by Davis, however, in *The Geographical Review* (1919D and 1923G). The student who reads between the lines of these reviews will soon detect the chief points and methods on which Davis and Passarge violently disagree.

Dr. Siegfried Passarge, known fifteen years ago for his explorations in South Africa, where he came upon the important idea of 'leveling without base-leveling', and subsequently professor of geography at the University of Hamburg, has frequently urged the importance of carefully studying the facts of the visible landscape before attempting to explain them. He now makes practical application of this manifest principle in a textbook, *Beschreibende Landschaftskunde*, (Hamburg, 1919, pp. 210) the first part of a contemplated four-volume work, *Die Grundlagen der Landschaftskunde*, which is as a whole intended to serve as a guide to the investigation and description of the visible features of geographical areas in their natural combinations. It is well said in the introduction to the first volume, now issued, that the study of the visible landscape is a branch of geography which has at last secured the place it should have long since occupied; for a knowledge of geographical spaces and their contents is the indispensable foundation on which all geographical descriptions and generalizations must be built; and it is further-more held to be a scientific necessity that the study of landscapes, including their land forms, atmospheric effects, plants, animals, and human inhabitants, as they are actually seen to occur together, should begin with an empirical description of facts of observation, uninfluenced by theoretical or explanatory preconceptions. The volume before us aims to assure the accomplishment of this empirical task.

(*Davis, 1919D, p. 266*)

The first part presents a systematic analysis of landscapes in the form of a compact list of the various features that compose landscapes. The meanings of these terms are then explained in the second and major part of the book.

The method here adopted may be illustrated by the treatment of land forms. These fundamental features are first classed as individual and as composite forms (*Grundformen und Gruppenformen*). The individual forms are mountains, valleys, plains, etc., each of which is composed of certain elements that cannot exist alone; thus a mountain has a summit, slopes, and base; a valley has sides, floor, and river channel. Composite forms, such as mountain ranges, are made up of many indivi-dual peaks, domes, and ridges; and composite forms are associated in larger regions and zones. The classification is thorough, sometimes to the point of being labored, as is indicated by the three lists of terms given above. The variety of forms treated is great, and a knowledge of them is profitable; but, as might be expected in any such text, certain items of equal rank with those included are not mentioned. (*Davis, 1919D, p. 267*)

Among the many omissions, Davis enumerates 'the frequently occurring "underfit" relation of stream curves to incised-valley curves, the cusps and concaves of terrace fronts', and so on. The definitions and descriptions of individual forms are as empirical and non-explanatory as the author could make them. But occasionally explanatory implications creep in and, accord-ing to Davis, 'illustrate the natural and constantly increasing habit of treat-ing geographical matters rationally in these modern days of evolutionary philosophy' (p. 269). The individual forms are illustrated with numerous outline drawings, many of which quite fail to match up to Davis' standards of expression and draughtsmanship. The last part of the volume contains nearly eighty empirical descriptions of actual land areas, which are meant to serve as examples. Most are based on Baedeker maps and pictures. The book, Davis suggests, will prove highly serviceable to travellers ignorant of

geography who wish to describe their journeys but the absence of explanatory matter makes its value as a university textbook very questionable.

> No one disputes the necessity for abundant observational study of geographical
> facts. No one proposes to describe landscapes by substituting theoretical explana-
> tions for direct observation. Even those geographers who wish to replace
> empirical description by explanatory description as far as it is reasonably possible
> to do so, fully agree that, when an actual landscape is to be described, the
> observation of its features must come first and their description, of whatever
> kind, second. This is logically inevitable; but it by no means follows that the
> preparatory indoor study, which is for the most part non-observational, must be
> even in its introductory stage limited to empirical definitions and descriptions,
> from which all rational, explanatory treatment is excluded. (*Davis, 1919D, p. 271*)

There is, Davis admits, great value in knowing a large variety of empirically-
defined types but a close association of rational explanation with observation
and description is desirable if geography is to remain alive and exciting. He
concludes the review with an all-out assault on the empirical method, and
then incidentally drags in G. K. Gilbert for the final charge.

> A textbook that imposes upon those who use it a prolonged empirical treatment,
> from which all the refreshing juice of explanation has been squeezed out, has a
> savor of heavy laboriousness. Such a book, unless in the hands of an experienced
> explorer who could not refrain from now and then adding an invigorating flavor
> of explanation, would make dull grinds out of its students and exacting
> Gradgrinds out of its teachers. It is difficult to understand how its arid method
> can be advocated by a man so experienced in exploration and so expert in
> explanation as Passarge.
>
> How different were the educational principles of Gilbert, our American master
> of geomorphology! He thought that the important thing in education is to train
> scientists rather than to teach science and that 'the practical questions for the
> teacher are whether it is possible by training to improve the guessing faculty and,
> if so, how is it to be done': he furthermore believed that the content of a science is
> often presented so abundantly as to obstruct the communication of its essence
> and that the teacher 'will do better to contract the phenomenal and to enlarge the
> logical side of his subject, so as to dwell on the philosophy of the science rather
> than on its material'.
>
> It would be imprudent to predict how far German professors and German
> students of geography will be content to accept so tediously prolonged an empiri-
> cal introduction to their science as the 'Beschreibende Landschaftskunde'
> imposes; but, except that it may have a certain value as a work of reference in
> association with texts of an explanatory nature, a book of this kind is not likely to
> be used in the United States. (*Davis, 1919D, pp. 272–3*)

In 1923 Davis published a long review on the third volume of Passarge's
Die Grundlagen der Landschaftskunde which, as *Die Oberflächengestaltung
der Erde*, had appeared in 1920. This volume first describes various types and
forms of crustal structures; then discusses in great detail sculpturing agen-

cies and their resultant surface forms; and concludes with a genetic classification of landforms. The contents, Davis admits, are exceptionally strong on
the formation and movement of soils, on wind action, and on the influence
of climatic effects and changes on landforms. However, he finds no difficulty
in detecting many errors and ambiguities in various parts of the book:

> ... it often seems as if more consideration were here given to form-producing
> processes than to process-produced forms; and, moreover, as these detailed
> chapters on soils are introduced earlier than the more general chapter on the
> sculpturing of land surfaces under the leadership of streams, the soil forms are
> insufficiently related to their physiographic environment and are almost wholly
> divorced from the various stages of the cycle of erosion in which they belong.
> That is unfortunate. More unfortunate still is the denial that soil creep is an
> effective process of degradation on forested slopes; and as a result of this denial
> the peneplanation of temperate forested regions is ascribed on a later page to the
> more effective degradational processes of a tundra or an arid climate, which are
> without further evidence assumed to have formerly operated in forested pene-
> plains. It is doubtful whether American observers of the Appalachian piedmont
> belt will accept this conclusion. (*Davis 1923G, p. 601*)

We, however, are more concerned here with Passarge's strong antagonism to
deductive methods. Type forms, he asserts, ought to be established by the inductive method, that is they must be based on field observations of actual forms
and of the processes working upon them. Davis shows that in spite of this condemnation of deduction Passarge does make use of it – occasionally incorrectly!
What is more, Passarge, even in his excellent chapter on wind action

> does not make sufficient mention of the interaction of wind and water in the
> production of plains of desert degradation; and that is curious, for this compound
> process is one concerning which Passarge himself announced a most original and
> important generalization on his return from South Africa about twenty years ago.
> But neither in this chapter nor in any other of Part II are plains of degradation
> duly treated: they are postponed to the final pages of the book, where the explana-
> tion proposed for them is presented quite apart from its proper context, as if to
> close the volume with a surprising climax. (*Davis, 1923G, pp. 604–5*)

The third part of Passarge's third volume attempts to classify land forms
'according to their manner of origin in a natural system'.

> The system proposed is extremely systematic. It includes types, classes, orders,
> families, genera, and subdivisions of genera and thus becomes so elaborate that its
> general adoption in landscape description is altogether improbable. There are two
> leading types, land forms and coast forms; then follow two classes, endogenetic
> and exogenetic. The first class is divided into these orders, tectonic, volcanic, and
> seismic; the order of tectonic forms is divided into three families according to
> their structure, flexed forms, faulted forms, and folded forms. In the second class,
> exogenetic forms, no attention is paid to the structure on which the external
> processes work but only to the processes; and the class is then divided into three
> orders, dissected or degraded forms, aggraded forms, and chemically modified

forms (chiefly soils). The orders of degraded and aggraded forms are each divided into ten families, which are named after the ten controlling agencies, beginning with lightning and ending with man. To illustrate the detail to which this scheme is carried, it may be noted that, under the order of aggraded forms, the family of lightning-made forms includes but one genus; namely slopes at the foot of cliffs which lightning has shattered! (*Davis, 1923G, p. 605*)

The book ends with a 'concluding consideration' embodying attacks on the cycle of erosion and on the widespread application of the notion of peneplains developed under 'normal conditions':

here at last Passarge presents, after a condemnation of another physiographic scheme than his and a justification of his own scheme, the postponed explanation of plains of degradation. It is printed in spaced letters as a sort of dénouement to the volume. The explanation first states that both at times of upheaval and at times of climatic change degradation is stimulated; but, while upheaval with increase of seaward slope results chiefly in greater linear erosion by streams, certain changes of climate result in greater areal degradation. In view of this a fundamental law, based on a suggestion made by Passarge's teacher, von Richthofen, in 1905 is formulated: 'Extensive and deeply penetrating planation is the consequence of repeated climatic changes.' For example, when a climatic change transforms a temperate, forested, hilly upland into either a sub-arid steppe or a sub-frigid tundra, the deep-weathered soils of the hill slopes that were – according to Passarge – practically stationary under the forest cover will be actively removed; but they will be formed again if a moist temperate climate returns. If such changes happen repeatedly, the hilly upland will be degraded to a low plain. Thus it is thought that all plains of degradation in temperate latitudes – but not plains of abrasion – are to be accounted for and not by soil creep under a forest cover in a persistently moist climate. Climatic alternations from arid to rainy in the torrid zone are similarly held responsible for the production of extensive degradational plains with their surviving 'Inselberge', such as those of Central Africa.

This novel principle is manifestly a deduction, and, in view of the mistrust of deduction previously expressed, the confidence with which the principle is here announced is somewhat surprising ... It is also easy to assume that, if a still-standing land mass should suffer changes of climate often enough, it would in the end be worn down to a plain; but it has not yet been shown that any worndown region really has suffered climatic changes of sufficient number to bring about its peneplanation in this way; nor has it been shown that a forest-covered region cannot be slowly degraded to lower and lower relief, largely by surface wash and soil creep, and thus eventually reach peneplanation. (*Davis, 1923G, pp. 605–6*)

Davis clearly resented this open attack on his own explanatory method, retorting that Passarge's idea on the formation of erosion surfaces was not yet tested sufficiently

... to warrant its acceptance as a replacement of the more generally accepted idea that peneplanation has ordinarily been accomplished chiefly by the long-continued action of the erosional processes of a single climate. (*Davis, 1923G, p. 606*)

To EDWARD and DOROTHY

MERRY CHRISTMAS
AND
HAPPY NEW YEAR

FROM THEIR FATHER

DEC. 25, 1926.

FIG. 97. Christmas card drawn by Davis for his son Edward and his wife in 1926
(Courtesy W. M. Davis II)

Davis concludes:

> It is certainly true that, in spite of all the work done by the earlier physical geographers, the careful description of land forms has been slowly developed as part of a geographer's duty. How far Passarge's method of treatment will increase such description remains to be seen; but it is to be feared that, in spite of the insistence he places on direct observation and empirical description in his first volume, a distraction from explanatory description toward analytical explanation will be the result of his third volume. In geomorphogeny, or the study of the development of land forms, such analytical explanation is appropriate enough, and it has been highly developed by German geographers as may be seen in the essays contributed to the Penck 'Festband' of a few years ago, but in the explanatory description of the visible landscape it is not the development of land forms but the developed forms that should stand in the foreground; and hence in such description the geographer should so contrive his phrases that the attention of his readers is held to the visible present instead of being left in the imagined past, as so often happens when process instead of product is given strong emphasis. In

this respect Passarge's 'Landschaftskunde', learnedly analytical as it is, leaves much to be desired. (*Davis, 1923G, p. 607*)

An interesting sidelight on the above review is provided by a letter from Kirk Bryan to Davis, who appears *inter alia* to have sought reassurance that peneplanation in a desert was not largely the work of the wind. Incidentally the reader may also glean some idea of the picture of Davis that existed in the minds of American geologists. Davis, who deduced a series of desert forms far beyond existing field studies, is subconsciously treated as a practical down-to-earth physiographer, while Passarge, who spent long periods in the deserts, is the one who 'made hypotheses and then rode them beyond the facts'!

I have delayed answering your letter of January 12, (1922), partly through pressure of work and partly because I am somewhat uncertain as to how to answer.

I am not much of a German scholar, and Passarge writes atrocious German. On this account I distrust myself as a critic of him, but not so much as I distrust him. It seems to me that he was a typical young German, very learned in books and very sure of the infallibility of science and rather fond of himself. He went to South Africa, found something different and thought that there must be an equally strange and different explanation. He made hypotheses and then rode them beyond the facts. He was also anxious to make phrases and terms, as for instance 'Zoogenous erosion' which seems quite unnecessary for his discussion and very unnecessary in general literature.

As for 'Inselgebirge', they undoubtedly occur; there are many examples in southern Arizona and other parts of the West; they are monadnocks or Umakas of a different type. The vital question is the character of the plains at their bases. Are these 'Rump flache' or 'denudation flache' sloping planes with gradients determined by stream erosion or are they wind cut?

Passarge does not give contour maps and it is difficult to judge, but since I have been unable to find any place in southern Arizona where wind erosion was able to destroy stream grades and very few places where wind deposits diverted streams, I am inclined to think that stream action is dominant in most deserts. South Africa can hardly be drier than Yuma with 3·5 inches of rainfall per year.

However, I believe, from observations in northern Arizona, that areas underlain by porous sandstones and therefore having a low water table will suffer from wind erosion even if the annual rainfall is as high as 10 inches. Thus the examples worked out by the Egyptian Survey and reviewed by Hobbs, in the Libyan desert may be valid examples of wind work. It does not follow that Hobbs' conclusions are correct. Wind erosion will, I believe, when properly evaluated be found quantitatively insufficient to reduce a region to a peneplain in any reasonable length of time.

As to alternations of climate: I have no data which indicates that southern Arizona, since Pliocene time, was ever so wet as to have the erosional forms of a humid climate or so dry as to allow wind work to be predominate over stream work.

As to the abrupt change in slope at the foot of mountains: I have analyzed the different factors of erosion on mountain slope and on the pediment for the Papago Country, a region of less than 11 inches of rainfall per year. This analysis will be published in Bulletin 730 B, now with the printer, which I will send you as soon as published. Briefly, on the mountain slopes, material moves by gravity and rain wash, the talus blocks by gravity and small fragments (in granite individual minerals) by rain wash; however, on the pediment material moves in streams which form by the concentration of rain wash. The grade of the mountain slope is determined by the size of the talus blocks, the grade of the pediment by the ordinary relation between volume of water and material to be carried. As, however, under conditions of extreme aridity the amount of material furnished to streams is small they have lower gradients than in some more humid but still arid regions.

East of Tucson, Arizona, the rainfall exceeds 11 inches and temperatures are lower. Perennial grass grows around the bases of many mountains and the break between mountain slope and pediment is much less sharp. I expect to analyze these factors in my San Pedro Valley paper on which I am now working. The Benson, Ariz. Quadrangle will give you the best notion of the appearance of pediments in a climate which is not extremely arid. With the exception of the trench of San Pedro River, the plains around the mountains which look like alluvial fans, have less than 50 feet of Pleistocene gravel and sand. They are erosion plains on granite, Mesozoic sandstones and shales, and Pliocene conglomerates and clays. With this in mind one can judge very easily, I think, of the utility of the word pediment for this variety of erosion plain. Coalescent pediments form the peneplain which is normal to this region. I have seen areas which have been so peneplained by the disappearance of the residual mountains (Inselgebirge, if you please). Such a peneplain differs from the normal type largely in the high grades of the component facets. It is of course theoretically possible that such a peneplain might be modified by wind erosion but I doubt if Passarge has valid examples.

In the light of our present knowledge of erosion in the driest part of the United States, due to the work of the Desert Watering Place Survey, it would be most interesting to check up Passarge in the field. He is, however, protected by distance. (*Letter: K. Bryan to Davis, 23 January 1922; Washington, D.C.*)

ALFRED HETTNER'S BLAST AGAINST DAVISIAN PRINCIPLES

Hettner was Professor of Geography at Heidelberg, where he had won a high reputation particularly for articles on the description of landforms published in the *Geographische Zeitschrift* of which he was editor. Ever since 1910 Hettner had been attacking Davis' notions in the *Geographische Zeitschrift* and in 1921 he collected these articles and revised and expanded them into a book of 250 pages entitled *Die Oberflächenformen des Festlandes*. The recent translation of the second edition of Hettner's work now permits English readers to appreciate the bases for his opposition to Davisian geomorphology, for example:

In Davis's school the concept of age has taken on a special form. The age of landforms is measured not against the geological time scale based upon the evolution of the plant and animal world, but on the degree to which they have been modified, the character of landforms, the physiognomy of the landscape. But since their degree of modification depends not only on how long has elapsed but also on rock resistance, landscape age is, as we have seen, not a purely chrono-logical term but one of degree of development. It is paradoxical but true that Davis's school, in which 'stage' is the third word of the trilogy structure, process and stage, in fact altogether neglects age. . . . Only when placed against the geological time scale can we assess the rate at which modification takes place and how it differs with rock type and climate. Climate and the plant and animal world have changed even in recent geological time, and these changes have exerted great influence on the surface modification of the earth's crust. We must therefore know how mountain building and changes in climate are chronologically related.

(*Hettner, 1928; 1972, p. 98*)

Davis's cycle theory expressed this approach in its fundamental theoretical form. His theory is not in fact the general theory of landscape development for which it sometimes passes, but is limited to the influence of uplift. It gains far-reaching significance only if we grant that uplift does play a major role, and that intervening periods of stillstand were of long duration. The theory maintains that the history of the land surface begins with its emergence from below sea level and elevation to an appreciable height. Thereupon, the flanks of stream-cut channels are flattened by weathering and mass-movement. Valleys are formed and develop progressively and continuously through different stages of life until they attain an end-state in which their bottoms display quite gentle gradients and their sides slope gently towards these. Their development to this stage was termed a cycle by Davis. After this, uplift may begin anew, whereupon erosion is re-invigorated and a new cycle begins. (A. von Bohm compares it with winding up a clock.) Such a development may be repeated again and again. . . .

Factually, this theory is not as original as has often been thought. It introduces only a novel form of expression, lays greater emphasis on the significance of uplift and in distinguishing periods of erosion, and considers planation during periods of still-stand more widespread than is commonly believed. But by its consistency and schematic nature it has led geomorphologists to devote more attention to such phenomena than they have hitherto. Distinguishing cycles has become a favourite aim of geomorphological studies. Indeed one can almost say that the setting-up of cycles has become a mania. . . .

If uplift is renewed before a broad valley floor has been formed, as is very often the case in small tributary valleys, the interruption in the progress of downcutting does not manifest itself in terraces.

. . . Just as zoogeographers once moved the earth to explain the distribution of a beetle. Davis's school has whole segments of the earth's surface moving up and down, being destroyed and levelled, as though they were stage scenery. But here again, the theory is usually not supported by the facts. Most 'peneplains' are postulated on slender evidence; and such planations as actually do exist may have to be explained in a different way.

The cycle of erosion concept was given special importance in this approach to the study of landscape, because it was thought that it provided a means of correlating what may be called landscape storeys with varying degrees of destruction; destruction at the lowest storey would form only narrow valleys, at an upper one, broad elevated plateaux. On the uppermost storey it would leave only isolated inselbergs. This may, but need not be the case. Such contrasts at different elevations can also result from differences in the resistence of strata. . . . To postulate cycles of erosion, and to explain differences of landforms at different elevations, are two-separate things . . .

. . . (Downward movements) are neglected by the cycle theory and in fact are considered only when subsidence has drowned the lower part of valleys to replace the normal cycle by a marine one. But land which has subsided may still be above sea level. When this happens a valley is displaced relative to the curve of equilibrium. If the river was still incising it can continue to do so only at a decreased tempo and with less energy. If it had more or less attained the curve of equilibrium it will now have to deposit material and build a gravel plain. But this explanation will hardly do for the gravel plains of major Alpine rivers. In their case, Penck's explanation, the infilling of overdeepened glacial valleys, is more likely to be correct, but it certainly appears true that in many other valleys, especially those of tropical mountains, gravel plains have been formed by subsidence or downwarping of the surface. In this way, too, a new cycle would begin, but it has the opposite effect. (*Hettner, 1928; 1972, pp. 103–6*)

To characterize landforms by their age leads to error. Whether old and senile valleys have ever been formed must remain an open question; Davis's 'old' valleys are, for the most part, valleys in weak rock; his senile valleys are either lowland valleys which have never had steeper or higher sides, or are not valleys at all but 'dells' (*dellen*), that is to say, smooth flat-floored hollows in benchlands or elevated surfaces. I will not dispute that commonplace fluvial degradation can bring about complete planation and the formation of a remnant surface – I developed the idea myself even before Davis did; but evidence that this has happened has not yet been produced. And it is questionable whether such an end-stage will be reached in nature. So-called senility can have quite different causes.

Thus the 'cycle' idea loses its real significance. That renewed uplift results in erosion being reinvigorated after periods of stillstand has long been known through the study of valley terraces; we have therefore long since spoken of periods of erosion. But a true cycle exists only when the life process is not broken in the stage of maturity or even youth, but progresses uninterrupted to the stage of senility, and only then, and not before, is reinvigorated. Such cases have not as yet been proven. The fact that geologically old remnant surfaces have been redissected in Tertiary and Quaternary times is something different.

Davis's purely geometrical approach is inseparable from its foundation on the deductive method. By deduction we can determine quantitatively the period during which processes have operated; but only observation will tell us the different kinds of process. While Davis talks a lot about 'life', his scheme lacks vitality, the landscape picture it gives has a moribund and dismal emptiness. The unending variety of rock types and the way they are arranged is submerged beneath the schematic contrast of 'hard' and 'weak'. Although he distinguishes a glacial and an arid cycle as well as the so-called 'normal' or fluvial one (in both of which his approach breaks down and deduction is seen to be an unusable tool), he considers the diversity of climatic influences in much too cursory a fashion. He takes no account of the differences in surface configuration within the fluvial cycle of polar regions, in the temperate zones, in steppe climates, in Mediterranean climate, or in the periodically and permanently moist tropics. Minor landforms have no place in the theory. It conveys nothing of the landscape's physiognomy or its structural style, and ignores many facets of its structural plan. Founded neither on rock type nor on climate, his description of the configuration of the earth's land surface is too superficial, and does not form part of the overall pattern of the earth's surface phenomena.

I can see Davis's approach only as an episode, not as a step forward in geomorphology. Its simplicity and the energy of its advocates has rapidly won for it a wide circle of adherents; it has enlivened geomorphological research and led to a number of correct results. But as a whole it has been abortive, and studies founded upon it have produced many fallacies. A lot of debris has to be cleared away to reopen the field to unrestricted research. As a whole, the earlier theory now condemned as backward was in fact on the right lines; it is upon this that we must build. (*Hettner, 1928; 1972, pp. 135–6*)

Davis reviewed the first edition in *The Geographical Review* for 1923(E) and put up a stout defence against Hettner's outright condemnation of the 'too hasty', too deductive methods of the cycle school. The review speaks for itself and wherever mention is made of 'other geographers' or 'an American geographer', 'Davisians' or 'Davis' should be substituted. At last – and how painful it must have been to Davis who was quite unrivalled in this field – someone has also criticized the deductive nature of his superb block diagrams.

The book addressed to advanced students and experts, purports to be a discussion of the fundamental principles involved in the genetic investigation and explana-tory description of land forms. The contents show a wide acquaintance with the literature of the subject and afford a large amount of profitable reading, often provocative of helpful reflection, this being especially true of the pages in which the history of land-form theories is traced; but the book as a whole is not so much characterized by novel and constructive suggestions as by homilies, truisms, hesitations, obstructive misunderstandings, and disputatious objections. It is a conscientious but reactionary protest against what its author regards as the too hasty methods of certain other geographers.

The following examples of homilies may be instanced. If a supposition as to the origin of a land form springs into an observer's mind, he must test it before accept-ing it (p. 14). Scientific investigation must be first inductive and analytic; that is it must set out from observed facts and, beginning with the simplest, test all possibili-ties of explanation before it can go so far as to establish an explanatory principle (p. 47). Theoretical deductions must be tested inductively by comparing them with observed facts (p. 32); this homily is held to be so important that it is stated twice again on later pages (pp. 167 and 214). Ideal block diagrams are to be recommended only when the ideas that they represent are correctly deduced; if the deductions are wrong, the diagrams are dangerous because they are so vivid (p. 246). Hettner solves this difficulty by using no block diagrams at all! (*Davis, 1923E, p. 318*)

Davis proceeds to give a long list of truisms and another of well-established matters on the verity of which Hettner still feels hesitant. These do not include the two-cycle explanation of certain mountainous highlands, worn to a peneplain and then broadly uplifted, which Hettner considers only an 'untenable speculation', a declaration of physiographic bankruptcy. No wonder Davis hits back, with these words:

Although the book is nominally devoted to the fundamental principles of land sculpture, it is in reality for the most part a diatribe against the scheme of the erosion cycle and against an American geographer who has for the past forty years interested himself in the development of that scheme. The name of this unfortunate individual is mentioned, usually with adverse comment, on the average rather oftener than on every other one of the 246 pages of the text; his method of thinking on physiographic problems is heartily condemned in nearly every respect. The condemnation is based partly on curiously insistent misunderstandings, some of which will be next instanced, partly on a disputatious preference for some other method than that of the erosion cycle for treating land forms, as will be shown in later paragraphs.

One of the most pervading misunderstandings results in the discussion of the 'age' of land forms instead of their 'stage' of development. 'Davis' genetic terminology is based primarily on the age of land forms' (p. 233). 'It sounds like a paradox, yet it is true, that the Davis school, in which age is the third term (in the description of land forms), makes no attempt whatever to determine the real age' (p. 141). As a matter of fact, not age but stage of development is the third term: and, although stage of development manifestly depends on the factor of time, it is practically determined not by measuring the geological periods during which a given feature has been undergoing erosion but simply by its visible form.

It is repeatedly insisted that the proper relation between induction and deduction, as expressed in the homilies above quoted, has not been observed in the exposition of the erosion cycle, which is said to be chiefly deductive (pp. 49 and 72); and this in spite of the abundant verifications by which nearly all the deductions of the cycle scheme have been inductively tested. For example, at the close of a chapter in my 'Erklärende Beschreibung der Landformen' which sets forth a series of elementary deductions – really for the most part based on inductions and given, as Hettner recognizes (p. 207), a deductive form only as a device of presentation – the explicit statement is made: 'Whether the deductions are correct cannot be known until they are confronted with facts.' In the next chapter the necessary confrontation is made, with the result of showing that while the deductions are as a whole correct they need modifications and extensions; and these are then proposed in a third chapter. But Hettner overlooks these verifications. In only one chapter of my book – the chapter on the arid cycle – was deduction carried far beyond observation; the reason therefore was explicitly announced to be the lack of sufficient descriptions of desert topography with which the deductions could be checked. Hettner, who has not traveled in desert regions, characterizes that chapter as 'schematic and without meaning' (p. 213). Erich Kaiser, professor of geology at Munich, who has made a close study of the coastal desert of southwest Africa, states: 'It is a region of the kind which Davis has deduced from the factors of an arid climate.' (*Davis, 1923E, pp. 319–20*)

The reader will not be surprised that Hettner also condemns the Davisian theory of peneplanation which rests, he asserts, on a 'geometrical construction' and not on the observation of actual processes (Hettner, 1921, pp. 94–5). But Davis retorts that he had formed the theory after seeing a 'peneplain' in eastern Montana in 1883, and that he had confirmed it by experience on the plains of Siberia in 1903. The theory was inductively originated and was not weakened because it is supported by deduction.

Hettner expresses dislike of the plan of treatment adopted in the exposition of the erosion-cycle scheme. It should, he insists, begin with the consideration of detailed forms and should pay more attention to structure or geological composition; it is too schematic; it is a 'theory' not a 'method'. Davis gladly accepts the charge of being schematic and stoutly affirms that 'structure, process and stage' *is*, in fact, a well-ordered method. He presents their main differences in presentation fairly to the English-speaking world.

> Hettner would seem to prefer an all-at-once plan of presentation, which should immediately set forth with fundamental principles, all sorts of complications in the way of varied rock composition, diversity of structure, irregular upheavals, alterations in the nature and rate of erosional processes with changes of climate, and so on. My preference is for a first entrance into the subject with simple examples and then a gradual advance to complex examples; on the firm foundation thus constructed one may later embroider as many elaborations as are desired.
> (*Davis, 1923E, pp. 320–1*)

It is interesting that Davis did not choose to answer in detail some of the most basic criticisms by Hettner, namely (1) that the cycle of erosion did not sufficiently admit the influences of structure on landforms; (2) that different initial conditions and processes are capable of leading to very similar end results; and (3) that certain landscape features are characteristic of particular climates, rather than particular stages of development within a given climate (Tilley, 1968, p. 266). Just as it was characteristic of Davis to react violently and disdainfully to criticism (particularly at this time to German criticism!), it was also characteristic of him that he chose to attack the most vulnerable points of criticism.

In short, the criticisms of both Passarge and Hettner were directed primarily at the theoretical model qualities of the cycle of erosion, and it was precisely these qualities which Davis thought most valuable. The whole argument was the now-familiar ideographic versus nomothetic one. It is probable that Passarge and Hettner did noticeably influence the Pencks, Albrecht and Walther (particularly the former), who were the leading

German physiographers, and they certainly influenced geographers, especially as Hettner went on to produce some notable textbooks. However, the most telling opposition to the cycle concept was to come from the Pencks themselves.

THE GROWING OPPOSITION OF ALBRECHT PENCK

Although, as we have shown, Albrecht Penck at one time adopted some of Davis' ideas there seems no doubt that the differences in their approach to geomorphology were always fundamental. The following letter sums up the divergence with a prophetic clarity.

Nov. 3rd was the big day. Taft was elected. Because of this, they made an incredible row in the streets. But in Berlin Davis held his Inaugural Lecture; the local 'Times' only reported this on the 8th, and it was plainly more interested in the Jewish faith of Alder than in the content of the speech of the geographer of purely American descent. But I hope to read in German papers about it. I am genuinely pleased that my students will be able to hear a lot from you and that you do not confine yourself to a mere 2 hours of lecturing. I expect nothing but good to come from this for, in my opinion, the advantage of the entire professional exchange resides in the very fact that, on the one hand, professors come to experience new circumstances and that, on the other, students come to know new teachers. In spite of your arguments I maintain that you are the deductive type and I the inductive one. Naturally, I know very well that you have done a lot of inductive work and that I, in the opinion of some people, have been far too active in the deductive field. Inductive and deductive are not, in my opinion, mutually exclusive but complement each other naturally, like one hand complements the other. But just as there are people who work predominantly with the right hand and others who make more use of the left, so one scientist is inclined to be deductive while the other is more inductive. I attach importance to the qualifying adjectives 'more or less', for if I teach here in Columbia more analytical morphology I think it equally important that my Berliners should now have the opportunity to listen to the representative of the synthetic school and I shall not think it in the least deleterious if as a result they become somewhat americanized. Such fears might for all I care move Passarge, who throughout his flabby [? illeg.] career has come to the idea that he is able to influence the more recent developments.

I have not yet been greatly pressed to attend dinner parties. Yesterday evening there was a reception for the first student tea party, and after shaking hands 300 times I admire Taft who did so 150,000 times. On Saturday I am going to Princeton of Libbey [? illeg.] in order to watch Yale play against Princeton. On Thanksgiving Day I should like to go to Boston, Cambridge and Newport. In Cambridge I shall not neglect to call on Frank Whitney but the real attraction will not be there, and the house between the Lockes and the Fowlers will be empty. Walther, who feels wonderfully well in Yale, will probably accompany me.

With my kindest regards for you and your dear wife.

(*Letter: A. Penck to 'my dear friend' Davis, 11 November 1908; Columbia University, New York*)

The long friendly association between Davis and the Pencks cooled after the appearance of Davis' works in German (*Grundzüge der Physiogeographie*, 1911; and *Die Erklärende Beschreibung der Landformen*, 1912) and, particularly, after the 1914–18 war in Europe. Albrecht Penck, increasingly influenced no doubt by his son Walther, believed ever more strongly in detailed field studies of structure, active tectonics and climatic change. His prolific writings included an excellent scheme of climatic geomorphology as well as superb studies of the earth's surface and, with E. Brückner, of *The Alps in the Glacial Period* (1909). This latter work was enthusiastically reviewed by Davis (1909H), although he took exception to some of the 'empirical terms' employed to describe the glacial landforms. Penck was a few years younger than Davis, whom he outlived, and the two of them, Penck in central Europe and Davis in New England, formed as it were a mighty geomorphological equipoise the like of which has never been seen again in the history of landform studies. True they often conjoined physically, especially in Europe, and in 1908–9 exchanged places at the end of the fulcrum as we have already described – at least from Davis' point of view.

By 1920 a notable lessening of warmth in the friendship and a widening in methodological differences becomes apparent when Davis chose to publish in *The Geographical Review* a derogatory account of a *Festband* presented to Albrecht Penck in 1918 on his sixtieth birthday. Perhaps we should add that Davis treated textbooks written by his own students in the same strict way if they did not support Davisian concepts. For example, he seriously offended Ellsworth Huntington for this reason. But, whereas a textbook criticized in private or public can be and should be improved, to condemn much of a *Festband*, an unalterable expression of goodwill, seems ruthless and more than ruthless if the condemnation is largely because the methods used are not Davisian.

The opening of Davis' review hardly prepares us for the later strictures.

Let the memories of four painful years be set aside as far as possible while this well-earned tribute to a great geographer is reviewed. The volume opens with an address from one hundred and sixty-three of his former students – the list would have been significantly longer if the events of those obtrusive years had not excluded a number of non-Teutons – an address glowing with gratitude for the instruction and inspiration received by the disciples from their master and with enthusiasm for the subject which he has, in their opinion, done so much to raise to the rank of an independent science. They admire the broad range of his studies, developed chiefly in different phases of physical geography and extended into human and political geography; and they properly take pride in his numerous activities as teacher, organizer, explorer at home and abroad, and prolific writer. The many geographers of all countries who have had to do with Penck at international congresses and on distant journeys will recognize the justice of this tribute; for during the last thirty years his has been one of the most active and attractive minds in every gathering that he has joined. His scientific originality and his genial personality have been equally enticing. He used to be liked as much as admired, but during the war some of his statements have lessened the esteem formerly felt for him: into that matter we do not enter farther here.

The Festband contains twenty-two essays by as many writers and thus covers a wide variety of subjects. Morphology of the lands takes the lead with eleven.

(Davis, 1920G, p. 249)

Davis selects for detailed review three regional physiographic studies of limited districts in central Europe, and strongly objects to their domination by the geological standpoint, which in some parts, he says, amounts to geology masquerading as geography.

A careful study of these three essays and a general examination of the others suggests the question whether full warrant can be found for the assertion in the dedicatory address to the effect that Penck has raised geography to the rank of an independent science; for, like his own work as exhibited in numerous publications, many of the essays in the Festband are by no means freed from domination by geology. Just as the masterly investigation, 'Die Alpen im Eiszeitalter', conducted by Penck jointly with Brückner of Vienna, is essentially a geological study, however interesting and valuable its geographical by-products are, so the work of his students as illustrated by several essays in the Festband appears to be concerned more with the evolution than with the description of land forms; and the evolution is treated largely from a geological standpoint rather than directed to a geographical end. Furthermore, not a single essay is devoted to full-fledged geography in the sense of giving a complete or regional account of a limited district, beginning with its physiography and climate and ending with its human inhabitants. *(Davis, 1920G, p. 250)*

Ignoring the fact that the Germans had experienced four years of a disastrous war, Davis condemns also the quality of presentation and publication whereas many Europeans would marvel at the production of a 450-page book in Germany in 1918.

The volume here reviewed is also disappointing in giving so much attention to the science of investigation as not to allow sufficient cultivation of the art of presentation. Even in the mere matter of book making the needs of readers are not enough considered; for most of the essays are not well divided into sections, page headings give only the author's name and the title of his essay, and there is no index. In so far as the morphological essays which occupy half of the volume are concerned, they are not notable for clearness of exposition. All the essays reflect an enthusiasm for research, and that is excellent; but their authors do not seem to have devoted much thought as to how their results could be best set forth or as to how the essentially geographical elements of their results should be given due prominence. This is shown by the greater attention frequently accorded to invisible past processes than to their visible present products, as well as by the scarcity of outline maps and diagrams and by the commonplace quality of such diagrams as are included. The volume is not so great a product, geographically measured, as might have been expected. (*Davis, 1920G, pp. 250–1*)

Later he chastises the authors for their 'imperial provincialism' in assuming that the place-names and local technical terms they use of their home area will be known to all the world. Their diagrams, needless to say, fall far below the coverage and standards required by Davis who illustrates this deficiency by drawing a truly superb example.

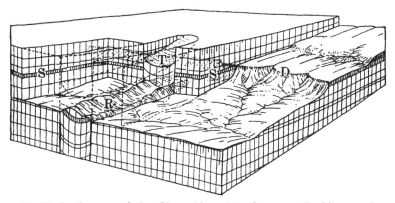

F I G. 98. Block diagram of the Ohmgebirge (D), Germany (looking south-east), showing restored extensions of the capping limestone (S) and of the structural depression (T), the latter giving rise to the Klien (R) – a down-faulted syncline (From Davis, 1920G, Fig. 1)

Even the greater labor of designing and drawing a diagram in explanation of so peculiar a ridge as the Klien would be warranted; for the conception and construction of such a diagram demand close attention to the facts and their meaning on the part of the author, and its interpretation yields a clear understanding to the reader. The diagram above presented [Fig. 98], modified from an original drawn on the ground ten years ago by the late Professor Grund of Vienna, Berlin, and Prague is reproduced to illustrate the point. In this case the diagram is the more worth while because it shows that each side slope of the residual Klien is one of those physiographic rarities known as an obsequent fault-line scarp. (*Davis, 1920G, p. 254*)

Davis constantly complains of the relative lack of explanatory descriptions of the visible forms and often repeats a salutary reminder that presentation in clear, intelligible statements and diagrams is very important.

... after the necessary field investigations are well advanced, let the geographical account of the region be freed from all irrelevant erudition so that single-minded attention shall be given to the facts of today; let these facts be presented according to some well-tested scheme, either empirical or explanatory; and let the presentation be easily intelligible, especially to readers who do not know the region, for in that class belong most of the geographers of the world. (*Davis, 1920G, p. 260*)

He draws on any weapon or analogy at his disposal and drags in astronomy and chemistry in a rather ambiguous way.

> The essay illustrates to perfection that exaggeration of pseudo-explanatory physiographic treatment which makes the origin of a thing more difficult to understand than the thing itself. Indeed, if such an essay as this is to be regarded as geographical, then those are justified who claim that geography is only geology masquerading under another name.
>
> A lesson may be here learned from the practice of astronomers, who in concisely announcing the elements of a comet's orbit do not include the elaborate mathematical calculations by which the elements were determined; or from the chemists, who in briefly stating the composition of a substance submitted for analysis say nothing of all the manipulations by which the composition was learned. Let morphogenetic geographers be as geologically analytical as they like and let them publish their geological analyses in appropriate journals if they wish; but, when it comes to presenting their geographical results, let their statement be directed to the manifestly geographical object of giving a helpful account of visible landscapes. (*Davis, 1920G, p. 257*)

Penck was obviously seriously upset by this Davisian onslaught on his students and replied to it at length by letter. Davis was reluctant to reply to his old comrade's criticisms, which included that the cycle of erosion was based on a false assumption and did not fit some observed facts. Penck, who made free use in his defensive letter of the works of Passarge and Hettner, seems to have lost faith in the general efficacy of the cycle method. In this respect, we must notice that Penck's *Morphology* referred to by Davis in the following letter is the fine two-volume *Morphologie der Erdoberfläche* published as long ago as 1894.

Surely most readers will regard Davis' reply as a masterly piece of literary browbeating. He concedes nothing and concusses his opponent with a welter of quotations and queries. The digression into chemists and dye-making seems peculiarly irrelevant for both parties. The letter is interspersed with frequent passages in German some of which are presumably copied from Penck's letter. These latter passages are marred by errors which seem to have occurred during their copying. In the text below these slips have been tacitly corrected where possible during the translation which is given in italics. A few of the German passages defied translation because of errors or omissions.

Your letter of Dec. 4 last was duly received; also your note of Feb. 13 with copy of December letter. Both have been on my desk over-long, waiting for answer; in the meantime I have been giving every available moment to the Gilbert memoir. It is growing very long, but I hope it may serve to tell our younger geologists something of Gilbert's nature and work.

It is pleasant news that your son, Walther, is established as professor in Leipzig where his father long ago studied. As he may have told you, I have enjoyed reading parts of his Argentine monograph, an able piece of work, and I have written asking him to specify the difficulties he finds in accepting the cycle theory. I am inclined to believe that he really does not know what that theory is any better than Jager did when he condemned it at the close of the Transcontal Excursion of 1912. J's idea seemed to be that the cycle is a rigid scheme, to which the facts of nature must be forcibly adjusted; and as the facts objected, the scheme must be given up. As a matter of fact, the reverse is true; the scheme is elastic, and must be continually modified and fitted to the facts. I will return to this topic later, in replying to your own misunderstandings.

Your comments on my review of the Festband show how easy it is to misunderstand the printed word; how easy it is to give an unmeant emphasis and thus develop differences instead of agreements. For example, the relations of investigation and presentation: 'Each sh'd be developed to best advantage' is my closing word. You can hardly feel otherwise. But you say: 'Eine Untersuchung verliert nichts von ihrem wissensenaftlichem endste Darstellung verfehlt ist, wenn ihr wissenschaftlicher Inhalt mangelhaft ist.' As to the second clause, of course. But what has that to do with the case? Who is talking about poor investigation well presented? Not I. There are four possible combinations. Bad investigation badly presented; Bad inv. well presented; good inv. badly presented; good inv. well presented. Why not strive for the last? Do not trouble to estimate the relative values of inv. and Pres'n, but give each its best value.

As to your second clause, I differ. A good investigation badly presented loses value, because its value cannot be apprehended in so far as it is badly presented. Of course, it is all right in the mind of the investigator, just as gold is all right in the chest of a miner; but it does no good in the world. Just in so far as an investigator fails to make his results clear, just so far does he fail of being as useful as he might be.

Scientific economy is surely served by careful presentation of good investigation. How can there be any question on that point? My idea in the review is that the authors of the Festband essays did not sufficiently attend to the art of presentation. That does not mean that they sh'd give less attention or care to the science of investigation; not for a moment.

You suggest that we, you and I, have perhaps a different conception of the art of presentation. Doubtless we differ in subordinate ways, but not in fundamentals. We both wish that the objective facts and the properly based inferences sh'd be clearly presented. We may differ because you have not given as much attention as I have to the contrasts between analytical presentation and descriptive presentation. For example, if any investigation involves principles and methods that are not fully established or not generally accepted, it sh'd be presented 'analytically' or argumentatively; and in problems relating to land forms, an analytical presentation must include much geological matter, as such; the article presentation thus assumes the quality of physiographic geology, in which the successive steps in the development of the existing land forms are properly presented, as a means of justifying the conclusion reached by the investigator, and of making clear his novel methods.

On the other hand, if an investigation of land forms employs generally accepted principles and methods, it is not necessary to repeat them in presenting the results reached; the results alone suffice; and much paper and ink saved. Also much time of readers. True, a young student may be required to indicate precisely the principles and methods that he follows, so as to insure his understanding of them; but a proficient geographer need not be required to do so. See my article on the Valdarno, in *Zft. Ges. f. Erdk. Berlin*, 1914. Of course, if an author's object is physiographic geology (the successive forms assumed by a land surface during passage of time), let his essay exhibit such succession; good examples of this kind are found in our Geol. Folios. But if the object is the description of the existing landscape, that is, physiography proper as a part of geography, then why drag in the past, unless in cases of problematic treatment. It is not worth while today to discuss whether the Rhine gorge is the work of the river; that is a past issue; settled long ago. Similarly, it is not worth while today to argue and demonstrate that the bench in the side of the gorge represents a former valley floor now uplifted and more or less deformed and dissected; that is already well proved. State the results, but don't delay further work by arguing that case over again.

As to my early articles on the Rivers and Valleys of Penna; and on the evolution of the Conn valley: both principles and methods therein employed were more or less novel; they needed explanation and justification; and I therefore attempted to analyse the investigation. But the analysis being once accepted, it is not necessary to go over it again. So with my paper on the Front Range of the Colorado Rockies; it is argumentative and explanatory; but as soon as readers generally understand what a morvan is, explanation and demonstration will be no more needed in describing a new morvan than in describing a new volcano. Of course, there will be difference of opinion as to whether demonstration is yet reached or not; thus some conservatives are still 'haggling' about glacial erosion. But there can be little difference of opinion over the principle here involved.

As to my analogies with astronomy and chemistry: It is easy to misinterpret them, if you wish, and easy to understand them. So long as the method of determining a comet's orbit was not well understood, it would need and deserve discussion; but after once being established as well as Gauss established it about 100 years ago, what is the use of going over his demonstration again. In a text book, repeat it, of course, but in a professional essay, no. Use it.

As to chemistry, the same. It is widely customary for chemists to present their innumerable results concisely, provided that the principles and manipulations by which the results are reached are well established and generally familiar. If not so, they ought to be explained, and they are explained in such cases, if a chemist proposes to make a scientific communication to his colleagues. Of course, if a chemist seeks to withhold some ingenious method of analysis or of synthesis for his own profit, he is perfectly free to do so; but such secrecy is not regarded as characteristic of scientific presentation.

German makers of dyes naturally keep their methods secret; so do American makers of gimlet-pointed screws and machine made watches; they have a perfect right to do; but what has that to do with the true principles of scientific presentation? Do you find any real likeness between such secrecy and my suggestion that geographical descriptions of well understood principles and methods should be concise?

As to the anxiety of American chemists to learn German methods of dye-making, three remarks. First, the value of imported dyes of German manufacture, of kinds that cannot at present be made in this country, but which are essential in certain of our industries, they amount to about $12,–15,000,000 a year; a small matter; Woolworth, who introduced a chain of 5-cent stores all over the U.S. some years ago and made a huge fortune out of them, sold about that value of candy in a single year. It is therefore not the money-cost of these finer dyes which we are concerned about. Second, a chemical corporation has lately been formed here, capital said to be $300,000,000, and it is now producing vast quantities of chemicals formerly imported; it is already making a larger percentage of high-grade indigo than is made in German factories. Third, look into this question 10 or 15 years hence, and see how it stands then.

But all this is aside from the point at issue; the point is that the presentation of familiar kinds of things like cuestas and incised meanders and many others, should be concise, without argument, but with sufficient direct description, qualitative and quantitative to enable the reader to understand what the writer is telling. As to secrecy – what has that to do with it?

I confess it is rather amusing to find you ranged with Friend Passarge in thinking that block diagrams are dangerous, because some readers may think they represent only observations and nothing of inferences. What sort of fools are such readers? Can they not see that all underground structures, such as you and Passarge freely introduce, and properly, in your sections, are nothing but inferences? Good inferences, to be sure, but inferences all the same. Then, is there really anything new in the idea that a line in a diagram expresses a more definite value than a word in the text? Certainly that idea has long been familiar hereabouts. Second, what is the remedy for this danger of block diagrams. One remedy would be better education of readers of geographical articles in which such diagrams are used. But for my own part I do not propose to use such diagrams less because some readers may be stupidly uneducated. Another remedy might be, dont use block diagrams; and that appears to be the remedy adopted by the authors of the Festband; but not truly so; the reason that they did not use block diagrams is not because they were afraid of misleading their readers; not at all. Ask the authors themselves and you will find it is because they did not know how to draw such diagrams. Try it and see if that is not the real cause of absence of diagrams.

The real need here, is to draw diagrams carefully; if there are doubtful points, indicate them by dotted lines, or by warnings in text. But the idea of giving up diagrams because they may be misunderstood is absurd. What will happen to a text on land forms if it has no diagrams. Is it to have maps only, purely superficial? Will the next edition of your *Morphology* (for which many readers are anxiously waiting) reject all such inferential diagrams as Figs. 11, 12, 13, 14, 15, in vol. II? No, if wisely advised, all those inferences will be retained; and all the figures will have the block-diagram quality faintly indicated in Fig. 15 but better drawn. Indeed, so drawn as to give emphasis to the visible and observed surface forms, and leave the underground inferred structures lightly outlined; just the reverse of the present style. Similarly, the next edition sh'd present much more pictorial views of Cirques (Figs. 17–20) so that the reader can more clearly apprehend the relation of preglacial form to glacial excavation and morainic deposition. The same with coasts Figs. 29–32; much improvement is here possible by adding surface perspective, and by indicating in successive diagrams or in successive blocks of one diagram, the successive stages of form development. Inference, to be sure; but can any one regard such inference as so dangerous as to exclude it? More diagrams like Fig. 36 will be helpful, still more so if a foreground structural section were added; and if the headlands were shown with increasing height inland; instead of in the very exceptional form of decreasing height inland; still better if several diagrams were included showing the differences between the ragged cliffs and narrow rock platforms of an early stage of cliff and platform cutting, with small bay-head deltas, and the intermediate stage of stronger cliffs and larger deltas; and especially the much later stage of far-retrograded cliffs and NO deltas (deltas cut away, like the former headlands). Inference, to be sure, but why not.

Why go back to a trite truism like *'Observation is the foundation of Geography'*. Why not advance, and say 'Observation is the foundation and inference is the upper structure'; both are essential in the completed edifice. Are we really to say only what we see? Must we not use the excellent inferential results regarding the effects of glaciation on Alpine form obtained thru the labors of Penck and Brückner? Must a volcano be described only as a conical mountain, not as a mountain formed by eruption? Of course, you may say, we see some volcanoes in eruption, of course we do; but all the same it is still an inference (an excellent inference!) to say that extinct volcanoes were, in their time, formed by eruption. Observe, of course; but why not think also? You may say, observation is safe and inference is dangerous. My answer is, observation without inference is stupid. Look at Passarge's first volume and see how he strives to suppress the natural spirit of inquiry that every right-minded, bright-minded student must feel. Infer carefully, of course, that is just as important as to observe carefully. But in my experience (never more clearly shown than in 1914 on the Pacific) thinking is an immense aid to observing. A Kodak can observe.

You say: *'What we need is not so much bold generalization and pregnant speculation but rather a rich accumulation of observations.'* The very way that I have found to promote observation, keen sharp critical observation, is precisely to think hard while you are observing hard. Neither mental process sh'd be suppressed; both sh'd be excited; and both sh'd be carefully trained. All over the Pacific, observation has been tried without thinking; and the result is a quantity of defective records. Plain matters of fact not recorded because not seen. I do not refer here merely to narratives of exploring sea- captains; but to records of scientific travellers. It is a melancholy study of poor observation. And the poverty of observation went largely with poverty of thinking and speculating.

As to the use of local village names as guides to physiographic features: Of course, introduce such names as freely as you wish if your object is to define the locality of some small outcrop; but how can you possibly defend the use of local names in Rudnyckyj's article on Podolia. You suggest that German readers resort to good atlases. I wish you would try your luck with Stieler or Debes, and see how many of R's local names you can identify. I found difficulty in identifying them even on the 1 : 200,000 and 1 : 75,000 sheets of the Austrian maps. What an amount of time he would have saved his readers if he had introduced a little outline map!

The principle to be observed here is clear enough. If you wish to indicate the location of some feature that is named on a good atlas, use the name of course. But if you wish to indicate something, such as a physiographic boundary, that is not shown on such an atlas, employ a better device than insignificant village names, most of which cannot be found in Stieler. If you wish to insist that a boundary runs close to a certain village, say so, after you have given its general course; but as such boundaries as R. referred to are not precise, the village names have little relation to them.

One thing I feel confident: It was not R's intention to show that his physiographic boundaries ran close to certain villages, so that his readers, when later in Podolia, could identify his lines. That was not his reason for using village names. The real reason was that he did not know any better! He followed an old-fashioned, traditional method, without asking himself whether a better method could not be found or devised. Scientific accuracy ('Wissenschaftliche Akkuratesse') was not his object; he simply did not know any other way; so he adopted that very awkward way. He was mentally clumsy. If he consciously had the relations of International and Local readers in mind when he wrote (very improbable) he could have easily satisfied the needs of both, a general outline for one; and local village names in text or an outline map for the other.

What is the attitude of high-class scholarship in this question. It is of course to give PROPER attention to investigation and to presentation. That is, make the investigation accurate, and the presentation intelligible. It is slovenly, not scholarly, to be accurate in investigation and careless in presentation. Accurate investigation shows ability, careless presentation shows selfishness. Consider Hassinger's very learned essay: No question of his learning, his ability. He knows a terrible lot. But how little he cares about his readers! He has evidently worked (not only observed; but thought; more thinking than observing I sh'd fancy) enormously; and he seems to think that his responsibility ceases if he simply throws his results together in some sort of fashion, and lets his readers struggle through them as best they may. That may be one kind of scholarship; but it is not the best kind.

Now let us turn to the scheme of the cycle. You say that I set out from a false assumption. 'You assume that a block (Scholle) is first uplifted and then eroded. The correct way of putting it is: the block will be subject to erosion from the moment that uplift begins.' I might have expected such a statement from Hettner or Passarge; but from Penck, no! It sadly exemplifies what I said at the beginning; that you do not know what the scheme of the cycle is; that your conception of it is imperfect and because that imperfect conception does not work, you will give it all up. Well, well, well! Where did you get your understanding of it?

Let me present the case as follows: First, my early statements of the scheme were rough outlines; the scheme was then incompletely conceived and therefore could not be fully stated. It has grown by continued attempts to adapt it to the complexities of nature. I have often wondered why its growth has been so slow; yet I find that even a Gilbert did not immediately develop all aspects of his Basin Range theory; and it is probably true that not even a Penck did at one stroke develop his present understanding of Alpine glaciation. However, let us look at some of my articles; and see how the question of uplift and erosion is stated.

Proc. Amer. Assoc. Adv. Sci. First article on subject. xxxviii, 1884. 'The channels (valleys) will be narrow and steep walled in regions of relatively rapid elevation, but broadly open in regions that have risen slowly, and I believe that rate of elevation is thus of greater importance than climatic conditions in giving the cañon form to a valley ... During maturity no vestige of plain surface remains (except in slow-rising, low plains that never acquire any marked relief).' That seems to me, as I now re-read it, pretty good for a starter.

In my Berlin paper, 1899, slow elevation is not explicitly considered; but possible complications are intimated in the statement regarding interruptions of a cycle: 'Only in this way can the theory of the cycle be made elastic enough to correspond to the variety of natural conditions', p. 288. Slow elevation not being considered there, I contributed a paper on 'Complications' at the Washington Congress, 1904, see p. 153. 'The elementary presentation of the ideal cycle usually postulates a rapid uplift of a land mass followed by a prolonged stand still. The uplift may be of any kind and rate, but the simplest is one of uniform amount and rapid rate ... In my own treatment of the problem, the postulate of rapid uplift is largely a matter of convenience, in order to gain ready entrance ... It is therefore preferable to speak of rapid uplift in the first presentation of the problem, and afterwards to modify that elementary and temporary view by a nearer approach to the probable truth.' On the next page, 154, the case of an open valley eroded during a slow uplift is specifically considered. 'In such a case there would have been no early stage of dissection in which the streams were enclosed in narrow valleys with steep and rocky walls.'

Geogr. Journ., 1899 'It should not be implied, as was done in Fig. 1, that the forces of uplift or deformation act so rapidly that no destructive changes occur during their operation. A more probable relation' (This is p. 7 of my reprint; in section headed Dev't of Conseq. Streams.) The figure referred to is first explained on provisional assumption of immediate uplift; then later on assumption of gradual uplift with accompanying erosion.

Erkl. Beschr. der Landformen., p. 146. 'Erosion during uplift. It is important to bear in mind that erosion does not in the least wait until uplift is completed before it starts to attack a land surface'.

On the whole the case seems clear enough. What puzzles me is why you sh'd have thought I assumed no erosion during uplift. That Hettner and Passarge might have thought so, yes; but that ...

Now your point as to the incompetence of the cycle scheme because you find that certain valley heads on flat uplands are not sharply incised in an early stage of renewed erosion, but are broadly opened; hence, according to such a succession, the cycle terminology would be old, young, old.

It is perfectly true that according to my original scheme, no such wide young valley heads were mentioned. I first came to know them in central France in the winter of 1898–99. I am sorry that, on returning home the following autumn I was so plunged into teaching that the only thing I wrote in detail was glacial erosion, the open young valleys remained in my note books. But the case is clearly enough presented in *Erkl. Beschr.* pp. 63, 259. On p. 63: 'Verwitterung und Gekriech sind an den Wanden eines langsam vertieften Tales – (von einem kleinen Quellfluss) – genau so wirksam wie an einem grossen Flusslauf, so dass das Tal eines kleinen Nebenflusses kraftig wahrend seiner Eintiefung ausgeweitet wird'. – 259. Die Hange der neueingeschnittenen Nebentaler werden im Oberlaufe so gestaltet sein, also ob sie bereits besser ausgeglichen waren als die Hange der Hauptschluchten'. etc. etc. [p. 63: *'Weathering and creep (solifluxion) are just as effective on the walls of a slowly deepening valley formed by a small headwater as with a large river bed, so that the valley of a small tributary becomes markedly widened during its incision. p. 259: The slopes of the upper courses of the newly-incised secondary valley will be formed in such a way as if they were already better adjusted (graded) than the slopes of the main gorges.'*]

The topic might easily have been expanded; but the book was already somewhat too large.

What interests me most in this connection is this: Suppose I had not understood this aspect of the problem at all; had said not a word about it. What then? Might not the special case be perfectly well added to the scheme when found; and the scheme, thus subordinately modified, continued in use, and with satisfaction because it is improved?

Personally I feel that it is still open to further improvement, to which I hope that you and others will contribute; and take all the credit therefore you wish; but because still improvable, I see no reason for rejecting the good parts of it. To be sure, if you insist, as Hettner seems to (see below) that the scheme once formulated in its earliest, simplest form, must not be modified for its improvement, it might be attacked in all sorts of ways. But why insist on anything so absurd?

Again, if you reject the scheme of the cycle, as you say you must (because of its deficiency as above) what are you going to do instead. I shall watch with interest to see what plan you introduce. Rovereto rejected the scheme because he did not know of any peneplains in Italy; therefore the scheme was false. (That is, he failed to understand that a cycle of erosion might be interrupted as often as need be; and failing to understand this simple idea, he gave the scheme up!). Passarge rejects it, because we do not yet know enough to use so venturesome a scheme. He seems to wish to wait till everything is known before anything is explained. It is lucky for the world that Newton did not wait till Einstein came along with relativity before formulating the law of gravitation, yet that would seem to be Passarge's idea. Penck proposes to give up the scheme of the cycle because he finds some valley heads that do not find explanation in the scheme, *as he knows it*. Well, well, well! And because the scheme assumes immediate uplift, without contemporary erosion. Go ahead; invent some other schemes. Do as Hettner does. Reject structure, process and stage, and say instead: *'Geographical description thus deals with three things: 1. facts concerning internal structure, 2. process of transformation, 3. the influence of the latter on the former, and the resulting surface forms and soil varieties.'* (*G.Z.* VII, 143.) As to the *'processes of trans-formation'*, he adds on the next page: *'Here we must take account of the duration of its effectiveness, but only as a secondary factor.'* I assume that Hettner intends to act honorably; but how he could be nearer to writing unfairly I can hardly imagine.

Other citations from Hettner are equally curious. I find it difficult to understand his querulous attitude. See what he says about Pidjin Deutsch and consequent, (I have forgotten the reference). See his misunderstanding about Alter (age) and Stadium (stage). *'It is true that the Davisian school, in which every third word is AGE, really does dispense with chronological determinations altogether'* (*G.Z.* XIX, 443). As a matter of fact, which he ought to have known, Alter is not the third word; it is Stadium. But Hettner and Passarge do not care to be accurate in a little thing like that; having once got it into their heads that die Davis'sche Schule uses Alter they stick to it, and abuse the Schule. (How about *'scientific accuracy'* in such a case?) See Hettner's diatribe against deduction, *G.Z.* XIX, 159, top of page ... *'because they, without any further examination ...'* What unfairness, what prejudice! And besides, he uses deduction as often as he wishes; but as a rule not thoroly, and therefore imperfectly. He used Stadium himself in one of his earlier articles. *'The forces of denudation (erosion?; Abtragung) at various (chronological) stages are followed by general planation (Einebung), but first of all they cause etc. etc.'* *G.Z.* IX, 202. On next page, bottom, he mentions 'the duration of external influences' but regards it as of subordinate importance. I attach greater value to it than he does.

Look at his apprehensive, his soliticous attitude, *G.Z.* XVII, 143, regarding the poor young morphologs who will be misled, etc. and then *'peacefully retire for the night'*. Note his plaint: *'The classification and terminology obtained in this way ... is only a scheme, which can only be considered valid by testing it against nature'*. What a marvellous discovery! Look at his complaint that the genetic scheme is no longer maintained in its 'purity' because it has been discovered that 'older' valleys in hard rocks have the same form as younger valleys in weak rocks. (all this, p. 142.)

Yes, I shall be greatly interested to see what plan you adopt instead of the pernicious cycle. But as you read all these pages, please understand that I am laughing, not scolding. It all amuses me immensely; to be sure, it is also exasperating at certain points, as when Hettner and Passarge so completely fail to apprehend my efforts. But why trouble myself about them. Let the time settle the matters in dispute. And while they are in process of settlement, please understand that I am laughing, not scolding. It may read like scolding, but that is because you cannot see or hear me while I write.

Do you not, yourself, think it rather amusing that you, of all persons, should have accused me of assuming that no erosion takes place during uplift? It strikes me so. But now, having given most of the morning to this over-long letter, I must close it. Write to me again, as fully as you like, if you can find the time'. (*Letter: Davis to A. Penck, 3 April 1921; Cambridge, Mass.*)

Some of the concluding remarks and irrelevances lead one to believe that Davis was genuinely disturbed when writing this epistle. To think of him 'laughing' in such a literary duel is quite beyond our belief unless he uses 'laughter' in the sense of the smile of a satiated tiger. As far as we know Penck never answered the letter in detail – who would? – but he now no longer needed to defend his views as his son Walther had already launched a vigorous frontal attack on the Davisians. Davis, however, kept up a correspondence with Walther at least for some months following.

In his 'Die Gipfelflur der Alpen' (*Sitzungsber. Preuss. Akad. Wissensch., Math.-Phys. Kl.*, XVII, 1919 (1), pp. 256–68) Albrecht Penck had been initially concerned with the height of the undulating plain connecting the summits of Alpine peaks (the *Gipfelflur*). Some thirty years previously he had thought this due to the existence of an upper denudational limit controlled by the rapid increase of erosional rates with elevation, and contrasting with the lower limit of denudation (i.e. peneplain, or *rumpffäche*). Subsequently the Davisian idea of an uplifted and dissected peneplain was invoked, but Penck believed that there were good reasons to support the view that the Alps had not been peneplained in pre-glacial times. Sharp peaks and ridges owed their dominance, Penck held, both to glacial backcutting and to the intersection of the steep valley-side slopes developing under the action of aggressive basal stream down-cutting. Under the latter conditions equal stream spacing would tend to produce summit accordance during progressive lowering, as Penck had previously suggested in his *Morphologie der Erdoberfläche* (Vol. 1, 1894, p. 365) and as Daly (1905) had supported. Penck was concerned with the identification and differentiation of active and relict slope forms, and considered the simultaneous association in the Alps of sharp and rounded slope crests to present a problem. The Davis cycle postulated that rounded crests develop from sharp ones, but when, Penck proposed, one considers 'the geographical cycle in a more comprehensive manner' other development sequences are possible:

> Uplift can begin quite slowly, so that low hills and shallow valleys result; if the movement becomes stronger, the valleys cut more deeply, and out of the low hills *Riedels* develop, rounded in shape, from which sharp edges result if the uplift continues. (*A. Penck, 1919; translated by Mrs G. Wills*)

Penck criticized Davis for having in his simple cyclic scheme separated uplift and erosion instead of considering their inevitable interactions. Penck considered the Davis cycle to represent the topographic response to only one possible pattern of uplift, and proposed three different 'series of transformations' the differences between which are largely determined by the intensity and duration of the uplift.

> The first transformation (*Umbildung*) sequence is characterised by strong, long-lasting uplift. Valleys are cut swiftly into the rising land; but they cannot incise the block as swiftly as it moves upwards, their floors (*Sohlen*) coming to lie above the original plain, gradually moving upwards with the land, although they become deeper and deeper. Between them sections of the elevated plain rise as

F I G. 99. A three-stage diagram to show the evolution of a hanging, fault-block valley
Background. A maturely-eroded fault block, with a large fan outwashed from its
chief valley into the adjoining down-faulted trough
Middle. The potential form of the landscape, after further up-faulting of the mountain block and depression of the trough
Foreground. The same landscape after some erosion and deposition consequent on
the renewed faulting

(From Davis 1930C, Fig. 4)

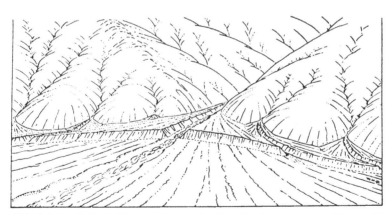

F I G. 100. A faulted alluvial fan at the mouth of a valley in a maturely-dissected fault-
block mountain, exhibiting an 'eyebrow' scarp (From Davis, 1930C, Fig. 5)

areas of *Riedel*. These become smaller and smaller because the valley sides (*Gehänge*) move laterally, until the *Riedel* disappears if the slopes of neighbouring valleys intersect in a sharp edge (*Schneide*). With prolonged uplift the edges do not rise as fast as the land, but only as the valley floors, from which they remain divided, . . . by an approximately equal difference in height. If the moment is finally reached when the animated river erosion has become strong enough to counteract the uplift, the upward-moving land does not further gain in height, but is levelled down by the rivers and by the destruction by the valley side slopes in proportion as the land rises. The upper uplift-limit has been reached. As long as the uplift continues, the ridges and peaks of the mountains formed maintain a constant height. Only when uplift abates are the rivers able to cut into the now high bases (*Sockel*) and depress the edges between them, until their vertical erosion slows down and the valley floors widen out. Then the edges are worn down and become rounded; from these rounded ridges (*Kämme*) result. Finally vertical erosion ceases, the valleys grow shallow and wide, and the ridges between them are levelled; in the end the land almost flat.

In this development sequence, the most noteworthy stage is that in which the edges remain at the same height over a long period. So long as this is the case, their level (*Flur*) indicates the upper limit of elevation, beyond which the land cannot rise in the given circumstances. We can then speak of a boundary peak-level (*Grenz-gipfel-flur*) as the final result of the uplift. This does not last as long as the edges, which achieve development both by approaching the upper levelling-limit (*Abtragungsgrenze*) and by sinking down beneath it. These sharp edges have slopes (*Abfälle*) of a youthful character, and between them lie youthful valleys, so long as glacial erosion has not interfered. This youthfulness of individual forms prevents us from describing the whole as 'mature', as W. M. Davis does; we prefer to speak of a fully-grown mountain group (*Gebirge*) at the edge-stage of development, which represents a counterpart to the ravine stage in valleys, although of shorter duration. Edges and ravines are corresponding solid and hollow shapes which are swiftly transformed.

The second transformation sequence, too, is characterised by a powerful uplift, but of limited duration. As with the first, to begin with *Riedels* are quickly formed, but before they can be destroyed through the development of very steep declivities, the uplift ceases. No edges are formed. The mountain group does not grow as high as the upper uplift-limit; it does not grow fully, but remains in mid-growth. Its heights are moderate, as are the differences in the heights. Its later variety of forms comes from the transformation of the *Riedels*; their rounding-off and levelling. Our second development sequence has initial and final stages similar to the first, but the characteristic middle stages are not found. They have been, in effect, missed out.

The third development sequence is associated with very slow uplift and lasts as long as the uplift. Rivers are never given the opportunity to quickly incise deeply. There is no formation of deep ravines, instead wide valleys develop and simultaneously the land between them lowers. The sequence again omits the middle stages of the last sequence described. Without there being a development of sharply-circumscribed or rounded-off *Riedels*, the very slowly heaving plain passes through the stage of flattened heights with shallow valleys into the *Rumpf*, never achieving great relief. The characteristic of this development is that the shallow-valley stage, which in both the other sequences comes at roughly the end, here appears close to the beginning in the uplift-phase, whereas in the other two sequences it develops only after the uplift is finished. The whole transformation runs its course in a single development stage, and this lasts only a little longer than the uplift. (*A. Penck, 1919, pp. 264–6; translated by Mrs G. Wills*)

Davis replied at length to Penck's article in the *Journal of Geology* for 1923 (J), repeating many of the arguments which he had made in private correspondence with the Pencks. The reply was divided into three parts.

The first part gave, with parenthetic comments, what Davis believed to be a fair exposition of Penck's view of the development of alpine landforms. As his title ('The cycle of erosion and the summit level of the Alps') suggested, however, Davis was incapable of viewing Penck's ideas except as elaborations of the cycle of erosion.

[Penck's article] proposes several refinements of the cycle of erosion; and it employs deduction to an extent that, while unquestionably helpful, has been decried by certain other German geographers ... [Penck] now extends his previous discussion, first by a more refined account of Alpine forms, in which he introduces an understanding of the effects of glacial erosion that had not been reached when his earlier article was written, and second by an elaborate deductive analysis of the cycle of erosion for mountains, in which the interaction of upheaval and degradation is more fully considered than has hitherto been done. (*Davis, 1923J, p. 2*)

Although the essay is the work of a geographer, the study as a whole is more largely concerned with conditions and processes of the past than with the forms of the present and hence its character is dominantly geological; and in this respect it is characteristic of German physiography in general. (*Davis, 1923J, p. 3*)

It would further seem as if a very special though accidental relation must exist between rate of upheaval, climate, drainage area, and rock resistance, in order that the good-sized streams here considered should maintain a graded flow, and that their valleys should maintain graded slopes, even while upheaval is in progress. For it must be remembered that as the streams are not yet deepening their valleys as fast as the mountain mass is rising, their fall must be increasing; and with increasing fall, their graded condition might be lost. Besides, with increase of mountain height, there must be increase of rainfall and of stream volume; and increase of stream volume would still further promote a return to a youthful, non-graded condition rather than a persistence in a mature or graded condition. On the other hand, the increase in valley depth and the accompanying increase in the area of wasting valley sides – part of this second increase being due to the development of many side ravines – will cause an increase in the detrital load that is to be swept away by the streams; and this may permit them to develop graded courses in spite of their increased fall; but none of these details are mentioned in Penck's deductive analysis. (*Davis, 1923J, p. 5*)

The most significant stage of this [i.e. Penck's] ideal cycle is held to be the intermediate one in which the sharp ridge crests maintain a constant absolute altitude as well as a constant relief, in consequence of a balance having been struck between the rate of upheaval and the rate of deepening the larger valleys; but the duration of this stage is shorter than that of the sharpened crests, which persist in the preceding stages while crest altitude is increasing, as well as in the following one while it is decreasing. For these three stages, in which the sharpness of ridge crests is maintained, my term 'mature' as originally defined, seems appropriate, in view of the maximum strength and variety of relief then acquired by the mountain mass with its steep but graded slopes; but Penck discards the German equivalent, reif, of that term which he had previously used, and speaks in paraphrase . . .

(Davis, 1923J, p. 7)

It may be here noted that valleys which, either by reason of slow upheaval or of weak rocks, are opened with graded slopes as fast as they are deepened, may be described as 'born mature'; also that a region which represents the third ideal cycle may be described as 'old from birth'. Furthermore, the late stage of Penck's third cycle will be very similar to the late stages not only of his second and first cycles, but also of a cycle of erosion introduced by an upheaval so rapid that little erosion is accomplished while it is taking place. The chief difference between the late stages of these four ideal cycles [note: including the Davis cycle] would seem to be that the streams would be best adjusted to weak structures by the process of river capture in the last mentioned cycle of very rapid upheaval, and least adjusted in Penck's third cycle of very slow uplift; this being for the reason that the more rapid the upheaval, the better the opportunity for large streams to cut down their valleys to a greater depth than that of small-stream valleys. *(Davis, 1923J, p. 8)*

It must, however, be a difficult matter to determine whether the conditions of uplift and erosion postulated in Penck's first ideal cycle really represent those under which the Alps gained their preglacial forms; for apart from the highly specialized relation of unrelated factors which, as pointed out above, is necessary for the persistence of graded streams and of graded ridge slopes during progressive upheaval, there remains the large uncertainty as to whether those parts of the high Alps in which no trace of earlier-cycle forms now survive may not have gained a considerably greater altitude than now at an early stage of the preglacial cycle in virtue of a rapid upheaval – such an upheaval, for example, as that which, acting on a much larger scale, gave the Himalayas their towering heights of today; and also an uncertainty as to whether the rough equality of present Alpine altitudes may not be competently explained chiefly by erosion after upheaval had ceased, as Penck formerly supposed, rather than by a delicate balance of erosion and upheaval as he now suggests. The idea of a constancy of height being imposed on mountain crests by a balance between upheaval and degradation is a beautiful one, and the ingenuity of the analysis by which the idea is carried out is not to be questioned, but its application to Alpine summits does not seem fully assured.

(Davis, 1923J, p. 9)

Let it be added that it is physiographically immaterial to which one of the three sharp-crested stages in Penck's first ideal cycle the preglacial Alps correspond; all three stages would look alike. Similarly, it is physiographically immaterial whether young mountains with flat interstream uplands and subdued mountains with rounded crests are to be explained by his first or his second ideal cycle, which in the early and late stages develop similar forms. Again, it is physiographically immaterial whether old mountains – that is, masses of deformed structure, once lofty but now reduced to undulating lowlands – are to be explained by his first, second, or third ideal cycle, or by a cycle introduced by a very rapid uplift, except in so far as a prevailing adjustment of drainage to weak structures would suggest the operation of a rapid uplift, or of the first instead of the third of Penck's cycles, as noted above. (*Davis, 1923ĭ, p. 10*)

The second part gives a very detailed and interesting account of the development of the cycle of erosion concept by Davis and of its extension to a variety of topographical conditions. He concludes:

One of the most common grounds for objecting to the scheme of the erosion cycle seems to be a general misunderstanding of its object. Several European geographers have misconceived it as a rigid scheme, to which the varied facts of nature must be forced to conform, instead of as an elastic scheme, readily modified to conform to the varied facts of nature. It has been misunderstood as always demanding a rapid or sudden upheaval, so rapid or sudden that practically no erosion could take place until the upheaval was accomplished. Yet slow upheaval movement with accompanying erosion is manifestly as easily postulated as rapid upheaval. Others seem to have supposed the scheme to present final and infallable conclusions; and on discovering an error or omission, they feel that the scheme must be discarded in its entirety. In my own case, at least, the scheme has been a growth, and its growth is by no means completed. Moreover, however many modifications, improvements, and extensions the scheme may now or later receive, it should be remembered that they are all based upon the valid principles of the scheme already established, and that but for the previous establishment of those principles the improvements of the scheme could not be made. Such modifications and extensions are like strengthened or reset rungs or newly added upper rungs in a physiographic ladder, on the lower rungs of which a good measure of ascent has already been made above the empirical level of the science fifty years ago. It is not to be questioned that various special cycles have yet to be worked out in order to develop form-sequences appropriate to peculiar structures and processes; and it is greatly to be desired that systematic studies of this kind should be combined with the observational studies of trained geographers in regions of unlike climates. And now after this long detour away from the Alps, return may be made there in order to show that Penck's explanation of the similar summit altitudes involves the elaboration of precisely such a special sequence of forms as contributes to the fuller development of the cycle scheme; but his study unfortunately includes an element of destructive criticism to which attention must also be called. (*Davis, 1923ĭ, pp. 24–5*)

Davis devotes most of the third section of his paper to demonstrating that he had, in many previous publications, allowed for the possibilities of different types of uplift affecting the landforms developed on them, and continually tries to play down his differences with Penck.

Whether the deductions from the postulate of long-continued upheaval are accepted as valid or not, the general sequence of forms that is traced out is a beautiful one. First, both the altitude and the relief of the young mountains are increasing, the increase of altitude, but not of relief, being as fast as the upheaval of the mountain mass; then while the ridges are sharpened, their altitude continues to increase, but now a little more slowly than the rate of upheaval, and the relief is held at a constant value; next, continued upheaval being balanced by degradation in full maturity, altitude reaches and is maintained at a constant maximum, while relief stands unchanged; later, upheaval ceases and altitude is slowly decreased, although for a time the ridges are still sharp and their relief is still unchanged; finally, the sharp ridges of the quiescent mass are rounded and lowered, and thus both altitude and relief are decreased to smaller and smaller values as old age is entered upon. There is elegance as well as originality in these deductions. (*Davis, 1923J, p. 26*)

Penck implies, however, on a number of his pages that his present views are corrections of earlier views. He states, for example (p. 263), that the sharp-crested preglacial Alpine ridges have been developed from pre-existent rounded ridges (dass manche Schneiden aus runden Formen hervorgegangen sind), and that such a sequence stands in opposition to a previously published scheme of a typical cycle of mountain erosion, in which a reversed sequence of forms is presented, the rounded or subdued ridges of late maturity being explained in that scheme as developed from the higher and sharper ridges of early maturity. But no real opposition occurs here, for in the earlier published statement the typical cycle of mountain erosion was supposed to begin after a previous cycle of erosion had reduced a region of deformed structure to a peneplain which, when upheaved, is dissected in such a manner that the sharp ridges of full maturity naturally enough precede the rounded and subdued ridges of later maturity in the same cycle. On the other hand, in the special case of the Alps, the sharp ridges of the pre-glacial stage of the present cycle were developed, as Penck clearly explains, out of the rounded ridges of an earlier cycle, which was interrupted by upheaval before peneplanation ensued. In other words, the rounded ridges were introduced, ready made, from an earlier cycle, and in that cycle they had presumably had sharp crests before they were rounded; it was in the following cycle that they were again sharpened. In this special case it is just as natural for the sharp preglacial ridges of the present Alpine cycle to have been developed out of the rounded crests of the earlier cycle, as it is for the rounded, late-mature forms of a single typical cycle which begins with an uplifted peneplain to develop out of the sharp forms of early maturity. (*Davis, 1923J, p. 27*)

Directly after his irrelevant statement that the actual sequence of forms in the Alps – rounded ridges converted into sharp-crested ridges – stands in opposition to the sequence that I have given as typical of an erosion cycle, he goes on to say that the scheme of the cycle should be treated, not as involving the action of erosion on an already upheaved mass, as Davis has done, but as involving the action of erosion during upheaval as well as afterwards (pp. 263, 264). It is perfectly true that I have frequently presented the scheme of the cycle as if introduced by upheaval and continued by erosion; the reason for such presenta-

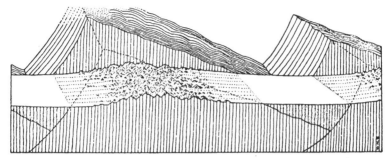

F I G. 101. The origin of worn-down rock floors around residual mountains
Background: Two up-faulted blocks in their potential (non-eroded) form.
Foreground: The inter-block depressions have been aggraded by detrital plains and
the uplifted blocks worn down to subdued mountains, flanked by smooth rock floors
(pediments) which grade imperceptibly into the detrital plains.
(From Davis, 1930C, Fig. 7)

tion being that it is the simplest way of placing the general idea before beginners; but it is also true that Penck has presented the scheme in the same simple
way.

... We have therefore both presented the scheme in a simple, elementary
fashion. But besides setting forth the scheme of the cycle in this simple and
elementary manner, I have quite as often extended the scheme by presenting it in
a more advanced manner, as opened by upheaval and erosion acting together and
completed by the continued action of erosion after upheaval ceases. Nevertheless,
the interaction of upheaval and erosion has never been presented by myself or by
anyone else in the beautiful manner deduced by Penck in his first ideal cycle of the
'Gipfelflur' essay; and for that reason his essay should be regarded as marking an
extension of the previous treatment of the cycle scheme. In order to justify the
opening statement of this paragraph I desire to cite a number of passages from my
earlier writings. (*Davis, 1923J, pp. 29–30*)

Davis closes, more in sorrow than in anger:

The composition of the three parts of this article has been attended with mixed
feelings. The analysis of Penck's 'Gipfelflur' essay in the first part was a pleasant
duty in so far as it was concerned with the constructive side of his study. The
general review and summary of the scheme of the erosion cycle in the second part
was also an agreeable task, as it brought to mind memories of work and progress
in association with many colleagues through forty years of busy life. The correction of Penck's corrections in the third part was a disagreeable necessity. It would
not have been undertaken but for his exceptional rank as a geographer and for the
high standing of the Academy in whose proceedings his essay is published. On
both those grounds it has been deemed desirable to show that his adverse criticisms are much less pertinent to my treatment of the erosion cycle than a reader
of his essay would be led to suppose, and that the real value of his essay, which is
unquestionably large, lies in the extension of the deductive treatment of the

erosion cycle with especial respect to a mature stage in which upheaval and erosion are balanced. (*Davis, 1923J, p. 41*)

Increasingly, however, Albrecht Penck, influenced by the views of his German colleagues and by the loyalty he felt to the work of his late son, reacted against Davisian ideas and by 1928 he had abandoned the idea of the sequential development of landforms.

CHAPTER TWENTY-THREE

Walther Penck and the First Breach of the Cycle

Walther Penck was born in Vienna in 1888 and soon became a fine mountaineer. He went with his father to the United States in 1908–9, when Davis was in Berlin, and met G. K. Gilbert who showed them among other features the great fissure caused by the San Francisco earthquake of 1906. The Pencks returned home via Hawaii where Walther became so interested in volcanism that he decided to study geology, which he did at Heidelberg and Vienna. In 1912 he was appointed geologist to the Dirección General de Minas in Buenos Aires and, aided by his mountaineering skill, acquired an authoritative knowledge of parts of the southern Andes. Three years later he became Professor of Mineralogy and Geology at the University of Constantinople (Istanbul) and eventually wrote several articles on this fractured landscape. In 1919 he was given the title of Professor at the University of Leipzig and worked steadily on the results of his researches in the Argentine Andes and Anatolia. He also continued active field work in southern Germany and the Eastern Alps, usually accompanied by his wife, in Davis' words 'the charming Ilse'. But he was already ill and on 29 September 1923, at the early age of thirty-five, died of cancer in the mouth.

Davis had always taken a kindly interest in Walther who, although a true academic disciple of his father whom he adored, always felt that he could turn to Davis for advice. We are most fortunate in having a full range of correspondence between Walther and Davis and in the history of geomorphology these letters are of great significance. The early letters by Walther hardly reveal the strong dislike of the simple Davisian cycle which we know he had.

Walther Penck was a fellow student of mine in Heidelberg and we spent much time together. While he was taking his Ph.D. degree in petrology (to prove that he could do first-rate work in a field wholly different from that of his illustrious father), he was then already thinking of devoting his life to geomorphology thinking along wholly new lines. When his book was published, in the telegram style of the frantic writing he used to beat his approaching death (by cancer), many of the chief new ideas sounded to me like an echo of the days when he expounded them with wonderful enthusiasm, dangling his legs from the bed on which he sat. Albrecht Penck was a thoroughly trained geologist and careful observer, thinking inductively. To Walther Penck he was still not thinking enough in terms of the dynamics of the processes involved. Davis' type of thinking was anathema to him. (*Letter: W. H. Bucher to R. J. Chorley, 22 February 1962, Houston, Texas*)

This could scarcely be otherwise when parental and environmental influences are considered. The Pencks grew up amid the Alpine fold system where earth movements had continued from at least early Tertiary geological times to the present. True they were also well acquainted with the worn-down stumps of the older mountain system, often called Hercynian or Armorican, which today form fractured upland blocks such as the Bohemian Massif and the Highland massifs near the middle and lower Rhine. But Walther also spent a considerable time in the southern Andes, one of the most unstable tracts in the world, where in parts of the west the whole landscape seems to be sinking rapidly beneath the sea and, inland, mountain chains appear to be rising, with concomitant fracturing into a remarkable array of faulted folds. Here any alpinist must be strongly impressed with the extent and continuity of crustal movements. In contrast Davis, let it be noticed, flourished amid the flat-topped Hercynian mountains and massifs of the eastern seaboard of the United States. When he crossed the southern Andes he was primarily a meteorologist and seemed remarkably imperceptive of landscape forces. When he went to the Alps, he was concerned mainly with glaciation and not with the landform effects of mountain-building. The mental attitudes of Walther Penck and Davis differed as widely as the characteristic scenery of the Andes and Appalachians. Davis subconsciously looked at landscapes as if they were tectonically stationary whereas Penck viewed them as eternally unstable. Paradoxically Davis was at heart a traditional catastrophist: if landforms moved he preferred them to move briefly with celerity. Penck, on the other hand, was a uniformitarian: the events may be catastrophic but in the past they went as indefinitely as they do today. The following letters and discussions will, we hope, reveal how these different mental attitudes were expressed as landscape concepts.

Walther Penck kept Davis supplied with his publications although their views on morphology differed considerably, at least in the opinion of the younger man.

Dear Professor,

I thank you for receipt of your kind letter in which you confirm the delivery of my book about the Permian, and about which you made kind remarks. This has pleased me for two reasons: on the one hand, it confirms my belief that scientific labours are still capable of bridging the gulf which arose during the last 5 years between our peoples, because it strives to be impersonal and is detached in its aims from the manifestations of life of both individuals and the communities to which we belong. On the other hand, I am pleased to receive your appreciation although my views on morphology as set out in my book considerably differ from yours. I hope to be able to talk with you about this at length at a later date. Today I am approaching you for some advice.

I cannot assume that sending my book to colleagues from the erstwhile enemy nations will always be regarded in so unpolitical and impersonal a manner as I would wish and as you, fortunately, have done. It has happened repeatedly that scientific communications sent abroad by Germans have been returned to the sender. I neither can nor wish to expose myself to this: if I send one of my works to a colleague beyond the German borders it will be for scientific reasons and I cannot wish that this should be misunderstood. This is what makes me hesitate to send my book to *R. A. Daly* whose sphere of activity is rather similar to my own, for I do not know how he would assess a work received by a German. If you believe that I can send a copy of my book to Daly without having to fear that he would consider this to be anything else but an attempt at a rapprochement for scientific purposes, I should be very grateful to you if you were to give me your opinion on the matter. Of course, I would send *R. A. Daly* my book in the hope that he makes available to me the works of *Jagger*, the Director of the Kilauea Observatory about Kilauea. These papers, particularly the last lengthy one, which I believe appeared in the *Amer. J. Sci.* in 1917 greatly interest me and I consider them so important for the furtherance of my own work that, frankly I would make a request for them. I would like your candid opinion, dear Professor, as to whether I can propose to Daly an exchange of my book for Jagger's Kilauea-papers without having to fear that this old-established custom would be interpreted in any other way than would have been customary before the war.

I hope you will have a happier Christmas than is possible for us Germans.

Your devoted

Walther Penck

(Letter: W. Penck to Davis, 24 December 1920; Leipzig. This and other letters of W. Penck in this chapter were translated by W. H. B. Simons, The Anatomy School, Cambridge)

Walther, now Professor of Geology in Leipzig, had sent Davis a copy of his detailed monograph on north-western Argentina (1920A). Davis told Albrecht Penck that he had enjoyed reading this Argentine monograph, which was 'an able piece of work' and added 'I have written asking him (Walther) to specify the difficulties he finds in accepting the cycle theory' (letter, 3 April 1921). Obviously Davis had not enjoyed those parts contrary to the Davisian system.

Walther Penck's reply (28 March 1921) suggests that Davis had blamed the Germans for starting the 1914–18 war. Walther rebuts this and then explains lucidly where his views differed fundamentally from those of the Davisian cycle of erosion. It is a difference of premise: landforms result from erosion *versus* uplift rather than erosion after uplift. All of this apparently is explained more fully in Walther's latest essay on morphological analysis (1920B), a copy of which was already on its way to Davis.

Dear Professor,

It is almost 2 months now since I received your letter of Jan. 17th, for which I owe you my thanks as well as a reply. The reason for this is that in order to answer your weighty letter I need a lengthy period of time, which I can only find during the Easter holidays. [*Trsl. Note*: Easter 1921 = 27 March.]

The frankness of your arguments enables me to answer you with equal candour. I shall try to deal with all your topics, except for the question about the origin of and blame for the late war. I could reply to your very frankly expressed opinions on page 2 of your letter, concerning the origins of the war, only by stating my own, equally personal, opinion, and I could then not remain silent about evidences of Western European Civilization which we have seen since the Armistice, consisting of an uninterrupted series of breaches of faith and atrocity after atrocity; to mention but a few examples of the latter: that in Upper Silesia Germans may, up to the present, be killed with impunity, or that in the occupied Rhineland it was possible for months on end for blacks to assault white women. All this happened in peacetime and not in war! All of us are still too close to the events of the past to be able to arrive at a reasoned verdict. This situation will only change with the passage of time. Then an impartial verdict can be given about the war, and will be delivered by humanity. We Germans look forward to this with the greatest equanimity.

I would greatly value it if you could inform me of the reasons which motivate the attitude of many American scientists towards their German colleagues. You tell me that personal dislike of us Germans is the main reason, and you substantiate this by giving a number of examples which I personally deeply deplore as constituting impertinences and tactlessnesses on the part of my compatriots. I am far from wishing to deny that amongst individual Germans there are those with bad manners, and even ruffians. But I should like to state with all possible emphasis that both types are representative of neither the general character of the German people nor of German scientists (as you yourself were obliged to concede), nor that they are typical of the German nation. Intellectual ruffians and discourteous people exist in all nations – to mention but a few examples which, amongst others, happen to come to mind: the Germans from the Cameroons [?illeg.] were, as you know, deported as early as 1914. In London the men were separated from the women, the former interned and the latter were sent to Germany. A woman who stood abandoned with her two children and who cried because of her harsh fate and her separation from her husband, was told by the delegate from the American Consulate who was present at her reception: 'When you return to Germany, take a rifle and shoot your Kaiser.' This to a woman who had just lost everything! My wife and I were greeted in Constantinople by an American Consul, in whose house we had lived, with the words 'let me see'. Certainly a polite form of greeting to a *lady*. Do you think that such things evoke sympathy?

However, neither myself nor most of those which you call 'reasonable types' would ever dream of judging the American people or American intellectuals by these slips, or make them responsible for the fact that some individuals have bad manners. The respect I feel for the American people will be influenced by their achievements, whether I personally like or dislike their manners. I can understand that a war may destroy friendships on both sides and may deepen existing dislikes, also that the late war has clouded judgements on all sides. But now Western European scholars proclaim in every possible strain that the fulfilment of their patriotic duty was a matter of course, but ours was a crime! As long as *such* arguments prevail a rapprochement will be out of the question in spite of the efforts of some sober mediators [?illeg.]. A rapprochement seems to me to be equally difficult to achieve as long as the tactlessnesses and blunders of some individuals and the dislike shown towards us Germans are used as a yardstick for the German people, rather than the values which it has created; as long as one requires us to respect others without showing any respect to ourselves. This, particularly at a time when the victor believes that the sole way of communicating with the vanquished is by force and disdain. I deplore these circumstances so greatly because *science* suffers from them. For its sake it is desirable that scholars should cooperate; but the establishment of relations which may lead to this will be made very difficult, even if – as is self-evident for us Germans from the above – this is undertaken *merely* for the sake of science. We must leave it to time to decide whether scientific relations will again provide a base for the resumption of personal relations. This can only be attempted when more reasonable views prevail on the other side.

I must now answer your enquiry as to how far I believe my results obtained in the Argentine differ from your morphological views. I can well understand your question, since your teachings have often been misunderstood and, as I am well aware, have therefore been unfairly criticized. However, whether I, too, have misunderstood your views, which I naturally only know from your writings, you will be able to judge from my essay about morphological analysis which I hope you have received in the meantime. The book on which I am engaged at present will also make this clear to you. I rate Passarge and Hettner as critics in the same way as you do. Neither will, I believe, be very pleased about my work, especially Hettner who fights against deduction because he has not recognized that morphology is a branch of science which formulates its problems in physical terms. This is the heart of the whole matter. As regards the differences which exist in my opinion between my results from the Argentine and your views, these appear on pp. 388–389 of my book: the shape [Gestalt] of landforms [Landformen] is a function of the ratio of the velocity of endogenous movement to intensity of erosion [Abtragung]. After the erosion cycle the – as *I* understand it – shape of land formations depends on the duration of erosion. This, after all, is something which is fundamentally different! Your view is based on the assumption that upheaval [Hebung] and erosion occur in succession. This view is held not only by you but, up to now, was generally accepted, also by your opponents and, at an earlier date, also by myself. Everybody merely knew that this view was incorrect, knew that upheaval and erosion act in the same way, but everybody merely believed that it was admissible to treat upheaval and erosion as if they acted successively. This gave rise to very crude simplifications. One of the results which I obtained from the Argentine is that it is inadmissible to treat upheaval and erosion as successive events. The exact argumentation is set out in my work on morphological analysis. Herein lies a further difference between us; it deals with the most important premise of morphological deduction. And, finally, I consider it no small deviation from your view when I found with certainty that the peneplains [Rumpfflächen] of the chains of the Andean [?illeg.] system of great folds [Grossfaltensystem] could not be the result of a degradation [Einebnung] of a previously-present mountain-range, but certainly represent the first, most primitive form of erosion which arose at the point where the crusts in question first formed chains and were exposed to erosion for the first time. The peneplain of the erosion cycle is found, on the other hand, not also at the start but only at the end of all erosion. As you can see, there exist large differences of opinion between us. If I take your views to be those which you express in your publications it becomes evident that my results differ fundamentally from them. My essay on morphological analysis brings out this point even more clearly than my book on the Permian. You will see from both that neither deduction as a method of investigation nor the developmental view (Entwicklungsgedanke), both of which I value as considerable acquisitions and advances for morphological science, form the point of departure of my critique, but rather the factual and methodical *premises* of the erosion cycle. In this resides the fundamental difference with Hettner, whose critique you rightly do not take too seriously, for he operates with the same premises as the erosion cycle and believes he can prove the inadequacy of deduction in morphology by stating that this method has been frequently applied by your followers in an irrelevant manner. That such endeavours are in vain you will see from their results: no *independent* mind shares Hettner's and Passarge's views. I do not make these statements without reasons: I am anxious not to be confused with Hettner or Passarge not even (or especially not) in the field of morphology.

If you find anything in my critique of the erosion cycle, particularly regarding my interpretation of your conception which I have drawn from your published writings, I would be sincerely grateful if you were to write to me about it. For the progress of science, which is my perpetual aim, can only benefit from an expression of differing opinions, as can also the elimination of misunderstandings through free discussions. I do not shrink from factual criticism, on the contrary, I welcome it.

I remain, very respectfully Yours,

Walther Penck

(*Letter: W. Penck to Davis, 28 March 1921; Leipzig*)

In the following May, although the article on morphological analysis had not yet arrived, Davis sent a detailed explanation of how Walther had misunderstood the cycle of erosion. Walther, however, was not convinced and in reply suggested that Davis was the cause of the misunderstandings. After rather tactlessly affirming that he was among the very few people who 'have carefully gone through almost all your works', he states in a blunt, clear way the fundamental differences in their methods. The cycle of erosion, Walther asserts, is not the norm – most landforms develop in a fundamentally different way from those postulated in the cycle. Landforms develop according to whether the rate of uplift is very slow, moderate or very rapid. He had not come to this conclusion through his father's influence but had on the contrary persuaded his father that there were not *three* main types of 'development possibilities' but an infinite number depending on whether upheaval was very slow (peneplain), more rapid (intermediate or ?mature) and very rapid (Alpine or ?young). All main types start with a peneplain (rumpffläche; primärrumpf) and end with a peneplain (endrumpf).

Dear Professor,

Thank you for your detailed letter of the 8th inst. I am glad to be able to reply so quickly because I am now fully satisfied that misunderstandings about morphological questions have arisen between us, which do not, however, seem to be due to me. When last I wrote to you I had already sent you an article on 'morphological analysis' beforehand. I assumed that you would certainly have received the reprint and, therefore, kept my letter short. Evidently, the article has not reached you. I deplore this greatly for two reasons: first I do not have available another copy to send to you, second, had you received the article it would certainly have prevented your understanding my letter and my views about the erosion cycle in a quite different sense from that which I intended.

As I wrote to you previously, I know your views on the development of erosion forms (Abtragungsformen)* only from your *published writings*. I believe I am among those very few who have carefully gone through almost all your works and I cannot find that I made an error about what you wrote. The situation, in my opinion, is as follows: you know, naturally, as well as I do, and have done so for a longer time, that upheaval (Hebung) and erosion (Abtragung) are simultaneously acting events, and I know of quite a number of passages in your works where you have stated this evident (natürliche) fact. On this point, therefore, there exists not a scintilla of doubt. But what of the *consequences* for the formational(?) development (Formenentwicklung)? The erosion cycle derives (ableiten) a developmental series (Entwicklungsreihe) of erosion forms, whose final member is the peneplain. In this, too, we agree completely. Which is the first member of the series? You start with some upthrust (gehobene) initial form (Urform) which has been dissected (zertalt). The valleys thus formed are to begin with narrow and have steep walls. You designate these alpine forms (steile Formen) as being young. They are the initial forms (Anfangsformen) of the genus (Formenreihe) which the erosion cycle derives. Before I proceed, I must ask you whether I have misinterpreted your views on this point, whether my statement of your line of argument is correct? – After some time the valleys become less steep and have become broader. You call this stage mature (reif), etc. Thus the genus of the erosion cycle is determined by the following characteristics: beginning with alpine formations (Steile Formen), narrow valleys, decrease of valley depth – a development proceeding via maturity to old age and the peneplain. Have I correctly reviewed your concepts? – I completely agree with you on this point, but I must make one essential reservation: this genus is only possible and only realized in nature if the crust does not undergo any movement. *This genus (young – mature – old, your definition) is only possible on a resting terrestrial crust, i.e. after cessation of upheaval.* In order to derive this genus, young – mature – old, you must presuppose that upheaval has ceased and that erosion acts quite by itself, undisturbed by tectonic processes. If before the end of the cycle young – Peneplain there is upheaval you assume that a new genus *again begins with steep, young forms*. If the

---

* Translator's parentheses.

upheaval takes place after the completed cycle, you assume that the *Peneplain* of the first cycle is dissected (zertalt) by the second cycle, *which again starts with the young, alpine (steep) forms.* Is there anything in the above with which you do not agree? Naturally, with such fundamental problems, to consider, we will disregard details. I see the situation as follows: according to your view, the development of erosion forms always starts with alpine forms, which you call young. Under the influence of erosion they develop into the milder forms of maturity and, finally, the peneplain, if there is no disturbance. If that is the case, then, at the very point where the old cycle broke off a new cycle arises, *which again has the alpine (steep) forms of youth.* This would be correct if upheaval were always rapid and sudden and if the upthrow (gehobene Scholle) were to remain stationary and gave time for erosion to produce the cycle young – mature – old (or as far as the cycle would precisely go). *The genus of the erosion cycle is a special case,* namely that case where a portion of land is rapidly upthrust, then remains stationary and erosion proceeds without hindrance until the next upheaval thrust (Hebungsruck). *In other words, this is the special case in which erosion, after cessation of upheaval, flattens on upthrow (gehobener Block) into a peneplain.* I do not know of any other genus (Formenreihe) occurring in an erosion cycle except that which starts with alpine (steep) formations, which are always called young, and ends with flattish forms (Flachformen) (type peneplain) which are called old (except for the influence exerted by the kind of rock which does not interest us here). True, you once mentioned the possibility of 'old-born' ('altgeborener') forms. But this is the only indication that you were aware that the course of events in nature might proceed otherwise than via the path young (= steep) – old (= flat). You will see that it is *not* without importance that during the derivation of form development (Formentwicklung) you always use the following theoretical pre-requisite, rough upheaval – tectonic rest – *then* erosion. I do not even consider to question the fact that with *this* pre-requisite the genus steep (= young) – flat (= old) comes into being, *which is why I do not by any means object to the erosion cycle.* But I state the following: such a genus, steep (young) – peneplain, i.e. the genus of the erosion cycle, *only* originates with the pre-requisite of upthrust and *then* erosion. My work in the Andes has shown that this pre-requisite is given in nature under quite specific circumstances. As a rule, one meets on earth other genera which fundamentally differ from those of the erosion cycle. While I was in the Argentine and Anatolia my father made similar observations in the Alps. I have now convinced him that there are on earth not one, nor *three,* but *infinitely many* genera. (This is why I do not quite understand your repeated remarks that I, influenced by my father had misunderstood your erosion cycle.) As you may see in my 'Morphological Analysis' all genera start with the *peneplain* [Rumpffläche] [Primärrumpf] and end with the peneplain (Endrumpf = peneplain). In between there lie an infinite number of developmental possibilities which are determined solely by the following law: very slow upheaval (Hebung): peneplain, more rapid upheaval: intermediate forms (Mittelformen) (what you call mature (reif)), very rapid upheaval: alpine forms (Steilform) (which you call young). In an upheaval which starts and rises to such an intense degree as occurs, e.g. in the Alps, the following series must necessarily arise (entsteht naturnotwendig): peneplain (Primärrumpf) – intermediate form (Mittelform) – alpine form (Steilform). This represents the generally distributed types in the mountain chains. This will look as follows in profile [Fig. 102]:

F I G. 102. Drawing by Walther Penck of the profile of an uplifted mountain chain
'3 = Primary peneplain, the first and oldest stage
2 = Intermediate forms (second form stage)
1 = Alpine (steil) forms (last, youngest form stage)'
(From Letter by W. Penck to Davis, 26 May 1921; Leipzig)

> where 3 = primary peneplain, the *first*, oldest erosion stage (your designation = old);
> 2 = intermediate forms (second form stage);
> 1 = alpine (steil) forms (last, youngest form stage).

The precursors of 1 are forms 2, that of 2 is the peneplain (Rumpffläche) 3, which latter has *no* precursor forms. In general, we may say that the terrestrial forms are the resultant (Ergebnis) of the intensity ration of upheaval (Hebung) and simultaneous erosion (Abtragung). The genus (Formreihe) of the erosion cycle is defined here as the resultant of the intensity of upheaval ($=\theta$) and the intensity of erosion of some maximal value, which decreases to value θ. This is where the peneplain (Endrumpf) lies. Examples of *this* series are found in the Continental Massifs, e.g., German East Africa, Uruguay, etc.

Naturally, I cannot go into such detail as in my article. Since you do not appear to have received it, I have been more explicit in this letter than would otherwise have been necessary, because I am anxious not to be misunderstood. To resume: If you admit that the genus (Formreihe) of the erosion cycle starts with alpine forms (steile Formen), and after every interruption by upheaval again starts with alpine forms, the cycle ending with the peneplain, then you must also admit that this is the one special case which can only occur with the pre-requisite: upheaval – and *then* erosion. The finding that the erosion cycle is a special theoretical case does not imply its rejection, nor does this in any way detract from the extraordinary advance effectively due to the erosion cycle. But research does not stand still, it progresses. I completely agree with you that research is better served if one collaborates. You can see from all my letters, and even more so from all my works, how much I subscribe to this view. But it is quite clear that a special case will be abandoned as soon as it becomes evident that, being an individual instance, it has no general validity, for which latter one will have to search. The generally valid case in the sphere of morphology has been found. You would hardly have considered this fact as being an attack on the erosion cycle if you had known the observations in North-West Argentina, if you had received and read my 'Morphological Analysis' before you wrote your last letter to me. Nor do I believe that you would then have reproached me with such positiveness that I had insufficient knowledge of your writings and that I had misunderstood the erosion cycle. I also believe that you would take a different opinion of my father's views if you knew his researches in the Alps (alpine peak plains) (Gipfelflur der Alpen). As for the rest, I do not know what he wrote you in the letter, nor do I know your answer. However, I shall ask him to let me see it, since you mention it.

I believe I have answered your various questions. But a recapitulation may be in order so as to avoid misunderstandings.

Do I believe that an upheaved mass can be eroded down to a low region (Land)? I believe so, provided that there is no further upheaval of the mass. I supplement this as follows: (a) if the mass is low it will bear a peneplain (Rumpffläche), if it continues to rise with sufficient slowness it will always be covered by a peneplain, even if upheaval and erosion are of unlimited duration; (b) if there is more rapid upheaval of the mass it will grow higher and is covered by ('mature' = 'reifen') intermediate relief (Mittelrelief). This persists as long as the upheaval takes place with equal velocity, even if upheaval and erosion are of unlimited duration. Such a mass can never decrease into a peneplain as long as upheaval occurs; (c) if upheaval is even more rapid, the mass grows higher, becomes deeply dissected (zertalt) and is covered by *steep* ('young') forms which persist as long as upheaval occurs with the same velocity. *In such a case* peneplanation (Einrumpfung) is completely excluded. Re (b) and (c): peneplanation only occurs when upheaval slows down and finally comes to a halt (erlahmt).

Do I believe that the peneplains with upheavals and dissections of varying intensity are the result of erosions of varying durations? No, I believe not. The peneplain of the Rhenish schist (Schiefergebirge) is a typical primary peneplain (Primärrumpf) and the gentle highlands (sanfte Berglandschaft) of the Harz mountains have evolved from a peneplain during a very old acceleration of the Harz-arching (upwarping) (Harzaufwölbung). This acceleration has increased up to the present, which is why alpine forms (steile Formen) have arisen along the increasingly *rapidly* deepening valleys which now attack (anfressen) the gentle highlands (Höhenlandschaft), thus separating them from their previous erosion base. It is precisely those examples which you give which, in the light of further research, certainly do not belong to the genus (Formreihe) of the erosion cycle. *They lead to an accelerated upheaval of the genus.* This represents one of many examples that the *form* of the erosion forms does not depend on the duration of erosion but rather on the intensity ratio of upheaval to erosion.

I send you my heartiest greetings

Yours sincerely,

Walther Penck.

(*Letter: W. Penck to Davis, 26 May 1921; Leipzig*)

The reader who is interested in what Davis said to Albrecht Penck will find the correspondence in full in Chapter 22. It is hard not to admire the way Walther stands up for his father and for himself. The simple cycle of erosion most used by Davis *was* a special case but, of course, we have yet to discover how far Walther's scheme falls outside the same category.

In July, on Independence Day, Davis replied in detail to Penck's letter, which he had much enjoyed. He had also examined with keen interest

Walther's article on morphological analysis (1920B). He addresses most of his remarks to both son and father but whereas his letters direct to Albrecht have, as we have shown in Chapter 22, been full of reproaches and scorn, this, and all others sent to Walther are friendly, polite and most encouraging. It is as if he welcomes the clever young student with open arms but cannot forgive the old contemporary who has, in his view, forsaken his camp. The length of the reply is formidable but it deserves preserving in print without mutilation. Davis welcomes both Walther's deductive approach, and 'the excellent improvements and advances and refinements' that Walther was introducing into landform analysis. But he points out that Walther has the great advantage of beginning about where he (Davis) stopped. Davis tells Penck that what he has done is to 'modify the scheme of the Cycle to its advantage; but *not to show* that it was altogether wrong and that a new scheme must be introduced in its place'. He advises Walther to keep up the good work of improving the cycle but to remember that he is only improving not destroying or replacing it. Is not, Davis asks, the Primärrumpf as special a case as the rapid uplift of the ideal cycle? As long ago as 1899(H) Davis briefly indicated that, on uplift, initial forms are immediately fashioned by denudational forces into segmental forms but this 'easily added idea' was never expanded. 'You have done so and deserve credit for it, but in doing so you have merely refined the scheme, it is still a relatively systematic sequence of forms that you are treating.'

Before we go further and assess the validity of Davis' suggestions we ask readers to peruse the letter for themselves.

> Your most interesting letter of May 26 was duly received several weeks ago and has been much enjoyed. Your pamphlet, Wesen und Grundlagen, came a few days after my previous letter to you; it has also been examined with much interest.
>
> I am glad to be corrected of my previous misunderstanding; but as far as I can see, it is now my turn to point out a misunderstanding that you – and I think, your father also – have entered upon.
>
> You refer to Geogr. Cycle, as I have usually treated it, with its young narrow valleys etc. as only a special case of the general problem of denudation. I shd say: the special case of rapid elevation of much slower erosion is a special case of the Geogr. Cycle. I have repeatedly indicated that the factors, structure, upheaval or deformation, erosion etc, are subject to all sorts of variations; as my articles and lectures have as a rule been directed to trying to explain the most simple case of the cycle (for when I began to understand the problem, even that most simple case was unrecognized), even you have failed to understand that all sorts of other cases fall under the general scheme. There is no question of that, as the originator of the term, I think that I ought to be allowed to state its meaning. Hence, while I welcome most heartily the many and excellent new ideas that you are bringing forth, I cannot agree that they involve the exclusion and rejection of the general idea of the Geogr. Cycle. They involve truly its improvement, its refinement, but not its rejection.

The first article I ever wrote on the subject (1884) contains the statement: 'The channels (river valleys) will be narrow and steep walled in regions of relatively rapid elevation, but broadly open in regions that have risen slowly, and I believe that rate of elevation is thus of greater importance than climatic conditions in giving the canon form to a valley.' (This was *à propos* of the idea then prevalent that canyons are cut only in arid regions.)

In a later article (Washington Internat. Geogr. Congress, 1904) I said: 'The elementary presentation of the ideal cycle usually postulates a rapid uplift of land mass, followed by a prolonged still stand ... the uplift may be of any kind and rate, but the simplest is one of uniform amount and rapid completion ... In my own treatment of the problem, the postulate of rapid uplift is largely a matter of convenience in order to gain ready entrance to the consideration of sequential processes ... Instead of rapid uplift, gradual uplift may be postulated with equal fairness to the scheme, but with less satisfaction to the student first learning it, for gradual uplift requires the consideration of erosion during uplift, etc. Then follows a case of a slow uplift and a valley which is widened as it is deepened; youth is eluded, etc.

Now it is perfectly clear, as you have pointed out, that my usual presentation touches chiefly upon cases of rapid uplift and hence of narrow young valleys; but I submit that the above quotation, in which the more general treatment of the cycle is considered, makes it clear that you are wrong in thinking that the Geogr. Cycle (as I have defined it) is only a special case of the general problem – a special case in which young valleys are narrow, etc. The presentations I have made have been merely special cases of the very general cycle scheme.

But as that matter, so is your case of a Primärrumpf. So indeed must every particularized case be; especially every case that is illustrated by diagrams of its successive stages, because diagrams are always relatively specific. By all means go on, multiply and elaborate these special cases, improve, refine them; but to think that they fall outside of the general scheme of the Geogr. Cycle because they are thus improved and refined is a mistake.

The case of the immediate modification of an uplifted Rumpf (away from its larger rivers) so that it at once becomes a Primärrumpf is a good example of your refinement of the scheme. I have somewhat briefly indicated that, when uplift takes place, initial forms endure only for a hypothetical moment and are immediately advanced, by slow processes of weathering, washing and soil-creeping into sequential forms; but I have never expanded this easily added idea. You have done so and deserve credit for it; but in doing so you have merely refined the scheme, it is still a relatively systematic sequence of forms that you are treating; and the very essence of the Geogr. Cycle is that it recognizes the systematic sequence of forms, as apart from the utterly irresponsible ideas of land sculpture that used to prevail.

Is not your Primärrumpf a special case, like my rapid uplifts? Suppose for example, that a mountainous region were uplifted, fast or slow, as you choose. You would not have a Primärrumpf thereupon produced; but simply a continuation of whatever stage of dissection had been attained before the uplift. That is therefore another special case. But all such cases fall into their appropriate places in the fully generalized scheme of the Cycle.

One phase of your work pleases me greatly; that is, the full value that you give to the deductive treatment of your problem. No one, after reading your essay, can doubt that deduction is an essential mental process, along with observation, etc. etc. in the rational study of landforms. In one respect, however, you do not seem to understand my use of deduction. You say, page 66, 'Hierzu bedient sich Davis der Deduktion' – as if I did not also use observation in making out the scheme of the cycle; that is, confirm my deductions by testing them with observed forms. I have frequently used deduction rather as a method of presentation than as a method of investigation; and when it is so used, it does not warrant a critic in saying that the investigation was deductive. As a matter of fact, all investigations of this sort are, or ought to be, both inductive and deductive. I do not see how they can escape being both but when it comes to presenting the results of the investigation, either the inductive or the deductive part may be given larger or smaller place as is desired. Thus when you say, p. 77, that your father reached the understanding of subaerial peneplanation by an inductive investigation. Also that he did not use the scheme of the cycle in reaching it. I believe both those statements are so incomplete as to give a wrong idea. He surely perceived that the worndown mountains he described had been gradually worn down; that they had once been lofty and rugged; that they were slowly reduced to fainter and fainter relief; how else could he have understood them? But such a succession of forms is essentially the basis of the scheme of the cycle. Further, while it is true that he did not explicitly state the deductions thru which he came to understand the origin of the worn-down mountains, he must inevitably have mentally conceived the deductive side of the explanation. Hence his investigation was partly deductive, partly inductive. His presentation was largely inductive; and to the extent that deduction was omitted, his readers must have had to supply it for themselves, if they were really to understand him. The difference between deduction in investigation and presentation is so clear that it ought not to be confused.

There is another aspect of your work to which I take the liberty, as a much older man than you are, to call your attention. It is that in the excellent improvements and advances and refinements you are introducing, you have had the advantage of beginning about where I have stopt. Had you begun where I did in 1884, I fancy that your progress would have been not very different from mine; and that some one else would have had to make the later steps which you are now making. I do not know how familiar you are with the extraordinary conservative not to say reactionary spirit that formerly prevailed among geographers. They objected, as many still object, to the use of explanatory methods of description, because of their danger, because of their use of deduction, etc. etc. In a word, they wished to remain purely observational, purely inductive. Indeed, many were so prejudiced against deduction that they decried it in others, even when they used it themselves.

Today things are different. You have the great advantage of beginning on the advance made by your predecessors; you grew up under the enlightening influences of your father's teachings; and very naturally, you are going on beyond your predecessors. I wish you great success in every such advance; but I hope that in making the advance you will not forget the advantages you had at the beginning of it.

As to your terms, fläch, mittel and steil; I have not the slightest objection. See

the closing sentence of the long paragraph, p. X, of Vorwort in my *Erkl. Beschr.*
The important thing is to recognize the systematic relations of forms in each
stage, and the normal succession of the stages. That, and not young, mature, and
old; not narrow young valleys etc. is the essence of the scheme of the Cycle. Go
on and improve it as far as you can, but do not forget that you are still only
improving it; not destroying or replacing it.

If you are writing to your father, please tell him that I have not yet received a
copy of his Gipfelflur der Alpen article; I sh'd like very much to see it.

I ought to have said, above, that there is plenty of warrant for the special case
of relatively rapid uplift and resulting narrow-rock-walled valleys; for examples of
such valleys may be found in many parts of the world. Hence, such a case is not
only the simplest special case for elementary presentation, but also corresponds to
many actual cases, of which I could give a long list. It shd also be noted that, even
if the main rivers of a rapidly uplifted region are narrow and steep-sided, the
upper courses of the smaller streams will have well opened valleys; see *Erkl.
Beschr.* 63, middle; 259 middle. See also, exercise on mountains in my Pract
Exercises, in which erosion of well opened valleys during upheaval is graphically
shown.

Note also *Erkl. Beschr.* 36, middle; that the embryo of your fully developed
idea of Primärrumpf is indicated in my kinsequente Abdachungen Ans; see just
below the case of a smaller stream in weak rocks having an open valley while a
larger river in resistant rocks cuts a steep-walled gorge. It is the expansion of
these briefly and often imperfectly suggested ideas that I am glad to see in your
essay. See also, *Erkl. Beschr.* 147, top. Also, exercise 3, p. 188.

On p. 2 of this letter, line 9, I said that I had somewhere intimated that initial
forms are immediately changed, etc. The citation is *Geogr. Journ.* 1899, 7th page
of reprint. 'The uplands also waste more or less during the period of disturbance,
and hence no absolutely unchanged initial surface should be found, even for some
time anterior to epoch 1. Instead of looking for initial divides separating initial
slopes that descend to initial troughs followed by initial streams ... we must
always expect to find some greater or less advance in the sequence of developmental changes, even in the youngest known land-forms. "Initial" is therefore a
term adapted to ideal rather than to actual cases, in treating which the term
"sequential" and its derivatives will be found more appropriate.'

It seems to me that such a statement as that fairly well anticipates the much
more explicit statement you have given under heading of Primärrumpf; and that it
therefore shows that what you have done is to *modify* the scheme of the Cycle to
its advantage; but *not to show* that it was altogether wrong and that a new scheme
must be introduced in its place.

Your reference to your father's priority in explaining the subaerial origin of
peneplains – in his lecture of Feb. 1887, tempts me to quote the following
passages from a report of mine, based on observations in Montana in 1883,
written in 1883–84, and published in 1886 in the XV vol. of the Tenth Census of
the United States, under statistics of coal mining – a place where no one would
ever look for it. (This was because the private survey for which the Montana work
was done encountered financial difficulties and could not publish the report; to
save it from being lost, it was handed over to the Census bureau. Had I then been

better experienced, I would have published extracts from the report in a scientific journal.)

In an account of the Great Plains, east of the Rocky mountains of Montana, certain very broad and shallow valleys are mentioned: 'These we consider the work of a river system now extinct, a system that, when in activity, cut and carried away an unknown thickness of overlying strata and brought the original surface down near its base level of drainage, thus producing the comparative smoothness of the existing plains.' (The amount of erosion of former overlying strata is estimated at from 3000 to 5000 feet as a minimum.) 'The plains are therefore to be regarded as plains of denudation, cut down as to an old base level of erosion' and afterwards broadly elevated and incised by the rivers.

'Their comparatively smooth surface is not a form of original deposition, but results from the denudation of a broad sheet of strata, as is shown by the attitude of the (volcanic) mesas and dikes (which surmount the plains). There is every reason to think that this denudation is the work of rain and rivers, but in order that these agents should produce a smooth surface they must perseveringly act until their valleys widen and consume the intervening hills and reduce the surface nearly to the base level of erosion.' As you will readily see, both induction of many observed facts and deduction of theoretical conclusions entered here, as they must have in your father's study. We appear to have been occupied with these subjects at about the same time (in *Proc. Amer. Assoc. Adv. Sci.*). The work in Montana was the basis of my first article, 1884, from which I quoted on p. 1 of this very long letter. (*Letter: Davis to W. Penck, 4 July 1921; Cambridge, Mass.*)

This friendly letter shows clearly enough that Davis either had a remarkable belief in the flexibility of the cycle scheme or did not appreciate fully that Walther Penck had tried to provide not an improvement but a replacement for it. When, for example, Davis writes 'Is not your Primärrumpf a special case, like my rapid uplifts? Suppose, for example, that a mountainous region were uplifted fast or slow, as you choose. You would not have a Primärrumpf thereupon produced; but simply a continuation of whatever stage of dissection had been attained before the uplift', he is still hidebound mentally with the idea of a beginning and an end. He cannot visualize that Penck was interested in the *rate of uplift* and that valley slopes would tend to change with it. The shape of valley slopes depended for Penck not on their age but on the prevailing relationship between erosion and uplift.

We have not yet found any manuscript reply to this letter from Davis but, as we have seen, Walther was ill and died in late September 1923. Albrecht, although dreadfully upset at the loss of his son, with typical courage and skill gathered up the threads of Walther's main unpublished writings on landforms. Among these posthumous publications was *Die Morphologische Analyse. Ein Kapitel der physikalischen Geologie* (1924), which consisted of the early and only completed chapters of a much longer projected work. It was written in Walther's hurried, telegrammatic style and proved incredibly difficult to translate. No English translation appeared until 1953 when Dr

Hella Czech and Miss Katherine Boswell presented the *Morphological Analysis of Land Forms* to a grateful, if mystified, public. However, in the first chapter of *Die morphologische Analyse* Penck clearly states his view that the Davis cycle was directed to a restricted combination of circumstances:

> What has found its way into morphological literature as the cycle of erosion is what Davis expressly defined as a special case of the general principle, one which was particularly suitable to demonstrate and to explain the ordered development of denudational forms. It is postulated that a block is rapidly uplifted; that, during this process, no denudation takes place; but that on the contrary, it set in only after the completion of the uplift, working upon the block which is from that time forward conceived to be at rest. The forms on this block then pass through successive stages which, with increase of the interval of time since they possessed their supposedly original form, i.e with increase of developmental age, are characterised by decrease in the gradient of their slopes.
>
> They are arranged in a *series of forms*, which is exclusively the work of denudation and ends with the peneplane. ... If a fresh uplift now occurs, the steady development, dependent solely upon the working of denudation, is interrupted; it begins afresh e.g. the peneplane is dissected. A new cycle has begun; the traces of the first are perceived in the uplifted, older forms of denudation. Thus it has become usual to deduce a number of crustal movements, having a discontinuous jerky course, from the arrangement by which more or less sharp breaks of gradient separate less steep forms above from steeper ones below.
>
> ... Both Davis himself and his followers have made and still make the tacit assumption that uplift and denudation are successive processes, whatever part of the earth is being considered; and investigation of the natural forms and their development has been and is being made with the same assumptions as underlie the special case distinguished above. There is, therefore, a contrast between the original formulation of the conception of a cycle of erosion and its application. Davis, in his definition, had in mind the variable conditions not only of denudation, but of the endogenetic processes; in the application – so far as we can see, without exception – use is made of the special case, with its fixed and definite, but of course arbitrarily chosen, endogenetic assumption. And criticism, with its justified reproach of schematising, is directed against the fact that the followers of the cycle theory have never looked for nor seen anything in the natural forms except the realisation of the special case which Davis had designated as such. ...
>
> As a method, the theory of the cycle of erosion introduces a completely new phase in morphology. Deduction, so far used only within the framework of inductive investigation, or as an excellent method of presentation, has become a means of research. Starting from an actual knowledge of exogenetic processes, the cycle theory attempts to deduce, by a mental process, the land-form stages which are being successively produced on a block that had been uplifted, is at rest, and is subject to denudation. Not only is the order of the morphological stages ascertained by deduction, but also the forms for each stage; and the ideal forms arrived in this way are compared with the forms found in nature. (*Penck, 1924; 1953, pp. 7–9*)

In the meanwhile Davis had not been idle. Working also with Walther's posthumous article on piedmont benches of the southern Black Forest (Penck, 1925), he tried to present Penck's ideas to the English-speaking world. Davis' article in 1932(G) on 'Piedmont benchlands and Primarrümpfe' was the first critical review of this article to be published in English. We will show later, in Chapter 28, that this assessment contained serious misunderstandings which long outlived both Davis and Albrecht Penck but the assessment itself still remained a remarkable feat for a man over eighty years old.

However, in a review shortly after Walther's death, Davis showed a rather benign attitude to his work, as well as considerable perception of the sedimentary evidence on which Walther ultimately based his notions of uplift.

> ... I recognize fully the important advance made by the lamented WALTHER PENCK, in which the progress of degradation on an uplifted area is measured by the nature of the sediments furnisht from it to a neighboring basin of deposition: but at the same time I question whether the occurrence of fine sediments in such a basin demonstrates the production of a Primär-Rumpf in the uplifted area as the first product of erosion of a previously formed peneplain: for a peneplain, before its uplift, may be covered with deep-weatherd soils; and in consequence of a small and gradual upheaval, the weatherd soils will be rather rapidly swept away from the shallow valleys excavated by the revived streams, and will thus furnish only fine sediments to a neighboring basin of deposition; but at the same time very little degradation will take place on the interstream areas of the peneplain, which will therefore as a whole lose the appearance of a Rumpf, and gain the appearance of a young hill-and-valley surface, to which the name Primär-Rumpf is not applicable; for such a surface results from the relatively activ deepening of the valleys at a faster rate than the much slower contemporaneous degradation of the hills, as is characteristic of a young stage of the cycle, and not from the gradual degradation of the hills at a very slow rate, and yet faster than the then greatly reduced degradation of the valleys, such as characterizes the production of a true Rumpf.
>
> On the other hand, there are many uplifted regions – for example, the Front Range of the Rocky Mountains in Colorado, and the western slope of the Sierra Nevada in California – in which the deep-cut valleys are so steep-sided and narrow-floored as to justify very well the assumption of a rapid but not instantaneous upheaval ... (*Davis, 1925F, p. 314*)

Mention of Davis' public antagonism to Walther Penck's ideas is looking nearly a decade ahead, and in the interval Walther's thoughts remained virtually shut off from the English-speaking world. When his ideas did begin to filter through the linguistic barrier and the stout Davisian ramparts, there arose a tremendous interest in the shape of slopes; and Davis' trilogy of explanatory landform description – structure, process, stage – was for some rapidly replaced by structure, process, stage and tectonic influence. For most physiographers an idealistic simplicity of downwasting of landscapes was now ousted by the realistic complexity of the backwasting of slopes. Neither Davis nor Albrecht Penck lived to see this first breach in the popularity of

FIG. 103. Sketch of Davis made by one of his students 'in a moment of boredom'
on 12 July 1928 (Courtesy H. F. Raup)

the erosional cycle. Symbolically the riper years of the two grand old men
illustrated perfectly Walther's premise – that outward form is an expression
of external decay versus interior rejuvenation. Both grew old in years but
neither experienced a Davisian stage of senility. We are, of course, in this
volume primarily concerned with Davis who as a septuagenarian had his ups
and downs, but cannot refrain from telling that he and Albrecht were
reunited by Walther's untimely death and later reappear together in the
American West.

CHAPTER TWENTY-FOUR
The Years of Depression

In July 1923 when Davis had refound marital contentment and much of his former buoyancy, his second wife died. On this occasion grief was not so prolonged. Molly Davis' death came at an age when it was less unexpected, and they had been married less than nine years. Yet Davis felt the loss deeply. Temporarily the rhythm of his life was broken and throughout the following years he struggled to regain some of the old productive pattern. He became sad and restless, and wandered from place to place trying to recapture his sense of fulfilment. Once more he first sought consolation in travel abroad and within three months was in the West Indies.

THE WEST INDIAN TOUR

Davis' travels in the West Indies are well documented as he wrote frequently to his family, who circulated his letters, and he looked forward eagerly to their replies. He chose the West Indies, and more particularly the smaller reef-bearing islands of the Lesser Antilles, because they were unknown to him and concerned his prolonged study of coral reefs.

He continued to write poems, including one, composed during a rainy afternoon on the minesweeper *Grebe*, which was later printed in *Harvard Graduates' Magazine* and as separate sheets to distribute to friends. This opens with the verse

> Look not alone upon the sunset glow
> Where red and gold, so gloriously blush
> Spread half across the cloudless firmament,
> But face the East as well, and there below
> The rosy twilight's rising arch descry
> A band of blue that broadens into sight
> The visible embodiment of night,
> The Shadow of the Earth upon the sky!

(*Poem: 'The Shadow of the Earth on the Sky' from a letter from Davis to his children, 21 October 1923; Barbados*)

Davis remained very proud of his poetry and was particularly fond of this one.

Davis visited the Lesser Antilles during October and November of 1923, after a detailed study of marine charts and topographic maps. His trip was a rapid one and it is clear that he had already made up his mind as to the postulated evolutionary sequence of the islands before he visited them (see Chapter 25). Of the more than twenty-five islands used in his scientific thesis, he set foot on ten and viewed eleven more briefly from passing steamers. He landed at St Kitts and 'drove around its younger cones'; took a 'short automobile trip into the interior' of Dominica; made an 'automobile excursion around the island' of Grenada: stayed overnight at St Croix; and paid 'a brief visit' to Barbados. He had longer stays on only three of the islands, spending nine days and ten days, respectively, on the 'first-cycle composite islands' of St Lucia and the Virgin Islands, at the latter stop being given the trip in the *Grebe*. 'The most instructive island I visited in the Lesser Antilles' was the 'second-cycle island' of Antigua where, during a ten-day stay, Davis 'received helpful courtesies from His Excellency the Governor'.

The letters to his family contain few geological and many personal references. Amid the rather tendentious descriptions of how he spent his time it is possible to discern one or two revealing facets of character. There is not the same depression as occurred on the death of his first wife but the constant search for companionship is repeated.

> Passengers mostly negroes and negresses – but as we were about to start a young white man came on board and I asked him to take the vacant place by me and tell me all about it as I was a stranger – 'Do you know me?' he asked in some surprise. 'Not at all' – 'I am inspector of agriculture and director of volcanic gardens at Choiseul' (further S. than Soufrière) – 'Then you can tell me many things I shd like to know – I am . . . etc.' – 'Won't you come and spend the night with me? – my name is Walters' etc. etc. So it was arranged I shd run down there next Tuesday and stay till Thursday, transportation being not obtained in three days. We had much talk on the two hours run and as the coast was more visible than the crops, I did most of the talking – and explained cliffs and bays and things. (*Letter: Davis to his children, 2 November 1923; Castries Club, St Lucia*)

Perhaps desire for an audience would be a more accurate way of describing it. Davis could listen to intelligent conversation but he had no time for small chatter.

> Mrs. D's breakfast table, at 12 o'c was well provided with good food – and was attended by a characteristic boarding-house gang. Mrs. McIntyre, a decided brunette, Mrs. Drinkwater, very inappropriately named I shd judge, Mr. Lambert, as above, Miss Douglas – and an old professor. Conversation was of the languishing kind and had to be stird to keep it going. However one mustn't expect too much in Grenada. Lucky the table was clean! (*Letter: Davis to his children, 26 October 1923; St George's Club, Grenada*)

A second attempt at company nearly ended unhappily.

> At Dominica, I took a car and invited two nurses, who are with me at the captain's table to ride up Roseau Valley, a fine sight of steep slopes and marvellous

tropical vegetation – but on the way back, the car balked at a hill, backt down, and unfortunately lodged in the inside gutter against a high bluff, instead of going over a precipice on the other side! My only adventure, but a pretty risky one. We had to walk a mile back to the town, where we consoled ourselves with ice-cream at the Home Industries Shop. (*Letter: Davis to his children, 13 November 1923; SS. 'Parmia' approaching Antigua*)

Desire for congenial conversation may have been his motive for introducing himself to the colonial governors of the islands. More likely he felt his professional status and reputation demanded it, yet when actually faced with attendance at an official reception he was always deterred by the excessive formality.

He missed most of all the clean, simple comforts of a home-life and disliked the rather impersonal politeness of a boarding-house routine. He appreciated good food, such as guava jelly, and a feeling of spaciousness, and although when necessary ready to suffer a little discomfort for travel, took great care of his own well-being. He strongly objected to excessive drinking and to swearing. The latter habit he seems to have associated with the imperial English whose manners and customs he regularly criticized. Not that we agree with his summary of the history of the great British Empire!

He [Davis] is facing the music bravely – doesn't want to join the Club here, because of the endless drinking. It is excessive . . .

Tuesday afternoon I hired a car – a very good one, with an excellent jet-black driver, invited a fellow guest at the hotel to come along and had a fine ride over good roads out and back. The guest was Mr. Clear of Demerara, a pleasant companion, who disresembled many colonial Englishmen by not swearing on the whole trip. He was four years in the war, was promised when he enlisted that his place with Mappin & Webb would be kept for him – but it wasn't – and after various tries, he has come out here, in the office of 'General Merchants' – His fiancee is soon to follow – and such is the history of the great British Empire. (*Letter: Davis to his children, 26 October 1923; St George's Club, Grenada*)

Davis, as we saw during his journeys in South Africa, was very quick to piece together isolated bits of information about a country's economy into an instructive analysis. But has he instinctively a New Englander's antipathy to colonialism?

The island [Grenada] is mountainous – a tangle of peaks, ridges, spurs, valleys, ravines and glens. The shore is more or less cleft where the spurs meet the sea, and embayed where the sea enters the valleys. Pasture everywhere altho the people are lamenting the lack of rain. Cocoa – or properly, cacao is the chief product – we past plantations of groves, in which the cacao pod could be seen budding from the trunk or larger branches. For a year or more after the war, the price for it was high. Then came the slump that seems to have hit most everything – and now the island is one large groan. Many of the plantations are run on credit – so the owners are in a sort of servitude to the merchants who have advanced money in future crops. Much of the trouble comes from London, where the Trinity – a combination of three large chocolate companies – Rowntree, Epps, and Cadbury? – hold down the purchase price on the producer and make it up on the consumer – slicing off a large profit for themselves in between. Dissatisfaction and complaint seem rather universal. No place where I have been has been satisfied with existing conditions. Everyone seems to think he is hard hit. In St. Thomas the dismantling of the distilleries at the sugar mills on the nearby island of St. Croix deprives them of the liquid which, with the bay-leaves of the other nearby island of St. John, they use in making bay rum – formerly an important product. – Rainfall was short there also. In fact, since the Virgin islands were bought by the U.S. nothing goes right. (*Letter: Davis to his children, 30 October 1923; St George's Club, Grenada*)

In November 1923 he began the journey back to the United States where he was soon wandering again. The main academic results of the West Indian tour consisted of one article and a sizeable book (1926D) on the Lesser Antilles, although the relevant material was also incorporated in *The Coral Reef Problem* (1928C).

FAREWELL TO CAMBRIDGE

After his second wife's death, Davis' feelings towards Cambridge changed for some reason. All his working academic life – nearly half a century – had been spent there and he knew the surrounding countryside intimately from numerous field excursions and summer vacation haunts. It seems a strange decision for an elderly man to forsake a town and society so familiar to him. Possibly he was no longer happy at Harvard, and certainly his resignation in 1912 could be attributed to this. The dominance of his ideas in physiography and his international academic renown should not beguile us into imagining that he was a revered leader on the Harvard campus. Some of his own references suggest the contrary and, of course, no one doubts that he always liked and expected a front seat.

> It almost made me think about that stone which the builders had rejected and which became ... what did it become? But the really most amusing aspect of all this is the way a shelved old professor, who generally had a back seat at his home town in the east, is given a front seat out here in the great and glorious west, where things are seen in their true relations. What is that old Latin proverb? Luna fulgeat inter stellae minores, or words to that effect; rather pat, isn't it. (*Letter: Davis to his children, 1 February 1928; Tucson, Arizona*)

For these and perhaps other causes, Davis on his return to the United States late in 1923 went not to Cambridge but to a club existence. He spent some time at the Oakley Country Club – a golf establishment in Belmont, and then moved on to the Cosmos Club in Washington, D.C.

The move to Washington undoubtedly saved Davis from stagnating, for the national capital was the place where he was most likely to be drawn back into society. So far since his return, his daily routine had been healthy but hardly exciting.

Breakfast about 8–9, generally with someone I know, for my acquaintance is growing fast – At work in my room till 11 or 12 – then a walk – unless I go earlier to Scarvey Library for books – Lunch at 1 – work in afternoon – again a walk – and about 5 a nap – supper at 7 – talk, reading, – occasionally writing in the evening – Bed by 10 or 11 – It is a fairly healthful round – and perhaps for that reason the cold I caught returning from Chattanooga was soon thrown off and I have been very well since. (*Letter: Davis to his children, 14 January 1924; Cosmos Club, Washington, D.C.*)

The opportunity to exchange intelligent conversation now provided a mental stimulus which had been lacking for some time. His letters become full of descriptions of interesting persons he had met, and of the responses of his table companions to his attempts at entertaining them.

First, as to my table companions; they continue to be varied and entertaining, not to say instructive; and surely it is hygienic, to talk during one's meals, instead of bolting them in silence. Among my recent companions are: Spinden of the Harvard Ethnol. Museum, on his way to Honduras for a few months; he gave me some good geographical points about a big delta that is growing out into the Caribbean. Macdougal, English professor of psychology at Harvard was another; and with him several of us talked spiritualism. (*Letter: Davis to his children, 12 February 1924; Cosmos Club, Washington, D.C.*)

Supper came along leisurely. I was deputed to escort Mrs. P., and did so without more ado than to offer her my arm and march her across the hall. She seemed to think we ought to have waited till a signal gun or something was fired. Alas, the supper was something of a disappointment; for I expected to have a chance to recite another sonnet to my companion between courses; but Mrs. McKim had M. Dupre; and that was the end of me. I shd have been a mourner at the feast, had not a kindly spirit, and a well embodied spirit too, been materialized on my other side in the form of Mrs. McKeon, lately of Baltimore, now resident in Washington. I was somewhat abasht at first, for she was rather impressively majestic in appearance; somewhat over the usual size, tho by no means so 'massive' as Lady Jane; she may be twenty years hence. But it is always an unfair advantage for a woman to take, when she sits higher than her neighbour. I shd have liked her better, had she been de la taille suivante plus petite.

However, after we got started, I liked her very much just as she was; for she was exceedingly fair to behold, and her smile, which was sympathetically facile, was illuminated by the most beautiful teeth ever. Really, she did smile several times, even to the point of throwing her head back, just a graceful little throw back, for the fuller enjoyment of some of my most delicate efforts. And Mrs. Posey was all the time so completely the prey of the dupe ... I mean so completely enwrapt in Dupre, that she never once turned her head to ask what bright little things I was saying. (*Letter: Davis to Isabel and May, 27 February 1924; Cosmos Club, Washington, D.C.*)

A patient listener, fond of travel and of show-offs in French, was a necessity for him on these occasions.

> Dr. Victor Vaughan of Ann Arbor, living here this winter, member Nat. Academy, proved to be a listener, so I told him a whole series of yarns and had him in my room to continue them after lunch; he did not seem bored, but left me saying he had been much entertained. (*Letter: Davis to his children, 12 February 1924; Cosmos Club, Washington, D.C.*)

However, his new-found tolerance of society, as we might have prophesied, did not extend to the professional failings of his academic contemporaries, and he was as quick as ever in pointing these out. Surely the kind Dr Vaughan had merited more charitable treatment?

> On Tuesday evg. to go backward, I heard Dr. Vaughan give his presidential address in Oceanography, etc. before the Washington Acad. Sci., a very inartistic performance, mixing commonplace and technicalities in a manner that showd no interest in 'form'. (*Letter: Davis to his children, 14 January 1924; Cosmos Club, Washington, D.C.*)

In such company, much of the familiar Davis begins to re-emerge and once again we see the pleasure with which he welcomes a former acquaintance. He was never happier than leading a serious discussion or talking to a close friend of the family. No doubt Mrs McKim had sized him up skilfully.

> But to whom do you suppose Mrs. McKim introduced me the very first thing on my entrance into her second-floor parlor? I am tempted to say, on my entrance, for it was none other than Mrs. Atherton, née Tilton, of Cambridge, who was formerly a girl of unusual charm, and who is now, as a woman grown, a charmer. Rather pleasant, was it not, to meet an acquaintance at the start; and all the better later on when she askt me to call, and said she would telephone to me at the Club here when she would be at home; and that is what I call real decent. Then I was presented to M. Dupre, the great organist, and as he seemed to have difficulty in pulling out the English stop of his vocal organ. I playd up a little leit-motif on my French flute – bien que je ne sois guere du bois dont on en fait and let him do the accompaniment. I think his eye glanced with approval at my petit ruban rouge, of which I ought not be so vain, se sert pour ses pourboires; but – ou peu s'en faut it is the fashion in Washington to wear such gauds, and I have adopted the fashion for formal shows. (*Letter: Davis to Isabel and May, 27 February 1924; Cosmos Club, Washington, D.C.*)

Davis' limited finances probably caused him concern and there are frequent indications that he was restricted by cost in the choice of what he did. However, there was plenty to do in Washington and he soon seems to be revived in spirits and geniality. His rather pedantic humour creeps back at more length into his letters but perhaps his family enjoyed what had been so long a substantial part of their literary diet. There is now even mention of sightseeing – the manuscript of Lincoln's immortal Gettysburg address of 1863 exhibited in the Congressional Library – and of a rekindled interest in golf and music, including some of Mrs Coolidge's magnificent concerts. Altogether a more relaxed and less dedicated character is emerging. Yet he still pushed himself hard, perhaps too hard for his age.

> Yesterday and today are taken up largely with packing up my things here, so as to vacate my room for about 10 days, while I am at Radnor and Pittsburgh for a vacation; and about time for one, too, for I have been rather steadily at work, perhaps too steadily since January; and sometimes, because of trying to get a special chore finished, working too many hours a day, so that I have been tireder than is prudent. (*Letter: Davis to his children, 4 April 1924; Cambridge, Mass.*)

Davis continued to concern himself with the aftermath of the European war and he makes occasional pleas in support of peace plans.

> What do you think of the Bok Peace Plan? It's certainly well advertised! I have read 4 copies – one from the Harvard Alumin. Assoc. one from the Phila. Geogr. Socy. – one from U.S. Playground Assoc. one from a western correspondent. Some persons here think that because it is so boosted it will produce a recoil against the League – Perhaps so among politicians, but not among the plain people. I hope you have all voted for it – as I have. Send in all the votes you can! (*Letter: Davis to his children, 14 January 1924; Cosmos Club, Washington, D.C.*)

> The H.O. has just sent out a letter to a hundred or more scientists, announcing a plan for Pacific exploration, and asking for suggestions. I call that very decent. The usual way for government departments, especially for Army and Navy men, is to tell the public, including the scientists, to mind their own business. The reason for the present geniality appears to be that the Navy wants to make itself solid and popular so as to get more support for big appropriations. I am willing to back up scientific re'h – that means, scientific work – but I do not commit myself to more battleships. (*Letter: Davis to his children, 13 March 1924; Cosmos Club, Washington, D.C.*)

Earlier dealings with the armed services had apparently made him a little cynical of their motives.

There is nothing new about these mental and behavioural patterns of Davis but it is interesting to speculate how all represent attempts to find himself a leading role. None really satisfied him and a solution did not come until he returned to teaching and field research. This study of coral reefs continued as slowly as ever.

> The Hydrographic Office has absorbed a good deal of time; for I am trying to finish up a long-delayed coral reef article. It is time to turn off on other things. The charts down there are innumerable; a good assistant gets them very quickly – indeed, when I hand a slip of paper with a lot of chart numbers on it, he says,

'Thank you.' (*Letter: Davis to his children, 13 March 1924; Cosmos Club, Washington, D.C.*)

Momentarily the assistance of the Hydrographic Office side tracked him into thoughts for an article on official maps.

> The next afternoon, he telephoned that if it were all right for me, his associate, Major Walker would call for me in his car this morning at 9; good. So I imagined an austere and haughty military officer, very set on the absolute correctness of everything the army does; with whom I have to be mighty careful not to tread on his toes. Quite the reverse; he had been a topographer on the Geol. Survey for over 25 years when the war took him to France; and he still retains his commission. No haughtiness at all; but every amiability. He showd me all sorts of things and gave me a lot of maps. The unfortunate effect of that is that I want to write an article about the maps at once; and away go coral reefs! (*Letter: Davis to his children, 13 March 1924; Cosmos Club, Washington, D.C.*)

The brief article appeared almost straightaway as 'Shaded topographical maps' in *Science* for that year (1924B).

Of more practical significance than this continued or quickened interest in publication was Davis' re-entry into the world of physiographic investigations. Contact at Washington with young geologists of the Survey inevitably led to discussions of current problems. Perhaps unconsciously at first, traces of his former enthusiasm for fluvial geomorphology began to come back. Once he saw that some of his suggestions were being accepted, he probably again began to hanker for a partial return to lecturing.

> Last Saturday evening was spent at Harry Ferguson's, where three Survey members were gathered to discuss a physiographic problem. It was rather good fun to be in it with the youth of the nation again. (*Letter: Davis to his children, 4 April 1924; Cambridge, Mass.*)

When it became known that Davis was willing to lecture there seems no doubt that he received numerous suggestions and several firm invitations. His career now entered an entirely new phase.

THE WESTERN CIRCUIT: CANADA

Davis' re-entry into the teaching world was probably assisted both mentally and practically by his attendance at the British Association meeting in Toronto in August 1924. This meeting also resulted in one of his few sizeable publications in the following year (1925C), which in quality was one of the poorest of his productive life. During the western excursion after the meeting he and several of the more geographically-minded members became interested in, among other landforms, the peculiar features of Lake Timiskaming, a long, narrow body of water upon the Laurentian plateau on the Quebec–Ontario boundary. The lake is bounded on the west by an almost rectilinear scarp of the crystalline uplands ... and is at least in part underlain by limestones gently inclined towards the scarp, and resting in strong unconformity on the rather even foundation surface of the crystallines beneath. The basin and its limiting scarp are evidently related to a fault, by which the limestones and their crystalline foundation have been locally depressed below the general upland level. (*Davis, 1925C, p. 65*)

Davis, as it happens, had always prided himself upon his classification of lakes. In 1882 he and Albrecht Penck had produced almost simultaneously their own classifications of lake basins and each had kept a sharp eye on later progress in the topic. Lake Timiskaming provided Davis with an opportunity which he was quick to seize. In a discursive article (1925C) he demonstrates that Roxen and many other lakes south and south-west of Stockholm appear to have much the same origin as Timiskaming. These lakes, he avers, are of second-cycle origin and therefore are only indirectly determined by a fault; in the first cycle uplift and fracture were followed by peneplanation; in the second cycle uplift, without fracturing, caused renewed erosion which, being greatest on the limestones (preserved in the previously down-faulted area), exposed the fault-scarp. In other words the present scarp is not a fault-scarp but a 'fault-line scarp' exposed by differential erosion.

Thus the present landscape consists of flat-topped crystalline uplands (peneplaned at the end of the first cycle) on either side of a belt where previous depression allowed limestones to persist and to protect an ancient peneplain. This depressed zone of softer rocks was excavated fairly rapidly in the second cycle so that it now consists of two smaller dissimilar belts.

The first is a gently slanting belt of the ancient peneplain, laid bare by the removal of the covering strata ... In the second belt, larger or smaller residuals of the limestones will remain well down in the 'safe angle' of the low-standing fault block, next to the fault line. The main river, which presumably ran near the fault line in Preglacial time, having been disorganized by glacial excavation and deposition, most of the low-lying limestone residuals and some of the stripped belt of the ancient peneplain will now be covered by a lake of the Timiskaming type.
(*Davis, 1925C, p. 67*)

Davis then discusses a more complex example where criss-cross fractures and uneven uplifts caused a jumble of prisms bordered by fault scarps, each overlooking a down-faulted patch of limestone which today is largely submerged beneath lakes such as Roxen. Most Swedish physiographers attributed these scarps and depressions directly to recent faulting whereas Davis favoured the scheme of two-cycle erosion described above and indeed stated in his *Die Erklärende Beschreibung der Landformen* (pp. 169–71) as long ago as 1912.

We must notice here that Davis' methods of analysing landforms had not changed in the last forty years. He studied the relevant topographic maps to detect a prevailingly even sky-line across the narrow fracture-guided trenches in southern Sweden and also made a casual inspection on the ground – the sort of 'unscientific' survey that German physiographers abhorred!

> ... but it is much more readily recognised by inspecting the uplands on the ground, as I had brief opportunity of doing in 1899. Hence the peneplain of the sky-line is of later date than the faulting on the fissures ... (*Davis, 1925C, p. 70*)

> First, is it worthwhile for geographers, whose studies are primarily concerned with the existing features of the earth's surface, to enter so far into the past geological history of lake basins or of any other land forms as has been done above? My own feeling is that it is well worth while to do so, because of the better understanding of many existing features that is thereby gained.
> The second question concerns terminology. If an explanatory description of a certain well-defined kind of lake basin is geographically helpful, should it be given a special name? (*Davis, 1925C, p. 73*)

> ... it is practically helpful to have a definite name for a definite kind of thing; ... Timiskaming is too long for a handy name. Roxen is more satisfactory because it is shorter and because the explanation above presented was applied to it a quarter-century before it was to Timiskaming. Hence a Roxen-type lake is one that occupies a basin that has been excavated in a second cycle of erosion after post-faulting peneplanation had been reached in a first cycle, and that is limited along one side by a fault-line scarp. (*Davis, 1925C, p. 74*)

FIG. 104. The origin of Lake Timiskaming in the Laurentian Uplands. A peneplain of deeply-weathered crystalline rocks has been stripped by a marine transgression over which limestone has been laid. Uplift and faulting (back) leads to the preservation of infaulted limestone during peneplanation (middle), renewed uplift results in the partial excavation of the weaker limestone and, finally, glaciation produces an irregular surface with lakes (front) (From Davis, 1925C, Fig. 1)

The infrequency with which the reader has heard, if ever, of Roxen-type lakes is a just commentary on Davis' typal nomenclature. Such a classification is not in itself instructive and cannot easily be reconciled with his strictures on dozens of authors who used little-known local names in descriptive geography. Probably some could find no real need for postulating a first-cycle peneplain post-dating the faulting, or disliked the idea that a surface buried beneath marine sediments would be a peneplain when it was more likely to be a platform of marine planation.

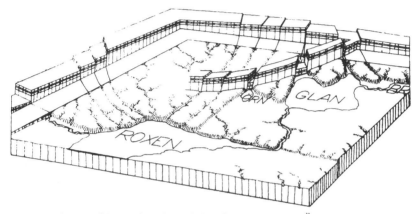

FIG. 105. Diagram illustrating the origin of Lakes Roxen, Örn and Glan, Sweden (From Davis, 1925C, Fig. 2)

We shall have to return later to Davis' interest in lake basins, for he wrote a fine paper on their classification when he was over eighty years old, but at the time of the Toronto conference although the old professional competence remained some of his flair and vision had slipped away. The next few years of his life would have tested the stamina and courage of a young man and it is one of his greatest achievements that he survived them with such honour and dignity.

THE WESTERN CIRCUIT: THE UNITED STATES

Davis' lecture tours in the American West are probably best described as they happened – a spasmodic succession of rather isolated events. According to his own account they were preceded by a rambling journey up and down the beautiful lakes of glacial erosion in the mountains of British Columbia in 1924 and opened

with a return to the southwest to lecture for short periods in several universities, and run about between times. There, except for relapses on Cape Cod in 1925 and '26, I have continued a peripatetic existence ever since, my chief resorts being the

University of Arizona at Tucson, the University of California, and Stanford University. (*Davis, 1930F, p. 397*)

He was a little uncertain about the venture and expected to miss the more civilized East.

It strikes me as a curious adventure I am embarking upon – I am wondering whether I can still hold an audience of students. The 'General Public' will be easy because the pictures of the Col. Canyon will keep them quiet – but perhaps a long lanky Texas boy will throw a lasso over my head when I am drawing chalk diagrams about the Lesser Antilles in a geography lecture and then what? . . . I explain to him that it is his business, not mine, to keep himself in order during the lecture . . . I hope the contingency will not arise . . .

Let me hear from all of you occasionally – I shall be very uninformed about most everything *Eastern* this winter except thru your letters. (*Letter: Davis to his children, 9 November 1924; Cosmos Club, Washington, D.C.*)

At Dallas, Texas he records in detail his first experience of traffic lights.

Perhaps the most novel feature is the method of regulating traffic by electric signals at important crossings. A green light means *Go* – a red light *stop*. After the red comes a yellow, and a bell ringing for 10 seconds, long eno for those on the crossing to get off – then green again. The curious thing is that all the signals are workt simultaneously from a central station – and that they keep on working in the night, when many of the streets are empty – but you must stop, quand. meme. In the city center, pedestrians as well as cars obey the signals. The time lost in waiting is said to be made by the full speed across cross streets, when the green light is on. When there is a fire alarm or an ambulance call, *all* the signals, all over town go *yellow*, the bell goes on ringing and *everything* stops, whether in the route of the engines or ambulances or not!

I had such a sample stop last night, but saw no engines. (*Letter: Davis to his children, 18 November 1924; Dallas, Texas*)

Many of his first impressions of the Western landscapes and townscapes were very favourable, and criticisms are minor ones.

Tucson was a great surprise, I had imagined it a rather desperate, last-chance sort of a place but it is quite otherwise a *beautiful* oasis in a *wonderful* desert basin with fine mountains around it. The University there was *far* beyond expectations – so good that I want to go there again for a month at least. (*Letter: Davis to his children, 26 November 1924; train from Tucson to Los Angeles*)

Tucson still amazes me . . . It was so wonderful to see so pretty a town in a desert. To be sure, *desert* here does not mean a waste of sand, for many bushes grow on it – but it is awful dry except where irrigated. What the town subsists on I can't imagine. Yet it seems thriving. It is as if an older town, with a well set population had been destroyed by conflagration or flood, and a wholly new town had sprung up in its place. Of course the streets are straight and wide, in squares . . . but the attractive houses and shade trees are *not* of course. I want to go there again and stay longer. (*Letter: Davis to his children, 28 November 1924; Santa Ana, California*)

– Needless to say, the Street is perfectly smooth and clean. They are all that – but to my surprise I have seen only one out of town road or boulevard with an in-slant at the curves – The effete East is ahead in that respect –. (*Letter: Davis to his children, 14 December 1924; Claremont, California*)

As the above correspondence shows, by November Davis had begun lecturing in California where he had many ex-students and other friends. Awkward in so many of his own human relationships, he was conscious of lack of sensitivity and restraint in others.

Wednesday was markt by a lunch at The Hollywood High School, a huge heavy building institution, of which Snyder is principal ... Remember him ... Davis & Snyder – Physical Geography, 1898. He was and is still an uncouth son of Maine – awkward, rough, nasal, countryfied – short, even stubbly – but a concentration of effective energy. He taught for a while, after graduating A.M.? at Harvard, in the Worcester Acady – then came here and captured his present position, where he is King. Real capacity, in spite of many handicaps ...

On Thursday I had a delightful lunch with Hoyt Gale and his wife at their bungalow home a few miles only from L.A. at Eagle Rock. He was one of my loyal students, 1900 – a very different fellow from Snyder – gentle, delicate in his manner. (*Letter: Davis to his children, 14 December 1924; Claremont, California*)

I have had Robert T. Hill of Texas to supper – a very able geologist and very queer fellow – Not the most restful companion, but *very* informing. (*Letter: Davis to his children, 28 December 1924; Los Angeles*)

We would like to say that his lectures in the West gathered a new quality from being delivered in a fresh environment, but this would not appear to be true. He had a repertoire of lectures which he shuffled around. For large public audiences there was the Colorado Canyon lecture with moving blackboard, the two religious themes and the volcanic district of Rome. Geographers were treated to coral reefs, the Basin Ranges or the geography of the United States.

The afternoon was quiet again – in preparation for the eveg. lecture on the Col. Can. Ransome gave me a *real* introduction, he had some real things to say – among others, that my lectures on the U.S. were the best lectures he had ever attended. I was glad to hear that – then began an hour of discomfort, for the lantern was out of filter – and the slides, after coming out properly would suddenly mount up the screen and disappear – unless the lantern man caught them and his efforts to do so made the audience *giggle*. It was rather fatiguing to keep my temper even and my talk going. A reception at Ransome's followed ... a whole string of professors and their wives. Appropriate remarks soon gave out and then I sincerely expressed my great pleasure ... (*Letter: Davis to his children, 28 November 1924; Santa Ana, California*)

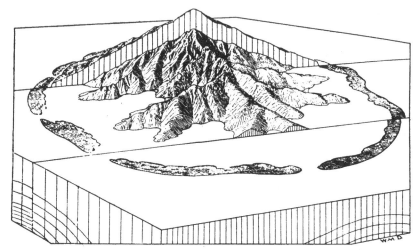

FIG. 106. Three-block diagram of a deduced subsiding island in a still-standing ocean, as postulated by Darwin. By this process a fringing reef (back) is converted into a barrier reef (middle) and then into an atoll (front) (From Davis, 1928C, Fig. 20)

FIG. 107. Three-block diagram of a deduced still-standing island in a rising ocean. A fringing reef in the front block grows into a barrier reef in the middle block, and into an almost-atoll in the back block (From Davis, 1928C, Fig. 21)

Davis' methods too had changed little but no doubt it was difficult to improve the lectures within their well-worn frameworks. Even good illustrations could not always offset factual out-of-dateness and, of course, Davis did not appeal to all students.

What qualifies me to claim (or admit) having been a student of Davis's is the fact that while I was a graduate student here at Berkeley the Department of Geology invited Davis to give a graduate seminar in that department. I had, of course, read much of Davis as a student, and eagerly seized the opportunity to enroll in the seminar. The result was a great disappointment to me. At one of the early meetings of the seminar I brought up a question about some statement Davis had made. Davis pointed a finger at me and said, 'If you had read (some article of his, with which I was familiar) you wouldn't ask that question.' It was evident that he did not want any discussion on the part of the students, and so I kept my mouth shut for the rest of the term. He spent his time expounding ideas he had already, in some instances long before, set forth in his writings. At the end I submitted some kind of paper, which I wrote without any enthusiasm, and wrote off the time spent as lost. Undoubtedly Davis was senile by this time (he was about 75 at the time), and should not have gone about pretending to instruct students. During his visit in Berkeley he delivered a public lecture, on the Grand Canyon, which he had probably been delivering for thirty years. He had nothing to add to what had been known since Dutton's day, and used illustrations on slides that were familiar from all textbooks. I have been told that in southern California he continued to lecture in public after he was so far gone that his wife . . . led him to the podium and put his notes in front of him when he was to begin. (*Letter: J. Leighly to R. J. Chorley, 18 October 1961; Berkeley, California*)

THE WESTERN CIRCUIT: DAVIS' SEVENTY-FIFTH YEAR

In 1925 Davis' lecturing tour began to acquire momentum. To a certain extent he himself created many of the invitations to lecture, although others occurred through ex-students, in particular Professor Ransome and his wife. John P. Buwalda of the California 'Tech' also proved a lasting friend as the following letters show:

Next is Lawson, of Canadian origin, but long head professor of geology here, and last year chairman of the Geol. Geogr. Division of the Nat. Research Council in Washington where I saw him rather often, thus reviving an earlier acquaintance: he gave the Great Geogr. Excursion of 1912 a fine trip to the earthquake rift of 1906. Louderback is somewhat younger – he and I have sympathetic views about the mountain ranges of the Great Basin and his latest MS on that subject is now in my hands for comment before publication. He was in Toronto last August and it was there we cookt up a plan for this 5 weeks sojourn, later officially ratified to the tune of $500 – so the problem before me is to be worth $100 a week to the University. Buwalda is a still younger fellow the finest looking of the three – delightfully frank countenance – and after a temporary try for a few years at Yale not long ago was called back here – I saw him at Los Angeles in Decr. and it was then he told me he would meet me on my arrival. (*Letter: Davis to his children, 21 January 1925; Berkeley, California*)

While there I remembered how closely confined I had been here last spring, in that single room and sleeping porch at Mrs. Lockwood's; remembered also how much I liked that little bungalow I had at Sierra Madre last September; and so wrote to young Prof. Brown here, would he see if a bungalette were available in the neighbourhood of the University? He mentioned the matter to Amy Ransome – do you know who she is? Born in Cordoba South America in 1871, her father was one of the assistants with me in Dr. Gould's observatory; – and Ransome was one of my Graduate students in 1897; afterward, geologist and chief geologist of the U.S. Survey, and treasurer of the National Academy of Sciences, until about 1922, when he came to be professor of geology here, but resigned dissatisfied with the university ructions, last spring and now has an excellent position in the Cal. Tech. at Pasadena: it was thru him I had my first stop and lecture here in 1924, and thru him also that I was invited here last spring. Well, Amy took the matter of the bungalette up with her customary energy; and here I am. (*Letter: Davis to his children, 1 February 1928; Tucson, Arizona*)

At first Davis seemed intent on re-establishing his confidence as a lecturer, which he did fairly quickly. He valued a responsive audience and if he had a bad evening he generally blamed them.

Queer contrast between my lecture at the La Jolla Community House last evening and at the San Diego Nat. Hist. Museum Sunday afternoon! The audience here seemed very indifferent and the lecture gave me no enjoyment at all – Some very under-aged children in the front row disconcerted me, not that they were disorderly but that they surely could not understand what I was driving at. It was a very cold sort of a show – At San Diego, the audience lookt unintelligent – but never did I speak to a crowd that gave closer attention – they actually leand forward as if to hang over my lips and at the end, a throng came up to say how much they had enjoyed the story ... a real worth-while Sunday afternoon pop. Sci. lecture – La Jolla was much more sophisticated – they bore the infliction patiently – having nothing else to do that would have bored them less – and when I stopt they probably said (as you may be sure I did) 'Well, now thats over' and went home to dinner. (*Letter: Davis to his children, 6 January 1925; La Jolla, California*)

Teaching was obviously a more exacting task yet Davis' early reactions at the University of California in Los Angeles are of self-congratulation, although it was twelve years since he had taught.

The first class meeting was occupied by brief reports made by the students on topics selected under Buwalda last week and by comments from the visiting professor. There will be six meetings, the last on Feb. 23, and the work done is to be regularly counted for the degree – that gives me much better standing than a visiting professor usually has – Consultation hours for personal talks were appointed and yesterday I had three of them – very satisfactory – It amused me to see how much I was like an old, circus horse, who all the time pranced about on his hind legs when the band played! – only I pranced rather better than ever, my present manner being more suave than it sometimes used to be. It was rather good fun, too, to find that I could still cite neat examples and give unread references on nearly every subject brought up and that, by good luck I knew the localities where the three consultants of yesterday had done their preparatory field work. One near Houston, Texas, where I had a day on the way out here – one in Colorado where my early explorations in 1869 stood me in good stead, to say nothing of later visits – one in S. California, when the recent automobile excursions enabled me to go the departing students one better, even to the point of local village names! Those were good base hits. (*Letter: Davis to his children, 21 January 1925; Berkeley, California*)

Again, some sleep and then at 10.30 Tuesday morning another class lecture, The Cycle of Erosion, treated first arithmetically, then algebraically – and it really amused me to see how much better that lecture was than the lecture I workt off on the defenseless Harvard undergraduate 40 years ago! (*Letter: Davis to his children, 14 March 1925; Anahecin, California*)

Although only a temporary tutor, he showed a real interest in the students' progress and was even considerate of the less talented.

It is now Wednesday p.m. At 2 o'c and the student conference – and the poor boy did so badly I am going to meet him again on Friday and try to boost him up to a proper level for a report on Monday. (*Letter: Davis to his children, 21 January 1925; Berkeley, California*)

Davis must have felt well as he always responded quickly to his state of health. Even fatigue was shrugged off. Perhaps he was buoyed up by a feeling of gratitude that at the age of seventy-five his efforts were still appreciated and that he still retained much of his old ability.

But what pleases me most is the way I can keep at it – To be sure, it doesn't mean so much work as it sounds, for I have lots of breathing spells – but all the same it is a rather active and a very varied performance – and it is going very smoothly. Again yesterday, after the evening lecture, a little throat rasping made me hoarse – but it is all gone this morning – everything smooth again – Long may it so continue. Now for a brief noon rest. (*Letter: Davis to his children, 6 November 1925; Ames, Iowa*)

After leaving California in the spring of 1925, Davis spent some time at Salt Lake City and repeated the visit about a year later. The comparisons between East and West are not unexpected.

It makes one realise what Harvard is, to visit a state university in the arid west. But Harvard would do well to make the visit, for there is a lot of earnestness and serious work, out here, even if some of it is a bit uncouth. One of the professors of geology is a very rough looking fellow, yet he knows his region pretty well. Of course, he is not very strong in physiography, that not being his special field; but he is active minded in catching on. The real difficulty with these men is as a rule, their inability to see beyond relatively small items. The students are a husky lot. If

I asked one of them to do something for me, he replies, You bet. But he does it all right. (*Letter: Davis to his children, 5 March 1925; Salt Lake City*)

I think his object in inviting me was to get my ideas about the University, regarding which he has pronounced views, including a general disappointment, both as to the standards of the professors and the indifference of other regents. I urged him to keep on trying, and not to expect any rapid change. It is clear enough that the general standards of work, both by professors and students, are not up to what Harvard takes as a matter of course. There is naturally, inevitably, a difference between the old East and the new West. (*Letter: Davis to his children, 27 April 1926; Salt Lake City*)

In April Davis was back in Washington, D.C., and had fixed up a lecturing circuit for the following winter. It was obviously to his advantage to have such visits made official and he did a certain amount of contriving and cajoling to achieve this.

A piece of very good news greets me here. A letter from the Harvard University Office asks me if I would like to take the 'Modern Circuit', Iowa, Colorado or California (Pomona) for next winter – and I have replied *yes* if I can arrange duties satisfactorily. So you may have another spell of me after all. *Every* hour I was about to write to you and others to see whether something of the sort could be arranged. Of course it may go all the better for being *official*. (*Letter: Davis to A. O. Woodford, 17 April 1925; Cosmos Club, Washington, D.C.*)

So in the fall he was off again, having earned enough from the previous lectures to keep his bank balance on the credit side.

Madison was left Wed. noon, Nov. 10, for Beloit, two hours away, where G. L. Collie, one of our students 30 years ago is now an advanced professor. He took me to his house for the night; and in the evening I gave another $50 worth of Col Can. That old lecture is much more than paying my way; especially as living expenses are reduced to a minimum by 'hospitality'. (*Letter: Davis to his children, 18 November 1925; Colorado Springs*)

When short of lectures he took up Society 'with a light catch; nothing elaborate' and had the advantage of a Harvard tie.

One of the things was a gathering of about 30 departmental heads and others to hear me talk; and that may remind you of the very true old saying that the moon shines among the lesser stars. However, even tho the moon is a waning moon in its last quarter, it gave out a few rays that afternoon, and provoked some discussion about a general course in science for those who study other subjects; that being a topic that I have been recently developing. There was also a neat little dinner of Harvard men one evening to whom I gossip for an hour or more; they are all hungry for news of the old place, even tho their attachment may be no more than a graduate year in one of our Schools. (*Letter: Davis to his children, 18 November 1925; Colorado Springs*)

We have left until last the most notable personal event for Davis in 1925, which was his seventy-fifth birthday on February 12th. The event was marked by a small dignified gathering.

> The arrival of the birthday itself, celebrated by a banquet especially laid out for me by a professional branch here, but given at the Bohemian Club 'one of the great clubs of the world' in San Francisco ... a magnificent building – a spacious private dining room, a great big oval table, with 14 or 15 around it – a delicious dinner – a single toast – some pleasant words by the Chairman, Prof. Louderback, who has done all sorts of nice things for me here, a fine address by President Campbell astronomer of the Obsy and pres't of the Univy here – & some rather random remarks & reminiscences by the tottering beneficiary. The handsomest thing of the kind that was ever done for, at & by me. – Then Prest. Campbell brought me over the ferry in his quiet car, most comfortably, right to the Club here – & as it was then 11.45 I went to bed without writing you about it! That was all of a Thursday. (*Letter: Davis to his children, 16 February 1925; Berkeley, California*)

By prearrangement several of his former students sent him congratulatory letters, two of which are printed below. Beneath their laudatory intent can be detected enough genuine recollections to show what a substantial impact Davis had made on each writer. Perhaps in the first letter, from Robert DeCourcy Ward, the comment on mathematical knowledge had already lost some of its justifications; and, of course, geography had ceased to be dominated by physiography, and Davis had long since ceased being a climatologist.

> Whatever I have accomplished in my science is due to you. You gave me the first inspiration, away back in the old undergraduate days of 'Natural History I', in the old Lawrence Scientific School building. Then, when I was abroad, after graduating, you wrote me about coming back here as your Assistant – the first one, I believe, that you had. You may recall that William Pickering at about the same time asked me to go to Arequipa with him, and that we had some correspondence in regard to which chance I should take. I decided to come back here, and I have never for a moment regretted that decision. Then I had those tremendously valuable years working with you, as Assistant, and as a graduate student, when we did the sea breeze monograph, and later the thunderstorms. I recall distinctly, when I was worried about my lack of mathematics, your saying that there was plenty of work for me to do without worrying about that matter. I thought you were wrong, but you proved to be right. I have rarely felt any real need of the higher mathematics, because, again at your suggestion, I turned to Climatology, and have made my job that, to me most absorbing side of Meteorology. So as I look back as your Birthday approaches, I cannot help feeling again the very great debt that I owe you. I want to send you that word on this occasion.
> The other thought which I have is about the part you played in the development of Physiography here at Harvard. As I look back to the years gone by, when you were giving all your splendid courses, and had those large numbers of students, I cannot help thinking of the disrupted and chaotic times we have had ever

since you stopped teaching. Nothing has ever been the same since, and it never can be. What now seems to go under the name of Geography is a very different thing. At Washington I was struck by the superficial kind of 'Geography' that was fed out to us, and the emphasis on the number of roads that go by farms, where so many cows are kept, which produce so many quarts of milk, which in turn gives so many pounds of butter, and so on. All very well, in its way, but it does not appeal to me. The Physiography which you taught was always interesting. You made the earth's surface live, even if you did not at that time say very much about Man. I find very little real Physiography nowadays. And I for one regret that, and realize how much you did, and how much we have lost here. (*Letter: Robert DeCourcy Ward to Davis, 6 February 1925; Cambridge, Mass.*)

Parts of the second letter, from Ellsworth Huntington, have already been used by us, as in them Huntington praised Davis' literary style and his activity in creating the Association of American Geographers. It was the kind of eulogy anyone would be delighted to get on his seventy-fifth birthday.

I am very glad that Dodge has reminded me that February twelfth is your seventy-fifth birthday. Many happy returns of the day. May each birthday make you feel more fully how much your students and the other geographers of America appreciate you and your work. Of late years I have regretted not seeing more of you, but I think frequently of the inspiration received from you during the past twenty-four years.

Few experiences have given me more profit and pleasure than the summer which we spent in Utah and Arizona in 1902, and the months when I was your assistant on the Pumpelly Expedition. Our long conversations on those two expeditions and your insistence on the importance of ontography largely determined the course of my life work. Your eagerness for the truth, your willingness to give every new idea fair consideration, your willingness to confess mistakes, your power of constructive criticism, and your ability to understand the other man's viewpoint have been an inspiration and a challenge all these years ... Thank you once more on this your birthday for all that you have done for me and for geography. In the fine old Biblical sense you, and you alone, have been my Master in geography. (*Letter: Ellsworth Huntington to Davis, 5 February 1925; New Haven, Conn.*)

Davis would hardly have been human had he not enjoyed such lavish praise, though his actual comment was a little cryptic.

The arrival of a flock of congratulation letters from geographers and others – for Dodge, a former student, gave out at the Xmas meetings in Washington that my 75th birthday was approaching – Some superfine letters too – and others subfine. (*Letter: Davis to his children, 16 February 1925; Berkeley, California*)

THE WESTERN CIRCUIT, 1926–7

By the beginning of 1926 the attractiveness of the West was influencing Davis. He was already considering it as a permanent home, his only fears being enforced idleness.

One of the piedmont villages that Wolff took me to see, Sierra Madre, about 10 miles east of here, is a most picturesque spot; the sort of place to which I may retire some years hence, when my lecturing days are over. Neat little bungalows where one could live quietly and economically, and perhaps contentedly; but that last is the most uncertain. An electric line connects it with Pasadena, and that is important; for libraries and things are to be found here. But the chance of me lecturing and sermonizing is still ahead of me; I have several letters out already for next autumn; but it is entirely possible that they may come to nothing but postage stamps. Time will show. (*Letter: Davis to his children, 4 January 1926; Pasadena*)

The winter climate of the West suited him well and he had ample time for rest.

This ten-days rest here has fixt me up splendidly for my next spell of lectures. Even bicarb. soda is no longer needed, tho it has been my constant companion ever since last summer. Drawing diagrams for a lecture on the Basin Ranges has been a placid employment, in my bright sunny room; reading silly novels has been a less useful use of time. Sleeping has been permissible I think, and I have lots of it, good and sound. Drives to the neighboring mountains have been entertaining; but they make me want to stop other things and make a regular study of the upheavals and erosions by which the mountains have been produced. It would be unwise to attempt that so I shall content myself with proposing a plan of work upon it to young Woodford, professor of geology, at Pomona, my February station. (*Letter: Davis to his children, 4 January 1926; Pasadena*)

Yet he took good care not to over exert himself in the hot summers and also to leave some time for completing his main work on coral reefs.

Did I write last week that I have felt it best to decline a $1000 invitation to give six-weeks course in the summer school at Berkeley, June 21 to end of July? It seemed wiser not to go on working so late into the summer, especially if I am to begin again in mid-September; besides I have a good piece of work to do on my big coral-reefs book, at Bass river, where Lucy and Carl again invite me to make myself at home this summer. Grace and the girls will all be there, so I shall have a delightful time of it. Work and play well mixt. (*Letter: Davis to his children, 31 January 1926; Stanford, California*)

One of the strongest impressions of his Western travels came from several visits to the Deep Springs School, California, where the isolation in a desolate basin, the unorthodox methods and careful selection of pupils on personal qualities greatly appealed to him.

A long letter from Pasadena shows a succession of changing enthusiasms and illustrates clearly how he planned ahead.

There seems to be a good deal of capacity to fall on his feet about this old professor; for here I am, after a fine week at Santa Barbara, beautifully establisht for the night in the handsome house of Prof. W. B. Munro, and as usual with a private bath and a typewriter at my service. A dinner party this evening is to give me the pleasant society of my former colleague, Wolff, with whom I stayd at Christmas time in the late '60s. Robert Gould Shaw, of the famous firstbaseman of the Harvard nine brother of George, who married your unknown cousin, Emily Mott and livd happily ever after, and Ernest Wright of Germantown by Philadelphia, with whose family I used to play in the interval between South America and Cambridge, that is, between 1873 and 1876. As you may perceive the only defect is the uneven spacing of these lines; and that is my own fault, because I do not yet understand how to adjust the spacing.

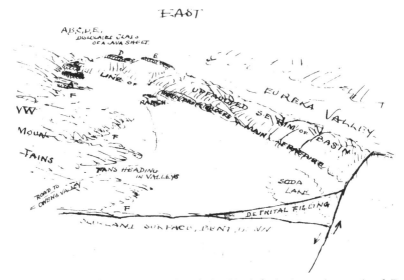

FIG. 108. Rough diagram by Davis of the block-faulted terrain south of Deep Springs, California, looking north-east

A letter sent you last Sunday from Claremont told of the week at Deep Spring School and my return, with repacking over a noon interval before going to Los Angeles and on to Santa Barbara (spacing is all right now) for a four day stand. In preparation for that trip I had charged Mr. Gibson, whom I had seen during my previous visit, when he acted as manager of the meeting of the Laymen's League there, with the opportunity, privilege, pleasure or duty as the case may be, of picking up a few paying engagements for my second visit; and he did it very nice. As the University Club there was full up, and no room for me available, he took good room, a really very nice room, for me in his private boarding house; and I was sufficiently comfortable, tho not luxuriously fixt. From his Church League he got me a $25 lecture on the Col. Canyon, which I generously made just as good as the usual $50 lecture; and then thru whom do you suppose he got me a lecture under the auspices of the local natural history Museum on any scientific subject I chose to treat. Ralph Hoffman, our old Brown and Nichols teacher and bird man, who is now among other things director of that same Museum, as well as teacher of Latin in Curtis Cate's school – Cate of '07 – remember him, Burt? – at Carpinteria, 12 miles east of Santa B. Ralph was just as nice and hearty as anyone could possibly be; had me out at his house over one night; took me to a very handsome dinner party one evening – and by good luck I had taken my dinner suit along, as a result of repacking at Claremont; and gave me a wonderful ride up to the crest of the Santa Inez mountains, so we could see into the valley on the north. Even better than that, he looks favorably on a scheme, to be told below, which may give me profitable, enjoyable and scientifically useful occupation for the winter of 1927–28. Why not for next winter, you may ask. Because letters found at Claremont fixt next winter in a fine and highly profitable fashion; October to December at University of Texas, Austin and January to May at University of Arizona, Tucson. Pretty fine that, isn't it? Well the following winter may be even better, especially if improvd by a suggestion just now made by Munro, to the effect that I shd give half a year to lectures in the Cal Tech. here and at Pomona College at Claremont (he is a trustee of the latter, and an influential friend of the former), and then the other half of the year might be applied to the scheme I cookt up with Hoffman; that I shd make a series of physiographic models for his Museum, and give occasional lectures at the State Teachers College, near by, and at the Cate School at Carpinteria, and at the Thatcher School in the Ojai valley (Call it O-High, please).

But this is a very rambling way of putting it; let me narrate events in proper order. Arrived at Santa B. Sunday evening and had a good supper at my boarding house. Monday morning, a call from Hoffman, making plans for later days. Errands afterwards with Gibson about a lantern for my evening lecture. Afternoon, a pretty drive with Mr. Gidney, president of the local chapter of the Laymen's League; evening lecture, to a good crowd, O.K. Fine sleep.

Tuesday morning, the eventful and fateful letters were written accepting engagements at Austin and Tucson for next winter; at noon a charming lunch, at the Paseo, in Spanish style, to which Hoffman as host had invited a choice group, including Lindgren, artist of the Colorado canyon, Sidebottom, an English doctor and amateur geologist, and several others, including Reginald Robbins, '92, who to my surprise proved very friendly and chummy. But I forgot one notable item, just before the lunch I went to the post office, and there at the stamp window found Russell Codman '83, who made an excursion with me to the uplifted delta of the Var near Cannes, in 1899, and whom I have hardly seen since; and then just as lunch was over, I saw a man darting toward me from across the court of the Paseo, and recognized him so I could greet him properly – Hello, Rollin Saltus; he was of our party to the Colorado Canyon in 1901 – remember him, Burt? – and is now an architect in Santa B. His charmingly beautiful wife was there too; but I had only a minute to speak to them. In the afternoon was the drive up to the mountain crest; and that ended the second day. Wednesday noon, a talk to the local Kiwanis club on the Nat. Hist. Goodness, which seemd to edify them – hold on, in the morning a drive along the mountain flank with R. Robbins, who was extremely entertaining and obliging and in the afternoon, what ho for Carpinteria with R. Hoffman to spend the night there, but before that, a return to Santa B. in evening togs, for the fine dinner at Mrs. Hazard's where I rattled and told stories in a most boyish manner; I wonder what they thot aild the old man.

(No wine). Then back to Carpinteria and a good night before an early Thursday breakfast, in time for an 8 oc lecture on that Interesting Volcanic District – Rome of course – which simply held the boys and Cate also captiv. There was some malice aforethot in my giving it to the school for nothing; it was as a sample of what the school might have in case the winter of 1927–28 is spent there. An anchor to windward, so to speak. I forgot to say that, Tuesday afternoon, besides the drive to the mountain crest, we took an elevated beach on the way out, and the Museum and Lundgres galley on the way back. No, the beach and the Museum – no, the Museum and the beach on the way out, and the galley on the way back. It was that sight of the Museum, which of course has no physiographic models, that suggested much talk to which Hoffman had to listen to on the way up the mountain; and he proved most receptiv. It may be that nothing will come of it, but then again it . . .

Well, Thursday morning, after the 8 oc lecture, Mr. George Bliss, whom I have met at the church on my previous visit, secured me for a drive to Santa Paula and the Ojai valley; and I may say emphatically that no visit to California shd be considered complete without that very fine drive. Bliss is a bright fellow; so is his wife; and much the same may be said of his children, two of whom, two small ones, went with us in an extremely comfortable Studebaker; much more comfortable than Hoffman's asthmatic Ford. Back in time to lunch with Cate and his school boys at 1.15; then with Hoffman back to Santa B. and up to the State Teachers College, for my afternoon lecture at 4 oc; subject, the new mountains of southern California in contrast to the old mountains of the Atlantic Coast. In spite of its length, Hoffman got it off correctly in introducing me; and thus did much better than the Kimanis president the day before; he askt twice what my subject was, just before setting me off; and then carelessly announced it as the Natural History of Education. I think he may not have been so quick witted as some presidents, for after much sport had been conducted under his discretion in the way of fining members for various imaginary offences, he took it very seriously when I told him that his club was just about the finest of any club I had ever seen. Well the Hoffman lecture went great; spell binder, in fact; and as four Museum trustees were present, my neatly introduced allusions to the importance of a physiographic exhibit may have some good effect for 1927–28. In order not to have to talk any more that day, I accepted Gibson's kind invitation to a play by an amateur company. Three Live Ghosts; well done and very amusing.

Up at 6.40 this morning; breakfast at 7.10; train at 8; Wolff met me at request of Munro, at Glendale, station outside of Los Angeles, and wafted me here, as Munro has a lecture at that hour. Nice lunch; and this afternoon some errands down town, and a call on George Hale, astronomer, wonderful fellow, at his private astronomical laboratory; very enjoyable, a really memorable call; for Hale is a great figure in American Science; he develop the great Mt. Wilson solar observatory up on the mountains here – a Carnegie Instn affair, but had to resign its direction by reason of ill health; now has his own fine s, which he showed to us most amiable-like. He is the moving spirit behind Cal. Tech. here; and it was he who braced up the National Academy of Sciences from its premature decrepitude organized the National Research Council, and was largely instrumental in securing the $5,000,000 endowment for the Academy and Council from the Carnegie

Corporation. He is a miracle worker. Munro is now doing his stint of afternoon work in his garden; hence I have his machine.

Too bad I am invited here for only one night. Tomorrow morning, to Claremont again, to repack before going to Noble's for the desert trip. Have I told you how I am going there? You see, Gale offerd, over a month ago, to take me a trip somewhere in March, and wanted to know where we had better go; so I suggested when I saw him that Sunday before going to Deep Springs, that we might run east to San Bernadino tomorrow, and then the next day, go up thr the Cajon pass, and on to Noble's at Valyermo on the north side of the mountains. He assented. Now an improvement is in my mind. He was going to pick me up tomorrow noon at Claremont but that involved my making my way alone, unprotected, in a mere public conveyance to Claremont from here. Why shdnt Gale call for me here at Munro's? It is almost directly on the way from his home in Eagle Rock to Claremont; so why not? He can wait, during the short half hour it will take me to repack; so I am going to call him up this evening on the telephone, and am ready to bet I can fix it without trouble. Rather cheeky of course; but these dear old boys seem really to enjoy wafting me about; so why not let them have their enjoyment. Another reason for stopping at Claremont is to pick up letters; none have been forwarded to Santa B. I will try to finish this sheet at San Bernadino tomorrow evening, or at Gale ranch, Sunday or Monday. Then the desert; and ve. likely no news, more than postal cards for ten days. I must be back about the end of the month, again to Claremont to pick up my trunk and move on to Salt Lake City, for all of April. My address there will be c/o the University Club. Now I will go out and look at Munro, farming. (*Letter: Davis to his children, 12 March 1926; Pasadena*)

Amidst all this contriving and lecturing, Davis now shows signs of his old zest for organizing groups. He persuaded the staff at Tucson to initiate a Shop Club on the lines of the one he had set up at Harvard. He talked Woodford into organizing intercollegiate excursions and into joining a new California Rift Club invented, of course, by himself.

Here's hoping that you will be free on Saturday, March 27, for on that day is to be initiated the southern California Rift Club, of which you are invited to be a member. Dr. Noble is to be president, and I am to be an honorary member, 'cause I invented the Club. Peters at whose ranch, over Devore in the lower part of Cajon pass I have just spent four delightful days, is to have the first meeting, March 27, beginning about 11 oc; and you are cordially and urgently invited to be there; also, to persuade Eckis to come along; he must be one of the Club, if he is going to work the rift along the mountain base. (*Letter: Davis to A. O. Woodford, 18 March 1926; Valyermo, California*)

We live on the great San Andreas Fault, and Davis had friends named Peters who lived on it, too, some fifty miles from here, so Davis thought up a club, calling it the Rift Club, for those who lived near or on it. It begun here with 17 people, and we all agreed to keep track of any possible movement and report it. We met once a month and had picnics, at which we reported 'No Movement', luckily, and then Davis would talk and talk. He was a splendid speaker, his choice of words and flashes of humor fascinated us all. His fame spread until other geologists asked to join, rift-dwellers or not, so it became a huge affair. (*Letter: Levi F. Noble to R. J. Chorley, 12 July 1961; Valyermo, California*)

In the late spring and early summer of 1926 Davis was at Salt Lake City and Logan, Utah, where he received with delight a copy of his book on the Lesser Antilles. The summer was spent in the East but he had already planned a circuit for the following winter months, beginning at Austin, Texas, and including, for January 1927, a lecture on 'Airplane views of the Alps' at 'Cal Tech' in California.

Letters from Texas and Arizona confirm the plans for next year; and as the pay is to be $1850 and $2000, it looks pretty good. (*Letter: Davis to his children, 1 April 1926; Salt Lake City*)

He has by now gained momentum and he is no longer satisfied just to lecture. He needed to feel that he was making a real impact on the audience. In some universities where he was a visiting lecturer presenting topics outside the general syllabus, the attendance was too small to stir up enthusiasm.

Work here is pleasant and light; so I am satisfied but in one respect it is disappointing. I have only five students. When Professors Simonds and Sel-rds invited me to come, last February, it was the understanding that qualified undergraduates as well as graduates should be in my class and since then the higher powers arbitrarily decided that undergraduates must not take graduate courses. As a result, my contributions to the education of the few students I have is costing the University rather high, but the Dean of the Graduate School, who is chagrind and apologetic in the matters begs me not to give that question another thought. Apart from the disappointment of addressing so small a party – I usually have from one to three instructors present also – I am contented; for with four students I have all the more time for other work, of which I have brought several packages along; the Natural History of Goodness and the Faith of Reverent Science are what I am trying to polish up just now. (*Letter: Davis to A. O. Woodford, 15 October 1926; Austin, Texas*)

Except for the fiasco of my little class, the time has been pleasantly spent; and the attentions of the University crowd have been agreeably bestowed. (*Letter: Davis to his children, 17 December 1926; Austin, Texas*)

Philip B. King, who attended Davis' courses at the University of Texas, is probably right when he avers that:

For the most part Davis' influence at Texas fell on sterile ground; students at Texas who took his courses were simply not prepared for him. One day after a long exposition of a region with cuesta topography, he asked, 'Where is it?' and one student said, 'It sounds like the country round Houston.' Actually, the region was south-eastern England, whose resemblance to anything in Texas was remote. His greatest influence at Texas, then, was on post-graduate geologists – myself, for one, and W. S. Adkins, for another, both of us at the time preparing geological reports on West Texas. Davis opened our eyes to the principles of arid erosion, not widely understood at the time, which bore directly on our work – although we were continually frustrated by his insistence on the form of concise 'geographical description', supposedly the ultimate objective of his whole system, but actually an absurd and unworkable formula, little used by Davis himself in his own writings. (*Letter: P. B. King to R. J. Chorley, 9 January 1963; United States Geological Survey, Pacific Coast Branch*)

There was an occasion outside the academic sphere when even Davis could not drag an enthusiastic question out of 'the best men in Pasadena'.

While at Wolff's Xmas time, Rev. Bradford Leavitt, Harvard '81? called on me and askt me to speak to the men's club in his neighborhood church, and Wednesday evening, March 24, was taken as the date. Leavitt spoke very highly of the members of the club, 'the best men in Pasadena', and all that; so I was rather keyd up about them, and expected a lot of fine questions at the end of my talk on the Faith of Reverent Science; and it fell *flatter* than a *pancake*. The chairman had hard work to get *anyone* to speak up; and with the exception of two or three reluctant speakers, *nothing* was said. To be sure, two of them said pretty nice things; but the general silence was oppressive, and I couldn't tell whether the fine-looking members – for they were really fine looking, middle aged and more so – thot my talk stale and stupid, or radical and shocking. After returning to Claremont, I wrote a note to Leavitt, expressing my little feelings of disappointment, and am wondering whether he will read the note at the next meeting of the club. Serve 'em right, if he does. (*Letter: Davis to his children, 2 March 1926; Deep Springs, California*)

With this succession of lectures and classes, Davis settled his financial worries. He was probably never actually short of money but was determined to balance his current expenditure. As we have already noticed, he had accepted a reduced share of his father's estate and the premature death of his second son Burt in 1926 must have placed some financial strain on the family. The royalties from Ginn & Co for the seven years ending in November 1922 had only averaged $800 annually but the new issue of his *Elementary Physical Geography* in 1926 must have restored his royalty earnings for a few years. He always considered himself poor, and took care not to incur unnecessary expenses, by staying when possible with friends, by eating sparingly (partly for health reasons) and by patronizing modest hotels and boarding houses. Woodford, a close friend at this time, certainly had the impression that Davis needed the money.

Then financial reverses in his family made it necessary for him to earn some money. So he went back to teaching and lecturing, acting as though he did it all just for fun. He was a very gallant old fellow. (*Letter: A. O. Woodford to R. J. Chorley, 4 January 1962; Claremont, California*)

However, finances must have been a worry as the money received during the winter months had to support him through the summer as well. The uncertainty in planning ahead and the need to accept so many unofficial appointments also raised many problems. Even religious audiences were charged a lecture fee and one secretary at least did not appreciate Davis' brand of humour in charging them extra for time spent in eating the supper they had provided free.

> I did a daring thing this morning: proposed to Revd Bradford Levitt that, after all, I would give him a sermon if he still wanted one; the Natural History of Goodness will be my subject, and I can promise a more consecutive treatment than he gave to his text in his sermon yesterday morning. Indeed, if you will promise not to tell, it was in good part his inconsecutiveness that persuaded me to try my own hand, and see if I couldnt keep more closely to my theme. Incidentally, it may mean $25, tho there may be some doubt about that, for I havent yet been paid what is owing me for two Chapter talks at Colo. Springs and Denver. Perhaps the Secretary of the Unitn Laymen's League to whom I sent a statement, is too serious a person to take a joke; for after making my proper charge of $50 for those two talks, I added $10 each for the two church suppers that I had to endure before the talks began; and the secretary, W. L. Barnard of Hingham, may not have thought that so funny as I did. Well if he wants more chapters cheered up by my style of eloquence, he had better send me a check pretty soon. (*Letter: Davis to his children, 4 January 1926; Pasadena*)

DAVIS' THIRD MARRIAGE

In the spring of 1927 Davis' financial and other troubles had been resolved sufficiently for him to decide that he would in future live in the West and would visit the East only on special occasions. With this in mind he tried to get himself reappointed at Tucson for the academic session of the following year and applied also for a position at the California Institute of Technology (Cal. Tech.) in Pasadena.

> It is hard to remember just how far my last report carried me; so I may repeat some items. One is that, altho no formal action as yet been taken on my application for re-appointment here next year, I am assured that it is to go thru. Another is the application I have made for appointment at the Cal. Tech. in Pasadena for the first half of the academic year has not yet brot a reply; so there is still hope. Dont imagine that my reason for suggesting a while ago that it was about time for me to stop these pedagogical antics is that I am worn out or tired or anything like that; no it is only because it is so unseemly for me I keep at it so over-long; and perhaps also because I am approaching the end of my resources in the way of chances. I doubt if any American professor has ever equald my cheek in asking for what I want and surely no one has ever got so much of what he askt for. So that's that. (*Letter: Davis to his children, 24 April 1927; Tucson, Arizona*)

Eventually he was reappointed for another academic year at Tucson but only after an hesitation which seemed to him an unforgivable affront, as the

Dean of the College of Mines had to take the case over the President's head to the Regents before the appointment was confirmed. At Cal. Tech. he was quickly successful due partly to the friendship of John P. Buwalda, to whom Davis had written.

> For the summer I am engaged at Berkeley but what I shall do between that summer course and the work here which will not begin till February, 1928, I do not yet know.
> In casting about for an occupation my inclinations lean strongly to your 'Cal. Tech' . . . The elementary course that I am now giving here is, in my opinion, both systematic and disciplinary; of good educational value. Much to my surprise, I believe that I am giving it as well as ever I did in earlier years. I shd. enjoy the opportunity of trying it on your students. The advanced course varies with the work of its takers, of whom there may not be many . . . (*Letter: Davis to J. P. Buwalda, 12 April 1927; Tucson, Arizona*)

The spring of 1927 at Tucson brought two pleasant events to Davis' later years. First, he resumes his friendship with Albrecht Penck although the old warmth never returned.

> The chief recent novelty is the arrival of Penck from Berlin with his young assistant, Dr. Haushofer. You may remember that I had something of a quarrel with Penck six or eight years ago; and that I have not been in active correspondence with him since then. But when he wrote he was coming to this country with a particular wish to see something of Arizona it seemd best to bury the hatchet and do what I could for him; and so he has been a sort of guest of mine for the last week, to the extent of relying chiefly on me for his chance of running about and seeing the country. (*Letter: Davis to his children, 16 May 1927; Tucson, Arizona*)

Davis in future made sure that when his friends, such as A. O. Woodford and R. J. Russell, went to Europe they called on Penck.

The second event was more remarkable. Having achieved a return to teaching, lecturing, and financial security, Davis decided that he must now procure a thing always dear to him – a patient and devoted wife. The reader may marvel at this seemingly impossible aim of a man of seventy-seven but to Davis all things were possible. He also needed such a woman as a kind of dignified background accompaniment, a listener, a foil, a provider of viands. He never ceased to be shocked at any lack of female subservience or of traditional propriety, as is shown in the following letter concerning what appears to be an early American version of a modern mini-skirt.

> We went up the front steps and rang the bell; no effect. Knockt; no effect; then turnd sadly away to the side gate again, and were about to disappear, when a curious figure a sort of nymph, made her appearance in a short, one-piece bathing suit and a sack, and said she would show us in. Had it been in Fiji she could not have been more naively unconscious of her costume, or rather of her lack of

costume. Her husband, the observer, was away; she talkt broken English and lookt like a Swede, or shd I say, Swedist, or perhaps Swedess. Her explanation of the recording instruments was surprisingly good; and her interest in our questions seemed genuine. But nothing was so genuine as her completely unembarrasst behaviour; never a word or a whisper of apology, never a sign of confusion. Perhaps bathing costumes are the rule at Balboa Palisades; but in winter I shd have expected something more extensiv. (*Letter: Davis to his children, 15 January 1926; Pasadena*)

F I G. 109. Davis and his third wife on a picnic near Tucson, Arizona, in the spring of 1930 (Courtesy P. B. King)

But to win a third wife, Davis had to unbend a little. We first hear of Lucy Tennant in April 1927.

Hoover and his wife came in to Phoenix, pickt up Lucy Tennant at the Episcopal cathedral (where they also are addicted to going) gathered me at a hotel; and then ran us out to a neat restaurant, the Casa Vieja, in Tempe, where I had ordered lunch for the party; in Spanish style, and very good, but mighty expensive; about double what I expected; however, I survive. (*Letter: Davis to his children, 24 April 1927; Tucson, Arizona*)

We are staggered to learn that Lucy smoked!

Death Valley is a marvelous place for alluvial fans and Davis whipped up a whole grist of descriptive names for their types. Some slopes were 'fan-bayed', some 'fan islanded', for instance, he named steep, narrow canyons 'Wine-glass canyons', which amused me, for he was an aggressive teetotaller. He was also bitterly opposed to smoking for women and told them so quite frankly, so we were amused when Lucy came to our house and lit a cigarette when she pleased, and no complaint came from Davis. (*Letter: Levi F. Noble to R. J. Chorley, 12 July 1961; Valyermo, California*)

Lucy Tennant, however, was an obvious candidate.

She had originally taught at Milton Academy in Massachusetts, was well known as a school teacher and was headmistress there for a time. Short, plump, later suffered from arthritis. Wore short hair even late in life. (*Interview: Ian Campbell with R. J. Chorley, August 1962; San Francisco*)

By June the friendship was already strong.

I used the vacant time to write up an epitome of the best chapter in the Coral Reef book, for a separate article; just to keep the interest of the scientific public awake. My friends there all let me alone, because I had written the time was to be spent in good part on my back, recovering from the activities of Tucson.

BUT, before I left Arizona it appeared that Lucy Tennant, of whom my previous letters may have made mention – Of course I told you how much I was cheered up for the last fortnight of my stay there by securing board at the house where she was staying, kept by two Canadian nurses, and So much pleasanter than the Copper Kettle restaurant where I had been mealing up that happier time.

Well ... Lucy T. said she would like awfully to attend the Rift Club meeting, the long-pland open meeting, to be held in the Cajon pass near San. Bo. on Sunday May 29, Why not, said I. So she arrived at Colton Saturday morning; and there in the station, waiting was Irwin Hayden of Riverside, and his car outside. I was also hanging round; and therefore when the train ran up, Hayden took Lucy's heavy suit case, and I myself took her lighter bag; and soon we were all five on the way to Riverside. You ought to remember Irwin Hayden as the Harvard man, 1902, who rescued me last January when Eggleston didnt meet me on my arrival one evening at Riverside, because I had forgotten to mail the letter informing him of my hour of coming there for an evening lecture.

Well ... we had a fine drive thru the beautiful residential part of Riverside, and then up Mt. Roubidoux, from which a glorious view is had over the district, largely occupied by orange groves. Then down again to Eggleston's house, where we pickt up him and his wife and carried them to the famous Mission Inn for a lunch; Unfortunately Hayden could not remain for that festivity; he had an appointment somewhere to meet a farmer. Too bad for him not to have all the fun after his nice attentions. The lunch was great sport; and I was especially glad to have the Egglestons there, for they have always been so nice to me on my previous visits. (*Letter: Davis to his children, 13 June 1927; Pasadena, California*)

As their friendship ripened, Davis' spirits rose and his whole attitude brightened by several degrees. He plays golf more regularly for morning exercise and seems bursting with self-satisfaction.

A typewriter on my table gave me indoor exercise; I polisht up a fine essay on a Migrating Anticline in Fiji, which is about ready now to go to the *American Journal of Science* in New Haven, where its appearance a few months hence will mark an epoch in coral reef literature; just remember that. It has a fine finale.

'Darwin's theory has such vigorous competence and irrepressible vitality that, in spite of the many obituaries written over it in the last forty years, it will regain in the coming half century the world-wide acceptance it enjoyd for a generation a hundred years earlier.' How's that for self-satisfied assertion! (*Letter: Davis to his children, 29 June 1927; Berkeley, California*)

He was equally confident that *The Coral Reef Problem* (1928C) would be 'a grand old book'.

In the mean time, the last of the paged proof of coral reefs has come and gone; to my great rejoicing. It is going to be a grand old book but too expensive to distribute generally. I shall have a copy sent to Mott and he can gradually spread it around, letting it pass via Hingham to Shirley, arriving at the latter place some time in the Spring, when Edward shall have returnd from Florida. You needn't read it all thru; not every page; but I think you may like to look at some of the pictures; for from my prejudiced point of view, a number of them are rather neat as well as effective. The conclusion reacht in favor of Darwin's old theory seems to me, and to a number of others here, whom I have belectured more or less on the subject, very solidly based; but at a meeting of the (geol) LeConte Club here last Saturday, Setchell, a seaweed botanist pitcht into Darwin almost savagely;

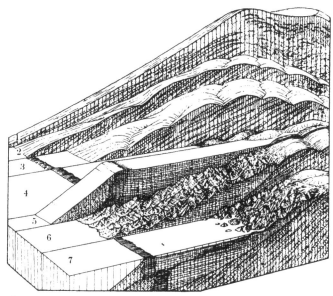

F I G. 110. Successive stages in the development of an uplifted and dissected reef, enclosed by a new barrier reef
1. Original volcanic island   2. Subaerial dissection
3 and 4. Growth of a barrier reef during subsidence
5. Rapid uplift of the island and reef 6. Renewed subsidence and subaerial dissection
7. Growth of new barrier reef on the irregularly-dissected previous reef
(From Davis, 1915I, Fig. 7)

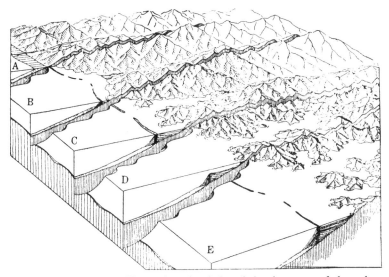

FIG. 111. Block diagrams illustrating the deduced development of the submerged
barrier reef of Palawan, Philippines
A. A maturely-dissected mountainous area with coastal detritus preventing cliffing
and reef building
B. Subsidence causes valley embaying and the upgrowth of a near-shore barrier reef
C. Increasing slow subsidence leads to accentuated embaying and reef growth
D. Rapid subsidence drowns the barrier reef
E. Renewed slow subsidence allows the renewed function of the barrier reef
(From Davis, 1928C, Fig. 215)

and thot he had overthrown the old subsidence theory. My idea is that he doesnt
understand the geol. evidence for subsidence; and his idea is that I dont understand the biological evidence against it. So there you are. Take your choice.
(Letter: Davis to his children, 15 December 1927; Stanford, California)

In May 1928 he finished his second academic year at Tucson, Arizona, and
was soon all set for Pasadena. His lightness had increased to a positive
sprightliness.

Well, here goes goodbye to Tucson for the second time. Lectures are over, exam.
books are read and marks handed in; trunk is nearly packt, various packages are
tied up and tomorrow evening I shall take train for Pasadena. But you may ask,
why not go tonight, if everything is so well prepared for departure. Why, tonight
is my Commencement Address, and it would never do to run away without
delivering it! Tucson is, as it were, all keyed up for me. And I am to hold my
audience of about 1000 spellbound for 42 minutes by the watch. The MS is all
markt up, with red border lines for paragraphs to be omitted, and with extra
commas to aid the reading emphasis. If I survive, you may hear a shout of joy over
the Rocky mountains. The Commt Exercises do not begin till 8 in the evening, as

the heat of the day will be somewhat lessend; and they are held outdoors; where we shall have all the advantage of lessend heat that is to be found. All the same I shall be thankful when it is over, and my work here is done. (*Letter: Davis to his children, 30 May 1928; Tucson, Arizona*)

In the following August Davis and Lucy Tennant were married, the ceremony being performed by a priest of the Episcopal Church. The old campaigner is back once more on the Davisian heights, with blue skies above and, miraculously, a large new comet in the East as *The Coral Reef Problem* makes its belated appearance.

CHAPTER TWENTY-FIVE

The Coral Reef Problem

It is rather ironic that when Davis retired prematurely from Harvard in 1912 he was immediately succeeded as Sturgis–Hooper Professor by Reginald Aldworth Daly who had previously (1905) written on the summit level of the Alps in anything but Davisian terms and was closely associated with the 'glacial control theory' of coral reef development, so much at variance with Davis' support for Darwin's theory of reef subsidence. Perhaps this antagonism was one reason why Davis, after his retirement, continued to play some part in Departmental activities!

During the weekly evening meetings of students and staff of the Department of Geology at Harvard from 1913–17 and again 1919–24 Davis sometimes showed up, always took part in the discussion, and occasionally gave an address. Daly (R. A.) was then fresh from his attacks on coral islands and his glacial control theory, bending every item brought to his attention toward substantiating this theory. He would talk volubly and overwhelmingly about his conclusions, and several times I remember Davis arising and with an exquisitely measured phrase knock down the whole house of cards.

I never saw Davis smile. J. B. Woodworth used to say 'Davis is always right. Nobody likes a man who is always right. That's why Davis has had but one graduate student in all his years at Harvard.' *

At one of our evening meetings someone (it may have been W. W. Atwood) had sketched a river-dissected mountain slope (with both hands!) with appropriate shading, but which failed to give an accurate impression of the relief. Davis marched up to the blackboard, drew a similar view, about the interpretation of which there could be no doubt, and pointing to the sketch asked 'What time is it!' The lesson we were to learn was that white chalk 'shading' on a blackboard might represent either light or shadow, hence morning or afternoon ... Woodworth, who respected Davis and tolerated Daly, was fond, whenever his lectures touched on coral reefs, of writing on the blackboard as follows:

DA RWIN
   NA
   VIS
   LY

accompanied by a semi-serious whispered aside that 'the first three were born in the same month – the fourth was not'. (*Letter: T. H. Clark to R. J. Chorley, 23 September 1970; Montreal*)

* F. P. Gulliver, *Shoreline Topography* (1896 and 1899), who applied the concept of the cycle to shoreline development, as Davis (1896G) advocated.

For the fifteen years immediately following his premature retirement Davis was primarily engaged on what he was to term 'the coral reef problem'. The task culminated in his major publication on the subject in 1928 and in this 'grand book' he gives a clear description of his introduction to, and treatment of, this fascinating problem.

> After forming an early acquaintance with Darwin's theory of coral reefs sixty years ago when I was a student in one of Shaler's classes in geology at Harvard University, those strange oceanic structures were relegated to the dim background of my thoughts; and there they remained ten years later when, on returning to Harvard as a teacher of geology at first and of physiography afterward, my attention was increasingly directed to the study and description of the forms of the lands. Not until that study had been continued some twenty years did I come accidentally upon the inference, while drawing an ideal block-diagram of a barrier reef for an elementary textbook, that if such reefs are formed by upgrowth around subsiding islands, as Darwin's theory proposed, the islands should have embayed shore lines. Then, on discovering from the charts of such islands that they do have embayed shore lines, the theory of subsidence seemed to me to be a true theory.
>
> But it was at about that time that Agassiz' report on the reefs of Fiji appeared; and his conclusions, reached after he had seen many well embayed islands, were so strongly opposed to Darwin's theory that my confidence in it was shaken; all the more so because of favorable references to several competitive coral reef theories invented twenty or thirty years before that time by Semper, Rein, Murray and Guppy, which I had met in reading and in none of which embayed shore lines were given any consideration. It was not until several years later that, on looking up Dana's report on the Geology of the Wilkes Expedition, I learned much to my surprise that he had clearly made application of the principle of shore-line embayments to the coral reef problem a long half-century before and had thus provided for Darwin's theory a support greatly needed by it. Yet on reading more recent authors I found to my still greater surprise that Dana's confirmation of Darwin's theory had been overlooked and neglected by nearly every one of them. I therefore called attention to this neglect in a special article on the centenary of Dana's birth [1913O].
>
> My interest in the old problem was thus aroused, and all the more strongly on perceiving that the studies of land forms to which I had been chiefly devoted might find novel application on reef-encircled oceanic islands. Hence, when, two years after resigning my professorship at Harvard in 1912, opportunity came to cross the Pacific, I made preparation therefore by reviewing all the coral reef theories then current [1914F] and then set out on a long voyage ... (*Davis, 1928C, pp. 546–7*)

> Cambridge, Massachusetts, was left on January 31, and after several stops on the way San Francisco was reached on February 9, whence the steamer *Wilhelmina* of the Matson line, sailing February 11, carried me to Honolulu, Oahu, on February 17; a week was then given to the elevated coralliferous limestones of that island. The fine steamer *Niagara* of the Canadian–Australasian line, running between Vancouver and Sydney, took me from Honolulu, February 25, to

Suva, on the chief Fijian island, Viti Levu, March 6. Seven weeks were spent in visiting eighteen Fiji islands, including Ovalau, Mango, Thithia, Vanua Mbalava, Taveuni, Rambi, and Vanua Levu, which were reached on a trading steamer, as well as Makongai, Wakaya, Kandavu, Ono, Nairai, Matuku, Ngau, Totoya, Moala, and Mbengha, which were reached by sailboats ...

On May 1 Suva was left on the *Makura*, also of the Canadian–Australasian line, and on May 5 arrival was made at Auckland, New Zealand. A month was spent there on North and South Islands, in making visits and excursions, for which free transportation was most generously given me on all railways, and excellent personal guidance was provided by the geologists of the colleges at Auckland, Wellington, Christchurch and Dunedin. Departure was made from Wellington, June 5, on the *Moeraki* of the Union Steamship Company of New Zealand, and Sydney, Australia was reached on June 9. Pleasant acquaintance was there made, through introduction by my most attentive friend Mr E. C. Andrews of the Department of Mines of New South Wales, with a number of members in other governmental bureaus. The French steamer *Néra* of the Messageries Maritimes, sailing thence on June 12, carried me to Nouméa, New Caledonia, on June 15; the greater part of a month was well used in making a circuit of that long island and in visiting the three Loyalty Islands to the north on trading steamers. A final week was spent in a cruise around the southeastern end of the island in a small sailboat under the pilotage of its owner, M. Cané, who, from his acquaintance with the natives and their very guttural language as well as with all the headlands and harbors of the coast, proved to be a most helpful guide ... A short trip was then made in company with Mr E. C. Andrews on the *Pacifique* of the Messageries Maritimes, which brought him from Sydney and on which we proceeded northward from Nouméa on July 18 to the New Hebrides, with stops at the islands of Efate, Epi, Ambrym, Malekula, and Espiritu Santo; we returned southward, touching at Nouméa on July 27, and continued to Sydney, where we arrived July 31 and had our first news of the outbreak of the World War from the pilot at the entrance to Sydney harbor.

The greater part of August was given to the meetings of the British Association for the Advancement of Science in Adelaide, Melbourne and Sydney. Time was then unexpectedly found for a journey northward by rail, August 27 and 28, and by steamboat August 29, as far as Cairns, September 1 and 2, on the Queensland coast, whereby opportunity was afforded me of seeing a long stretch of the shore line inside of the Great Barrier Reef and of visiting one of the reef islands. After a return to Sydney on September 10, that beautiful city was left on the 11th, and the return voyage to San Francisco then begun was accomplished on three steamers of the Union Steamship Company: the *Manuka* carried me to Wellington, New Zealand, September 15; from Wellington the *Moana*, sailing on the 17th and touching at Rarotonga in the Cook group on the 22nd, conveyed me to Papeete, Tahiti, the chief island of the Society group, on the 24th two days after that quiet little French colonial capital had suffered the excitement of a brief bombardment by the German cruisers *Scharnhorst* and *Gneisenau*, then on their way from China to the coast of Chile. Four weeks were allotted to that superb island and five others, Moorea, Huaheine, Raiatea, Tahaa, and Barabora which, were reached by launches ... Sailing from Papeete on October 23, a placid voyage on the *Marama* brought me to San Francisco on November 5 ... (*Davis, 1928C, pp. 143–5*)

It was my intention on returning home at once to prepare a narrative of the voyage and a statement of the conclusions I had been led to adopt; but, on continuing the review of the problem which I had previously begun (1914), it seemed advisable to enlarge the scope of my work so that it should attempt to cover the whole subject. Hence only a condensed summary of the results gained on the Pacific was published in 1915, and the present volume was postponed. Progress in the enlarged discussion was slow because of its many complications, and my work upon it was much interrupted by the distractions of the World War. The delay thus occasioned was, however, by no means disadvantageous, for it gave me time to gather and to reflect upon an increasing store of pertinent information from many books and articles and charts. It also enabled me gradually to reach an understanding of various matters which had before been obscure or unnoted. Among these are: the conditions under which young volcanic islands remain reef-free and exposed to the attack of the surf, and the change from those conditions which later permits the islands to become reef-encircled; the testimony furnished by the visible forms of certain volcanic islands as to the great subsidence that they have suffered; the compelling evidence for subsidence given by the disappearance of large volumes of detritus from maturely dissected, reef-encircled islands; and, following a principle embodied in Daly's Glacial-control theory although departing widely from the postulates on which that theory is based and from the conclusions to which it leads, the occurrence of marginal belts around the coral seas, where reef growth appears to have been interrupted during the Glacial period. These and various other aspects of the problem have been set forth in a preliminary way in a number of articles published during the last twelve years and listed in the Bibliography at the end of this book: the results thus gained are here all brought together.

Shortly after the occurrence of the marginal belts of the coral seas was recognized, opportunity came in 1923 to visit the Lesser Antilles, where the peculiar features expectable on marginal belt islands were found to be developed in the most striking manner. For there the reefs, mere timid Postglacial novices standing well back from the outer margin of their banks, were found to be strongly unlike the stalwart veterans of the true coral seas in the Pacific. (*Davis, 1928C, p. 2*)

Davis stressed that the coral reef problem could only be solved by studying the physiography of the whole coastal area and not just of the coral reefs.

It should here be emphasized that independent evidence as to the value of competing theories of reef origin can be best obtained not from the reefs themselves, but chiefly from the physiographic features of the coasts, either insular or continental, that are bordered by fringing reefs or fronted by barrier reefs; and also from the structure of elevated reefs. It is for this reason that the following pages are so largely occupied with the physiography of land forms rather than with the biology of coral reefs. The biology and especially the symbiosis of the reefs are unquestionably important subjects in themselves, but the opportunity for the establishment and growth of reefs is so largely determined by the physiographic conditions of insular and continental coasts, over which reef-building organisms have no control, that those physiographic conditions necessarily assume the leading role in the problem under discussion. (*Davis 1928C, p. 3*)

He strongly favoured Darwin's theory of upgrowing reefs on intermittently-subsiding foundations and considered that changes of sea level and of sea temperatures in the glacial period, of which Darwin had no knowledge, supplement and enlarge, rather than invalidate, that theory.

Work upon (this book) has brought me many enjoyable experiences, among which are to be counted the recognition of a delay in the establishment of fringing reefs on a young volcanic island until after both its dissection and its subsidence are somewhat well advanced; also the interpretation of the evidence for continued subsidence as furnished by the disappearance of large volumes of island detritus; to say nothing of the detection of the likeness and the contrast between the formerly cliffed and now reefless Madras coast of India and the formerly cliffed and now reef-fronted northeastern coast of New Caledonia; or of the evidence given by the small thickness of many uplifted fringing reefs for instead against a certain phase of Darwin's theory. Greater enjoyment still came from the discovery, prompted by the climatic factor of the Glacial-control theory, of the marginal belts of the coral seas, for that seems to me a long step forward in the home study of coral reefs; and also from the confirmation of the occurrence of a marginal belt in the Atlantic as secured from a visit to the Lesser Antilles in 1923. The same may be said of the illuminating contrast between the island-free and therefore never reef-protected southern half of the coast of eastern Australia and the island-swarming and therefore long reef-protected northern half; for in this contrast seems to lie a complete refutation of the idea that the Great Barrier Reef is of recent formation on a northward prolongation of the southern continental shelf of that far continent. The latest items of these many enjoyments came during the final revision of the preceding chapter: they are the evidence given by minute oceanic reefs for the subsidence of their foundations and therefore for the subsidence of the foundation of larger atolls also, and the evidence given by nullipores for the shallow origin of barrier and atoll reefs. But the most gratifying experience during the progress of this investigation has been that of seeing many complicated facts of apparently disorderly occurrence – those of eastern Fiji, for example – arrange themselves in orderly sequence as soon as their explanation came to be guided by the clue of Darwin's theory.

Through all this period of observation and study I have therefore felt a growing admiration for the keen insight of the young naturalist of the *Beagle* and have thus come to appreciate better than ever before the calm fairmindedness with which he, in his maturer years, taught a new philosophy to an unwilling world. It has been a great pleasure to try to secure for his early-framed theory of coral reefs, which after world-wide adoption between 1840 and 1870 was so strongly objected to and even rejected by a number of later observers between 1870 and 1910, the broader consideration that it so fully deserves. The theory was only an outline in the form conceived by Darwin; for the facts of the coral reef problem were then imperfectly known, and various phases of the theory itself were left undefined.

Yet, as new facts have been brought to light, it has been most impressive to note the ease with which they have been explained by inevitable extensions of the old theory. The explanations were, indeed, latent in the theory from the beginning. The only serious exception to this statement concerns the islands of the marginal belt of the coal seas, for the explanation of which Darwin's theory must be supplemented by the wholly new process of low-level abrasion, first brought into the problem in recent years by Daly. (*Davis, 1928C, pp. 547–8*)

The main conclusion of my many pages is that Darwin's theory of upgrowing reefs on intermittently subsiding foundations, extended by introducing the effects of Glacial changes of sea level and temperature, as proposed by Daly, and by adding such minor modifications as may be called for in special cases, will deservedly regain in the present century the general acceptance which it enjoyed through the middle of the past century. (*Davis, 1928C, p. 3*)

There is little doubt that Davis considered his work on coral reefs to represent one of his most important contributions to physiography and certainly it provided, during some of the darkest periods of his life, a source of genuine pleasure and satisfaction.

Have I told you that my little book on the Lesser Antilles and their coral reefs is out? Yes, the devoted Bowman of the American Geographical Society in New York has made it look very neat. A copy reacht me at S.L. City as I came up from Provo. Now if the uninformed students of coral reefs will only pay heed to what I have there stated and explaind, they will do better work in the future than they have done in the past. One of my special pleasures, in looking the book over, is to see how nicely I have treated Dr T. W. Vaughan, formerly of the U.S. Geological Survey now of the Scripps Oceanographical Institute, at La Jolla, near San Diego, Calif., where I visited him a year ago last December. He went to the Lesser Antilles some years ago; found the little, imperfect reefs there developed and mistook them for likenesses of the much greater stalwart veteran reefs of the Pacific, which he had not seen; and explained the Pacific reefs by the same processes that he applied to the L.A. reefs. I was doubtful of his conclusions before going to the L.A.; and after going there, was convinced that the Pacific reefs were of a totally different order. Now I have set forth my way of thinking by contrasting the two sets of reefs in what seems to me a very fair and a very convincing manner; and I am wondering what Vaughan thinks about it after reading my views. Anyhow, I have seen a great many more Pacific reefs than he has; so there. (*Letter: Davis to his children, 10 May 1926; Logan, Utah*)

Fortunately for Davis his great efforts were well received by his contemporaries and even Daly, his chief critic, summarized the results with guarded generosity:

In 1923 he added to his field experiences by travel among the reef-bearing islands of the Lesser Antilles. For twelve years his time was largely spent on the study of his own observations, of the multitude of island charts issued by the hydrographic offices of the world, and on the voluminous literature on the controversial subject of reef origin. At intervals, he published the results of his correlations, producing twenty-eight papers and a book on the Antilles. In 1928 there appeared his weighty monograph, entitled 'The Coral-Reef Problem', giving his complete views concerning the relative merits of the many hypotheses which have been offered as solutions to the reef problem.

Davis was fascinated by the beauty and apparent cogency of the Darwin–Dana view that atolls and barrier reefs are best regarded as the products of slow subsidence of the foundations on which these structures are built, and at first (1915) thought the subsidence hypothesis to be alone competent in explanation. Later he accepted the idea of 'Glacial controls' as useful in accounting for the 'platform foundations' and crowning reefs in the marginal areas of the earth's coral-reef zone. His treatment of the problem was dominated by the double principle of deduction and verification, but in the opinion of the present writer Davis failed to give adequate consideration to some of his premises, including the geological dates when the reef foundations were prepared and when the wave-resisting species of corals became abundant in the tropical ocean. Nor was sufficient attention paid to the relatively enormous areas and remarkable flatness of the lagoons inside atoll and barrier reefs – features which are almost universal and not to be expected on the Darwin–Dana hypothesis. It may further be remarked that this hypothesis is not supported by the findings at test boreholes in Bermuda and at Michaelmas Cay and Heron Island inside the Australian Great Barrier Reef.

Notwithstanding such failure to secure the premises on which the author of 'The Coral-Reef Problem' based his own conclusions, this book will long remain the Bible for geologists and geographers who need a richly illustrated handbook summarizing the facts known about these marvelous structures of the coral seas, or are interested in the relation of the reef controversy to the fundamental question as to the strength and stability of the earth's crust. (*Daly, 1945, pp. 278–9*)

However, Daly was less guarded when someone praised one of Davis' coral reef works at a Harvard dinner in the 1920's, and he is reported to have replied: 'Excellent illustrations, but not a word of truth in it!'

It is in the light of these comments that we now give full details of Davis' long incursion into the coral reef problem.

CORAL REEF THEORIES AND DAVIS' EARLIEST WRITINGS ON THEM

Darwin's theory of coral reef development, expressed in *The Structure and Distribution of Coral Reefs* (London, 1842), was based on the premises that coral growth ceases at a comparatively small depth, and that the subsidence of islands must have caused fringing reefs to develop by stages first into barrier reefs and then into atolls, as subsidence progressively decreased the size of the island and was accompanied by upward and outward coral growth (Davis, 1928C, pp. 22–40). He pictured subsidence as the dominant mechanism promoting continued reef growth, but regarded it as essentially intermittent. Such subsidence would cause thin fringing reef deposits (Darwin, 1842, p. 98) to develop into thick barrier reefs and complex atolls on the subsiding foundations. According to this theory lagoon depths inside the reefs should be related both to the amount of local subsidence and the amount of local deposition, allowing for a loss of sediments seaward between the reefs.

This theory appealed to Davis for several reasons. It was evolutionary and contained well-defined stages into which observed features could be placed in time sequence; it had a wonderful simplicity leading to a large number of theoretically-deduced consequences which could be checked in the field; and, finally, it was obviously itself the product of theoretical deduction, so in tune with Davis' own *modus operandi*:

> No other work of mine was begun in so deductive a spirit as this; for the whole theory was thought out on the west coast of S. America before I had seen a true coral reef. I had therefore only to verify and extend my views by a careful examination of living reefs. But it should be observed that I had during the two previous years been incessantly attending to the effects on the shores of S. America of the intermittent elevation of the land, together with the denudation and the deposition of sediment. This necessarily led me to reflect much on the effects of subsidence, and it was easy to replace in imagination the continued deposition of sediment by the upward growth of coral. To do this was to form my theory of the formation of barrier-reefs and atolls. (*Barlow, 1958, pp. 98–9. Referred to by Stoddart, 1962, p. 1*).

Darwin's subsidence theory found strong support from James Dwight Dana (*On Coral Reefs and Islands*, New York, 1853; *Corals and Coral Islands*, New York, 1872), who pointed to the coastal embayments of submerged subaerial valleys in reef-fringed islands as physiographic evidence of sinking of the land. But workers in the late nineteenth century tended to rely on other theories of reef formation, most of which postulated that reefs were thin veneers growing on suitable marine platforms cut with reference to an unchanging sea-level. Murray, for example, proposed that reefs grew outwards from truncated or built-up sea mounts and that the lagoons had been formed by solution of the less active reef. Agassiz also postulated coral reef growth on marine abrasion platforms, but this view was made less tenable by the relative absence of marine cliffs along tropical coasts fringed with reefs and by the presence of small islands within many lagoons. Various modifications of the marine planation theory of reef formation were presented by Semper, Le Conte, Sluiter, Guppy, Vaughan and Gardiner.

It was inevitable that the late nineteenth-century growth of knowledge relating to the Pleistocene glaciations should soon make an impact on theories of coral reef development, from the viewpoint of both sea-level and sea temperature changes. At about the time Davis retired from Harvard his successor, Reginald Aldworth Daly, began to develop his glacial-control theory of coral reefs (1910 and 1915). Daly assumed that during the glacial period sea-level would have fallen universally by some 33 to 38 fathoms and sea temperature would have dropped so that over large areas corals would have been unable to grow. At the time of lowest sea-level the waves would have bevelled off the dead pre-Glacial reefs and cut benches across the adjacent lands and islands. During deglaciation new colonies of corals would have grown in the rising and warming sea on the outer edges of the platforms forming barrier reefs, or, where the central island had been completely bevelled off, atolls. Thus post-Glacial coral growths would be expected to consist merely of shallow veneers on glacial platforms of marine abrasion. Daly believed that one of the strongest physiographic evidences in favour of his theory was the postulated accordance of depth and flatness of lagoon floors, a condition held to have been inherited from the glacial low-level abrasion platforms but this accordance, as we shall see, was strongly challenged by Davis and others. Another consequence of Daly's theory was that coastal valleys should exhibit features of recent drowning; but here too Davis was quick to point out that the amount of drowning to be inferred from the bedrock form of coastal valleys backing tropical reefs is far in excess of that required by the glacial-control theory. A further consequence of Daly's theory was that cliffs should be strikingly evident on the landward sides of lagoons, although masked by some post-Glacial drowning. It was pointed out, however, that in tropical areas prominent cliffs are commonly absent, and that the shores of reef-protected lagoons generally have minor cliffs only, such as could well have been produced by waves generated within the lagoon. Although Daly's theory did not demand absolute stability of the reef foundations, it was clearly opposed to Darwin's concept of reef upgrowth on subsiding foundations in which coral deposits could be expected to occur unconformably on a bedrock basement, rather than on a bevelled-off pre-Glacial coralline marine platform.

Davis soon discounted the general application of the glacial-control theory to tropical oceans, although he admitted that it could be applied significantly to the development of reefs in marginal climatic belts such as Hawaii and the Lesser Antilles. He concluded:

Daly's several presentations of the Glacial-control theory have introduced an important process that is, I believe, destined to be of permanent value in the coral reef problem; namely, the production of platforms by the low-level abrasion of circuminsular masses of coral reef and lagoon origin as well as by the slower abrasion of insular masses themselves, the encircling reefs of which were killed by the lowered temperature of the ocean in the Glacial epochs. But the area over which this process is thought to have acted appears to have been assigned too great a value in his theory, and the availability of the process has been limited in his discussion by its application only to prevailingly stable islands. On the other hand, the process of low-level abrasion is here regarded as having been inoperative over the greater part of the coral seas; for the absence of plunging cliffs on the great majority of barrier-reef and almost-atoll islands shows that their reefs were not killed. But the process is believed to have been very successfully operative on certain islands in the marginal belts of the coral seas . . . although those islands are thought to have been about as unstable as their more numerous neighbors in the coral seas.

The moderate and fairly accordant depths of most reef-enclosed lagoons, upon which the Glacial-control theory is based, is here explained chiefly as a result of effective lagoon-infilling during the slow upgrowth of the encircling reefs on subsiding foundations. The occurrence in subsidence is independently proved in many cases, and the rate of subsidence is shown to have been slow by the fact that reefs have been able to grow up while it was in progress.

The absence of deep moats between barrier reefs and their central islands and the corresponding absence of deep basins within atoll reefs of pelagic regions may properly be taken to indicate that lagoon infilling is a much more active process than it is assumed to be in the Glacial-control theory, and indeed not much slower than reef upgrowth. The crustal stability demanded by the Glacial-control theory must therefore be regarded as of vastly longer duration than the brief 'stationary period' which Darwin postulated for reef upgrowth and lagoon aggradation around certain islands. Indeed, the very slow subsidence which will fully satisfy Darwin's theory may perhaps be the equivalent of the prevailing crustal stability that is postulated in the Glacial-control theory for most of the coral seas of the world. And it is conceivable that the 'renovating agency', namely subsidence, which Darwin thought was necessary to prevent the filling of lagoons, was helpfully supplemented by the exportation of silt from shallowed lagoons in non-Glacial times and by the degradation of emerged lagoon floors during the Glacial epochs.

The occurrence of atolls with ordinary lagoon depths, as far as they have been sounded, in the neighborhood of notably unstable islands in the western Pacific is held to be of prime significance. For if lagoon infilling within upgrowing reefs has been successful in giving those lagoons ordinary depths where instability is well proved, the occurrence of ordinary lagoon depths in the atoll groups of the mid-Pacific, where nothing is directly known about the behavior of the earth's crust, cannot be held to prove its long-enduring stability.

Of all the conclusions above reached, the one that seems to me most convincingly opposed to the Glacial-control theory is that which excludes from the coral seas the assumed process of low-level abrasion by which the supposed sub-lagoon platforms of those seas were produced in the Glacial period. Had this process been so effective as to provide platforms for the numerous atolls of the coral seas by the complete truncation of old, well degraded and deeply weathered volcanic islands, there ought to be a moderate number, at least, of less old, less degraded, and less deeply weathered islands which would be incompletely truncated; and as such they would now be characterized by plunging cliffs surmounting their barrier-reef lagoons. Although a number of islands show precisely such features over the banks of the marginal belt adjoining the coral seas, the barrier-reef islands of the coral seas are as a rule not thus characterized. The few barrier-reef islands that are cliffed are relatively young and owe their cliffs to a lack of defending reefs as determined by local, individual conditions provided by the islands themselves, and not to the widespread climatic conditions of the Glacial epochs. (*Davis, 1928C, pp. 118–20*)

The conflicting theories of coral reef development provided for Davis an ideal proving ground for the application of his version of the multiple-working hypothesis. Like Darwin, he began by speculating about reefs before he had actively studied them in the field (Davis, 1913O) and conducted a 'Home study of coral reefs' (Davis, 1914F) based completely on previous writings and published maps. Davis' method was, allegedly, to take each theory and to deduce its logical consequences, which could then be tested against observational evidence, although he had clearly made up his mind at the outset of the general correctness of Darwin's subsidence theory. The following extracts show his concern for identifying evolutionary sequences, for throwing the emphasis of his work on physiographic rather than stratigraphic considerations, and for use of block diagrams not only as a means of presentation but of analysis.

[Speaking of Wharton's truncation theory.] The ordinary relation of fringing and barrier reefs to their central island suffices to show that the work of marine truncation would be arrested by the growth of reefs, just as soon as the abraded platform became broad enough to afford a foundation for coral growth a moderate distance away from the outwash of fresh water and its detritus; and when such a reef is once established the further truncation of the island by wave work is practically stopped; the waves of most lagoons are too weak to be effective agents in cutting away the land (*Davis, 1913O, p. 176*)

Murray's theory of outward growth and solution, whereby fringing reefs are converted into barrier reefs during a prolonged still-stand of a volcanic island, is also a manifest possibility; but it involves several consequences, easily deducible from the theory but not usually stated with it. For example, if the central island be several miles in diameter and a thousand or more feet in height, its streams will wash down abundant detritus upon the fringing reef. By the time the reef has grown outwards far enough to be called a barrier, the stream-borne detritus will have formed deltas fronting the mouth of each valley; and with farther outgrowth of the reef the deltas will become laterally confluent, so as to form a low alluvial plain round the original shore line of the island; and this original shore line should not exhibit sinuosities of the kind that are produced by subsidence . . . During the outgrowth of the reef its superficial parts will be more or less dissolved away in the excavation of the lagoon; hence if such a reef is uplifted, so as to form a terrace around its mountain center, its stratification will be inclined at a significantly steeper angle than the slope of the foundation on which it is built; while in the case of a reef formed during subsidence a large part of the reef may consist of horizontal strata . . . The second deduced consequence of the theory of outward growth appears to be seldom supported by the facts, for the strata of uplifted reefs, as far as I have read, are usually horizontal. (*Davis, 1913O, pp. 177–8*)

The diagrams by which Darwin illustrated the consequences of his theory are simple transverse sections, and as such they do not sufficiently represent the deductive side of the problem. If they be transformed into block diagrams, in which surface and section are both exhibited, their value in this respect is increased;

for, as Passarge has pointed out, block diagrams show what an author thinks rather than what he sees; indeed, it may be claimed for them that they help their author to think, even compel him to think. Therein lies half their value; while the other half lies in the ease with which the reader can, with their aid, understand the thought of the author. (*Davis, 1913O, p. 180*)

But the most significant feature is yet to be mentioned, although the diagram has probably already suggested it. Darwin recognized the diminishing size and final disappearance of the subsiding island as an essential consequence of his theory: but another equally essential consequence, of which no mention appears in his writings, is the transformation of the relatively simple shore line of the initial island into an indented shore line, as the sea enters the valleys, between the ridges of the dissected and subsiding island-mass. As long as subsidence continues, no large deltas can be built forward in the bays by the decreasing streams, no strong cliffs can be cut in the ridge-ends by the relatively quiet lagoon waters. Here are several consequences of the theory of subsidence which were entirely unsuspected by its inventor. (*Davis, 1913O, pp. 180–1*)

The alternative theories of outward growth on still-standing islands, of veneering barrier reefs on sea-cut platforms, and of upward growth on up-built submarine banks, certainly deserve to be considered wherever they can find application. The down-wearing of uplifted reefs and atolls is evidently pertinent in certain cases of complicated history. Every case must, as Agassiz repeatedly insisted, be independently investigated. Nevertheless, it may now be fairly said that the theory of subsidence deserves for a number of well studied examples the acceptance that it long enjoyed, and that it for a time and in part lost. It affords not merely a possible explanation; it exposes a well-supported explanation of many barrier reefs, and hence probably also of many atolls. The postulate of subsidence on which Darwin's theory is based, is justified and established by the independent evidence brought forward by Dana. (*Davis, 1913O, p. 188*)

In 1915 Davis began a long series of publications based on his Pacific tour of the previous year, again subsidized by the Shaler Memorial Fund together with a grant from the British Association whose annual meeting in Australia he attended in August 1914. By concentrating attention on the physiographic features of island coasts associated with reefs and on the structural features of uplifted reefs, Davis (1915A; 1915C; 1915I) propagated the subsidence theory and attacked the glacial-control theory. In this, of course, he was aided by the flexibility of the local subsidence theory as distinct from the relative inflexibility of Daly's eustatic mechanism. The lack of spur-end cliffs and valley-in-valley forms consequent on the glacial-control theory, and the magnitude of the drowned island valleys, the variation in lagoon depths and the unconformable resting of elevated reef deposits on their bedrock foundations (Davis, 1915C, p. 456) were all evoked to support Darwin and Dana over the theories of outgrowing reefs on still-standing islands and the glacial-control theory.

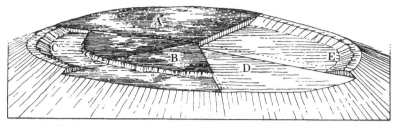

F I G. 112. Deduced stages of an atoll, according to the glacial-control theory
A. The pre-glacial reef-limestone plain
B. The cutting of an abraded lagoon floor a mile or so wide by the lower glacial sea level
C. The present-day barrier reef developed on the outer edge of the abraded lagoon floor surrounding a central tabular part
D. Shows complete glacial marine abrasion
E. A recent barrier reef grown on a glacial abrasion platform
(From Davis, 1915I, Fig. 4)

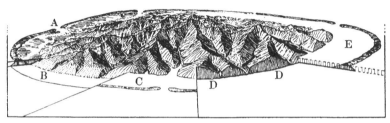

F I G. 113. Deduced stages of a barrier reef, according to the glacial-control theory
A. A pre-glacial barrier reef
B. The marine abrasion of the dead reef during the glacial lower sea-level period
D. This abrasion should produce spur-end cliffs
E. The post-glacial rise in sea-level should lead to the growth of a new barrier reef on the outer edge of the glacial abrasion platform
C. The usual features of a barrier reef coast (Contrast with E)
(From Davis, 1915I, Fig. 5)

Lagoon Floors are not Wave-cut Platforms. – Consider the case of a narrow-lagoon barrier reef, C, [Fig. 113]. If the floor of such a lagoon, half a mile or a mile wide, represents as much work as marine abrasion can accomplish in reducing a dead, preglacial reef, A, to a smooth platform, B, during the lowered sea stand of the glacial period, it follows that the lagoon-floor of a large reef, such as that of Hogoleu (or Truk) in the Carolines, 34 miles in diameter, cannot be of the same origin; still less could any large part of the vast lagoon of the Great Barrier reef of Australia, sometimes 70 or more miles wide, be due to marine abrasion during a lowered sea-stand, for it could have been attacked by the waves only on one side. So large an atoll as Hogoleu) should preserve the central tabular part, B, [Fig. 112] (more dissected than here drawn), of its preglacial reef-limestone plain, A, standing practically at sea level and surrounded by an abraded lagoon-floor a mile or so wide, outside of which a barrier reef, C, should rise to-day. No such central limestone table is known in any atoll. If, on the other hand, the floor, D, [Fig. 112], of a wide lagoon, E, like that of Hogoleu, or of the great Barrier reef of Australia, represents the abrasive work of the lowered and chilled sea acting on a preglacial reef-plain, A, during the glacial period, then not only the whole width of a narrow preglacial encircling reef, A, [Fig. 113], ought to be cut away, as at B, but cliffs, D, D, truncating the spur ends, should have been cut all around the margin of the central volcanic island as well, and after the barrier reef is built up in the rising sea the spur-end cliffs should rise from the postglacial lagoon, E ... But cliff-rimmed central islands are not known, with the exception of New Caledonia and Tahiti, as stated above, and Tahiti is a 'lava-formed island', which Daly places with those that should 'stoutly resist abrasion'. [No account is here taken of certain islands in which two or three spurs are cut off in cliffs, while twenty or thirty are not cut off: these will be considered in my detailed report.] Hence it appears impossible to explain both wide and narrow lagoons by the theory here under consideration. Escape from this dilemma may be found by assuming that the corals of narrow-lagoon barrier reefs were not killed by lowered ocean temperature so soon as those of wide-lagoon atolls; but this assumption is too arbitrary to be acceptable. Hence, as far as these two lines of evidence go, it must be concluded that the corals were generally not killed during the glacial period, and that lagoon floors are generally not abraded platforms. (*Davis, 1915I, pp. 239–40*)

All this seems to show that if the glacial lowering of sea level endured long enough for an island, then undefended by living reefs, to suffer mature dissection by small streams, its shore must at the same time have been cut back in mature cliffs by the waves. Wave attack would be much stronger than valley weathering; hence if drowned valleys are one or two miles wide, the spur ends between them ought to be cut far back in high cliffs; this must be true whether the island rocks are hard or soft. But, as has already been said, sea cliffs are of exceptional occurrence around the shores of barrier-reef islands; the spur ends are as a rule very little cut back, and where low cliffs are seen, they are usually fronted by a visible rock platform, showing that the cliffs were cut at present sea level. Hence as far as these two additional lines of evidence go, it must be concluded, as before, that reef-building corals were generally not killed during the glacial period, and that the flanks of preglacial reefs were protected by growing corals at whatever level the waves beat upon them. Sea waves should not, therefore, be appealed to as of dominating importance in the abrasion of the lagoon floors now enclosed by barrier reefs or by atolls. (*Davis, 1915I, pp. 241–2*)

In judging this matter, it must be borne in mind that the valleys of small streams are steep-sided only during the early or immature stages of their whole cycle of erosion, while downward corrasion is still rather active; when downward corrasion practically ceases the slower process of lateral corrasion in the mature stage of the cycle allows the valley sides to weather to gentler slopes. Hence if a maturely open, flat-floored valley be partly submerged, its depth will not be indicated by prolonging its side slopes downward until they meet; but if a steep-sided immature valley be partly submerged, its depth may be fairly estimated in that way. Now the existing embayments of certain barrier-reef islands are enclosed by spurs of relatively steep slopes, which must be interpreted as the sides of relatively immature valleys, not so far advanced in their pre-submergence development as then to have had flat floors; hence their depth near the bay mouths may be fairly determined by the depth at which the downward prolongation of the bay sides intersect; and this is often as much as 600, 800 or perhaps 1000 feet. So great a depth of erosion cannot be reasonably ascribed to the revival of the streams during the glacial period. (*Davis, 1915I, p. 243*)

[The theory of outgrowing reefs on still-standing islands] is inapplicable to any of the barrier reefs that I have seen, (1) because the embayments of the central island, as in sector O, [Fig. 114], prove that the relative level of land and sea has

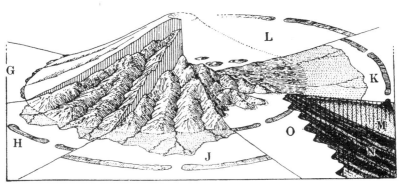

FIG. 114. Deduced sector diagram of an outgrowing reef enclosing a dissolved-out lagoon around a still-standing volcanic island (G, H, J, K, L). Sector O shows the comparative effect of subsidence resulting in barrier-reef upgrowth (M) on the volcanic basement (N) (From Davis, 1915I, Fig. 2 and 1928C, Fig. 28)

changed, as already stated; (2) because deltas advancing beyond the outer margin of the non-embayed central island into the lagoon, sectors H, J, K, as they should if the island had stood still, are conspicuously absent in all but two of the barrier-reef islands I have visited, although deltas of moderate size are always to be found in the bay heads; (3) because the lagoons, far from suffering excavation from solution, seem to be receiving new deposits by overwash from the outer reef, by outgrowth of inner fringing reefs, by inwash from stream deltas, and by organic growth on the lagoon floor; (4) because uplifted reefs, as far as known, do not show, as at M, the steeply inclined talus-stratification resting on non-eroded volcanic beds, as they should under this theory; (5) because no barrier reefs are known in which the central volcanic island is worn down to a lowland within a delta plain, as in sector K, and because no almost-atolls are known in which the residual central islands are vanishing lowland remnants, as in sector L; although both these forms should occur if, as Murray briefly suggested, barrier reefs are converted into atolls by the wearing away of the central islands.

The two islands which have deltas advancing into their lagoons are both large: one is Viti Levu, the largest of the Fiji group, where certain rivers of a considerable volume and of greatly increased flow after heavy rains have built their deltas forward, outside of the general outline of the island; the other is Tahiti, the largest of the Society group, where a present still-stand, recognized by Darwin as well as by Murray, is attested by a belt of alluvial flats and deltas that is more or less continuous around the island; but recent subsidence is well proved for both these islands by the form of their delta plains, which head between advancing spurs in strongly reëntrant spaces that were assuredly drowned-valley embayments before the deltas filled them. New Caledonia might be added as a third example of a large barrier-reef island which possesses good-sized deltas, but the deltas there are as a rule contained in the embayments and do not yet extend beyond the partly drowned headlands by which the embayments are bordered. Large deltas are similarly situated between the headlands of the embayed Queensland coast. Submergence unquestionably preceded delta formation in all four of these examples.

Reefs as Veneers on Wave-cut Platforms. – The explanation of barrier reefs as relatively thin veneers on the outer edge of wave-cut platforms around still-standing islands, was long ago suggested by Tyerman and Bennet (quoted by Darwin) and afterwards advocated by Guppy and Agassiz; it was extended by Wharton to the explanation of atolls, by supposing that in such cases the volcanic islands were completely truncated before corals were established on the resulting platform. The explanation is not satisfactory (1), because, as already stated, the occurrence of drowned-valley embayments in barrier-reef islands shows conclusively, as in sector O, [Fig. 115], that the relation of land and sea level has changed; (2) because the prevailing absence of sea cliffs, sectors H, J, around central islands, negatives, as Darwin long ago pointed out, the occurrence of a sea-cut platform of significant width; (3) because the depth of many lagoons is greater than that of wave-cut platforms of the same width; (4) because no almost-atolls

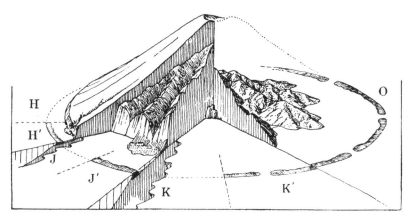

FIG. 115. Deduced sector diagram showing the stages (H, J, K) of marine abrasion followed by reef growth around a still-standing island (Guppy's theory). This theory requires deltas, cliffs, hanging valleys and barrier reefs (H′, J′) – a combination of phenomena which Davis believed to be practically unknown (From Davis, 1915I, Fig. 3, and 1928C, Fig. 31)

are known in which the residual central island is a cliff-rimmed stack, as in sector K'; (5) because no sufficient reason has been found to explain the delay in the establishment of reefs, as here required, while a broad platform was abraded; (6) because no uplifted reefs have been found in the form of relatively thin veneers on the outer edge of broadly abraded rock benches, as in Sections H and J. (*Davis, 1915I, pp. 231–2*)

After careful consideration I have had to discard (Daly's) theory, except insofar as it may have produced small results that are altogether subordinate to the larger effects of some more efficient cause. My reasons are in brief that, if the lagoons of large atolls have been abraded across their whole diameter of 20 or 30 miles, the central volcanic islands within narrow-lagoon barrier reefs should have been strongly cliffed by the lowered sea all around their shores, and their lagoon waters should now rise on the cliffed spur ends; but this is not the case; the spurs generally dip gently into the lagoon with small cliffs or none. Further, if the embayments of the central islands within barrier reefs occupy new-cut valleys that were eroded during the lowered sea-stand of the glacial period, the up-stream parts of such new valleys should be visible beyond their embayed parts, and should there appear as incisions beneath the floors of preglacial valleys, producing a valley-in-valley landscape; but in the hundreds of embayments that I saw, no such composite valleys occurred. Finally, it is doubtful if the lowering of sea temperature generally sufficed to kill the corals and expose the reef flanks unprotected from sea attack; for on the atolls of the Paumotus Agassiz found many instances of slightly uplifted reef limestones, which he regarded as 'Tertiary,' and hence as preglacial, on the inner border of the present encircling reefs; and in such cases the lagoon floors could not be the result of marine abrasion in glacial time. (*Davis, 1915A, p. 149*)

The Merits of Darwin's Theory. – The chief merits of Darwin's theory of subsidence are its simplicity, its breadth, its capacity to explain critical facts – the drowned valleys of barrier-reefs islands, as in sectors J, K, [Fig. 116], and the

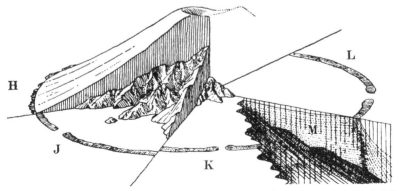

FIG. 116. Deduced stages of an upgrowing barrier reef (J, K) around a subsiding volcanic island (H), producing an atoll (L) in which the thick reef limestone rests unconformably on the volcanic rocks (M) (From Davis, 1915I, Fig. 9)

unconformable contacts of elevated reefs on eroded foundations, as in section M
– that were not known when it was invented; also the ease with which it may be
modified by the addition of supplementary processes without invalidating its own
essential process. It contains no postulate so arbitrary as that of a still-standing
reef foundation on a fixed ocean bottom, of which the rigidity is relaxed only to
permit elevation but not subsidence; it avoids the assumption of great uplifts in
other parts of the ocean to produce less great submergence in coral-reef regions,
and explains local submergence by local subsidence of the same amount; but it
easily accepts the possibility of some submergence or emergence being caused at
any time by a general change of ocean level from whatever cause; it understands
that other organisms than corals contribute to the formation of coral reefs; it
freely recognizes that subsidence may be interrupted by still-standing pauses or
reversed by elevation; it considers and accepts the possibility of the outward
growth of reefs during still-stands, but regards such outward growth as subordinate to upward growth, because existing sea-level reefs are generally of moderate
breadth and because the deltas which occupy the bay-heads of central islands
within barrier reefs are generally too small to project into the lagoon; it considers
and rejects the explanation of barrier reefs as veneers on abraded rock platforms,
because the central islands are not cliffed as they should be if this theory were
correct; it perceives the possibility of reefs being established on submarine banks,
built up to the shallowness required for coral growth by other kinds of organic
deposits, but it accepts this idea, to which Murray later gave wide application,
only for rapidly-growing banks near continental borders and rejects it for pelagic
banks, where the very slow up-building would be easily overcome by a less slow
submergence or outstripped by a less slow upheaval, and where wave-action might
sweep away fine deposits in greater depths than twenty fathoms.

Dana's Confirmation of Darwin's Theory. – The chief deficiencies in Darwin's
statement of his theory are: the failure to deduce the embayment of central
islands within barrier reefs and the unconformable contact of the reef mass with
its volcanic foundation as essential consequences of subsidence, and the omission
of the possibility of submergence, due to a rising ocean surface, as an alternative
to submergence due to a subsiding ocean bottom. The theory as first stated
involved a greater uniformity of subsidence over certain large areas than now
seems necessary, in view of later discovered facts ... (*Davis, 1915l, pp. 266–7*)

However, despite his opposition to the glacial-control theory, even at this
early date he was prepared to allow its possible efficacy in marginal belts.

Certain islands in the South Pacific, especially Norfolk island, east of Australia,
rise in bold cliffs from extensive shallow platforms, on which a good number of
soundings in depths 20 or 30 fathoms record 'crl' (coral) on the bottom. Certain northwestern members of the Hawaiian group, notably Midway, Lisianski
and Nihoa islands, rise from similar platforms. It is possible that these extensive
platforms may be to a greater or less degree the result of marine abrasion by the
lowered and chilled sea of the glacial period, and that the imperfect development
of reefs upon them to-day may be due to rapid submergence that would result, as
will be shown in a later section, from the combined action of island subsidence

and rising ocean level at the close of the glacial period. But it is the Marquesas
group at the eastern limits of the coral seas and to-day without barrier reefs, that,
according to the scanty accounts available, should present the effects of the
glacial-control theory in their most interesting development; for detailed charts
show that the embayments of these islands are separated by truncated spurs, such
as are demanded by the glacial-control theory, but such as are prevailingly absent
in the warmer seas. A special study of that group with particular reference to the
several rival theories of coral reefs would be highly instructive. (*Davis, 19151,
p. 247*)

Between 1915 and his trip to the Lesser Antilles in 1923 Davis, although
hampered by his preoccupations with the War, published some twenty
papers employing his observations in the Pacific in support of the subsidence
theory. His lines of attack are identical with those already mentioned,
although often much more detailed (e.g. 1920H; 1922N), but an important
feature of this period was the exploitation of the local flexibility inherent in
the subsidence theory, often in a most theoretical manner, relying heavily on
geometrical drawings.

Let a sea-level fringing reef be formed around a young volcanic cone at D,
[Fig. 117], and let the horizontal lines on the right side of the figure represent

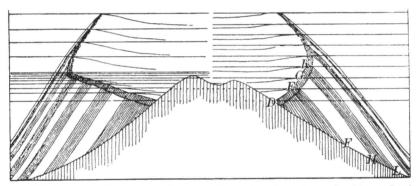

FIG. 117. Features of reef growth at varying rates of sea-level rise (From Davis,
1916C, Fig. 1)

successive levels of the sea with respect to a uniformly subsiding cone. Let the
subsidence be so slow that the reef at first grows upward and outward, DE, on its
own talus, which lengthens as subsidence progresses. The talus material consists
chiefly of fragments broken by waves from the corals and other organisms which
grow on the outer face of the reefs. In a given period of time a growing reef face
of given perimeter and 20 fathoms deep cannot produce more than a certain
maximum volume of new growth; and as some of the new growth stands firm to
build the reef upward, while some is broken off and washed inward to form the
flat reef and the lagoon shoals, only the remainder can be sacrificed for the talus.

Let this remainder be represented in section by DEF. If equal remainders are applied to talus building in equal time intervals, they may be represented by the shaded and unshaded quadrilaterals, EFHG, GHLK, etc., in which the breadth diminishes as the length increases.

So long as the reef growth is inclined outward, its increase in perimeter will, by providing a larger volume of coral growth, aid in supplying the demand for more talus material due to increase of depth; it is indeed conceivable that, if a reef foundation subsides very slowly and of the talus slants down to a level ocean floor of moderate depth, increase of perimeter may more than compensate for increase of talus length in the deepening water. In this case, outgrowth will continue at a constant or a lessening instead of at a steepening angle, as on the left side of figure [117]. (*Davis, 1916C, pp. 467–8*)

The six sectors, G to M, of [Fig. 118] graphically represent the successive changes thus inferred. Sector G shows part of a group of confluent volcanoes, the area of

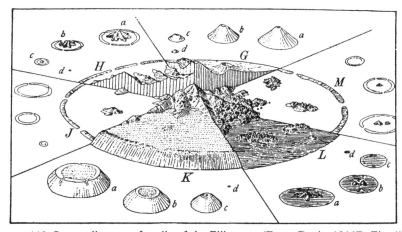

F I G. 118. Sector diagram of atolls of the Fiji group (From Davis, 1916D, Fig. 1)

which is about the same as that of the present Exploring Isles lagoon; several smaller cones rise near by; all together they resembled the present Lipari islands of the Mediterranean. After suffering prolonged dissection and partial submergence, the resulting embayed islands encircled by an upgrowing barrier reef are represented in sector H; this stage corresponds to that of Kandavu and its neighbors in southwestern Fiji. Down-sinking of the volcanic islands and upgrowth of the barrier reef continued until the total subsidence measured x + 600 + y feet, when but few volcanic hills survived, as in sector J; the small Gambier islands in a large lagoon southeast of the Paumotus, or the small islands of Budd reef in northeastern Fiji represent this stage. An uplift of 600 + y + > 100 feet then occurred, as in sector K: the resulting limestone plateau is typified by the uplifted atolls of the Loyalty group. The compound mass thus exposed to erosion was reduced over most of its limestone area to low relief surmounted here and there by residual hills, as in sector L; the hills of volcanic rock have smooth soil-covered slopes, those of limestone have steep cliffs and ragged crags. A recent submergence of 100 feet or more introduced the present conditions, as in sector M, where the lagoon floor has been smoothed by renewed deposition.

The exterior volcanic islands, a, b, c, d, of sector G, must have suffered essentially the same series of changes as the larger central volcanic islands. The atoll built on the largest cone, a, may have shown no volcanic knob in the stage of sector J, but one is afterwards laid bare in sector L, and its summit remains visible, along with a limestone knob, in the largest reef ring of sector M. Island b having a less initial height in sector G, it now shows only limestone knobs, sector M. Island c, beginning as a small volcanic cone, sector G, was deeply covered with limestone in sector J, and reduced to a low surface without high limestone knobs in sector L; and this is reasonable enough, for its area is not so large as any one of several uninterrupted lagoon areas within the adjacent great barrier reef. Its present reef is the result of upgrowth during the subsidence which transformed sector L into sector M.

Seven of the twelve outlying reefs near the Exploring Isles are true atolls, like the true atoll of sector M. The foregoing discussion gives, I believe, a nearer approach to a demonstration of the origin of these sea-level atolls by upgrowth during sub-recent subsidence than has been provided for any other sea-level atolls, except Funafuti which has been penetrated by a deep boring. The smallest reef in sector M represents mere ledges of coral rock on the Admiralty chart: the 100-fathom line around each of three such reef rocks near the Exploring Isles is less than a mile in diameter, and the rocks are mere points; hence these minute reefs, beginning on small volcanic cones in the stage of sector G, must have been extinguished in sector J, resurgent in sector K, a little enlarged by outward growth in sector L, and almost extinguished again in sector M. (*Davis, 1916D, pp. 473–5*)

Various combinations of diverse conditions may be imagined. For example, the succession of events may be as follows: (1) Moderate cliff-cutting during a still-stand period before reefs are developed; (2) moderate submergence and reef-upgrowth; (3) a second still-stand period, resulting in the smothering of reefs by outwashed detritus, and renewal of cliff-cutting; (4) further subsidence and renewed reef-growth. Tahiti seems now to be approaching the third phase of this succession, for, if the present still-stand that is attested by the alluvial lowland around the island border endures as long as the earlier reefless period of valley- and cliff-cutting, the lagoon will be overfilled, the smothered reefs will be abraded, and a new attack will be made by the waves on the cliffs at a higher level than before.

In any event, the only way of developing a barrier reef around a deeply dissected and non-cliffed volcanic island seems to be either to allow it to subside rapidly to a great depth while its reef is growing up, or to allow it to subside to a less depth after strong cliffs have been cut around its shore. And inasmuch as Réunion, Tutuila and the Marquesas, and Tahiti exemplify the second of these alternatives, the first alternative is regarded as the less probable of the two.

(*Davis, 1919G, p. 22*)

He even elaborated his own variation of the marine planation theory – the reef-plain theory – when commenting on Vaughan's idea that the Great Barrier reef was built on a Pleistocene marine platform:

> Imagine a recently uplifted and reef-free coast, or for convenience of graphic illustration imagine a recently constructed, reef-free volcanic cone, of which an undissected sector is shown at D, [Fig. 119]. The shore is attacked by waves

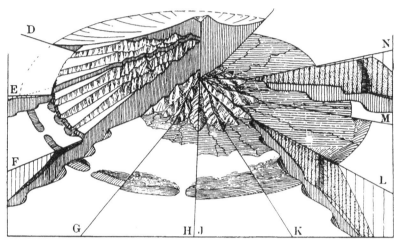

F I G. 119. Deduced stages in the formation of a reef-plain on a recent volcanic cone (D). Subsidence sets in (E) and, when the embayed valleys enable near-shore deposition, a narrow young reef develops (F). This is succeeded by a long period of still-stand during which the lagoon is filled by deposition and reef growth (G, H, J, K), resulting in the formation of a mature reef-plain (L). Eventually the detritus will smother and kill the reef and a marine platform will be cut across reef and lagoon deposits (M, N) (From Davis, 1917D, Fig. 5)

and a cliff and platform are developed, sector E, on which no reefs can be formed until submergence embays the radial valleys, sector F, when a narrow young reef will make its appearance, enclosing a lagoon where aggradation takes place. Tahiti, I believe, exemplifies this sequence of development. Let it be supposed that the reef thus initiated is maintained in a narrow or youthful stage with a widening lagoon by continued subsidence, until in sector G a long enduring still-stand period sets in. The youthful reef will thereupon widen to more mature form by outgrowth and overwash; the embayed valleys will be filled by deltas, which will advance farther and farther into the lagoon, as in sectors H, J and K; and the lagoon will be at last converted into a mature reef-plain, as in sector L.

The mature reef-plain is not, however, the limit of orderly change if the island suffers no variation of level. The detritus from the central island will be delivered by the streams to the outer face of the mature reef, and the growing corals may thus be smothered. When that eventuality is reached the waves will cut the dead reef farther and farther back, reducing it to a shallow submarine platform, as in sectors M and N. Finally the central island itself will be again attacked, unless before that goal is reached subsidence sets in anew, whereupon a rejuvenated reef will make its appearance on the outer edge of the widening platform, or on the outer edge of the narrowed plain, or on the island border, according as the submergence is large and rapid or not. (*Davis, 1917D, pp. 346–7*)

The Great Barrier reef of today does not in all parts of its extraordinary length rise from the outer margin of its platform, and this has been taken to show that the platform is an inorganic continental shelf and not an antecedent reef-plain. But no reason has been adduced to show why a reef should grow up at a moderate distance back from the margin of a smooth continental shelf, and why it should not grow up at a similar distance back from the margin of submerged and almost equally smooth mature reef-plain. On the other hand, reasons for the upgrowth of a young barrier reef at a certain distance back from the margin of a submerged reef-plain are suggested in sectors M and N of [Fig. 119]: a partly abraded reef-plain might, after rapid submergence, be too deep for coral growth at the outer margin of its abraded platform, but not too deep at the margin of the unconsumed part of the plain. Hence the control of the location of the present barrier reef may well have been the result of other conditions and factors than those which formed the platform that serves as its foundation.

There is one way, to which attention has not hitherto been called, in which the inorganic processes now at work in developing the continental shelf of New South Wales may be in the future, and may have been in the past, unfavorable to the development of a mature reef-plain along the Queensland coast. In that part of the coast where the Great Barrier reef of Queensland and the great continental shelf of New South Wales adjoin, the long-shore currents are at present engaged in forming extensive sand reefs, which appear to be extending northward. The sand reefs have already, in the relatively short interval since the latest flexure by which the coast was embayed, gained lengths of scores of miles. Hence in an earlier and much longer interval, sufficient for the partial peneplanation of an upflexed coastal belt, similar sand reefs might have extended much farther northward along the coast than they now reach; as fast as they advanced they would certainly kill all the reef-building organisms, and thereafter the waves might be able to cut away the reef. (*Davis, 1917D, pp. 349–50*)

Pausing occasionally to give massive reviews of the coral reef problem (1916I; 1919G; 1923C), Davis bent every effort on the one hand to explain local conditions not in keeping with the subsidence theory (e.g. the cliffs of Tahiti, New Caledonia and Réunion; Davis, 1916A) and on the other in showing the general applicability of the theory. He employed contemporary gravity determinations to support the isostatic subsidence of coral islands (1917B), marshalled evidence for the general subsidence of the western and central Pacific (1918G), and pointed to the thickness of uplifted Fijian reefs in support of subsidence during reef building:

The newest discussion of this problem is by Molengraaf, who calls attention to the results of recent gravity determinations, from which it appears that the volcanic islands of the Pacific 'as far as they have been studied are not isostatically compensated, and, without exception, show a larger or smaller positive anomaly of gravity ... These volcanic islands, rising ... as cones or groups of cones of considerable bulk, cannot always remain in existence; under the influence of gravity they will without exception yield and sink down slowly ... The yielding and slow sinking of the volcanic islands under the influence of gravity must be regarded as the cause of the downward movement of large amount and long duration which must be assumed in order to explain the formation of barrier-reefs and atolls in true oceanic regions'.

This appears to me an important suggestion, and one that is likely to remove the objections to Darwin's theory of coral reefs in so far as they are directed against a great subsidence of broad ocean-floor areas: for although Darwin himself was led by his coral-reef theory to infer the subsidence of such areas, it is clear from the original exposition of his theory that local subsidence of reef foundations will serve all its needs. It may be added that the accumulation of the great limestone masses of atolls upon slowly sinking volcanic foundations must aid and prolong their sinking; also that no comparable sinking of volcanic cones upon continents need be expected, not only because of the differences supposed to exist between the earth's crust in continental and oceanic areas, to which Molengraaf calls attention, but also because continental volcanoes suffer erosion, whereby their waste is carried away and widely distributed, while oceanic volcanoes, even if they rise for a time above sea level and suffer erosion, retain the waste from their summits on their flanks. (*Davis, 1917B, pp. 652–3*)

It is not, however, to be supposed that general warpings and deformations of the ocean floor, upward and downward, should be left out of consideration; such movements have surely taken place to a less or greater degree, particularly in the western Pacific, where coral reefs border continental islands. The integrated effect of all these causes of change in the level of the ocean surface cannot now be determined, because so little is known regarding the various factors of the problem: but nothing in the little that is known and in the much more that may be fairly inferred should be regarded as discountenancing the theory of upgrowing reefs on subsiding foundations, essentially as Darwin supposed. His primary theory of coral reefs holds good, although his supplementary theory of broad ocean-floor subsidence needs modification. (*Davis, 1917B, p. 654*)

The chief value of the ingenious Glacial-control theory may therefore be found not so much in its postulate of the prevalent stability of reef-bearing islands, or in its assumption that reef corals were killed and that reef-bearing islands were abraded while the ocean was chilled and lowered in the Glacial period, but in the emphasis that it gives to changes of sea level from climatic causes as a factor in the coral-reef problem; for it is manifest that if the post-glacial rise of sea level coincide in time with the subsidence of an island, the resulting submergence will be at an accelerated rate and of an increased amount; while if the fall of sea level occasioned by the oncoming of a glacial epoch coincide with a subsidence, the resulting submergence will be at a retarded rate and of a decreased amount. It cannot however be supposed that two processes so unlike in cause as external climatic changes and internal crustal deformations should be closely related in

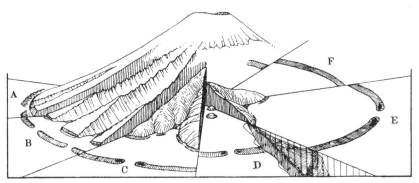

FIG. 120. Deduced sector diagram showing the upgrowth of a reef around a slowly-subsiding island. The fringing reef (A) becomes a barrier reef (B, C & D) with new fringing reefs along the shoreline. The latter seldom develop into inner barrier reefs during continued subsidence. The great reef thickness is shown in E, and in F an atoll has been formed (From Davis, 1928C, Fig. 16)

time; their coincidences must be fortuitous. Throughout the central Pacific the rate and amount of recent submergence have not been as a rule too great to be compensated by reef upgrowth; witness the abundant atolls and barrier reefs. But in the region of the Australasian archipelagoes compensation of submergence by reef upgrowth has frequently been unsuccessful; witness the rarity of well developed barrier reefs and the almost entire absence of atolls. As the climatic changes of ocean level must have been everywhere the same, the factors which have determined the success or the failure of reef upgrowth would appear to be the rate, the amount and the date of subsidence. (*Davis, 1918G, p. 203*)

The surviving islands are pertinent in the present connection because several of them show volcanic rocks unconformably covered by eroded limestones, remnants of the uplifted almost-atoll: the contact of the two kinds of rock is not a level platform at a depth of about 240 feet below the highest limestones; on the contrary, the contact exhibits rounded forms and moderate slopes such as characterize volcanic islands maturely dissected by subaerial erosion; and the occurrence of such forms beneath heavy limestones, 600 feet or more in thickness, clearly demonstrates the submergence of a previously eroded volcanic mass by over 600 feet, while the limestones were forming. Thus not only the recent history of the present barrier reef around Vanua Mbalavu, but also the Pleistocene history of the now dissected almost-atoll, of which Vanua Mbalavu is a remnant, testifies unqualifiedly in favor of Darwin's theory of coral reef and against all other theories. (*Davis, 1917A, pp. 478–9*)

In virtually every one of his publications during this period Davis continued to attack the glacial-control theory, devoting a number of papers specifically to this purpose (Davis, 1918C; 1919G; 1923C). His criticisms are those which have already been mentioned:

A. The great amount of assumed submergence of coastal valleys in tropical islands.

By constructing longitudinal sections and cross sections of the great valleys from the topographical maps, I have inferred that the amount of submergence since the valleys were cut down and the cliffs were cut back is some 800 feet. But whether submerged or not, the island of Hawaii ought not to be instanced in proof of the belief that the absence of spur-end cliffs on the central islands of barrier reefs elsewhere in the Pacific is due to the resistance of their lavas: to my reading Hawaii supports the other side of the discussion, namely, that volcanic islands which are indented by drowned-valley embayments half a mile or a mile wide must, like northeastern Hawaii, have had strong cliffs cut in their spur ends if they were exposed to abrasion during the considerable period required for the erosion of the now drowned valleys; and as such cliffs are prevailingly absent on reef-encircled islands the islands must have been protected from abrasion – that is, their reefs must have been clothed with growing organisms of some sort – while the now drowned valleys were eroded. (*Davis, 1918C, pp. 291–2*)

B. The absence of general evidence of spur-end cliffs cut at a lower Pleistocene sea level.

The lesson to be learned here is that the time which sufficed for the erosion of the steep-sided valleys of Tahiti sufficed also for the cutting of its great cliffs; hence, far from proving the power of fresh lavas to resist the breakers, Tahiti proves that, resistant as its lavas may be – and they certainly look very resistant where ledges are exposed in the cliff faces and on the valley sides – the waves of the trade-wind sea can strongly abrade the coastal slope of a volcanic island in the same period of time that is needed for the erosion of submature or mature valleys; and that the spur-end cliffs thus formed may be of such height that they will still rise hundreds of feet above sea-level after the cliff base and the platform in front of it and the valley ends between the cliff spur ends are submerged hundreds of feet below sea-level. (*Davis, 1918C, p. 293*)

C. His disbelief in tropical reef destruction during the Pleistocene.

It is not without careful consideration that I have been constrained to reject the assumption that reef-building organisms were so completely killed during the glacial period as to leave the reefs an easy prey to the waves. The assumption is, as already noted, certainly a plausible one at first hearing, and it merits careful examination; but as the result of the best examination that I have been able to devise it proves to be erroneous ... Perhaps the organisms were killed and the reefs were cut away on islands near the borders of the coral zone, as will be further considered later. Perhaps the corals were very generally killed, but the nullipores were not; if so, the nullipores, although unable to construct a reef alone, might cover and protect the exposed flanks of an already constructed reef for a geologically brief epoch until the corals could again establish themselves upon it. (*Davis, 1918C, pp. 297–8*)

D. The assumption that the form of lagoon floors is due mainly to deposition and not to low-level Pleistocene marine plantation.

> ... if the subsidence of an atoll or a barrier-reef island is relatively rapid, the reef will be somewhat submerged, the inwash of detritus from the reef will be active, and the increase of lagoon depth will be retarded; if subsidence is, on the other hand, relatively slow, the reef will be maintained at sea-level and will broaden its surface, sand islands will be formed along its edge, and inwash of detritus into the lagoon will practically cease; hence, in spite of slow subsidence the lagoon will not be rapidly shoaled. Thus, unless subsidence be unusually rapid, there appears to be a series of spontaneous reactions which tend to prevent lagoon depths from varying by large measures. Wherever unusually rapid subsidence occurs, the atoll would be drowned and converted into a submarine bank. The scarcity of such banks in the Pacific, as far as it is now explored, suggests very strongly that subsidence has rarely been unusually rapid; and slow, equable, or intermittent subsidence, being a near approach to long-continued stability, does not appear to be particularly incredible. (*Davis, 1918C, p. 302*)

Davis, in fact, devoted a considerable paper to lagoon floor evidence (Davis, 1923C), suggesting that the mountain-top islets of 'almost-atolls' might offer a quantitative measure of rates of subsidence when compared with rates of fluvial erosion of the remaining island.

E. The belief that the form of existing reefs can be interpreted with reference to present processes and present sea-level.

> First, regarding the exterior profile of continuous reefs: In view of the evidence already given, from which it appears that the production of platforms by abrasion during the glacial period is improbable if not impossible, and in view of the fair accordance between the depths at the outer margin of the uninclosed sector of a lagoon floor and the depth on the outer face of a reef where the change takes place from a gentle slope to a steep pitch, I am persuaded that the change of slope at 40 fathoms is not an inheritance from a time when that part of the reef lay at or close to the surface of the ocean, but, as already stated, a consequence of adjustment between the detritus to be transported and the agents of transportation with respect to present sea-level. Like the reef itself, the two elements of its exterior profile – namely, the gentler slope down to 40 fathoms and the steeper pitch below – have been brought by organic growth and by inorganic processes into a normal relation to sea-level. Secondly, regarding the free border of an uninclosed lagoon-floor sector of atolls and barrier reefs: In view of the alternate retardation and acceleration of submergence by the combination of prevalent subsidence with the periodic changes of ocean-level during the glacial period – for this is, in my mind, the chief value of the glacial-control theory – I am strongly inclined to regard any 'platforms' that may exist, now more or less aggraded, beneath incompletely or completely inclosed lagoon floors as nothing more than the

surface of earlier reefs normally broadened while submergence was so slow that narrow reefs were transformed into mature reef plains. Thus interpreted the present reefs are merely new, still young, and relatively narrow growths above their mature predecessors; narrow, because they have been developed while submergence has been accelerated. (*Davis, 1918C, pp. 307–8*)

Davis' support for the subsidence theory continued to flourish and he tended to lean increasingly upon the geological considerations of the sedimentary environments of reefs (1918C), reef structures (1919G) and reef foundations (1920H), although he generally avoided questions relating to rate of reef growth (except for 1926A):

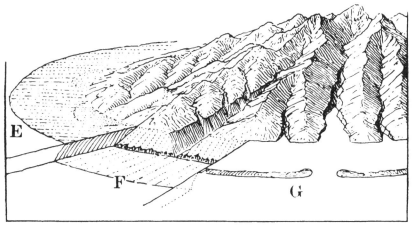

F I G. 121. Deduced sector diagram showing the effects (G) of a short (glacial) epoch of low-level marine abrasion (F) on a still-standing island with an outgrowing reef (E) (From Davis, 1928C, Fig. 39)

It is important that the observer who has opportunity of examining a dissected reef should locate the structural details that he may discover with respect to the total reef-mass; it is also important that he should bear in mind the expectable structure of reefs formed according to the several chief theories of reef-origin, . . . for he will thus be led to make special search for critical structures in their appropriate locations.

Thus if the great body of an elevated reef consist, . . . of steeply sloping layers of reef detritus mostly free from admixture with volcanic sands and gravels, resting conformably upon a non-eroded volcanic slope, T, and more or less complicated by slides, the reef should be explained as a product of outgrowth during a prolonged still-stand period. Darwin clearly recognized the possibility of reef-formation in this manner, but regarded it as seldom occurring, because it would not result in the formation of a reef-enclosed lagoon from 20 to 40 fathoms in depth. Murray attempted to overcome this difficulty by assuming, as Semper had before him, that the lagoon-cavity would be excavated by solution; but the assumption is not supported by the features of lagoons, as has been noted above.

On the other hand, an elevated reef may show a three-part structure, . . . The steeply dipping, exterior strata, T, may be formed of detritus chiefly derived from the reef, but with some fine sands and silts from the central island. The slanting layers may be sometimes complicated by slide-structure as in the preceding case; they may rest on a heavy deposit of volcanic detritus, . . . which should be associated with a buried cliff. The intermediate wall-like structure, . . . should contain much coral in place, as well as large and small fragments. The outward or inward slant of the wall appears to be dependent on the rate of subsidence during its formation (Davis, 1916C). The nearly horizontal interior strata, . . . may contain coarse sand near the reef-wall, fine lagoon deposits in the middle, and volcanic sands and gravels near the central island; and the whole mass, with the exception of some of the outer slanting layers, may lie unconformably on a rock surface of subaerial erosion. In such a case reef-upgrowth during prolonged submergence probably due to subsidence would be inferred. Irregularities in the reef-wall . . . would indicate changes in the rate of submergence. A horizontal outgrowth, . . . would occur during a long still-stand period, when delta plains, . . . might almost fill the lagoon. (*Davis, 1919G, pp. 27–8*)

This aspect of the coral-reef problem has a geological flavour and therefore will not be treated at length here; but it should be said that, excepting the small elevated reef at Walu bay, Suva, which lies conformably between inclined delta-like beds of volcanic muds containing marine fossils and locally known as 'soapstone', all the other elevated reefs as well as practically all the sea-level fringing reefs of Fiji lie in unconformable contact on foundation surfaces of subaerial erosion. The same is true of nearly all the elevated and sea-level reefs elsewhere of which I have found detailed records. Hence submergence, probably due to local subsidence of the foundation, must have, as a rule, preceded or accompanied reef formation. It may be noted in passing that the above-mentioned 'soapstones' are the uplifted marine beds which Guppy regarded as contradicting the supposed subsidence of Viti Levu; but in reality they seem to give evidence of two periods of subsidence; for as well as I could make out their structural relations they lie unconformably on an eroded slope of older volcanic rocks, and they are without question overlaid unconformably by the younger sea-level reef-limestones of the southern coast. (*Davis, 1920H, p. 212*)

Despite these geological sallies, based largely on his own fieldwork, Davis still continued to rely heavily on deduction.

In any case, the best opportunity for reef upgrowth will be found if, as above suggested, the pre-subsidence still stand or the very slow subsidence of the island is continued long enough to witness a great diminution of its original size by the subaerial erosion of deep and broadly matured valleys before any persistent reefs are formed, and therefore to witness also the abrasion of a broad platform, outside of which a great volume of detritus from the valleys and cliffs is deposited in deeper water. If subsidence then begins and an upgrowing reef is formed on or near the outer margin of such a platform, it will enclose a broad lagoon, . . . that cannot be filled with discharged detritus until after a very long stationary period; and before it is filled, subsidence may be renewed and continued, . . . whereby the embayments will be deepened again and the cliffs eventually submerged. In any case the discharge of detritus from the shortened valley-heads of a deeply dissected and greatly submerged island will be slower than it was from the same valleys when the island was larger and higher and the valleys were younger, longer and steeper. It thus appears that an upgrowing barrier reef, once well established, has an increasing chance of escape from being smothered by detritus; for be it remembered that continued subsidence diminishes the area and the height of the central island and increases the breadth of the surrounding lagoon. (*Davis, 1920H, pp. 216–17*)

Several reasons for the neglect of these essential considerations [for the solution of the coral-reef problem] may be suggested. One is that the investigators of coral reefs have often been zoologists untrained in geological inquiry. Another is that the physiographic principles which are involved in a critical study of the reef problem are not always familiar even to geological observers. A third and perhaps the most important reason is that few investigators of coral reefs appear to have taken the time necessary to think out the essential consequences of the several leading theories of reef-origin in order to discover which of the consequences are best supported by the facts. A fourth, as important as the third, is that observers have too often given their chief attention to the reefs, and have not attended sufficiently to the islands that they encircle. A fifth is that the origin of coral reefs is a very complicated matter, because many different factors may have a share in it, and many different solutions therefore appear possible. (*Davis, 1919G, p. 29*)

It is typical of the breadth and elasticity of Davis' conceptual frameworks, even in his later years, that, at the same time as he was attempting to show that Penck's uplift and erosional sequence was simply a special case of his own cycle of erosion, he was suggesting that the glacial-control theory could be incorporated in his reef subsidence theory for areas marginal to the tropics. Not only did Davis so skilfully exploit the inherent flexibility of the submergence theory that he was able to produce a 'custom built' explanation for the evolution of any coral island, but, even before his visit to the marginal belt of the Lesser Antilles, he was set on assimilating Daly's theory into his own extended theory of coral-reef development.

THE INVESTIGATION OF REEFS IN THE LESSER ANTILLES: 1923–26

Before visiting the Lesser Antilles in October and November of 1923 Davis published a general article on the marginal belts of coral seas (1923K) which he pictured as having reefs which grew up to ocean level in warmer non-glacial climates, but in which reef growth was inhibited, reefs cut away and the backing islands attacked by the waves of the lowered and chilled ocean during the glacial stages (1923K, pp. 181–2). The broad submarine banks of these marginal belts he still attributed to accretion rather than marine planation, but they were obviously a source of difficulty for him. His trip to the West Indies provided a period of concentrated theorizing the fruits of which

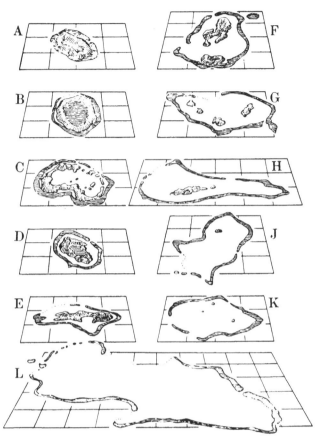

FIG. 122. Diagrams of elevated and dissected atolls in eastern Fiji (squares are 2 miles a side) arranged in order of increasing amount of erosion (A to L), showing the absence of a platform at 30–40 fathoms below the limestone crest, as required by Wharton's theory A – Naiau; B – Kambara; C – Fulanga; D – Tuvuthá; E – Namuka; F – Ongea; G – Yangasá; H – Oneata; J – North Argo; K – Reid; L – Great Argo (From Davis, 1928C, Fig. 32)

were immediately apparent in the first publication following his return which proposed a complex history for these marginal belts, although still dominated by subsidence.

Subsidence of Islands and Building of Banks. – All the maturely dissected volcanic islands have been more or less submerged since their eruptive growth was completed and while their dissection was advancing. This is shown by the embayments which enter the valleys or by the delta plains which replace the embayments. The submergence is ascribed to island subsidence and not to Postglacial ocean rise, because the embayed valleys are so maturely widened and, in a number of instances, their estimated rock-bottom depth at the bay mouth is so great that it is believed they could not have been excavated only during the Glacial epochs of lowered ocean level. Hence the banks around these islands have been built up by aggradational processes upon subsiding foundations. This makes it probable that, in the case of banks that bear no islands, the island foundations have been completely submerged.

The Islands Were Reef-Protected while the Banks Were Aggraded. – Coral reefs are at present imperfectly developed on the submarine banks, chiefly as discontinuous fringing reefs well back from the bank borders; but inasmuch as the interbay headlands have been only moderately and recently clift, it is believed that, during most of the period of island dissection and subsidence and hence during most of the period of bank aggradation, the islands have been protected from cliff-cutting by vigorous barrier reefs. It is only lately, and probably during the Glacial epochs when the temperature and level of the ocean were somewhat lowered, that such reef protection failed and that the headlands were clift. The fact that many of the headland cliffs plunge to a small depth below present sea level favors this view.

The Banks Represent Reef-Enclosed Lagoon Floors, Formed According to Darwin's Theory and Modified by the Processes of the Glacial-Control Theory. – The Lesser Antillean banks are therefore interpreted as having been formed by the aggradation of lagoon floors that were enclosed by up-growing barrier reefs over slowly subsiding volcanic foundations, essentially according to Darwin's theory of coral reefs; but the barrier reefs are thought to have been cut off by low-level abrasion, the lagoon floors are suspected of having been somewhat planed down, and the headlands are believed to have been moderately clift, when the corals of the protecting reefs were weakened or killed by the slightly reduced temperature of the lowered ocean in the Glacial epochs, as postulated in Daly's Glacial-control theory. The reef foundations, however, are not supposed to have been stable, as is also postulated in that theory; and the 30-fathom measure there accepted for the lowering of the Glacial ocean has not been confirmed by observation.

The Lesser Antilles Lie in the Marginal Belt of the Atlantic Coral Seas. – The absence of vigorous, bank-border barrier reefs around the island-bearing banks of today and of vigorous atoll reefs around the island-free banks is ascribed to the position of the Lesser Antilles in the marginal belt of the Atlantic coral seas, where the temperature of Postglacial time is thought to be not quite so high as that of Interglacial and Preglacial time. Some of the banks have been shown by Vaughan to be slightly benched at several levels, as if by abrasion in the rising Postglacial ocean.

Contrast between the Lesser Antillean Bank Reefs and the Typical Reefs of the Pacific Coral Seas. – The existing bank reefs of the Lesser Antilles, as here interpreted, are novices of recent establishment, and therefore not the direct successors of the inferred bank-border, lagoon-enclosing barrier reefs of Preglacial time. The two sets of reefs – the earlier ones of large depth, the later ones of very small depth – are probably separated by a surface of abrasion which truncated the earlier established reefs, and which serves as a foundation for the island-fringing novices of today. In this respect, the novice reefs of the Lesser Antilles are unlike the veteran barrier and atoll reefs of the Pacific coral seas; for there the reefs of earlier date do not appear to have been cut away in the Glacial epochs; and the present reefs, with their exterior slopes descending into deep water, are the direct successors of the earlier ones. It is only in the marginal belts of the Pacific coral seas that novice reefs of new establishment are found, comparable to the novice bank reefs of the Lesser Antilles.

Second-Cycle Islands. – The Lesser Antillean islands which are composed of volcanic and calcareous rocks, or of calcareous rocks alone, have evidently suffered uplift; they and their submarine banks are therefore at present in a second cycle of development. They appear, however, to have been formed, before their uplift, in the same manner as the islands and banks which have not been uplifted. Insofar as their calcareous areas have subsequently been worn down to lowlands and submerged, or abraded to low-level platforms in the Glacial epochs, the resulting banks may be treated as of a second generation. Several of these second-cycle islands merit brief description [e.g. Antigua and Barbuda; Davis, 1924A].

(*Davis, 1924J, pp. 206–7*)

An ideal scheme of island development may now be outlined in terms of which the islands of the Lesser Antilles may be genetically classified. The scheme includes one complete cycle of island history and part of a second cycle. The first cycle opens with the eruptive upgrowth of a volcanic island, presumably on a slowly subsiding foundation; as the cycle advances the island may be increased in size by new eruptions, which for a time more than make good the loss of size caused by erosion and subsidence; then after eruptions cease, the submarine slopes of the slowly subsiding and diminishing island are more and more encroached upon by accumulating lagoon deposits within an upgrowing barrier reef; and when the island is wholly submerged the barrier becomes an atoll reef enclosing an island free lagoon. A small episode of reef abrasion and of headland cliffing occurs at whatever stage of the cycle is attained when the Glacial period is encountered. The second cycle, introduced by upheaval or uptilting, may interrupt the first cycle at any stage in its progress. The second-cycle island thus exposed to developmental changes is not a young and growing volcano, such as was formed at the beginning of the first cycle, but an uplifted barrier reef with a central island or an uplifted atoll without a central island, and is composed of calcareous strata on a volcanic foundation; it then, with or without renewed volcanic activity, runs through a sequence of erosional changes associated with subsidence and renewed reef growth, again with an episode of abrasion when the Glacial period is encountered. Evidently, an island now well advanced in the second cycle was originally formed by volcanic eruption at a much earlier geological date than an island now at the beginning of its first cycle: but as the island in the second cycle is, after its time of upheaval, again subject to subsidence, it is inferred that young, first-cycle islands are also as a rule formed during the subsidence of the ocean floor. (*Davis, 1924J, pp. 208–9*)

Two years later Davis (1926D) produced his book on *The Lesser Antilles* in which he set out to show how the features of islands in the marginal belts (apparently between latitudes 25° and 30°) could be explained by a complex fusion of the modified glacial-control theory, with that of subsidence.

Attention must now be called again to the contrast, already briefly noted in an early paragraph of this essay, between the novice reefs on circuminsular banks in the marginal belts of the coral seas and the veteran reefs in the true coral seas; for without a clear understanding of that contrast, the new light thrown by the Lesser Antillean reefs upon the old coral-reef problem cannot be appreciated. The contrast will be first stated for the Pacific ocean where it is best developed. There the novice reefs of the marginal-belt islands and banks – namely, the smaller north-western members of the long Hawaiian chain and certain southern members of the Bonin and Riu-kiu groups in the North Pacific; Norfolk and Lord Howe Islands, and probably Raoul Island in the South Pacific – are discontinuous, one might almost say timid structures, which stand well back from their bank border; and the islands with which these reefs are associated have cliffed shores, sometimes embayed, sometimes not. But the veteran reefs that abound in the vast area of the Pacific coral seas are strongly developed, stalwart structures, which rise bravely, not to say boldly, from the outer border of their lagoon floors next to deep water; and, if the lagoons include islands, their shores are not cliffed unless the islands are young. These veteran reefs appear to have been growing continuously since middle or late Tertiary time and are therefore the direct successors of ancient ancestors, without more interruption in the succession than has been involved in a small downward migration of reef-building corals on the outer reef slope and a solutional erosion of the emerged reef crest with each lowering of the ocean, and a similar upward migration and rebuilding of the reef crest with each rise of the ocean in successive epochs of the Glacial period. On the other hand, the novice reefs are not the direct successors of ancient ancestors; their predecessors appear to have been cut away to platforms of low-level abrasion in the Glacial epochs, and the present reefs are therefore only of Postglacial date. They thus exemplify, within the narrow limits of their belts, certain processes of the Glacial-control theory of coral reefs, as set forth by Daly, although their islands do not exemplify the stability of reef foundations postulated in that theory, and their banks do not show the accordance of depth that is supposed in that theory to have been produced by low-level abrasion, little modified by Postglacial aggradation.

Thus the islands have gradually come to take their place in my mind as members of a developing sequence; so that, the sequence once being known, the general features of any island may be concisely learned from the stage it occupies in the sequence; and, the general features being thus learned, the special features can be easily apprehended as individual modifications of the general features.

The sequence in its simplest form begins with a volcanic island, built up from a slowly subsiding sea floor by a succession of active eruptions; after eruptive growth ceases, the island gradually diminishes by erosion as well as by subsidence, while a lagoon-enclosing barrier reef is built up around it; and finally, when the island is completely submerged, it is succeeded by an atoll, essentially as imagined by Darwin in his theory of coral reefs. But this simple sequence is affected by complications of three kinds. In the first place, an island instead of being completed in small size by a relatively short series of eruptions from a single vent, may, after subaerial erosion is more or less advanced on the earliest-formed cone, resume its eruptive growth from the same or from a new vent and become a larger composite island. In the second place, a simple or a composite island and its reef system may be uplifted or uptilted at any stage of its development by deformational forces, which thereupon interrupt the first-cycle sequence and introduce a second cycle; and the island thus upraised will then be again characterized by erosion, subsidence, and reef growth during the new cycle of development upon which it thus enters.

In the third place, on recognizing that the initial eruptive formation of different islands took place at various times from early Tertiary almost to Recent, it will be understood that, whatever the stage of either cycle reached by an island when the Glacial period intervenes, the waves of the lowered and chilled ocean, exemplifying certain postulates of the Glacial-control theory of coral reefs as stated by Daly, first truncate the encircling reef, because its corals are weakened or killed by the chilled waters; then continue their abrasive work by planing down the lagoon floor wherever it is laid bare, thus converting it into a rimless calcareous platform; and eventually, though for a short time compared to that needed for the mature dissection of an island, attack its headlands and cut them back in cliffs. Next, as the ocean rises to its normal level in Postglacial time, the abraded calcareous platform will be more or less aggraded, chiefly by organic detritus, and converted into a shoaling ban; but, inasmuch as the Lesser Antilles are believed to stand in the marginal belt of the Atlantic coral seas where, as will be explained in a later section, the ocean temperature has not been very favourable to reef growth in Postglacial time, no new bank-border reefs but only discontinuous, island-bordering reefs are developed. Coral growth in the Lesser Antillean chain is therefore now limited for the most part to discontinuous reefs along or near the island shores, well back from the bank border. The bank border, therefore, usually remains as rimless as when it was abraded. These relatively new and discontinuous bank reefs must be regarded as feeble novices compared to the stalwart veterans of the Pacific coral seas, which are now still vigorously continuing the growth that was begun in Preglacial time. They were not truncated in the Glacial

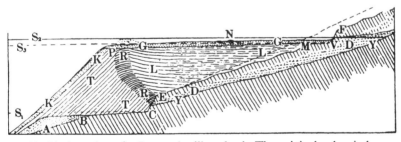

FIG. 123. Ideal section of a Lesser Antillean bank. The original volcanic basement slope (AF) is made up of an original submarine slope (AB), an original marine terrace (BCE) cut when sea level was at S_1 and a subaerial slope (EM). Subsidence of the island (giving sea level S_2) resulted in barrier reef (RR) growth, with talus (TT) and lagoon (LL) deposits; the latter extending as delta deposits (DD) up the embayed valleys (YY) cut with reference to the embayed valleys (YY) cut with reference to sea level S_1. The glacial drop in sea level (to S_3) led to the cutting of a platform (PMV) and to the deposition of a detrital embankment (KK). The post-glacial rise of sea level led to the growth of novice reefs (N) and of detrital deposits (GG) on the glacial marine platform (From Davis, 1926D, Fig. 1)

period, although they were probably more or less dissected by solvent erosion when the ocean was lowered. It is only in the marginal belt of the Pacific coral seas that feeble novice reefs are found, like those of the Lesser Antilles.

Thus interpreted, the circum-insular banks of the Lesser Antilles are complex structures, as shown in [Fig. 123]. Their foundation must be in nearly all cases a volcanic mass, AF, which should include a part of the originally submarine slope, AB, below sea level, S_1, at the time when the volcano became extinct, and a part of the originally subaerial slope, EM, which was more or less eroded before it was submerged and possibly also an early-cut cliff and platform, ECB. The greater part of the terrace-like mass of the banks will consist of a barrier reef, RR, with its external talus deposits, TT, and its enclosed lagoon deposits, LL, the latter being extended inwards by the delta deposits, DD, in embayed valleys YY. But a surface fraction of the reef and its lagoon floor will have been cut away by low-level abrasion in the Glacial epochs, with respect to the lowered ocean, S_3, and thus a gently inclined calcareous platform, PM, will have been produced, widened inward by a cliff-base belt of abraded volcanic rocks, MV, and outward by a detrital embankment, KK. In Postglacial time the abraded platform will have been aggraded by novice reefs, N, and detrital deposits, GG, into the present bank, over which the spurs of the island will be cut off in the plunging cliffs, FV. The bank alone is a matter of observation by sounding; the abraded platform at a small depth below the bank surface and the sloping volcanic foundation at a greater depth are matters of inference. The three terms, foundation, platform and bank will be used as thus defined throughout this essay. It is easily conceivable that platforms of low-level abrasion may have been produced at successive levels in successive Glacial epochs around a progressively subsiding island and that then they may have been aggraded in successive non-Glacial epochs; but this complication of the problem is not further considered. (*Davis, 1926D, pp. 12–16*)

Despite the complexities of Davis' theories it is clear that they are in the main purely deductive ones unrelated to actual field observations except of the most casual kind supported by the study of hydrographic charts. Davis is now in his element! A first-cycle sequence of forms is recognized – Saba, The Saints, Redonda and submarine banks (Davis, 1926D, pp. 35–40), and of second-cycle forms – Marie Galante and Sombrero.

This interpretation is necessarily in large measure hypothetical, for most of the conditions and processes it involves are lost in the past. Nevertheless, in so far as the interpretation is correct, the past conditions and processes involved in the interpretation were facts in their own time just as truly as are the observable conditions and processes of today. That the hypothetical interpretation may be accepted as essentially correct appears from the simple manner in which it assembles and correlates a variety of insular features which at first sight seem to have little in common. (*Davis, 1926D, p. 44*)

He speculates on the limiting depth of submarine bank deposition (pp. 58–61), compares the cliffs of Dominica with those of Tahiti and associates cliffed headland coasts with the inhibition of recent reef growth by valley mudflows. Davis found Antigua of particular interest as here his geological observations seemed to support his physiographic conclusions.

Antigua was the most instructive island I visited in the Lesser Antilles: it not only exhibits in its physiographic features an interesting stage in a second-cycle sequence of island development but presents in its geological structure a most encouraging confirmation of the scheme of a first-cycle sequence. Indeed, it reveals the deep under-structure of coral-reef lagoon deposits better than any of the 35 reef-encircled islands seen by me in the Pacific in 1914, for most of those islands had not been elevated, and those which had been elevated gave no such exhibition of their under-structure as Antigua affords. Its well displayed series of volcanic and calcareous strata is far more demonstrative of the conditions under which atolls are formed than is the famous Royal Society boring in the reef of Funafuti, an atoll in the Ellice group of the Pacific; for while that boring, which was but little more than 1000 feet deep and did not reach a volcanic foundation, yielded only a slender rock core whose interpretation has given rise rather to dispute than to agreement, the beveled section of the Antigua monocline is broadly open to deliberate observation in many outcrops, road cuts, and quarries in its surface of scores of square miles, as well as in many headland cliffs around its shores; and the evidence that its tuffs and limestones give for long continued and great subsidence during this deposition is not to be questioned. . . . Its upper 1500 feet of strata show conclusively, by their fossils as well as by their structure, that they were built up as persistently shallow-water deposits on a bank or lagoon floor by the slow accumulation of organic detritus on a slowly subsiding foundation; and the presence of a reef around the bank while it was building is made so highly probable by various converging lines of evidence that I am persuaded the bank will, in time, come to be accepted as representing the deposits of a former barrier-reef lagoon and later of an atoll lagoon, and indeed as the most instructive section of an atoll thus far known in the oceans. (*Davis, 1926D, pp. 164–5*)

Finally Davis turned to the broader aspects of the speculations on the Lesser Antilles. He assumed, firstly, that the first-cycle islands of the inner island arc are younger than the second-cycle ones of the outer arc, a line of volcanic activity having migrated westward from an earlier position on the outer curve to a later position on the inner curve (Davis, 1926D, p. 177. This was paralleled by his rather similar thesis on the Fiji islands: Davis, 1927B) and, secondly, that the islands of the Lesser Antilles had been built up from a subsiding foundation (a previous possible land connection with South America). Davis' theory, leaning heavily on reef upgrowth and lagoon infilling was set in opposition to the earlier work of Vaughan because Davis (pp. 182–5) believed:

1. That reefs, now submerged and buried were already in existence in mid-Tertiary time.
2. That the thick reef deposits rest uncomformably on their foundations.
3. Barrier reefs developed from fringing reefs.
4. Lagoon infilling was associated with Tertiary reef upgrowth and that

detritus had accumulated on various submarine banks since they were prepared by low-level Pleistocene marine abrasion on the 'pre-existent circuminsular, terrace-like masses'.

5. That low-level marine abrasion 'has recently acted to transform great barrier-reef, terrace-like masses into circuminsular islands'.

The same contrast is found in the Atlantic, but it is less marked, because the true coral seas of that ocean, as indicated by the occurrence of stalwart reefs bordering deep water, are imperfectly developed; they are limited to the Caribbean sea and to parts of the Gulf of Mexico. Yet in the Atlantic the contrast is of precisely the same kind as in the Pacific; the novice reefs of the Lesser Antilles appear to be of Postglacial date and are associated with cliffed islands; they are discontinuous and weak as compared with the bank-border reefs of Cuba. The Cuban reefs have probably grown up from a considerable depth, as there is no indication that their predecessors were cut away by low-level abrasion; the Lesser Antillean novices seem to be based on pre-existent platforms produced by low-level abrasion and afterwards shoaled by aggradation; and, as Vaughan has pointed out, in the production of these platforms the present-day reefs have not been concerned. . . .

. . . The novice reefs of the two oceans are closely alike; and the novices of one ocean may be well believed to have the same kind of abraded platforms beneath them and the same Postglacial origin as those of the other; but the novices should not be accepted as exemplifying the origin of the veterans. In both oceans the novices of the marginal belt are strikingly unlike the veterans of the coral seas. It is therefore important that these two classes of reefs should be distinguished. (*Davis, 1926D, pp. 194–7*)

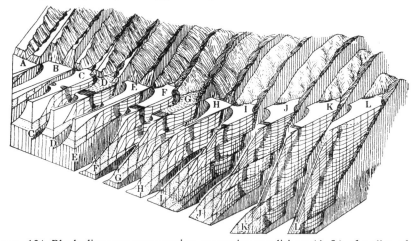

F I G. 124. Block diagrams representing successive conditions (A–L) of a slice of a subsiding volcanic island. The sections show volcanic rocks (close vertical lines), abraded detritus (dotted), coral reefs (close horizontal lines), lagoon deposits (spaced near-horizontal lines) and reef-face deposits (spaced slanting lines). A–D represent slow subsidence, E rapid subsidence and abrasion, F–G slow subsidence, H rapid subsidence and abrasion, and I–L slow subsidence (From Davis, 1920H, Fig. 9)

Davis provided a penetrating analysis of his own work:

> It is felt that these several lines of evidence converge upon the conclusion above announced and that they are well enough argued to show that the conclusion has not been adopted hastily or without good warrant. It may, indeed, be declared that this conclusion has the merit of a successful theory in that it brings order out of confusion by providing for so many apparently independent items of observation a systematic arrangement in a well coordinated whole. When such a theory has been framed, the facts that pertain to it are no longer regarded only as so many isolated, individual occurrences; they fall into their proper relative positions so naturally, become the essential elements of a reasonable entity so spontaneously and so unconstrainedly, and in these positions supplement and support one another so helpfully that one wonders why they ever seemed unrelated. Yet in my own experience an incubation period of five months elapsed after observational exposure to the facts above detailed before their coordinated meaning 'broke out' upon me. Since then the islands, banks, and reefs of the Lesser Antillean chain have seemed almost transparent, so great has been the aid given by the adopted explanation of their origin in penetrating their structures and in recognizing the past conditions and processes through which they were brought into being. Yet the large share of the adopted explanation, which, apart from the facts of direct observation, consists of mental inferences, is after all only a figment of the imagination. Perhaps it is incorrect. Time will show. (*Davis, 1926D, pp. 184–5*)

There is one simple observation we would like to add to the above. Most of the 'novice reefs' and 'marginal reef areas' mentioned by Davis in the Pacific lie outside the tropics, while the 'master reefs' lie near the equator. Yet his postulated 'novice reefs' and 'marginal-belt islands' in the Atlantic Ocean are in latitudes 12°–18°N and actually lie well equatorward of Cuba (20°–23°N), the area given as the ancient master region of grand continuous reef-building. It seems to us that in this very important respect, and in its climatic corollaries, that the novice reefs of the two oceans are *not* 'closely-alike'.

THE GRAND SUMMARY

At the age of seventy-eight Davis (1928C) published a large volume reiterating and elaborating his views on the coral reef problem, beginning with what he considered to be the 'observable features' of coral reefs, including a new emphasis on the direction of marine currents:

> The facts above presented regarding coral reefs, most of which have been known for the greater part of a century, warrant a number of generalizations. First, coral reefs are the product of congeries of lime-secreting organisms, among which polyps known as corals and algae known as nullipores are the most abundant; the former presumably provide the greatest volume of reef-building material, while the latter are believed to be of essential importance in binding the material together on the reef face so that it shall better withstand the assault of storm waves. The growth of these reef-

building organisms is possible only in pure sea water of a relatively high temperature and the corals are usually limited to depths less than 25 or 20 fathoms.

Second, reefs may be initiated on shores of firm rock, free from detritus outwashed from the land; they may also grow up from small depths on submarine shoals, provided firm rock or little-disturbed shells, gravels, or cobbles are present on which the floating embryos of reef-building organisms can attach themselves.

Third, the face of a fringing reef on the shore of a reef-enclosed lagoon is veneered by growing corals for only a few fathoms below the lagoon surface, because its lower slope is cluttered over with detritus which the lagoon waves do not remove in spite of the violence which such waves occasionally attain; but fringing reefs that front the open ocean resemble barrier and atoll reefs in that their exterior growing face extends downward to a greater depth.

Fourth, it is from the growing face of barrier and atoll reefs that detritus is supplied to be swept by waves outward, especially at times of gales and storms, down the gentle slope to the steep pitch, where it is believed to form a talus that descends to great depths; and also inward upon the reef flat and across it into the lagoon. The profile of the gentle slope and the steep pitch is believed to be in adjustment to the action of waves and currents. The growth of a reef as a whole must be much slower than the growth of corals and other organisms on the reef face.

Fifth, the presence of fine detritus, mostly calcareous, on lagoon floors indicates that the large volume of sea water which, surging over the reef flat, drifts slowly through the lagoon and finds exit chiefly through the leeward passes, is not able to dissolve all the detritus and carry it away in solution. Hence lagoon floors suffer aggradation rather than excavation. The occasional disturbance of the lagoon waters by storm winds appears to cause a fairly equable distribution of fine detritus over lagoon floors, as well as to drift the finer sediments toward the leeward breaches in the reef and to export the finest sediments to deep water.

Sixth, reefs are usually of more continuous growth on the windward than on the leeward arc of their circuits; this believed to be because the reef-building organisms on the windward side are constantly receiving a good supply of food and of oxygen in the on-drifting ocean water, while on the leeward side the supply of food and oxygen is somewhat depleted; also because only clear ocean water comes to the reef on the windward side, while much fine detritus is drifted toward the leeward side.

Seventh, it is to be inferred from the above generalizations that new fringing reefs will be formed on any new shore line where fitting conditions for reef attachment and growth are provided; that a reef thus formed will increase in breadth by outward growth on a stationary coast; that it will increase in thickness by upgrowth on a slowly subsiding coast; and that, if the upgrowth be vertical or outward, the fringe will separate itself from the receding shore and become a barrier reef. On the other hand, the growing face of a reef will migrate down the slope of a slowly rising coast. But it is also to be inferred that a rapid or sudden subsidence may submerge a reef to such a depth as to 'drown' the reef-building organisms, whereupon a new fringing reef may be formed at the level where the shore line thereafter rests; and, on the other hand, that a rapid or sudden upheaval will cause the emergence of a reef, after which a new fringing reef may be formed on the exterior slope of the emerged reef. (*Davis, 1928C, pp. 20–1*)

Discarding alternative theories (1928C, p. 88), Davis restates Darwin's theory of upgrowing reefs on intermittently subsiding foundations and deduces its extensions, with particular stress on the removal of island debris:

It thus appears that the long continuation of the processes ordinarily at work upon a coral reef should, if it stands stationary, result not only in closing passes and breaches but also in broadening the reef flat by outgrowth on its exterior slope and by deposition on its interior slope. After the passes and breaches are closed so that no exportation of fine sediments is possible, the lagoon of a still-standing reef would be shoaled and eventually filled. If the original reef were a barrier, the advancing deltas of the central island, the mountainous volume of which is often many times greater than that of the lagoon waters, would aid in converting the lagoon into a plain. Guppy calculated, on the basis of observations made during a visit to Keeling Atoll in the eastern Indian Ocean – the only atoll that Darwin studied – that 5000 tons of detritus are annually washed across the reef flat from the outer face into the lagoon and that at this rate the lagoon will be filled in about 3000 years ...; but no account was taken of silt exportation. (*Davis, 1928C, p. 18*)

Darwin's theory of coral reefs was based on a well assured condition; namely, on the limitation imposed on reef formation by the small depth at which .reef-building corals can live. Its leading postulate, island subsidence during reef up-growth, was wholly reasonable; but the theory as the young naturalist presented it included only a few of its deducible consequences. After its invention he evidently searched for other of its consequences than the manifest one of the transforma-tion of fringing reefs into barrier reefs and of barrier reefs into atolls, which the theory was invented to explain; but only three additional consequences were announced, and two of them were not confronted with the appropriate facts, because those facts were not known to him. One of these two was the great thickness that barrier and atoll reefs should possess if they have been formed by upgrowth during the deep subsidence of their foundations; the other was the internal structure that reefs thus formed should possess. A third consequence, namely, the drowning of reefs by rapid subsidence, was thought to be verified by the submerged Chagos Atoll in the Indian Ocean. Several other deducible conse-quences will be pointed out in the following chapter. Darwin's theory therefore did little more, apart from correlating many dissimilar facts, than explain the phenomena that it was invented to explain; but it did this so simply, so beauti-fully, that it was widely accepted as giving a counterpart of invisible past pro-cesses, in so far as the formation of coral reefs is concerned. That the theory was not criticized as incompletely developed is probably because the training of those who accepted it had not been especially directed to the importance of deduction and confrontation in geological theorizing. (*Davis, 1928C, pp: 39–40*)

... it is evident that a maturely dissected island, exhibiting radial ridge-and-valley forms such as are so generally seen within barrier reefs but possessing no wave-cut cliffs, can be explained only by supposing that a great volume of rock detritus has been removed by subaerial erosion; and the disappearance of the detritus therefore demands reasonable explanation. Rough estimates that I have made of the volume of detritus removed from certain well dissected Pacific islands in the coral seas indicate that it is from 50 to 100 times greater than that of the lagoon waters within the near-by barrier reefs and from 500 to 1000 times greater than that of the bay-head deltas now visible . . . The disappearance of large volumes of detritus from maturely dissected islands and the development of barrier reefs around their non-cliffed but embayed shores must therefore be accounted for in some reasonable manner in any competent theory of coral reefs . . .

The only reasonable manner of accounting for the disappearance of the detritus that has been discharged from the valleys of a maturely dissected, non-cliffed, well embayed and now reef-encircled island is to suppose, as I have elsewhere explained . . . that the island has been subsiding during the detrital discharge . . .

Under this view of the case the larger part of the discharged detritus must be supposed to lie now at a considerable depth on the non-eroded, submarine slopes of the island; and the smaller remainder must be found in the delta and lagoon deposits that have been accumulated at rising levels during subsidence, after a lagoon was enclosed by an upgrowing reef. Manifestly necessary as this explanation is, it has hitherto had no place in the discussions of the coral reef problem. It clearly gives good reason for accepting a decidedly greater measure of island subsidence than that indicated by the inferred rock-bottom depth of valley-mouth embayments. Indeed, when an observer views a maturely dissected and well embayed island from its barrier reef and realizes the vastly greater volume of discharge detritus than of the lagoon waters, he must gain a strong impression of the evidence for subsidence thus afforded. To my regret, that impression in the presence of such an island was not included among the experiences of my Pacific voyage, for this particular verification of Darwin's theory did not occur to me until several years after my return home. (*Davis, 1928C, pp. 51–2*)

An extended criticism of the general application of the glacial-control theory is tempered by its limited acceptance in marginal belts, and Davis devotes three highly-deductive chapters to the latter topic, noting the rarity of uplifted and truncated stationary coral islands.

Yet, inasmuch as the margin of the coral seas is determined by a limiting temperature of ocean water it can hardly be questioned that the reefs situated in a marginal belt were killed and cut away in the Glacial epochs. Thus restricted, this active element of the Glacial-control theory is, in my opinion, the most original and valuable contribution to the coral reef problem that has been made since Darwin's early time. It remains therefore to evaluate, if possible, the breadth of the marginal belt of the coral seas over which low-level abrasion of dead reefs actually took place. (*Davis, 1928C, p. 112*)

The chief conclusions of the preceding sections are as follows: Stationary islands in the cooler seas should, if of recent origin, have sea cliffs rising from narrow platforms of normal abrasion. If of earlier origin, they should have higher cliffs rising from broader platforms, around which low-level platforms are to be expected. If of ancient origin, the islands should be completely truncated; the larger ones should have normal platforms rimmed with low-level platforms, while the smaller ones might be wholly reduced to low-level platforms. So far as any cliffed islands survive, they should be but slightly embayed, if embayed at all.

Stationary islands in the coral seas should, if young, be cliffed, slightly embayed, and encircled with narrow reefs. If of earlier origin, their cliffs should be weathered and somewhat dissected, their embayments should be better developed although not of great inward-reaching length so long as the island is mountainous, and their reef-enclosed lagoons might be partly occupied by normal platforms. If of ancient origin, they might be today represented by almost-atolls, in which the surviving islets should be cliffed on the outer side; or by true atolls, in the center of which the surviving part of a normal platform should stand as a shoal, unless the island was small enough for such a platform to be completely cut away by low-level abrasion. The reef flats might have acquired a considerable width.

Stationary islands in the marginal belts should, if they have not been completely cut away, be cliffed; and the cliffs should usually plunge below normal sea level. In so far as such islands survive in good size, they would be of rather massive pattern and moderately embayed or not embayed at all. Instead of having plunging cliffs, the islands may be surrounded by a normal, cliff-backed platform, the outer part of which may be cut back and rimmed by the submarine bluff of a low-level platform. The reefs of Postglacial growth will probably be slender and incomplete novices of the bank-barrier or bank-atoll kind. In all three regions, the expected platforms, either at normal or at low level, will not be directly observable, because they will have been converted into banks by Postglacial aggradation of unknown thickness.

The features of rising islands are so simple that they need not be summarized here.

Subsiding islands in the cooler seas should be cliffed, in so far as they survive;

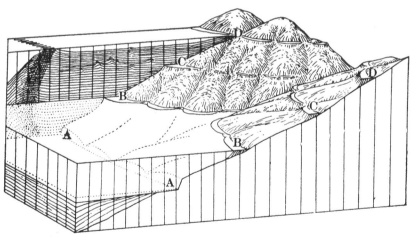

FIG. 125. Block diagram to show the effects of rapid coastal subsidence and elevation, compared with the barrier reef developed during slow subsidence (back block). BB represents the present fringing reef, CC and DD elevated unconformable fringing reefs, and AA the possible original shoreline with reference to which the embayed valleys were eroded (From Davis, 1928C, Fig. 19)

their submarine profiles should be more or less distinctly benched. After their submergence, continued subsidence would lower them to depths where their discovery by sounding would be unlikely. Hence as a rule only the younger subsiding islands of these seas would be known.

Subsiding islands in the coral seas would, if very young, be cliffed and reefless. If somewhat less young, they should be cliffed, moderately embayed, and reef-encircled. If of earlier origin, well embayed with non-cliffed spurs and encircled by barrier reefs of variable breadth enclosing lagoons of ordinary depth. If of ancient origin, they might be represented by almost-atolls with round-topped, non-cliffed islets or by atolls which would memorialize vanished islands.

Subsiding islands in the marginal belts might, if they have subsided slowly and slightly, resemble stationary islands; for they would have plunging cliffs surrounded by a bank of ordinary depth; their reefs would be of the bank-barrier kind. But if they have lately subsided more rapidly, they might be of skeleton outline, rather strongly embayed and with cliffed spur ends; their surrounding banks might be benched at various depths.

It is not to be questioned that the discussion of this chapter is largely speculative, especially in its sections concerning the marginal belts of the coral seas. It is intentionally so. But the more clearly the characteristics of imagined stationary and subsiding islands in the several seas are deduced, the more definitely the characteristics of actual islands and banks in the several seas can be studied . . . In so far as the interpretation is true, it gives a meaning to the peculiar features of marginal belt islands and to their singular bank reefs that has not been heretofore perceived; and it gives also a quantitative or areal value to the influence of Glacial changes in ocean level and temperature upon the growth of coral reefs, a value that is found to be very different from that assigned to it under the Glacial-control theory. The interpretation is, moreover, fundamentally unlike the interpretations of that theory in that it associates the influence of Glacial changes of ocean level and temperature with subsiding instead of with stationary islands; but in justice to the inventor of that theory it is a pleasure to state that my own understanding of the possible entrance of Glacial influences into the coral reef problem was wholly derived from his discussions of them.

It will be later shown that certain banks in the marginal belts are of so large a size as to demand a considerable lapse of geological time for their production by the combined action of reef upgrowth and lagoon-floor aggradation on the one hand and of low-level abrasion on the other; and this would seem to indicate that barrier reefs had been well developed there in Preglacial time. Thus countenance is given to the belief that the atolls of the coral seas also were probably represented by atolls or by barrier reefs in Preglacial time; and this is another respect in which the present interpretation differs from the Glacial-control theory.

(*Davis, 1928C, pp. 139–41*)

The preceding survey of the marginal belts in the Pacific, Indian, and Atlantic Oceans leads to several novel results. First, the belts certainly exist, although their boundaries are regrettably indefinite over broad oceanic regions, because of the absence of islands. Second, the embayed islands in these belts give good evidence of instability with prevalent subsidence; the islands that bear elevated reef limestones also testify for instability. Among these, Oahu is a strong witness for subsidence which, before a recent elevation of small measure, was in progress even during the eruptive growth of its younger volcanic dome. Third, the cliffed islands and stacks give no direct evidence of subsidence, but the presence of banks around them as well as around the embayed and the limestone-bearing islands, while similar banks are practically wanting in the cooler seas, leads to the belief that the banks represent great, terrace-like, coral reef masses, formed by reef upgrowth and lagoon infilling during the subsidence of their foundations and recently modified in a superficial and subordinate manner by low-level abrasion in the Glacial epochs and by aggradation in Postglacial time.

Fourth, the cliffs by which the marginal-belt islands are so generally characterized show clearly that the low-level abrasion to which the cliffs are due was not similarly operative in the coral seas, where, except on a small number of young volcanic islands, cliffed shores are wanting. Fifth, it is disappointing to find that in these marginal belts, precisely where low-level abrasion acted most effectively, the level at which it acted cannot be determined, chiefly because of the Postglacial aggradation of the platforms that it is believed to have produced and partly because of possible Postglacial changes of island level. Sixth, inasmuch as the bank atolls of the marginal belts are believed to crown islands that have been more or less abraded in the course of their subsidence, they may differ from the typical atolls of the coral seas in having, beneath their banks, moderate-sized platforms of volcanic rocks instead of the rounded mountain tops such as are supposed, according to the subsidence theory, to occur under atoll lagoon floors in the coral seas. (*Davis, 1928C, p. 217*)

The study of the Atlantic marginal belt in its course through the Lesser Antilles has been particularly gratifying in several respects. Each member of that island chain proves to be worthy of special investigation for itself alone, because divers problems of eruption and erosion, of emergence and submergence, of reef growth and reef abrasion are more or less completely represented there. But each member gains an added interest when a comparison of all shows them to exemplify various stages of development in a simple but comprehensive scheme of island evolution. And, when the island chain as a whole comes thus to be understood, it is found to have a still larger value because it is so clearly characterized by certain peculiar features found to characterize also the islands in the marginal belts of the Pacific; namely, features that indicate, first, a continuance of successful reef growth on subsiding islands in Preglacial time; second, a relatively short-lived inhibition of reef growth with resultant low-level abrasion in the Glacial epochs; and third, a more or less successful revival of reef upgrowth in Postglacial time . . .

Thus understood, the marginal belts not only add a most interesting complexity to the coral reef problem of Darwin's time; they provide also a most salutary correction for the excesses of the Glacial-control theory. The cliffed islands of the marginal belts truly validate the occurrence of low-level abrasion, as postulated in that theory, and thus substantiate the most significant process by which the study of coral reefs has been enlarged since the subsidence theory was formulated. But the cliffed islands at the same time restrict the area over which such abrasion has acted, and many of them discredit the associated postulate of insular stability by which the Glacial-control theory is unfortunately limited.

It is certainly significant that the islands and banks here described verify the expectations deduced . . . regarding subsiding islands in a marginal belt where reef growth has been alternately favored and inhibited in the Glacial period. And it is no less significant that the circuminsular banks of these cliffed islands, the like of which are practically absent from the cooler seas, become relatively prevalent as soon as a climatic boundary is crossed. This fact, not previously considered in the study of coral reefs, is believed to give strong support to Darwin's theory of upgrowing reefs on subsiding foundations, subordinately modified by the addition of changes of ocean level and temperature and therefore by changes in the marginal belts from reef growth to reef death as proposed in the Glacial-control theory. (*Davis, 1928C, pp. 218–19*)

Following treatments of such special cases as 'reefless coasts in coral seas' (due to unconsolidated sediments or excessive detritus) and 'volcanic islands with plunging cliffs and embayed shorelines', Davis treats in a long chapter the characteristic islands of the coral seas – the 'embayed, non-cliffed islands with barrier reefs'. Special treatment is given to the Society Islands where

. . . the 12 members of the group exhibit so precisely the systematic sequence of island and reef forms that is called for under the subsidence theory, it is then legitimate not only to regard the theory as well supported but also to use it in interpreting the history of the group. It is therefore here concluded that the volcanic rocks seen in eight islands and inferred to exist under the others represent islands, the eruptive construction of which migrated eastward with the passage of time, and which have, since their eruptive construction, suffered erosion and subsidence proportional to their age. (*Davis, 1928C, p. 307*)

Stressing the equilibrium conditions which may be achieved under continued subsidence, Davis restates his support for his elaboration of Darwin's theory.

Under this theory the unrimmed part of the lagoon floor is supposed to have been kept free from reef growth by the outdrift of sediments, largely supplied by inwash over the windward reef flat. Thus the floor there has now, as it has supposedly had in the past, a depth of about 40 fathoms, because at that depth it is in adjustment to the waves and currents acting upon it. As long as inwash from the windward reef flat and outwash from the open arc of the lagoon floor continue, the leeward breach in the reef may be maintained; except that it may be slowly encroached upon by coral growth on the reef ends at either side. The lagoon floor may as persistently have its open margin built up to about 40 fathoms depth, provided that subsidence continues at a not too rapid rate. Thus explained, the even margin of the lagoon floor in a reef breach gives no indication whatever of the form or the depth of the rock foundation below it. (*Davis, 1928C, pp. 333–4*)

. . . it now appears that [Darwin's] theory succeeds equally well in explaining also certain additional facts that it was not invented to explain; namely the preliminary establishment of fringing reefs, as told in an earlier chapter, and the embayment of reef-encircled islands together with the disposal of the great amount of detritus that such islands have lost during their dissection, as told in this chapter.

These three additional facts should be pondered over, in order to appreciate the altogether unexpected and yet wholly natural, indeed inevitable manner in which, in spite of their dissimilarity, they take their places as essential elements of Darwin's theory. While the original theory took explicit account of the transformation of fringing reefs into barriers and of barriers into atolls as a consequence of reef upgrowth during island subsidence, it gave little attention to the share that subsidence might have in providing opportunity for the initiation of fringing reefs. Yet it now appears that a vigorous fringing reef can hardly be formed around a young volcanic island until subsidence begins after the island has been more or less dissected and cliffed; also, that after a fringing reef has been once initiated, its escape from being overwhelmed and smothered by downwashed detritus as well as its gradual transformation into a barrier reef depends on the continued subsidence of its island.

Similarly, while the original theory recognized that a subsiding island should diminish in size in consequence of its subsidence and recognized also that supposedly subsided islands have shore-line embayments, it did not find any proof for diminution of size or offer any explanation of shoreline embayments. Yet it now appears that the shore-line embayments of an island give good proof that its size actually has been diminished by subsidence; and it appears also that both diminution of size and embayment of shore line are unescapably associated with the upgrowth of a barrier reef around an island that is slowly subsiding.

Furthermore, the original theory took no account of the disappearance of a great volume of detritus from an island that becomes maturely dissected as it slowly subsides; but as soon as it is perceived that the vanished detritus must be reasonably disposed of, behold, instead of our having to complicate the old theory by introducing new conditions and processes into it as a means of disposing of the detritus, its disposal is easily and completely accounted for by a previously unnoticed consequence of the fundamental postulate of the theory in its first-announced form.

A theory that is so unexpectedly successful in meeting new requirements can hardly be wrong. Yet there are still other new requirements, quite as significant as those here brought forward, to be set forth in the following chapters, of which Darwin took no account but for which his theory provides a completely satisfactory explanation as soon as attention is consciously directed to them.

Still another matter which has been presented in this chapter deserves further emphasis. This is the occurrence of broad and smoothly aggraded lagoon floors in association with embayed and uncliffed coasts; for example, the broad lagoon-floor west of Vanua Levu in Fiji. It would certainly be difficult to account for the presence of embayments and the absence of shore cliffs back of such lagoon floors unless the floors were enclosed by barrier reefs while their aggradation was in progress. The Mbengha lagoon floor and the lagoon floor back of Rambi, both in Fiji, are of significance in this connection . . .

It must not be overlooked that, although good reasons have been now given for believing that barrier reef islands have subsided during the upgrowth of their reefs, the enclosed lagoons are of ordinary depth; hence the processes of lagoon aggradation must be regarded as competent to fill up the space back to the reefs. Island stability, as postulated in the Glacial-control theory, is not a necessary condition of reef formation. (*Davis, 1928C, pp. 373–4*)

Drawing on evidence relating to 'almost atolls' and atolls, together with that furnished by elevated reefs of all kinds, Davis concludes that atolls represent the last stage of coral island development due to coral upgrowth on subsiding foundations.

Throughout the remainder of his life Davis (1929B; 1933F; 1934D) maintained his interest in relative sea-level changes and in the origin of coral reefs. Davis and the 'coral reef problem' had served each other well. The former had dissected and analysed the problem as far as was possible without the detailed field evidence which was only just beginning to become available in his declining years; the latter had provided the master with perhaps the most convincing stage for the demonstration of his deductive genius. In conception, method and execution the attack on the coral reef problem exemplified Davis' scholarly characteristics more effectively than almost any other phase of his work.

Recent researches have supported Davis' belief in the veracity of Darwin's subsidence theory of atoll formation, although for different reasons from those evinced by Davis. Stoddart's (1969, pp. 436–41) excellent and long review of coral reef research has shown that, although surface features of existing coral reefs are ambiguous guides to geologic history, the evidence of deep drilling (particularly on Bikini and Eniwetok atolls in 1947 and 1951, respectively) through hundreds of metres of shallow-water limestones, together with the existence of drowned and bevelled volcanic islands ('guyots'), has amply demonstrated the correctness of the subsidence theory when applied to widespread open-ocean atolls. Dating indicates average subsidence rates of up to 50 metres/10^6 years, and that this subsidence has operated intermittently during much of the Tertiary. Barrier reefs along continental margins also broadly conform with the subsidence theory, although their structures indicate rather complex diastrophic histories. The fact that some reefs have been identified which undoubtedly formed in areas of uplift does not invalidate the general applicability of the subsidence theory. The status of glacial control is less clear, but support for it has been weakening, partly because of the slow rates of operation of erosional processes observed on exposed reefs.

PART FOUR

Final Rejuvenation

FIG. 126. Davis in California, 1932 (Courtesy Dr Edward Sampson, Princeton)

CHAPTER TWENTY-SIX

Western Rejuvenation:
The Desert Environment

After remarriage Davis' reactions were stronger still. The environment might have been transformed. It suddenly assumed a set of bright new colours and everything was now to his satisfaction. The same euphoria was reflected in his work and it was as if he had experienced a marked rejuvenation. He had aged and most of his ideas had aged with him, but he was now happy and capable of functioning with much of his former efficiency. His letters from Stanford University in the following October exude cheerfulness.

It is, I fear, some time since a general letter has gone to all of you, because there are so many distractions to interrupt leisure hours (and before I forget it, let me ask if you know what Al. Smith's first words will be, when he sees the White House next spring? If you *dont* know, you will find them farther on). The most impressive change I have to report is the civilising influence of my dear Lucy. I had relapsed for once into barbarism, living alone really alone . . . it wasnt worth while to keep things in order – or to set the table for a lone meal. But naturally I rise to the refinements of normal life! You ought to see how neatly we live in this very comfortable little apartment! Our living room and bedroom open widely into each other with sliding doors that move to a touch. Patent vanishing beds are in both rooms, turn up into the walls thru the day, open out thru the night. The little kitchen is only a slice of a room, but that makes for few steps, and Lucy proves to be a talented cook. We patronise an excellent grocer, telephoning our orders in the morning and getting the goods in the afternoon. Fresh green peas and beans, delicious strawberries, all sorts of grapes are abundant and cheap. We avoid meat most of the time but occasionally fall back on bacon and eggs. My share is not much more, in the work of preparation than setting and clearing the table and wiping the dishes if I haven't gone off for a lecture. Lucy plans the bill of fare with tasteful skill and it is mighty different, all this, from my life here last Fall in one room.

My friends of last year are standing by most generously I've had lots of calls and a fair number of invitations . . . Yesterday, Prof. Blackwelder and his wife carried us in their car 50 m. to Berkeley, for a geological club meeting (poorly conducted) and I showed Lucy the familiar Faculty Club, where I had spent two summers.

In the meantime my classes are going on nicely – 35 beginners, as against 27 last year and 7 or 10 advanced as against *none* last year. One of the latter exclaimed 'We have all been waiting for you to come back' which was hearty, wasn't it? Thus all my hours are occupied, ex Sat or Sunday – and some conference hours besides – pleasant work. A good library supplies me with plenty of reading and abstracting.

I am becoming better informed on various regions than ever before! There is still the plan of a College Physiography in my head – and it is increasingly transferred to paper – both in written and drawn form and I believe that many of my pen outlines will be more informing than the poor half tone photo pictures now so commonly – too commonly used. But this *completion* of a book is a far ahead prospect. In the meantime work on it keeps me out of mischief – or out of *other* mischief. (*Letter: Davis to his children, 14 October, 1928; Stanford, California*)

Davis' return to a more civilized life was happily not too extreme as his wife retained some of the forced economy and abstemious habits that seemed to suit him best.

It is then time to get up, so we rise from our twin beds, not exactly simultan-eously, but in good order, have a warm tub and a cold shower, and then take a light breakfast off a card table in our sitting room, Lucy being the table setter. I do like light breakfasts. For lunches we usually go up the roof garden on 16th floor; plentiful lunch may be had for fifty cents, which we think is very reasonable. Dinner is seldom taken there unless we have guests. It costs more and is far too filling. (*Letter: Davis to his children, 13 October, 1931; New York*)

His lecturing engagements continued as before with the aid of several official appointments in California. When at Stanford University, in the Fall of 1928, he writes:

Plans are already forming for January [1929]. Between here and Tucson, we shall probably be at Cate's School again, or near enough to it to find lessons or lectures there for about 3 weeks – so that means we shall stay here till Dec. 31 – instead of only till Dec. 20. But no final decision is yet reacht. Plans are also in the making for next summer which may be spent at the Los Angeles Section of the University, Calif., so as to avoid the chilly winds of Berkeley. (*Letter: Davis to his children, 14 October 1928; Stanford, California*)

In his last years Davis lectured mainly at the University of California at Berkeley, the University of Oregon at Eugene, Stanford University at Palo Alto, the University of Arizona at Tucson, the California Institute of Technology at Pasadena and Columbia University in New York, and infre-quently at more than a score of other places. In academic circles he went as low as schools and in social circles willingly descended to ladies' civic groups, with whom at times he virtually reached the stage of lecturing on the topic chosen by them. Occasionally he was well paid for his talks.

Our rooms are remarkably quiet, considering. We see the 8th Ave. elvd. trains a few blocks to the east, but hardly hear them, except faintly at night. Just imagine, instead of looking across a narrow court against a blank wall or into someone's rooms, we look out to half the sky. The weather has been fine most of the time.

Air is deliciously fresh today. Wasnt I wise in telling Johnson last spring that I would come on a *good salary*, but would not come if I had to economize. How in the world he managed to get Columbia to cough up $3500 for less than half a year, after the budget has been settled and closed, I do not yet know. But he did. (*Letter: Davis to his children, 31 October 1931; New York*)

When over eighty years old Davis actually improved his blackboard technique by means of a system of pulleys.

All Friday I was occupied with preparations for an evening lecture on the Colorado Canyon, with new and improved apparatus. It fetcht out a large audience and went beautifully; at least Lucy said it did and the audience gave it a big glad hand. (*Letter: Davis to his children, 31 March 1931; California*)

Do you remember that day we spent at the UCLA last spring, when one of your colleags spoke to me about my repeating the Col. Can. lectures at Pomona College? I have heard nothing from him since than and wonder why he got what are called, by the irreverent, 'cold feet'. If you happen to meet Dean Smith of Scripp's perhaps you could suggest to her that my lecture would be a good thing for the girls. Even you have not heard that lecture in its present polisht form, with a fixt sea level represented by a stretcht level tape: behind which the blackboard rises or sinks, tilts one way or the other. It is very edifying. (*Letter: Davis to A. O. Woodford, 10 November 1932; Pasadena, California*)

However, there seems no doubt that Davis' greatest successes were with mature students and young practising geologists.

Although his great system of geomorphology (or physical geography) had been evolved during his Harvard years, in many ways the period after his retirement was the most influential of his career.

Whereas at Harvard he was, to his students, a crochety and unhelpful taskmaster (so G. R. Mansfield told me), during his career in the western schools he lectured with few inhibitions regarding his philosophy; these lectures, and his many personal contacts, profoundly influenced a whole generation of younger geologists. All these younger geologists knew instinctively that Davis was a great man and teacher, at whose feet they should sit and worship. Davis was never trivial or petty! All his faults and virtues had epic proportions! (*Letter: P. B. King to R. J. Chorley, 9 January 1963; Menlo Park, California*)

We have space only for a few examples of the great impact Davis made on some members of his audiences. At Eugene, University of Oregon, in the summer of 1930, John Allen was one of the undergraduate geologists who attended Davis' courses.

It turned out to be one of the most stimulating experiences in my college career, and was largely responsible for my life-long interest in geomorphic problems . . . One of my greatest treasures is the class notebook I kept during these weeks, in which I was able to copy in color the beautiful diagrams presented each day to us on the blackboard. It records about 150 maps, cross-sections, and block diagrams.

> Even at the age of 80, WMD presented rigorous courses, and I was proud to have received A's in both of them . . . WMD was a little man, and he would come briskly into the classroom in his old-fashioned black serge suit, with high starched Herbert Hoover collar and a black silk skull cap; he would turn immediately to the blackboard, and with colored chalk delineate surely and quickly with almost no erasures, the most complicated of block diagrams. (*Letter: J. E. Allen to R. J. Chorley, 17 March 1971; Portland State University, Oregon*)

Davis succeeded equally well at the California Institute of Technology where Robert Webb attended his courses in the early 1930s.

> At that time he was in his early eighties and was as peppy and as vigorous, so I was told, as he had ever been. Dr. Davis was a very articulate person, but basically a rather poor teacher because his command of the English language was so perfect that the method he used in his classroom made it impossible for his students to satisfy him. If one could persuade Professor Davis by some device to deliver a formal lecture, then the delivery was magnificent and as a teacher he couldn't be surpassed. But he insisted on class discussion in all of his courses . . .
>
> Dr. Davis would stay in a classroom as long as any student remained who wanted to discuss geology . . . the conversations went on sometimes for two hours. He was without exception, I think, the most stimulating professor I ever knew.
>
> I learned that he loved to take field trips and I always volunteered my car and he would ride with me. I remember one field trip into Death Valley from Pasadena, California . . . We proceeded into Mojave Basin to enter the southern end of Death Valley along what was then the Cave Springs Road through the Avawatz Mountains. Professor Davis was very interested in the pediments which typically surround many of the Basin-Ranges . . . We camped that evening on the pediment surface. Our field party was spread out asleep under granitic knobs along and near the single track desert road. (*Letter: R. W. Webb to R. J. Chorley, 14 January 1971; University of California, Santa Barbara*)

FIELDWORK AND DESERT EXCURSIONS

In between lecturing and formal gatherings Davis and his wife relied for relaxation mainly on the hospitality of friends such as the Buwaldas, Ransomes, Finlays, Peters, Nobles, Engels and others. At times a small party would make an expedition into the desert or mountains and would spend a short while under canvas. The following recollection, which makes Davis a few years older than he actually was, sums up these excursions admirably. The writer, a consulting engineer and geologist, was a student under Davis at both the California Institute of Technology and the University of Arizona during the late 1920s.

> Dr. Davis was an unbelievably robust man even in his sunset years. I well remember an excursion into the Galiuro Mountains of Central Arizona in the late spring of 1929. By mid May, Pinal County, Arizona, sometimes enjoys temperatures in the mid 90's. At this time the Arizona chaparral is literally alive with

desert fauna. Myriad types of insect and avian life literally and continually agitate the foliage, while the short-lived bunch grass and wild flowers seem ever to be in constant movement with the passage of the small reptilean, rodent and mammalian animal forms.

Dr. Davis, at this time in his early 80's as I recall, favored hobnail boots, a long staff as protection against rattlesnakes, and three Mexican sombreros all worn at the same time. Thus outfitted, he was in fair condition to outpace students in their early 20's.

Perhaps this explains to some extent his success with a bride a quarter of a century his junior, who at the time was in residence on the Arizona campus. Dr. Davis seemed to be a bundle of tireless energy. His small white goatee was accentuated against a complexion made florid by his exertions at these elevations. He was not only an inspiring lecturer but seemed to soak up each new experience and observation with an insatiable appetite.

My last memory of Dr. Davis was that of his rotund but active figure vanishing upstream along the Kern River of California hot on the scent of 'kernbuts and kerncols', i.e. fault incised trenches and truncated remnant spurs. (*Letter: G. Austin Schroter to R. J. Chorley, 2 April 1963; Los Angeles*)

An account of some of Davis' activities when living at Pasadena in March 1931 will exemplify the sort of life Lucy and he led. On the 5th he lectured to the Geological Society and the Branner Club of Los Angeles on the 'Nature of Geological Proof, or How do you know you are right?' (Davis, 1931D). Apparently he had seldom had a better audience. A few days later he and Lucy gave a well-planned dinner at the Athenaeum to about twenty of their best friends. From the 13th to 15th the Rift Club had a trip to southernmost California. About fourteen cars and forty passengers made the excursion, Davis and Lucy travelling with Professor and Mrs Engel. On the way back they stopped at a date farm kept by Bartlett Hayes, one of Davis' students in 1898.

The big show of the month was a weeks' trip to Death Valley with Sam Storrow '86 and Wolff '79. It began by a lovely ride with Amy Ransome and her and our friend, Mrs. Prof. Kemp of Columbia, all the way to Redlands, about 80 miles, on the 17th. There we stayed with our excellent friends, the Finlays, overnight, so I might be ready for a lecture at 8.50 the next morning at Redlands Univy. on the site of Rome. That over, a ride into the mountains with the local geologist, and then with Lucy up to Peters Ranch, for a two days stay . . . They gave us a fine dinner party, and had an afternoon tea for Lucy.

Friday morning, March 20 Storrow and Wolff turned up in their redoubtable roadsters about 9'oc, and soon we were off for the wilds. You all know Wolff as my former Harvard Colleag, professor of mineralogy and such. Storrow was one of my students in the middle 80's, and appears to have conservd good memories of that period; for we were his guests the whole week we were away . . . he had made a career and a name for himself as an engineer and had retired on a fair fortune . . . the contracts which he put thru for his clients had totald for them about $250,000,000.

The contrast between his life and mine has amused me. He is an agressively activ doer. I hav been a thinker. (*Letter: Davis to his children, 31 March 1931; Pasadena, California*)

The trip included a descent into a remarkable depression, the Saline valley, where the bottom 'has dropt out, about a mile deep and ten or twenty miles across'. On Friday, 27th, the day after returning, Davis lectured on the Colorado Canyon at Pasadena. On 31st they packed and left for Palo Alto where a lecture had been arranged for 10 a.m. on 1 April.

It is perhaps no wonder that the desert became Davis' favourite environment. Undoubtedly he enjoyed the works of nature more than those of man, an attitude stimulated by his religious belief. Worldly pleasures were allied to the flesh and immoderation and had to be avoided; the delights of nature were the creations of God and therefore to be enjoyed. His writings, his taste for music, his dislike of war – all mark him as a man of strong emotion. This feeling had to have some outlet and it is particularly evident in his desert descriptions.

> But much better than a dry camp on a desert plain is a camp in a valley mouth at the base of a desert mountain range, where 'the streams come down at night', as I have found them to do at the House Range in western Utah. Truly, many a little stream, so scantily fed by its valley-head springs that it withers away on its steep course through the heat of the day, ventures farther down-slope after nightfall. Indeed, in the chilly desert dawn, such a stream may trickle out from its rock-floored mountain valley upon the piedmont gravel slope, there soon to be lost by sinking into the sieve-like detritus. But when the unclouded sun blazes forth a few hours later, the stream actually evaporates upward faster than it flows downward, and so it retreats to the shade of the quivering aspens in its upper glens. If, after a weary day in the saddle crossing an arid intermont plain, the traveler ascends the long detrital slope to a mountain valley-mouth, where one of these nymph-like streams is known to descend at sunset, the desert becomes a pleasant land to rest and sleep in. I spent two nights at such a valley-mouth in 1904, and the stream came down with clock-like regularity both evenings and trickled cheerfully till morning. One can hardly help imagining that, like the barking coyotes and the jumping mice of the dry plains, the gnomes and elves of the mountains as well as the nymphs of the valley-head springs come forth in the dark for their nocturnal revels: at least, so they must have done before prospectors and cowboys, who care nothing for such little creatures, frightened them away. (*Davis, 1930F, p. 402*)

He was conscious that the desert possessed an inexhaustible source of spiritual contemplation which the modern trappings of civilization have not.

> Yet much as there is to see in daylight the desert is at its best in summer nights, always provided that it is mastered. The cooling air is so refreshing, the absolute stillness is so comforting, the clearness of the entire sky is so inspiring! What a refuge one may find there from the turmoil of the over-busy world from the over-crowded streets of big cities, and above all from that climax of vulgar display, that

glaring midnight horror, known as the Great White Way, with all its false standards and its metropolitan provincialism. While one is at rest in the desert silence the starry universe, slowly turning overhead, seems to draw nigh, so that one may more easily place himself in reverent and prayerful accord with it. A night thus passed in the desert is an unforgettable experience, which no traveler, especially no city resident in the humid East, should fail to enjoy when he visits the arid West; and he will be of dull mentality if a night in the desert does not move him to deep contemplation. (*Davis, 1930F, p. 403*)

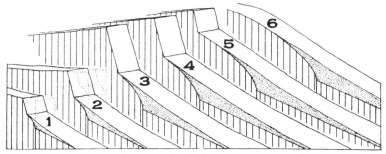

F I G. 127. Block diagrams of a retreating fault scarp over a growing, detritus-covered rock floor
1. Parallel retreat of the inclined slope developed from the original fault
2. Renewed uplift and resulting erosion
3. Renewed uplift and resulting erosion
4–6. Successive stages of retreat after 3
(From Davis, 1938A, Fig. 8)

We, however, will have no difficulty in showing that professionally Davis was more interested in the mountains surrounding and surmounting deserts than in the sand-covered basin floors. Even in the description above the term 'piedmont gravel slope' hints at a flattening at the foot of the mountains and of the work of water upon a piedmont bench. It happens that when Davis saw any landscape he soon put his ideas on it – and other people's ideas – into print. During his western rejuvenation he wrote almost as fluently and as voluminously as ever.

DAVIS' WESTERN PUBLICATIONS, 1928–32

The early years of Davis' western circuit produced, as we have seen, relatively little literature except *The Coral Reef Problem* and several articles on the same theme. Considering the burden that travel and lecturing must have been to a widower, it is not surprising that 1929 was his least productive literary year since 1882. After the rejuvenation inspired by his third marriage Davis regained much of his old progressive tempo and 1930–2 proved fruitful years. The western fieldwork is reflected in long, interesting articles on the Galiuro Mountains, the Peacock Range and the Santa Catalina Mountains in Arizona, and on the Santa Monica Mountains in California.

The origin of limestone caverns and of rock floors or piedmonts in arid and humid climates formed other fruitful themes. At the same time *Elementary Physical Geography* appeared in Japanese and Davis began a public attack on Walther Penck's ideas on landform analysis.

Before summarizing this varied work it seems important to notice that during his western rejuvenation Davis experienced a small but significant mental change towards landform description. Not only had the physical environment changed from humid to semi-arid and desert, the social and academic environment was equally different.

In the East Davis had become cock of the roost; he led and dictated. In the more egalitarian West the aged physiographer was usually one of many; he talked and discussed with younger geologists and was prepared within limits to modify his views. Always fieldwork played a large or dominant role and, apart from a continued lack of measurements, his findings are based on reality. No longer do glimpses from a train or steamer suffice for landscape suggestions; inspection is on foot in the manner approved by all geomorphologists. Deduction and the normal cycle inevitably begin to play a less dominant role in his thoughts.

> You asked for a personal opinion of Davis' approach to geomorphology. Certainly his views dominated the American attitude toward geography for most of a generation. His concept of the subject was so well organized in his mind that he was able to communicate his logic to many followers; at the same time his inability to adjust his thinking to possible alternative approaches or to changes provided something of a strait-jacket for other geographers. Only under pressure would he alter his views; the arid cycle of erosion illustrates my point. (*Letter: H. F. Raup to R. J. Chorley, 5 January 1962; Kent, Ohio*)

What A. O. Woodford says in the following comment on the cycle of erosion is what Davis himself averred from at least 1902 onwards. The new ideas, however, provide welcome evidence of the stimulus of fieldwork in the intensely fractured cordillera of the West.

> Davis never changed his attitude concerning the cycle of erosion. But he did say, again and again, that it was a mental aid only, just one model. He fully realized, and said that he always had realized, that initial uplift was not followed by stillstand that lasted throughout the erosion cycle. But in his teaching and in his studies of California land forms, he usually decided that uplift mostly came first. Then a lot of erosion, with the development of a set of land forms. Then some more uplift, and rejuvenation.
>
> Davis had some new ideas in California. He supervised a paper by Rollin Eckis, published in the *Journal of Geology* in 1928, that described alluvial fanhead benches and midfan mesas as normal-erosion-cycle features after block faulting. He concluded that complex fracturing of crystalline massifs may result in surface forms (or subsurface forms) shaped like anticlines. Mason L. Hill demonstrated such structures a little later (*Univ. Calif. Publ. Geol. Sci.*, 1930). (*Letter: A. O. Woodford to R. J. Chorley, 4 January 1962; Claremont, California*)

Carl Sauer, who also knew Davis well in California, sees a slight but significant weakening in his attitude to the cycle.

> To some extent he did see after retirement that the cyclic system was not enough. In the Santa Monica Mts he became somewhat aware of sea level changes, of the irregularity of time that called into question the concept of base level. (*Letter: C. O. Sauer to R. J. Chorley, 18 October 1961; Berkeley, California*)

In fact there is a great deal to be said for the following idea:

> I have the impression that the three main parts of Davis life consist of: The 'essay days', followed by 'defence of essay days' and finally 'emeritus days', particularly during the Tucson–Pasadena phase, when he began to get some new ideas, really from induction. (*Letter: R. J. Russell to R. J. Chorley, 5 January 1962; Baton Rouge, Louisiana*)

But, having made our point, we can as usual let the master speak for himself.

> . . . the oldest mountain-making disturbances along the up-starting Pacific coast are less ancient than the youngest of such disturbances along the settled-down Atlantic coast; and the scale on which deposition, deformation and denudation have gone on by thousands and thousands of feet in this new-made country is ten-or-twenty-fold greater than that of corresponding processes in my old tramping ground. On shifting residence from one side of the continent to the other, a geologist must learn his alphabet over again in an order appropriate to his new surroundings. So this spring I am going to Arizona for yet another try at my new lessons. (*Davis, 1930F, p. 404*)

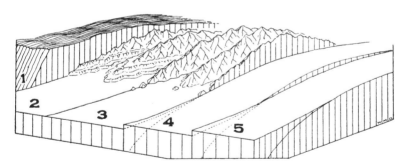

F I G. 128. Reduction of a fault block, first, to an irregularly-embayed mountain and, finally, to a smoothly-arched dome

1. Original block
2. Block with a benched face resulting from intermittent uplift (Not observed by Davis)
3. An embayed mountain front and pediment slope
4. A smooth convex dome
5. A faint dome resulting from prolonged erosion

(From Davis, 1938A, Fig. 6)

Davis' attitude to this reorientation was expressed in a long publication 'Physiographic contrasts, East and West' (1930C). In this he contrasts the continuity and adjustment of the valley systems of the north-eastern United States with the discontinuity of those of the south-west. Davis also contrasts the broad upheavals which produced a three-cycle landscape in the Appalachians with the irregular faulted uplift of the Basin Ranges. He refers to 'a newer idea' of drainage diversion resulting from the belief 'that mountain ranges rise slowly instead of rapidly' (Davis, 1930C, p. 405) and confirms that the Basin Ranges, like the fractured and tilted Appalachian blocks, 'appear to be the work of secondary forms, acting superficially, in consequence of deep-seated deformation due to primary forces' (Davis, 1930C, p. 409). Most of the remainder of the paper is taken up with a description of the formation of fault-block topography, the rejuvenation of fault-line features and the production of pediments from the destruction of desert mountains.

Yet during this period of Davis' life in the West there is evidence of something more than the influences of the physical and of the social environment. The octogenarian has lost some of his personal aggression. Instead of ruthlessly taking other people's ideas and expanding and refashioning them into a Davisian mould, he now co-operates with friends, and many of his western fieldwork studies, were produced in collaboration with practising geologists, such as Baylor Brooks, William C. Putnam and George L. Richards, Jr. In our opinion, the general result is highly beneficial. However, those collaborators did not attack the cyclic theory and when that was under pressure Davis could still defend it stoutly. So during this period his publications fall into three main categories: landform observations; anti-cyclic retaliations; and, inevitably for a man who has been producing for nearly fifty years, reprints or resurrections.

LANDFORM OBSERVATIONS IN THE WEST

Davis' publications during and after his eightieth year, and therefore mainly written after his third marriage, amount to thirty-three articles exclusive of separate summaries repeated in different journals and of new editions or reprints of books. Another three articles written mainly by him were published posthumously. So as an octogenarian Davis produced for the geological world a total of well over 800 pages of text and about 250 line drawings. When seen together, this four-year output forms a bulky corpus of knowledge that most modern geomorphologists would not exceed in a lifetime. Together it represents four pages of text and more than one line drawing for every week of his last four years. At the same time he did extensive fieldwork, lectured, attended conferences and kept up a wide correspondence. What he did not do was to miss opportunities of publication.

With his remarkable fluency and keen deductive approach he could complete the writing up of fieldwork with a facility and ease of mind lacking in lesser men. He kept publication almost in step with investigation which is a major reason for his stature.

These octogenarian writings, apart from a religious theme and an important defence of the cycle each of which we will deal with separately in later chapters, are concerned largely with problems of the analysis and classification of landforms characteristic of the American West. These articles may be grouped into seven main categories: Basin Range problems; mountain forms and slopes in deserts; older desert mountain forms: rock floors in arid and humid climates; marine benches on coastal mountains; the origin of caverns; and the classification of lakes.

Basin Range problems
Davis had been interested in faults since long before 1900 when he discussed at some length fractures and block mountains in his *Physical Geography*, as well as publishing a separate article on a fault scarp in the Lepini Mountains, Italy. After an excursion in 1902 he wrote more than half a dozen articles on the Basin Ranges of Utah, Arizona and Nevada. However, the Colorado Canyon proved a powerful counter-attraction here and it was not until 1913(L) that he brought typical Davisian methods to the topic when he attempted a 'Nomenclature of surface forms on faulted structures'. Davis' interest in Great Basin landforms was heightened by his close friendship with Grove Karl Gilbert (1843–1918) of whom he compiled a fine biography which was not published until 1927(F). In the meanwhile, as we have seen, Davis wrote on African rift valleys (1920A), and on 'Faults, underdrag and landslides of the Great Basin Ranges' (1922H). In 1925(E) Davis wrote more extensively on 'The Basin Range problem' in which he combined Clarence King's ideas on folding, Powell's on the reduction of the region to a post-Jurassic peneplain (perhaps preserved as a 'Louderback' by lava flows) and Gilbert's on block-faulting possibly related to considerable horizontal extension. Davis suggested that the low inclination of Basin Range faults (30°–40°) might suggest that 'the recently revealed part of the fault faces now visible along the base of certain ranges may be 5000 to 10,000 feet below the Powell land surface of the time when the Gilbert faulting was initiated' (1925E, p. 389) in that the angle of normal faulting probably decreased with depth. Reasoning from the observed post-Bonneville erosion, Davis estimated that the destruction of a fault-block like the Oquirrh Range would take a cycle of erosion of roughly 2×10^7 or 2×10^8 years (1925E, p. 391). Davis' address at the first meeting of the Rift Club on 27 March 1926 on 'The rifts of Southern California' (1927A) was much concerned with the topographic recognition of the position and angle of normal faults, the lateral movement of the San Andreas – which, he gave the impression, was subsidiary to the

vertical movements – and to the erosional and depositional modification of topographic fault lines. The last general contributions to the Basin Range problem in 1932(F) and 1933(H) consisted of a reiteration of his previously published ideas and reference to his more detailed studies of the Peacock Range (1930A), the Galiuro Mountains (1930H) and the Santa Catalina Mountains (1931F).

The detailed articles on the mountains of Arizona were based on field trips from Tucson. The account of the Galiuro Mountains (1930H) was written in

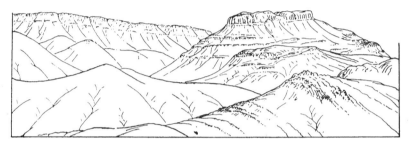

F I G. 129. Sombrero Butte, Arizona (looking south-east), a lava butte separated from the main front of the Galiuro Range (left background) (From Davis, 1930H, Fig. 7)

collaboration with Baylor Brooks a graduate student at the University of Arizona, Tucson, when Davis lectured there in the spring of 1929. They interpreted this mountain range as a tilted, dissected fault-block, or a typical Basin Range, upstanding above down-faulted or down-tilted adjoining troughs. The upland surface consists largely of a great thickness of Tertiary lavas (mainly andesites) which present a bold escarpment on the west where the underlying crystalline and Palaeozoic rocks outcrop in parts. Observations showed that the buried landscape was hilly or sub-mountainous. It was considered that earth movements fractured the lava-covered mass into an uplifted, tilted range block (the present Galiuro Mountains) and a depressed trough, now drained by the San Pedro river. The reason for claiming the main western scarp as a fault is that it maintains its steepness and simple, rectilinear continuity for fifty miles although it transects various rock structures. Similar physiographic, as distinct from geological, evidence was used, Davis says, by Henri Baulig (*Le Plateau Central de la France* . . ., Paris, 1928) who thereby showed the existence in mainly massive crystalline rocks of several scarp-producing faults which the geologists had not noticed. In the Galiuro Mountains dissection by ravines had removed visible parts of a western fault but the great scarp could only have been fashioned out of a fault face. What is more the sediments in the adjacent San Pedro trough also showed signs of deformation and uplift.

Davis' article on the Peacock Range, Arizona, was presented before the

Cordilleran Section of the Geological Society of America in February 1930(A). It first traces the various theories proposed to explain the nature of the isolated mountain ranges in the Great Basin of the American West. Clarence King in 1870 considered these ranges to consist of conformable, stratified beds compressed in late Jurassic times into vast mountain folds and uplifted bodily as a wide, huge corrugated plateau. The synclines were later filled with detritus eroded from the anticlinal ridges. G. K. Gilbert (1875) could not find evidence of normal anticlinal structures nor of the expected depth of stream incision on the range margins, so he suggested that each range was a fault-block, uplifted, without lateral compression, on near-vertical fissures. King (1878) incorporated these ideas into his own theory by postulating compressional folding in the late Jurassic period and vertical fracturing and uplift in late Tertiary times. Davis points out that this involved an important physiographic principle.

> Whenever a crustal movement of relatively recent date enters a regional problem, the pre-movement form of the region, the effect of the movement and the work of erosion or deposition in consequence of the movement must all be stated. (*Davis, 1930A, p. 295*)

This principle, however, had already been applied to the Basin Range province by Powell (1876, pp. 32–3) who believed that the region had been, before its block-faulting, worn down to a surface of low relief. What is more, Dutton (1880, p. 47) went further and suggested that the denudation to a state of low relief occurred between the Jurassic folding and the Tertiary fracturing. Thus the dominant idea was that each tilted fault-block in the Great Basin had a back slope or dip slope that was formerly part of the worn-down relief, and a steep front, and perhaps one or more other sides, that were newly-formed fault scarps.

Davis in his studies of the region insisted that erosion would inevitably accompany faulting and would continue after the movement ceased. He also devised what he considered a helpful terminology to the various elements of the landform development. The 'King mountains' of the Jurassic folding were worn down to a 'Powell surface' of low relief in Tertiary time before being uplifted and fractured into 'Gilbert fault-blocks'. However, to those, a fourth term – a 'Louderback' or lava-capping – must be added after the geologist who in 1902 discovered that the back slopes of the two east-tilted fault-blocks investigated in the Humboldt Lake ranges in north-west Nevada were covered with an originally continuous lava sheet that had preserved the 'Powell' peneplain and permitted the subsequent tilted dislocation of the ranges to be assessed (Louderback, 1904). Subsequently many other more or less lava-covered ranges had been found and Davis himself had seen nine successive fault-blocks, all more or less lava-covered, all tilted to the east and fairly well eroded, in the neighbourhood of Death Valley, south-eastern

California. Where these Louderbacks occur it is easy to demonstrate fault-block origin. But where Louderbacks are absent, as in many of the Basin ranges, erosion has gone so far that it has removed traces of the Powell surface on the back slope and of spur-end facets (blunt ends) along the main fault-scarp. In this event, it is difficult to demonstrate block fracturing.

Davis (1930A) describes the Peacock Range in north-west Arizona as intermediate between the Louderback and non-Louderback type. Its lava-cover is today restricted to several small cappings on the spur ends on the lower edge of its back slope. These are tilted and, since lava-cappings on a nearby isolated block to which they were once attached at the surface are horizontal, fault-block origin can be postulated for the Peacock Range. In

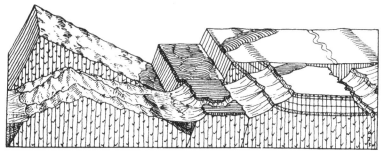

FIG. 130. Block diagram of the Peacock Range, Arizona
Back block. A reconstruction of the partly lava-covered and faulted blocks
Front block. The present eroded forms of the fault blocks
(From Davis, 1930A, Fig. 5)

this article, Davis agrees with Gilbert's latest idea (published in 1928 in 'Studies of Basin-Range structure': *U.S.G.S. Prof. Paper*, 153) that faults are not vertical but extensional. As the local dip of the fault planes often averaged between 30° and 70° but seldom became much steeper it was impossible to postulate a simple vertical movement. Davis ends this account in a way characteristic of his approach to all problems since at least 1900 – a way, it must be emphasized, that most of his disciples never took to heart! He shows clearly and cleverly the complications of the Basin Range problem. In some parts of the Great Basin the folding continued into Tertiary time. The fracturing of the Powell surface into Gilbert fault-blocks was distributed over a time-scale sufficient for some to be almost worn away while others, especially in the north, are only moderately eroded. The worn-down Powell surface itself was irregularly warped so that some parts of it were covered with stratified sediments and perhaps also with volcanic ash and dust before block-faulting occurred. In that event, when these deposits were elevated with a fault-block, they were relatively quickly worn down to low relief as inter-range plains, so forming, as Eliot Blackwelder had suggested in 1928, degradational plains in a region characterized by plains of aggradational origin.

In October 1931(F), Davis described the Santa Catalina Mountains, Arizona, and included in it his best resumé of the Basin Range problem. Here, after the primary upfaulting along its south and west border, a minor secondary uparching of several hundred feet occurred. The idea suggested itself independently to J. B. Tenney of the State Bureau of Mines and to Davis when they were returning separately from a picnic excursion there in 1930. Davis apparently raced Tenney to the pen and benefited from the latter's revision and helpful suggestions. We cannot here go into details of the Santa Catalina range but the general gist of the argument was that the northern parts have at the base of the main scarp a rock pediment up to 1000 feet higher than that on the southern parts where eventually it disappears southward beneath detrital deposits [Fig. 131]. Davis then gives a long account of

FIG. 131. Condensed diagram (with some restoration) of the Santa Catalina Mountains, Arizona, surmounted by the dome of Mt Lemmon. At the northern end there are even (unshaded) pediment compartments above a steep rocky slope (vertical shading); at the southern end ten steep valleys are occupied by detrital bahadas (dotted). The horizontal scale divisions are miles and the vertical scale divisions are in 1000's feet above sea level. The numbers 1–4 indicate the locations of the cross-sections shown in Fig. 132 (From Davis, 1931F, Fig. 2)

the special features of the Basin Ranges. The normal or best-certified simple type are the two Humboldt Lake ranges of north-western Nevada described by Louderback. They clearly exemplify Gilbert's theory of 1928. The (King) mountains were reduced to a (Powell) surface of low relief when lava flows (Louderbacks) covered it unconformably. The compound mass was then fractured and uptilted in two main (Gilbert) fault-blocks each of which had a lava flow (Louderback) slanting down its back slope at a moderate angle from a newly-exposed, steep fault-scarp.

Variations from this simple scheme of a linear fault-block are regarded as departures from the norm. The Galiuro Mountains have an exceptionally uneven buried Powell surface, an exceptionally thick Louderback, and raised outwashed detritus. The Peacock Range has retained lava-patches only on the spur ends of its lower back slope; in contrast the Argus range has large patches of lava on its back slope. Some ranges have experienced recent secondary faulting or, as in the Santa Catalina block, secondary tilting of the whole mass. Others, covered before faulting not with lava but with relatively soft volcanic ash or sediments, have been fashioned by differential erosion into inter-range 'plains'. The mind of the reader, seeing endless combinations of differences in uplift, fracture, tilting, erosion and deposition, may boggle at the potential variety within a large territory. Similarly, Davis at the age of eighty-one had no delusions and quotes his favourite author, G. K. Gilbert: 'all our results in geology are tainted by the tacit assumption of simplicity that does not exist'. The personal, 'pathfinder' terms suggested by Davis for the elements of the development of linear fault-blocks in the American West never caught on and went into limbo with so many other of the technical terms that he invented.

If the reader wonders what has happened to the popularity of the fault-block hypothesis of the basin-and-range province devised by Gilbert and upheld by Davis, he should turn to 'Pediments in Southeastern Arizona' by Yi Fu Tuan (1959) which includes a contour map of the mountains near Tucson. The Gilbert–Davis idea of 'regional peneplanation followed by block-faulting in late Tertiary or early Pleistocene times is inadequate and incompatible with the cumulative evidence for repeated orogenic activity' (Tuan, 1959, p. 15). Although the tectonic history is not yet clearly understood

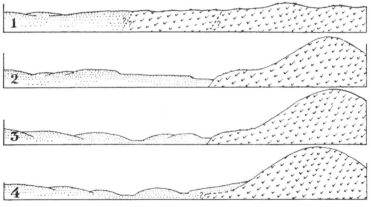

FIG. 132. Four schematic profiles across the western base of the Santa Catalina
Mountains, Arizona, at the locations shown in Fig. 131

1. Small fault-block uplift reduced to a peneplain, together with the adjacent granitic area to the west (left). The peneplain has been uplifted but is, as yet, undissected

2. The sixth pediment compartment. The area having been raised and partly redissected

3. The second pediment compartment. Here the uplift has been smaller, and the erosion shallower, than in profile 2

4. Depression and burial of the inferred pediment near the south-western corner of the mountains

(From Davis, 1931F, Fig. 6)

enough is known to throw doubt on the adequacy of the hypothesis of block faulting considered simply as rapid uplift along marginal faults to the extent of producing mountain ranges from a former regional peneplain. The great accumulation of basin deposits . . . suggests that crustal disturbances have taken place, perhaps sporadically, throughout the later part of the Cenozoic era. Tilting of the older basin beds and their extreme dissection in comparison with the younger beds at the center of the basins indicate that the relative uplift of the mountain ranges may not yet have come to an end. Whether this uplift occurs primarily along major fault lines cannot always be determined from the available evidence. The nature of the uplift (block faulting, tilting, doming, or a combination of these) and the rate of uplift probably vary from place to place. (*Tuan, 1959, pp. 21–2*)

There is no doubt that both Gilbert and Davis would have agreed with this modern assessment as they were convinced of the complexity of the problem and of the need for further fieldwork.

Mountain forms and slopes in deserts
Davis spent four spring terms (1927–30) at the University of Arizona at Tucson, and most of his time thereafter at the Balch Graduate School of the Geological Sciences at the California Institute of Technology, Pasadena. Here he modified his ideas on the cycle of arid erosion as well as on the interrelated problem of Basin Range relief. It was clear that the cycle of erosion devised by him in 1905(L) applied particularly to a region with block-faulted ranges and many enclosed basins. Maturity was said to be marked by reduction of relief and an appreciable degree of integration of adjacent drainage basins. Old age was associated with a general lowering because of the loss of finer debris by deflation. More resistant rocks might be left upstanding as were the inselbergs of the Kalahari. The ultimate result was a kind of desert peneplain.

This scheme was recognized as highly speculative and soon needed modification in regard to its ideas on baselevel and on the processes at work and the nature of the slopes resulting from them. However, much of the detailed fieldwork came now from the American West, and Davis revised and expanded his own concepts at the same time as he acted as a mouthpiece for those of others. His main summaries are 'Rock floors in arid and in humid Climates' (1930E); 'Granitic domes in the Mojave Desert, California' (1933D); 'Geomorphology of mountainous deserts' (1936A) and 'Sheetfloods and streamfloods' (1938A). As the first also concerns the normal cycle of erosion we will deal with it separately. The second was published in the *Transactions of the San Diego Natural History Society* (Vol. 7, No. 20, pp. 211–58) and contained 34 figures and 7 photographs. The two posthumous publications (1936A and 1938A) are of high quality, that on 'Sheetfloods and streamfloods' being an enduring masterpiece of geological literature.

These various articles show that Davis' ideas on the baselevel of arid cycle erosion are now more complicated, being divisable into three main possibilities:

First, the development might proceed with normal baselevel control where the drainage reaches the ocean. Such desert regions are eventually reduced, for the most part, by 'differentially degrading sheetfloods until they eventually become bare rock floors of faintly concave slope' (1938A, p. 1402). McGee as long ago as 1897 had described the general effect of sheetflooding in the Sonoran region of north-western Mexico where nearly half the baselevel plains are exposed granite and the remainder are only covered by a veneer of detritus in transit.

Second, if desert mountains drain to playas or basins which are being lowered by the deflation of fine detritus, the surrounding area is attacked by differentially degrading sheetfloods which lower its convex domes and arches, and wear away the thin edges of detrital covers. Thus the extent of graded rock floor increases. Examples occur in south-western California but extensively degraded rock floors are probably best developed in the great 'gobis' of the Mongolian desert.

Third, the development of arid and semi-arid landforms might proceed under the influence of a slowly rising local baselevel due to the aggradation of an enclosed playa. This is common in the Mojave Desert and elsewhere in the American West and is associated with extensive rock floors (pediments) and decidedly convex uplands, as will be described later.

Davis in his old age also shows a much greater interest in the influence of rock structure and of erosional processes on the shape of slopes. In deserts he distinguishes between 'granitic' and non-granitic landforms. In the former and in certain hard volcanic rocks that also have widely-spaced joints, the large angular joint blocks weather into rounded boulders, fine grains and dust 'without scraps of intermediate size'. As the fine material is soon removed by sheetfloods these boulder-covered slopes develop with little or no talus at their base. Consequently the break of slope or angle between the mountain front and its base that is usually present in desert topography becomes especially pronounced in granitic and granitic-type weathering rocks because the large boulders maintain a steep gradient on the mountain face and its foot is not normally cloaked with deep accumulation of more gently-sloping debris. Davis considered that other rocks broke down for the most part into detritus of intermediate size which 'blends by gracefully concave basal profiles into the graded piedmont slope' (1938, p. 1359). He was in this respect following the work of A. C. Lawson (1915) who had shown that in uplifted hard rocks the slopes were usually steeper than were gravity slopes related to size of particle, and of Kirk Bryan (1922) who had observed in southern Arizona gentle to moderate slopes of under 20° on softer materials weathered into fine waste, steep slopes on bedded lavas and granite (30°–35° being characteristic) and, in rare instances, very steep slopes on a few hard rocks, mainly igneous, with extremely wide-spaced jointing.

The boulders of steep granitic and granitic-type weathering slopes were considered by Davis to orginate, as Blackwelder had pointed out in 1925, as rounded, residual cores due to

> underground chemical decomposition of certain minerals, which appears to be promoted by small quantities of percolating water initially entering along joint planes and gradually penetrating the pores of the joint blocks. This is proved by the occurrence of disintegrated rock well below the surface, as shown in artificial cuts. Decomposition penetrates not only all round the subsoil boulders but also below them ... Thus prepared, the boulders are laid bare by the removal of the soil from above, around, and beneath them by wash and creep. When a higher boulder is thus left unsupported it falls or rolls down upon a lower one. Only when thus exposed, may physical disintegration and surface flaking be promoted by diurnal changes of temperatures. (*Davis, 1938A, p. 1360*)

When exposed the boulder is reduced to smaller and smaller size by a continuance of the chemical decomposition already begun underground aided by physical disintegration. When reduced to a diameter of a foot or less it was considered to fall to pieces in grains and dust, except where sizeable scraps were supplied from quartz veins or aplite dikes.

> So interpreted, the presence of joints is the pre-requisite for boulder production. Close-set joints determine small boulders; wide-spaced joints determine large

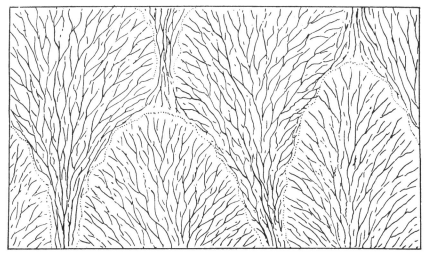

F I G. 133. Ideal diagram of the enmeshed lines of sheetflood flow forming fan-like units commonly measuring 300 to 500 feet in length (From Davis, 1938A, Fig. 2)

boulders; but the latter are eventually reduced to small size. (*Davis, 1938A, p. 1361*)

We may interrupt the argument here to point out the great change in Davis' attitude to chemical weathering in deserts. In 1905 and even in 1930 he thought desert processes were characterized by a predominance of rock disintegration by physical weathering, there being little chemical weathering. By 1933 when this article was first written, he has relegated physical weathering of rocks to a minor role. However, his attitude has changed also towards slope retreat.

The initial fault-scarps of granitic desert mountains, usually homogeneous in large masses and weathered chiefly by subsoil decomposition, are soon worn back to 'roughly graded, boulder-clad slopes of 35° or 40°; which, as Lawson and Bryan have shown, maintain that declivity while the mountains are reduced to mounts and knobs' (1938A, p. 1413). This parallel slope retreat or backwearing is due to weathering and rillwash on the whole scarp and is not aided by the lateral erosion of sheetfloods or streamfloods at the mountain base. On the other hand the initial scarps of non-granitic mountains (more heterogeneous in structure and weathering into grains of all sizes) tend to wear back into irregular forms, being ungraded where the hardest rocks are exposed and graded where covered with detritus. The youthful general steepness is not maintained as in granitic mountains and the scarp slopes are slowly reduced to lower and lower declivity.

Davis inevitably attempted to fit these and other new findings into his cycle of desert erosion. His observations relate specifically to the Mojave Desert of south-eastern California where for six years he had gone on excursions with Dr L. F. Noble, of the United States Geological Survey; Mr Samuel Storrow, of Davis' Harvard class of 1887, now a retired engineer in Los Angeles; Mr Myron Hunt, architect of Pasadena and lover of all outdoors; and several students at California Institute of Technology at Pasadena, all of whom provided 'stimulating encouragement, many photographs, and comfortable transportation' (Davis, 1938A, p. 1342). These observations were supplemented by fieldwork during his long stays at Tucson in southern Arizona, where 'uplifted fault blocks are also a characteristic initial form of a cycle of desert erosion'.

Thus this cycle consists essentially of uplift, usually as fault-blocks, followed by dissection and degradation under the influence of weathering and surface runoff. In his 1938 article Davis pays particular attention to the relative importance of sheetfloods and streamfloods – or conversely to the shape and retreat of slopes – during the various stages of the desert erosion cycle. He uses extensively the writings of McGee who from experience considered that streams differed greatly from sheetfloods.

McGee (1897, p. 100) had the good fortune to witness a flood that came roaring out of a canyon in the mountains, 'thick with mud, slimy with foam and loaded with twigs, dead leaflets and other flotsam ... The torrent advanced at race-horse speed at first, but, slowing rapidly, died out in irregular lobes not more than a quarter of a mile below the road; yet, though so broad and tumultuous, it was nowhere more than about 18 inches and generally only 8 to 12 inches in depth ... The front of the flood was commonly a low, lobate wall of water, 6 to 12 inches high. ... Within the flood transverse waves arose constantly, forming breakers with such frequency as to churn the mud-laden torrent into mud-tinted foam; and even when breakers were not formed it was evident that the viscid mass rolled rather than slid down the diminishing slope.'

The flow was generally uniform, though locally retarded by bushes, 'but now and then a part of the sheet ... began to move more rapidly, when almost immediately the flotsam would shoot forward at twice or thrice the ordinary rate, the flood surface would sink toward the upper end and swell toward the lower part of the rush line, while the roar would rise above the rustling tumult of the more sluggish waters ... then the waters would diverge and slacken ... The whole process of gathering and respreading of the waters commonly lasted but a few seconds, or perhaps a minute or two ... So common were these rushes that two or three or even half a dozen might be within the field of vision at the same time – some just starting, some dying away' (McGee, 1897, p. 101). The flood lasted about ten minutes; it was over a mile wide, and nearly blended with others from adjacent canyons. In half an hour the ground was drying.

'A highly significant effect was found on examining the track of one of the more violent rushes within the flood. At the upper end this was a gully reaching two feet in depth and one or two yards in width, newly gouged in the gravelly and sandy silt of the plain; at the lower end it was an elongated delta or fan.' (McGee, 1897, p. 102). (*Davis, 1938A, p. 1343*)

Davis elaborates the detailed characteristics of sheetfloods; shortness of overland flow; briefness of duration; ill-defined lateral limits; extremely long time-interval between successive floods, and so on.

The condition essential to the formation of a sheetflood appears to be a heavy rain draining to or falling directly on a barren, graded detrital slope, where, immediately loaded to capacity, the flood may flow freely. A secondary condition appears to be the absence of a low-water stream during the long interval between sheetfloods; for such a stream would develop a channel in the detrital slope, and the channel would delay, if not prevent, the outspreading of the next sheetflood.
(*Davis, 1938A, p. 1346*)

The work of sheetfloods differs widely from that of streamfloods so-called by Davis because they are spasmodic, short-lived and impetuous unlike the flow of a normal river.

If the rain falls on a barren mountain mass which drains to a graded detrital slope, the slanting surfaces of the mountain quickly become covered with a multitude of audibly rippling rills and streamlets, which unites in rushing torrents or streamfloods on the narrow beds of many confluent, steep-pitching ravines and thus reached a main valley. So far, probably because of high velocity by reason of concentrated flow, the flood is somewhat underloaded and therefore performs much corrasive and erosive work; but after issuing from its narrow valley upon the piedmont slope it quickly spreads into a sheet-flood which, running slower and at the same time picking up whatever additional detritus it can carry, promptly becomes fully loaded. Thereafter, the stream losing volume by in-soaking, the load is gradually laid down.

The sheetflood drainage of a graded piedmont slope below a mountain face of back-weathering contrasts strongly with the inbranching torrential or stream-flood drainage in the dissected mountain mass above and back of the slope. There the stream lines are narrow and well separated by high, rocky divides which they cannot submerge. But as soon as a torrential mountain streamflood issues from its valley upon the piedmont slope, it is transformed into a broadening sheetflood, an essential undivided, interwoven complex of drainage lines. (*Davis, 1938A, p. 1346*)

It is possible for the 'balanced conditions of a sheetflood flow on a graded slope' to be modified so that the floods begin to erode and then the sheet of water is transformed into 'individualized, confluent streamfloods, which excavate systems of close-set, inbranching valleys along the nearly parallel down-slope lines' (p. 1347). Such flow soon begins to dissect an alluvial fan or rock floor. Probably the chief local cause of transformation from sheet- to streamflood is increase of slope due to differential uplift, though climatic change is a possible factor in some cases.

The relative importance of sheetflow and channel- or streamflood will vary also with the stage of the cycle, initiated by uplifted fault blocks. Davis summarizes the general development of the typical landforms.

While the highland surface of an uplifted fault block is more or less dissected by streamfloods, the bold scarp face is caused to retreat by back-weathering; and the face, as it retreats, leaves in front of it a smooth rock floor of diminishingly convex profile, buried under an aggrading detrital cover which slopes with a faintly concave profile to a central aggrading playa. As the erosion cycle develops, deep embayments are formed in the mountain fronts and grow until those on the opposite sides of an uplifted mass meet and coalesce. Eventually the whole mass is reduced to one or more low domes, which are characteristic of the old-age stage of the cycle. (*Davis, 1938A, p. 1341*)

In the youthful stage, on the uplifted areas, streamfloods accomplish erosion and transportation, aided by rain and rill wash on the steep slopes. When streamfloods emerge from canyons or valleys at the mountain front they spread out into sheetfloods which in time build up the surface of the bahada (waste-floored plain) and carry the finer particles to the aggrading playa.

In the intermediate or mature stage of the cycle of desert erosion in the American West both the valley-side slopes and sides of embayments in the main scarp faces show increasingly strong back-wearing. Davis pays close attention to the development of embayments. A main scarp, he writes, retreats especially fast where it is incised by streamfloods 'which trench the undulating highland, back of the original mountain face'. At this exit or incision the face will be 'worn and weathered back' into a triangular embayment. In time adjoining embayments may consume the intervening mountain face and so replace it successively by sharp spurs, a series of isolated mounts and knobs and finally by a gently-sloping rock floor – called by others a pediment. Davis examines the various theories advanced to explain the production of embayments, and so also of rock floors and pediments, in great detail.

All agree broadly on the backweathering of the adjacent mountain faces but one school of thought considers it is at least accelerated by lateral erosion by floodwater and the other that lateral stream erosion is absent because mountain floodwater escaping from a narrow canyon is transformed into a non-eroding sheetflood. The lateral erosionists included Sidney Paige (1912), Kirk Bryan (1922) and Eliot Blackwelder (1931). Similarly D. W. Johnson (1932), although particularly concerned with rock fans or rock planes, postulated for arid mountain blocks, an inner upland zone of dissection and degradation, a marginal zone of lateral corrasion and pediment formation, and beyond it an outer, lower zone of bahadas and playas.

Since the episodic nature of the runoff made observation virtually impossible, Davis examines the 'method of reasoning' at great length. Lateral erosion by streamfloods should be revealed, he affirms, by some or all of the following minor landform details. Well-marked water channels on the alluvial fans flooring the embayments; basal trimming and steepening of the mountain base near the embayment head where the stream has swung laterally or shifted its course; a relatively large apical angle in a slightly receded bay in the same spot for the same reason; no surviving residual rock knobs or nubbins rising above the fan surface; and differences in the heights of adjacent embayment floors, due to difference in the gradients and sizes of the streams.

Davis now becomes exasperatingly prolix, or cautious, for an exposition in a crack geological journal. It would need a genius to misunderstand his arguments for the absence of lateral corrasion at the edges of the embayments and upon their floors.

Peculiar consequences of back-wearing. – The left part of Figure [134] . . . show[s] the forms expectable if a mountain stream, on issuing from its narrow canyon upon a widening and detritus-covered embayment, is promptly converted into a broad sheetflood which exerts practically no lateral erosion and therefore leaves the mountain sides free to retreat under the action of back-wearing alone. These features are: First, the presence of innumerable enmeshed flow lines; and in association therewith, smaller or larger, fresher or staler, delta-like floodsheets of

F I G. 134. Contrasted large-scale features of embayments produced by back-wearing
(left) and lateral streamflood erosion (right) (From Davis, 1938A, Fig. 9)

detritus, irregularly overlapping each other. (The presence of 'wash' channels is by
no means excluded here, particularly near the bay head; but they are not essential
and they should not be of great length). Second, the absence of steepening in basal
slopes along the bay sides and around the isolated and outstanding spur ends, all
of which should have more or less frayed-out margins. Third, . . . the possible
occurrence of rock knobs and nubbins, rising from the fan floor in front of the bay-
side mountain faces, or in irregular association with isolated spur ends. Fourth, the
sharpness of the apical angle at the head of a bay which receives a large stream
from its mountain canyon, because in such a bay the corrasional recession of the
bay head will be much faster than the back-wearing of the bay-side mountain
faces; conversely, the openness of the angle in a bay which receives a small stream.
Fifth, a prevailingly close accordance of level between the floors of adjoining
embayments where their separating spurs are worn away, provided they drain to
the same playa. (*Davis, 1938A, p. 1372*)

He proceeds to give details of many embayments in granitic and non-
granitic mountain faces which lack basal trimming or undercutting by
streams and, indeed, fail to reveal any other evidence of lateral erosion. He
concludes

It would be significant to discover a desert in which all or most of the mountain
bays had undercut forms along their sides, thus proving their origin by lateral
erosion. (*Davis, 1938A, p. 1387*)

It is interesting to notice that in 1935, Kirk Bryan after further detailed
fieldwork, rejected the idea of lateral stream corrasion being a, or *the*, general
factor in embayment-side retreat. Yet he still considered that where large,
though ephemeral, streams debouch into embayments, the surfaces of the
latter will be slightly convex (fan-shaped) in cross-profile and some lateral
corrasion will take place. However, Davis had already argued (1933A but
printed in 1938A, p. 1368) that convexity and fan-shape are not decisive as

they can be formed by water issuing from its mountain canyon either dominantly as a sheetflood or dominantly as a streamflood.

Older desert mountain forms

The desert mountains so far discussed are relatively youthful as they have been worn back only a moderate distance from the assumed initial fault-scarp and still supply an abundant amount of debris to the piedmont areas. But as maturity develops the embayments are enlarged by back-weathering (aided by rain and rill-wash) of their sides and relatively rapid erosion at their heads. At the same time, valleys on the upland usually widen and become flat-floored and steep-sided. The floors of some may eventually join up with the extensions of the rock floors or pediments that aline the foot of the retreating embayment and mountain faces. Davis in his 1933(D) article, with acknowledgments to Kirk Bryan, described in detail how retreating embayments on opposite sides of a range will meet and coalesce, leaving flanking residual masses separated by 'a completely graded arch'. These large residual masses are in the course of time reduced to increasingly smaller remnants until they are completely removed. In the cycle parlance the landforms have then gone through maturity to old age, or at least to the penultimate stage of the sequence.

A comprehensive term for the equivalent of a peneplain in the normal cycle of erosion had long been sought for in the arid cycle. A. C. Lawson in a classic paper on 'The epigene profiles of the desert' (1915), had suggested *panfan* which he made clear was, as with the peneplain, the penultimate rather than the ultimate stage of degradation. But the term, meaning 'all fan', did not seem appropriate as fan indicates deposits whereas so much of the late-stage area is exposed rock. In 1933(D) Davis thought he had found the penultimate stage of desert erosion in certain *granitic domes* in the Mojave Desert. The much more appropriate term *pediplain* was not coined until the year after his death and he himself always preferred *rock floor* to *pediment* although his friend Carl Sauer talked of pediment passes and pediment gaps for the coalescent features of retreating embayments in the late mature stage of the arid cycle.

However, Davis soon realized that *domes* may rarely also be developed on non-granitic rocks so he called these features *desert domes*. Usually, however, as in the earlier stages of the arid cycle, the penultimate forms of granite and non-granitic mountains were dissimilar. The common penultimate form on rocks of appreciably heterogeneous structure was 'a cluster of small, smoothly graded mounts of gently declivity' (1938A, p. 1388). Each individual mount was too small to develop a good-sized stream and on all mounts 'cloudburst' sheetfloods dominated degradation processes. In contrast, on homogeneous granitic-type mountains the penultimate form was a convex arch, or dome, of narrow convexity if the original mass was very high and of broad convexity if derived from an initial mass of moderate height. The convexity decreased with the delay in the introduction of the change of the sheetfloods from aggrading to degrading agents.

Fine examples of the penultimate form of granitic desert topography occurred, according to Davis, in the broad Cima Dome and in the great Cuddeback Arch in the Mojave Desert. The central convex area forms the upper 700 feet of the Cima Dome and has a diameter of $2\frac{1}{2}$ miles. As already noticed, Davis described the 'granite domes' in the Mojave Desert, at great length in the *Transactions of the San Diego Natural History Society* for 1933(D). This elaborate article would probably have received relatively little notice had not Charles A. Cotton summarized it in his *Climatic Accidents in Landscape Making* (1942) and so gave it and several of its many line-drawings wide international advertisement.

Moreover, much of its content is repeated by Davis in his posthumous paper of 1938(A). Yet the earlier paper happens to be important to the geomorphologist and to the historian because of the high praise and advertisement given by Davis to A. C. Lawson's 'keenly analytical essay' on the 'Epigene profiles of the desert' (1915). In this 'indispensible introduction' to desert landforms, Lawson assumed the upfaulting of a mountain mass to be completed before its recessional degradation began.

> he begins with ideally simplified conditions and successively introduces a large variety of complications by which actual conditions are approached. Close study is needed to appreciate his thorough deductive analysis, which deserves to be ranked as a worthy supplement, for arid regions, to Gilbert's famous essay on Land Sculpture in his 'Report on the Henry Mountains . . .' (*Davis, 1933D, p. 217*)

The cynical reader will not be surprised at Davis' admiration of such an approach to landform analysis but he may be more surprised at the way in which the octogenarian geomorphologist accepts Lawson's idea of parallel slope retreat.

> where the original slope of an upheaved mass is steeper than the slope of repose of its weathered detritus 'hard rocks present persistently steep slopes throughout the entire period of their degradation'. (Lawson, 1915, p. 27) . . . Hence after the original slopes are weathered back to steep faces of proper declivity, commonly about 35° for granitic rocks, they will retreat at a uniform rate parallel to themselves . . . until each side of the upheaved mass is completely reduced to a systematically inclined rock floor of hyperbolic profile, over-spread with a detrital cover, which thickens at a diminishing rate and which usually slants forward, with gently concave profile to an aggrading playa. (*Davis, 1933D, pp. 216–17*)

In Lawson's scheme the retreat of the steep mountain face would in time be prolonged beyond the thin edge of the detrital cover so as to form a bare

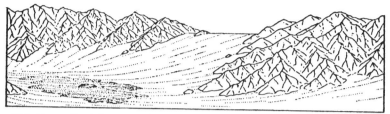

F I G. 135. Generalized features of the Mohave Desert. Some of the ranges still reach far down towards the flat axial floors of the aggraded intermount troughs, where they are wrapped around by floodsheets of detritus. Other mountain fronts have receded to produce sharp ridges separating intermont troughs, or to vanish altogether leaving a smooth skyline arch or dome (From Davis, 1933D, Fig. 9)

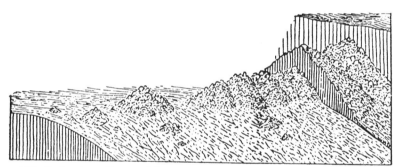

F I G. 136. Reduction of a granitic mountain (right to left) first into indented and embayed margins; later narrowing to an irregular ridge with a serrate crest; then worn through in graded passes; and eventually subdivided into isolated mounts, knobs and nubbins (From Davis, 1933D, Fig. 4)

F I G. 137. Granitic bouldery nubbins scattered near the summit of the Cima Dome, representing the last residuals of the eroded mountains (From Davis, 1933D, Fig. 19)

rock surface with much the same gradient as the detritus. Eventually the extending rock floor would extinguish the surmounting rock residual by meeting a similar rock surface extending up the other side of the uplifted mountain mass. The degraded mass would be a low rock ridge with symmetrical slopes of essentially uniform declivity in their upper parts. However, if the original uplifted mass was circular or ovoid, rather than linear, in outline the eventual shape would be a blunt-angled cone.

Davis, however, differed from Lawson on the question of the penultimate shape of granitic landforms in deserts, and considered that an upheaved mass which did not increase in height towards its mid-line would develop convex rather than concave (hyperbolic) slopes. The idea of the development of convex profiles first took conscious form in Davis' mind during the night of 30 April 1932 when he was camping where the Cave Spring road passes near the base of a large rock knob on the dissected part of the rock-floor piedmont of the Granitic Mountains, nearly fifty miles north-east of Barstow. He soon put the concept into print.

> If an upheaved mass had an originally nearly level surface [as assumed by Lawson] . . . its retreating face must become shorter and shorter and the load of detritus delivered from it must become smaller and smaller as time goes on. A time must therefore come when the delivered detritus will not provide a sufficient load for the piedmont sheetfloods and they will no longer act simply as transporting agencies. They must then begin to rob the cover of some of its detritus or the rock floor of some of its disintegrated grains . . . As soon as such rock-floor robbing sets in, the up-slope extension of the already thin detrital cover will cease, and the rock floor thereafter developed will be of less and less acclivity as it is retrogressively extended farther and farther towards its eventual summit. For after the robbing of the floor has once begun, it must go on at an increasing rate so long as the detritus delivered from the retreating mountain face continues to decrease; and in consequence of such increase of floor robbing the floor must acquire a convex profile instead of maintaining an essential uniform acclivity to its head. (*Davis, 1933D, pp. 219–20*)

In postulating true domes and broad arches as the penultimate form in senescent granite mountains, Davis realized that his views differed significantly both from those of Lawson (who preferred a slightly ridged form that he called a *panfan*) and of Kirk Bryan who favoured the development of a blunt cone on circular masses and of a tent-shaped ridge with a half cone at each end on elongated masses. He could see clearly the difficulty or paradox of postulating, for the continued wearing down of desert domes (draining to aggrading playas), a more rapid degradation at the top of the dome where it is flattest and less rapid degradation a short distance away where the slopes are steeper. One of the main reasons for writing 'Sheetfloods and streamfloods' in 1933 (not published until 1938) was to expound this problem, the answer to which lies in the nature of sheetfloods.

In seeking to understand this apparent paradox, it must be borne in mind that sheetfloods, the main agents of degradation on dome tops, behave very differently from the rivers of humid regions. Besides their great breadth of sheet-like flow, they are short-lived in time and patchy in area; they take up, carry, and deposit their load, all within moderate distances compared to the radius of the dome slope from its summit to its playa, and the deposit laid down by an earlier flood may wait long before it is carried farther by a later one. Insofar as their cloudburst rainfall is provoked by the height of the dome over the surrounding lower land, their average volume should decrease from dome summit to playa; and near the summit, where the divergence of slope lines is pronounced, the depth of a flood should be decreased as it advances. It is probable, however, that they are, in the long run, evenly distributed; and if the extended periods of inactivity are elided and the brief periods of activity are run together, the floods may be conceived as acting almost continuously; and they will be so treated in what here follows.

(*Davis, 1938A, p. 1391*)

The mechanism of the process depends on the relationship between length, depth and velocity of surface flow and load.

A sheetflood forming on the top of a dome may apply all its small measure of energy to the removal of summit detritus, because no energy is there needed for the transportation of detritus received from a still higher source. The flood therefore at once picks up as much load as its small carrying power permits. On moving away from the summit and increasing in velocity with increase of slope, the flood will take up more and more detritus and will thus be kept continually loaded to capacity; but the gain of velocity will be lessened by thinning as the flow spreads over greater and greater breadth on the rapidly diverging slope lines; and the additional load taken up from any given area will be less than that taken up from a similar area nearer the summit, because so much load had already been taken up. As soon as insoaking causes diminution of volume deposition must begin; before long, all the lifted load is laid down again. When another flood begins where a previous flood stopped, the load then taken up will consist largely of detritus previously deposited. As the new flood advances, it will be accelerated as the slope increases, and more load will be lifted and carried; and so on, as far down the slope as the declivity continues; yet at each area where load is increased, the increase will not be so great as the load taken up from a similar area at the flood source.

On reaching the contour or belt of inflexion, where gravitative acceleration remains constant for a brief space, a condition of steady motion will obtain, because all the acceleration is there applied to overcoming resistance and no load can be added. To be sure, a flood beginning at that belt will lift up load and thereby degrade the surface; but the degradation thus accomplished will correspond rather closely to the aggradation caused by a similar earlier flood which stopped there. On passing this belt and coming to the concave slope, flood flow will be progressively retarded because acceleration there falls below resistances. The flood must thenceforward not only cease to take up new load but must deposit some of that already taken up. Inasmuch as all these changes take place differentially, we may conclude that effective degradation is greatest at the dome summit, gradually decreases from a maximum there to zero at the belt of in-flexion, and that the maximum of deposition is on the playa. (*Davis, 1938A, pp. 1391–2*)

The reader will immediately recognize certain difficulties in the above eloquent description. First, Davis assumes increase in slope (and so of velo-city) outward from the dome summit, whereas we are not told here of the *direct cause* of this convexity. Second, he talks only of sheetflood transporta-tion which seems puzzling since he is describing how domes form and once formed how they are lowered. However, these discrepancies are completely removed by the reference to his earlier 1933 article on granite domes and rather incidental references elsewhere:

> bays and domes respectively represent earlier and later stages of sheetflood grada-tion; but in the earlier stages, sheetfloods act chiefly as transporting agencies with a faint tendency to aggradation, and in the later stages they still act chiefly as transporting agencies, but with a faint tendency to degradation on the higher, convex slopes. (*Davis, 1938A, p. 1391*)

Thus Davis recognizes that sheetfloods on bare rock surfaces, granite and presumably non-granitic also, rock floors (pediments) and domes alike, do cause a certain amount of what McGee called 'sheetflood erosion'. The dominant process is the mere removal of the detritus of weathering or to use Davis' term 'rock-floor robbing'. Such 'robbing' by sheetfloods will become relatively more effective as the upstanding mountain mass, and the supply of debris furnished by it, diminishes. In other words, as the rock floors, bare domes and arches increase in extent, the decrease in available detritus allows the load to be carried by sheetfloods flowing down smaller gradients. But that Davis should also give some recognition to slight erosion during rock-floor robbing seems inevitable. Since he also writes at length (1938A, pp. 1402–11) on *sheetflood degradation* in deserts draining to the ocean or to degrading (as distinct from aggrading) playas, and on streamflood channel erosion on rock floors (1938A, p. 1414), he incorporates almost all the best contemporary hypotheses on rock floor or pediment formation into his own analysis.

Davis was well aware that the smooth granitic dome and arch normally did not occur on non-granitic rocks, presumably because they did not weather solely or largely into fine debris. However, he was equally aware that such smooth convexities did not always develop on granite. In fact even his closest followers admit that domed and arched summits are 'rather excep-tional forms in the senescent peneplain of arid and semi-arid erosion'. In 'non-granitic' terrain the penultimate form often consisted of groupings of low ridged and conical mounts, or round-topped hills of gentle slope, whereas in granite areas he had found several examples of assemblies of low, irregular hillocks, crowned with ledges, knobs and boulder-heaps. Davis, being unable to associate these irregularities with his back-wearing, scarp-retreat theory, postulated for origin a form of 'down-degradation' due to streamflood erosion. He assumed that they were the intermediate or late stages of an erosion cycle dominated by streamfloods or even the youthful stage succeeding or accompanying a slow or moderate uplift. They probably had been etched, by stream erosion, out of either the gentle back-slope of a moderately uplifted fault-block or an upheaved mass that lacked a fault-scarp. Davis leaves us in no doubt that he believes in the eventual develop-ment of a smooth dome-like mass.

> Where upheaved masses of low relief are not raised in fault blocks, exposing steep scarps as previously specified, but in broad swells, the small streams which run irregularly away from the crest of the swells proceed to erode branching valley systems; but the headwater valleys are deepened less rapidly than they are widened. The upraised surface thus becomes diversified by many hills of moderate height and slope. The hills are slowly worn down to lower and lower relief, until the resulting domelike mass becomes so smooth that broadly enmeshed sheet-floods replace the branching streamfloods. Examples of this process are believed to have been found in various stages of degradation. (*Davis, 1938A, p. 1414*)

When Davis died he left two unfinished manuscripts (1936A and 1938A) both on arid geomorphology. In his 'Geomorphology of mountainous deserts' (1936A) he gave a brief sketch of his modified arid cycle in which fault blocks are back-weathered, the basal debris removed by sheetfloods to the playa basins causing a rise in baselevel. The slopes are dissected into re-entrants and the thinly-veneered pediment edge encroaches on the moun-tains until they are replaced by desert domes surmounted by steep residuals. These domes are slowly worn down by 'sheet-flood robbing' to a fainter and fainter convexity and the residuals shrink to piles of rocks or 'nubbins'. Allowing that the above scheme applies best to granite terrain where the rock breakup is in two stages (joint blocks then individual mineral grains), Davis points to the greater concavity and lack of basal breaks of slope with other rock types.

Rock floors in arid and in humid climates

In his work on desert landforms discussed above Davis made it clear that his findings were largely concerned with American deserts and should be tested against observations in the rest of the world. Probably the last suggestions he ever made for geomorphological research were for further investigations in the Death Valley area and in the Sonoran desert of northwestern Mexico; and for the exploration of sub-arid regions generally to disclose landforms 'intermediate between those described for the desert and the more-familar forms of humid regions'. He had already attempted a direct comparison between arid and humid landforms in his article on 'Rock floors in arid and in humid climates' (1930E). This is important because with typical opportunism and caution he recapitulates his ideas on both the normal and the desert cycles of erosion. Some readers, noticing the subtle introduction of what may appear to them to be new adjectives and innuendos, may wonder if this is really the geographical cycle on which they were nurtured. But Davis gives chapter and verse from his 'Geographic classification, illustrated by a study of plains, plateaus and their derivatives' (1885D). Here, he wrote that as a body of horizontal strata emerges from its parental ocean a smooth unbroken plain is revealed. As 'effective elevation' is gained, 'rivers establish their courses, the smooth plain is trenched . . . The channels [valleys] will be narrow and steep walled in regions of relatively rapid elevation, but broadly open in regions that have risen slowly . . . Rate of elevation is thus of greater importance than climatic conditions in giving the cañon form to a valley . . .' He then describes, as we have already discussed, maturity and simple old age. Nearly fifty years later he still suggests that the first study of a cyclic sequence is helped by assuming, temporarily, that the cycle of erosion is introduced by a relatively simple upheaval of a lowland underlain by homogeneous rocks. Then all the various complications and elaborations can be studied, as he did in 1905(J) and more fully in his *Die Erklärende Beschreibung der Landformen* (1912J) and as D. W. Johnson had done for a special process in a detailed analysis of the cycle of marine action (1919).

The idea of the cycle of erosion that the octogenarian Davis wished to propagate in 1930 may be judged from his outline of its three essential principles. The reader is especially asked to notice the suggestions of slow or rapid uplift and of 'sequential stages *during* uplift'.

First, the forms produced by the action of erosional processes upon uplifted masses of earth crust vary with the nature of those processes, as well as with the under-structure of the crustal mass, its previous surface form, and the attitude in which it is slowly or rapidly placed by uplift; second, the various members of a group of forms produced by a given process at any stage in its action on a given structural mass are reasonably related to one another; and third, the groups of forms produced in successive stages of an undisturbed cycle follow in systematic order from the initial stage, when uplift begins, through many sequential stages during uplift and after it ceases, to the long-delayed ultimate stage when erosion is essentially completed. (*Davis, 1930E, pp. 1–2*)

After this partly irrelevant introduction, Davis turns his attention to rock floors in arid regions. He praises the descriptions of these features given by Gilbert, McGee, I. G. Ogilvie, Paige, Lawson and by Bryan, who proposed in 1925 the term *mountain pediments*. He then passes on to the description of extensive rock floors, surmounted in parts by abrupt-sided *Inselberge*, in Africa.

They were first described by Bornhardt, who proposed an elaborate explanation of them in terms of successive changes of level with alternations of erosion and deposition. They were afterward examined in detail by Passarge who some years later followed an informal suggestion of von Richthofen's in explaining them as the product of repeatedly changing climates, the deep weathered soils of humid epochs being removed in the following arid epochs. Hence, he concluded that *Inselberge* are *Vorzeitsformen*, the work of past conditions and not now in process of formation.

Several other German observers have adopted less specialized explanations. Thus Jaeger treats rock floors and their surviving island mounts as representing a far-advanced stage of erosional processes; earlier stages include the cutting of steep-walled river valleys and the retreat of the valley walls under attack of the weather. Similarly, von Stapf regards the *Inselberge* of East Africa as resulting from an extreme yet fully normal operation of destructive processes in a late stage of a cycle of erosion and believes that as such they may be formed in any climate.

(Davis, 1930E, p. 6)

Davis considered the two latter more simple views a decided advance on the elaborate schemes of Bornhardt and Passarge but that they did not discriminate sufficiently between the differences in the processes and products of humid and of arid erosion. Most later German workers, such as E. Obst, Leo Waibel and W. Penck, also accepted normal erosion as adequate to explain the production of island-mount landscapes (*Inselberge Landschaften*).

W. Penck believed that the weakening of river erosion to stillstand is decisive for the production of *Inselberge*, but denied the occurrence of an angular change of slope at their base; and like von Stapf, he concluded that *Inselberge* are perfectly normal members in the series of degradational forms which a region assumes in consequence of a sufficiently long, undisturbed relation to baselevel, and that *Inselberglandschaften* may occur in all climates as the result of general destructional processes. But his first statement overlooks the slow degradation that rivers continue to perform after they are graded while their load slowly decreases; the second contradicts the testimony of many observers; and the last statements are questionable . . . (*Davis, 1930E, p. 7*)

Davis, after also describing recent observations by geologists on desert landscapes in east and west Africa, Australia, Mongolia, and Arizona, then attempts to show, first, that the cycle of arid erosion may be reasonably treated as a variant of the cycle of humid erosion, and, second, that in spite of many resemblances arid erosion and humid erosion differ so much in process and product that 'they cannot be clearly understood if they are briefly brought together as examples of normal erosion, as has been done by several of the German writers above cited' (1930E, p. 14).

He proceeds to discuss approvingly the origin of pediments as analysed by American geologists from G. K. Gilbert to A. C. Lawson and Kirk Bryan. Special emphasis is laid on the peculiarity of the comparatively abrupt transition from the pediments or rock floors to the faces of the mountain masses or inselberge. Here the almost angular change may be from a pediment or rock plain with a slope of only 5° or less to a bold mountain front rising at a gradient of 30° or 35°.

To make a fair comparison between the forms produced by a cycle of humid erosion and a cycle of arid erosion Davis imagines a series of granitic fault-block mountains to exist in a humid climate, and compares these with known examples of granitic fault-block mountains in deserts. With typical Davisian method, he discusses the pre-uplift surface of low relief and makes it quite clear that when block-faulting or uplift is initiated erosional processes at once set to work, gaining force as altitude and slopes are increased during the uplift. For our purposes, we hardly need to go into the elaborate details of the early contrasts and resemblances of humid and arid drainage on such fault-blocks. Valley forms, valley floors and their aggradation, dissection and degradation of the mountain mass and arid peneplanation are discussed fully for humid regions, thus giving Davis a chance to restate emphatically his cyclic ideas. New emphases and new impacts abound.

> It may be noted in passing that, while the power of running water to carry theoretically increases with the sixth power of velocity, the carrying power of an actual stream will not increase faster than the fifth power of its velocity, because its cross-section will diminish as much as its velocity increases. That is, if a given stream runs twice as fast as before, it will be only half as large, and it will therefore have only half as many threads of current available for the work of transportation. On the other hand, inasmuch as a stream in flood increases in size as well as in velocity its carrying power will be increased by more than the sixth power of its velocity! (*Davis, 1930E, p. 140, footnote*)

This praiseworthy hydraulic notice does not prevent Davis from ruthlessly elevating load into a powerful hydraulic control of hypothetical floodplain development.

> Furthermore, after reaching a maximum of aggradation in the maturity of a normal cycle, when the discharge of waste from the valley sides is believed to reach its greatest measure, a graded river will have a decreasing load, because of decrease of valley-side height and slant; the river will then resume its degradational work and will slowly wear away the flood-plain deposits; yet in so doing it will not terrace them because their surface will be worn down *pari passu* with the slow wearing down of the entire valley floor. The river will thus in its later maturity sweep away all the earlier-formed flood plain and return to the long-postponed task of slowly wearing down the rock basement of its first graded course, and will gradually reduce it to the fainter and fainter gradients of old age. In the meantime the previously formed and temporarily buried lateral strips of valley floor will be laid bare and will gain increased breadth by the recession of valley-side spur ends! (*Davis, 1930C, pp. 140–1*)

Further on he admits that it is today 'extremely rare to find even a peneplain, much less a plain of degradation, still holding the attitude with respect to baselevel in which it was degraded' (p. 143).

When it comes to comparing the erosional processes and the forms of humid and of arid regions Davis finds that the differences are a matter of degree rather than of kind, that the resemblances are much more striking than their dissimilarities. For example, although the processes of weathering are more physical than chemical in arid regions* and more chemical than physical in humid regions, on bare-rock in the latter, physical forces dominate and beneath detrital deposits containing ground water in deserts weathering may be largely chemical.

Similarly, the 'maturity graded slope' of a forested granitic mountain in a humid climate is really about as well boulder-clad as its barren counterpart in a desert but the former is hidden beneath soil and vegetation. Indeed rock decomposition penetrates to a lesser depth in arid than in warm humid regions where underlying granites may be decomposed in mass 'to depths of 30, 50 or more feet below the sub-surface boulders'. In each, the boulders are for the most part destroyed by disintegration while they are resting on or slowly descending the slope.

A striking difference between granite mountains in humid and arid climates is the development of slope angles with age. In humid areas the slopes flatten as the erosion cycle advances, whereas in deserts they retain a constant angle. As a result, the residual mounds on humid peneplains and island-mounts on arid pediments, although strictly homologous, are not essentially identical.

> The first are pale forms of weakening convexity and lessening slopes which merge, without any sharp lines of demarcation, into the surrounding lower surface, and which gradually fade away as time passes. The second are of vigorous form and are, especially in granitic areas, sharply separated from the surrounding rock floor; they retain their vigour, in spite of losing size, even to the moment of their extinction. (*Davis, 1930E, p. 150*)

Davis did not pursue the 'true homology' which could be drawn between the slow degradation of a humid, soil-covered peneplain of 'hardly perceptible declivity' after its mounts were worn away, and of an arid, thinly-veneered pediment of 'very visible declivity' on which parallel-slope retreat had consumed all its 'peaks and pinnacles'.

The whole of this long article is, however, of immense interest to the student of Davisian geomorphology as it illustrates the octogenarian in his most expansive and all-embracing mood. He begins the breakdown of the

---

* Davis had changed his mind on this relationship before 1933 (see pp. 657–8). This may however be a rare slip of the pen, 'weathering' being used as degradation or all forms of erosion.

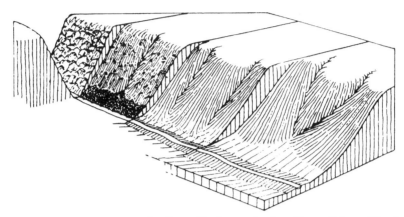

FIG. 138. Four early stages of valley-widening under humid conditions. The final block shows the elements of floodplain, 'lateral valley-floor strips', valley-side slope and intervalley ridge (From Davis, 1930E, Fig. 5)

distinction which he himself proposed between the humid and arid cycles, which is so much a feature of modern geomorphology, by identifying 'homologous features' in the two environments; he is almost Penckian in his identification of three stream segments on the active fault-blocks, with the intermediate course (possessing for a time an ungraded torrent and steep valley sides) merging headward into an open shallow upper valley; humid valley sides are illustrated as possessing narrow basal 'strips' in maturity (Fig. 138) which look suspiciously like embryonic *haldenhange*; humid and arid slopes, although apparently so different, are both identified as 'graded'; and even humid 'pediments', largely masked by soil, are identified.

Throughout this discussion we see many signs of his high deductive powers and of his acute knowledge of the contemporary literature. He has

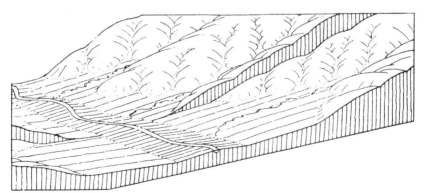

FIG. 139. Two advanced stages of valley widening under humid conditions following those shown in Fig. 138 (From Davis, 1930E, Fig. 6)

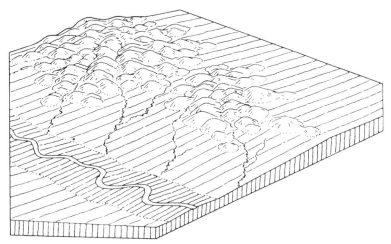

F I G. 140. A late stage of valley-widening under humid conditions (following those shown in Fig. 139) showing the dissection of the intervalley ridge and the isolation of its severed parts (From Davis, 1930E, Fig. 7)

regained his old pawky humour: only a poet from a humid region could have written of a brook that ran 'on forever'; only a migrant from a rainy clime could have exclaimed, on seeing clouds of dust whirled up from a dried-up wadi floor, 'I never saw a river whose bed was so well aired!'. Nor has Davis lost any of his determination to offer advice on further research:

> A fifth, and to me a very baffling, problem is concerned with the relative rates of erosion and degradation in humid and in arid regions. It seems as if humid stream erosion must – other things being equal – be more rapid than arid stream erosion in the early stages of an erosional cycle; also that, in a much later stage, degradation may be more rapid on the bare slopes of an arid region than on the plant-covered slopes of a humid region. It may even happen that an inclined but smooth rock-floor plain of degradation is sooner prepared under an arid climate than a nearly level, soil- and plant-covered plain of correspondingly faint relief can be produced under a humid climate; the reason for this being found in the persistence with which the steady-flowing humid rivers continue to degrade their courses and their valley floors to fainter and fainter gradients until they are very nearly level, thus leaving the fading intervalley hills always in relief, while the intermittent rivers of arid climates seem content to permit the slanting planation of their basins while their gradients are still rather strong; but no definite pronouncement on these problematic matters is here intended. Evidently enough, abundant physiographic problems await the observer who has opportunity for research in arid region. (*Davis, 1930E, p. 158*)

It says much for Davis that these incidental speculations, made almost on his eightieth birthday, have in recent years been proved more right than wrong.

CHAPTER TWENTY-SEVEN

Western Rejuvenation: Coastal Benches, Limestone Caverns and Lakes

Outside the fault blocks and aggraded basins, Davis found many other geomorphological attractions in the arid and semi-arid American West. He had always liked coastal strips particularly where mountains looked on the sea and also took a lifelong delight in lakes, where a liquid contour bisects and reflects the adjacent slopes. The intricacies of limestone topography had also been an earlier interest especially after his visit to the Adriatic karst in company with Albrecht Penck and students from the University of Vienna in 1899. Now, in his old age, his spirit stirred him to undertake elaborate summaries of these topics, a task which his friends and colleagues greatly encouraged him to tackle in a true Davisian vein with utter disregard to length and publication space.

So, Davis returned to the study of lakes, on the classification of which he had written so long before (1882A; 1887B). This revival of interest was heralded by an address on the landslide-dammed Clear Lake in the northern California Coast Ranges (1931C) and resulted in his extensive account of 'The lakes of California' (1933E). This added virtually nothing to the fundamental principles and classifications of limnology, but as a regional summary of a land exceptionally rich in its variety of lakes it is valuable for its wide range of examples. Indeed, it would have been difficult to improve greatly on A. Delebecque's *Les Lacs Française* (1898), and as long ago as 1894 Israel C. Russell had published *The Lakes of North America*.

LIMESTONE CAVERNS

Davis' summary of his own and other people's ideas on limestone caverns was on an even grander scale. It was presented to the Cordilleran Section of the Geological Society of America in February 1930 and appeared in their *Bulletin* for that year as an article of 154 pages, including 8 plates and 62 line-drawings. The bibliography contains nearly 100 separate items. Although rather over-deductive (a weakness lamented by Davis himself), in method, presentation, visual illustration and tone it shows the mature master at his best. We will attempt the unenviable task of reducing the article to a few intelligible pages.

In many parts of the world limestones have been riddled with elongated passages or caverns which are often irregularly interconnected and exhibit occasional expansions in great chambers or domes. The cavern spaces may be partly occupied by dripstones (stalagmites and stalactites) which are considered by Davis for the most part younger than the caverns because supplementary cavern enlargements subsequent to the formation of the dripstones are rarely seen except where the roof has collapsed for lack of support. Thus two chief epochs are recorded in cavern history; an earlier epoch of solutional or corrasional excavation, and a later of 'depositional replenishment'.

This aspect of the cavern problem was given little or no attention in standard textbooks on geology, which generally attributed the formation of limestone caverns to the solutional and corrasional activities of subsurface (vadose) water above the water table, when the limestone tracts have assumed their present attitude with respect to baselevel.

In place of this one-cycle theory Davis proposes a two-cycle theory;

> that caverns may be produced largely by ground-water solution below the water table and then, after regional elevation, drained of their previous water-filling and, on thus becoming filled with ground air made ready for dripstone deposition.
> (*Davis, 1930B, p. 480*)

Current theories for the origin of limestone caverns provide no adequate reason for the reversal of processes involved in cavern excavation by solution and in cavern replenishment by dripstone deposition. It is here proposed to explain this reversal first by ascribing solutional excavation to an earlier time when the cavern region stood so low that the cavern rock was below the local water table and therefore permeated with ground water; and next, by ascribing depositional replacement to a later time when, in consequence of regional elevation, the excavated cavern was raised above the water table.

Some solution of calcite undoubtedly takes place in carbonated ground water as it descends through limestone crevices toward the water table; but as soon as the cavities thus excavated become large enough to be partly filled with ground air, allowing the escape of CO_2, a replenishing deposition of calcite in the form of dripstones is more probable than a continuation of solutional excavation. But below the water table, no such escape of CO_2 is possible; and even if descending ground water becomes saturated with calcite before reaching the water table (which need not always be the case), it will, as it sinks below the water table, gain increased solvent power by reason of increasing pressure. Thus conditioned its water-logged paths of movement may be slowly dissolved out and enlarged to cavernous size at considerable depths . . .

The time element is held to be important in subwater-table solution; for the limestone of most caverns has spent by far the greater part of its existence below the water table. (*Davis and Killingsworth, 1931G, pp. 308–9*)

The sinking of the water-table in consequence of regional uplift or some other effective cause is thus the essential feature in dividing cavern history into an earlier period of excavation and a later of deposition.

Davis had practically completed his article (1931A) before he found out that his main thesis had been anticipated by Alfred Grund in two fine accounts in *Penck's Geographische Abhandlungen* (1904 and 1910). However, Davis had not been able to review the extensive European literature that arose from Grund's ideas. At the same time a few American students of caverns, such as Matson (1909), Lee (1925) and Weller (1927) all had a two-cycle concept, in whole or part. None, however, appears to have paid attention to the significance of the 'network rather than branchwork arrangement of large cavern galleries' which Davis does in his paper.

The discussion, he says regretfully, is academic rather than based on his own fieldwork, as although he visited a few caves in the United States, Europe and Australia in his earlier years he had 'now passed the age when underground exploration is advisable' (1930B, p. 484). First, he considers the terminology and adopts that of Meinzer (1923) as far as possible. However, he coins the term *dripstone* to replace the 'cumbersome pair stalactite and stalagmite', and adds *flowstone* for calcareous deposits formed by laterally flowing water and *rimstone* for those formed around the rims of overflowing basins. Fortunately the two latter terms did not last long.

The long list of acknowledgements for help includes many members of the State Geological Surveys, Mr W. D. Johnson, Jr, of the U.S. Geological Survey, Dr Isaiah Bowman, and Mr Cecil Killingsworth, student of Stanford University, California, 'who in the autumn of 1929 wrote a thesis on the origin of caverns under my direction and this brought the subject, which had been in the background of my mind for several years, to more conscious attention, as a result of which the preparation of this essay has taken most of my time during the spring of 1930' (1930B, p. 486).

The solvent action of subsurface water is discussed first.

> The calcite of limestone rock is soluble in about 30,000 parts by weight of pure water, but it is seven time more soluble in water charged with carbon dioxide under ordinary atmospheric pressure. This gas may be absorbed in small amount by raindrops as they fall through the air; but in subsurface water it is believed to be derived chiefly from the decomposition of organic matter in the soil, probably promoted by soil bacteria, while rain water is descending from the surface as vadose water to the water table. There the vadose water joins the curiously skeletonized body of ground water which so generally and so intricately occupies all the accessible joints, partings and pores of the rock mass below the water table, and which, although fairly well defined at its upper surface, is very vaguely limited downward. (*Davis, 1930B, p. 486*)

The argument now proceeds on typically Davisian lines. The two main epochs in the two-cycle theory of cavern formation will be related to the stages of lesser-importance in the proper sequence of a cycle of erosion during which an uplifted limestone region is worn down to its ultimate planation. The processes will necessarily include the more ordinary work of surface weathering and washing as well as the solvent action of sub-surface waters. These processes are first imagined to act upon a body of 'modern marine limestones, imperfectly indurated, rather porous in texture and with moderate diversity of density in its horizontal strata, elevated as a broad upland 300 or 400 feet above sealevel' with some warping and moderate jointing. When degraded to a lower level this limestone mass is assumed to resemble the lake-dotted upland of central and northern Florida. Here the floors of some of the lakes descend to below baselevel and strong submarine freshwater springs bursting out from the sea-bed offshore, as happens also off Cuba and Yucatan, suggest that deep-seated solution below the water-table occurs in such regions.

Having discussed the evolution of cavernous passages in porous limestone, Davis now proceeds to treat the developmental changes experienced during the first erosion cycle by a broad upland of generally dense, impervious, level-bedded limestones with fairly open bedding planes and well-developed nearly vertical joints. Here sink-holes and caves evolve to their maximum perfection and there is a gradual concentration of vadose water into the low-level galleries. This concept had already been stated by J. M. Weller (1927) and others in the United States but Davis gives fuller recognition to the presence of deep-lying ground-water galleries and to the transformation of a network system of underground passages at the early stage of the erosion cycle into a branchwork system at a later stage.

> maturing cavern streamlets in level strata develop their low-level stretches chiefly by the development of new shafts, and of new galleries at lower and lower levels. It would seem to be largely in this early production of small and local network systems with respect to galleries at various levels and their gradual integration into larger systems based on low-level galleries that the underground dissection of dense limestones differs from that of porous limestones. (*Davis, 1930B, p. 503*)

Davis attempts to determine the relative importance of corrasional and solutional excavation by vadose water and to outline the pattern of passages characteristic of each. He assumes that a subterranean stream with a free upper surface behaves essentially the same as sub-aerial streams, and also that in the dense limestone block normal river-valleys deepen at an appreciable rate.

> The most significant result of the corrasional deepening of low-level galleries should be the transformation of the network system of minute or slender inter-connecting passages with occasional ascending paths, which earlier prevailed under the action of solution, into a persistently down-grade branchwork system of larger passages, with many incoming but no outgoing branches. Accompanying the transformation should go a rather orderly increase in the size of the passages from their heads near the underground drainage divide to the mouth of the concentrated trunk-stream gallery, where it opens into a valley of normal erosion. But no such transformation is to be looked for in galleries at higher levels, from which the streams have been withdrawn. (*Davis, 1930B, pp. 509-10*)

There are some local exceptions to the rule that mature galleries of vadose corrasion should have a branchwork pattern – as for example where the dip of a relatively-impervious stratum has determined a network system of minute passages with a gradient and altitude about the same as that of the graded profiles of the streams that may concentrate there later as they are enlarged. But most dominantly corrasional systems will be free of elaborate loops, especially at much the same level, and of intricate channel-splittings.

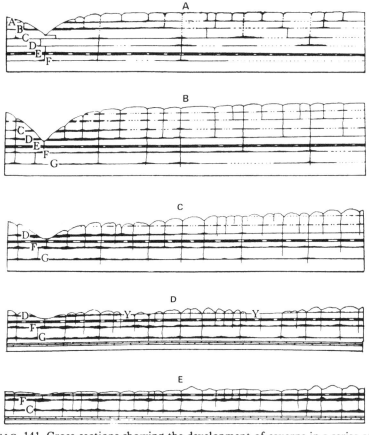

FIG. 141. Cross-sections showing the development of caverns in a series of horizontal limestone rocks (A–G)

A. Youth
B. Early maturity
C. Maturity
D. Late maturity (Showing blind valleys Y, Y)
E. Peneplanation

(From Davis, 1930B, Figs. 4, 5, 22, 23 and 30)

The lower-level galleries will undergo lateral corrasion at bends and restrictions and, after grade is established, will suffer a general widening of their floors.

The relation between solution and corrasion is complex. During early stages of the cycle in a dense limestone mass, solution efficiently dissolves passages along joints and bedding planes; in later stages trickling water, if not yet saturated with lime, may dissolve runnels in the walls of galleries and so slowly widen them. But most slowly-percolating (trickling and dropping) water will be saturated and will deposit rather than dissolve when it enters a large cavern. An exception to lack of solvent power of vadose water at a considerable depth will occur when surface water quickly descends underground down sinkholes and shafts and so is not saturated with lime on arriving at the lower levels. The occurrence of travertine deposits at the exits of springs in many limestone areas is not taken by Davis to be evidence of the solvent efficiency of vadose water. Rather it postulates the solution of a certain amount of calcite from the walls of water-filled cavities below the water table where the carbon dioxide contained in the water does not escape and, as a result, calcite is not deposited until the ground water rises and emerges at a surface spring.

THE MATURE OR KARST STAGE IN THE FIRST CYCLE IN DENSE LIMESTONES

As is illustrated from the karst of the north-east Adriatic coastlands, at the mature stage of limestone development the sinkholes extend, greatly aided by the collapse of underground channels, and streams become intermittent or mainly subterranean. Eventually the sinkholes coalesce into large flat-floored 'blind valleys'; the residual hills lessen in size and are riddled with cavernous passages; the drainage becomes increasingly on the surface and 'the broadened floors' of the rivers 'foreshadow the peneplain to which an entire limestone district will, at a still later stage, be reduced when all its caverns above the water table are destroyed' (1930B, p. 531). Even upon a stationary upland, the valleys may, after their streams have reached grade, undergo first aggradation and then a slow degradation of the valley floors, which continues indefinitely into old age. This idea, which is explained by Davis in his article on 'Baselevel, grade and peneplain' (1902A, p. 96) and again in his 'Rock floors in arid and humid climates' (1930E, pp. 139–41), is based on a decrease of available load. This valley degradation may lower the local water table and so drain some of the deeper-lying galleries of their water, and so transfer them to the vadose water zone. Thus a late-mature one-cycle cavern may imitate the nature of a two-cycle cavern due to regional elevation; this is demonstrated by two underground streams and two drained ground-water galleries in Indiana (1930B, pp. 536–46) which seem to be one-cycle products but, in fact, are 'in an advanced stage of an (n + 1)th cycle of erosion', their region having been peneplained at least once. Large first-cycle caverns in dense limestone are unknown.

FIRST CYCLE STAGE OF PENEPLANATION

At this (penultimate) stage the residual ridges have been worn down (mainly by subaerial erosion) to faint relief and in the process practically all the cavernous passages excavated by vadose water and the shallower ones formed by ground-water solution will have been unroofed and exposed to the sky. Only the solution caverns at deeper levels will still be subterranean and water-filled. Indeed, because the supply of non-saturated water sinking below the water-table may now be increased, the conditions are favourable for the enlargement of deep-lying galleries by ground-water solution. A good share of the rainfall soaks into the inter-stream areas and, because its journey to the water-table is relatively short, may retain a good proportion of its solvent power to well below the water-table. In addition this penultimate stage will, provided no uplift occurs, endure much longer than any of the previous stages.

> Hence the solutional enlargement of a deep-lying network of interconnecting shafts and galleries by ground water, even to the point of widening the galleries irregularly into great chambers in the most soluble layers, may during this long-lasting stage of the erosion cycle come to be of large importance. It is possible that some corrasional enlargement also may then take place when vadose streams descend below the earlier level of the water table. (*Davis, 1930B, p. 551*)

There is also the possibility of the solvent enlargement both of galleries through which sub-surface water movements may be concentrated and of more stagnant deep-lying galleries in which 'a slow circulation of ground water will be set up in consequence of its taking limestone into solution from the walls and roof' (1930B, p. 554).

The long time period involved in the solutional excavation of deep-lying galleries by ground water is important. Davis actually enters into a quantitative estimate.

> The excavation of even the largest cavern by ground-water solution requires only that its water-filling shall, after becoming saturated with calcite, be changed 30,000 times; and if an average of 33 years be allowed for a filling to become saturated . . ., the time required for the total excavation would be hardly more than 1,000,000 years, even without the aid of carbon dioxide . . . If a water-filling is already somewhat charged with calcite when it enters the growing cavern, the time required for total excavation will be prolonged; if the entering water-filling contains some unemployed carbon dioxide the time required will be shortened. (*Davis, 1930B, p. 560*)

At this point, Davis recapitulates the contrasts between limestone cavern and passage forms due to faster-moving vadose water and to slow-moving ground water.

First, percolating vadose water should almost cease to be a dissolving agent once the passage it follows has become appreciably larger than the thread of water, whereas ground-water solution, if it occurs at all, has theoretically no limits to its deep-lying channel enlargement.

Second, high-level galleries initiated by vadose-water solution are likely to remain small with angular networks, whereas low-level galleries minimally above the water-table may be enlarged by vadose water corrasion of their floors which will be graded to some external control, usually a nearby valley floor. In contrast, ground-water solution may open and enlarge deep-lying galleries along bedding planes at various levels because being below the water-table lower galleries do not drain dry the upper galleries. In addition, the channels will tend to be oval and not widest at their floors as in those affected by vadose-water corrasion.

Third, marks of vadose-water corrasion will be common on the perimeters of small galleries and on the floors of large galleries but ground-water solution will tend to cause 'smooth, rounded forms' on the perimeters of all caverns excavated by it.

Fourth, low-level galleries widened by vadose-water corrasion will have a branchwork network with downstream gradients, whereas ground-water solution galleries will be characterized by many loops and outgoing channels, and by irregular interconnections with various irregularities of profile. In other words, both vadose and ground water develop initially ungraded network systems of interconnecting passages, but the ungraded network is maintained by the latter while the former (the vadose) ordinarily converts its channel-pattern into a graded branchwork.

TWO-CYCLE CAVERNS

As no example is known of a maturely-dissected upland of dense level-bedded limestone with a karst surface that has been uplifted by crustal movement, Davis passes quickly on to the regional uplift, or second cycle of erosion, of limestone peneplains.

> If a rapidly and recently uplifted limestone peneplain be somewhere found, trenched by narrow and steep-sided valleys, the caverns to which the young valleys give entrance may be found so extensively developed as at once to suggest their origin before uplift occurred, especially if the largest galleries are not at the lowest levels and if they possess a network instead of a branchwork pattern!
> (Davis, 1930B, p. 562)

Observers who accept the one-cycle theory of cavern origin will regard all caverns in an uplifted peneplain as having been formed by vadose water since upheaval occurred, whereas it is possible that most of them were excavated by ground-water solution prior to uplift, and were modified later by vadose-water action. In fact nearly all the chief dripstone-clogged caverns in the United States are in regions of uplifted and dissected peneplains of dense limestone, such as the horizontal Palaeozoic limestones of Kentucky, Indiana, Tennessee and Missouri; the tilted Palaeozoic limestones of the Appalachian valleys; and, for the most part, Carlsbad and other caverns in the Permian limestones of New Mexico and Texas.

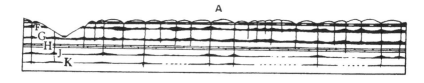

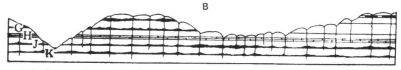

FIG. 142. Cavern formation in an uplifted and trenched mass of limestone layers
(F–K)
A. Soon after uplift
B. Maturely-dissected in the second cycle
(From Davis, 1930B, Figs 35 and 36)

Davis now expends much space in showing that if at an early stage of a second cycle the caverns below an uplifted limestone peneplain are wholly the work of vadose water then they will have a branchwork pattern with the lowest gallery as the only large one. If large galleries occur at several levels below an uplifted limestone peneplain, they are probably associated with pauses in the regional uplift and should occur at much the same vertical interval throughout a region and should bear some correlation to various features of sub-aerial degradation on the adjacent topography. However, if the caverns beneath the uplifted peneplain are mainly due to ground-water solution in the preceding cycle, a three-dimensional network of shafts and galleries would occur; the largest galleries need not be the lowest; smooth, rounded surfaces, intricate loops and many other minor forms characteristic of solution in water-filled cavities would abound. Many examples are described from the Blue Grass and Mammoth Cave districts of Kentucky, central Tennessee, southern Indiana, and the Carlsbad cavern in the Guadalupe Mountains of New Mexico.

Davis finds that in these regions most of the caves are best explained by the two-cycle theory. This even includes the well-explored Mammoth Cave.

So much has been gained from Weller's and Lobeck's reports that it is with regret I am constrained to dissent from their conclusions. The one-cycle theory, demanding that the larger galleries of Mammoth Cave have been excavated chiefly by graded water-table streams during pauses in the elevation of the region, is far from being proved correct. But the alternative two-cycle theory that the many galleries of that great cavern have been excavated in their curiously nonstream-like patterns by ground-water solution during the peneplanation of the region before its last elevation took place, and only subordinately modified by stream action during and since that elevation, is also far from being proved correct; yet it seems to be worthy of consideration in competition with its nonproved rival by those who are fortunate enough to study the problem on the underground. (*Davis, 1930B, p. 603*)

The account proceeds with a description of caverns in dense, inclined limestones where the evidence contradicts the theory of ground-water solution. The problem of dripstone formation is then discussed at some length. In Davis' opinion the evaporation of percolating water is as important a cause of dripstone formation as is the escape of carbon dioxide. Most of the larger galleries experience two epochs, one of solutional excavation and the other of 'depositional replenishment'. Probably the excavation of large caverns, like Mammoth, Wyandotte, Endless and Carlsbad, had been essen-ially completed, except for roof falls, before the dripstone replenishment began. The change-over occurred with the replacement of water with dry air, the withdrawal of the ground water being due either to 'the gradual lowering of the water table as a result of continued valley-deepening in a late stage of an erosion cycle' or to regional uplift of the limestone mass. Even in the first cycle of limestone erosion the withdrawal of streamlets from high to low galleries might also lead to dripstone formation in the higher galleries but such caverns would be small. On this reasoning, stalactites are not a safe index to the age of a cavern but only a measure of the time lapse since dripstone deposition began. The peculiar case of caverns lined with calcite crystals strongly supports the idea of water-filling but here some of the liquid may have been ascending juvenile water. The octogenarian Davis ends this literary marathon with beguiling sagacity.

Favourable consideration has been given, in view of their network patterns, to the ground-water solution of large, two-cycle caverns in dense level-bedded lime-stones. It is not, however, my intention to insist on the proved correctness of the two-cycle theory, in view of its lack of success in accounting for the several levels of certain caverns in inclined limestones; but rather to urge that, before it is either accepted as true or rejected as untrue, much more observational study of caverns, especially in inclined strata, should be made by observers who shall bear in mind all proposed theories of cavern origin and all the consequences deducible from each theory. Care should be taken not to be distracted from the study of cave excavation by the fascinations of dripstone ornamentation ... Whatever theory is under consideration, its relation to the general physiographic evolution of a cave district should be clearly defined and a provisional place should be found under it for every item of observed cave form. Doubtless many items may be, for a time, incorrectly explained by false processes and erroneously assigned to false places; but in the end, after many studies of many caverns have been made by many observers, errors will be ruled out and a successful theory of cave origin will survive. (*Davis, 1930B, p. 623*)

In the third volume of our *History of the Study of Landforms* we shall trace the progress in the knowledge of limestone terrain in the Davis era, as well as the European work which preceded it. Here it must suffice to notice that Davis' article stimulated a spate of calcareous literature in the United States, some for and some against his views. For example, Swinnerton (1932) placed more emphasis on solution by vadose water and on lateral movement at the top of the water-table, as had Alfred Grund and Albrecht Penck. On rather different lines, Gardner (1935) invoked the aid of the solvent action of static ground water accumulated in structural underground aquifers. C. A. Malott, in various articles 1932–52, emphasized that many large caverns were formed largely by surface water diverted underground. However, Davis found a true champion in J. H. Bretz, who in many articles from 1938 onwards postulated a two-cycle theory of limestone erosion, wherein most of the caves were formed in the mature stages of the first cycle by ground water circulating under a hydrostatic head; whereas after uplift vadose water began to modify these caverns and to clear out much of their sedimentary accumulations.

Thus Davis remains the effective introducer to American geologists of the two-cycle theory of limestone cavern formation and this remarkable literary effort of an old man stimulated undoubted progress in the knowledge of karst terrains.

COASTAL PLATFORMS AND GLACIAL SEA-LEVEL CHANGES

During many periods of residence at Pasadena from 1927 to 1932 Davis and his friends made twenty or more trips to a thirty-mile stretch of the southern California coast near the Santa Monica Mountains, west of Los Angeles. Many aspects of the landforms were discussed on the spot with these friends, among them Professor J. E. Wolff (former colleague at Harvard), Samuel Storrow, various local geologists, three recent students from the California Institute of Technology and another three from Stanford University, including W. C. Putnam. It appears that Davis slept none too well in his old age, for we know that although he had recognized on the Santa Monica coast two marine abrasion platforms in 1929, the possible role of climatic changes during the Glacial Period in causing rises and falls of sea-level, did not occur to him until the 'watches of the night' on 27 April 1931 when he was staying in Palo Alto while lecturing at Stanford University.

It is a measure of Davis' aversion to invoking the eustatic mechanism that

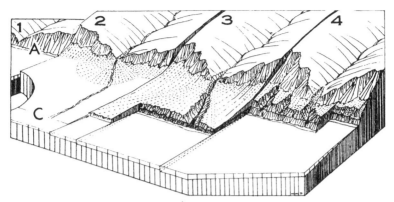

FIG. 143. Development and abrasion of a detrital cover on an emerged marine
platform
1. Cutting of marine platform by cliff retreat (A)
2. Uplift of the coast and fan building on the raised platform (C)
3. Stream dissection and marine abrasion of the raised platform
4. Further fluvial and marine dissection of the raised platform
(From Davis, 1933B, Fig. 5)

it was not until so late in life that he applied it to explain non-coral coastal
forms. Davis was well acquainted with the effect of glacial epochs on the
topography of non-glaciated regions from the writings of G. K. Gilbert
(especially on Lake Bonneville) and of Albrecht Penck, who in 1913 had
published a scheme of the shift of climatic zones in Glacial and post-Glacial
times based on landform evidence (Penck 1913 and 1914). Ellsworth
Huntington had done much the same thing for Mexico and arid North
America generally (1914). Also, R. A. Daly had since 1910 repeatedly demonstrated that coral reefs throughout the warmer seas showed the effect of sea-
level lowering in the Glacial Period. As we have seen, Davis, after his Pacific
voyage of 1914, accepted the occurrence of coral platforms of low-level
abrasion but limited them to narrow marginal belts of the coral seas largely
because to postulate their existence throughout tropical oceans would
demand 'a much greater lowering of terrestrial temperatures than is
indicated by evidence from the Alps, as worked out by Penck and Brückner'
(Davis, 1933B, p. 1115). He was also well aware of the writings of R. W.
Sayles who in 1931 had analysed the probable correlation between the alternations of weathered soils and dune sands in Bermuda and the Pleistocene
glacial epochs of north-eastern America.

He read a paper on the topic to the Cordilleran Section of the Geological
Society of America in March 1931 and to the general meeting at Tulsa in the
following December. The script, entitled 'Glacial epochs of the Santa
Monica Mountains, California', was edited for printing by Professor Douglas
Johnson of Columbia University who, according to Davis, greatly clarified

F I G. 144. Condensed diagram of marine terraces on the northern coast of the Santa Monica Mountains, California. Cape Viscaino is in the background (From Davis, 1933B, Fig. 26)

and improved it. It appeared in full, after a shorter report (Davis, 1932C), in the *Bulletin of the Geological Society of America* in October 1933(B) resplendent with 38 fine photographs and 26 excellent line-drawings. Some of the latter, such as the condensed diagram of the marine terraces towards the north of the Santa Monica Mountains (Fig. 144), demonstrate the characteristic features much more clearly than they can be seen on the thirty-mile stretch of coast where they are very discontinuous.

The coastlands facing the Pacific show strong remnants of cliff-backed marine abrasional platforms at 200 feet and 100 feet above present sea-level which is assumed to be the third and most recent abrasional level. The two earlier-cut cliffs are much modified by weathering and their basal platforms have been largely covered with detritus from the cliffs and the higher ground behind them.

These raised platforms and cliffs, Davis admits, could be partly explained by local elevatory land movements but in any event some account would have to be taken of sea-level changes of 200 or 300 feet during periods of agglaciation and deglaciation in colder areas of the world as discussed by R. A. Daly in his 'Swinging sea level of the Ice Age' (1929). Thus, it seemed probable that the successive epochs of marine abrasion on the coast of the Santa Monica Mountains could be associated with both glacial changes of sea-level and regional changes of land level. The latter would increase, decrease, or neutralize the former.

Davis assumed that the three local records of abrasional advance of the sea coincided with the on-coming, or deglaciation, of three non-glacial epochs and that the two intermediate epochs of sea-withdrawal were glacial epochs or times of agglaciation. The rising sea-level fashioned abraded platforms backed by cliffs and during the ensuring low sea-level subaerial detritus deposits covered the emerged platforms. It also seemed possible to detect a coastal upheaval of about 100 feet during the first of these glacial epochs; and another upheaval, which increases from small measures near the west end of the mountains to 300 feet or more near their eastern end, during the second. (*Davis, 1933B, p. 1046*)

But continued study of the coast convinced Davis that the above simple scheme of three main sea advances and two main withdrawals was inadequate for two reasons.

First, the vast amount of subaerial deposition on the second platform seemed far too large for a simple agglacial epoch; and second, the amount of abrasion in the present Santa Monica Bay, as well as the amount of valley erosion in the Monica plain inland, seemed excessive for one deglacial period.

The present sea floor has a cliffed shore practically all along the Santa Monica Coast; and in the eastern part of the coast, which is adjoined by Santa Monica Bay, the shore cliffs have heights up to 200 feet or more. It may be inferred from this that the Bay floor is a platform of abrasion, more or less cloaked with marine deposits, and that the excavation of the Bay was an immense piece of abrasional work. (*Davis, 1933B, p. 1048*)

Therefore it seemed reasonable to introduce an additional 'pair' (one up-and-down oscillation) of sea-level change between the second and third abrasional epochs and so to assign to the lowest platform two epochs of deglaciation, namely the last glacial and the post-glacial. During these two latest phases it is also assumed that there was no significant land upheaval. Thus on this scheme, the coastal features of the Santa Monica Mountains are interpreted as representing an alternation of three agglacial lowerings of sea-level and four deglacial rises of sea-level (Fig. 145), imposed upon intermittent local uplift and tilting.

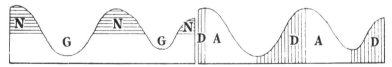

F I G. 145. Diagrams illustrating the distinction between (Left) non-glacial and glacial epochs, and (Right) deglacial and agglacial epochs (From Davis, 1933B, Figs. 1 and 2)

In this elaborate account, Davis explains diagrammatically how, in his terminology, deglacial and agglacial epochs differ from non-glacial and glacial and makes it clear that whenever he uses the latter, more familiar, terms, the former meaning should be understood.

We hardly need to summarize the topographical details and geological sections of the seven segments into which the thirty-mile coastline is divisible linearly. Among the interesting features is the remnantal nature of the highest platform which, however, is perfectly represented in the salient triangle of Point Dume where its cliff base stands 200 feet above present sea-level. It is attributed to the second from last non-glacial epoch. As the second platform on Point Dume is about 100 feet lower than the highest platform, Davis assumed (allowing about 200 feet for a world-wide glacial drop in sea-level) that a land uplift of about 100 feet occurred in the interval between the abrasion of the two platforms. However, he wisely adds

> but the uplift need not have been precisely of the same measure as the difference of altitude between the two platforms, because part of that difference may be due to a difference of sea level resulting from an unlikeness of climate in the two nonglacial epochs. (*Davis, 1933B, p. 1051*)

Yet it seems strange that he did not also consider the possibility that the various glacial or non-glacial epochs might also vary in length.

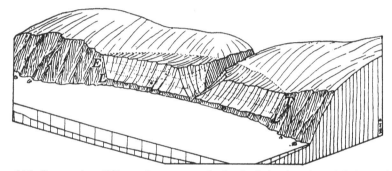

F I G. 146. Composite cliff, partly composed of raised detritus (ELLT), in a bight between the simple cliffs of the two headlands (From Davis, 1933B, Fig. 7)

The second marine platform, which ranges from heights of 40 to 130 feet at Dume Point, sinks to below present sea-level at the western part of the Santa Monica Mountains and rises to 300 feet at their eastern end. This uneven tilting and uplift obviously occurred after the abrading of the marine platform, which is attributed to the oncoming of the penultimate non-glacial epoch. Thus the highest platform must have experienced two uplifts.

Davis stresses the difficulty of recognizing the altitude to which an emerged platform has been raised since its abrasion was completed. This altitude must be measured at the inner border, or cliff-base, of the platform as was stated by Lawson (1897, p. 127) and re-emphasized more recently by Johnson (1932). Thus, when an emerged platform is revealed in a complete cliff-section, its true upheaval is not its present altitude above sea-level but that altitude added to the altitudinal difference between it and the platform at the cliff-base at its inner margin. However, the result requires a further correction because, 'In view of the apparently well grounded belief that the nonglacial epochs were warmer than the present Postglacial Epoch', its value needs to be reduced by 10 to 20 feet.

The analytical value of small surviving remnants of detritus cover of the base of the cliff-line of an emerged abrasional platform is discussed at some length. The penultimate stage of cliff retreat, in which small 'cover-heads' survive, is occasionally represented on the Santa Monica coast. Once a 'clinging cover-head' is completely destroyed, the new cliff loses its composite character and no longer bears any traces of an earlier abraded platform. Then the cliff usually retreats inland gaining in height in the process.

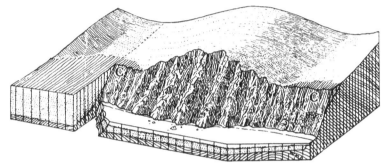

F I G. 147. Small cover-head remnants (CC) on a strongly-cliffed coast. A partial restored section of the original volume of cover is given in the left block, and a section of existing cover on the right (From Davis, 1933B, Fig. 10)

The effect of an isolated ledge of resistant rock in detaining locally for a time the fairly regular retreat of a cliff in a belt of weak rocks is discussed and illustrated.

> Several successive stages in the development of a salient thus caused . . . may be described as faint, strong, peninsular, shrinking reduced, and vanishing . . . I have elsewhere proposed the term cobark, made up like cabal and Anzac from the initials of its defining words, 'cut-out-behind-a-rock-knob', for the shrinking, reduced and vanishing stages of this class of forms, but it has not been well received. (*Davis, 1933B, p. 1062*)

Among the offshore topographic features noticed by Davis are 'submarine valleys' which are common off the Californian coast and 'have been interpreted by some geologists as the work of stream erosion during a time of sea-floor emergence' (1933B, p. 1049). Davis, however, considered that along the Santa Monica coast the lowest marine platform has been heavily masked by marine deposition and that these 'mock valleys' should be ascribed not to river-action on an emerged platform but to the scouring action of marine agencies 'during the recent rise and advance of the sea when the present cliffs were cut back'. He proceeded later to expand this idea in brief contributions on 'Submarine mock valleys' (1933F and 1934B).

Throughout the article discussed above, on the south Californian coast, the reader can hardly fail to notice the complexity and flexibility of approach shown by the octogenarian Davis. He may seem at times rather too didactic but he is in fine fettle and, for example, the nice distinction between glacial and deglacial epochs is typical of his subtle analysis. We share his regret that he could not continue the coastal work and his hope that others might take it up. He never lacked followers and three years later his former student W. C. Putnam described (1937) the probable shoreline cycle on a steeply-sloping emergent coastline as typified by the vicinity of Ventura just north of the Santa Monica Mountains where sun, sea, mountains and active friends had maintained Davis in what seemed to be a prolonged stage of maturity.

CHAPTER TWENTY-EIGHT

Davis' Last Defence of the Cycle

In Chapter 23 we printed in full Davis' long private correspondence with Walther Penck on landform analysis. He always treated Penck courteously and had a great respect for his deductive ability. Indeed at times we can detect an almost paternal warmth for the brilliant youngster. 'The case of Davis versus Penck' is so important to modern geomorphology and has been so grossly mishandled by historians that we intend to give a full account of it. As individuals, both suffered at the hands of critics on the other side of the Atlantic. Davis certainly had his fair share of misjudgments and misunderstandings largely because he outwrote most of his readers. Walther Penck was still more harshly dealt with but misinterpretations of his concepts were partly linguistic. When Germans criticized Davis they had his fine *Die Erklärende Beschreibung der Landformen* (1912J and 1924L) to consult, whereas with Walther Penck's *Die morphologische Analyse* (1924) the English-speaking world lacked any generally available translation until 1953. Penck's German was enigmatically telegrammatic; Davis' English was the last word in endless explication. You need insight to read Penck and stamina to master Davis.

Quite obviously Davis was the person to interpret Penck's message to the English-speaking world. He had already revealed to Americans the thoughts of many German authors on landforms but until 1932 Walther Penck was not among them. The delay may have been due to their private exchanges, or to the untimely death of Penck; but, on the other hand, it may have arisen because a public reply in defence of the Davisian system might look stronger coming from a neutral observer. And who could be better for the purpose than Dr Isaiah Bowman, director of the American Geographical Society of New York and well known in Europe as a peace-consultant and boundary-arbitrator? Davis and Bowman were old friends. In 1903 Bowman as a Junior at the Lawrence Scientific School, Harvard, attended Davis' course on the 'Physiography of the United States'.

In 1904, their relationship gives the lie to the oft-quoted rumour that Davis was cold and aloof from his students.

Dear Mr. Bowman,
 Will you not come with me to the Department dinner for Sir John Murray and Prof. A. Penck at the Colonial Club, Monday evening (informal) at 6:30 – after the conference on Oceanography at the Museum?
 (*Letter: Davis to I. Bowman, 10 November 1904; Cambridge, Mass; quoted in Knadler, 1958*)

Within a fortnight, the invitation is more cordial.

Dear Mr. Bowman,
    This is a late note – but if you are free for Thanksgiving Dinner come take it
with us, at 2 o'clock. Your note about New Haven excites my curiosity.
    Prof. Penck will be with us, and some 'family',
    (*Letter: Davis to I. Bowman, 24 November 1904; Cambridge, Mass; quoted in
    Knadler, 1958*)

In 1905, Bowman graduated from Harvard and joined the faculty at Yale.
Seven years later he acted, with Mark Jefferson and R. E. Dodge, as a
Marshal on the Transcontinental excursion. In most physiographic matters
Davis and Bowman spoke as one voice; the former supplies most of the
evidence without acting as the advocate for his own cause.

ISAIAH BOWMAN ON WALTHER PENCK

Bowman's (1926) article on 'The analysis of landforms: Walther Penck on
the topographic cycle' opens with a brief history of the growth of the study
of landforms from an American point of view. We are told that, following on
the work of Powell, Gilbert and Dutton, Davis 'systematized the sequence of
forms through an ideal cycle and provided a terminology'.* By 1885 these
ideas were being applied to specific landscapes and soon the accepted technique of landform description was expressed in the tripartite evolutionary
theme: nature of pre-uplift relief; agencies of post-uplift erosion, their intensity and observable results; and stage in the cycle.

    Foreign geomorphologists, and especially Passarge, attacked the cycle
concept on the grounds that its alleged simplicity was illusory and because
many topographical features did not appear to fit into its postulated
sequence. An inexplicably persistent misunderstanding was of *stage* to mean
age and the refusal to describe two parts of the same valley as 'mature' where
excavated in weak rock and 'young' where cut in more resistant rock.

    Having outlined the Davisian concepts, Bowman proceeds to review
Walther Penck's *Die morphologische Analyse*, which, with its fine illustrations, he commends for 'thorough reading to every American student of
physiography' (Bowman, 1926, p. 124). Its outstanding quality, says
Bowman, is its explanatory method which applied ingeniously to fieldwork
studies in northwestern Argentina, Anatolia and central Europe, had
produced the clearest and most profound book to criticize 'the American
school of physiography'. As it attempts to rewrite the systematic study of
landforms and to place it on a new foundation, its arguments will eventually
have to be met in detail. Bowman, however, intends to indicate its general
concepts and the chief difficulties in their application.

---

* Bowman's manuscript was, in fact, submitted to Davis for editing before publication, and
this phrase was added by Davis (see Bowman, 1934, p. 184).

He criticizes Penck for believing that in the Davisian cycle progress takes place *always in a definite sequence*, whereas Bowman and Davis considered that the cycle was an orderly progress subject to interruption and complication at any time. Rapid uplift, followed by prolonged stand-still, was to Penck a *special case* and, says Bowman, the uninterrupted cycle continued to its completion would indeed be a *special case*: 'but there is nothing new in that'.

Penck considered that the relative rapidity of uplift, erosion and denudation must be taken into account from the start of the crustal movement. The principle of the *mobility of the crust of the earth* and its corollaries are the bases of his geomorphological scheme. Yet, says Bowman, although Penck thinks he has devised 'a new and entirely different formulation of the cycle' he has really only expounded in detail *complications* of the Davisian cycle.

However, Bowman greatly admires Penck's elaborate analysis of the formation and development of slopes, which is a contribution of the first order, far excelling anything similar done in the United States except perhaps the analytical work of the late Joseph Barrell of Yale University. Such studies would 'put a large portion of weakly-descriptive physiography upon a rational basis'. Penck closely relates uplift and erosion. Thus the gentler slopes of the crest and upper flanks of the front range of the Andean Cordillera in north-western Argentina are to him evidence not of an uplifted Davisian peneplain but of the results, during prolonged uplift, of erosion which also produced the thick sediments that flank the mountains and form the floors of the adjacent basins.

> Certainly he has brought into the literature a number of useful terms and has challenged prevailing explanations at a number of quite critical points. His recognition of the piedmont step; the brow of the uplifted block upon which an old erosion surface may be identified; the sharp topographic unconformity that exists between the old erosion surface and the residuals that rise above it and the remarkable localization of this last feature on a line almost as definite as a strand line; the importance of studying the effects of piedmont stripping upon an upland border where accumulation of sediments was made before an old erosion surface was developed upon the adjacent uplifted mass; the agencies which effect a change in the form of a landscape from concave to convex or from convex to concave; the constant challenge with which he meets a new grouping of forms – these are among the most important contributions of his book. A far-reaching influence is attributed to *intensity and degree of uplift* (as opposed to climate) in the modeling of slopes. If convex, the slopes are a response to rapid uplift; if concave, they bespeak a slower rate. (*Bowman, 1926, p. 128*)

These and other qualities ensure that if the book were translated for the use of American students 'it would certainly lead to a profound stimulation in physiographic fieldwork' in the United States as was, indeed, needed.

Bowman raises four main criticisms of *Die morphologische Analyse*. First, except the central German mountains, most of the illustrations are for border ranges of cordillera where the specialized or concentrated phenomena are not 'a key to systematic physiography as a whole'.

Second, Penck almost ignores American physiographic literature and fieldwork. For example, he does not use G. D. Louderback's (1904) work on the Basin Range structure of the Humboldt region. Here on the present faulted mountains (see our Chapter 26) cappings of lava flows have preserved the pre-existing peneplain and the relative lack of erosion on these cappings shows that uplift was rapid enough to 'put them in their new position before significant erosion could take place'. In other words, rapid uplift was more important than the amount of erosion during uplift, a conclusion supported by the work of W. H. Emmons (1910), and Alfred Knopf (1918) during Penck's lifetime, by Davis after his death, and most recently by Schumm (1963).

Third, Penck overlooked the recognition of 'varying rates of uplift in the scheme of the cycle' and emphasized 'precisely the wrong things in his exposition'. Bowman (1926, p. 126) says that Davis made a special point of the mobility of the earth's crust and of pre-uplift relief and of 'subyouthful stages during uplift', yet Penck insists that Davis was still a slave to the simple rapid-uplift followed by prolonged stand-still idea, which was rather an instructional device for beginners. Here we must interpolate that in fact Penck (1924, pp. 8 and 12) gave Davis credit for keeping well in mind 'the importance of concurrent uplift and denudation' although 'he scarcely ever made use of the notion, and his followers never'. This was a fairly just remark in the early 1920s and it is difficult to see how Bowman could have missed it. The comment was, of course, not true of Davis' later writings after his migration to the American West. Bowman goes on to show that Penck had found in the mid-German mountains, the rounded, smoother higher surfaces of a dome (continuously uplifted and expanding) flattening and progressing towards old age while the lower margins of the dome were acquiring youthful forms which worked headward and so replaced the older forms of the interior upland. In other words, the Mittelgebirge were advancing first to old age and then to youth. But Bowman could not agree that this reversal of the normal cycle was 'a discovery that requires the revision of physiographic science'. Davis had already shown in various uplifted uplands that during the renewed dissection in headwater streams, 'youth' in the sense of steep-sided ravines may be completely by-passed as the valleys deepen slowly and widen as they deepen.

Bowman's fourth major criticism is that the relation between isostasy and physiography is not known with sufficient accuracy to base any fine argument on it. Assuming that he means by isostasy, crustal movements, few would disagree with his suggestion and least of all Penck who wished to use visible surface forms as a guide to the intensity of such movements.

The review ends on a prophetic note. According to Bowman eventually the Davisian topographic cycle will be 'the most important part of interpretative generalization' in landform analysis but it will be combined with the 'fundamentals of isostasy'. It is of note that some years before Davis (1910L) had attacked the strict application of the isostatic theory.

A copy of Bowman's review was sent to Albrecht Penck, now sixty-eight years old, whose reply contains the following points.

> I must agree fully with you ... that we must conceive every form to be primarily a link in a definite developmental series. That is the permanent contribution of Davis that he has brought this point of view into modern physiography.
>
> But a land form can also be taken as the point of departure of other observations ... My son set up an entirely different problem from Davis. It was not in the least his aim to reform Davis' cycle or even to establish a new system on the basis of special cases. Davis wants to explain the manifold forms of a landscape, and Walther Penck wants to deduce from them the intensity of crustal movements ... While Davis lays the main emphasis on the factor time, Walther places in the foreground the intensity relation of the processes, viz. the effect of processes in a given time unit. Davis thereby is naturally led to the terms young, mature, old; Walther to the recognition of the fundamental difference between ascending and descending [waxing and waning] development. (*A. Penck, 1926, pp. 350–1*)

Concerning this letter, Bowman points out that Walther Penck's approach – essentially geological and undesirably non-climatic – has limitations of value and scope not inherent in Davis' cycle which, because 'his expression of the form may be refined *at any stage or at all stages* by the closer analysis of causes' and is 'a growing principle ... just as evolution is a growing principle, and not a dead system rigid in concept and precisely final in form' (Bowman, 1926, p. 352).

We are tempted to criticize the logic and scientific application of this assessment but must refrain. From our point of view the essential fact is that later American ideas on Walther Penck can only be understood with reference to Isaiah Bowman's long review.

It seems clear to us that Bowman for some reason, perhaps lack of space, omitted the fact that Walther Penck was trying to devise some general types of crustal movement that were reflected in general types of landform assemblages. Penck was concerned with a general scheme and not with a 'special case', although he only discussed a few special areas in detail. He claimed that outside the very large tracts of the continents that were stable, were unstable areas liable mainly to local lateral (compressive) folding and/or to regional uplift or subsidence of a vertical nature. Most great fold ranges (geoanticlines?) such as the Alps which had experienced severe lateral compression owed, he said, their present altitude to regional dome-like uplift which accompanied or followed the folding. But dome-like upheaval could cause high tracts without folding and could affect areas outside the mountain belts.

Thus the earth's surface exhibited three main types of tectogenetic units: cordillera or great fold ranges affected by regional doming; regions arched up by doming; and regions of stability. In each there was a certain 'linking of slope, form associations and sets of land forms'. In *Die morphologische Analyse* Penck wrote at length on 'broad folds', but said relatively little on piedmont flats and benchlands (regional doming), and less still on *inselberg* landscapes which were to him distinctive not of any one climate but of continental masses where erosional intensity was low. He never developed his ideas on regions of stability but continued his field studies on Piedmontflächen in the southern Black Forest (Penck, 1925) and it is this work which Davis later reviewed. Thus Davis' views on Penck discussed below were concerned primarily with one special aspect and, although typically he did digress, he must not be blamed for not including what Bowman omitted to say on the universality of the Penckian system.

During the first three decades of the twentieth century most geomorphological ideas relating to the character of the vertical uplift of mobile areas derived from inferences which had been drawn from the associated stratigraphic record. Following the development of the facies concept, it was believed that the distribution of sedimentary facies provided the key to an understanding of the pattern of vertical uplift of the source areas. Thus, some workers (e.g. A. Penck, 1919) recognized the possibility of a wide range of such patterns, whereas others preferred to interpret the stratigraphic record as indicating mainly short periods of rapid uplift. There can be no doubt that much of Walther Penck's geomorphic work was an attempt to provide physiographic support for a general pattern of uplift which he had previously inferred from stratigraphical evidence. The importance which Penck placed on the ability to identify the movements of the source area from the record of sedimentation is clearly stated in Chapter 1 of his *Die morphologische Analyse* (1953, p. 5) and resulted largely from his studies in the Andes (Penck, 1953, pp. 277–9). Where he employed physiographical reasoning regarding uplift history, it was often in the acknowledged absence of strata which might be correlated with the uplift (this was particularly true of the Black Forest; Penck, 1925, p. 86: see also Penck, 1953, p. 86). It is interesting that nowadays few geologists would attempt more than to postulate the occurrence of some rather generalized uplift from the sedimentary record alone, and not infer the pattern of uplift in any great detail. This reticence grew largely from the development of studies of the relations between tectonics and sedimentation – i.e. of 'sedimentary tectonics'. In 1917, in a most important paper, Barrell showed that much of the character of the sedimentary record is controlled by the nature of the subsidence of the basin of sedimentation, as distinct from the behaviour of the adjacent source area. Although these behaviours are often so closely linked that it is difficult to distinguish between the two, the work of Barrell began to cast doubt on the simple association between the nature of sedimentation and the pattern of uplift of the source area. This doubt, however, did not really become universal until after 1930 when geomorphologists were forced to adopt much less doctrinaire views regarding the interpretation of patterns of uplift from the sedimentary record (Chorley, 1963, p. 961).

DAVIS' VIEWS ON WALTHER PENCK

The posthumous publications of Walther Penck's researches were constantly brought to Davis' notice by friends and no doubt by Albrecht Penck's visit to him at Tucson in 1927. Yet he remained silent in print, apart from incidental comments in his various articles, on Penckian conclusions that did not support his own views, such as Penck's denial of an angular change of slope at the base of *Inselberge* quoted in our Chapter 26. However, we are left in no doubt that Davis found much to admire in *Die morphologische Analyse* although dissenting *inter alia* from its new non-cyclic terminology. In 'Rock floors in arid and in humid climates' he writes

> A more general discussion of land sculpture is to be found in Walther Penck's *Morphological Analysis*, written shortly before and published shortly following his lamented death, which occurred soon after he had been appointed professor of geology at the University of Leipzig. His volume contains a restatement of various established principles as well as the announcement of certain new ones, the latter being based in part on the author's observational experience in Argentina and elsewhere, and in part on his well-developed deductive faculty as well as upon his wide reading; and thus constituted the whole is profitable although sometimes difficult reading. (*Davis, 1930E, pp. 3–4*)

In a footnote Davis explains that Penck's 'morphological analysis' may be defined as

> a procedure by which the nature of crustal movements is to be inferred from external processes and surface forms; hence physical geology is its object and physiography is only a means to that end . . . Walther Penck's critical discrimination in the use of terms is illustrated in connection with recently uplifted peneplains, which most physiographers have been content to describe as such while fully recognizing that they have suffered more or less areal degradation as well as linear dissection in consequence of their uplift. W. Penck, however, distinguishes by name between a lowlying peneplain [Endrumpf], still in the penultimate stage of a long-undisturbed cycle of erosion, and the slightly modified surface form [Primärrumpf] of the same mass as soon as it begins to suffer elevation. (*Davis, 1930E, p. 5*)

It appears that sometime in 1931 Davis decided to write a critique of Walther Penck's posthumously-published description of 'The piedmont benches (flats) of the southern Black Forest' (1925).

> I think Dr. Davis shared my appreciation of the views of the younger Penck and at my instigation he read the book with interest and even wrote a paper on Penck's

ideas. (*Letter: E. B. Knopf to R. J. Chorley, 27 January 1962; Stanford, California*)

News of Davis' essay on Penck filters through to his letters where he is typically much more forthright than in print.

It is pleasant to have your letter of the 20th telling of your visit to [Albrecht] Penck during a summer abroad. I am glad to know that he is revising his *Morphology*, a belated task. It will be interesting to see how he treats the cycle of erosion, and how far he follows his son's schemes. My article will probably exasperate him, because he has explicitly stated that his son was right where I believe he was wrong; and all American opinion I have consulted is on my side.

(*Letter: Davis to R. J. Russell, 26 April 1932; Pasadena, California*)

The passage in W. Penck's book on convex profiles is not satisfactory to my reading, because he explains convexity of profile chiefly by increase in rate of upheaval and consequent increase in rate of stream-downcutting. In doing so he overlooks the plain possibility of the production of a convex profile by the rounding off of the 'shoulder', after uniform uplift and uniform slope of valley-side, by action of weather and creep. Similarly, concave profile he usually ascribes to diminution of rate of uplift; but it may be produced by retreat of a uniform slope. He on some other page assumes or demonstrates that valley sides retreat parallel to themselves, and thus brings about his extraordinary idea that the slopes of a peneplain are all concave. He also confuses an ultimate plain of degradation with a peneplain. And so on and on. Some or much of his best work is only elaboration of previous ideas, in unncessary prolixity.

In view of his Black Forest essay, written during his illness, I can't help thinking his illness led him to erroneous deductions; affected his mental working.

(*Letter: Davis to A. O. Woodford, 10 November 1932; Pasadena, California*)

Before embarking on Davis' article we ought in all fairness remark that had Penck seen the comment on 'valley sides retreat parallel to themselves' he might well have written of Davis 'I can't help thinking his old age led him to erroneous deductions . . .!' But Walther was long since dead, his father was well over seventy and Davis was so strongly dominant in the English-speaking world that there was no one of sufficient stature to challenge this Olympian oracle. However, Davis was eighty-two years old and the task was excessively difficult and it would be hard to decide which should be the greater, the praise for the remarkable effort or the blame for its defects. The point is that his comments and diagrams were eagerly and widely accepted by the English-speaking world as models of accuracy.

PIEDMONT BENCHLANDS

The general principles of the piedmont benchland problem are discussed in *Die morphologische Analyse* where it is shown that various mountains in central Germany are not surmounted by a common single erosional level as was previously assumed but reveal a stepped series of such flattenings.

'Young Penck', to use Davis' term, develops this idea of 'Piedmontflächen' and 'treppen' more fully in his 1925 paper on the southern Black Forest. The main thesis is that a domed upland experiencing continually-accelerated uplift and continually-expanding area will develop a series of stepped or terrace-like flattenings on its flanks and of downward-broadening valley-in-valley forms along its watercourses, as may be seen in the Black Forest. Davis and most American physiographers consider that these features are better explained as the result of erosion during successive pauses in an intermittent upheaval, a view explicitly rejected by Penck. The explanation of the observed landforms is based on certain general principles but before discussing these in detail Davis draws attention to some of their character-istics. First, they are highly deductive as Walther Penck was 'an avowed advocate of the careful use of this essential mental process' (Davies, 1932G, p. 404). Secondly, the principles range almost from mere truisms to elaborate ideas, some of which are 'very hazardous, if not altogether incorrect'. Thirdly, the explanation of the observed landforms of the Black Forest is based on the potential form of a dome (gewölbe) undergoing continuous expansion and acceleration of uplift, as distinct from that of a 'broad' or 'great' fold (Grossfalte) which does not expand or may actually contract in width during compressive up-arching. Regional doming, exemplified by the Black Forest, was the tectonic assumption which Davis attacked in 1932(G); whereas broad folding, producing thrusted and superficially-faulted 'basin and range' structures, was mainly treated in *Die morphologische Analyse*.

F I G. 148. Davis' view of Penck's benchlands on a domelike highland (From Davis, 1932G, Fig. 1)

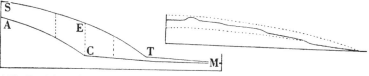

F I G. 149. Davis' profile of an expanding dome with a basal angle. This attempts to show that it is impossible to maintain a steep marginal angle by uplift which is consistently greatest over the dome crest and decreases towards the margin. To produce angle STM, the uplift (EC) of the previous marginal angle ACM must be greater than that towards the dome centre (AS) (From Davis, 1932G, Fig. 2)

Where both sorts of movements were associated, Penck believed, as in the Alps, the most complex structures were produced.

Davis examines the various kinds of crustal upheaval movements and illustrates them diagrammatically (Fig. 150). He begins with two highly deductive and most improbable cases: (A) the essentially instantaneous uplift of a landmass on which no change of form occurs during upheaval:

> This is clearly an impossible case, except for small movements in earthquakes; but it may be temporarily considered to advantage as a hypothetical case in elementary presentation. (*Davis, 1932G, p. 406*)

and (C) the uplift of a smooth surface, such as a sea-floor, that emerges so slowly that its rise is counterbalanced by degradation. In both, 'most of the time demanded for their degradation' would be spent after their soon-reached maturity. Davis proceeds to add many possible types of uplift, such as

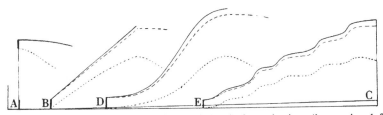

FIG. 150. Varying patterns of uplift (solid lines) through time (increasing left to right) and of the accompanying altitudes of the main divides (dashed) and of the main streams (dotted). The thick vertical lines show the general relief when upheaval begins

A. Instantaneous uplift
B. Uniform uplift, suddenly started and finished.
C. Very slow uniform uplift
D. Accelerating and then decelerating uplift
E. Uplift complicated by pulses and pauses

(From Davis, 1932G, Fig. 3)

(B) uniform upheaval, starting and stopping suddenly; (D) uplift that starts slowly, accelerates at a fairly rapid rate and then ends about as slowly as it began; or (E) intermittent upheaval by impulse and pauses. But he considers that the establishment of such a curve for any particular region, for example the Massif Central of France, would be very difficult. However, Penck attempts this for the Black Forest and chooses type D, accelerated uplift, combined with lateral expansion of the dome, to explain the marginal benchlands. Such uplift is simply one of the 'innumerable kinds of upheaval' that a landmass may suffer and, says Davis, Penck's choice is merely a special case under the comprehensive scheme commonly known in America as the cycle of erosion.

Having somewhat biased the reader in his own favour, Davis now examines the various principles used by Penck to explain the observed landforms of the southern Black Forest. These, for clarity, Davis (1932G, pp. 401–4) had interpreted as 24 main statements, and we will retain his divisions and translation but will precede each of our major groupings with relevant quotations from Penck's Black Forest paper.

Development of valley-sides

The amount of denudation which takes place while a stream cuts down to a given depth is greater if the rate of erosion is smaller, and conversely. In other words, assuming uniformity of the rocks, the more swiftly a river cuts down, the steeper do the valley slopes become, and conversely. If a river's rate of erosion increases, so the valley slopes growing up from it become steeper in successive units of time. *Convex slope profiles result.* As erosion intensity weakens, the reverse is found, *concave* profiles result. . . . A valley slope which first experiences increase, then decrease and standstill of erosion, has one after the other convex and later concave forms. Every convex and concave break of gradient has, therefore, a definite meaning in denudation relief, the first indicating increase, the second decrease of the rate of erosion. (*Penck, 1925, pp. 88–9, trans. by Martin Simons. See also the quotation on slopes from the same source later in this chapter.*)

1. If the stream's erosive power is progressively increased, its valley sides will steepen from top to bottom; that is, will have convex slopes. If the erosive power is decreased, the side slopes will become concave.
2. When valley deepening ceases, the degradational retreat of the valley side goes on parallel to itself and leaves a surface of less declivity [a valley floor] below it as it withdraws.
3. A valley which was first deepened somewhat rapidly while its stream was increasing in erosive power and later deepened more slowly while the stream's erosive power was decreasing, will have its side-slopes convex above and concave below. Hence every convex valley-side slope means an increase of erosive power, and every concave slope a decrease.
4. A stream's erosive power depends, for a given volume, wholly on its gradient . . . A stream will in time develop a graded course, which is the steeper the smaller the stream, and along which the down-cutting is exactly as much as the rise of the land mass.
5. But if upheaval is accelerated, the graded stretches are disturbed . . . and at the same time the erosive power is temporarily increased by increase of gradient.
6. Under otherwise equal conditions, convex valley-side profiles therefore indicate acceleration of upheaval.

Davis, saying that (1) called for no comment, wrote at length only on principles (2) and (3) (Davis, 1932G, pp. 408–10). The idea that a valley side retreats parallel to itself seems, he says, to be erroneous because that retreat is usually accompanied by a lessening of slope as well as by the development of a convex profile at its top or shoulder and of a concave profile at its base. The upper convexity is due to weathering and creep on its two faces and not necessarily to increase of the stream's erosive power. Moreover, 'the principle of parallel retreat' would lead to sharp ridges on the penultimate hills of a peneplain (Fig. 151A) such as could hardly form in the Black Forest unless

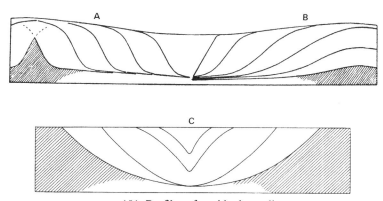

FIG. 151. Profiles of a widening valley

A. The parallel-retreat of a double-curved valley-side slope producing sharp penultimate hills which 'may have been Penck's view of the case'

(From Davis, 1932G, Fig. 4)

B. Retreat of a valley-side slope accompanied by a decrease of steepness and the development of a convex profile above and a concave profile below, favoured by Davis

(From Davis, 1932G, Fig. 4)

C. Formation of a concave valley-side slope by the parallel retreat of slope elements, postulated by W. Penck

(After W. Penck, 1925; From Tuan, 1958)

a capping of local hard rock held up the erosion of the summits. In certain cases nearly parallel retreat of slopes can occur as in granite mountains in an arid climate where boulder-covered slopes can (according to Lawson and Bryan) maintain their steepness as mountain masses diminish to inselbergs. But, says Davis, these are in arid climates and ordinarily penultimate hills are low, gentle convex swells.

In any case, convex upper side-slopes of a well-opened valley are not reliable evidence of accelerated downward erosion unless other independent evidence supports that conclusion. As the same kind of convexity may result from different kinds of uplift it seems hardly logical to use landforms to determine the nature of crustal upheavals.

Development of stream profiles

... the volume of water in every river increases from the upper to the lower course. The lower course is thus always a more powerful agent of erosion than the upper. The lower course is consequently always the first to be able to counteract the increase of gradient produced by the accelerated upheaval. Thus, if the upheaving, which in time units 1–3 was uniform (thus, itself uniform, the same amount of incision of the long profile), becomes now more rapid, so inevitably the water-rich lower course immediately erodes deeper than before, in each time unit, but not the feebler upper course. A *convex nick* is formed in the long profile of the stream. This nick eats back in accordance with the laws of headward erosion, and forms an erosion base for the upper reach of the course. This, therefore, is no longer tributary to the erosion base at the edge of the rising mass, but to a point in the valley course which is not sinking relatively, but has become upheaved and, moreover, moved up valley. The upper reach of the river is from now on withdrawn from the renewal of gradients caused by the upheaval of the mass: the nick, local base of erosion for the upper course, has become lifted up and, in addition, moved up valley, and thus rises relatively to the upper course section. In this, consequently, erosion intensity weakens and concave slope profiles come to be formed there. Continuous acceleration of upheaval leaves in the long profile of the stream one convex nick after the other, all move up valley, below each one there begins a narrow, steep course reach, with convex valley slopes, above each there is a broader reach with concave slope profiles. (*Penck, 1925, pp. 89–90, trans. by Martin Simons*)

In his interpretation it is interesting that Davis insists on imposing the term 'grade' on Penck!

7. The lower course of the river, having a larger volume and therefore also a greater erosive power than the upper course, will be the first to be able to overcome the increase of its gradient due to accelerated upheaval.
8. Hence, as upheaval becomes faster, the larger lower course cuts down faster than before, but the weaker upper course does not. Thus a convex nick is formed in the river profile [at some unspecified point] separating an upper and a lower segment of its graded course.

Davis says little on these statements except (8) which 'is of the highest significance in Penck's scheme' and needs therefore careful examination (Davis, 1932G, pp. 411–15). It is, he believes, 'wholly erroneous'. A radial stream on an actively upheaved dome may consist of three parts (viz., the upper, small, feeble, headwaters with weak gradients; the actively-corrading, non-graded, medial torrent course; and the more voluminous, graded lower course) but these, Davis asserts, merge into each other and are not sharply and abruptly separated.

The gradient of the headwaters gradually steepens as the stream passes along a convexity of its profile into the beginning of the medial torrent, where valley depth is actively increasing. Conversely, the lower end of the torrent gradually increases its gradient and passes along a concavity of its profile into the graded lowest course ... (*Davis, 1932G, p. 412*)

If the dome is uplifted slowly the graded course will occupy the greater part of the stream length and will accomplish more work than 'the headwater torrent'. This, says Davis, appears to be the case assumed by Penck for the southern Black Forest. If in a later state the uplift accelerates, the torrent course (presumably the medial torrent) will extend retrogressively upon the feeble headwaters and progressively upon the graded lower course. But will this extension produce nick points?

In an advanced stage of uplift the 'torrent head' works back to the centre of the dome crest and develops there a sharp divide downstream from which the steepened torrent may be minutely irregular but would retain an 'almost continually concave profile' with no possibility of important nick points if the rocks were homogeneous. The progressive encroachment of the torrent on the graded lower course also poses serious problems and even Davis' most ardent admirers must admit that his own arguments now become obscure. He admits the truth of statement (5) but thinks that 'at the stage of development considered in Penck's statements (6), (7) and (8) there is nothing new in this offering of added power in the lower course' because it has been made continuously and increasingly ever since uplift began. With accelerated uplift:

> The graded lower part of a stream at once expends, in maintaining its graded flow, all the added power that is offered to it. Although the upheaving forces wish to upheave the stream, it does not wish to be upheaved and it has power enough to satisfy its wish. As it thus expends what it receives, it becomes no more powerful than it was before. (*Davis, 1932G, p. 413*)

Apart from the mode of expression, which may not appeal to most scientists, this conclusion seems to us meaningless unless it is changed to 'after it has expended what it received, it becomes no more powerful than before'. And surely, as Penck has stated that the dome also constantly expands, the outer margin is in fact a newly-elevated one. However, from his own scheme of stream development, Davis proceeds to state that no nick can be produced under the condition of continuously accelerated upheaval.

> The torrent must always merge into the graded course, and the graded course must maintain its even profile, except in so much of its uppermost part as it slowly cedes to the torrent. This is inevitable if the upheaval is continuously, not intermittently, accelerated because, in addition to the stream being thereby offered more erosive power in each successive time-unit, it is at the same time given more work to do; and as the work added in each successive unit is greater than that added in the preceding unit, the graded course can be maintained as successive time-units only by the more and more powerful part of the river; that is, by that downstream part where the river is larger and larger. But in that part of

the course a graded profile *must* be maintained. It is impossible to produce a nick there. (*Davis, 1932G, p. 414*)

Davis reiterates several times that in his opinion a 'gradually accelerated upheaval can not raise a graded stream more than the stream can cut down' and he quotes as proof Penck's own statement (4), 'down-cutting is exactly as much as the rise of the land-mass'. But the reader who consults the German original will see immediately that Penck was in fact discussing 'the uniform upward movement of a land mass that does not alter the gradients of rivers flowing on it'. He was merely defining a *graded stream* and not saying that a stream *must* remain graded during uplift. So in our view the problem becomes 'was there an increase of gradient?'. Davis thinks 'a temporary steepening impossible although exception must be made of the uppermost part [of the lower course] that is ceded to the torrent'. The separation of the competent and incompetent parts of the stream should be,

as I see it, located at the lower end of the slightly lengthened torrent; it is implied by Penck [statement 8] to lie somewhere in the lower graded course ... Such nick production (here) seems to me quite impossible; and I must therefore conclude that under the postulated condition of continuously accelerating upheaval, the lower stream course can not be divided into two segments by a nick ... (*Davis, 1932G, p. 415*)

Having disproved the possibility of nick formation to his own satisfaction and shown that certain German physiographers also dissented from Penck's view on this point, Davis has no difficulty in refuting the remaining theories on the development of the stream profile, given below.

9. The top of this nick, retrogressively eroded, serves as a local base-level for the upper segment of the graded course, which is therefore no longer controlled by the more general baselevel at the margin of the [expanding] dome. Moreover, the local baselevel is, while working upstream, raised with the rise of the dome and therefore rises in relation to the upper segment of the graded course. Hence there the erosive power of the upper stream is weakened, and concave basal side-slopes are developed below the higher convex slopes.

10. Continued acceleration of upheaval causes the production of a series of nicks, all working headward, in the stream profile. Next below each nick is a narrow valley with steep fall along its stream-line and convex side slopes; farther downstream is the deepening and widening valley of a graded lower segment. Upstream from a nick is a widening floor with concave [basal] side slopes.

11 and 12. This type of valley-form predominates in the Black Forest, which is therefore assumed to be a land-mass experiencing continuously accelerated uplift.

Davis found these statements curious and dubious. The changes in the river profile would, he says, be gradual, and no graded upper segment could be withdrawn from its relation to the general baselevel of the graded lower segment in such a way as to be controlled by its own baselevel at the nick-top. Here he is undoubtedly right. An erosional nickpoint has no influence whatsoever upon the stream course *above* it until it has worked into and through the segment in question. Also, having disproved the production of one nick, he waxes eloquent on the still greater impossibility of producing on radial streams a tier of five or six marked nicks, accordant over a wide area, on a dome undergoing continuously-accelerated uplift, and concludes that in 'so nicely defined a combination of conditions as those adopted by Penck for the case of the Black Forest', the only part of the stream profile that could be markedly convex would be where 'the weak headwaters pass into the active torrent'; and that convexity would be a gradual curve, not an angular nick.

The production of marginal benchlands

A dome is a formation which, during its growth, suffers the greatest amount of upheaval in each time unit, at the divide; on the other hand the margin, generally speaking, suffers no uplift. Near the margin the intensity equals a very small rate of upheaval, towards the divide it increases. Accordingly the landforms are arranged in orderly fashion: near the margin a peneplane forms, closely similar to the Primärrumpf type, which, towards the divide, passes over into a dissected mountainland. Such a peneplane we will call a *piedmont flat* (Piedmontfläche) . . .
    When the rate of upheaval increases and the dome broadens, so the piedmont flat comes into the sphere of livelier upward movement and becomes dissected, while the out-pushing edge of the dome now finds itself in the zone of slowest upheaval in which, outside the edge of the old dissected edge of the piedmont flat, a newer lower one is formed. The two flats are separated from one another by steeper gradients, always by a concave (lower) and a convex (upper) break of gradient. Such gradients and the general arrangement of piedmont flats one above the other signify not in the slightest that the upheaval was intermittent, but only becoming continually more rapid! The younger piedmont flat is the erosion and denudation base, with reference to which the advancing dissection and denudation of the fore-edge of the surface above proceeds. In it one sees the lower, flatter and broader valleys which grow narrower upwards, and the interfluves become finally reduced to isolated hills which one finds sitting on the lower piedmont flat in front of the downstep like Inselbergs. (*Penck, 1925, p. 91, trans. by Martin Simons*)

Davis (1932G, pp. 403–4) interpreted Penck's view of benchland production summarily as follows:

13. If a land-mass of low relief suffers a slow, domelike upheaval, all its streams will be impelled to gentle erosion, and the shallow valleys that they then excavate will have faint side slopes.

14. The initial surface being a Rumpffläche of low relief with streams of faint fall, it will, after its streams have excavated their shallow valleys, become a Primärrumpf.

15. The remnants of the uplifted peneplain will be seen in its central upheaved area between its radial valleys.

16. The Black Forest does not, however, show only one Rumpffläche with the steepest inclination in it its remnants around the margin of the dome, but many out-slanting and down-stepping benches which have less and less outward slant the nearer they are to the dome margin and the lower they lie.

17. An expanding dome of accelerating upheaval suffers, in each time-unit, the greatest rise at its crest and the least rise around its outer border. Hence its slowly rising border will be degraded to a surface of low relief, a Piedmontfläche; and at the same time the central part of the rising dome will be transformed into a dissected upland.

18. As the expanding domelike upheaval becomes more rapid and embraces a larger area, the first-formed marginal Piedmontfläche is uplifted and dissected, while a newer one is developed outside of and below it.

19. The successively formed Piedmontflächen or benchlands will be separated from each other and from surviving parts of the central highland by steplike convexo-concave slopes of moderate height.

20. The steplike succession of benchlands does not indicate in the least that the upheaval proceeds intermittently, but that it is continually accelerated.

21. Each bench serves as a local baselevel with respect to which the next higher bench and eventually the central highlands are retrogressively dissected and consumed. Residuals of an upper bench may stand forth on the next lower one like Inselberge, unrelated to rock structure.

22. Each marginal bench advances into the highland along one of the broadened and gently ascending valley floors.

23. A bench can not be safely identified by its altitude [because it rises inwards, a caution to which American students may well give heed] nor from non-detailed maps. Every bench surface is limited between a concave slope which ascends to the next higher bench and a convex slope which descends to the next lower one.

24. The benchlands are everywhere developed without the slightest regard to rock structure.

Davis begins his criticisms of the above principles in a devious way by gratuitously devising a scheme that would ensure radial drainage from a dome. He then embarks on a 14-page detour on the nature and superfluity of the Primärrumpf concept (Davis, 1932G, pp. 417–30). With the acceleration of the slow uplift of the Rumpffläche the valley-side slopes deepen and become convex, characteristic of waxing development (*aufsteigende Entwicklung*) and a Primärrumpf has been formed. In Davisian terminology it is a slightly uplifted peneplain slightly modified in the infantile stage of the new cycle by the onset of slow uplift. The term was introduced by Penck to distinguish between the beginning of a new cycle and the end or senile form of the older cycle (or normal Davisian peneplain) which he called *Endrumpf* and which was characterized by concave slopes.

Davis shows that the so-called Primärrumpf could also develop – but most probably does not – by complete degradation of a 'slowly rising and writhing crustal mass' (line AC, Fig. 152) and by erosion on the central part of a large

F I G. 152. Davis' view of main valley-cutting in a peneplain (ADC) subjected to accelerated uplift giving rise initially to a shallow valley (EGF) and then to the convex form (RTS) (From Davis, 1932G, Fig. 5)

peneplain that was rapidly and evenly uplifted and protected for a long time from much stream dissection by its interior situation. However, examples of present and past Primärrümpfe appear to be extremely rare or non-existent, although certain German observers believe fully in their reality. In recognizing 'that highly specialized land form', W. Penck has abandoned young, mature and old stages of the cycle of erosion and replaced them by waxing and by waning development (*aufsteigende* and *absteigende Entwicklung*), the former characterized by increasing rate of uplift and convex valley-side slopes and the latter by decreasing rate of uplift and valley-side slopes which are, 'at least near their base', concave.

This replacement, says Davis, seems regrettable. To postulate that uplift begins slowly and to adapt a terminology for that belief, seems to set as much restriction upon the scheme of the cycle of erosion as did the Davisian

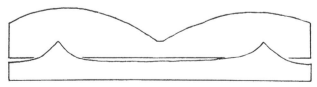

F I G. 153. Generalized profiles of (above) a Primärrumpf and (below) an Endrumpf from Priem (*Geogr. Anzeiger*, 1927) (From Davis, 1932G, Fig. 6)

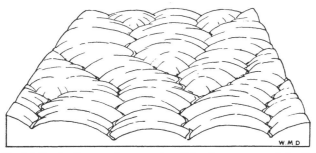

FIG. 154. Davis' block diagram interpreting Priem's concept of the Primärrumpf
(From Davis, 1932G, Fig. 7)

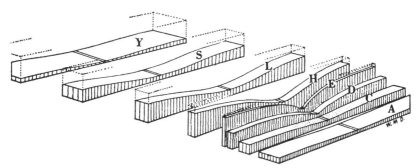

FIG. 155. Davis' stages in the erosion of an uplifted peneplain (A) during an early accelerated phase (C, D, E), followed by conventional maturity and old age (H, L, S, Y)
(From Davis, 1932G, Fig. 9)

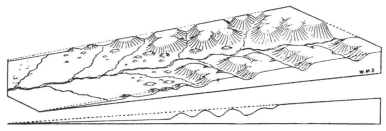

FIG. 156. Davis' interpretation of a dissected benchland by block diagram and
profile. The frayed-out border of the benchland overlooks the marginal lowland
(From Davis, 1932G, Fig. 10)

idea of rapid uplift followed by a long stand-still which the Germans condemned. Would it not be wiser to leave the scheme open, as it always has
been, so that it may be adjusted to any kind of uplift? Moreover, to assume
that the pre-uplift form must always be a peneplain (Penck in fact said
Rumpffläche) is artificial as the cycle is often interrupted at any stage as well
as in old age. Thus a slowly-uplifted mature-stage landmass will develop
forms resembling the mature forms from which they are derived.

Davis now returns to Penck's error in assuming that the final (old age) peneplain will be characterized by concave profiles. Rather the normal peneplain will have 'broadly convex swells between broadly concave swales; the convexities being due chiefly to the processes of surface wash and soil creep on the divides.' Had Davis referred to *Die morphologische Analyse* (p. 116) he would have seen that Penck there admits 'the flattening of the upper, better exposed parts of the slope unit' to a limited extent, the less the gradient of the slope unit, the greater the flattening of its upper part and vice versa.

This long digression by Davis on Primärrumpfe, is followed by a wordy discussion on the production of piedmont benches or flattenings (statements (13) to (24)). Davis admits that Penck's fieldwork observations are not in doubt but considers that no adequate explanation is given of the process by which successive marginal benches are differentiated. Indeed Penck's explanation of the 'intermittent development of the lowlands during non-intermittent upheaval must be rejected'. Davis prefers an explanation based on the intermittent uplift of an expanding dome, because it alone *demands* a systematic correlation of certain observed features around the dome margin. But if this is accepted each 'lowland bench', or Piedmontfläche, is a true peneplain (Endrumpf) and not a Primärrumpf. The visible surface forms may be the same but only the hypothesis of intermittent upheaval *must* include the production of downstepping benches.

> If it be thought that an overlong argument is here presented in the discussion of an apparently simple problem, the answer is that I have not ventured to consider the problem simple, in view of the fact that a brilliant and devoted student of physiography, as well as a number of his disciples, have adopted another solution for it than the one that seems to me inevitable: and I should not regard it as fair to the disciples to reject their leader's solution without first giving it careful consideration. As a result of such consideration, the most difficult aspect of the problem remains unsolved: How can serious students of physiography ever have persuaded themselves that continuous upheaval would or could cause intermittent erosion? (*Davis, 1932G, pp. 439–40*)

Perhaps we should warn the reader that the few sentences above condense nearly ten pages of Davis' text, which includes homilies on multiple working hypotheses and on the treatment of alternative hypotheses as well as a detailed exposition of how intermittent expanding upheaval would affect surface forms.

THE QUALITY OF DAVIS' ASSESSMENT OF PENCKIAN LANDFORM ANALYSIS

Davis' German was good and his methodology as near faultless as non-mathematical models can be, so not surprisingly this essay on the piedmont benchlands of the Black Forest is a remarkable feat of advocacy for a man of eighty-two. But in spite of its general fairness it does contain a few ambiguities, some mistranslations and misrepresentations of the German and, inevitably, a bias towards Davisian principles. More startling – for him – are one or two printing errors, among the very few that we have found in his vast output of impeccable print.

The article does not always do full credit to Penck's aims and originality. Thus he was propounding a new tectonic theory on the formation of the mid-German mountains in attributing them to domal uplift and not to block-faulting (horsts). Davis not unnaturally concentrates on the weakest parts of Penck's arguments – such as the production of nicks and discontinuities in a continuously expanding and uprising dome – and rather ignores the fuller details given on these features in *Die morphologische Analyse*. Indeed he says he deliberately avoids reference to the major work as far as possible, for the sake of brevity. Further, Penck's article and his main book have diagrams which Davis did not use direct nor study deeply. Those in the Black Forest article are so crude and those in *Die morphologische Analyse* so geometrical and complicated that they probably revolted Davis' three-dimensional artistic spirit. They seem to have goaded him to draw something better and subconsciously the improvement turned toward his own interpretation. His devotees, having long ago decided – quite irrationally and against his repeated warnings to the contrary – that all his drawings were of photographic accuracy, welcomed these interpretations as Penck's own ideas.

At least three aspects of Davis' detailed study of Penck's account of the southern Black Forest have met serious criticism.

First, it seems to some geomorphologists that Davis assumed that Penck postulated a simple relationship between rate of uplift and rate of river erosion, whereas in fact Penck insisted that the nicks and convexo-concave valley-in-valley forms of the Black Forest were associated with a continuously accelerated uplift and not with waxing and waning of the upheaval. Thus he considered the slope concavities here to have developed during increasing uplift. There seems no doubt that Penck was wrong in his interpretation of the various Piedmonttreppen landform associations as Davis so determinedly pointed out. Moreover, we must admit that in his *general principles*, Penck does give the general impression that, stream discharge being fixed, increased rate of uplift leads to increased gradient and so to increase of river erosion and presumably to convexity of valley-side slope.

Second, Davis, because of the same erroneous assumption of a simple relationship between rates of uplift and erosion, wrongly credited Penck with the assertion that all convex slopes are formed during accelerating uplift and concave slopes during decelerating uplift. Thus Davis says 'waxing development' is 'applied to forms during upheaval of increasing rate and hence characterized by convex valley-side slopes' (Davis, 1932G, p. 427).

Whereas Penck states of waxing development; 'The occurrence of convex breaks of gradient and of convex slope profiles is ... necessarily bound up with increasing intensity of erosion' (*Penck, 1953, p. 155*).

Third, the greatest disservice Davis did to Walther Penck was to attribute to him a belief in the parallel retreat of valley-side slopes. Davis' translation of this aspect ran as follows: 'When valley deepening ceases, the degradational retreat of the valley side goes on parallel to itself and leaves a surface of less declivity (a valley floor) below it as it withdraws' (Davis, 1932G, p. 402). What Penck actually wrote may be more correctly and more fully translated as:

> The denudation on all inclined flats (geneigten Flächen) now proceeds in such a way that the flats (slope units) retreat parallel to themselves at a constant gradient, and at their foot a flat of smaller gradient grows upward increasingly, as long as no river erodes deeper there, hindering the formation of the gentler foot-flats. On every rock wall one can observe the operation of this law of the denudation process. The retreating flats finally become crowded out and replaced by the upward growing lower foot flats [Fig. 151C]. After that, the flattening and lowering of the land continues as long as there is no further erosion in depth. But after that also *the slope profile is preserved* as long as there is no newly growing slope unit eating through, from below upwards. (*Penck, 1925, pp. 88–9, trans. by Martin Simons*)

Davis' substitution of 'valley sides' for Penck's 'inclined flats' (geneigten Flächen), used in the sense of 'slope units', was a gross misrepresentation. It was all the more inexcusable as Penck showed clearly in a crude diagram in the Black Forest article (Penck, 1925, Fig. 4: see our Fig. 151C) that he believed that valley-sides flattened with age when valley deepening ceased; then each slope unit could be said to retreat at a constant gradient but the slope *as a whole* through time becomes less steep. Thus Davis' mistake simply ignored Penck's 'law' that intensity of denudation, which is equal to the rate of development of the slope units, varies with their gradient. Davis drew his own diagram (Fig. 151A) to illustrate what he thought were Penck's ideas on parallel slope retreat. This was directly opposed to Penck's 'law' quoted above as it showed all parts of the slope irrespective of their gradient being denuded at the same rate. Penck was in fact an *opponent* of the hypothesis of parallel slope retreat and for him an Endrumpf, or peneplain, had mainly concave forms.

We can only suggest two reasons for Davis' serious misrepresentation. He was an octogenarian and he had just been advocating parallel slope retreat of boulder-strewn mountain faces in arid climates. May he not subconsciously have jumped to the wrong translation? But strange to say in his ideas on arid slopes also he had disagreed with Penck who denied the occurrence of an angular change of slope at the base of Inselbergs.

Davis' ideas on Penck spread rapidly throughout the English-speaking world, as we shall show in detail in Volume Three of this *History*. In December 1939, nearly six years after Davis' death, the American geographers held a symposium on the work of Walther Penck, its contributions to geomorphology and its relation to the Davisian system of land form analysis (Symposium, 1940). The extent to which American students of geomorphology depended on Davis' interpretation may be judged from the replies to a questionnaire prepared by O. D. von Engeln of Cornell University and circulated before the meeting. This included questions such as:

W. Penck held, that assuming homogeneous structure, the profile of slopes, convex, plane, concave, is determined by the circumstances of the uplifting action. If the rate of uplift is constantly accelerated [*aufsteigende Entwicklung*], the slopes will be convex; if uniform [*gleichförmige Entwicklung*], plane; if declining [*absteigende Entwicklung*] concave.

W. Penck maintained that once established with reference to an unvarying altitude of base, the further history of slopes was retreat in planes parallel to the original degree of declivity, and extinction when this plane is finally intercepted along divide planes by the progressive growth and rise of *Haldenhange* [basal rock floor beneath talus at foot of main slope]. Does this concept of parallel retreat of slopes have verity, if so under what conditions? (*Symposium, 1940, p. 222*)

Among the participants of the conference there was fairly general agreement that Penck's ideas had been stimulating, especially his details on the development of slopes, which ought to encourage further study of processes, an aspect largely neglected by Davis.

The impact of the radical doctrines of Walther Penck (1924) on the American school of geomorphology has all the effect of a cold shower on a complaisant reveller. Slightly bemused by long, though mild intoxication of the limpid prose of Davis's remarkable essays, he wakes with a gasp to realize that in considering the important question of slope he has always substituted words for knowledge, phrases for critical observation. (*K. Bryan in Symposium, 1940, p. 254*)

Penck discussed essentially only one set of processes, namely those associated with the development of slopes; but the same kind of attention to process will remake the rest of geomorphological science. Davis's great mistake was the assumption that we knew the processes involved in the development of land forms. We don't; and until we do we shall be ignorant of the general course of their development. In his eagerness to set up a general system, Davis jumped over the preliminary, necessarily painfully slow study of processes, and so left his system with an inadequate foundation. (*J. Leighly in Symposium, 1940, p. 225*)

Upon the tectonic aspect, Douglas Johnson, student, colleague and friend of Davis for thirty years referred approvingly to Davis' 'matured views' on crustal movements, which included slow uplift accompanied with concomitant extensive erosion.

On the other hand, Howard A. Meyerhoff of Smith College thought:

> As American geomorphologists we all have upon us the mark of the American school of thought, developed so convincingly by William Morris Davis. If we adhere to the precepts of this school, we are promptly handicapped by a tectonic premise which ... assumes very long still-stands of the land and the sea ... (*Symposium, 1940, p. 247*)

Meyerhoff, with regard to tectonics, wanted to discard the tectonic premises of both Davis and Penck (which he thought diametrically opposed) and to replace them with a concept 'of a restive earth in differential but intermittent motion'.

Opinions on the relation of the form of slopes to crustal movements included Douglas Johnson's forthright comment:

> Penck's conception that slope profiles are convex, plane, or concave, according to the circumstances of the uplifting action, is in my judgment one of the most fantastic errors ever introduced into geomorphology. (*Symposium, 1940, p. 231*)

Ideas on the verity of the parallel retreat of slopes showed Davis' baleful influence but reveal some independent trends. Leighly thought Penck's concept of the parallel retreat of slopes 'is probably not to be taken in an absolute sense'. However, he also considered 'Davis' thesis of a progressive flattening of slopes was an illusion' and gave examples of parallel retreat of slopes from the arid American West (as indeed Davis had also done!). On the other hand Douglas Johnson did not accept the ideas of A. C. Lawson and Davis on the parallel retreat of mountain fronts due to weathering in arid climates, as he thought basal sapping of the slope by stream action was essential. Johnson also found Penck's analysis of the cause of parallel retreat of slopes unconvincing. George B. Cressey of Syracuse University considered that 'the fundamental difference between Davis and Walther Penck concerns the flattening of slopes with time' (p. 235). Kirk Bryan of Harvard approached Penck's concept more closely in stating that, once formed, slope units 'persist in their inclination as they retreat'; they disappear only 'when all the volume of rocks above the encroaching foot-slopes or pediments has been consumed' (p. 266). Bryan discussed textbook diagrams of the retreat of slopes but reproduces (p. 258) only one – that drawn by Davis (1932) to show how his views differed from Penck's. Bryan did not indicate how erroneous was this Davisian representation of Penck's ideas. In fact, none of the participants in the conference seemed fully aware that Penck had made it quite clear in *Die morphologische Analyse*, that if denudation or general weathering is considered, as distinct from erosion due to rivers and streams, the gradient of slopes would lessen with age.

> No part of any surface on the earth, no matter how denudation works upon it, can ever thereby become as a whole steeper. It can only become less steep. The most important law obeyed during the development of denudational forms is this principle of decrease of gradient. (*Penck, 1953, p. 121*)

However, as to the conflict between the two main systems of landform analysis, it seemed that many of the geomorphologists present agreed that they were not mutually exclusive. Leighly summarizes the contrast succinctly. Davis was more comprehensive and Penck merely a fragment, but sufficient 'to constitute nearly an entire system by itself'. The ideal complete theory of the future will include part of each 'but not all of either' (p. 224).

Whatever this Symposium contributed to the understanding of the work of Walther Penck, it certainly pointed to a degeneration of the application of Davisian methods to the study of landforms. Whatever had been the early advantages of the heavily theoretical and semantic quality of Davis' work, by the Second World War in the hands of many of his followers it had developed into almost a legalistic exercise in which a kind of 'research by debate' was conducted by reference to texts and precedents, rather than to observations of landforms themselves.

DAVIS VERSUS PENCK: THE LATEST INTERACTIONS

It seems obvious from the comments of American geographers in 1940 that several had looked at Penck's *Die morphologische Analyse* but few had studied his article on the piedmont benchlands of the southern Black Forest, except Hellmut de Terra who in a wide scholarly survey showed how unlikely it was 'that piedmont benchlands express in all instances a lawful evolution of mountain growth even though intermittent uplift took place' (*Symposium, 1940, p. 246*).

In France, Henri Baulig in his 'Sur les Gradins de Piedmont' (1939) took much the same point of view as Davis, rejecting Penck's explanation of piedmont benchlands, and writing of Penck's theory of slope development:

> But the radical error in this conception consists in believing that a graded valley side is made up of a number of distinct elements *successively developed*. In fact, a graded slope, like a graded river profile or any other profile of equilibrium, cannot exist and persist except through the action of a *loose mass* (in this case, the debris) *in motion*: which amounts to saying that all parts of the profile are mutually dependent, all in constant though imperceptibly slow change, *all adjusted to present conditions and hence totally independent of past events.* (*Baulig, 1939, p. 303*)

But for concepts on slope retreat, most non-German geomorphologists, whether they had read *Die morphologische Analyse* or not, took Penck's ideas from the convenient visual summary in Davis' mistaken diagram. This flashed through the English-speaking world like a radio message and gained momentous advertisement. It was reproduced in C. A. Cotton's world-popular *Landscape* (1942, second ed. 1948, p. 232), and was referred to in Von Engeln's *Geomorphology* (1942) and in L. C. King's important article on 'Canons of landscape evolution' (1953).

Yet readers of Penck's major work should have been left in no doubt that he postulated a valley-side flattening with age. A few did realize this. For example, at Berkeley, J. E. Kesseli had lectured and issued notes on this subject previous to 1940 but most other geomorphologists refused to struggle with the difficult German text. Eventually in 1953 Hella Czech and Katherine Boswell produced a scholarly translation of *Morphological Analysis of Land Forms. A Contribution to Physical Geology*. The intricate diagrams were now placed before a wider public and Davis' error was exposed in print. Yi Fu Tuan, who had studied at Berkeley under Kesseli, led the way with a succinct and fully-illustrated analysis on 'The misleading antithesis of Penckian and Davisian concepts of slope retreat in waning development' (1957). Unfortunately this necessary exposé, although reiterated at Oxford and perhaps elsewhere in England, took some time to permeate geomorphological thought in western Europe and also apparently in the eastern United States. In 1962, Martin Simons published for British geographers a comprehensive review of the inadequate and misleading commentaries in English on the landform concepts of Walther Penck, concluding that they present much of value.

> In particular it seems very possible that the argument he used to demonstrate the dependence of stream erosion on crustal movements, though highly artificial, is capable of some limited applications. It may be true that the character of the uplift of a land mass is stamped on the landforms during the early stages of the erosion cycle at least, and may, thereafter be recognizable. Similarly, Penck's insistence that slope forms are a reliable record of erosional history does not seem unreasonable and certainly cannot be dismissed out of hand. In this respect, the validity of Penck's law of denudation that, other things being equal, the rate of retreat of part of a slope is proportional to its steepness, remains to be tested. It may even be said that quantitative investigation of this principle is of first importance to geomorphology. (*Simons, 1962, pp. 12–13*)

Davis, had he lived, would we feel sure have also come round to this appreciation. None would have regretted more than he the comment:

> Penck's early death, his involved style, and above all the carelessness of his chief critic, W. M. Davis, have delayed the proper understanding of his ideas for almost forty years, and have thus restricted the development of the study of landforms.
> (*Simons, 1962, p. 13*)

We hope to say more on this problem, especially on peneplains and Primärrumpfe, in the third volume of our *History of the Study of Landforms*. There was, we must reiterate, much that was correct and admirable in Davis' pioneer translation and critique of Penck's Black Forest article. No doubt his startling misrepresentation of slope retreat, by expressing a major contradiction in ideas, itself stimulated the investigation of slopes and processes. Although we cannot condone Davis' error, we cannot allow it to invalidate completely this remarkable effort by an octogenarian to fill a major gap in English geomorphic literature.

CHAPTER TWENTY-NINE

Ultimate Base Level

INCREASING STABILITY

After his eighty-first birthday Davis began to travel less. His global wanderings were over and his major lecture tours virtually completed. His venture to Columbia University, New York, was specially arranged by his dear friend Douglas Johnson and was his last prolonged lecture tour away from California, although he continued to attend conferences and give single lectures in the American East. The warm winters of southern California suited him well and there seems no doubt that he had now to be more careful of his health and of catching cold.

Even Mark Jefferson, always a favourite of his, could not entice him to Ypsilanti.

> On reading your letter of the 8th again, which I acknowledged by a hasty and rather brief postal card as is perhaps too often my habit, I see now that it deserved a fuller reply.
>
> The luxurious room with twin beds and private bath that you promise me all in the same building with the meeting, exhibits and meals, would tempt a less timid old duffer than I am to throw prudence to the wintry winds and stop over at Ypsilanti with the rest of you. But unhappily for all concerned and especially for me, I am not a less timid but a more timid old duffer, because any small, even slight lowering of temperature sets me barking, and a strong change would probably knock me out from lecturing for a time, and I must avoid that dangerous chance at all hazards. Everything depends on my being able to meet the engagements that have been most generously offered to me; and that is why I feel compelled to decline all extras. One exception is made to this rule, because it is part of the Columbia engagement. I have agreed to go with Johnson on his Pennsylvania trip in a motor-bus next week, and must not back out; but I will confess privately to you that I am shivering with apprehension of it. Should you happen to meet the bus and see a heavily muffled figure groaning in a corner, that will be ME, as I now picture myself. (*Letter: Davis to Mark Jefferson, 11 October 1931*)

We now have rare references to fatigue, a condition in the past he always induced in others, young and old alike, without knowing the meaning of it. In thanking J. K. Wright for a copy of his (edited) book on New England, he wrote:

> I am sorry not to have had time to read your book properly; but at present I am so busy trying to finish two of my own articles that it is impossible to look at any

others except superficially. Time to stop as my machine is getting tired. (*Letter: Davis to J. K. Wright, 2 June 1933; Pasadena, California*)

During the last few years he and Lucy had changed house several times. On Christmas Day 1930 he wrote from Palo Alto to J. P. Buwalda telling him

F I G. 157. Davis and Dr La Motte at Cashmere, Washington, on 16 August 1930. The photograph was taken by G. T. Renner, who appended the following comment: 'I took this picture just after Dr Davis and I had concluded arguing for half an hour over how to spell "peneplane", and buying a watermelon.' (Photograph by G. T. Renner; Courtesy A. N. Strahler)

that they were planning to move southward on the night of 31 December to their house, 359 S. Wilson Ave., Pasadena, which had been secured with the help and approval of Mrs Ransome.

> I am especially gratified by the arrangement you have made for the students to be ready to begin work promptly ... I shd prefer *not* to begin at 8 oc., because of the early start it demands for Mrs Davis. We hav had that hour for my lectures here three days a week, hence it can be taken if necessary; but a somewhat later hour is preferred ... Both Mrs Davis and I are looking forward with great pleasure to our sojourn with you.

Later in 1931–2 the Davises lived at 441, South Chester Avenue, Pasadena, and on 1 July 1932 moved to 656, South Mentor Avenue, a 'new and better house' in the same town. Here he wrote continuously and in his own – and our – opinion produced some of his finest essays (already discussed in Chapters 26 and 27). His personal correspondence also maintained its old abundance amid a growing tendency to reminiscence and expressions of real tenderness. On his eighty-third birthday he writes to his son Edward at Winter Park, Florida.

> Thy Aunt Lucy regards the above date as so exceptional that she has put a little rosebud in my button hole, and if the etymological connection that I suspect between bud and button is correct, the conjunction of the botanical and sartorial variants of the word are eminently appropriate . . .
>
> Besides all that, 83 has a sort of dyspeptic quality that bothers me. Pangs and qualms come along in a semi-periodic wave-like fashion; they interfere with my work, of which big piles remain to be done. To be sure I hav been lucky enough to send off one article lately and hope to hav it in print in a month or so; and over 60 pages of MS went to be copied yesterday. But they have been very slow in ripening . . .
>
> My recent family letter told of our plans to spend three weeks at Curtis Cate's school, where our address will be Carpinteria, Calif. from Feb. 26 to March 22. Talks to the boys. I hope it wont be too much for those qualms and pangs.
>
> Now I must turn to censoring a MS for the Geol. Soc. Amer., a chore by which the standard of the Society's publication is held up pretty high.
>
> With much love and nice messages to all of you.
>
> Thy loving OLD Father
>
> (*Letter: Davis to his son Edward, 12 February 1933; Pasadena, California*)

Davis adds a footnote, 'I am glad thee enjoy'd that $100', from which we infer that he realized how difficult things were for his son with five children and a depression hitting the fruit-growing business.

On 29 July Davis tells Edward,

> Thy readable letters are always welcome; and the one that brings good news of a bequest from Dorothy's aunt, is especially prized . . . Would that I had a lot of spare cash to hand over to thee; but that pleasure is denied me . . .
>
> We are doing all we can here in the way of hoping to fend off hailstorms from your orchards. But just now our hoping hasnt had any effect in preventing very hot weather here; 104, 105 reported officially. The heat rather knocks me out; makes me feel good for nothing; but a preceding cooler spell allowd me to work a good number of hours a day.
>
> Work just now is copying off, for the nth. time, a long MS, 80 big pages, on Sheetfloods and Streamfloods; it tells of the processes of land sculpture in the desert and is pretty good as far as I can see . . . (*Letter: Davis to his son Edward, 29 July 1933; Pasadena, California*)

LATE MATURITY

With increasing age and stability Davis began to round off and complete certain aspects of his landform research. We have already discussed these fine works written in the East. Thus on 30 July 1933, he wrote to the secretary of the American Geological Society

> The emotions excited in your secretarial breast on the arrival of a long paper I am sending you on Sheetfloods and Streamfloods ought to be of various kinds. In the first place, on learning that this is the last paper I have on hand with which to burden the G.S.A., you ought to rejoice, secretary fashion. But on reflecting that I may not contribute further to a Bulletin that has long been my refuge, you ought to grieve ... Let me say also that, in sending you two rather long papers [the other was 'Glacial epochs of the Santa Monica Mountains, California', 1933B] I am earnestly trying to redeem the promise I made to the G.S.A., when it presented me with that superb Penrose Medal at Tulsa; to go on and do some more work. In my own perhaps erroneous opinion, both these papers are as good as any I have ever done. Others may have different opinions of them; but they don't know them as well as I do. (*Davis, 1938A, pp. 1339–40*)

The Secretary states that he and other members of the Society

> felt that the paper ought to be somewhat expanded in order to contain some of the philosophy which Professor Davis had evolved in his long years of work and study, and he [Davis] had, accordingly, asked for the return of the manuscript at Christmas time, 1933. Apparently he had begun to reconstruct it in accordance with the new plan and was in the midst of making the re-arrangement, for he left it in an incompletely organized condition. The necessary revision and preparation for publication was undertaken by Hoyt Rodney Gale, son of Hoyt S. Gale, a friend and former student of Davis at Harvard. The younger Gale was also a personal friend and admirer of Professor Davis and is familiar with many of Professor Davis' ideas ... When the text was revised, care was taken to avoid as far as possible alterations in the ideas of the original manuscript. (*Editorial note in Davis, 1938A, pp. 1339–40*)

The result was the remarkable 80-page article that we have appraised at length.

Davis, however, in these years had other irons in the fire. He was working with John Haviland Maxson on an account of the Panamint Mountains, the first draft of which was finished in the autumn of 1933, and the last was published in 1935(A). He wrote a long review of J. S. Gardiner's *Coral Reefs and Atolls* (1934D) and a description of the Long Beach earthquake of 10 March 1933. This earthquake, which cost 120 lives and about $50 million in damage, was not, Davis shows, the rare and unrepeatable catastrophe that public and private authorities were making it out to be. One might regard this article as a forerunner of those on 'environmental perception':

> Life is full of hazards, and we must take our chances among them. The chances of an enjoyable life in southern California are, in spite of its occasional earthquakes, undeniably excellent. (*Davis, 1934A, p. 11*)

He also completed the promised article on 'Submarine mock valleys' (1933F and 1934B). In this he stresses that he remains a confirmed advocate of Chamberlin's multiple working hypothesis.

A submarine valley should be regarded as merely an element in the geological history of its coastal region, and its explanation should therefore be consistent with that history. Moreover, the larger part of the history must be based on the visible surface of the coastal land area ... while the sea floor is, as a rule, imperfectly charted, its structure is unknown except for a veneer of surface sediments, and its processes are still uncertain. (*Davis, 1934B, p. 297*)

Therefore, Davis suggests tentatively that the 'mock submarine valleys' of the inner coastal waters of the south Californian coasts were cut by localized marine bottom currents (later to be termed 'turbidity currents') in coastal offshore sediments. He gives examples of the submarine degradation of the sea-floor in other parts of the world. An earlier contribution on 'undertow and rip tides' (1931B) foreshadows this interest in bottom currents.

In his article on mock valleys Davis referred to his 'value of outrageous geological hypotheses' (1926B). In this he lamented the prosaic and restrained character of contemporary scientific meetings and the passing of outrageous geological speculation and of 'rudely polemical dissension'.

To encourage our patience, let me recall another outrageous idea of recent introduction, which in itself is only a sort of reaction from an outrage of somewhat earlier invention and a return toward a more primitive view; namely, the recent idea that those topographical features which we call mountains owe their leading feature, namely, their height, not as has been until lately supposed to a vertical movement of escape from the horizontal thrust by which their rocks have been crowded together, but to an uplifting force which acted long after the rocks were crowded together, and in which, as was thought when the view of a mobile earth crust was first promulgated, no component of horizontal thrusting is necessarily involved. A chief difference between that primitive view and its revival in the recent outrage is that the first view took little account of erosion and implied that each individual ridge and peak was the result of an individual or localized uplift; while the second view takes great account of erosion, not only in ascribing the present intermont valleys to the long and slow action of that patient process during and after recent uplift, but still more in ascribing the destruction of the surface inequalities, that must have been earlier produced when horizontal thrusting forces crowded the mountain rocks together, to a vastly longer action of erosion before the recent uplift of the worn-down mass was begun; for where in the whole world can we find mountains that today owe their height to an upward escape from horizontal thrusting; in other words, where in the world can we find any existing mountains that are still in the cycle of erosion which was introduced by an upward escape from the horizontal thrusting that deformed their rocks, and not in a later cycle of erosion which was introduced by uplift alone after the inequality of surface form due to earlier thrusting had been greatly reduced, if not practically obliterated! (*Davis, 1926B, p. 466*)

Two other articles at about this period show Davis' continuing concern with the 'nature of geological truth' (1931D) and the 'preparation of scientific articles' (1930D). The latter gives an insight into the painstaking manner in which the master constructed his own articles.

In addition to these completed literary projects, Davis always had before him his major philosophical–theological theme, *The Faith of Reverent*

F I G. 158. Davis examining a road-cut at Beverly Boulevard, Los Angeles, on 27 February 1931 (Courtesy J. H. Maxson)

Science (1934C). He presented and polished this oration scores of times, and it was usually received with great enthusiasm.

> Davis spent a week with us at the University of Nebraska about 1932. He was as alert and enthusiastic on our field trips as a neophyte. However, when Dr. Bengston, our department chairman, undertook to talk shop with him in a private conversation, Davis shifted to philosophical themes. After discovering Bengston's liberal views in religion and that he was a Methodist, Davis told Bengston that he should be ashamed of himself – that in reality he should be a Unitarian. Bengston soon became a Unitarian.
>
> If you haven't read 'The Faith of Reverent Science', you surely do not know William Morris Davis. I think the article is a masterpiece. I had an off-set copy made for my personal use. (*Letter: Earl E. Lackey to R. J. Chorley, 5 February 1962; Natick, Mass.*)

For our many readers who are interested in this lecture, which was published in the May edition of *The Scientific Monthly* for 1934, we have reprinted it in full in an appendix. Unfortunately we have not the space to comment fully on it but ask for it to be read aloud – with more than a little verbal emphasis here and there. It is not startling but it shows how easy it is to underestimate Davis. It exhibits an understanding and power of foresight of twentieth-century 'spiritual' needs well above the average. His insistence on a faith built of a few simple ideas which could be comprehended and put into practice by the bulk of common humanity is the nub of the problem with which contemporary theologians of all denominations are just beginning to grapple. As there was a progression of landforms, so he envisaged a

F I G. 159. A Rift Club outing. Davis appended the following comment: 'Here is a shady sycamore grove, near Peters' house, where we luncht on Feb. 21; a party of 20 or more brought together by Eggleston for my benefit. It was here that I played the part of a polygon or paragon of wisdom; see general letter. The grove lies or stands in a gully, worn by wet-weather streams from the mountains down the alluvial slope.'
(Courtesy R. Mott Davis)

comparable advance towards higher forms of civilization. But formal religion has failed to keep pace with social changes and 'the laymen are advancing beyond their ministers in recognizing the universal essentials and in discarding the denominational unessentials of religion'. With which, and more, being recorders of the history of geomorphology not theology, we must refer our readers to the Appendix.

Towards the close of 1933, Davis was still hard at work.

We spent last Sat–Sunday in the mountains with the Hunts and Peterses. Wonderful roads up to 6000 or 7000 feet; then ordinary trails to this and that cabin or cabin group. Instead of finding the night cold, it was surprisingly mild ... The

Tech gets to work next week, the 25th. Then I shall hav 3 lectures and some extra hours a week. Also some excursions with the class, in case I hav a class. I am a great deal prouder of my appointment than I have told you all.

With my love to every member of the family,

Thy affectionate old Father.

(Letter: Davis to his son Edward, 19 September 1933; Pasadena, California)

Before setting out on a dash to Boston, leaving on the Christmas evening and returning the morning of 3 January, he wrote a long letter to J. P. Buwalda on behalf of Curry, one of their students who had done an excellent geology paper for Davis but had failed in mathematics. Most geologists, says Davis, make very little use of mathematics except the few who specialize in geophysics.

THE END

At the end of December 1933 Davis delivered what proved to be his valedictory at the meeting of the American Association for the Advancement of Science at Cambridge, Mass. In affirming the 'Faith of Reverent Science' he was hailed as a philosopher and a scientist. On his return he contracted 'flu but by late January 1934 had recovered and was his old self. On the 18th he wrote to Robert Webb criticizing his tentative use of 'parasitic' and 'mimic' for subordinate or secondary fault blocks in an essay he wished to publish. But, says Davis, 'Go on take your chances!; this feeling on my part may be largely due to mental inertia.' He also busied himself with the reorganization of 'Sheetfloods and streamfloods' and made time to write to his youngest son Edward. This letter is Davis at his best: who could guess from it that he was within a few days of his eighty-fourth birthday?

A series of entertaining letters from thee keeps us informed of Floridian affairs, which seem to be going well in general, although unduly retarded in connection with the College Museum. It looks as if the honourable President didn't know how to subdivide his work and assign it to assistants. Perhaps the universal shortage of money is at the bottom of it.

'Universal shortage' is a good joke, tho, when it comes to borrowings by the Government against our grandchildren. What a time they will have, paying taxes. Also, what an awful good time there must be behind the scenes, in the distribution of the governmental billions to the right and proper persons. My belief is that it is actually impossible, under present conditions, to distribute so much money honestly. No wonder the democratic majority vote to support the President as long as he distributes lots of money thru democratic committees and such-like.

Now that I am up again and able to do a moderate amount of work, I have given a final polish to my Faith of Reverent Science, and shall soon have a clean copy of it to send to the Scientific Monthly, in which it shd appear in March or April. There has been some talk of bringing it out later in book form: But I do not favor the idea. Too many books are coming out already. I have had lots of letters, based on the brief abstract in the Lit. Digest, which was good as far as it went. They are of three kinds. The first and simplest ask where they can see the whole story, as if they were interested. The second pitch into me for being so rank a heretic. The third work off some of their own stuff, mostly rather awful, on me. Some are absolutely crazy: one quotes a Russian who has been for forty years counting the letters in the words of the O. and N. Testaments, and thus proving a great mathematical system at the base of the whole revelation.

What queer things we are, to be sure.

Thy enterprise in the way of bird talk is excellent; the more the better during the winter. I shd like to hear one of them, to see how thee develops thy subject. I recall that my father once gave me some good points about speaking: Plan your start and your finish, and fill in between among a general order of topics. Pitch your voice as if talking to some one in the back of the room or hall. I have since added other points, of which the chief is: Don't make excuses for lack of preparation or any other short-comings; go ahead and do your best without apologies. I seem to have a good reputation as a speaker here, for I am not infrequently called upon. My last chance was at the Men's Club of the Neighbourhood Church which we attend irregularly; topic, The Eastern United States: the influence of their geology, thru their geography, upon settlement, boundaries, industries, history and politics. One of my best points is the relation of the Glacial Period to the Civil War. That talk was put at the head of the list when the teachers of the region were gathered here just before Christmas, for conferences and addresses. In case I am on the list for Cal. Tech. Lectures in other cities again this spring, the E.U.S. will be my topic, as the Colorado Canyon, my crack piece, was given last spring.

Here are a few details about the Boston trip. We made out a program for every day before starting; and by holding to it and refusing many extras, we survived; also by cutting nearly all the scientific meetings. They were excessive in number. Two things amused me. I had been appointed by the French Academy of Science, of which I am a correspondent, to represent them as their delegate. On arriving at Boston I sent a note to Secretary Ward, informing him of my appointment. He replied, he knew of it already; and in virtue of it I had the right to attend the Council meetings of the Association; but as they were held at 9 am in Cambridge, in below zero weather nary one did I attend. The joke was that that was all my being a delegate amounted to. So I propose to report back to the French Academy that appointment of a delegate is a useless formality, more honord in the breach than in the observance.

The other amusement was the neglect of the Maiben Lecturer (n.M.D.) by the officials of the Association. Except for meeting a few of them by chance, not one approacht me or spoke to me, except the president, H. N. Russell, astronomer of Princeton, a long-time friend, who voluntarily attended the lecture. Now if I had been a stranger in Boston, what opinion must I have formed of being a Maiben Lecturer? He is supposed to be a distinguished person, and the lecture is supposed to be one of the things of the meeting. The trouble is, that the meeting is a sixty-six ring circus; and the officials are overwhelmed with work. I have suggested to Russell that a special committee shd be appointed, to perform the proper courtesies, as the regular officers are too busy to perform them.

Two visits to Cambridge gave me chances of seeing the new Institute of Geographical Exploration on Divinity Avenue, which I had seen briefly in 1931;

this time an exhibition of geographical maps and drawings was set up in one of the rooms; and I was quite overcome with the number and real beauty of some of my old work. I mean, with the number of really beautiful drawings that I did years and years ago and that were somehow scraped up for the show.

The other visit included a lunch in Tom Barbour's room in the old Museum, where I met several good old friends. Barbour is an immense success there; he was elected to the National Academy of Sciences last year; well deserved honor.

On Saturday afternoon reception, 4 to 5, in the second floor front of the New Faculty Club, replacing the old Colonial Club on Quincy St. went off admirably. It was followed by a delightful dinner at Mardie Wyman's on Craigie St; Lucy Davis of Cambridge, Katherine Francke, Helen Davis (Carl's eldest daughter) were the others. Quite like old times.

Our Sunday–Monday visit to Lucy's relatives in Holyoke went off well and our Tuesday–Wednesday stop in New York with one of her oldest friends was very restful. Then we took train and returned to our sunshine and flu.

(*Letter: Davis to his son Edward, 27 January 1934; Pasadena, California*)

This letter is especially significant because it must have been the last important one which he wrote and the subject matter so exactly comprehends all the attitudes and interests he had carefully acquired throughout his life. If one attempted to summarize Davis' character, emotions and sentiments it would be impossible to do it as economically. There was no expectation of death and he was still planning ahead to get a little extra out of life. On this occasion he had only a few days more to live, for on 1 February 1934 he suffered a heart attack which he survived four days, dying on the 5th. As well as the collapse of the heart an autopsy revealed a healed duodenal ulcer.

J. P. Buwalda, a devoted friend, has given the following account of Davis' final days in California:

When he decided to go East to the Boston meeting he understood full well that it was probably an even and odd chance that he would never return, in view of travelling conditions and cold weather. He made a new will in which among other things he specified quite precisely what should be done in case of his death. He had weakened somewhat during the Fall and I think he felt that it was probably the last good opportunity for him to revisit Cambridge. At the time of the meetings he and Mrs Davis gave a tea to all his old friends, I believe in one of the rooms at the Harvard Union, and he enjoyed seeing many of his old acquaintances once more. On returning to Pasadena he was ill with influenza the last day or two on the train, and went immediately to bed on arrival here. I visited him at that time and cautioned him to get a thorough rest, but in two or three days he insisted on being up and about again. On my next visit two days after the first, he was up and working at his table, and I urged him once more to live a very lazy life for two or three weeks, knowing the insidious effects of the toxins retained from influenza attacks. He insisted that he had a speaking engagement the following evening and he kept it. He continued working and soon after spent a long evening with the Harvard Club in Los Angeles, which fatigued him greatly. He worked vigorously until Thursday, February 1st, when he suffered some pain in his heart; he had previously felt this slightly during the influenza attack. On Friday he was quite ill but still himself. On Saturday and Sunday he was conscious only part of the time and not rational during those periods, the difficulty being that the pulse rate had dropped down to about 30, due to partial paralysis of one portion of his heart. On Monday he was continuously in coma and his physician told me that he would not last through the night. Mrs Buwalda and I were at their home during the after-noon and evening, and he died about eight o'clock, as a consequence of a gradual lowering of his pulse rate and circulation. Mrs Davis was a very courageous woman, in spite of much loss of sleep and fatigue.

Dr. Davis specified in his will that he should be cremated, which was done three days after his death. He asked that there should be no funeral or other ostentation. Mrs. Davis, Dr. Soares, the ministers at the non-sectarian Neighborhood Church at which Dr. Davis had preached a time or two when Dr. Soares was absent, and I agreed that it would be appropriate and in accordance with Dr. Davis' wishes if a brief memorial service were held following the usual morning service at the Neighborhood Church, and that was done last Sunday. The service was very fitting, Dr. Soares having acquainted himself quite thoroughly with Dr. Davis' philosophy as expressed in the manuscript of his address on 'The Faith of Reverent Science', given at Cambridge, and with Dr. Davis' memoir on G. K. Gilbert. A multitude of his friends and former students of the last fifty years attended the memorial service. (*Letter: J. P. Buwalda to D. W. Johnson, 13 February 1934; Pasadena, California*)

Just over two years later a memorial gate was dedicated to Davis at the California Institute of Technology, together with a bronze plate bearing the inscription:

IN MEMORIAM

WILLIAM MORRIS DAVIS

1850–1934

A FOUNDER OF PHYSIOGRAPHIC GEOLOGY

FOREMOST INVESTIGATOR AND INSPIRING TEACHER

PROFESSOR IN HARVARD UNIVERSITY

1879–1912

VISITING PROFESSOR AT CALIFORNIA INSTITUTE OF TECHNOLOGY

1930–1934

In 1932 Davis had been asked by the National Academy of Sciences to place on record some brief comments on himself. He wrote:

William Morris Davis 'Scientific investigation'.

CHAPTER THIRTY

Epilogue

There seems little purpose in attempting to summarize the achievements of Davis which we have spent a large volume in presenting. But there are a few aspects that we would like to emphasize and many judgments written about him after his death that must be reassembled. As a person he was a social crusader who preferred the pen to the platform in the market square.

> As a Quaker he abhorred physical strife as strongly as he liked to strive mentally with his fellow men. He was a true grandson of Lucretia Mott, ardent emancipationist. In his chapter on the United States in Hugh Robert Mill's 'The International Geography' (1899) Davis wrote: 'Better than the plain (Southern Coastal Plain) should never have grown a pound of cotton, better that its fertile strata should never have emerged from the waters of the sea, than that slavery and its dire long-lasting consequences should have come upon the United States'. *(Bowman, 1934, p. 178)*

As a lecturer and teacher he improved rapidly and became excellent at the former and highly successful with gifted students at the latter. The comments speak with a clear-cut unanimity.

> Students came to him from all parts of the country and from abroad. Mountains were living things ever afterwards to the person who once heard Davis explain their forms in terms of 'process, structure, and stage'. He accomplished his purpose not only by fascinating expositions of fact and meaning but also largely by confining the attention to 'relevant detail', and he underscored the phrase. It was a singularly dull man whose candle was not lit by the end of the first exercise!
>
> There was a relentless quality to the Davis discipline. He looked for steel and a spark, and his classes not infrequently declined in numbers as the year advanced. 'Great stuff if you can stand it', was the comment. Complacent, slipshod, and uncritical work did not receive from him the coddling approval that self-expression demands in some of the schools of a later day. *(Bowman, 1934, p. 177)*
>
> The seminar was most stimulating, but very exhausting on the students. Each of us presented at least one paper a week, and sometimes two. You can well imagine how some of our other courses suffered. Davis, as you probably know, was quite caustic at times in his criticisms; so much so that it was often difficult to control one's vexation. Looking back on the experience, however, it was one of the best things that could have happened to us. To avoid his sharp comments, we made a super-human effort to sharpen up our reasoning and to improve our oral presentations. Believe me, we became painfully aware of inductive and deductive reason-

ing, of the importance of considering all possible explanations of a particular set of facts, of the inexcusability of confounding fact and inference, and of the utter lack of consistency in following a careful scientific investigation with a slip-shod written or oral presentation. (*Letter: A. D. Howard to R. J. Chorley, 13 February 1963; Stanford, California*)

His lectures were definite, logical, precise, and appealing. He lacked the width of interest and the breadth of sympathy of Shaler; but he, above all others I came under, was scientific in manner of work and in methods of exposition ... Any skill I have for clear thinking, or in making concise, accurate statements of fact or theory, I owe to Davis in particular, for he was a master of exposition equalled by few men I have known. (*Interview: R. E. Dodge with P. F. Griffin, 31 July 1951: quoted in Griffin, 1952*)

DAVIS' WORK ON BIOGRAPHY

During his long life Davis produced a number of biographies and obituaries which help to shed light both upon his own personality and upon his views on geomorphology and physical geography. These writings fall conveniently into three groups: those on his teachers and Harvard colleagues – James Henry Chapin (1893A), Nathaniel Southgate Shaler (1906F and 1906P), Raphael Pumpelly (1919H), and Josiah Dwight Whitney (1896N); on his students – Frederick Putnam Gulliver (1921E) and Albert Perry Brigham (1932E); and on other eminent American geologists and geographers – James Dwight Dana (1895F), George Perkins Marsh (1906L), John Wesley Powell (1915F), John Peter Lesley (1915G) and Grove Karl Gilbert (1918F and 1927F).

Most of these biographies are brief and factual except that Davis occasionally pauses to approve an extension of the cyclic theory to marine erosion (1921E, p. 114), or to applaud the scholarly social life of a Harvard Professor (1896N, pp. 206–7), and the 'pure and guileless heart' and clear, sagacious head of a tearaway, turned academic (1919H, p. 61). Indeed, in some respects, they are as illuminating for their omissions as for their content. For example, despite Davis' (1906L, pp. 79–80) vague approval of Marsh's geographical view that 'whereas Ritter and Guyot think that the earth made man, man in fact made the earth', he certainly did not seem to grasp its real significance and his own work in physical geography made few concessions to it.

It is in the two major biographical essays on Powell (1915F) and, especially, on Gilbert (1927F) that Davis' method of treatment is most informative. Where the work of these two famous geomorphologists lays the foundations for the concept of the cycle of erosion Davis deals at length with it in approving terms, as in respect of the concepts of baselevel (1915F, pp. 32–4), planation (1915F, pp. 34–6 and 1927F, pp. 48–9), the laws of erosion (1927F, pp. 49–50), subsequent valleys (1927F, pp. 104–6) and the idea that in the teaching of science, philosophy should be stressed more than the physical content (1927F, pp. 98 and 146).

The famous chapter on 'Land sculpture', which is expanded from a briefer statement of the same problem in the essay on the 'Colorado plateau province as a field for geological study', above analyzed, offers an illustration of Gilbert's manner of thought that is both pleasing and edifying; and it is particularly with reference to the revelation of his inner nature thus afforded that the chapter is here reviewed. It exemplifies a principle which he announced later, that in the teaching of a science more attention should be given to its philosophy than to its material content; for having recognized that an understanding of sculpturing processes would lead to the understanding of sculptured forms, he expounded the philosophy of the elementary processes of erosion as well as philosophy of the evolution of land forms, and set forth both doctrines in the most genial and competent manner thus elevating them to the grade of serious studies. (*Davis, 1927F, p. 98*)

Where the ideas of Gilbert and Powell do not extend sufficiently to give direct support to those of Davis, he tends to read implications into them, as with Powell's implied but 'undeveloped' idea of stages of erosion (1915F, p. 30).

Occasionally Davis disagrees with the conclusions of these two giants of geomorphology; for example, he points to the inadequacies of Gilbert's theory of basin ranges. Davis' own prejudices are also apparent, as when he questions the antecedence of the Green River across the Uinta Mountains and proposes instead a less dramatic type example of the Meuse in the Ardennes (1915F, pp. 23–7). A telling example of Davis' inability to digest ideas which could not be squared with his own is given by his complete failure to recognize the functional equilibrium basis of Gilbert's approach to the analysis of landforms:

The absence of the important physiographic factor, time, from Gilbert's reports is more perplexing. He must have known perfectly well that the existing conditions of drainage systems as well as the existing forms of the land surfaces are the product of erosional processes acting upon structural masses through longer or shorter periods of time; yet his account of streams and of land forms is much more concerned with their existing status than with their evolutionary development from an earlier or initial status into the present status. It is only by reading between the lines that the idea of systematic change with the passage of time is to be gathered, and even then but incompletely. The passage about stream volumes and grades quoted in the second preceding paragraph concerns only a maturely developed river system; nevertheless the law of increasing steepness upstream is, without qualification, said to apply 'to every tributary and even to the slopes over which the freshly fallen rain flows in a sheet before it is gathered into rills. The nearer the watershed or divide the steeper the slope; the farther away the less the slope'. Yet this evidently holds good only for ready-made, full-grown drainage systems, neither young nor old. It is true that the need of much erosional work in the production of a systematic increase of river grade from mouth to source is intimated a few lines later, when it is said that such an arrangement 'is purely a matter of sculpture, the uplifts from which mountains are carved rarely if ever assuming this form'; but the idea of development here intimated is not fully carried out. Consideration of the time factor is exceptional all through the chapter on 'Land sculpture'. Even the possibility that rivers may grow old and that mountains may be worn down is presented only as an unrealizable tendency...

(*Davis, 1927F, p. 107*)

But the reader may ask, if not exclaim: Why point out these shortcomings in a study that had so many excellencies? The answer is: Partly to reinforce the lesson of the preceding section that progress is not made all at once; but even more to spur on those discouraged physiographers of to-day who seem to fear that their science is now completely developed, and that no new progress is to be expected.

(*Davis, 1927F, p. 108*)

In terms of personality Davis clearly preferred Gilbert's sensitivity to Powell's muscularity, and in writing of the former one feels that Davis is voicing his own aspirations.

He was always ambitious to do good work, but he never strove for office or for position. The nearer one lived to him and the longer one knew him the clearer it became that his personal nature was as exceptional as his scientific capacity; for in his private life as in his geological tasks he was fair-minded, self-controlled, serene, gentle in his manner, simple in his ways, uncomplaining under trials and disappointments, loyal to his duties, steadfast in his friendships. Little wonder that those already old when he was young should have recognized in him one who would continue the work they had begun and carry it forward into regions of space and of thought they had never entered; or that those still young when he was old should have looked upon him with respect akin to awe... (*Davis, 1927F, p. 1*)

Gilbert was essentially and consistently a rationalist. He stood by himself, thoughtful and independent, in all religious matters. He was a member of no sect; he rarely went to any church, and when he did go it was to listen to the preacher as he would listen to a lecturer. His beliefs were the product of his reason, not of his emotions; they were essentially ethical rather than theological. He saw clearly that man's advance in natural knowledge, based on observable evidence and logical inference, has long been accompanied by a decrease in his supernatural beliefs, derived from revelation or inspiration, so-called. He therefore trusted wholly to natural knowledge, and let the decrease of the supernatural take its course. He was much interested in the history of the successive steps in this decrease and knew that dangers were imputed to each step by those unwilling to take them; but he knew also that others who have taken some of the successive steps have found the imputed dangers to be only imagined; and he was therefore not dissuaded from taking his own further steps. His intellect was his guide, and he abandoned all views for which it gave no support.

Men like Gilbert would of course be called 'disbelievers' by the modern conservative who, apparently unaware that the supernatural elements in his ancestors' faith during the Dark Ages were many as compared to the few that he accepts to-day, regards his own belief as a standard that has been fixed, permanent and constant through the Christian centuries; and who, ignorant of the great share that the intellect of great men has had in dispelling the superstitions of earlier times, looks upon any intellect that may lead to a less belief than the one he holds as a cold and self-willed guide. Yet Gilbert's nature was not cold, nor was his life selfish. On the contrary, he was warm-hearted; his sympathy was easily aroused, and his generosity was always responsive; his disposition was gentle, kindly, loyal, and helpful. Religiously independent as he was, he never sought to disturb the religious dependence of his friends. If his judgments were sometimes stern, they were always sincere; and they were applied with much more severity to himself than to anyone else. His will was strong, but it was controlled by a tender conscience. 'Virtue is its own reward' was to him no empty platitude; it was the rule of his life; further reward he neither asked nor expected. (*Davis, 1927F, p. 203*)

Davis' final paragraph refers with obvious pride to his own association with Grove Karl Gilbert.

Gilbert's death occurred five days before his seventy-fifth birthday, in celebration of which, as has already been told, a host of his geological colleagues had written him congratulatory letters that he never received. It is sad to think that he could not read the many messages of esteem and affection that these letters contained, and it is sad to know that, great as was his accomplished work, beneficent as his influence had been, his final wish to review and complete one of his earliest and greatest studies was not realized. Nor was his intention to establish himself near San Francisco brought to completion. All plans had been made for the journey across the continent in the spring; a chosen friend was to join him on the way so that he should not go alone to the new home he was to make in California, a home not so far as Washington from the field in which he hoped to work. It was as if, even at the age of three-score and fifteen, he still looked forward to finding a new life beyond the mountains he had so often crossed before. Alas, the companionable journey across the Great Divide was not made; he crossed the Greater Divide alone. (*Davis, 1927F, p. 303*)

We cannot refrain from commenting on the greater good fortune of Davis. At the time of writing this he too was about to cross the Great Divide and to find there not only a new paradise but also a congenial companion.

DAVIS' WORK FOR GEOGRAPHY

Being primarily concerned with landform analaysis we have not been able to expand in great detail Davis' outstanding contribution to American geography as a whole. Fortunately this was well done by Dr Vera Rigdon who actually submitted her text to Davis for approval shortly before he died. She stresses his pioneer work in educational geography, including the preparation of textbooks for the elementary field. From 1875 to his death he had a part in every important movement for the improvement of geography. He both organized and systematized geography in the United States and won recognition for the subject as a mature science and as an academic discipline.

Through various agencies his influence has permeated the whole field of geography in America and spread to every inhabited continent, thus justifying the title awarded to him by common consent – the dean of American geographers.
(*Rigdon, 1933*)

The following comments show clearly how much he understood the need for geography as such and, while devoting himself to physiography, unwittingly advanced the organic side of the subject. We shall begin our analysis with the important influence which Davis exerted on his fellow-members of the sub-committee for geography of the Committee of Ten on Secondary School Studies, whose report under the chairmanship of T. C. Chamberlin was published in 1894. This report perpetuated a system of geography teaching in which 'the earth was approached from the viewpoint of the physical and natural sciences: it was studied in its own terms and for its own sake' (Leighly, 1955, p. 309). Although subsequently Davis paid greater lip service to human geography, by modern standards his attitude to the discipline was almost crudely deterministic, owing far more to Ritter (Davis, 1900O, p. 158) than to the subject of one of his biographical essays – George Perkins Marsh (1906L). Davis termed the 'organic half' of geography *ontography* (1902E, p. 240), to distinguish it from Huxley's *physiography* (Davis, 1900O, pp. 161–2), but for him the former was always secondary to, and derivative from, the latter. Writing of school instruction based on the state map of Massachusetts, he gave as its aim the inculcation of

a clear understanding of the manner in which our mode of living is related to the earth on which we live . . . From the time of the first occupation of the State its people have been constantly influenced . . . by the geographical features that they observed about them. (*Davis, 1897L, p. 496*)

The evolution of the earth and the evolution of organic forms are doctrines that have reinforced each other; the full meaning of both is gained only when one is seen to furnish the inorganic environment, and the other to exemplify the organic response. (*Davis, 1904I, p. 675*)

When regarded objectively, the geography of today is nothing more nor less than a thin section at the top of geology, cut across the grain of time; and all the other thin sections are so much more like the geography of today than they are like anything else, that to call them by another name – except perhaps 'paleogeography' – would be adding confusion to the earth's past history instead of bringing order out of it. (*Davis, 1904I, p. 681*)

the essential in geography is a relation between the elements of terrestrial environment and the items of organic response; this being only a modernized extension of Ritter's view . . . the reasonable principle of continuity will guide us to include under geography every other example in which the way that organic forms have of doing things is conditioned by their inorganic environment. (*Davis, 1904G, p. 167*)

If Davis' own work in physical geography and his influence on his contemporaries had attempted to put this form of environmental determinism into operation things might have been very different, but this was far from the case. Apart from a handful of instances, for example such essays as those on the 'Geographical factors in the development of South Africa' (1911O) and on Lower California (1921D), his work on physical geography consisted almost entirely in a discussion of physical forms and processes, dominated by the 'geographical cycle' (Davis, 1899J, p. 473), and his fleeting reference to man's activities were usually restricted to banalities about the eskimos, bushmen and bedouins (e.g. Davis, 1898E).

> Savage tribes must, I believe, be studied first; because only among savages do simple and immediate relations between physical environment and human habits control the greater part of the social economy. A valid reason for introducing an early account of more civilized peoples is that our schools are situated in the midst of such a people, and hence in so far as this side of geography is based on local observation, it must make a start on a complex example. But on the other hand I question whether school children can to advantage become so introspective as to open the study of mankind with a study of themselves and their neighbors. Moreover, the savage, although distant, never fails to excite interest in young classes; probably from the unconscious kinship between the rude conditions of the human race and the undeveloped conditions of childhood. My own habit in his division of the subject is to begin with such examples as the dwarfs of the equatorial forests of Africa and the Eskimos of Greenland, and from these primitive peoples to illustrate one of the consequences of the globular form of the earth. Later I would take such examples as the wandering Bedouin of the Sahara and the Indians of the Amazon forests, for illustration of the consequences entailed by the systematic planetary circulation of the atmosphere, in producing arid deserts in one region and abundant rain in another. Still further on come illustrations of the terrestrial elements of climate; that is, those elements which depend on the oblique attitude of the earth's axis to the plane of its orbit; such as the floods of the Nile, the necessity of irrigation in southern California, and the habits of thrift engendered in climates having a winter that can be survived only by gathering a summer harvest. It is too commonly the habit to separate these causes and consequences by much irrelevant matter. (*Davis, 1898J, p. 87*)

We know that Davis was very much concerned with the teaching structure of geography and of its relationship with allied disciplines.

> The quality of geography as a whole may be presented and emphasized by a simple graphic device that experience has shown to have some value. Imagine the four frameworks, E, A, O, L, [Fig. 160A], to stand in a vertical plane over the line EL, and to contain compartments for all the topics of systematic physiography under the larger heading of the Earth as a globe, the atmosphere, the oceans, and the lands. The compartments may be taken to represent types with respect to which actual examples are classified ... The framework on the further side of [Fig. 160B], by which a classification of organic responses is here roughly indicated must therefore be taken for the present as suggesting what may yet be done rather than as representing what has already been done in this direction ...

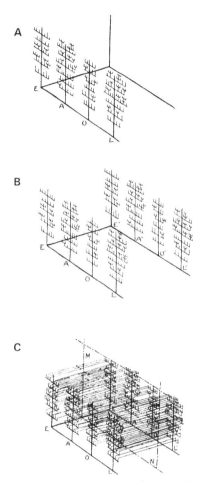

FIG. 160. Davis' scheme of geography

A. The major compartments of physical geography; the earth as a globe (E), the atmosphere (A), the oceans (O) and the lands (L)

B. The organic responses to physical geography (E', A', O', L')

C. The intersections of the cross-lines of correlation with the moveable plane MN represents the content of systematic geography which may tend to move either towards the physical or towards the biological environment. The mid-way position shown here represents the attitude of the 'well-balanced geographer' (From Davis, 1903H, Figs. 1, 2 and 3)

Imagine now a vertical plane, MN, parallel to the planes of the two frameworks and midway between them [Fig. 160C]. This new plane will be intersected by all the cross-lines of correlation, and the summary of all the points of intersection may be taken to exhibit the whole content of systematic geography. The plane may be shifted towards one or the other framework, if it is desired to give greater emphasis to one or to the other part of geography, but the well-balanced geographer will doubtless prefer to hold the plane in a midway position, as here indicated. In case the biological classification of ontographic items were adopted, the lines of correlation would not be parallel; a group of lines would radiate from each kind of physiographic control in the nearer framework to all of its ontographic responses, and various lines from numerous kinds of controls would be focussed on the kind of organism in the further framework that responded to them. (*Davis, 1903H, pp. 416 and 420*)

In 1909 A. P. Brigham cleverly summarized Davis' attitude to geography as a whole and, in particular, to the views of Hettner (Davis, 1907F):

Davis feels the need of careful definition of the aim of geography, so that effort may be focussed, and to this end he devoted his address, as president of the Association of American Geographers, on 'An Inductive Study of the Content of Geography' (*Bull. Amer. Geogr. Soc.* 1906), in which he emphasized the importance of treating the many items which go to make up the whole of the subject, in view of the mutual relations which they have come to have through the long and slow processes of evolution during recent geological time, and with special attention to the relations established between inorganic environment and environned organism. He sympathizes fully with Hettner in regarding geography in its regional aspects as embodying a description of the material filling of terrestrial spaces (die dingliche Erfüllung der Erdräume), but gives more emphasis to the explanatory aspects of systematic geography, in which things of a kind are studied together, wherever found, and thus made ready for more appreciative treatment in their regional associations. He is disposed to extend the boundaries of geography very far, and would include phenomena of language and religion, as well as phenomena of commerce and other more material matters, so far as they reflect the influence of environment, under this wide-embracing science; but to do him justice in this respect one must read and consider his various papers.

While thus cordially recognizing the great breadth of geography as a whole, while constantly pointing out the relation of inorganic and organic elements in his writing and teaching, and while greatly regretting the small representation of the organic phases of geography among the members of the Association of Geographers, Davis' chief interest in recent years has centered in the physiography of the lands, and in the search for means of clear, definite, systematic and comprehensible description of land forms. (*Brigham, 1909, pp. 52 and 54*)

As we have seen, Davis was always encouraging his students to explore human geography in a deterministic or rational way (Davis, 1906I, p. 112) but when they did so too assiduously he was curiously dissatisfied with their efforts. Bowman, Huntington, Dodge, Jefferson and many others had unhappy experiences in this respect.

Professor Davis' great contribution to geography has been, as we have seen, the systematizing of the science of physiography, but that is not his only claim to our gratitude. So far as the other half of geography is concerned – the half which he has called ontography – he has avowedly done but little. He has made many most interesting and valuable suggestions as to the relation of physical features to the activity of living beings, but he has not attempted to frame the laws of the subject or to elucidate its principles as he has in the case of physiography. This is not because he does not realize the importance of the subject for he is keenly alive to it. More than once I have heard him say to his students, 'You younger men must go on and work out the relation of man and other living things to their geographical environment. You must get at the laws which govern this matter, and must find out how to present the subject. That I cannot do. It is too late now; my work has been to develop the subject of physiography. That was the need when I began; the next need is a fuller study of the relation of life to geographical environment . . .' (*Huntington, 1912, pp. 32–3*)

When the first annual meeting of the American Geographers' Union was held at the University of Pennsylvania in 1905 no less than 14 of the 19 papers presented dealt with landforms (Colby, 1936). By 1911 when the first issue of the *Annals* appeared, the needs of what was coming to be called 'commercial geography' were leading to a significant change of emphasis. In 1895 the universities of Wisconsin, Cornell, Chicago, Yale and Harvard together offered 8 courses in physical geography and none in human geography, whereas in 1910 the numbers were 18 physical and 27 human (Whitbeck, 1910).

However Davis seemed to become more and more confused regarding the relationships of the various branches of the discipline as human geography developed in the United States during the first three decades of the century, particularly with respect to the relationship between physical geography and geology. Despite the great play he made for the exclusion of geologic age descriptions from geographical work, physical geography for him was always concerned with the last part of the last chapter of earth history.

The essential content of geographical science is so large that the successful cultivation of the whole of it demands all the energies of many experts. Those who are earnestly engaged in cultivating geography proper should treat non-geographic problems in the same way that a careful farmer would treat blades of grass in his cornfield: he would treat them as weeds and cut them out, for, however useful grass is in its own place, its growth in the cornfield will weaken the growth of the corn. So in the field of geographical study, there is no room for both geography and history; geography and geology; geography and astronomy. Geography will never gain the disciplinary quality that is so profitable in other subjects until it is as jealously guarded from the intrusion of irrelevant items as is physics or geometry or Latin. Indeed the analogy of the blades of grass in the cornfield is hardly strong enough. It is well known that Ritter, the originator of the causal notion in geography, and therefore the greatest benefactor of geography in the nineteenth century was so hospitable in his treatment of history that his pupils grew up in large number to be historians and his own subject was in a way lost sight of by many of his students who became professors of geography, so-called in the German universities, until Peschel revolted and turned attention again to the essential features of geography proper. (*Davis, 1904G, p. 166*)

. . . all geography belongs under geology, since geography is neither more nor less than the geology of today, and since all geology is essentially the sum of a long succession of past geographies. The separation of the geographical part from the geological whole was a natural consequence of the opinions that prevailed a century ago, when most geographers were merely observational empiricists and most geologists were in large measure fanciful catastrophists; but such a separation is, systematically considered, an absurdity at the beginning of the twentieth century, when we all are convinced that the flow of the past into the present has been without a break . . . (*Davis, 1912H, p. 121*)

In much the same way, Davis' attempts to preach the subordination of history to human geography have a rather hollow ring; although, like James Thurber's grandfather, he does show disconcerting flashes of sanity!

In any case I believe that the treatment should be rational, explanatory, and even evolutionary, but not historical; that is, that all pertinent features should be treated in view of their present stage in the whole series of changes that they suffer during a city's growth, from its beginning in a village to its most flourishing development and perhaps to its end in vanishing ruins. The presentation of the growth and development of a city in historical order should be avoided as carefully as the presentation of the erosion and development of land forms in geological order; what is needed is a statement of the existing condition and appearance of the city in terms of its historical growth, provided that the general sequence of changes in the growth of cities can be analyzed and systematized in some such manner as the evolution of land forms has been. Different kinds of villages and cities in different stages of development would then be recognized.
(*Davis, 1915K, pp. 98–9*)

The geographical treatment of historical events requires that they should be classified not historically but geographically, that is, with relation to localities; and that they should be described geographically, that is, with relation to the inorganic factors that have influenced them. For example, all the occasions in which the passage of a military force along a certain valley, or through a certain notch in a ridge, has been fraught with historical consequences should be brought together, whenever they happened; and their relations to the valley or the notch should be set forth. Thus, all important movements along the subsequent valley of Ape dale, between the axial Caradoc ridge and the uniclinal Wenlock Edge, should be set forth not merely as historical events, but as events in which history has been guided by topography. All the historical events relating to the gorge of the Severn in the Wenlock limestones, or to its open straths up or down stream from the gorge, ought to be set forth and compared in order to learn and to teach how the influences of these physiographic features were exerted. (*Davis, 1920K, pp. 289–90*)

A village is an admirable subject for observation of human conditions. Notice the increasing closeness of the houses towards the center, around the stores and offices; note the larger open spaces about the border of the village. See how the roads converge towards it from the surrounding country. Consider the traffic on the roads, inward and outward. All of this should be taken, not merely as local fact, but as an example of a way in which some of the people of a certain country in the world live. After direct observation comes simple explanation. The post office is near the center of the village, because it there best serves general convenience. A single road leads out for a mile and then forks into two; because it is cheaper to reach two districts in this way than by two independent roads. Some of the villagers work in shops, others are employed in a bank, or at a railroad station; thus diversity of occupation is first observed, then accounted for. The growth of the village may be explained, story-fashion. Then, like the hill, it is seen to have a life-history. (*Davis, 1905C, pp. 3–4*)

Again, Davis increasingly preached the virtues of 'regional geography', especially in his splendid article of 1915(K) which we have already quoted in Chapter 17, but his conception of this, in so far as it is comprehensible, seems poorly developed even in his old age.

Let me make it clear why I lay so much emphasis on regional geography in contrast to systematic geography. However important the different divisions of systematic geography are, they yield only a discontinuous sort of knowledge. Under the division of land forms we may study about volcanoes, and learn as much as possible about all of them wherever they are; but the items of knowledge that we thus gain compel us to leap about all over the world in order to locate them. Similarly, we learn under human geography all about fishing villages where-ever situated; but there again we must leap about to locate the items that we learn. On the other hand, the study of regional geography gives us a continuous or areal sort of knowledge. If we study the regional geography of Japan, for example, we shall learn not only the distribution of its many volcanoes and its many fishing villages, but of all its other geographical features; and it is this continuous or areal knowledge that is usually and properly implied when we speak of geography, in a general way. The study of the various divisions of systematic geography has, indeed, its chief value as a preparation for the study of regional geography; hence no one should consider himself a geographer until he has become expert in the regional geography of at least one large area, preferably a continent. (*Davis, 1922C, p. 123*)

Indeed, even when properly limited, regional geography is so complex that some critics assert it cannot be successfully mastered and treated by one man. There is no question that it can not successfully be treated by a geographically uninformed, untrained, undisciplined man; or by a trained man who gives part of his attention to other sciences; and it may perhaps be true that it cannot be so well treated by one informed, trained and disciplined geographer working alone as by several informed, trained and disciplined geographers each responsible for a part of the total subject and all working together in a team. No decision need be made on that question now, because the problem has never been given study long enough and serious enough to answer it. (*Davis, 1922C, p. 127*)

Even after the developing concept of 'geography as human ecology' (Barrows, 1923) had effectively reversed the relative significance of physiography and ontography, Davis continued to propound much the same old philosophy.

* Moreover, when properly applied in a direct explanatory description instead of in a roundabout argumentativ analysis, the scheme of the cycle saves physiography from the reproach of being only maskt geology, because it enables one to state the results reacht by explanatory analysis in terms of physiographic types, without going thru the argumentativ demonstration that the analysis involves; and it thus directs attention to the visible facts of present landscapes instead of diverting attention to the invisible conditions and processes of past time. (*Davis, 1924G, p. 190*)

There are studies of certain states or regions in which the effort is made to treat the inorganic or physiographic aspects of geography as thoroly as the organic or ontographic side ... But there is also evidence that geography, as conceivd by some of our geographers, has been humanized to such an extent that its physiographic side is reduced to a minimum, and analytical study of land forms being relegated to geology. (*Davis, 1924G, p. 200*)

Thus our progress had led to the apparent existence of two schools, each of which thinks that the other gives too much consideration to a favorite factor. Those who cultivate chiefly the human side of the subject believe that physiography as they conceive it is too geological to be retaind; those who cultivate chiefly the physiographic side believe that human geography as they have encountered it goes too far into history and economics. Every freedom should of course be given to the development of individual views; but for the development of geographical science as a whole, it would seem that the next step of progress should include a meeting of these two extremes for the purpose of discussion, with the wish of reaching a better understanding of each other's views, and perhaps even the hope of at least lessening their differences, if not of approaching an agreement.

The possibility of reaching such an agreement may be materially increast if two preparatory steps are taken. One is a step recommended to physiographers to the effect that they should separate the analytical investigation of land forms, which ought to be regarded as a phase of geology, from the non-argumentativ statement of the results reacht, which must be accepted as good geography. (*Davis, 1924G, pp. 200–1*)

Douglas Johnson held much more liberal views regarding the relationships between physical and human geography than these traditional ideas of Davis but his attempt to put new wine into an old bottle was rebuffed by the master in his eighty-second year.

* Davis is here using the 'simplified' phonetic spelling with which he experimented during his later years.

Thus while Professor Davis's writings, considered as a whole, may appear to support the view that geomorphology is geography rather than geology he has in later years moved much closer to the conception of geography advocated in the present essay. I understand that he would not now consider most of his writings as strictly geographic in character, and that he would classify himself as a physiographer or geomorphologist rather than as a geographer. (*Johnson, 1929, p. 209*)

Perhaps as a means of lightening the geographical load, Johnson places the study of Land-forms or Geomorphology with Geology; but this seems to me as geographically regrettable as it would be geologically regrettable to place Mineralogy with Chemistry, or Optics with Geometry. Landforms may, to be sure, be studied from a geological point of view if one desires to trace their historical development; but they may also be studied from a geographical point of view if one wishes to describe their present appearance. (*Davis, 1932B, p. 217*)

Today's part of the play is Geography; the integration of all its days forever is Geology. (*Davis, 1932B, p. 230*)

Davis, however, was obviously an advocate rather than an exponent of the more human branches of geography and in his explanatory method unwittingly over-stressed the influence of the natural environment on organic life, although not unaware of psychological factors.

The delay of the subject to reach mature treatment did not surprise Davis, who regarded it as 'perhaps the most complex of all sciences'. Although he did not mention it, not the least of the complications in human geography is man's free will, so often obscuring his responses to physiographic controls. Thus for more than one reason Davis himself did comparatively little in illustrating his fundamental principle of relationship between organisms and environment. He wisely restricted himself to spade work on the inorganic side of the vast subject. (*Daly, 1945, p. 272*)

In 1950, at the symposium held by the Association of American Geographers in honour of the centenary of Davis' birth, Kirk Bryan of Harvard revealed how far the dominance of physiography in geographical studies had waned.

He preached the gospel of geography on all occasions during his long and active life: geography in the elementary schools, in the high schools, in the colleges, and in the universities. As an advocate he had no peer. Devoted and dedicated to the analysis of land forms, he also spent much time and thought on the methods of teaching [all] the elements of physical geography ... Moreover, he continuously advocated investigation of the relations of all the primary factors of physical geography with the life of man. He was consistently friendly to the cause of 'human geography' ... Davis' advocacy of the study of the interrelation of man and his environment has triumphed. Except as propaganda, his own contribution was small, but the missionary zeal of Davis and his contemporaries has borne fruit. Today an overwhelming majority of our members are principally concerned with various elaborations of 'human geography'. (*Bryan, 1950, p. 197*)

John Leighly's following comment seems particularly appropriate:

> Davis's theory of the intellectual function of geography thus rested on the presumption of an unbroken chain of causation linking the physical phenomena of the earth's surface, the organic realm, and human society. This monistic and mechanistic interpretation has deep roots in Western thought, but acquired special emphasis in the late nineteenth century through the extension of the Darwinian concept of evolution through natural selection to the intellectual and social realm ... When the link of assumed causation between these sets of incommensurable phenomena was finally recognized as being hopelessly weak, the two halves of Davis's structure of geography fell apart, and the two sets of phenomena toward which it was directed retained only their empirical association in space. But the concatenation assumed earlier had a lasting effect through the selection of the individuals who were to carry American geography into its post-Davisian stage. To uncritical minds the linkage of physical cause with cultural effect offered an explanation of cultural phenomena, which were included in the body of facts at the end of the Davisian chain of causation. The impression that geography offered an explanation of matters relating to human beings attracted to it more adherents with a primary interest in finding explanations of historical events than with a curiosity about the physical earth. The copious writings of Miss Semple, for example, whose interests lay wholly within the 'human' part of Davis's 'ontography', attracted many who would have been repelled by the soberer investigations a Davis or a Salisbury pursued in his original work. (*Leighly, 1955, pp. 312–13*)

Davis himself always thought the pendulum had swung too far away from physiography towards the human side. In his 'Retrospect of geography' (1932B, p. 229) he lamented the fact that organic topics had 'almost uprooted the ge from geography!' And surely nobody had a greater right to complain.

DAVIS' CONTRIBUTION TO PHYSICAL GEOGRAPHY

Davis dropped his meteorological studies relatively early in his academic career but he had a great influence on the teaching and spread of meteorology in the United States. In 1906, Cleveland Abbe when discussing the present condition in American schools and colleges of the study of climatology as a branch of geography and meteorology, wrote

> But the great pioneer in school and college work was our colleague Professor William Morris Davis, who made his elective courses for Harvard freshmen so thorough and so attractive, they became the ideal model for all others. The influence of the Harvard school of Meteorology has been felt everywhere. (*Abbe, 1906, p. 121*)

In 1906, a thousand graded schools in the United States were teaching the elements of climate and attempting simple forecasting with the aid of daily weather maps. In most normal schools meteorology or climatology was included in the syllabus and no less than 144 colleges and universities had some meteorological instruction (Rigdon, 1933, p. 187). We have, of course, no direct evidence of the exact part Davis played in this florescence, and his popular *Physical Geography* had only a short section on the atmosphere.

On the study of landforms he soon became, as we have shown at length, strongly dominant at least in America, much of western Europe and Australasia. We hope that our patient readers will think that we have already done him justice in this respect. But consider again the length, breadth and depth of his limpid prose. Consider his method of presentation and analysis – too lacking, it is true, in mathematical formulation – but for clarity a model for all time. He rarely failed to keep to his own precepts and invariably took incredible care in presentation.

> He was always an analyst and a generalizer ... To him the method was, if anything, more important than the result. Thus many of his papers consist of two parts; an analysis of the probable results of processes, the deduction, and then a citation of the field facts which prove that one or another of the deductions is the correct explanation. This method he has bequeathed to us and whatever the awkwardness and ineptitude of the system in incapable hands, it still remains a standard both in the preparation and in the presentation of geographic results.
>
> (*Bryan, 1935, pp. 24 and 31*)

For length, we can but point out that having published nothing until he was thirty years old, Davis wrote steadily and lucidly for half a century. 'If it's worth writing it's worth printing' was his motto. Eventually he published well over 500 sizeable articles and books on aspects of geography and geology, exclusive of short reviews, brief notes and of a few more esoteric themes such as 'The song of the singing mouse' (1889G). Some of his articles in periodicals are of book length, just as many of his private letters were of article length. They all, without being great literature, have a competence and lucidity outside the common range. He was determined not to be misunderstood, and so high was his ranking and so powerful his editorial connections that he was encouraged to be at the same time explicit, discursive and comprehensive. Today financial costs alone would prohibit many of his performances, which include essays on the history of geography; biographical memoirs of famous geologists, meteorologists and geographers; philosophical treatises on the content of geography; and several hundred articles on landforms analysis alone. He published in at least forty-eight different periodicals and in several the first issue of volume one carried an article by him. At various times he was associate editor of *Science, American Naturalist, Journal of Geography, American Journal of Science* and the *American Meteorological Journal* as well as being a corresponding member of many learned societies, nine of whom awarded him a medal. For his scientific writings he was also created a Chevalier de la Légion d'Honneur.

Many people find certain of his essays particularly pleasing. Henri Baulig, a stout Davisian, thought the 'Disciplinary value of geography' (viz. of geomorphology) (1911P) 'a great classic of scientific method' that should be familiar to everyone engaged in geological or allied researches. For us half a dozen other of his methodological essays equal or excel it, and particularly 'The principles of geographical description' (1915K). His biography of Grove Karl Gilbert is worthy of that great geomorphologist. His later longer essays on landforms are remarkable expositions which of their kind are not likely to be excelled.

But his pen had other qualities. As an illustrator of ideas on landforms he was outstanding. At times his critics say he was far too good, far too convincing but when again shall we see among professional geographers his diagrammatic versatility, compactness and verisimilitude? His three-dimensional concepts or block diagrams have a surprising reality. He said just before his death that he was quite overcome with the number of really beautiful drawings that he had done years earlier, and we have in this volume tried to give a fair sample of his ability. Familiarity does not lessen the entrancing first impressions of his visual concepts from which sprang a whole school of vivid geomorphic artists including Armin K. Lobeck and Emmanuel de Martonne.

However, our chief purpose is to assess the main contributions made by Davis to the study of landforms. Henri Baulig, of the University of Strasbourg, wrote in his 'William Morris Davis: Master of method' (1950)

> Davis's contributions to particular problems of land sculpture are numerous and important. It seems, however, fair to say that his greatest service to geomorphology was to raise it finally to the status of an autonomous science by clearly defining its object, fundamental principles, and methods.
>
> For Davis, the ultimate aim of geomorphology is and should be 'the explanatory description of landforms'. (*Baulig, 1950, p. 188*)

In 1925, Davis himself thought his main contributions had been to provide a terminology and to systematize the sequence of forms through an ideal cycle. Each calls for some kind of summary.

THE DAVISIAN TERMINOLOGY

Nothing epitomizes more Davis' nineteenth-century scholarly roots than his concern for precision in the use of words. If one of his seminar students was injudicious enough to use the word 'stuff', Davis would suspend the proceedings while he delivered a lecture on woollen fabrics. In this respect he exemplified Ruskin's dictum that the essential difference between education and non-education resides in the accuracy with which one can interpret the meaning of words. Thus, whatever scholarly project Davis tackled, he tried to sharpen his precision of thought and expression by developing a terminology of explanatory technical terms. During the first few years of his career as a lecturer at Harvard he introduced a few new terms and continued to do so for over half a century. He recognized how handicapped the study of physical geography was by the lack of an adequate vocabulary and deliberately set out to remedy the deficiency. In meteorology his efforts, for example, to obviate the common misuse of the terms *cyclone* and *tornado*, had little effect, but in landform analysis he was so dominant in the English-speaking world that he, and his disciples, met with little competition and much success. He coined, concocted and advertised for new terms with his usual vigour. As early as 1890, having found in northern New Jersey an old baselevelled region (either a peneplain or a truly ultimate plain) slightly uplifted so that youthful valleys were being cut into its simple, featureless surface, Davis could think of no better terms for it than 'past-plain' or 'past-peneplain'. So he solicited assistance.

> Wanted: a name applicable to those broken rugged regions that have been developed by the normal processes of denudation from the once continuous surface of a plain or peneplain. The name should be if possible homologous with plain, peneplain, and pastplain; it should be of simple, convenient, and euphonius form; it must be satisfactory to many other persons than its inventor, and its etymological construction should not be embarrassed by the attempt to crowd too much meaning into it. The mere suggestion that it was once a plain and that it is now maturely diversified will suffice. (*Davis, 1890K, p. 89*)

As far as we know no one supplied the solution and we are quite sure that Walther Penck, then aged two, had not yet exclaimed 'Primärrumpf'!

Davis was often criticized at home and abroad for his reckless invention of terms, more so 'than seems warranted by the small numbers of terms introduced'.

> For example, at the Geological Congress in Zurich in 1894, Richthofen told Davis, with courteous frankness, that in his opinion the new terms had been too freely used; but on January 23, 1896, Richthofen wrote (in English) stating a change in his opinion with generous candor, as follows:
>
> 'I have read in the winter with much interest and pleasure your papers on the basins of the Seine and Thames, the importance of which is being generally acknowledged. Owing to the sagacity of your mind and to your exact studies, you have contrived to acquire a mastership in the analysis of the forms of the land and the history of rivers ... Both papers, which are but prominent examples of your acquired art, have been reported on in my geographical institute, and I have recommended them to my pupils as masterpieces of geographical analysis. I will not enter upon your numerous other papers which testify to your wonderful capability of work. But it gives me pleasure to tell you that I now fully value the use of the exact terminology which you have introduced, and I beg you kindly to excuse the remarks which I have been too prone to make at Zürich against the introduction of new terms ...' (*Brigham, 1909, pp. 26 and 28*)

Nothing would give him more pleasure than winning over so eminent an opponent. Yet, strange to say, in his early days Davis sometimes indicates that his critics over-emphasized his terminological enrichment of physiography. In 1903 he stated that he had added at most twelve new terms (1903I, p. 485) whereas some geomorphologists consider that he added more than that number in 'Rivers and valleys of Pennsylvania' (1889D) alone. Thus it seems that when he gave a technical meaning to a common word – such as subsequent, mature, beheaded etc. – he did not consider that he was introducing a new term.

For five decades he harried his colleagues, students and co-travellers for suitable technical terms. Some of his own introductions proved highly successful. He was right, for example, in suggesting that 'monadnock' (a remnant hill of more resistant rock surmounting a peneplain) had 'a whole paragraph of explanation' packed into it. He served on a committee of the Geological Society of America to investigate and standardize the nomenclature of faults and the findings were published in 1913(L). A great many of his coinings never got into circulation; 'skiou' was intended to be temporary; 'morvan' proved unpopular; COBARK (cut-out-behind-a-rock-knob) was mercifully stillborn. On the other hand, the extension of geomorphic terminology by the use of similes and of adjectives with a human quality met with widespread approval although they may nauseate modern scientists. 'River-pirate' and 'river-piracy', introduced more in jest than earnest, became immediately popular as also did phrases such as 'stream-capture' and 'elbow of capture'.

In 1919 Charles R. Dryer, president of the Association of American geographers, in complaining bitterly of the inadequacy of the technical language of geography, which he compared to 'a white-haired centenarian with the speech of a child of ten!', says:

> Davis has succeeded in 'putting over', at least on American geographers, peneplain, monadnock and cuesta, but only with the help of the geologists. We may hope he will succeed with his latest, 'hermatapelago' for a sunken reef sea. (*Dryer, 1920, p. 14*)

Some may not regret that 'hermatapelago' has passed into limbo but against it and numerous other failures must be set the very successful use, especially in Davis' later physiographic essays, of participal adjectives such as 'inverted', 'beheaded', 'extended', 'betrunked' and 'dismembered' streams. Inevitably he is often credited with the first introduction of terms that were in fact introduced by others. Powell invented 'consequent' for rivers but Davis added 'subsequent', 'insequent' and 'obsequent'. 'Young' and 'old age' were being used in England some time before Chamberlin and Salisbury took them over and bequeathed them to Davis, who added 'mature', 'infancy' and 'adolescence'. Davis really revelled in coining and concocting technical terms when he moved to the American arid West. In describing the mountain bases in the Basin and Range Province his adjectives include 'fan-bayed', 'fan-frayed', 'fan-dented' and 'fan-wrapped' (1925E, pp. 391–2). In his essay on limestone caverns he coined 'dripstone', 'flowstone' and 'rimstone' but only the first was welcomed and withstood competition.

The total number of popular technical terms introduced by Davis is still uncertain. In 1933 Marius R. Campbell of the United States Geological Survey gave him credit for the introduction of over 150 terms or phrases. But to these, as Vera Rigdon points out, must be added the new terms introduced by his own students. These probably exceed 100, including at least 32 by Douglas Wilson Johnson alone.

Henri Baulig, in describing 'the procuring of a terminology' as one of Davis' main contributions to geomorphology, presents an adequate modern appreciation.

> His terms are ingenious, pleasant, expressive, easily understood, easily transferred to other languages. Partly borrowed from common speech, partly artificial, his terminology serves well its purpose, though it cannot have the rigorous precision of the vocabulary of exact sciences: but geomorphology is no exact science. It may, however, be objected that his terminology is entirely genetic, forms being classified and named after their assumed origin. It seems that provision should have been made for a distinct set of terms, denoting, without any implications as to genesis, some simple and obvious relationship, as for instance between structure and direction of drainage: monoclinal, cataclinal, and anaclinal as applied to valley, are rarely synonymous with subsequent, consequent, and obsequent, because first-cycle landscapes in which they exist, appear as investigation progresses to be more and more exceptional. (*Baulig, 1950B, pp. 191–2*)

Against this must be weighed the detrimental effects of Davis' use of loaded and figurative metaphors for physical phenomena. Dury (1963, p. 74), for example, lists some of the terms used by Davis to describe streams: robust, bravely, loyal, uncertain, impoverished, bewildered and many more!

THE DAVISIAN EXPLANATORY DESCRIPTION

Davis set out to replace empirical description with explanatory or causal description and to systematize the hitherto multifarious methods of describing landforms. We have described in adequate detail, for we believe the first time, how it all came about and have presented sufficient original evidence to allow the reader to form an independent judgment. It seems to us that the struggling Davis rapidly slipped into the cycle concept and enthusiastically became enmeshed in its possibilities. However, he soon became aware that in the 'ideal cycle' he had saddled himself with an old man of the sea. Unfortunately he had in his nature a stubborn streak and a deductive urge which for didactic purposes constantly led him to call attention to a model which he himself later admitted publicly – as he had, in our opinion, long held privately – to be most improbable or impossible or a 'special case'. However, it was his disciples and detractors, not himself, who would not forget the

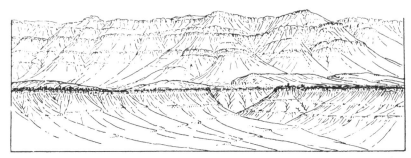

F I G. 161. Escarpment of thick lava flows half-way along the Galiuro Mountain front, looking north-east. Low lava slabs rise in the middle distance. In the foreground the valley has been eroded in detrital deposits consisting of gravel-covered silts (From Davis, 1930H, Fig. 3)

F I G. 162. The Argus Range, south-eastern California
Above. The western faulted face of the northern Argus Range, showing step faults in the lava-covered rock mass
Below. The eastern dip slope of the northern Argus Range, showing considerable erosion of the surface where not lava-covered
(From Davis, 1930A, Figs. 1 and 2)

'ideal cycle'. He went over very early (1884N) to the systematic description of landforms by means of structure, process and stage but in the popular mind he was, and still is, credited mainly with *stage*. Even in 1970 a new geomorphological textbook stated that *age* or *stage* denoted *length of time* whereas it never meant more than the relation between the amount of work already done compared to the amount yet to be done.

The landform-analysis aims of Davis became steadily more involved with systematic explanatory description than with the cycle concept in itself. What Brigham wrote in 1909 was reasonably true of the early Davis.

> It is largely in order to make the results of investigation thoroughly intelligible that Davis has in recent years so often urged the advisability of at least an approach to uniformity of method in the treatment of land forms by geographers who have to do with this aspect of the science; and for this purpose, following the outline indicated in his paper on 'Geographic Classification' (1884), he has formulated the method of 'structure, process and stage'. . . . Following this plan, land forms are to be treated, first, according to the structure and attitude of the crustal mass of which they are the surface; second, according to the process, – normal, glacial, arid, solvent, or marine – by which they have been carved; third, according to the stage reached in the work of the process on the mass, in view of the whole sequence of possible stages from the initial form produced by uplift to the ultimate form produced by indefinitely long erosion. Moreover, under this method actual land forms studied by observation are to be described in terms of an elaborate and systematic series of ideal land forms with which the geographer must become familiar by conscious and careful deduction; the process of deduction, always held apart from that of observation, being guided by the scheme of structure, process and stage, in which all manner of possible combinations lead to the deduction of a vast variety of ideal types. Thus observation and deduction, seeing and thinking, are held to be indispensable complements, one of the other, in modern geography. One of the most helpful results of this method is the increased recognition that it gives to the essential correlation of the different elements of a landscape so that the mention of one part suggests the most concisely described in terms of a corresponding ideal and generalized type with the addition of quantitative and local details as they are needed.
>
> The adoption of this method of treatment, made elastic by the permissible variation and combination of its various elements has repeatedly been shown to have practical value in field studies; but it should be candidly stated that there is no magic in it; it may aid and guide, but it cannot replace serious field study; its real value is found in giving clear and systematic expression to well ascertained facts.
>
> (*Brigham, 1909, pp. 56 and 58*)

Long before Brigham's appraisal, Davis' main themes had begun to be not the cycle but interruptions and complications of it. All landscapes were proving two-cycle or polycyclic. By 1910 and after, the 'ideal cycle' has slipped almost out of sight as 'a special case' of high educational value for beginners. The emphasis was henceforward on the nature of relief before

FIG. 163. Christmas card sent by Davis and his wife in 1932 (Courtesy J. K. Wright)

uplift and the development of relief after uplift; or before glaciation, during, and after glaciation; and so on. Eventually under Walther Penck's influence, the effect of erosion during uplift begins to be more patently included.

From correspondence during his eighty-third year we are left in no doubt what Davis himself wished the world to think that he stood for. He stood as an advocate and disciple of the multiple-working hypothesis; and he stood for something more than structure, process and stage.

In his later teaching Davis gave emphasis to a corollary of the method of structure, process and stage, which he found helpful in physiographic description, but for which no special name has been devised. It applies to any region which at any stage of any process suffers a change which does not significantly affect its structure, but which entails a change in the further progress of its erosional processes. For example after a given land mass suffers a simple uplift at any stage of its erosional development, its description involves three statements: first what it was before uplift (this is to be given as usual in terms of structure, process, stage), second what the uplift did to it, and third what has happened since ... In case what has happened since involves a *great* measure of change, terms one and two are not needed. (*Letter: Davis to V. E. Rigdon, 1933; quoted in Rigdon, 1933*)

MODERN REACTIONS

As a progressive and powerful geomorphologist Davis faced a great difficulty always inherent in an inexact science. In an exact science *laws* provide stepping-stones for progress whereas in geomorphology, an inexact science, hypotheses provide themes for further investigation. Unfortunately when a person such as Davis becomes an *authority* his hypotheses tend to be uncritically accepted as laws by his lesser disciples, and so may lead to a lessening of scientific investigation.

There were many weaknesses in the Davisian armour; for example, he never really believed in, or largely ignored, non-glacial eustatism, mathematical equations and scientific formulae, laboratory physical experiments on erosional processes, and elaborate measurements and intensive instrumentation of mechanisms and processes. His idea of look, think, deduce, describe and draw explanatorily did not include sufficient exact measurement of component elements such as slopes.

By 1950 the revolt against certain weaknesses in Davis' method had become important in the United States. This revolt is well exemplified by the remarks of Arthur N. Strahler of Columbia University, a student of Douglas Johnson's. He objected to Davis' qualitative approach, to his lack of precise measurements both of slopes and of the mechanics of processes.

> Davis' treatment appealed then, as it does now, to persons who have had little training in basic physical sciences, but who like scenery and outdoor life. As a cultural pursuit, Davis' method of analysis of landscapes is excellent; as a part of the basis for the understanding of human geography it is entirely adequate. As a branch of natural science it seems superficial and inadequate. (*Strahler, 1950, p. 209*)

Strahler considers that had G. K. Gilbert's quantitative-dynamic method of geomorphic study, as revealed in his classic report, 'The Transportation of debris by running water' (1914), prevailed among students of landforms, the analysis of slopes would not have been so long delayed.

> As it was, apparently Davis won over a large following and the study of landforms became dominated by such figures as Douglas Johnson, C. A. Cotton, N. M. Fenneman and A. K. Lobeck, whose contributions to descriptive and regional geomorphology have provided a sound basis for studies in human geography ... Even as recently as 1943, Douglas Johnson presented his graduate classes in geomorphology with subject matter faithfully reproducing the principles and details as written by Davis 45 years earlier. (*Strahler, 1950, p. 210*)

Thus to some the Davisian method had become a stranglehold or at least a sedative. The school of quantitative geomorphologists wished to pay full attention to the dynamics of the erosional processes. In doing so they, to some extent, confirmed Davis' concept of declining slopes with age but could not comfortably adopt his cycle scheme.

> Instead the concept of a steady state in an open system seems a logical replacement for the idea of 'maturity' while the stage of 'old age' may well be abandoned. This change will bring the theory of the erosion-transportation process into line with systems of flow of fluids and heat and various other dynamic systems which reach and maintain steady states. (*Strahler, 1950, p. 213*)

We must, however, point out that Davis – a skilled geologist – aimed at the explanatory description of landforms and not at the explanation of landform processes. Davis and his disciples wished to move towards the understanding of physical forms; today we wish to grope towards their formative processes, as a means of predicting their transmutations.

This revolt in favour of quantitative geomorphology had grown rapidly by 1970 and within a few decades it may be hard for people to realize the intense loyalty of Davis' disciples throughout the world, particularly as their then leader, Sir Charles Cotton, died in 1970. Fortunately we get a glimpse of the loyalty – or as some would say 'fanaticism' – of the true Davisians from a review of Davis' *Geographical Essays* (on its republication in 1954) by S. W. Wooldridge, a redoubtable buccaneer of physiography and a power in British geography who had long declaimed against 'taking the Ge out of geography' and so 'throwing away the baby with the bathwater'. Wooldridge welcomes the reprint as notable, timely and important especially in view of the recent active, and at times almost bitter discussion about the internal organization, aims and methods of geomorphology and its place among the Earth Sciences. He notices the strong reaction against some of Davis' work in the United States.

> We must distinguish, however, between the 'slope question' which gave the ground for dispute to his chief critic Walther Penck and the more general evaluation of his system which has failed latterly to find admirers or supporters in the United States. Thus Sauer in the Bowman Memorial Lectures for 1952 . . . goes out of his way to claim that W. M. Davis delayed somewhat our learning about the physical earth by his system of attractive but unreal cycles of erosion. In these references there is a distinctly bitter or rancorous note which many of us, still heavily in debt to Davis, must resent and, indeed, flatly repudiate. (*Wooldridge, 1955, pp. 89–90*)

Wooldridge continues in a forthright way:

> What seems to have been lost sight of both in Britain and America is the superbly 'clean' and intellectually attractive quality of land form study as Davis taught it. In its quasimathematical form and style of reasoning it has something of the quality of economic analysis. It conveys a point of view which can permeate one's

thinking and make each and every landscape live . . . Would that these essays could be made compulsory reading, not so much for students, as for university teachers of geography . . . Human or social geography has yet to find anyone of the stature of W. M. Davis to give it coherence and organization and until this is achieved, it cannot be a wonder that appreciable numbers of geographers, not content either with amassed information as such, or the pedantic quest for mere 'areal differentiation', finds intellectual stimulus and sustenance in geomorphology. All such look to W. M. Davis as a potent factor in the fashioning of their minds and they will take down this reprinted volume from their shelves with deep respect, gratitude and affection and the prospect of much renewed pleasure. (*Wooldridge, 1955, p. 90*)

To us, a main feature of this enthusiastic review is that Davis' pre-1909 essays are still considered 'true classics, as alive today as when they were written'; whereas we, admitting some of their enduring qualities, consider that the true Davis evolved *after* they were written. We hope that by drawing attention to his later works, we have rendered an even greater service to Davis than that performed by D. W. Johnson or any other of his devoted disciples. We have endeavoured to portray without bias his mature ability, his weaknesses and his strength. This is what he himself would have wished in his riper, richer years. He was an outstanding protagonist and practitioner of landform study with a deductive power and range of topics of extraordinary dimensions. He may well be always revered as a master of method and of presentation but we hope he will also be given some small credit for positive contributions to field knowledge. For the late-nineteenth century his aims were high but for the late-twentieth century they may well be too low, too out of tune with scientific progress. In the meanwhile few will disagree with William D. Thornbury's (1954, pp. 10–11) assessment that

Despite objections which have arisen to some of Davis' ideas, it can hardly be denied that geomorphology will probably retain his stamp longer than that of any other single person.

Appendices

The Faith of Reverent Science*

PART I

Outline

The object of this address is, first, to direct attention to the enormous
service of science in liberating our minds from their century-long subjection
to ancient dogmas, thus enabling us freely to enjoy the modern understanding of the world and of our place in it; and second, to consider certain
responsibilities that are placed upon us in consequence of our liberation.
With this object the contents of the address are as follows: It will begin with
an introductory statement concerning two great products of the human
mind – religion and science. Then will follow several retrospects, the first
of which will recall the harmonious relation existing between primitive
religion and primitive science among primitive peoples; the second will
touch upon the unhappy conflict which arose between struggling science
and theologically dominated religion in the later centuries of European
history; and, after a brief interlude, the third retrospect will tell of the
victory of science over various theological elements of our religion in a
recent era which might well, for that reason *alone*, be called the Victorian. In this short review it will not be so much the events of the several
retrospects as the state of mind behind the events that we shall try to
consider.

Then will follow a sketch of the actively growing reconciliation of
Christian theology with modern science, which has so fortunately occupied
the short period from the Victorian era to the present. After that, an outline
will be given of the faith toward which multitudes of our population, who
have become more or less conscious of the reconciliation just mentioned, are
now advancing. There will then come some reflections on our new responsibilities in consequence of that reconciliation, and also a prognostication of
the manner in which our new responsibilities can best be met, this being the
main theme of my address.

---

* The Second Hector Maiben Lecture presented at the meeting of the American
Association for the Advancement of Science on 28 December 1933 at Cambridge,
Massachusetts. Reprinted by permission of the Editor of *The Scientific Monthly*.

Introduction

Two of the greatest products of the human mind are the various systems of faith known as religions and the various bodies of knowledge known as sciences. Both have slowly evolved from ancient and simple beginnings to their present complications; and the evolution of both appears to have been largely guided by outstanding leaders who have arisen from time to time in the two lines of thought. Religion is believed to have begun as a form of magic, in which the 'medicine man' or shaman used certain incantations to compel the 'powers' to act as he wished them to act. Later, when the powers were personified as deities, they were no longer compelled by incantations but implored by prayers and sacrifices to act as their priests wished. In its simpler stages religion was largely concerned with relations of humanity to the gods; in its more advanced stages it comes to be much concerned also with the relation of human beings to one another. Thus we have the division of religion into its two parts, theology and ethics. Science must have begun among primitive peoples as simple observation, soon followed by elementary induction; but, as we now know it, science involves mental processes far too elaborate for primitive peoples. Like religion, it must have had a gradual beginning, and the beginning may have been as little like its present state as a seed is like a full-grown plant.

At present religions are systems of faith which their devotees believe have been made known in smaller or greater part by inspiration or revelation from the supernatural world. Sciences are, on the other hand, bodies of demonstrable knowledge regarding the natural world, as far as it is open to observation and to inference based on observation. Both these lines of thought have come to be largely under the direction of two groups of specialists; the priesthoods of the churches and the professorhoods of the universities; but the priesthoods are supplemented by learned men among the laity, and the professorhoods are reinforced by many skilled experts in the arts and the industries.

Before going farther, let me repeat a significant word – the word 'human'. I have said the systems of religious faith and the bodies of scientific knowledge are both the products of the human mind. That rejects supernatural revelation on which the followers of various religions believe them to be based. This statement may arouse dissent. Yet every one present will doubtless agree that all the sciences are purely human achievements. Most of those present will, I presume, accept also the purely human origin of the ruder religions of primitive humanity, perhaps including the hard religion of fear and punishment that was developed several thousand years ago among the barbarous Israelites. Many of those present may perhaps accept the wholly human origin of all the modern religions of today, excepting Christianity. And some may believe, as I do, that all religions, ancient and modern, including Christianity, are, like all the sciences, wholly of human origin. The grounds for this belief will be presented in the review of the recent reconciliation of Christian theology and science.

First Retrospect

Our first retrospect is brief. It recalls the harmonious relation between the crude science and the rude religion of primitive peoples. If scientific explanation of natural phenomena is there attempted it is often wrong, largely because it is so commonly given a supernatural nature. For example, dwellers in arid regions do not ascribe the rains of their thunderstorms to the adiabatic cooling of ascending air currents, but to their rain god, to whom they therefore address prayers when rain is wanted. The priesthood has charge of such supernatural beliefs, and they develop them in accord with the rest of their religious system. Conflict between their science and their religion is thus avoided.

Second Retrospect

Our second retrospect deals summarily with the later European centuries. We there find at first one and later several priesthoods, each in charge of its own phase of Christianity. Let us remember that, as then taught, Christianity was little concerned with the simple ethical principles preached by its founder; principles phrased in easily understood words and addressed to an uneducated population. Christianity of the Middle Ages and the next following centuries was largely concerned with highly theological creeds which had been adopted in earlier centuries by majority votes of humanly, not to say politically appointed delegates at the councils of a well-organized and dominant church. Moreover, some of the councils were held at times of violent metaphysical disputes within the church; in one instance about a matter so recondite as the difference expressed, as Gibbon puts it, by the vowel of homo-ouzion and the diphthong of homoi-ouzion.

It is truly singular that a great religion should be, not only then but almost to this present day, so largely concerned with the theology of complicated, majority-vote creeds rather than with the simple ethics of its founder; but such is clearly the case.* The explanation of this singularity is clear also; it is the result of ecclesiastical organization; and it has been through such organization that the peoples of Europe and of European stock have been so long held subject to a theologically dominated religion. And here let me state a very curious feature of that subjection. By reason of a strange concatenation of ancient events, the theological doctrines which have so long dominated European Christianity are the outgrowth of a series of crude,

* Edwin Hatch, 'The Influence of Greek Ideas and Usages upon the Christian Church' (The Hibbert Lectures, 1888). Ed. by A. M. Fairbairn, Williams and Norgate, London and Edinburgh, 1890.

superstitious beliefs which originated several thousand years ago among the barbarous, ignorant, credulous peoples of southwestern Asia; a body of beliefs which was recorded, in the form commonly known to us, by a people who believed themselves 'chosen' by their god from among all other peoples; a body of beliefs which was introduced into Europe in close association with a new gospel which that 'chosen' people rejected, a rejection which later caused three centuries of cruel persecution. And yet we – that is, all peoples of European stock – adopted both the old beliefs and the new gospel as the infallible 'Word of God'. Does history record anything more extraordinary?

Is there any wonder that a conflict arose when European mentality began to assert itself? These old beliefs, especially the cosmological myth standing at their beginning, were fitting enough for the race which, in the childhood of humanity, invented them. But how impossible it was that they should accord with the discoveries of science made by another race in another region under altogether different conditions!

Specifically, it was the heliocentric arrangement of the solar system, an old Greek view revived by Copernicus, which conflicted with the traditional position of a fixed earth in a geocentric system, as revealed in Genesis. Science, more or less aware of our many ignorances, tried by observation and reflection to establish a safe beginning of knowledge. Theology insisted that a finality of knowledge had been given us on this, as on so many other subjects. Hence the Copernican system was condemned as heretical, and under the strongly organized Christian priesthood of that time, heretics had a terrible time of it. Even the protesting Luther said: 'The fool' – Copernicus – 'wishes to reverse the entire science of astronomy, but sacred Scripture tells us that Joshua commanded the sun to stand still, and not the earth'.

Let it here be emphasized that the real significance of the conflict which thus arose did not lie so much in the difference of the two systems of astronomy as in the unlikeness of the reasons for adopting them and in the unlikeness of the minds which accepted those reasons. From the theological point of view, the relation of earth and sun was not to be studied; it was infallibly known and settled by revelation. From a scientific point of view that relation was, like any other problem, open to unprejudiced discussion. The scientific mind would be inclined to say that, if the revealed relation of earth and sun were true, it could only be confirmed by unprejudiced investigation, while if it were not true it would be thereby corrected. But the theological mind refused to take the risk of alternative conclusions, and it therefore opposed all such investigation. Thus the right of the human mind to think was infringed upon, its practise in thinking was lessened, and its progress in the difficult art of learning how to think correctly was retarded. In our freedom today, when investigation is aided and encouraged in every direction, we can hardly imagine how persistently it was hampered and opposed in earlier centuries. Experimentation in physics and chemistry was frowned upon if not suppressed. Philological study was retarded by the belief that Hebrew, as spoken by Adam and Eve, as well as by Jehovah and the Serpent, in the Garden of Eden, was the original language of mankind. The humane care and cure of the insane was long delayed by the theological belief that insanity was due to 'demoniacal possession'. Even in the nineteenth century the use of an anaesthetic in childbirth was opposed by the orthodox on the ground that it avoided 'part of the primeval curse upon women', although it was at the same time ingeniously defended by pointing out that in preparation for 'the first surgical operation ever performed' – that of taking a rib from Adam's side for the creation of Eve – a 'deep sleep' was caused to fall upon the rib-loser.

All this opposition was epitomized centuries earlier; for when the Franciscan friar, Roger Bacon, wished to increase his knowledge by experimentation instead of by deducing all truth from sacred texts, he was attacked on the ground that 'he did not believe that philosophy had become complete and that nothing more was to be learned'. He was forbidden by the 'general' of his order to lecture at Oxford; his brilliant studies of the refraction of light in the production of rainbows were condemned because the rainbow was theologically believed not to be a result of natural laws but a 'sign' supernaturally placed in the heavens to assure mankind that another universal deluge was not to be feared. And in his old age Bacon was imprisoned for fourteen years because he was 'dangerous'. How glorious is our mental freedom compared to the enslavement of those earlier times! When we realize the long struggle that our freedom cost our predecessors, should we not strive to use it worthily?

Let me close this retrospect by acknowledging my deep indebtedness to White's two masterly volumes on 'The Warfare of Science and Theology in Christendom', from which the above statements are taken, and in which one may find the following general conclusion: 'The establishment of Christianity . . . arrested the normal development of the physical sciences for over fifteen hundred years' (vol. 1, p. 375). Those who have not read this great work should do so without delay, for it will teach them many wonderful stories; for example, that of John Wesley, surely a man of no mean intellect, who nevertheless believed that spiders did not eat flies until after Adam sinned.

Interlude

Several eventful centuries elapse between our second and third retrospects, during which the spirit of scientific rationalism gained much strength; for this was an era in which the minds of Europeans were slowly learning to work more and more independently of preconceived opinions. With this growth of rationalism the number of heretical unbelievers – that is, unbelievers of various theological complications – greatly increased. But that same era witnessed also some growth of the spirit of Christian ethics, and although unbelievers were still branded as heretics, the branding was only verbal; they were no longer burned at the stake.

This era witnessed another manifestation of mental independence in the multiplication of religious sects, each one having its peculiar creed or organization based on Biblical texts which were still taken as the 'Word of God', and elaborated in the mind of some able leader. At the risk of repeating what is already well known to some of you, I wish to give a few examples of the theological doctrines which were then still in force, in order that you may contrast them with the relative freedom of today. Nearly 300 years ago a famous 'Confession of Faith' was formulated by a body of 150 learned and godly men who were summoned by an act of Cromwell's Long Parliament and were in close consultation for five years. Two articles of this confession may be here cited. One of them sets forth again the old doctrine of original sin, which many a piously taught child of earlier times than now learned in its condensed, rhythmical form: 'In Adam's fall we sinnèd all'; but which, phrased in more sententious form by those godly men, is as follows: 'From this original corruption' – that is, from Adam's sin – 'whereby we are utterly indisposed, disabled and made opposed to all good and wholly inclined to all evil, do proceed all actual transgressions'. There can be no question that this extraordinary article was authentically based on Scriptural texts; but there can also be no question that its doctrine is now so completely opposed by the teachings of evolutionary ethnology as to be scientifically absurd. Can there be a more striking example of the manner in which ancient error was perpetuated?

But that is not all. That now absurd article was supplemented by another which taught the perdition of most of the human race, because of a baneful clause found near the end of the Gospel of St Mark but not in the other Gospels; a clause now suspected by competent students to be spurious, for it is given as a saying of Jesus in spite of its violent contradiction of the spirit of his more authentic teachings; a clause which 'has cost the world more innocent blood than any other' (White, II, 387); a little clause of only eight words, reading: 'And he that believeth not shall be damned' (Mark 16: 16). Chiefly upon that clause a monstrous conclusion was stated in another article of the above-mentioned 'Confession of Faith', as follows: 'No men, not professing the Christian religion, can be saved in any way whatsoever than by becoming Christians, be they ever so diligent to frame their lives according to the light of nature and the law of the religion that they profess; and to assert and maintain that they may be is very pernicious and to be detested.'

Had Jesus been present when that detestable article was adopted – he who taught us: 'Be ye therefore merciful, as your Father also is merciful' – would he not have said once more: 'Father, forgive them, for they know not what they do.' Or if that article had been found in a distant land, expressing the will of some pagan Juggernaut, would not all godly Christians have joined in condemning it as altogether opposed to the will of a loving and forgiving Father in Heaven? And yet, under abject submission to ancient myths and to very fallible records of later date, those 150 earnest, devoted, conscientious men compelled themselves to believe that monstrous theological doctrine, with its imagined alternatives of perdition and salvation!

Third Retrospect

The conflict between the theological elements of Christianity and science reached a climax in the Victorian era, to which our third retrospect is directed, because science had by that time gained immensely in strength and independence. Not long before that era opened, the past history and the future continuity of the earth had been shown to be vastly longer than the Bible taught. As the Scotch geologist, Hutton, put it: 'In the history of the earth there is no trace of a beginning and no prospect of an end.' Then, shortly after the era opened, the antiquity of man was shown to be much greater than the few thousand years which the Biblical chronology allowed it. And but little later, Darwin, after holding his growing ideas under 'study and meditation for nearly twenty years', announced his revolutionary, evolutionary theory, including man's development from some anthropoid mammal, and thus presented a new philosophy to an unwilling, a very unwilling world.

It must not be imagined that the makers of these great discoveries had been working with the object of unsettling religious opinions; not in the least. They were devout men, possessed of vigorous intellectual curiosity, who were objectively pursuing geological and biological studies simply with the wish to learn the truth, whatever the truth might be. They rarely if ever precipitated a conflict by attacking the theological beliefs which their discoveries traversed. The attack was made by the theologians, most of whom were wholly untrained in scientific habits of thought. When the above discoveries, particularly Darwin's, were announced, the violence of the protests they aroused among the orthodox can hardly be credited today. Men learned in Christian theology then asserted, 'Darwin requires us to disbelieve the authoritative word of the Creator'; and again, 'If the Darwinian theory is true, Genesis is a lie, the whole framework of the Book of Life falls to pieces, and the revelation of God to man, as we know it, is a delusion and a snare.' Not twenty years ago, a preacher told his congregation that God had made monkeys look like men so that heretics should be led into error! In a word, the clergy of the Victorian era, as well as some of their successors, were still as much blinded to the grandeur of the evolutionary scheme of the varying organic world as their predecessors had been to the grandeur of the Copernican scheme of the solar system. It is therefore gratifying to know that, some twenty-five years after Darwin's 'Origin of Species' appeared, a broader view of the world was taken by Bishop Temple of London who wrote: 'It seems something more majestic, more befitting Him to whom a thousand years are as one day, thus to impress His will once for all on His creation, and to provide for all the countless varieties [of organic forms] by this one impress, than by special acts of creation to be perpetually modifying what he had previously made.'

A curious feature of the objection to Darwinism was that, while it strained at the little gnat of specific evolution, it unhesitatingly swallowed the huge camel of individual development. Of these two processes the latter is immensely the more marvelous. It involves for every individual the change from a slightly differentiated ovum into an elaborately differentiated adult, closely resembling its parents; and during this change there is, first, a rapid recapitulation of many ancestral antiquities, and later a more gradual addition of certain evolutionary novelties. On the other hand, the appearance of a new species involves only the addition of a few extra-novelties, highly significant for the species then appearing, but nothing like so wonderful as the long sequence of inherited antiquities which precedes them. But the huge camel of individual development was, like the magnificent phenomena of sunrise and sunset, a familiar commonplace which the Mosaic record took as a matter of course, while the little Darwinian gnat of specific evolution was a choking innovation, which the Mosaic record explained by supernatural acts. And these supernatural acts were believed in by the multitude, because the unscientific mind desires a supernatural power to enter frequently into mundane affairs, where the scientific mind sees only the orderly working of nature.

The story is told of an African chieftain who, thirty years ago, watched the construction of the steel-arched railway bridge across the gorge of the Zambezi. At its beginning he told the engineers: 'You can never build it across.' When it was built across he said: 'It can not bear the weight of a train.' Finally, when a train was safely driven over it, he said: 'The finger of God holds it up.' But the engineers said: 'The bridge stands and bears the weight of the train because of the strength of its steel.' This story has a long moral. Many devout Victorians, sixty or seventy years ago, could not believe that anything so marvelously intricate as the development of successive organic forms could be accounted for by natural processes. Like the African chieftain they could not leave out the 'finger of God'. But Darwin said: 'I believe in the doctrine of descent with modification, not withstanding that this and that particular change of structure can not be accounted for, because this doctrine groups together and explains . . . many general phenomena of nature.' And the grandchildren of the devout Victorians now, with no less of devoutness, follow the belief of Darwin rather than that of their grandparents. Such is the order of human progress.

But why repeat these old stories of the Victorian era? There may perhaps be present some persons who are so completely modernized that they look upon that era as already too much talked about, because it was so fatiguingly prim and platitudinous. Truly, its earlier years were rather prim, for church-going two or three times every Sunday was then a matter of course for the piously respectable part of our population: but it is that very primness which must be understood if we, much less prim, are to measure our indebtedness to science for having led us out to a more vigorous life.

It must be remembered that hell and the devil were vivid realities to devout Christians in those recent days; and that some very orthodox parents then taught their children an awful catechism. The loving parent would ask: 'Doth original sin wholly defile you, and is it sufficient to send you to hell, though you had no other sin?' and the little child was to answer, 'Yes,' though no child could possibly have any adequate conception of that shock-ing doctrine. Again, when asked: 'What is your natural state?' the little one was to say: 'I am an enemy of God, a child of Satan and an heir to hell.' Could anything be more incredible? Yes, more incredible still, parents were informed by the authorities of their church that it was proper, even before the little ones had learned to read, to teach them those frightful answers and others equally far beyond childish understanding. There are actually men now living who were taught that sort of thing in their early years by their fathers. Do any of those men, I wonder, now teach the same catechism to their sons? Very few of them do; and it is to scientific rationalism that they owe their deliverance from such folly. If the question is again asked, Why repeat this old story of the Victorian era? the further answer may be given: It is repeated in order to show the younger generation of today, who are now growing up in ignorance of such stories, how recent is the escape of many of us from the theology of medievalism, based on ancient myths.

It is truly difficult to believe that such theological beliefs were persisted in by devoted, conscientious men, who therefore violently opposed anything so heretical as Darwinism. On the other hand, there were also many liberalized rationalists who welcomed the new discoveries and rejoiced in the better understanding of the world that they provided. But there was an inter-mediate and very important class of Victorians; for between the unyielding conservatives and the open-minded rationalists was a large number of some-what mobile-minded yet still orthodox believers, who could not altogether reject the new discoveries, although they had a hard time accepting them. Many of those earnest men suffered great mental distress on finding that certain elements of the creed which they had been taught as essential verities were so inconsistent with the findings of science that they must be regarded as erroneous. And the more logical of them perceived, with still further distress, that if the Bible, which they had been taught was infallible from cover to cover, were shown to be fallible in one or more of its parts, the infallibility of the rest was thereby made questionable, to say the least. To what guide should they turn if the guide that they had so devotedly trusted proved to be not wholly trustworthy? Moreover, if they gave up the infallibility of the Bible, would they not thereby forfeit the right to call themselves Christians? That fear caused profound unhappiness to many sincere men before they found their new bearings. But some of them at least were wisely shown that if they followed the example of their orthodox predecessors in studying and interpreting Christ's teachings under the best advice they could get, and if they then still experienced a joyful exaltation over the profound ethical value of the teachings and a warm desire to live in accordance with them, they would have just as good a right to count themselves among his followers as any one else. In that happy conviction, thousands and thousands of former conservatives, liberalized by the discoveries of science, are living today.

It is as if the mobile-minded orthodox believers of recent time had been near the top of a long flight of stairs, each step of which represented an article of faith, and on or near the top step of which their great-grandfathers had stood. Wherever the mobile-minded orthodox found themselves, they at first insisted they would never take a single step downward. Then in a few years, finding themselves, in consequence of more or less unconscious thinking, several steps lower on the long flight, they again insisted they would never descend any more. But they did, step after step; and many of them are now not far above the solid ground of rationalism at the bottom of the flight. If they should turn and look back at the long flight, they would marvel at the insecurity of the scaffolding that holds it up! And if they there conferred with one another, they would say: 'How few of us today hold all the beliefs that our fathers so ardently held!'

Scientific Study of Religions

One of the most potent causes of disconcertment to orthodox theologians, not yet mentioned, was actively at work all through the last century; namely, the application of the objective, scientific method, not alone to the study of plants and animals, not alone to the study of the structure and history of the earth and of the distance and composition of the stars, but to the study of religions. In spite of their wide diversities the religions of the world were thus found to show a broad community of content and an inherently human origin. They very generally make more or less definite claim of intercourse between their founders and their gods, or of revelations *to* their founders *from* their gods, or of miracles which attest the power of their gods or the authority of their founders. Some religions are found to be more, some less refined in their ethics; but whatever form they take, they reflect much of their natural environment. Among the factors of environment must be included the characteristics of the people who develop them, as well as the climate and the topography of the region in which the people live. In an open country where occasional mountains rise over lowlands, mountain tops have repeatedly been chosen as residences for the gods, or as sites for altars where sacrifices to the gods should be most fittingly made. It is on the sea coast that a god of the sea is worshipped; it is in the desert that a god of rain is worshipped. That is, the gods vary with the homes and the needs of the people who pray to them; they vary also with the ethical standards of their worshippers.

But whatever the quality that a religion acquires, its believers, knowing little of any religion but their own, naturally take it to be superior to all others; its gods are the greatest gods; its people are divinely favored. This self-centered belief has made many an ardent young Christian missionary much harder-hearted and harder-minded than he should have been with respect to the beliefs of the heathen whom he was sent out to convert to what he earnestly believed was the only true faith. In contract to that self-centered belief, the more sympathetic belief of the student of religions is such that he comes to see in every religion the striving of humanity toward a fuller understanding of man's place, man's opportunity and man's duty in the world. And this surely makes for brotherhood and good will among men. The change that has thus been made desirable in missionary work is most beautifully and understandingly set forth in a remarkable report * made last year by an interdenominational 'Commission of Appraisal' of fifteen members, who visited India, Burma, China and Japan. Further reference will be made to it later.

Great benefit has been conferred upon Christianity from scientific study of another kind, closely related to the scientific study of religions; namely, from the scientific study of the Bible, the so-called higher criticism. This study has applied a most helpful correction to the beliefs of irrationally convinced credulity by the use of impartial rationalism. One may now, guided by the illuminating results of such study, appreciate the great value of the Bible as a venerable human work; manifestly human in its naïve record of various scandalous stories about its leaders as well as in the candid record of successive stages in a great advance from savagery to barbarism. Moreover, as long as the Bible was regarded as the infallible 'Word of God', there were persons who, with more sense than sympathy, scoffed at it because so many of its passages are ungodly. But the ground is taken from under the feet of scoffers when the human origin of the Bible is recognized; for then its cruder passages are seen to be only the inevitable imperfections of struggling humanity, with which we, still struggling *today*, must warmly sympathize.

If some passages in the Old Testament are fatiguing in their enumeration of the generations of Adam or of the places where the Israelites 'pitched' during their traverse of the Wilderness, many of its passages are inspiring. How grand is the solemnity of the first four words: 'In the beginning God.'

* 'Re-thinking Missions', by the Commission of Appraisal, W. E. Hocking, *chairman*. New York, 1932.

Even a rationalist must respect the devoutness of the nineteenth Psalm: 'The heavens declare the glory of God; and the firmament showeth his handiwork.' Where can be found a finer gem of reverent thought, even though it lies amid much dross, than Micah's declaration: 'For what does the Lord require of thee but to do justly, and to love mercy, and to walk humbly with thy God.' And if some passages in the New Testament, especially in Revelations, are extravagant, who can fail to marvel at the words of Jesus, uttered in an age of violence: 'Love ye your enemies, and do good . . . and ye shall be children of the Highest; for He is kind to the unthankful and the evil . . . Be ye therefore merciful as your Father is also merciful.' His reputed miracles in the way of casting our devils are as nothing compared to his clear perception of the value of gentle goodness.

The Recent Period of Reconciliation

The transition from the Victorian era to the present time is of particular interest to those of us who have lived through it, not only because it has witnessed an unprecedented advance of science, but even more because it has been characterized by an amazing reconciliation of Christian theology with science. To be sure, there are still many earnest preachers who, like the conservative hold-fasts and die-hards of 80 years or more ago, can not give up the beliefs of their fathers. Instead of being the leaders of their congregations into the newer understanding of the world; they are vainly striving to hold them back in the older misunderstanding. This is because they were treated when young very much as the children of the Flathead Indians are – or used to be – treated. Stiff boards were bound to the Indian children's heads, so as to flatten their skulls as their parents thought they ought to be flattened. Similarly, the children of the ultra-orthodox Christians have had stiff theological creeds and catechisms bound to their minds, so as to shape them as their elders devoutly thought they ought to be shaped. It is therefore most natural that, when such children grow up, their minds remain so shaped; they can not reshape them if they would, especially not if they become ministers pledged to maintain the stiff beliefs to which their minds were bound in their growing years.

It was a mind-bound conservative of this kind who not long ago exclaimed: 'If I could not believe that Joshua made the sun stand still in the heavens, I should lose faith in the Bible and in God.' He had probably been educated in one of those sectarian colleges, to which orthodox parents send their sons to safeguard them from scientific error. It was as a teacher in such a college that a young Harvard graduate was some years ago, as he later told me himself, instructed on his arrival there by the president of the institution: 'You may teach as you like about minerals and rocks, but if the students ask you about the age of the earth, refer them to me.'

Theologians would have found their change of belief easier had they been scientifically educated, but they have not been. They have presumably studied 'apologetics', in which they learned how to argue out a problem; but that method of argumentation is not free; it is directed to the support of a predetermined conclusion and is therefore absolutely unscientific. The essence of scientific method is that it gives equally impartial consideration to all conceivable solutions of a problem, and adopts the one which pragmatically works best, after it has been tested by many men in many places through many years. But it is recognized also that the human mind is not infallible, and therefore all adopted solutions are open to revision whenever new evidence bearing upon them is discovered. It should be understood, however, that scientific method thus characterized is a modern development, the result of centuries of experience; for although the science of logic is of more ancient origin, its unprejudiced application to scientific problems is difficult to learn, even in scientific schools; and it is probably never learned in theological schools.

In spite of all obstructions, liberation from outworn theological creeds seems to have been repeatedly accomplished during the period of reconciliation, even among conservatives, by a gradual and insensible shift of belief. The shift has rarely been caused by direct argument, and it has nothing whatever in common with the hysterical conversion to religion of old-fashioned camp-meetings. The shift is in great part a consequence of the simple process of growing up in the rationalized atmosphere of modern times. The present generation has become habituated to accepting the results of scientific discoveries. It has at the same time become increasingly incredulous about miracles, which are now more and more overlooked, neglected, forgotten. Along with the rationalizing influence of scientific progress has been a marked lessening of the antagonism which used to prevail between religious sects; and a falling-off in the number of severely doctrinal sermons, the kind that used to expound 'hell-fire religion'.

There has also been a marked relaxation regarding matters of belief. One of the most striking instances of this kind was that by which the implied damnation of non-elect infants was omitted from the confession of faith of one of our most important denominations. Many of the laity and clergy had practically ceased to believe it, and finally at a General Assembly of the church in Philadelphia about thirty years ago, the phrase 'Elect infants dying in infancy are saved' came up for discussion. A motion was made to strike out the word 'elect', and the Conservatives objected that there was no specific Scriptural evidence that all infants were saved. In the vote which followed, however, they were decisively beaten and it was then officially declared in the faith of that denomination that 'Infants dying in infancy are saved'. When the motion was declared carried, an elder arose – he must have had a keen sense of the humor of the situation, in which a majority vote of ministers and elders could be taken as deciding how their God would act – he arose and said: 'I move that vote be made retroactive.' The solemn assembly was overcome with laughter. Let it be noted that that antiquated doctrine was not argued away; it was simply sloughed off by the weight of its inherent absurdity and soon forgotten. Should we not be very grateful to the growth of scientific rationalism by which so monstrous a belief was crowded out of acceptance?

The rapid advance of rationalism during these recent years may be seen by comparing the beliefs held at successive ten-year periods in the life of a single individual. It has been written by an experienced evangelical clergyman that 'the recognition of the orderly working of nature weakened confidence in the miraculous, and as years passed theologians . . . interpreted scenes, teachings and epistles of the Bible in a way that would have shocked themselves ten years before'. We may tell further of a man of altogether admirable character who had been taught as a boy the standard Christian myths, such as that of the devilish serpent which talked with Eve in the Garden of Eden; and who, after giving up those childish beliefs in his early manhood, found that still further liberalization was experienced in his mature years. He thus in a single lifetime came down many steps from near the top of the long flight where his excellent father had stood, but he is not yet at the bottom. However, if he lives a few decades longer he may, after interpreting the Bible's teachings in a way that would shock himself today, come farther down, nearer the solid ground of wholly rational belief at the bottom of the flight. Many examples of this rationalizing process may be found among one's friends, if not in one's own experience.

There is one disappointing element in this story. It has often been claimed that the marvelous progress of civilization in Europe has resulted from the adoption of the Christian religion; and it would be indeed gratifying to know that the refined ethics of Christianity have really brought about that progress. But in view of the dependence of European civilization largely on the advance of science, which did not begin until the revival of learning, centuries after the adoption of Christianity, and in further view of the persistent opposition with which the organized forces of Christianity so generally resisted the advance of science even in so beneficent a study as medicine, and in still further view of the dominance even today of many non-Christian principles of behavior among peoples of European stock, the claim that European civilization is a result of Christianity can be allowed only in part, perhaps only in small part. Modern European civilization is the result of mental rather than of moral achievements, and as such it is more closely associated with the mentality of European peoples than with the religion which they adopted nearly 2000 years ago, the ethics of which they have not yet learned to practise. Had they done so, the claim would be more true. It may, indeed, be seriously doubted whether Christian ethics had much to do with the adoption of organized Christianity by early Europeans. They were an ignorant, superstitious, credulous, warlike and cruel people, and it may be well believed that they made professions of Christianity less in order to practise, during their lives on earth, the lofty ethical principles preached by its Founder, than to secure for themselves through a vicarious atonement for their natural sinfulness a blissful future life in heaven promised by propagandizing theologians, and thus to escape a threatened eternity of torture in a horrible hell.

Human Origin of All Religions

It was said at the beginning of this address that all religions, like all sciences, should be regarded as the natural products of the human mind. It was added that the grounds for that belief would be stated in this part of the address. The grounds are these. First, as to Christianity: Many theological elements of our religion, based on Scriptural texts, have been shown by scientific research to be erroneous; they can therefore be no longer taken as infallible revelations; they are only humanly invented beliefs, accepted by the ancient Asiatic people who recorded them. But various other elements of our religion are still held by conservatives to be of supernatural origin. They are so held on the evidence of other Scriptural texts; and these other texts, being of essentially the same nature as those formerly held to demonstrate the supernatural origin of the abandoned beliefs, can no longer be taken as competent witnesses to the truth of the remaining supernatural beliefs. Hence these remaining beliefs should also be interpreted as humanly invented. The upshot of this is that, while the Bible unquestionably gives us an invaluable record of the beliefs of an ancient people, the competence of its texts as witnesses for the supernatural is invalidated. The same argument applies to all other religions, but with less directness because they have not come into conflict with science so much as Christianity has.

A corollary of this conclusion is that the precepts of our religion should be accepted not on authority but on merit. They should be judged, just as we judge the precepts of other religions, by their appeal to our moral sense. In support of this conclusion let us recall the various traditional beliefs which, passed down to us from an ignorant and credulous people, were for centuries held to be essential elements of the theological part of our religion: recall the proposed replacement of those beliefs by the findings of science, won from nature by patient, truth-loving research; recall the vehement protests conscientiously made against such replacements because they contradicted the 'Word of God'; recall finally the gradual fading away of the protests and the acceptance of the protested replacements. What a striking resemblance there is in all these stories; the same sequence over and over again. Should we not learn from so uniform a repetition of the same experience that the Bible is, as just said, not a competent witness to the occurrence of supernatural events in human history, and that it should be studied as other human records are studied?

But let another matter be recalled. All the discarded elements of Christianity thus far noted are of a theological, not of an ethical nature. The ethical elements of Christianity have not been disturbed; they have not been discarded, even by those who believe in the humanity of their great preacher. Nor has the value of the ethical elements been lessened in the least by the abandonment of the theological elements. The great principle, 'Whatsoever ye would that men should do unto you, even so do ye also unto them', is still as valid as ever. Indeed, its validity has been strengthened by the comparative study of the races and the religions of mankind. In a word, the whole drift of scientific opinion through the centuries since the revival of learning has been away from theological superstition toward rationalism; and during the same centuries, Christian ethics have been gaining ground. This is true, notwithstanding the fact that the principles of Christian ethics are, like those of the ruder ethics which they have displaced, increasingly regarded as of purely human origin, quite as much so as tables of logarithms or of chemical elements.

It is, of course, to be expected that many persons will dissent from the untraditional views just expressed and will continue to believe that, however human other religions may be, our own Christian religion is truly based on supernatural revelations; and they will thus resemble the African chieftain who believed that the finger of God gave an otherwise incredible strength to the Zambezi bridge. To them should be told the story of George Fox, founder of Quakerism, and his convert, William Penn. George, meeting William not long after his conversion, said: 'William, I see thou art still wearing thy sword.' 'Yes,' said William, 'it seems best to wear it.' 'Wear it as long as thou canst,' replied George; and the next time they met, William wore no sword. The moral of this story is that those who still devoutly believe in the occasional interruption of natural processes by supernatural processes should continue to hold that belief as long as it is helpful to them. Only when it ceases to be helpful should it be set aside. But that such beliefs have repeatedly ceased to be helpful is shown by the vast number of modern William Penns who have rationally given up wearing their theological swords in the last seventy or fifty or thirty years.

The change of mental habit thus indicated has been astonishingly rapid; more rapid, indeed, than corresponding changes have been made in the creeds of rationalized church members. Indeed, many churchmen have been led by orthodox preaching to think that theological beliefs constitute so essential a part of Christianity that, when they found those beliefs no longer tenable, they thought they no longer had any religion. This is greatly to be regretted, for the majority of those men really still hold the essentials of Christian ethics in their hearts as their ideal.

Before turning to my next topic, let me point out one feature of the growing reconciliation of Christian theology with science that is of especial significance. The reconciliation is not a compromise, in which each side has yielded something. It has been brought about wholly by the modification of theological views so as to bring them into accord with scientific views. Not a single one of the various scientific discoveries which, at the time of their announcement, proved to be so disconcerting to the orthodox, has been reversed. On the contrary, the estimates of the earth's age and of the antiquity of man are now greater than when they were first calculated; and the evidence in favor of organic evolution is immensely stronger than it was in Darwin's time. True, there was a flurry some twenty years ago, when an eminent English biologist, Bateson, declared in an address at Toronto that 'Darwinism is dead', or words to that effect. The half-informed public misunderstood him to mean that evolution is dead. He meant nothing of the sort, for he was a pronounced evolutionist! What he meant was that the process of natural selection, which Darwin had thought was the mainspring of evolution, was in his opinion inadequate, and that evolution had been brought about in some other way.

PART II

Clearing the Ground

Before taking up the 'Faith of Reverent Science', to which the foregoing pages may serve as a preface, let a word of warning be introduced. Thus far much has been said of various outworn theological beliefs which have been displaced by rational beliefs. Perhaps that has given the impression that the object of my address is destructive. Not so! The object of the address is constructive, in that it places great value on the sounder beliefs which have replaced the outworn beliefs. The case is like that of our pioneer forefathers, who had to clear away the forests before they could plant their productive crops. Similarly, science has had to clear away old errors before it could implant new truths; but that is a wholesome, constructive process. It is, however, because of the abandonment of the outworn beliefs by scientific rationalists that the orthodox have so often given them the derogatory name of unbelievers, although the rationalists felt themselves to be just as ardent believers of the faith that they maintained as the orthodox were of theirs.

What, then, are the beliefs of scientific rationalists? No one can say definitely what they are, because rationalists have not organized themselves into a religious body and therefore have not as a body formulated a creed. No one is authorized to speak for them; they are individualists; like St Paul, each one holds his own gospel (II Tim. 2: 8). This is, in one respect, unfortunate, because, standing alone, they lose the great advantage of concerted action; but, on the other hand, they conserve a freedom of thought, which organization is likely to limit. They might, however, unite in neighborhood churches, where each member would be 'free to follow truth as he sees it to its uttermost bounds', yet at the same time 'enjoy religious fellowship with others and work together for the common good'.

But besides being unorganized there are, among scientific rationalists, as among other groups of our population, persons of unlike temperament: some of them take religious matters lightly, inattentively; others of them take such matters seriously and reverently. It is with the latter sub-group, whose faith may therefore be called the faith of reverent science, that we are here concerned. It is highly significant that, as far as I can judge from a considerable acquaintance among them, they hold many of the essential principles of Christian ethics. Those principles are, as already told, fortified by scientific research, not only because they harmonize with the great modern truths which have supplanted Biblical myths, but also because so many of them are represented in other great religions, also of human origin like ours.

The Faith of Reverent Science

The chief articles of the faith of reverent science, thus understood, may be summarized as follows:

(1) Reverent science devoutly refrains from assuming to know the nature, the thoughts, the acts of a Supreme Being by imputing even the best of human acts and thoughts and nature to him. It stands humbly silent before the ever-expanding mystery of the universe. In the sincere agnosticism of profound ignorance regarding the supernatural, no satisfaction is found in the limitations of a Supreme Being which invariably accompany the attempt to define him. Let no one make the blunder of confusing this agnostic attitude with atheism: inability to form an estimate of a quantity is no ground for saying its measure is zero. Recognition of our ignorance regarding supernatural matters goes with the growth of our knowledge regarding natural matters. It was appropriate enough in Noah's time, when little was known of the world, that a legend should describe a creator who repented of his creation. Somewhat later, Moses still knew so little of the world around him that he did not hesitate to define his chief deity in very human terms, even to the point of conceiving him as being shamed out of his wrathful threats by human reproaches. From then till now a long series of concepts has been promulgated, generally becoming more refined as time went on, although one of the most popular included a malevolent Devil working in conflict with a beneficent God. The humanly inconceivable quality of immanence came to be attributed to Deity by metaphysical theologians, but that quality must be terribly strained, even to metaphysicians, now that the universe is found to extend, along only a part of one of its dimensions, for 300,000,000 light-years. It is on reviewing these various unsatisfactory hypotheses that the reverent scientist retreats to humble silence.

But by no means are all scientists silent agnostics in this matter. One of our leading physicists is reported to have said: 'It is through science that man has discovered that his own soul is God's greatest purpose in the universe': but it is not explained how the discovery was made. Another defines the 'God of science' as the 'spirit of rational order'; but without going on to specify the place that is occupied in rational order by the terrible griefs caused to humanity by irresistible hurricanes, devastating floods, incurable plagues and above all by barbarously irrational wars.

Agnostics are baffled by the cruelties and miseries of the world; always a marvelous and often a beautiful world, but also for ages and ages a merciless world, on which, even while the gentle rain fell from heaven, carnivora remorselessly devoured herbivora; and even now sometimes a terrifying world, as when an unforeseen earthquake wave sweeps away a whole village of simple fisherfolk; and too often a cruel world, in which, until a few years ago when medical science came to their relief, countless innocent and devoted young mothers have had, through no fault of their own, to endure the agony of seeing their little ones, whom they had no power to cure, strangled to death before their eyes by diphtheria. The mystery of such a world is too profound for our solution. The easy invention of a devil gives no acceptable aid in understanding it. Hence in no spirit of irreverence but only in the sincere humility of acknowledged ignorance does the agnostic refrain from making assertions regarding a Creator.

(2) Reverent science has a secure faith in the persistence of the order of nature through time and space, because such persistence has repeatedly been shown to be in the highest degree probable; but it is not absolutely proved; science has no means of reaching absolute proof in matters of such magnitude. Yet in view of this faith, certain reported events, known as miracles, which interrupt the order of nature, are discredited because there is no sufficient evidence forthcoming that they have ever taken place. Such disbelief is confirmed by the historical fact that alleged miracles are usually reported from ignorant and uncritical communities. With the vanishing of belief in miracles, legions of elves, gnomes, imps, witches, demons, devils, fairies, spirits and angels have also vanished from among us.

It is not alone among rationalists that miracles are disbelieved today. Many of our orthodox churches now discredit certain Biblical miracles which, a generation ago, were literally believed. Consider, for example, the miracle performed by Christ in calling forth devils from a demoniac, and at their entreaty giving them leave to enter a herd of swine. Is there today any one who believes that the demoniac was possessed by actual devils; or that when they came forth from him they orally besought Christ to let them go into the swine; or that, on being given leave, they actually entered the swine? Does not every one now take the central episode of the story to be an example of more or less hypnotic healing, and then interpret its miraculous elements as legendary attachments demanded by the ancient belief that certain diseases were due to possession by devils; the swine being introduced as the best means of disposing of the devils when they came forth from the demoniac.* And when thus interpreted where is the miracle?

The orderly wonders of science would have been taken for miracles by peoples of ancient and medieval times, as they still are by modern savages. Some of them may well seem miraculous, even to intelligent moderns. Consider, for example, the computation of a comet's orbit, as worked out by Gauss on the basis of Newton's laws. A new comet is ordinarily, when first seen, only a small and faint nebulous wisp, with neither head nor tail. Its distance from us and its direction and velocity of motion are unknown; yet if three observations of its apparent position among the stars are made at intervals of about a week, as it is seen from our revolving, rotating earth, it is then possible to calculate not only its distance from us and the direction and velocity of its motion at the times of those observations, but also where it will be seen among the stars and how it will be moving at any desired date for a considerable period of time in the future. This suffices to show how enormous is the range of human mentality, a conclusion that should not be forgotten when we are tempted to regard the announcement of new ethical principles as necessarily indicative of superhuman powers.

(3) Reverent science believes that various communities or tribes or peoples have, through their purely human efforts, gradually formulated not only their theological beliefs but also such rules of behavior, or codes of morals, or principles of ethics as seemed fitted for their needs in the successive stages of savagery, barbarism, civilization and enlightenment. It is therefore concluded that, like all other codes, the Christian code has been humanly formulated instead of supernaturally revealed. A strong support for this conclusion is that many of the so-called revelations of the Bible do not seem beyond man's own discovery; also, that the essential equivalents of many Biblical revelations are found in other religions.

Moreover, the human origin of the Biblical systems of ethics, as well as of other systems, is strongly indicated by their gradual improvement with the passage of time and with the progress of the peoples who formulated them. Let us not forget in this connection that, if we go back far enough, lying, stealing and slaying have not always been 'wrong'. Such acts were instinctively 'right' among our semi-brutal ancestors, and they did not become wrong until, after thousands and thousands of years of experience, they were first condemned by moral leaders, and later condemned by the local public. Similarly, some acts which we consider right may be in time condemned by our improved descendants. It is truly difficult to believe that 'right', as we know it, is not a matter of objective reality but only of subjective human opinion, subject to change as humanity advances. Many persons may refuse to believe that acts which we hold to be wrong were ever right; yet such acts

---

* The long-continued opposition of theologians to medical science is broadly treated in White's 'Warfare . . .', II, Ch. xii.

had, in their time, just the same claim to be right as our gentler and more merciful acts have today; namely, the general approval of their communities as guided by their leaders: for example, Moses, the great law-giver, boasted of how he had taken all the cities of Sihon, 'and utterly destroyed the men, and the women, and the little ones, of every city'; and of how he had likewise taken all the cities of Og, king of Bashan, 'utterly destroying the men, women, and children of every city' (Deut. 2: 34; 3: 6). Centuries later, the gentle little Samuel thought, after he had outgrown his childish gentleness, that it was his Lord's wish – and therefore surely 'right' – that Saul should smite the Amalekites and 'slay both man and woman, infant and suckling'. And therefore Saul 'utterly destroyed all the people with the edge of the sword', sparing only Agag, the king; but Samuel made up for this sinful omission, for he himself 'hewed Agag to pieces before the Lord' (Sam. I, 14).

It is truly a difficult proposition to believe that what we think is 'wrong' could ever have been 'right'. Many persons may shrink from accepting it; but what other conclusion can we reach if we face the facts; that is, if we study open-mindedly the evolution of savage man from brutes and of civilized man from savage man. Thus explained, an evil act of today, whether it be a sin against a religious code or a crime against a civil code, is generally nothing more than an act that was permitted and condoned in an earlier era, but that has come to be condemned and prohibited in a later era. The torture of prisoners of war was once a matter of course; now it is as a rule no longer sanctioned. Human nature has changed for the better.

(4) Reverent science preserves an earnest faith in the value of sacrificing one's own selfish preferences for the common good and of prayerfully consecrating one's best efforts to the betterment of humanity; but it has very generally given up the belief in the possibility of turning the course of events, which are beyond one's own control, into a desired direction either by the sacrifice of animals or by the prayer of humans. A waning belief in the efficacy of prayer is one of the most marked effects of the waxing influence of rationalism. It goes with a fuller realization of the persistence of the order of nature. More than any other phase of rationalism it varies with personal temperament. Many persons no longer pray for rain during a drought, because they have come to understand that rainfall results from natural movements of the atmosphere; but the same persons may pray for the recovery of a friend from illness, perhaps because of a feeling that human affairs are less physical, more spiritual than atmospheric movements. Yet science has shown that human illnesses involve conditions and processes that are quite as natural as those which control rainfall, even though organic instead of inorganic. But what shall be said of a hospital which announces on advertising cards in London buses that it is supported wholly by prayer? If it be so supported, why should such an announcement be posted?

(5) Reverent science accepts, without asking that it shall be revealed to us, whatever fate is in store after death, be it immortality or annihilation, in the complete trust that it is a fate fitting the part we have to play in the unfathomable mystery of existence. It leaves aside all transcendental questions about the imagined regions of heaven and hell, and it rejects absolutely the monstrous doctrine that most of mankind have long been and still are condemned to horrible torment after death; also the specious doctrine that salvation from such a fate has been and is still granted to but a small minority of mankind, and to them only through vicarious atonement for sin they never committed, the original sin of a fictitious Adam. The jealous Jehovah visited, according to Moses, 'the sins of the fathers upon the children unto the third and fourth generations', but Christ's merciful Father in Heaven has been made by some would-be Christians unforgiving to the children of countless generations of those that never had opportunity of hearing of him. The persistence of this doctrine of damnation into our times must do serious injury to Christianity in the minds of intelligent 'heathens'.

Let me draw special attention to the first two lines of this article: 'Reverent science accepts, without asking that it shall be revealed unto us, whatever fate is in store after death.' This acceptance of ignorance of the unknown is characteristic of an honest scientific attitude with regard to an unsolved problem. A suspended judgment is maintained, which refuses to settle down on an unwarranted solution. That attitude is distasteful to the scientifically untrained; they do not wish to suspend judgment; they wish to adopt some preferred solution instead of remaining agnostic in the absence of a demonstrable solution. It sometimes seems as if not only persons of untrained minds, but also certain highly trained philosophers and metaphysicians were likewise unwilling to suspend their judgment; for, after discussing problems that are far beyond solution by scientific methods, they appear to settle down on one or another solution of such problems, their choice being guided more by subjective preference than by objective proof.

(6) Reverent science is much concerned with making our life on earth as good, as unselfish and as helpful to others as possible, not in order to receive posthumous reward for doing so, nor in fear of posthumous punishment for not doing so, but in the convinced belief, based on long human experience, that in a life so conducted – a simple, kindly, helpful life – man finds his highest and deepest satisfactions and his fewest regrets; a convinced belief that doing good in a sincere and unselfish spirit is man's best means of maintaining the progress of the past and of contributing to the progress of the future; a convinced belief that in so conducting his life he is playing the best part accessible to him; and that while so conducting it he will find his truest happiness in his home, with his neighbors, among his countrymen and over the whole world.

Summary

The foregoing articles of faith suffice to show that no one who holds them should be called an unbeliever; for they represent the essence of Christianity, after many mythical, miraculous and theological elements have been withdrawn from it. Alas, that it is so much easier to state these articles than to practise them! Alas and alas, that even while earnestly believing them one may fail and fail again to live up to them. Without making that confession I should not be willing to write these pages.

One of the large merits of such a faith as that above outlined is that it may easily keep pace with the growing knowledge of the world. No one who holds it would be content to use, as his guide, a creed formulated centuries ago by majority votes of credulous and disputatious theologians and based on texts several centuries older, or a body of documents of doubtful authenticity older still by many more centuries. Yet certain well-organized religious denominations still profess to be guided by such creeds and texts and documents. In such profession we find an unfortunate consequence of the very rapid reconciliation of theology and science, because the change of religious thought involved in the reconciliation has taken place so fast that the change of formal expressions of religious belief has not kept pace with it. The laity of such denominations, therefore, find themselves in the dilemma of either sitting under hold-fast conservatives who are as out of date as the creed they have solemnly promised to preach, or under liberalized progressives who have as solemnly promised to preach a creed which they no longer believe; and the worst of it is that the laity are so indifferent to their dilemma that they take no efficient steps to get out of it.

It may, perhaps, be thought that what has been said thus far about the reconciliation of Christian theology and science goes too far. Remember, therefore, that the reconciliation has not been described as by any means complete, but only as rapidly growing and as being in some quarters already far advanced. In support of this conclusion, let me here read some extracts from the above-cited report, 'Re-thinking Missions', by a many-denominational Committee of Appraisal. They unanimously agree on the following statement: 'Only a religion whose first principles are capable of the simplest formulation can become a religion for the modern man ... The religion which assumes too much knowledge of the supernatural realm, its system of heavens and hells, or its inner mechanisms of eternal justice, can no longer be a living issue.'

Or again: 'Western Christianity has in the main shifted its stress from the negative to the affirmative side of its message; it is less a religion of fear and more a religion of beneficence. It has passed through and beyond the stage of bitter conflict with the scientific consciousness of the race over details of the mode of creation, the age of the earth, the descent of man, miracle and law, to the stage of maturity in which a free religion and a free science become inseparable and complementary elements in a complete world view. Whatever its present conception of the future life, there is little disposition to believe that sincere and aspiring seekers after God in other religions are to be damned; it has become less concerned in any land to save men from eternal punishment than from the danger of losing the supreme good.'

The following passage concerning the larger Christian denominations is also instructive: 'They were formed at a time when a precise and definite theological system of doctrine was generally stressed as vitally important, and this theological emphasis has remained up to the present time a dominant feature of these conservative churches. This excessive occupation with theological doctrine has kept such churches out of touch with trends of thought and intellectual problems in the world around them. Churches of this sort appeal only to a certain type of mind. Students in the main leave them coldly alone and are apt to be turned against Christianity if this is the only kind of Christianity which they know. It seems to them too often a complicated religion of words and phrases, dealing with the issues of a former age, not a living force for the moral transformation of the world and for the remaking of the present social order.'

Strong support is thus given to the above-stated faith of reverent science, but its various omissions may make it unsatisfying to many devout Christians who, although now standing much lower down on the long flight of theological steps than their grandfathers did, have not yet descended to the foot of the flight. They surely have the same right to adopt and maintain their beliefs as we have to form and maintain ours. As long as their beliefs help them to lead good and happy lives, let them be held fast; for in addition to the above-recited articles of the faith of reverent science, there is another: 'Give up no belief until it may be replaced by a more helpful one.' Thus may be avoided the unhappy fate of those who, over-rigidly brought up, think that the abandonment of their early faith leaves them with none.

Reflections

Let us now turn to some reflections on the statement made at the beginning of the address, that all religions are, like all sciences, of human origin; in other words, that religion and science are both examples of the natural evolution of human thought. Note particularly that this does not involve any change in the articles of belief of any religion. They stand unchanged. True, the attribution of religious beliefs to human sources may cause a change in the attitude of those believers who are accustomed to taking authority for truth; they may feel differently toward the religion they profess if they come to regard it of human origin. But others who are trained to take truth for authority will not change their attitude; the evidence which had previously convinced them of their religion's verity will still convince them. Dissent from the view that all religions are of human origin will of course be expressed by the conservatives of today, just as their grandfathers two generations ago expressed dissent from the view that all plants and animals are examples of the natural evolution of organic forms. The present-day conservatives will urge that at least their own religion is based on supernatural revelation; and the same claim would be made by the conservatives of other races, if the question ever arose among them. The present conservative belief in the supernatural revelation of various religions is therefore a parallel to the orthodox belief of our grandfathers in the supernatural creation of organic species. But this latter belief has been given up by many intelligent persons in the era of reconciliation. Hence the belief in supernatural revelations as the basis of religions may also be increasingly given up as time passes. It certainly seems to be losing ground at present. What will our grandchildren think about it?

We may here profitably quote again what was said fifty years ago by Bishop Temple, of London, regarding Darwinism. He came to think that it was more fitting for a Supreme Creator 'to impress His will once for all on His creation, and to provide for all the countless varieties [of organic forms] by this one impress, than by special acts of creation to be perpetually modifying what he had previously made'. This view of organic evolution may be rephrased so as to apply to it religious evolution, as follows: 'It is more consistent with the modern understanding of the universe to suppose that moral progress is inherent in human nature, and that all such progress is therefore the result of a continuous natural evolution, than to ascribe it to supernatural revelations, those of later date modifying those of earlier date.' Thus rationally interpreted, a new faith would arise when a prophet makes an acceptable modification of an earlier faith. The new faith would first appear in some limited area of the earlier faith; once established there, it would spread as far as it could into larger and larger areas; but as it spreads, new modifications, which we know as sects, would branch off from it; and they in turn would spread as far as they could. The faith of reverent science is one such.

The spread of new species of plants and animals from the areas in which they arise is controlled largely by geographical and climatic factors; the spread of new religious faiths – that is, of new species of belief – appears to be largely determined by racial mentalities, for the greater religions of today are rather closely related to the races of man. In view of this parallel between species of organisms and species of religious faith, the natural instead of the supernatural origin of religions may come to be more and more accepted by our children and theirs, just as we have come more and more to accept the natural instead of the supernatural origin of organic species.

How instructive it would be if we could obtain some definite measure of the rate of change of these opinions. A measure might be secured if a good number of churches of various denominations would cooperate, first, in preparing questionnaires on a variety of articles of Christian belief, and second, in asking their members to indicate their opinions of those articles. Replies would probably be of four kinds: Some church members might not have definite beliefs; let them remain in their uncertainty. Others might have definite beliefs but be disinclined to express them; let them remain silent. But the rest, and probably a good majority, would know and be willing to express their views. If such a census were taken every five or ten years for a century, we should be able to measure fairly well the drift of religious opinion in the census area. We could then learn something as to the rate at which descent is made from the long flight of theological steps, far up on which our grandfathers stood, and something also of the number of those who have reached the solid ground at the bottom of the flight, where they may hold the faith of reverent science. The value which such censuses would have in the future may be estimated by the value we should attach to them today if they had been taken during the last hundred years.

Another reflection concerns the value that will be placed by future historians on the enormous change of religious beliefs which has been brought about by scientific study during and since the Victorian era. They will look back on this period as one in which many have liberated themselves from the centuries-long enslavement to the crude myths of an ancient and ignorant Asiatic people. The historians will see, as we also may see, in the long period during which the enslavement lasted, a measure of the astounding credulity of the human mind and of its incapacity to think out its problems by a reasonable, scientific method. They will see, as we indeed may see already, that confidence of belief and certainty of conviction do not suffice to prove the objective truth of the opinions so earnestly held. The historians will moreover recall the self-satisfied assurance with which the fathers of the church, while grossly ignorant of the natural world around them, believed they could solve the profoundest mysteries of the supernatural world, and they will contrast that assurance, as we may too, with the increasing mistrust which we moderns feel concerning the supernatural solutions reached by the fathers, in spite of our amazing increase of natural knowledge.

Those future historians will say our liberation from that centuries-long enslavement was tantamount to a declaration of independence, of as great moment in the spiritual world as the declaration of independence, which we Americans made in 1776 was in the political world. But the two declarations are unlike, in that the earlier one was definite in form, place and date, while the later one is indefinite in those dimensions. We can not, therefore, celebrate it by parades and orations and fireworks on a certain date every year, but it is nevertheless worthy of celebration, and I wish to outline a method of celebrating that spiritual declaration in a manner befitting its importance. This is the prognostication mentioned in my introduction.

Prognostication

Our declaration of spiritual independence can be best celebrated by a long festival of cooperation between the organized forces of religion and science – the priesthood and the professorhood – directed to the object which the priesthood has always, but the professorhood has not always held in view; namely, the betterment of humanity; and the festival must be continued into the future at least as long as the period of antagonism which so unhappily divided the two forces in the past. They may be ushered in during an era of brotherly love, vastly more truly Christian than the unhappy medieval era from which we have escaped, and therefore an era of enormous importance in human history; for the festival of cooperation should enlist all those members of the priesthood who, recognizing the victory of modern science over ancient theology, desire to replace a good share of their study of theological apologetics by a scientific study of the nature of modern man and of the methods by which he can be ethically moved; and it should enlist also all those members of the professorhood, who, still preserving their scientific methods but impelled by the faith of reverent science, wish to turn their studies toward humanity rather than to the further investigation of the non-human universe.

Let the attack on all those vast non-human problems be of course continued by such members of the professorhood as are not attracted to human problems; let the attack go on until all the island universes 300,000,000 light-years or more distant shall have been discovered and catalogued; let it go on until the almost infinitely minute electrons have been resolved, if they are composite, into their infinitely more minute constituents, as atoms have been. Similarly, let all the more conservative members of the priesthood, who do not accept the results of modern science, continue their efforts toward human betterment by orthodox theological means. But it will be from this festival of cooperation, as it is entered into by more and more priestly and professorial members, that we shall expect the greatest progress in bettering humanity. Largely by such cooperation will appropriate scientific methods be applied, more than ever before, to the heavy task on which a theologically dominated religion has labored so long; for on that heavy task, if we may judge by the unscrupulous greed of our growing criminal class, and by the cheap complacency with which a still larger class seems to look upon the successes of the criminals, the long lasting labor has not been expended with great success. How can science *dare* to stand aside any longer, instead of taking a more active part *with* religion in directing its best efforts to the accomplishment of that greatest of all tasks? A reproach has been directed against Nero because he fiddled while Rome was burning. What will our descendants say of us of the professorhood who, assured of our salaries or our pensions, continue to work upon our recondite, non-human problems, while our neighbors suffer and our nation is criminally demoralized.

I am not unmindful of the efforts already made for betterment, nor would I overlook whatever measure of success those efforts have reached. In particular would I recall the historic fact that, even during the era of antagonism between theology and science, efforts for human betterment were made more largely by the priesthood than by the professorhood. Those efforts were, to be sure, guided by what we now consider mistaken beliefs, and they were too often directed by a wish to secure a verbal profession of faith rather than an actual performance of good works; but they had the high merit of being closely associated, all through the Christian era, with correction of error, relief from distress and consolation in affliction, as they are still. But as far as efforts of the priesthood toward ethical betterment are concerned, they have been, as a rule, too narrowly limited to precept and exhortation, without a sufficient use of what would, in scientific teaching, be called laboratory exercises. It is for that reason that we may have confidence in the attainment of a greater measure of ethical success in the future, when our efforts toward that end are guided less by theological than by scientific principles; and when those better guided efforts are applied not only through exhortation to the young people who go to church on Sunday, but also in a more practical manner to the larger numbers of them at school and college all through the week. A great change must therefore be made in our educational methods during the festival of cooperation. There was a time not long ago when the heads of colleges were, almost as a matter of course, chosen from the priesthood; and when the morals of the students were cared for by the required attendance at morning prayers every day and at church services once or twice every Sunday; in other words, morals were then taught by precept. That was a time when physics and chemistry, botany and zoology were also taught by precept. We have now come to a time when the heads of colleges are seldom drawn from the priesthood but more generally from the professorhood; and when the sciences are increasingly taught by laboratory exercises, in which actual performance is required of every student. Yet in spite of these scientific advances, *less* rather than *more* attention is now given in colleges to the inculcation of morals, perhaps because of an increasing consciousness that the old method of inculcating them by precept was so ineffective. To be sure, elective courses are offered in ethics; but they are not very largely taken and they treat chiefly the ethics of peoples and races; or if individuals are mentioned they are usually famous persons, not mere boys who call themselves *men*.

This inattention to the local individual must be corrected. Just as physical health and bodily strength are increasingly secured for each student by the performance of appropriate bodily exercises in gymnasiums or on playgrounds, so ethical health and the building of finer characters must in the future be secured by the performance of appropriate exercises in what may come to be called an ethical laboratory; and such exercises will gradually penetrate downward from the colleges where they are developed into the schools. There is today a not unnatural unwillingness to have any one religious faith taught in the schools; there will be no such unwillingness shown regarding the practical inculcation of ethical principles, for ethical principles are substantially alike in all the religions professed among us.

But it is not only by downward penetration from colleges into schools that better methods of ethical education will be extended. They may also penetrate upward from schools into colleges. The efforts of an important movement, already developed among school-teachers and known as 'progressive education', are directed toward discovering the best means not only of teaching children but of developing their finer qualities. What career, therefore, can be nobler than one in which men and women of real ability are thus not instructing children only in the rudiments of knowledge but are developing their loyalty to high ideals; for in that career those consecrated to it are not merely *earning* a *living* out of teaching, but are *spending* their *lives* on it. The discouragements in work of this kind will be many; assured successes will be long delayed and few. No immediate and definitive results, such as those which commonly attend investigations in physics and chemistry, are to be expected. Experiments in human betterment must last a lifetime. Hence only those of the professorhood should undertake betterment problems who desire, as the best of the priesthood have for centuries desired, to consecrate their lives to a task of uncertain rewards, yet a task on which the future of humanity must largely depend.

This proposed festival of cooperation may perhaps be decried as absurd, and its practical exercises in ethics may be deemed fantastic by doubting Thomases; but the cooperation is eminently feasible, and the practical exercises in ethics can and must be developed. Nowhere are they more needed than by the rising generation in our own country, where the lawless greed which so abounds in the grown-up generation excites astonishment, to say the least, among our friends in Europe, while our own inefficiency in correcting lawlessness, not to say our indifference to it, mortifies us so profoundly at home; for we are living in an era when professional politicians are increasingly governing us to their selfish satisfaction and to our disgrace, and when organized criminals, ingeniously sheltered from the law by expert legal advice, are to our shame conducting a more profitable business than any one else. We must surely during the festival of cooperation make a vastly more active and effective educational effort toward ethical betterment than we are now doing.

Our universities must come to recognize the need of such effort. They have grown enormously in the last century. Opportunity for study within them has been immensely broadened by enlarging the equipment of faculties, laboratories and libraries. Direction and supervision of study is so much improved that scholarship has risen gratifyingly in spite of the many distractions that tend to lower it. Let the next step in advance be that of character-building. Let courses in practical ethics be introduced, tentatively at first, with more confidence later on. Perhaps some such courses are already established, but they are not yet usual, as they must come to be.

Imagine a general introductory course of 100 or more students. At its close let the professor in charge invite a promising group of its members to go on with him for another year in an avowedly experimental course, in which all shall agree to study each other, all to estimate the strength of various ethical qualities possessed by each, and then jointly to devise practical exercises which shall strengthen the weaker qualities. Something valuable would surely be learned after ten or twenty years of such experimentation in ten or twenty colleges; and from that beginning further steps might be taken. For remember, our festival of cooperation is to go on for centuries. When the planning of these exercises is under discussion, all persons should be excluded who say at the beginning that such exercises can not be successful. Plans should be made only by those who insist that at least some exercises *can* be and *shall* be successful. Above all, let no one say, 'Human nature can not be changed.' It has already been enormously changed and it is going to be changed more still. Every ardent disciple of Christ must believe that. Could any university president have a higher ambition than to see, during his administration, the successful development of wholesome courses on character-building, as a fitting supplement to the broadening of opportunity and the raising of scholarship by his predecessors?

Mistakes will be made, of course, but mistakes can and will be corrected. The most manifest will be the development of self-conscious priggishness; that must be avoided by the cultivation of a truly Christian humility. It is already known that profitable elementary exercises in ethics may be introduced unconsciously in games and sports; witness the expression, 'It isn't cricket', by which meanness and under-handedness are condemned in England, the home-land of fair play. We may build much farther on that beginning. Let the ability, the inventiveness, the perseverance which have characterized progress in non-human sciences be applied practically in ethical science and it will then advance as it never has yet; but it must not be expected that the advance will be rapid. No one may know today the most effective methods of forming habits of self-control and self-sacrifice, up to a truly Christian standard; but better methods will surely be discovered by scientific study and experimentation. Such study may even teach us how to do unto others as we would that others should do unto us; and when we learn how, let us call that great principle 'the rule of unselfish happiness', not the 'golden rule'. Golden is not a good enough name for it.

The development of school and college courses in practical ethics should not, however, be by any means the only object of the festival of cooperation. The festival should be so conducted that it will attract large endowments for the investigation of the best methods of advancing human betterment, in order to aid in the development of school and college courses. We have such endowments already for various objects of a somewhat like nature. One of them is endeavoring to stamp out certain diseases at their source. Several endowments are studying the means of promoting public health. We have various excellent organizations for practical betterment, such as eugenic societies, institutes of family relations, and so on. Some investigators are striving to discover the cause and cure of cancer. But most of these are working collectively on the human body; few are working on individual human natures. May we not therefore hope that future endowments shall be more directly applied to the discovery and cure of the mental disorder known as selfishness, which lies at the very root of so many of our disorders. Surely that mean and degrading quality is more damaging to humanity than cancer is! The wonder should be that a concerted, scientific attack on it has not been already made. Is the delay perhaps due to an aloofness of the professorhood because of a semi-conscious but mistaken feeling that the correction of moral faults is the business of the priesthood? If so, let us hope that the mistake will be soon corrected; not that the priesthood shall cease their efforts but that the professorhood shall helpfully cooperate with them. How wholesome it will be to bury the acrimonious disputes which separated those two great forces in the past, under the cordial relations of the future. Can there possibly be a nobler crusade than one formed by the union of those forces marching unitedly against the defects of humanity? If our defects are due to our glands, then let the crusade march against the glands!

But let it be understood that the crusade is not to be begun by proclamations calling attention either to it or to its object. It will begin inconspicuously; it has undoubtedly begun already in various quarters. But it is at present everybody's and therefore nobody's business. What it now needs is organization and direction; and it is in supplying that need that a long-lasting endowment for coordination will be immensely useful; an endowment directed by men of constructive and persuasive wisdom, who would give their whole time to it and who would, for years and years to come, encourage and focus on a central object the many lines of religious and scientific effort which are now pursued too independently for the greatest efficiency. Can this association, which is pledged to the advancement of science, possibly do better work than promote the establishment of such an endowment? Is there anything in the world more important and more difficult to work upon than the improvement of human nature? Can science anywhere find a more worthy subject for investigation than the methods of accomplishing such improvement? As an encouragement to persevere against difficulties, let us remember again that we live in a world which was merciless for ages and ages, but a world in which the quality of mercy was developed, not many centuries ago, by unconscious evolution; a world in which a sense of justice has similarly arisen, telling us that might is no longer right, although it truly used to be right. If so much has been gained by unconscious evolution in the last few thousand years, how much more may we hope to gain from conscious evolution during the plenitude of time to come!

Reference to science has often been made in this address. Let us be careful to understand that there is no mystery or magic about it, in spite of its enormous power. Science is merely *well-attested knowledge*. Let no one make the mistake of imagining that it is hard or unfeeling. In tasks where vigorous action is needed, it works with active vigor; in those which call for gentle sympathy, it is gentle and sympathetic. Its faithful handmaid, logic, will be deemed cold only by those who like to repose on mossy banks of outgrown beliefs in shady groves of tradition, for naturally enough such slumberers do not wish their repose disturbed by the clear light of truth. The fields of science include everything that can be observed; for example, human conduct: human conduct being merely one of the many problems which finds a better explanation under the scientific philosophy of natural evolution than under the credulous theology of supernatural creation. The results of science are not final or absolute; they are always open to correction and extension. The essence of its spirit is a search for truth, by which, as an old philosopher long ago said, *no man has ever yet been harmed*. That search has been wonderfully successful in the physical world. Great renown has come to members of the professorhood through their discoveries, even if what they have discovered is as utterly remote, as utterly beyond application to the betterment of humanity as the dimensions of the orbits of a binary stellar system, the two members of which have never been seen apart even in the most powerful telescopes. True, such discoveries *do* teach us more of the mystery of the universe as well as of the possibilities of achievement by the human mind; but unfortunately, the privileged few who have the time and the ability to achieve such discoveries and to appreciate their meaning constitute almost a secret order, far above and regrettably apart from the ordinary run of mankind. By all means let those distinguished members of the great professorhood go on and discover new marvels in the physical world and the astronomical universe, but may we not hope to see, alongside of them and equally honored with them, another group of the professorhood who shall apply themselves with equal ability and assiduity, and in harmonious cooperation with a liberalized and growing group of the priesthood, to solving the everyday terrestrial problems of selfish humanity? Many difficulties lie ahead. Progress will be slow at best. But surely we may look forward with confidence to bettering days when our festival of cooperation is well under way. For my own part, and I trust for many others also, the ground for that forward-looking confidence is an optimism which is based on a study of the past and which springs from a firm belief in the philosophy of evolution and the faith of reverent science.

Postscript. In reply to a religious questionnaire from the National Academy of Sciences, Davis wrote in 1911:

When I was twelve years old, my father, having enlisted as quartermaster under Fremont, was expelled from the Society of Friends in Philadelphia. My mother, having approved his action, thereupon resigned. When I was afterwards old enough to maintain membership in the Society, and to understand the case, I withdrew, stating the above facts as sufficient reason. . . . My early home influences were extremely liberal and rationalistic; Sundays were holidays, not days of formality. But all my surroundings were of high moral motives and actions, being in a family devoted to the abolition of slavery, when that meant ostracism by 'respectable' society, and equally devoted to woman suffrage, then usually ridiculed, and to peace, almost before any peace societies were organized. With these influences all thru the week, Sundays might well be holidays from church. . . . A strong individualistic tendency was given me by the protestant attitude of my elders during my youth. This was probably intensified by my scientific education, in which inquiry before acceptance has been the leading principle. This tendency makes it difficult for me to join in any formal organization for religious purposes. . . . My studies have led me to discount so much of what is ordinarily taught in 'the church' that in that sense they have probably discouraged my interest in it; but so many sermons are occupied with matters irrelevant to modern life, and are given by persons so incompetent to treat the large subjects that they touch, that a sermon more often arouses my opposition than my consent. . . . [I have acquired] 'advanced' views regarding the history of the church, and the manifestly human origin of what is ordinarily regarded as divine. On this matter I find myself in close sympathy with Emerton's *Unitarian Thought* [See p. 256] . . . the share that my family takes in social service is in larger part performed by the wife than by the husband, and I see no reason why it should not be so. My time is largely taken in earning a salary; but I reserve a small portion for my duties at the Avon Home [for destitute children, where Davis was a Trustee and Vice-President]. . . . Mrs Davis has for some years past . . . been active in several movements.

The Davis Family Tree

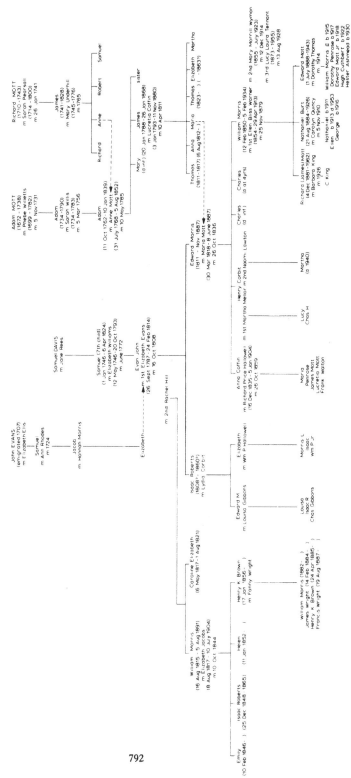

The Bibliography of William Morris Davis

1880

A. Banded amygdules of the Brighton Amygdaloid, *Proceedings of the Boston Society of Natural History*, Vol. 20, pp. 426–8.

1881

A. Illustrations of the Earth's Surface: Glaciers (with N. S. Shaler), *Science*, Vol. 2, pp. 581–4 and 624–30.

B. Remarks on the geology of Mt. Desert, Maine, *Proceedings of the Boston Society of Natural History*, Vol. 21, pp. 117–18.

1882

A. On the classification of lake basins, *Proceedings of the Boston Society of Natural History*, Vol. 21, pp. 315–81. (Abst. *American Naturalist*, Vol. 16, pp. 1028–9.)

B. Glacial erosion, *Proceedings of the Boston Society of Natural History*, Vol. 22, pp. 19–58.

C. The structural value of the trap ridges of the Connecticut Valley, *Proceedings of the Boston Society of Natural History*, Vol. 22, pp. 116–24.

D. Brief notice of observations on the Triassic trap rocks of Massachusetts, Connecticut and New Jersey, *American Journal of Science*, Third Series, Vol. 24, pp. 345–9.

E. The little mountains east of the Catskills (New York), *Appalachia*, Vol. 3, pp. 20–33.

1883

A. The Cachai earthquake, *Science*, Vol. 1, p. 67.

B. An early statement of the effect of the earth's rotation, *Science*, Vol. 1, p. 98.

C. Lakes and valleys of northeastern Pennsylvania, *Science*, Vol. 1, pp. 304–5.

D. The origin of cross valleys, *Science*, Vol. 1, pp. 325–7 and 356–7.
E. Temperature and ice of the Bavarian lakes, *Science*, Vol. 1, p. 393.
F. Classification of islands, *Science*, Vol. 1, p. 484.
G. Lake Bonneville, *Science*, Vol. 1, p. 570.
H. Meteorological charts of the North Atlantic, *Science*, Vol. 2, pp. 654–7.
I. Becraft's Mountain, *American Journal of Science*, Third Series, Vol. 26, pp. 381–9.
J. The nonconformity at Rondout, New York, *American Journal of Science*, Third Series, Vol. 26, pp. 389–95.
K. On the conversion of chlorine into hydrochloric acid, *Proceedings of the American Philosophical Society*, Vol. 2, pp. 102–9.
L. On the relations of the Triassic traps and sandstones of the eastern United States, *Bulletin of the Museum of Comparative Zoology, Harvard College*, Vol. 7, pp. 281–309.
M. The folded Helderberg Limestones east of the Catskills, *Bulletin of the Museum of Comparative Zoology, Harvard College*, Vol. 7, pp. 311–29.
N. Deflective effect of the earth's rotation, *Van Nostrand's Engineering Magazine*, Vol. 28, pp. 297–8.
O. Charcoal as applied to the deposition of gold from copper and other impurities, *Franklin Institute Journal*, Vol. 85, pp. 274–87.

1884

A. Whirlwinds, cyclones, and tornadoes, *Science*, Vol. 3, pp. 40, 63 and 93.
B. How do the winds blow within the storm-disk?, *Science*, Vol. 3, pp. 402–3.
C. Paleozoic high tides, *Science*, Vol. 3, pp. 473–4.
D. The older wind charts of the North Atlantic, *Science*, Vol. 3, pp. 593–7.
E. Light in the deep sea, *Science*, Vol. 4, p. 94.
F. Drumlins, *Science*, Vol. 4, pp. 418–20.
G. Temperature and its changes in the United States, *Science*, Vol. 4, pp. 569–70.
H. Tornadoes and how to escape them, *Science*, Vol. 4, pp. 572–3.
I. The winds and currents of the equatorial Atlantic, *American Meteorological Journal*, Vol. 1, pp. 48–56.
J. The relation of tornadoes to cyclones, *American Meteorological Journal*, Vol. 1, pp. 121–7.
K. On the definition of a tornado, *American Meteorological Journal*, Vol. 1, pp. 159–60.
L. Local and tropical weather cards, *American Meteorological Journal*, Vol. 1, pp. 245–7.
M. Rainfall maps, *American Meteorological Journal*, Vol. 1, pp. 302–3.

N. Gorges and waterfalls, *American Journal of Science*, Third Series, Vol. 28, pp. 123–32.

O. The distribution and origin of drumlins, *American Journal of Science*, Third Series, Vol. 28, pp. 407–16.

P. *Whirlwinds, Cyclones and Tornadoes* (Lee and Shepard, Boston), 90 pp.

Q. Ueber Samun und Böen, *Meteorologische Zeitschrift*, Vol. 1, pp. 243–5.

R. Five cent cyclones (note), *American Meteorological Journal*, Vol. 1, p. 158.

S. On the connection between meteorology and seismology (note), *American Meteorological Journal*, Vol. 1, p. 296.

1885

A. Work of the Swiss Earthquake Commission, *Science*, Vol. 5, pp. 196–8.

B. The Meteorological Observatory on Blue Hill, *Science*, Vol. 5, p. 440.

C. The reddish brown ring around the sun, *Science*, Vol. 5, pp. 455–6.

D. Geographic classification, illustrated by a study of plains, plateaus, and their derivatives, *Proceedings of the American Association for the Advancement of Science*; Vol. 33, pp. 428–32.

E. Earthquakes in New England, *Appalachia*, Vol. 4, pp. 190–4.

F. Reduction of barometer readings to latitude 45°, *American Meteorological Journal*, Vol. 1, pp. 510–11.

G. The deflective effect of the earth's rotation, *American Meteorological Journal*, Vol. 1, pp. 516–24.

H. Terminology of atmospheric vapour, *American Meteorological Journal*, Vol. 2, pp. 6–7.

I. The New England Meteorological Society, *American Meteorological Journal*, Vol. 2, pp. 383–6.

J. Geographic evolution (Abst.), *Proceedings of the Boston Society of Natural History*, Vol. 23, p. 223.

1886

A. Winter on Mount Washington, *Science*, Vol. 7, pp. 40–2.

B. Chinook winds, *Science*, Vol. 7, pp. 55–6.

C. The festoon cloud, *Science*, Vol. 7, pp. 57–8.

D. The recent cold wave, *Science*, Vol. 7, pp. 70–1.

E. A recent ice-storm (Letter), *Science*, Vol. 7, p. 190.

F. Bishop's Ring during solar eclipses, *Science*, Vol. 7, pp. 239–40.

G. A thunder squall in New England (Letter), *Science*, Vol. 7, pp. 436–7.

H. Temperature diagrams, *American Meteorological Journal*, Vol. 2, pp. 169–75.

I. Foreign studies of thunder-storms, *American Meteorological Journal*, Vols. 2–3, pp. 289–99, 489–99, 40–8, 65–6 and 69–79.

J. The temperature of Mediterranean Seas (Letter), *American Meteorological Journal*, Vol. 3, p. 49.

K. Weather prediction in New Zealand, *American Meteorological Journal*, Vol. 3, pp. 103–5.

L. Derivation of the term 'Trade Wind', *American Meteorological Journal*, Vol. 3, pp. 111–12.

M. Cyclones, anticyclones and pericyclones, *American Meteorological Journal*, Vol. 3, pp. 117–18.

N. Notes on studies of thunderstorms in Europe. First Papers, *American Meteorological Journal*, Vol. 3, 11 pp. (Reprinted.)

O. Brief notices of papers read before the Geological Section of the American Association, *American Journal of Science*, Third Series, Vol. 32, pp. 319–24.

P. The structure of the Triassic Formation of the Connecticut Valley, *American Journal of Science*, Third Series, Vol. 32, pp. 342–52. (Abst. p. 321.)

Q. Bishop's Ring around the sun, *Popular Science Monthly*, Vol. 28, pp. 466–74.

R. On the methods of study of thunder-storms, *Proceedings of the American Academy of Arts and Sciences*, Vol. 21, pp. 336–47.

S. Thunderstorms in New England in the summer of 1885, *Proceedings of the American Academy of Arts and Sciences*, Vol. 22, pp. 14–58.

T. Relation of the coal of Montana to the older rocks, *Tenth Census of the United States, 1880*, Vol. 15, pp. 697–712.

U. American contributions to meteorology (Philadelphia, Nov. 19), reprinted from the *Journal of the Franklin Institute*, Vol. 127, 27 pp.

V. Earthquakes in New England, *Appalachia*, Vol. 4, pp. 190–4.

W. Mountain meteorology, *Appalachia*, Vol. 4, pp. 225–44 and 327–50.

X. Height and velocity of clouds (with A. McAdie), *Annual Report of the Director of the Harvard Astronomical Observatory*, Vol. 40, p. 10.

Y. *A series of twenty-five colored geological models and twenty-five photographs of important geological objects, each accompanied by a letter-press description* (with N. S. Shaler and T. W. Harris) (D. C. Heath & Co., Boston).

Z. The chinook winds of the Northwest (Abst.), *Proceedings of the Boston Society of Natural History*, Vol. 23, pp. 249–50.

A′. Relations of the atmosphere to organic life, *American Meteorological Journal*, Vol. 3, pp. 57–8.

B′. Observations on thunderstorms, *American Meteorological Journal*, Vol. 3, pp. 105–6.

1887

A. Advances in meteorology, *Science*, Vol. 9, pp. 539–41.
B. The classification of lakes, *Science*, Vol. 10, pp. 142–3.
C. Snow as a cause of cold weather, *American Meteorological Journal*, Vol. 3, pp. 389–90.
D. Water-vapour and radiation, *American Meteorological Journal*, Vol. 3, pp. 443–5.
E. The height of cumulus clouds, *American Meteorological Journal*, Vol. 3, pp. 492–4.
F. The foehn in the Andes, *American Meteorological Journal*, Vol. 3, p. 507.
G. The foehn in New Zealand, *American Meteorological Journal*, Vol. 3, pp. 442–3.
H. Land and sea breezes, *American Meteorological Journal*, Vol. 4, pp. 59–61.
I. The foehn, *American Meteorological Journal*, Vol. 4, pp. 182–91 and 224–7.
J. Mechanical origin of the Triassic monoclinal in the Connecticut Valley, *Proceedings of the American Academy of Arts and Sciences*, Vol. 35, pp. 224–7 and *Proceedings of the Boston Society of Natural History*, Vol. 23, pp. 339–41. (Abst. *American Journal of Science*, Third Series, Vol. 32, 1886, p. 321.)
K. Instruction in geological investigation, *American Naturalist*, Vol. 21, pp. 810–25.

1888

A. Wasp-stings, *Science*, Vol. 11, p. 50.
B. The topographic map of New Jersey, *Science*, Vol. 12, pp. 206–7.
C. Synclinal mountains and anticlinal valleys, *Science*, Vol. 12, p. 320.
D. Local weather predictions, *American Meteorological Journal*, Vol. 4, pp. 409–12.
E. On the use of meteorological maps in schools, *American Meteorological Journal*, Vol. 4, pp. 489–92.
F. A classification of the winds, *American Meteorological Journal*, Vol. 4, pp. 512–19.
G. The structure of the Triassic Formation of the Connecticut Valley, *7th Annual Report of the United States Geological Survey*, 1885–6, pp. 455–90.
H. *Two Chapters on the Physical Geography and Climate of New England* (Cambridge, Mass.), 10 pp.
I. Blanford on the relation of rainfall to forests, *American Meteorological Journal*, Vol. 5, pp. 93–6.

J. Hill on anomalies of winds in Northern India, *American Meteorological Journal*, Vol. 5, pp. 87–9.

1889

A. A river-pirate (Dear Run, Pa.), *Science*, Vol. 13, pp. 108–9.
B. The contoured map of Massachusetts, *Science*, Vol. 14, pp. 422–3.
C. Geographic methods in geologic investigations, *National Geographic Magazine*, Vol. 1, pp. 11–26.
D. The rivers and valleys of Pennsylvania, *National Geographic Magazine*, Vol. 1, pp. 183–253. (Abst. *American Geologist*, Vol. 6, 1890, pp. 60–1.)
E. The faults in the Triassic Formation near Meriden, Connecticut: A week's work in the Harvard Summer School of Geology, *Bulletin of the Museum of Comparative Zoology, Harvard College*, Vol. 16, pp. 61–87.
F. The intrusive and extrusive Triassic trap sheets of the Connecticut Valley (with C. L. Whittle), *Bulletin of the Museum of Comparative Zoology, Harvard College*, Vol. 16, pp. 99–138.
G. The song of the singing mouse, *American Naturalist*, Vol. 23, pp. 481–4.
H. Methods and models in geographical teaching, *American Naturalist*, Vol. 23, pp. 566–83.
I. Some American contributions to meteorology, *Journal of the Franklin Institute*, Vol. 127, pp. 104–15 and 176–91.
J. The geographic development of northern New Jersey (with J. Walter Wood Jr.), *Proceedings of the Boston Society of Natural History*, Vol. 24, pp. 365–423. (Abst. *American Geologist*, Vol. 6, 1890, pp. 195–6.)
K. The glacial origin of cliffs, *American Geologist*, Vol. 3, pp. 14–18.
L. Report on the investigation of the sea-breeze, *American Meteorological Journal*, Vol. 6, pp. 4–6.
M. Topographic development of the Triassic Formations of the Connecticut Valley, *American Journal of Science*, Third Series, Vol. 37, pp. 423–34.
N. The ash bed at Meriden and its structural relations, *Transaction of the Meriden Scientific Association*, Vol. 3, pp. 23–30.

1890

A. Oscillations of lakes (Seiches) (Letter), *Science*, Vol. 15, p. 117.
B. Dr Hann's studies on cyclones and anticyclones, I, *Science*, Vol. 15, pp. 332–3.
C. Vertical components of motion in cyclones and anticyclones, (Letter), *Science*, Vol. 15, p. 388.
D. Ferrel's convectional theory of tornadoes, *American Meteorological Journal*, Vol. 6, pp. 337–49 and 418–63.

E. Secular changes in climate, *American Meteorological Journal*, Vol. 7, pp. 67–81.

F. The features of tornadoes and their distinction from other storms, *American Meteorological Journal*, Vol. 7, pp. 433–6.

G. The level of no strain, *American Geologist*, Vol. 5, pp. 190–1.

H. The Iroquois Beach, *American Geologist*, Vol. 6, p. 400.

I. Investigation of the sea breeze (with R. DeC. Ward), *Annals of the Astronomical Observatory of Harvard College*, Vol. 21, Part 1, pp. 215–65.

J. Structure and origin of glacial sand plains, *Bulletin of the Geological Society of America*, Vol. 1, pp. 195–202.

K. The rivers of northern New Jersey with notes on the classification of rivers in general, *National Geographic Magazine*, Vol. 2, pp. 81–110.

L. An outline of meteorology, *Johns Hopkins University Circular*, Vol. 9, pp. 71–2.

1891

A. The Lawrence tornado (with H. F. Mills and H. H. Clayton), *American Meteorological Journal*, Vol. 7, pp. 433–43.

B. Cumulus clouds over islands, *American Meteorological Journal*, Vol. 7, pp. 563–4.

C. European weather predictions, *American Meteorological Journal*, Vol. 8, pp. 53–8.

D. The physical geography of southern New England, *Johns Hopkins University Circular*, Vol. 10, pp. 78–9.

E. Tornadoes: A story of long inheritance (Abst.), *Johns Hopkins University Circular*, Vol. 10, p. 78.

F. The Triassic Sandstone of the Connecticut Valley, *Johns Hopkins University Circular*, Vol. 10, p. 79.

G. Dr Hann's studies on cyclones and anticyclones, II, *Science*, Vol. 16, pp. 4–5.

H. The geological dates of origin of certain topographic forms on the Atlantic slope of the United States, *Bulletin of the American Geographical Society*, Vol. 2, pp. 545–84.

I. The lost volcanoes of Connecticut, *Popular Science Monthly*, Vol. 40, pp. 221–35.

J. Two belts of fossiliferous black shale in the Triassic Formation of Connecticut (with S. W. Loper), *Bulletin of the Geological Society of America*, Vol. 2, pp. 415–30.

K. The story of a long inheritance, *Atlantic Monthly*, Vol. 68, pp. 68–78.

L. Was Lake Iroquois an arm of the sea?, *American Geologist*, Vol. 7, pp. 139–40.

M. Tornadoes and their distinction from other storms, *American Meteorological Journal*, Vol. 7, pp. 433–43.

1892

A. The Loup Rivers in Nebraska, *Science*, Vol. 19, pp. 107–8 and 220–1.
B. The convex profile of bad-land divides, *Science*, Vol. 20, p. 245.
C. Ferrel's contributions to meteorology, *American Meteorological Journal*, Vol. 8, pp. 348–59.
D. Theories of artificial and natural rainfall, *American Meteorological Journal*, Vol. 8, pp. 493–502.
E. Mirage on a wall (Letter), *American Meteorological Journal*, Vol. 8, pp. 525–6.
F. Meteorology in the schools, *American Meteorological Journal*, Vol. 9, pp. 1–21.
G. Note on winter thunderstorms (Letter), *American Meteorological Journal*, Vol. 9, pp. 164–70.
H. The teaching of geography (The Physical Basis of Descriptive Geography), *Educational Review*, Vol. 3, pp. 417–27.
I. The teaching of geography (What to avoid in teaching geography), *Educational Review*, Vol. 4, pp. 6–15.
J. The Catskill Delta in the Post-glacial Hudson Estuary, *Proceedings of the Boston Society of Natural History*, Vol. 25, pp. 318–34. (Abst. *Journal of Geology*, Vol. 1, 1893, pp. 97–8.)
K. On the drainage of the Pennsylvania Appalachians, *Proceedings of the Boston Society of Natural History*, Vol. 25, pp. 418–20.
L. The subglacial origin of certain eskers, *Proceedings of the Boston Society of Natural History*, Vol. 25, pp. 477–99. (Abst. *Journal of Geology*, Vol. 1, 1893, pp. 95–6.)
M. Remarks on drumlins, *Proceedings of the Boston Society of Natural History*, Vol. 26, pp. 17–23.
N. The ancient shore-lines of Lake Bonneville, *Goldthwaite's Geographical Magazine*, Vol. 3, pp. 1–5.
O. The Cañon of the Colorado, *Goldthwaite's Geographical Magazine*, Vol. 3, pp. 98–102.
P. The folds of the Appalachians, *Goldthwaite's Geographical Magazine*, Vol. 3, pp. 251–5.
Q. The Appalachian Mountains of Pennsylvania, *Goldthwaite's Geographical Magazine*, Vol. 3, pp. 343–50.
R. Observations of the New England Meteorological Society in the years 1890–91, *Annual Report of the Astronomical Observatory of Harvard College*, Vol. 31, pp. 1–93 and 161–2.

S. The Lawrence tornado of July 26, 1890, *Annual Report of the Astronomical Observatory of Harvard College*, Vol. 31, Pt 1, pp. 119–37.
T. *Outline of Elementary Meteorology* (Cambridge, Mass.), 13 pp.
U. Sketch of William Ferrel, *Popular Science Monthly*, Vol. 40, pp. 686–95.
V. The extension of physical geography in elementary teaching, *School and College*, Vol. 1, pp. 599–608.
W. Clouds in thunderstorms, *American Meteorological Journal*, Vol. 9, pp. 231–2.

1893

A. Memorial of James Henry Chapin, *Bulletin of the Geological Society of America*, Vol. 4, pp. 406–8.
B. Geographical work for State Geological Surveys, *Bulletin of the Geological Society of America*, Vol. 5, pp. 604–8. (Abst. *American Geologist*, Vol. 13, p. 146.)
C. Note on winter thunderstorms, *American Meteorological Journal*, Vol. 9, pp. 164–70.
D. Winter thunderstorms, *American Meteorological Journal*, Vol. 9, pp. 238–9.
E. The general winds of the Atlantic Ocean, *American Meteorological Journal*, Vol. 9, pp. 476–88.
F. On the difficulties in weather forecasting, *American Meteorological Journal*, Vol. 9, pp. 550–3.
G. Proposed subjects for correlated study by State Weather Services, *American Meteorological Journal*, Vol. 10, pp. 68–74.
H. The deflective effect of the earth's rotation, *American Meteorological Journal*, Vol. 10. pp. 195–8.
I. The theory of cyclones, *American Meteorological Journal*, Vol. 10, pp. 319–21.
J. The winds of the Indian Ocean, *American Meteorological Journal*, Vol. 10, pp. 333–43.
K. *Geographical Illustrations: Suggestions for teaching physical geography based on the physical features of southern New England* (Cambridge, Mass.), 46 pp.
L. Facetted pebbles on Cape Cod, Massachusetts, *Proceedings of the Boston Society of Natural History*, Vol. 26, pp. 166–75. (Abst. *American Geologist*, Vol. 13, pp. 146–7.)
M. Geography in grammar and primary schools, *School Review*, Vol. 1, pp. 327–39.
N. The abandonment of entrenched meanders: Wye, Evenlode, Cherwell, Thames, *Proceedings of the Geologists' Association*, Vol. 34, pp. 81–96.

O. The improvement of geographical teaching, *National Geographic Magazine*, Vol. 5, pp. 68–75.

P. Artificial and natural rainfall, *Boston Commonwealth*, March 26.

Q. The Osage River and the Ozark Uplift, *Science*, Vol. 22, pp. 276–9.

R. The winds and disease (Letter), *American Meteorological Journal*, Vol. 9, p. 433.

S. Katechismus der Meteorologie by W. J. Van Bebber (Review), *American Meteorological Journal*; Vol. 10, pp. 326–7.

T. Notice of H. F. Blanford, *American Meteorological Journal*, Vol. 10, pp. 74–6.

U. Mountain and valley winds (Letter), *American Meteorological Journal*, Vol. 10, p. 143.

V. Cloud measurements at Blue Hill Observatory, *American Meteorological Journal*, Vol. 10, pp. 107–9.

1894

A. A speculation on topographic climatology, *American Meteorological Journal*, Vol. 10, pp. 333–43.

B. Note on diffusion of water vapor and on atmospheric absorption of terrestrial radiation, *American Meteorological Journal*, Vol. 11, pp. 147–51.

C. Festooned mammiforms and pocky clouds, *American Meteorological Journal*, Vol. 11, pp. 151–3.

D. *Elementary Meteorology* (Ginn and Co., Boston), 355 pp.

E. Physical geography in the university, *Journal of Geology*, Vol. 2, pp. 66–100.

F. The ancient outlet of Lake Michigan, *Popular Science Monthly*, Vol. 46, pp. 217–29.

G. *Report on Government Maps for Use in Schools* (with C. F. King and G. L. Collie), Prepared by a Committee of the Conference on Geography Held in Chicago, Ill., December, 1892, (New York), 65 pp.

H. Eastern boundary of the Connecticut Triassic (with L. S. Griswold), *Bulletin of the Geological Society of America*, Vol. 5, pp. 515–30. (Abst. *American Geologist*, Vol. 13, pp. 145–6; *American Journal of Science*, Third Series, Vol. 47, pp. 136–7.)

I. *List of Geographical Lantern Slides* (Cambridge, Mass.), 17 pp.

J. *Papers from the Physical and Geographic Laboratory of Harvard University*, Reprinted from Ann. Rep. School Comm. of the City of Cambridge for 1893.

K. Meteorology in the schools, *School Review*, Vol. 2, pp. 529–39.

L. Note on Croll's Glacial Theory, *Transactions of the Edinburgh Geological Society*, Vol. 7, pp. 77–80.

M. An outline of the geology of Mount Desert, in *Flora of Mount Desert Island, Maine: A preliminary catalogue of the plants growing on Mount Desert and the adjacent islands,* by E. L. Rand and J. H. Redfield (Cambridge, Mass.), pp. 43–71.

N. A step towards improvement in teaching geography, *Harvard Teachers Association Leaflet No. 11.*

O. The Redfield and Espy Period, 1830–1855, *Report of the International Meteorological Congress, Chicago, August 1893, U.S. Department of Agriculture, Weather Bureau Bulletin,* No. 11, pp. 305–16.

1895

A. Winds and ocean currents (Letter), *Science,* Vol. 2, pp. 342–3.

B. The absorption of terrestrial radiation by the atmosphere, *Science,* Vol. 2, pp. 485–7.

C. Notes on geological excursions (Abst.), *Science,* Vol. 2, p. 744.

D. Theories of ocean currents, *Science,* Vol. 2, p. 824.

E. The development of certain English rivers, *Geographical Journal,* Vol. 5, pp. 127–46.

F. Professor J. D. Dana, *Geographical Journal,* Vol. 5, pp. 599–600.

G. La Seine, la Meuse, et la Moselle, *Annales de Géographie,* Year 5, pp. 25–49.

H. The need of geography in the university, *Educational Review,* Vol. 10, pp. 22–41.

I. Bearing of physiography on uniformitarianism (Abst.), *Bulletin of the Geological Society of America,* Vol. 7, pp. 8–11; *American Geologist,* Vol. 16, pp. 243–4; *Science,* Vol. 2, p. 280.

J. *The New England States,* Supplement to Frye's 'Complete Geography' (Boston), 31 pp. (Also 1902 edition.)

K. The physical geography of southern New England, *National Geographic Society, Monograph 1, No. 9,* pp. 269–304.

L. *Physiography as an Alternative Subject for Admission to College,* Official Report of the 10th Annual Meeting of the New England Association of Colleges and Preparatory Schools, pp. 38–46. (Also *School Review,* Vol. 3, pp. 632–40.)

M. A note on Croll's Glacial Theory, *American Meteorological Journal,* Vol. 11, pp. 441–4.

1896

A. An elementary presentation of the tides, *Science,* Vol. 3, pp. 569–70.

B. Physiographic features of the middle Susquehanna region, Pa, *Science,* Vol. 3, pp. 786–7.

C. The State Map of Connecticut as an aid to the study of geography in grammar and high schools, *Connecticut School Document*, No. 6, 14 pp.

D. A speculation in topographical climatology, *American Meteorological Journal*, Vol. 12, pp. 372–81.

E. Large-scale maps as geographical illustrations, *Journal of Geology*, Vol. 4, pp. 484–513.

F. Plains of marine and subaerial denudation, *Bulletin of the Geological Society of America*, Vol. 7, pp. 377–98. (Abst. *American Geologist*, Vol. 7, pp. 96–7; *Science*, Vol. 3, pp. 50–1.)

G. The outline of Cape Cod, *Proceedings of the American Academy of Arts and Sciences*, Vol. 31, pp. 303–32. (Abst. *American Geologist*, Vol. 17, pp. 95–6; *Science*, Vol. 3, pp. 49–50.)

H. The peneplain of the Scotch Highlands, *Geological Magazine*, n.s., Decade 4, Vol. 3, pp. 525–8.

I. The quarries in the lava beds at Meriden, Conn, *American Journal of Science*, Fourth Series, Vol. 1, pp. 1–13.

J. The Seine, the Meuse, and the Moselle, *National Geographic Magazine*, Vol. 7, pp. 189–202 and 228–38.

K. The soaring of birds and currents of air, *Auk*, Vol. 13, pp. 92–3.

L. The State Map of New York as an aid to the study of geography in grammar and high schools and academies, *University of the State of New York, Examination Bulletin No. 11*, pp. 503–26.

M. The State Map of Rhode Island as an aid to the study of geography in grammar and high schools, *Rhode Island Educational Publication*, 15 pp.

N. Josiah Dwight Whitney, *Harvard Graduates' Magazine*, Vol. 5, pp. 206–9.

1897

A. Home geography, *Journal of School Geography*, Vol. 1, pp. 2–7.

B. Field work in physical geography, *Journal of School Geography*, Vol. 1, pp. 17–24 and 62–9.

C. The use of geographical periodicals, *Journal of School Geography*, Vol. 1, pp. 81–5.

D. Temperate zones, *Journal of School Geography*, Vol. 1, pp. 139–43.

E. Topographic maps of the United States, *Journal of School Geography*, Vol. 1, pp. 200–4.

F. The Harvard geographical models (with G. C. Curtis), *Proceedings of the Boston Society of Natural History*, Vol. 28, pp. 85–110. (Abst. *Science*, Vol. 7, 1898, p. 81.)

G. Winds and ocean currents, *Scottish Geographical Magazine*, Vol. 13, pp. 515–23.

H. The coastal plain of Maine, *Report of the British Association for the Advancement of Science*, pp. 719–29.

I. Is the Denver Formation lacustrine or fluviatile?, *Science*, Vol. 6, pp. 619–21.

J. The present trend in geography, *A Paper Delivered at the 35th Convocation, University of the State of New York, Senate Chamber, Albany, New York*, 29 June, pp. 192–201.

K. Science in the schools, *Educational Review*, Vol. 13, pp. 429–39.

L. The State Map of Massachusetts as an aid to the study of geography in grammar and high schools, *Massachusetts State Board of Education, 60th Annual Report*, 18 pp.

M. Current notes on physiography, *Science*, Vol. 6, pp. 834–5. (See also other notes by Davis in *Science* at about this time.)

1898

A. Winds and ocean currents, *Journal of School Geography*, Vol. 2, pp. 16–20.

B. Waves and tides, *Journal of School Geography*, Vol. 2, pp. 122–32.

C. The equipment of a geographical laboratory, *Journal of School Geography*, Vol. 2, pp. 170–81.

D. The selection of topographical maps for school, *Journal of School Geography*, Vol. 2, pp. 240–45.

E. *Physical Geography* (assisted by W. H. Snyder) (Ginn and Co., Boston), 428 pp.

F. *Outline of Requirements in Meteorology* (with R. DeC. Ward) (Harvard University, Cambridge, Mass.), 16 pp.

G. The Triassic Formation of Connecticut, *United States Geological Survey Eighteenth Annual Report 1896–97*, Pt. 2, pp. 1–192.

H. Geography as a university subject, *Scottish Geographical Magazine*, Vol. 14, pp. 24–9.

I. The grading of mountain slopes (Abst.), *Science*, Vol. 7, p. 81.

J. Systematic geography, *4th Yearbook of the National Herbart Society*, pp. 81–91.

1899

A. The continent of North America, In *International Geography* (by H. R. Mill) (Newnes, London), pp. 664–78.

B. The United States, In *International Geography* (by H. R. Mill) (Newnes, London), pp. 710–73.

C. Un exemple de plaine côtière: La plaine du Maine, *Annales de Géographie*, Vol. 8, pp. 1–5.

D. Vallées à méandres, *Annales de Géographie*, Vol. 8, pp. 170–2.

E. La Peneplaine, *Annales de Géographie*, Vol. 8, pp. 289–303 and 385–404.

F. Balze per Faglia nei Monti Lepini, *Bollettino della Società Geografisca Italiana*, Fasc. 12, pp. 3–17 (translated by Fr. M. Pasanisi).

G. Consequent and subsequent streams, *Scottish Geographical Magazine*, Vol. 15, p. 436.

H. The geographical cycle, *Geographical Journal*, Vol. 14, pp. 481–504.

I. The peneplain, *American Geologist*, Vol. 23, pp. 207–39.

J. The rational element in geography, *National Geographic Magazine*, Vol. 10, pp. 466–73.

K. The circulation of the atmosphere, *Quarterly Journal of the Royal Meteorological Society*, Vol. 25, pp. 160–9.

L. Die Cirkulation der Atmosphäre, *Das Wetter*, Year 16, pp. 201–3, 228–32 and 253–9.

M. The drainage of cuestas, *Proceedings of the Geologists' Association of London*, Vol. 16, pp. 75–93.

N. 'Helm-Wind' Beobachtet In Den Cevennen, *Meteorologische Zeitschrift*, Vol. 16, pp. 124–5.

O. The system of the winds, *School World*, Vol. 1, pp. 244–7.

1900

A. Physiographic terminology with special reference to land forms, *Science*, Vol. 11, p. 99.

B. The basin deposits of the Rocky Mountain Region (Abst.), *Science*, Vol. 11, p. 144.

C. The conditions of formation of conglomerates, and criteria for distinguishing between lacustrine and fluviatile beds (Abst.), *Science*, Vol. 11, p. 429.

D. Current notes on physiography, *Science*, Vol. 11, pp. 790–1.

E. Fault scarp in the Lepini Mountains, Italy, *Bulletin of the Geological Society of America*, Vol. 11, pp. 207–16.

F. Continental deposits of the Rocky Mountain Region, *Bulletin of the Geological Society of America*, Vol. 11, pp. 569–601. (Abst. *Science*, Vol. 11, p. 144.)

G. History of the Cincinnati Anticline (discussion), *Bulletin of the Geological Society of America*, Vol. 11, pp. 604–5.

H. Glacial erosion in France, Switzerland, and Norway, *Proceedings of the Boston Society of Natural History*, Vol. 29, pp. 273–322.

I. Glacial erosion in the Valley of the Ticino, *Appalachia*, Vol. 9, pp. 136–56.

J. Local illustrations of distant lands: I. A Temporary Sahara, *Journal of School Geography*, Vol. 4, pp. 171–5.

K. Notes on the Colorado Canyon District, *American Journal of Science*, Fourth Series, Vol. 10, pp. 251–9.

L. Physical geography in the high school, *School Review*, Vol. 8, pp. 388–404 and 449–56.

M. Practical exercises in geography, *National Geographic Magazine*, Vol. 11, pp. 62–78.

N. The freshwater Tertiary Formations of the Rocky Mountain Regions, *Proceedings of the American Academy of Arts and Sciences*, Vol. 35, pp. 345–73.

O. The physical geography of the lands, *Popular Science Monthly*, Vol. 57, pp. 157–70.

1901

A. Schleswig-Holstein, *Science*, Vol. 14, p. 224.

B. The ranges of the Great Basin: Physiographic evidence of faulting, *Science*, Vol. 14, pp. 457–9.

C. Local illustrations of distant lands: II. The Lakes and Rivers of the Laurentian Highlands, *Journal of School Geography*, Vol. 5, pp. 85–8.

D. Maps of the Mississippi River, *Journal of School Geography*, Vol. 5, pp. 379–82.

E. Practical exercises in physical geography, *Proceedings of the Fifth Annual Conference of the New York State Science Teachers' Association, 28–29 December 1900, Albany, New York*, 11 pp.

F. Note on river terraces in New England (Abst.), *Bulletin of the Geological Society of America*, Vol. 12, pp. 483–4.

G. An excursion in Bosnia, Hercegovina, and Dalmatia, *Bulletin of the Geographical Society of Philadelphia*, Vol. 3, pp. 21–50.

H. An excursion to the Grand Canyon of the Colorado, *Bulletin of the Museum of Comparative Zoology, Harvard College*, Vol. 38, pp. 107–201. (Abst. *Bulletin of the Geological Society of America*, Vol. 12, p. 483; *Geological Magazine*, Vol. 8, p. 324; *Science*, Vol. 13, p. 188.)

I. The geographical cycle, *Verhandlungen Siebenten International Geographical Kongresses 1899*, Vol. 2, pp. 221–31.

J. The causes of rainfall, *Journal of the New England Waterworks Association*, Vol. 15, pp. 338–50.

K. Peneplains of central France and Brittany (Abst.), *Bulletin of the Geological Society of America*, Vol. 12, pp. 481–3.

L. Les enseignements du Grand Canyon du Colorado, *La Géographie*, Vol. 4, pp. 339–51.

1902

A. Base level, grade, and peneplain, *Journal of Geology*, Vol. 10, pp. 77–111.

B. Field work in physical geography, *Journal of Geography*, Vol. 1, pp. 17–24 and 62–9.

C. Progress of geography in the schools, in *First Yearbook of the National Society for the Scientific Study of Education*, Part II (Chicago), pp. 7–49.

D. River terraces in New England, *Bulletin of the Museum of Comparative Zoology, Harvard College*, Vol. 38, Geological Series V, pp. 281–346.

E. Systematic geography, *Proceedings of the American Philosophical Society*, Vol. 41, pp. 235–59.

F. The terraces of the Westfield River, Mass., *American Journal of Science*, Fourth Series, Vol. 14, pp. 77–94.

G. *Elementary Physical Geography* (Ginn and Co., Boston), 401 pp. (Also 1926 printing.)

1903

A. Effect of shore line on waves (Abst.), *Science*, Vol. 15, p. 88; also *Bulletin of the Geological Society of America*, Vol. 13, 1901, p. 528.

B. Walls of the Colorado Canyon (Abst.), *Science*, Vol. 15, p. 87; also *Bulletin of the Geological Society of America*, Vol. 13, 1901, p. 528.

C. The Blue Ridge of North Carolina (Abst.), *Science*, Vol. 17, p. 220.

D. The fresh-water Tertiaries at Green River, Wyo. (Abst.), *Science*, Vol. 17, pp. 220–1; also *Bulletin of the Geological Society of America*, Vol. 14, p. 544; also *Journal of Geology*, Vol. 11, p. 120.

E. Block mountains of the Basin-Range Province (Abst.), *Science*, Vol. 17, p. 301; also *Bulletin of the Geological Society of America*, Vol. 14, p. 551; also *Engineers' Monthly Journal*, Vol. 75, p. 153.

F. An excursion to the plateau province of Utah and Arizona, *Bulletin of the Museum of Comparative Zoology, Harvard College*, Vol. 42, pp. 1–50.

G. The mountain ranges of the Great Basin, *Bulletin of the Museum of Comparative Zoology, Harvard College*, Vol. 42, Geological Series VI, No. 3, pp. 127–78.

H. A scheme of geography, *Geographical Journal*, Vol. 22, pp. 413–23.

I. Geography in the United States, *Proceedings of the American Association for the Advancement of Science*, Vol. 53, pp. 471–502.

J. Practical exercises in physiography, *Journal of Geography*, Vol. 2, pp. 516–20.

K. The development of river meanders, *Geological Magazine*, n.s., Decade 4, Vol. 10, pp. 10, 145–8.

L.   The question of seminars, *Harvard Graduates' Magazine*, Vol. 11, pp. 363–70.
M.   The stream contest along the Blue Ridge, *Bulletin of the Geographical Society of Philadelphia*, Vol. 3, pp. 213–44.
N.   *The Teachers' Guide to Accompany the Text Book of Elementary Physical Geography* (Ginn and Co., Boston), 80 pp.
O.   The basin ranges of Utah and Nevada (Abst.), *Journal of Geology*, Vol. 11, p. 120.
P.   The Blue Ridge in southern Virginia and North Carolina (Abst.), *Journal of Geology*, Vol. 11, p. 121.

1904

A.   The scheme of geography, *Journal of Geography*, Vol. 3, pp. 20–31.
B.   The scope of geography, *Nature*, Vol. 69, pp. 403–4.
C.   A flat-topped range in the Tian-Shan, *Appalachia*, Vol. 10, pp. 277–84.
D.   Glacial erosion in the Sawatch Range, Colo., *Appalachia*, Vol. 10, pp. 392–404.
E.   A summer in Turkestan, *Bulletin of the American Geographical Society*, Vol. 36, pp. 217–28.
F.   The Hudson River described, *Bulletin of the American Geographical Society*, Vol. 36, pp. 557–9.
G.   Geography in the United States, *American Geologist*, Vol. 33, pp. 156–85.
H.   Geography in the United States, *Science*, Vol. 19, pp. 121–32 and 178–86.
I.   The relations of the earth sciences in view of their progress in the nineteenth century, *Journal of Geology*, Vol. 12, pp. 669–87.
J.   Physiography and glaciation of the western Tian-Shan Mountains, Turkestan (Abst.) (with E. Huntington), *Bulletin of the Geological Society of America*, Vol. 15, p. 554.

1905

A.   Glaciation of the Sawatch Range, Colorado, *Bulletin of the Museum of Comparative Zoology, Harvard College*, Vol. 49, Geological Series VIII, No. 1, pp. 1–11.
B.   The Wasatch Canyon and House Ranges, Utah, *Bulletin of the Museum of Comparative Zoology, Harvard College*, Vol. 49, Geological Series VIII, pp. 15–56.
C.   Home geography, *Journal of Geography*, Vol. 4, pp. 1–5 and 32.
D.   Illustration of tides by waves, *Journal of Geography*, Vol. 4, pp. 290–4.
E.   Levelling without baseleveling, *Science*, Vol. 21, pp. 825–8.

F. The Colorado Canyon (Abst.), *Science*, Vol. 21, p. 860.

G. A journey across Turkestan, in *Explorations in Turkestan*, edited by Raphael Pumpelly (Carnegie Institute, Washington, D.C.), pp. 23–119.

H. Africa as seen by the British Association, *Nation*, Vol. 81, pp. 397–9 and 419–20.

I. An opportunity for the Association of American Geographers, *Bulletin of the American Geographical Society*, Vol. 37, pp. 84–6.

J. Complications of the geographical cycle, *Report of the Eighth International Geographical Congress, Washington, 1904*, pp. 150–63.

K. The bearing of physiography upon Suess' theories, *American Journal of Science*, Fourth Series, Vol. 19, pp. 265–73. (Abst. *Report of the Eighth International Geographical Congress, Washington, 1904*, p. 164.)

L. The geographical cycle in an arid climate, *Journal of Geology*, Vol. 13, pp. 381–407.

M. A day in the Cévennes, *Appalachia*, Vol. 11, pp. 110–14.

N. Tides in the Bay of Fundy, *National Geographic Magazine*, Vol. 16, pp. 71–6.

O. College entrance examination in physiography, *Report of the Eighth International Geographical Congress, Washington, 1904*, p. 956.

1906

A. An inductive study of the content of geography, *Bulletin of the American Geographical Society*, Vol. 38, pp. 67–84.

B. Physiographic notes on South Africa, *Bulletin of the American Geographical Society*, Vol. 38, pp. 88–9.

C. The mountains of southernmost Africa, *Bulletin of the American Geographical Society*, Vol. 38, pp. 593–623.

D. Incised meandering valleys, *Bulletin of the Geographical Society of Philadelphia*, Vol. 4, pp. 182–92.

E. Observations in South Africa, *Bulletin of the Geological Society of America*, Vol. 17, pp. 377–450.

F. Professor Shaler and the Lawrence Scientific School, *Harvard Engineering Journal*, Vol. 5, pp. 129–38.

G. The content of geography, an instructive study, *Journal of Geography*, Vol. 5, pp. 145–60.

H. The geographical cycle in an arid climate, *Geographical Journal*, Vol. 27, pp. 70–3.

I. The physical factor in general geography, *Educational Bi-Monthly*, Vol. 1, pp. 112–22.

J. The sculpture of mountains by glaciers, *Scottish Geographical Magazine*, Vol. 22, pp. 76–89. (Abst. *Report of the British Association for the Advancement of Science*, Vol. 75, pp. 393–4.)

K. Was Lewis Evans or Benjamin Franklin the first to recognize that our northeast storms come from the southwest?, *Proceedings of the American Philosophical Society*, Vol. 45, pp. 129–30.

L. *Biographical Memoir of George Perkins Marsh, 1801–82* (Washington), 10 pp.; also *Biographical Memoirs of the National Academy of Sciences*, Vol. 6, 1909, pp. 71–80.

M. The Colorado Canyon and its lessons, *Proceedings of the Liverpool Geological Society*, Vol. 10, pp. 98–102.

N. The physiography of the Adirondacks (Formation of scarps), *Science*, Vol. 23, pp. 630–1.

O. The relations of the earth sciences in view of their progress in the nineteenth century, *Congress of Arts and Sciences (St. Louis, 1904)*, Vol. 4, pp. 488–503.

P. Professor Nathaniel S. Shaler, *American Journal of Science*, Fourth Series, Vol. 21, pp. 480–1.

1907

A. Current notes on land forms (with D. W. Johnson and I. Bowman), *Science*, Vol. 25, pp. 70–3, 229–32, 294–396, 508–10, 833–6 and 946–9; Vol. 26, pp. 90–3, 152–4, 226–8, 353–6, 450–3 and 837–9; Vol. 27, pp. 31–3.

B. Postglacial aggradation of Himalayan valleys, *Science*, Vol. 25, p. 231.

C. The terraces of the Maryland coastal plain, *Science*, Vol. 25, pp. 701–7.

D. Hanging valleys, *Science*, Vol. 25, pp. 835–6.

E. The place of coastal plains in systematic physiography, *Journal of Geography*, Vol. 6, pp. 8–15.

F. Hettner's conception of geography, *Journal of Geography*, Vol. 6, pp. 49–53.

1908

A. Die Methoden der Amerikanischen Geographischen Forschung, *Internationale Wochenschrift für Wissenschaft, Kunst, und Technik* (Berlin), Nov. 14.

B. The prairies of North America, *Internationale Wochenschrift für Wissenschaft, Kunst, und Technik* (Berlin), Vol. 2, pp. 1011–18 and 1045–50.

C. The physiographic subdivisions of the Appalachian Mountain system, and their effects upon settlement and history (Abst.), *Report of the British Association for the Advancement of Science*, Vol. 78, pp. 761–2.

D. *Practical exercises in physical geography* (Ginn and Co., Boston), 148 pp.; Atlas, 50 pp.

E. Causes of Permo-Carboniferous glaciation, *Journal of Geology*, Vol. 16, pp. 79–82.

1909

A. Der grosse Canyon des Colorado, *Gesellschaft Deutscher Naturforscher und Ärtze* (1908), Vol. 1, pp. 157–69.

B. *Die Erklärende Beschreibung der Landformen* (English manuscript circulated: see 1912J).

C. *Geographical Essays* (edited by D. W. Johnson) (Ginn and Co., Boston), 777 pp.

D. Der Grosse Cañon des Colorado-Flusses, *Zeitschrift Gesellschaft Erdkunde* (Berlin), Vol. 3, pp. 164–72.

E. Glacial erosion in North Wales, *Quarterly Journal of the Geological Society*, Vol. 65, pp. 281–350.

F. The Colorado Canyon: Some of its lessons, *Geographical Journal*, Vol. 33, pp. 535–40. (Abst. *Report of the British Association for the Advancement of Science*, Vol. 78, pp. 948–49.)

G. The systematic description of land forms, *Geographical Journal*, Vol. 34, pp. 300–18; with subsequent discussion pp. 318–26.

H. The Alps in the Glacial Period, *Geographical Journal*, Vol. 34, pp. 650–59.

I. The lessons of the Colorado Canyon, *Bulletin of the American Geographical Society*, Vol. 41, pp. 345–54.

J. The valleys of the Cotswold Hills, *Proceedings of the Geologists' Association*, Vol. 21, pp. 150–2.

K. Der Grosse Cañon des Colorado, *Himmel und Erde*, Vol. 22, pp. 22–41.

L. The Rocky Mountains, *Internationale Wochenschrift für Wissenschaft, Kunst, und Technik* (Berlin), 16 pp.

1910

A. Antarctic geology and polar climates, *Proceedings of the American Philosophical Society*, Vol. 49, pp. 200–3.

B. Experiments in geographical description, *Bulletin of the American Geographical Society*, Vol. 42, pp. 401–35.

C. The Kamerun Region by K. Hasseb, *Bulletin of the American Geographical Society*, Vol. 42, pp. 670–2.

D. Explorations in Bolivia by P. H. Fawcett, *Bulletin of the American Geographical Society*, Vol. 42, pp. 673–5.

E. Le Relief du Limousin by A. Demangeon, *Bulletin of the American Geographical Society*, Vol. 42, pp. 840–2.

F. Schleswig-Holstein by K. Olbricht, *Bulletin of the American Geographical Society*, Vol. 42, pp. 842–4.

G. Experiments in geographical description, *Science*, Vol. 31, pp. 921–46.

H. Experiments in geographical description, *Scottish Geographical Magazine*, Vol. 26, pp. 561–86.

I. Topographical maps of the United States, *Nation*, Vol. 91, pp. 359–60.

J. Deutsche und Romanische Flussterminologie; *Geographischer Anzeiger*, pp. 121–3.

K. Practical exercises in physical geography (Abst.), *Proceedings of the Ninth International Geographical Congress*, Vol. 2, pp. 169–70.

L. The theory of isostasy (Abst.), *Bulletin of the Geological Society of America*, Vol. 21, p. 777.

M. Die Umgestaltung der Gebirgsformen durch die Gletscher (Abst.), *Verein für Erdkunde Leipzig Mitteilungen*, pp. 28–9.

1911

A. Morphologie des Böhmerwaldes by D. H. Mayr, *Bulletin of the American Geographical Society*, Vol. 43, pp. 46–51.

B. Wellington Harbor, New Zealand by J. M. Bell, *Bulletin of the American Geographical Society*, Vol. 43, pp. 190–4.

C. The place of deduction in the description of land forms, *Bulletin of the American Geographical Society*, Vol. 43, pp. 598–602.

D. Mountain passes by Johann Sölch, *Bulletin of the American Geographical Society*, Vol. 43, pp. 602–3.

E. Terraces in south-central Italy by A. Galdteri, *Bulletin of the American Geographical Society*, Vol. 43, pp. 603–4.

F. A cuesta in Middle Germany, *Bulletin of the American Geographical Society*, Vol. 43, pp. 679–84.

G. The Argentine Cordillera by F. Kühn, *Bulletin of the American Geographical Society*, Vol. 43, pp. 847–50.

H. Grundzüge der Oberflächengestaltung Cornwallis by H. Spethmenn, *Bulletin of the American Geographical Society*, Vol. 43, pp. 851–3.

I. The Colorado Front Range, *Annals of the Association of American Geographers*, Vol. 1, pp. 21–84. (Abst. *Science*, Vol. 33, p. 906.)

J. Short studies abroad: The seven hills of Rome, *Journal of Geography*, Vol. 9, pp. 197–202 and 230–3.

K. Repeating patterns in the relief and in the structure of the land, *Bulletin of the Geological Society of America*, Vol. 22, p. 717.

L. Geographical descriptions in the folios of the Geologic Atlas of the United States (Abst.), *Bulletin of the Geological Society of America*, Vol. 22, p. 736.

M. *Grundzüge der Physiogeographie* (with G. Braun) (Teubner, Leipzig and Berlin), 322 pp. (Also edition of 1915–17.)

N. New England States, in Frye's *Geography*, State Supplement (Ginn and Co., Boston), 32 pp.

O. Geographical factors in the development of South Africa, *Journal of Race Development*, Vol. 2, pp. 131–46.

P. The disciplinary value of geography, *Popular Science Monthly*, Vol. 78, pp. 105–19 and 223–40.

Q. An item for commercial geography, *Journal of Geography*, Vol. 9, pp. 157–8.

1912

A. L'ésprit explicatif dans la géographie moderne, *Annales de Géographie*, Vol. 21, pp. 1–19.

B. La vallée de Armançon: 8ᵉ excursion interuniversitaire (Mars 1912), *Annales de Géographie*, Vol. 21, pp. 312–22. (Translated by F. Herbette.)

C. The Tertiary Gravels of the Sierra Nevada of California by Waldemar Lindgren, *Bulletin of the American Geographical Society*, Vol. 44, pp. 908–10.

D. Begleitworte zu den '40 Blättern der Karte des Deutschen Reiches' by Walter Behrmann, *Bulletin of the American Geographical Society*, Vol. 44, pp. 911–12.

E. The Geography of Godavari by S. W. Cushing, *Bulletin of the American Geographical Society*, Vol. 44, pp. 912–13.

F. *Guide Book for the Transcontinental Excursion of 1912 of the American Geographical Society* (Editor) (Ginn and Co., New York), 144 pp.

G. A geographical pilgrimage from Ireland to Italy, *Annals of the Association of American Geographers*, Vol. 2, pp. 73–100.

H. Relation of geography to geology, *Bulletin of the Geological Society of America*, Vol. 23, pp. 93–124.

I. American studies on glacial erosion, *Proceedings of the Eleventh Geological Congress, Stockholm* (1910), pp. 419–27.

J. *Die Erklärende Beschreibung der Landformen* (translated by A. Rühl) (Teubner, Leipzig), 565 pp.

1913

A. The Rhine Gorge and the Bosporus, *Journal of Geography*, Vol. 11, pp. 209–15 and 276.

B. Grand Canyon of the Colorado, *Journal of Geography*, Vol. 11, pp. 310–14.

C. Age of the earth, *Journal of Geography*, Vol. 12, pp. 70–4.

D. Southern Vancouver Island by Charles H. Clapp, *Bulletin of the American Geographical Society*, Vol. 45, pp. 360–1.

E. Southern Nigeria by A. E. Kitson, *Bulletin of the American Geographical Society*, Vol. 45, pp. 361–2.

F. Morphologie des Moselgebietes zwischen Trier und Alf, by Bruno Dietrich, *Bulletin of the American Geographical Society*, Vol. 45, pp. 362–4.

G. Notes on the Physiography of the Southern Tableland of New South Wales by C. A. Süssmilch, *Bulletin of the American Geographical Society*, Vol. 45, p. 364.

H. Brevi note sui sette colli di Roma, *Bollettino Reale Società Geografica* (Rome), Series 5, Vol. 2, pp. 163–75.

I. Valli Consequenti e Subsequenti, *Bollettino Reale Società Geografica* (Rome), Vol. 12, pp. 1429–32.

J. Submerged valleys and barrier reefs, *Nature*, Vol. 91, pp. 423–4.

K. Dana's proof of Darwin's theory of coral reefs, *Nature*, Vol. 90, pp. 632–4.

L. Nomenclature of surface forms on faulted structures, *Bulletin of the Geological Society of America*, Vol. 24, pp. 163–216.

M. Speculative nature of geology (Abst.), *Bulletin of the Geological Society of America*, Vol. 24, pp. 686–7.

N. Dana's proof of Darwin's theory of coral reefs, *Scientific American Supplement*, Vol. 75, pp. 214–15.

O. Dana's confirmation of Darwin's theory of coral reefs, *American Journal of Science*, Fourth Series, Vol. 35, pp. 173–88. (Abst. *Science*, Vol. 37, p. 724.)

P. Human response to geographical environment, *Bulletin of the Geographical Society of Philadelphia*, Vol. 11, pp. 63–102.

Q. Meandering valleys and underfit rivers, *Annals of the Association of American Geographers*, Vol. 3, pp. 3–28.

R. Kelvin on 'Light' and 'The Tides', *Science*, Vol. 3, pp. 29–32.

1914

A. Le Maroc Physique by Louis Gentil, *Bulletin of the American Geographical Society*, Vol. 46, pp. 36–9.

B. La Théorie de Bloc-Diagram by Paul Castlenan, *Bulletin of the American Geographical Society*, Vol. 46, pp. 39–40.

C. The Balkan Peninsula by D. G. Hogarth, *Bulletin of the American Geographical Society*, Vol. 46, pp. 41–2.

D. Glacial origin of fiords of the South Island of New Zealand, *Bulletin of the American Geographical Society*, Vol. 46, p. 285.

E.   Lakes of the Balkan Peninsula, *Bulletin of the American Geographical Society*, Vol. 46, pp. 525–7.

F.   The home study of coral reefs, *Bulletin of the American Geographical Society*, Vol. 46, pp. 561–77, 641–54 and 721–39.

G.   The Coast of New Caledonia, *Bulletin of the American Geographical Society*, Vol. 46, pp. 609–10.

H.   The publication of the Memorial Volume of the Transcontinental Excursion of 1912, *Bulletin of the American Geographical Society*, Vol. 46, pp. 684–5.

I.   Physiography of arid lands, *Report of the British Association for the Advancement of Science*, Year 84, pp. 365–6.

J.   Sublacustrine glacial erosion in Montana (Abst.), *Bulletin of the Geological Society of America*, Vol. 25, p. 86.

K.   Der Valdarno: eine Darstellungstudie, *Zeitschrift Gesellschaft für Erdkunde, Berlin*, 68 pp.

1915

A.   The origin of coral reefs, *Proceedings of the National Academy of Sciences*, Vol. 1, pp. 146–52.

B.   The Mission Range, Montana, *Proceedings of the National Academy of Sciences*, Vol. 1, pp. 626–8.

C.   Preliminary report on a Shaler Memorial study of coral reefs, *Science*, Vol. 41, pp. 455–8.

D.   Problems associated with the origin of coral reefs suggested by a Shaler Memorial study of the reefs (Abst.), *Science*, Vol. 41, p. 569.

E.   Sculpture of the Mission Range, Mont. (Abst.), *Science*, Vol. 42, p. 685.

F.   Biographical memoir of John Wesley Powell, 1834–1902, *Biographical Memoirs, National Academy of Sciences*, Vol. 8, pp. 11–83.

G.   Biographical memoir of John Peter Lesley, 1819–1903, *Biographical Memoirs, National Academy of Sciences*, Vol. 8, pp. 155–240.

H.   The development of the Transcontinental Excursion of 1912, in *Memorial Volume of the Transcontinental Excursion of 1912* (American Geographical Society, New York), pp. 3–7.

I.   A Shaler Memorial study of coral reefs, *American Journal of Science*, Fourth Series, Vol. 40, pp. 223–71.

J.   Preliminary report on a Shaler Memorial study of coral reefs, *Nature*, Vol. 95, pp. 189–91.

K.   The principles of geographical description, *Annals of the Association of American Geographers*, Vol. 5, pp. 61–105.

1916

A. Clift islands in the coral seas, *Proceedings of the National Academy of Sciences*, Vol. 2, pp. 284–8.

B. Symposium on the exploration of the Pacific (with others), *Proceedings of the National Academy of Sciences*, Vol. 2, pp. 391–437. (See especially pp. 391–4.)

C. Extinguished and resurgent coral reefs, *Proceedings of the National Academy of Sciences*, Vol. 2, pp. 466–71.

D. The origin of certain Fiji Atolls, *Proceedings of the National Academy of Sciences*, Vol. 2, pp. 471–5.

E. Sinking islands versus a rising ocean in the coral-reef problem (Abst.), *Science*, Vol. 43, p. 721.

F. Expedite the map, *Science*, Vol. 44, pp. 525–6.

G. Mission Range, Montana, *Geographical Review*, Vol. 2, pp. 267–88. (Abst. *Annals of the Association of American Geographers*, Vol. 4, pp. 135–36.)

H. Practical exercises on topographic maps, *Journal of Geography*, Vol. 15, pp. 33–41.

I. Problems associated with the study of coral reefs, *Scientific Monthly*, Vol. 2, pp. 313–33, 479–501 and 557–72.

J. Coral reef problem (Abst.), *Bulletin of the Geological Society of America*, Vol. 27, p. 46.

K. Marcellus Hartley Memorial Medal, *Monthly Weather Review*, Vol. 44, pp. 205–7.

1917

A. The structure of high-standing atolls, *Proceedings of the National Academy of Sciences*, Vol. 3, pp. 473–9.

B. The isostatic subsidence of volcanic islands, *Proceedings of the National Academy of Sciences*, Vol. 3, pp. 649–54.

C. Sublacustrine glacial erosion in Montana, *Proceedings of the National Academy of Sciences*, Vol. 3, pp. 696–702.

D. The Great Barrier Reef of Australia, *American Journal of Science*, Fourth Series, Vol. 44, pp. 339–50.

E. *Handbook of Travel*, 'Geography', pp. 423–8; 'Geology', pp. 439–50 (Harvard University Press, Cambridge, Mass.).

F. *Excursions around Aix-les-Bains*, Published for the Y.M.C.A. National War Work Council by the Appalachian Mountain Club of Boston (Cambridge, Mass.), 27 pp.

G. *Grundzüge der Physiogeographie*, Pts 1–2, Ed. 2 (Teubner, Leipzig).

H. Topographic maps of the United States, *National Highways Association, Division of Physical Geography, Physiographic Bulletin No. 1*, 15 pp.

1918

A. *Handbook of Northern France* (Harvard University Press, Cambridge, Mass.), 174 pp.

B. *Practische Übungen in physischer Geographie* (Übertragen und neu bearbeitet von Karl Oestreich, hierzu ein Atlas mit 38 Tafeln) (Teubner, Leipzig), 115 pp.

C. Coral reefs and submarine banks, *Journal of Geology*, Vol. 26, pp. 198–223, 289–309 and 385–411.

D. Geological terms in geographical description, *Science*, Vol. 48, pp. 81–4.

E. The Cedar Mountain trap ridge near Hartford, *American Journal of Science*, Fourth Series, Vol. 46, pp. 476–7.

F. Grove Karl Gilbert, *American Journal of Science*, Fourth Series, Vol. 46, pp. 669–81.

G. Fringing reefs of the Philippine Islands, *Proceedings of the National Academy of Sciences*, Vol. 4, pp. 197–204.

H. Metalliferous laterite in New Caledonia, *Proceedings of the National Academy of Sciences*, Vol. 4, pp. 275–80.

I. Subsidence of reef-encircled islands, *Bulletin of the Geological Society of America*, Vol. 29, pp. 489–574. (Abst. pp. 71–2.)

J. The reef-encircled islands of the Pacific, *Journal of Geography*, Vol. 17, pp. 1–8, 58–68 and 102–7.

1919

A. The alleged journey of James White through the Grand Canyon, 1867, *Geographical Review*, Vol. 6, pp. 355–6.

B. The young coasts of Annam and Northern Spain, *Geographical Review*, Vol. 7, pp. 176–80.

C. Drainage evolution on the Yünnan–Tibet frontier, *Geographical Review*, Vol. 7, pp. 413–15.

D. Passarge's Principles of Landscape Description, *Geographical Review*, Vol. 8, pp. 266–73.

E. Fringing reefs of the Philippine Islands, *Scientific American Supplement*, Vol. 87, pp. 58–9.

F. The geological aspects of the coral-reef problem, *Science Progress in the Twentieth Century*, Vol. 13, pp. 420–44.

G. The significant features of reef-bordered coasts, *Transactions and Proceedings of the New Zealand Institute*, Vol. 51, pp. 6–30.

H. Pumpelly's Reminiscences, *Science*, Vol. 49, pp. 61–3.

1920

A. African rift valley, *Science*, Vol. 52, pp. 456–8.

B. Changes of ocean levels, *Scientific American*, Vol. 87, p. 294.

C. Features of glacial origin in Montana and Idaho (A Shaler Memorial Study), *Annals of the Association of American Geographers*, Vol. 10, pp. 75–147.

D. Glacial erosion of Snowdon, *Geological Magazine*, Vol. 57, pp. 381–2.

E. Physiographic relations of laterite, *Geological Magazine*, Vol. 57, pp. 429–31.

F. Geography at Cambridge University, England, *Journal of Geography*, Vol. 19, pp. 207–10.

G. The Penck Festband: A Review, *Geographical Review*, Vol. 10, pp. 249–61.

H. The islands and coral reefs of Fiji, *Geographical Journal*, Vol. 55, pp. 34–45, 200–20 and 377–88.

I. The small islands of Almost-Atolls, *Nature*, Vol. 105, pp. 292–3.

J. The framework of the earth, *American Journal of Science*, Fourth Series, Vol. 48, pp. 225–41. (Abst. *Bulletin of the Geological Society of America*, Vol. 31, p. 110.)

K. The function of geography, *Geographical Teacher*, Vol. 10, pp. 286–91.

1921

A. Airplane views of the Alps, *Journal of Geography*, Vol. 20, pp. 36–9.

B. Union of Geographers at the University of Leipzig, *Journal of Geography*, Vol. 20, pp. 116–17.

C. Coral reefs of Tutuila Samoa, *Science*, Vol. 53, pp. 559–65.

D. Lower California and its natural resources: A review, *Geographical Review*, Vol. 11, pp. 551–62.

E. Memoir of Frederic Putnam Gulliver, *Annals of the Association of American Geographers*, Vol. 11, pp. 112–16.

1922

A. Deflection of streams by earth rotation, *Science*, Vol. 55, pp. 478–9.

B. Geological overthrusts and underdrags (Abst.), *Science*, Vol. 55, p. 493.

C. A graduate school of geography, *Science*, Vol. 56, pp. 121–34.

D. Home geography, *Journal of Geography*, Vol. 21, pp. 28–32.

E. Geographic orientation, *Journal of Geography*, Vol. 21, pp. 316–19.

F. The reasonableness of science, *Scientific Monthly*, Vol. 15, pp. 193–214.

G. Topographical maps of the United States, *Scientific Monthly*, Vol. 15, pp. 557–60.

H. Faults, underdrag and landslides of the Great Basin Ranges, *Bulletin of the Geological Society of America*, Vol. 33, pp. 92–6.

I. Peneplains and the geographical cycle, *Bulletin of the Geological Society of America*, Vol. 33, pp. 587–98.

J. *A Graduate School of Geography* (Clark University Library, Worcester, Mass.), 26 pp.

K. Coral reefs of the Louisiade Archipelago, *Nature*, Vol. 110, pp. 56–8.

L. Coral reefs of the Louisiade Archipelago, *Proceedings of the National Academy of Sciences*, Vol. 8, pp. 7–13.

M. Dixey's Physiography of Sierra Leone, *Bulletin of the Geographical Society of Philadelphia*, Vol. 20, pp. 131–41.

N. The barrier reef of Tagula, New Guinea, *Annals of the Association of American Geographers*, Vol. 12, pp. 97–151.

O. The eustatic theory (Discussion), *Bulletin of the Geological Society of America*, Vol. 33, pp. 483–5.

1923

A. Drowned coral reefs south of Japan, *Proceedings of the National Academy of Sciences*, Vol. 9, pp. 58–62.

B. The marginal belts of the coral seas, *Proceedings of the National Academy of Sciences*, Vol. 9, pp. 292–6.

C. The depth of coral reef lagoons, *Proceedings of the National Academy of Sciences*, Vol. 9, pp. 296–301.

D. The Halligs, vanishing islands of the North Sea, *Geographical Review*, Vol. 13, pp. 99–106.

E. The Explanatory Description of Land Forms (A Review of Alfred Hettner's 'Die Oberflächenformen des Festlands') *Geographical Review*, Vol. 13, pp. 318–21.

F. New Zealand Land Forms (A Review of C. A. Cotton's 'Geomorphology of New Zealand. Pt. 1'), *Geographical Review*, Vol. 13, pp. 321–2.

G. The Shaping of the Earth's Surface (A Review of S. Passarge's 'Die grundlagen der Landschaftskunde'), *Geographical Review*, Vol. 13, pp. 599–607.

H. A working model of the tides, *Scientific Monthly*, Vol. 16, pp. 561–72.

I. The island of Oahu, *Journal of Geography*, Vol. 22, pp. 354–7.

J. The cycle of erosion and the summit level of the Alps, *Journal of Geology*, Vol. 31, pp. 1–41.

K. The marginal belts of the coral seas, *American Journal of Science*, Fifth Series, Vol. 6, pp. 181–95.

1924

A. A tilted-up bevelled-off atoll, *Science*, Vol. 60, pp. 51–6. (Abst. *Science*, Vol. 59, p. 544; *Pan-American Geologist*, Vol. 42, p. 74.)
B. Shaded topographic maps, *Science*, Vol. 60, pp. 325–7.
C. Modification of Darwin's theory of coral reefs by the glacial-control theory (Abst.), *Pan-American Geologist*, Vol. 42, pp. 73–4. (Also *Report of the British Association for the Advancement of Science*, 92 Meeting, 1924, pp. 384–5.)
D. Classification of oceanic islands (Abst.), *Pan-American Geologist*, Vol. 42, p. 319. (Also *Bulletin of the Geological Society of America*, Vol. 37, pp. 216–17.)
E. Oceanic problems related to coral reefs (Abst.), *Journal of the Washington Academy of Sciences*, Vol. 14, pp. 348–9.
F. Gilbert's theory of laccoliths (Abst.), *Journal of the Washington Academy of Sciences*, Vol. 14, p. 375.
G. The progress of geography in the United States, *Annals of the Association of American Geographers*, Vol. 14, pp. 159–215.
H. Notes on coral reefs, *Proceedings of the Second Pan-Pacific Scientific Congress* (Melbourne, 1923), Vol. 2, pp. 1161–3.
I. The oceans, *Natural History*, Vol. 24, pp. 554–65.
J. The formation of the Lesser Antilles, *Proceedings of the National Academy of Sciences*, Vol. 10, pp. 205–11.
K. The Explanatory Description of Land Forms, in *Recueil de travaux à Jovan Cvijić*, (Belgrade), pp. 287–336.
L. *Die Erklärende Beschreibung der Landformen* (translated by A. Rühl) (Teubner, Leipzig), 565 pp. (Second Edition).

1925

A. Undertow myth, *Science*, Vol. 61, pp. 206–8.
B. The Stewart Bank in the China Sea, *Science*, Vol. 62, pp. 401–3.
C. A Roxen lake in Canada (Lake Timiskaming), *Scottish Geographical Magazine*, Vol. 41, pp. 65–74.
D. Les Côtes et les Récifs Coralliens de la Nouvelle-Calédonie, *Annales de Géographie*, Vol. 34, pp. 244–69, 332–59, 423–41 and 521–58.
E. The Basin Range problem, *Proceedings of the National Academy of Sciences*, Vol. 11, pp. 387–92.
F. Comment on Dr C. W. Kochel's 'Abstraktionen in der Geologie', *Geologische Rundschau*, Vol. 16, pp. 313–14.
G. Laccoliths and sills (Abst.), *Journal of the Washington Academy of Sciences*, Vol. 15, pp. 414–15. (Also *Bulletin Volcanologique*, Year 2, pp. 323–4.)

1926

A. Subsidence rate of reef-encircled islands, *Proceedings of the National Academy of Sciences*, Vol. 12, pp. 99–105.
B. Value of outrageous geological hypotheses, *Science*, Vol. 63, pp. 463–8.
C. *Elementary Physical Geography* (Ginn and Co., Boston), 401 pp.
D. The Lesser Antilles, *American Geographical Society, Map of Hispanic America, Publication No. 2*, 207 pp.
E. Origin of the Lesser Antilles (Abst.), *Bulletin of the Geological Society of America*, Vol. 37, pp. 220–1.

1927

A. The rifts of Southern California, *American Journal of Science*, Fifth Series, Vol. 13, pp. 57–72.
B. A migrating anticline in Fiji, *American Journal of Science*, Fifth Series, Vol. 14, pp. 333–51.
C. Fortunate failure of philosophy, *Science*, Vol. 65, pp. 362–3.
D. Channels, valleys, and intermount detrital plains, *Science*, Vol. 66, pp. 272–4.
E. Barrier reefs of Tahiti and Moorea, *Nature*, Vol. 120, pp. 330–1.
F. Biographical Memoir of Grove Karl Gilbert, 1843–1918, *Biographical Memoirs, National Academy of Sciences*, Vol. 21, 5th Memoir, 303 pp.

1928

A. Fractures and fiords in the Faroes, *Science*, Vol. 68, pp. 419–20.
B. The formation of coral reefs, *Scientific Monthly*, Vol. 27, pp. 289–300.
C. *The Coral Reef Problem*, American Geographical Society, Special Publication, No. 9, 596 pp.
D. Die Entstehung von Korallenriffen, *Gesellschaft Erdkunde Berlin Zeitschrift*, Nos. 9–10, pp. 359–91.

1929

A. Geological map of New Mexico, *Science*, Vol. 70, pp. 68–70.
B. Wharton's and Darwin's theories of coral reefs, *Science Progress in the Twentieth Century*, Vol. 24, pp. 42–56.

1930

A. The Peacock Range, Arizona, *Bulletin of the Geological Society of America*, Vol. 41, pp. 293–313.

B. Origin of limestone caverns, *Bulletin of the Geological Society of America*, Vol. 41, pp. 475–628.

C. Physiographic contrasts, East and West, *Scientific Monthly*, Vol. 30, pp. 394–415 and 501–19.

D. Preparation of scientific articles, *Science*, Vol. 72, pp. 131–4.

E. Rock floors in arid and in humid climates, *Journal of Geology*, Vol. 38, pp. 1–27 and 136–58.

F. The desert of the great Southwest, *Harvard Graduates' Magazine*, Vol. 38, pp. 395–404.

G. The earth as a globe, *Journal of Geography*, Vol. 29, pp. 330–44.

H. The Galiuro Mountains, Arizona (with B. Brooks), *American Journal of Science*, Fifth Series, Vol. 19, pp. 89–115.

I. *Elementary Physical Geography* (Japanese translation), 413 pp.

J. *Practical Exercises in Physical Geography* (Japanese translation), 193 pp.

K. Geology and Geography 1858–1929 (with R. A. Daly), in S. E. Morrison (Ed.), *The Development of Harvard University Since the Inauguration of President Eliot* (Harvard University Press, Cambridge, Mass.), pp. 307–31.

L. Periodicity in desert physiography (Abst.), *Pan-American Geologist*, Vol. 53, p. 320.

1931

A. Origin of limestone caverns, *Science*, Vol. 73, pp. 327–31.

B. Undertow and rip tides, *Science*, Vol. 73, pp. 526–7.

C. Clear Lake, California (Abst.), *Science*, Vol. 74, pp. 572–3.

D. Nature of geological proof, or how do you know you are right? (Abst.), *Pan-American Geologist*, Vol. 55, pp. 357–8.

E. Shore lines of the Santa Monica Mountains, California (Abst.), *Pan-American Geologist*, Vol. 55, pp. 362–3; also *Bulletin of the Geological Society of America*, Vol. 43, 1932, p. 227.

F. The Santa Catalina Mountains, Arizona, *American Journal of Science*, Fifth Series, Vol. 22, pp. 289–317. (Abst. *Pan-American Geologist*, Vol. 55, pp. 372–3; also *Bulletin of the Geological Society of America*, Vol. 43, 1932, p. 235.)

G. Origin of caverns (Abst.) (with C. Killingsworth), *Bulletin of the Geological Society of America*, Vol. 42, pp. 308–9.

H. Elevated shore lines of Santa Monica Mountains (Abst.) (with W. C. Putnam and G. L. Richards), *Bulletin of the Geological Society of America*, Vol. 42, pp. 309–10; also *Pan-American Geologist*, Vol. 54, p. 154.

1932

A. The college life of Robert DeCourcy Ward, *Annals of the Association of American Geographers*, Vol. 22, pp. 29–32.
B. A retrospect of geography, *Annals of the Association of American Geographers*, Vol. 22, pp. 211–30.
C. Glacial epochs of the Santa Monica Mountains, California, *Proceedings of the National Academy of Sciences*, Vol. 18, pp. 659–65.
D. Remarks on arid pediments (Abst.), *Pan-American Geologist*, Vol. 56, p. 236.
E. Albert Perry Brigham (A Memorial), *Journal of Geography*, Vol. 31, pp. 265–6.
F. Basin Range types, *Science*, Vol. 76, pp. 241–5.
G. Piedmont benchlands and Primärrumpfe, *Bulletin of the Geological Society of America*, Vol. 43, pp. 399–440. (Abst. *Pan-American Geologist*, Vol. 58, p. 68; also *Bulletin of the Geological Society of America*, Vol. 44, p. 154.)

1933

A. Work of sheetfloods (Abst.), *Bulletin of the Geological Society of America*, Vol. 44, p. 83.
B. Glacial epochs of the Santa Monica Mountains, California, *Bulletin of the Geological Society of America*, Vol. 44, pp. 1041–133. (Abst. *Proceedings of the Geological Society of America*, 1933, pp. 304–5; also *Pan-American Geologist*, Vol. 59, pp. 306–7.)
C. Geomorphogeny of the desert (Abst.), *Pan-American Geologist*, Vol. 57, pp. 374–5.
D. Granitic domes of the Mojave Desert, California, *Transactions of the San Diego Society of Natural History*, Vol. 7, pp. 211–58.
E. The lakes of California, *California Journal of Mines and Geology*, Vol. 29, pp. 175–236.
F. Submarine mock valleys, *Transactions of the American Geophysical Union, 14th Annual Meeting*, pp. 231–4. (Abst. *Pan-American Geologist*, Vol. 59, pp. 307–8.)
G. San Francisco Bay, in 'Middle California and Western Nevada', *Sixteenth International Geological Congress*, Guidebook 16, pp. 16–21 (United States Government Printing Office, Washington).
H. Geomorphology of the Basin and Range province, in 'The Salt Region', *Sixteenth International Geological Congress*, Guidebook 17, pp. 6–14 (United States Government Printing Office, Washington).

1934

A. The Long Beach earthquake, *Geographical Review*, Vol. 24, pp. 1–11.
B. Submarine mock valleys, *Geographical Review*, Vol. 24, pp. 297–308. (Abst. *Proceedings of the Geological Society of America*, 1933, p. 306.)
C. The faith of reverent science, *Scientific Monthly*, Vol. 38, pp. 395–421.
D. Gardiner on 'Coral Reefs and Atolls', *Journal of Geology*, Vol. 42, pp. 200–17.

1935

A. Valleys of the Panamint Mountains, California (Abst.) (with J. H. Maxson), *Proceedings of the Geological Society of America*, 1934, p. 339.

1936

A. Geomorphology of mountainous deserts, *Report of the 16th International Geological Congress*, 1933, Vol. 2, pp. 703–14.

1938

A. Sheetfloods and streamfloods, *Bulletin of the Geological Society of America*, Vol. 49, pp. 1337–416.

An Analysis of Davis' Publications*

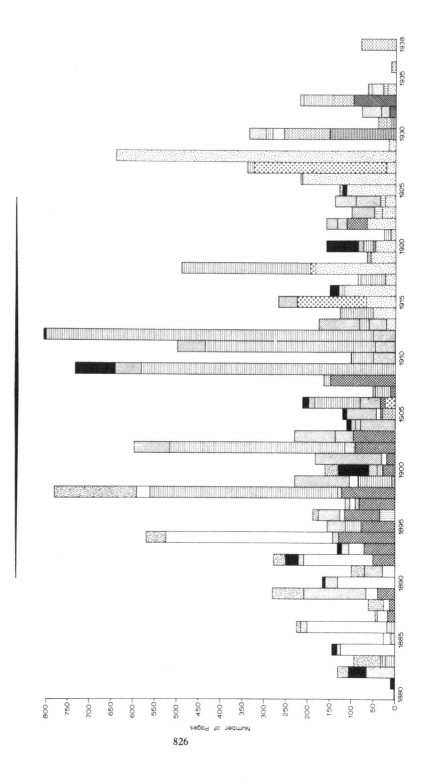

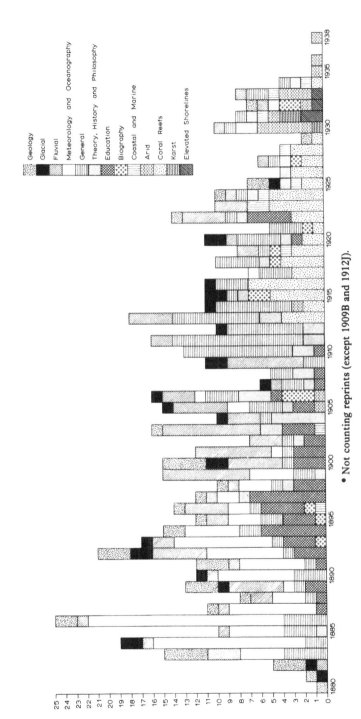

• Not counting reprints (except 1909B and 1912]).

APPENDIX V

General References

ABBE, C. (1906) The present condition in our schools and colleges of the study of climatology as a branch of geography and meteorology, *Bulletin of the American Geographical Society*, Vol. 38, pp. 121–3.

American Men of Science, 1933, William Morris Davis, 5th edn.

ANON (1934) Dr William M. Davis, *New York Times*, 7 February, p. 20.

ANON (1936) Memorial to Professor W. M. Davis, *Harvard Alumni Bulletin*, Vol. 38, No. 35, pp. 1219–20. (Contains a photograph taken at the unveiling of the Davis Memorial Gateway in Pasadena.)

BARLOW, N. (Ed.) (1958) *The Autobiography of Charles Darwin 1809–1882* (Collins, London), 253 pp.

BARRELL, J. (1917) Rhythms and the measurement of geologic time, *Bulletin of the Geological Society of America*, Vol. 28, pp. 745–904.

BARROWS, H. H. (1923) Geography as human ecology, *Annals of the Association of American Geographers*, Vol. 13, pp. 1–14.

BAULIG, H. (1928) *Le Plateau Central de la France et sa bordure Méditerranéenne* (Armand Colin, Paris), 590 pp.

BAULIG, H. (1939) Sur les 'Gradins de Piedmont', *Journal of Geomorphology*, Vol. 2, pp. 281–304.

BAULIG, H. (1950A) L'oeuvre de William Morris Davis, in *Essais de Géomorphologie* (Publications de la Faculté des Lettres de Université de Strasbourg, Paris), pp. 13–29. (Also in *Information Géographique*, 1948, No. 3, pp. 101–8.)

BAULIG, H. (1950B) William Morris Davis: Master of method, *Annals of the Association of American Geographers*, Vol. 40, pp. 188–95.

BECKINSALE, R. P. (1970) Physical problems of Cotswold rivers and valleys, *Proceedings of the Cotteswold Naturalists' Field Club*, Vol. 35, pp. 194–205.

Biographical Encyclopaedia of Pennsylvania of the Nineteenth Century (1874) Edward M. Davis (Galaxy, Philadelphia), 672 pp.

BLACKWELDER, E. (1925) Exfoliation as a phase of rock weathering, *Journal of Geology*, Vol. 33, pp. 789–806.

BLACKWELDER, E. (1931) Desert plains, *Journal of Geology*, Vol. 39, pp. 133–40.

BORNHARDT, W. (1900) *Zur Oberflächengestaltung und Geologie Deutch-Ostafrikas* (Dietrich Reimer, Berlin), 595 pp.

BOWMAN, I. (1926) The analysis of land forms; W. Penck on the topographic cycle, *Geographical Review*, Vol. 16, pp. 122–32.

BOWMAN, I. (1926) Reply to the Letter by A. Penck, *Geographical Review*, Vol. 16, pp. 351–2.

BOWMAN, I. (1934) William Morris Davis, *Geographical Review*, Vol. 24, pp. 177–81.

BRIGHAM, A. P. (1909) William Morris Davis, *Geographen Kalender*, Year 7, (Justus Perthes, Gotha), pp. 1–73.

BRIGHAM, A. P. (1915) History of the Excursion, *Amer. Geog. Soc. Mem. Vol. of the Transcontinental Excursion of 1912* (New York), pp. 9–45; 407 pp.

BRIGHAM, A. P. (1929) An appreciation of William Morris Davis, *Annals of the Association of American Geographers*, Vol. 19, pp. 61–2.

BROOKS, C. F. (1934) William Morris Davis, *Bulletin of the American Meteorological Society*, Vol. 15, p. 56.

BRYAN, K. (1922) Erosion and sedimentation in the Papago Country, Arizona, *U.S. Geological Survey, Bulletin 730B*, pp. 19–90.

BRYAN, K. (1925) The Papago Country, Arizona, *U.S. Geological Survey, Water Supply Paper 499*, 436 pp.

BRYAN, K. (1935A) William Morris Davis, *Annals of the Association of American Geographers*, Vol. 25, pp. 23–31.

BRYAN, K. (1935B) Processes of formation of pediments at Granite Gap, *Zeitschrift für Geomorphologie*, Vol. 9, pp. 125–35.

BRYAN, K. (1950) The place of geomorphology in the geographic sciences, *Annals of the Association of American Geographers*, Vol. 40, pp. 196–208.

BUCHER, W. H. (1947) Douglas Wilson Johnson (1878–1944), *National Academy of Sciences, Biographical Memoir*, Vol. 24, No. 5, pp. 197–230.

BUWALDA, J. P. (1934A) W. M. Davis: An appreciation, *Scientific Monthly*, Vol. 38, pp. 385–7.

BUWALDA, J. P. (1934B) A tribute to William Morris Davis, *Science*, Vol. 80, p. 46.

CALVERT, P. P. (1934) Obituary notice of W. M. Davis, *Entomological News*, Vol. 45, p. 84.

CHAMBERLAIN, R. W. (1936) *There is no Truce: The life of Thomas Mott Osborne, Prison Reformer* (Routledge, London), 420 pp.

830   APPENDICES

CHAMBERLIN, T. C. (1885) Administrative Report 1884–5, *Sixth Annual Report of the U.S. Geological Survey, 1884–5*, pp. 33–40.

CHAMBERLIN, T. C. and SALISBURY, R. D. (1885) Driftless Area of the Upper Mississippi Valley, *Sixth Annual Report of the U.S. Geological Survey, 1884–5*, pp. 199–322.

CHAMBERLIN, T. C. (1897) The method of multiple working hypotheses, *Journal of Geology*, Vol. 5, pp. 837–48.

CHAMBERLIN, T. C. and SALISBURY, R. D. (1909) *Geology*, 2nd edn, 3 vols (Murray, London), 684 pp.

CHORLEY, R. J. (1963) Diastrophic background to twentieth-century geomorphological thought, *Bulletin of the Geological Society of America*, Vol. 74, pp. 953–70.

CHORLEY, R. J., DUNN, A. J. and BECKINSALE, R. P. (1964) *The History of the Study of Landforms: Volume 1 Geomorphology before Davis* (Methuen, London), 678 pp.

CHORLEY, R. J. (1965) A re-evaluation of the geomorphic system of W. M. Davis, in R. J. Chorley and P. Haggett (Eds), *Frontiers in Geographical Teaching* (Methuen, London), pp. 21–38.

COLBY, C. C. (1936) Changing currents of geographic thought in America, *Annals of the Association of American Geographers*, Vol. 26, pp. 1–37.

COMSTOCK, G. A. (1924) Biographical Memoir of Benjamin Apthorp Gould (1824–1896), *National Academy of Sciences, Biographical Memoirs*, Vol. 17, pp. 155–80.

COTTON, C. A. (1922) *Geomorphology of New Zealand: Part I Systematic* (Dominion Museum, Wellington), 462 pp.

COTTON, C. A. (1942) *Climatic Accidents in Landscape-Making* (Whitcombe and Tombs, New Zealand), 354 pp.

COTTON, C. A. (1948) *Landscape*, 2nd edn (Cambridge University Press), 509 pp.

CROMWELL, O. (1958) *Lucretia Mott* (Harvard), 241 pp.

CVIJIĆ, J. (1909) Bildung und Dislozierung der Dinarischen Rumpffläche, *Petermanns Geographische Mitteilungen*, Vol. 55, pp. 121–7, 156–63 and 177–81.

DALY, R. A. (1905) The accordance of summit levels among alpine mountains, *Journal of Geology*, Vol. 13, pp. 195–205.

DALY, R. A. (1910) Pleistocene glaciation and the coral reef problem, *American Journal of Science*, Ser. 4, Vol. 30, pp. 297–308.

DALY, R. A. (1915) The glacial-control theory of coral reefs, *Proceedings of the American Academy of Arts and Sciences*, Vol. 51, pp. 155–251.

DALY, R. A. (1929) Swinging sea level of the Ice Age, *Bulletin of the Geological Society of America*, Vol. 40, pp. 721–34.

DALY, R. A., PALACHE, C. and PARKER, G. H. (1934) Minute on the life and services of William Morris Davis, *Harvard University Gazette*, Vol. 29, pp. 149–50.

DALY, R. A. (1945) Biographical Memoir of William Morris Davis: 1850–1934, *National Academy of Sciences, Biographical Memoirs*, Vol. XXIII, Eleventh Memoir, pp. 263–303.

DANA, J. D. (1853) *On Coral Reefs and Islands* (Putnam, New York), 144 pp.

DANA, J. D. (1872) *Corals and Coral Islands* (New York), 398 pp.

DANA, J. D. (1890) Archaean axes of eastern North America, *American Journal of Science*, Vol. 39, pp. 378–83.

DARWIN, C. (1842) *The Structure and Distribution of Coral Reefs* (Smith, Elder and Co., London), 214 pp.

Defence of Dr Gould by the Scientific Council of Dudley Observatory (1858), 3rd edn (Albany, New York), 93 pp.

DELEBECQUE, A. (1898) *Les Lacs Française* (Paris), 436 pp.

DE MARTONNE, E. (1929) La morphologie du Plateau Central de la France et l'hypothèse eustatique, *Annales de Géographie*, Vol. 38, pp. 113–32.

DICKINSON, R. E. (1969) *The Makers of Modern Geography* (Routledge and Kegan Paul, London), 305 pp.

Dictionary of American Biography (1944) William Morris Davis, Vol. XXI, Supp. 1 (Oxford), pp. 229–31.

Dictionary of American Biography (1932) Richard Price Hallowell, Vol. VIII, (Scribner's, New York), p. 160.

DODGE, R. E. (1934) William Morris Davis, An appreciation, *Journal of Geography*, Vol. 33, pp. 148–50.

DRYER, C. R. (1920) Genetic geography; *Annals of the Association of American Geographers*, Vol. 10, pp. 3–16.

DURY, G. H. (1963A) Underfit streams in relation to capture: A reassessment of the ideas of W. M. Davis, *Transactions of the Institute of British Geographers*, No. 32, pp. 83–94.

DURY, G. H. (1963B) Geographical description: An essay in criticism, *The Australian Geographer*, Vol. 9 (2), pp. 67–78.

DURY, G. H. (1964) Principles of underfit streams, *U.S. Geological Survey, Professional Paper* 452-A, 67 pp.

DUTTON, C. E. (1880) Report on the Geology of the High Plateaus of Utah, *U.S. Geographical and Geological Survey of the Rocky Mountain Region* (Washington), 307 pp.

EMMONS, W. H. (1910) A reconnaissance of some mining camps in Elko, Lander, and Eureka Counties, Nevada, *U.S. Geological Survey, Bulletin 408*, 130 pp.

FENNEMAN, N. M. (1936) Cyclic and non-cyclic aspects of erosion, *Science*, Vol. 83, pp. 87–94.

FLEMAL, R. C. (1971) The attack on the Davisian system of geomorphology: A synopsis, *Journal of Geological Education*, Vol. 19, pp. 3–13.

GARDINER, J. S. (1931) *Coral Reefs and Atolls* (Macmillan, London), 181 pp.

GARDINER, J. S. (1934) Professor W. M. Davis, *Nature*, Vol. 133, pp. 973–4.

GARDNER, J. H. (1935) Origin and development of limestone caverns, *Bulletin of the Geological Society of America*, Vol. 46, pp. 1255–74.

GILBERT, G. K. (1875) Report on the Geology of portions of Nevada, Utah, California and Arizona (1871–72), in *Report on Geographical and Geological Explorations and Surveys West of the One Hundredth Meridian*, in charge of First Lieut. G. M. Wheeler (Washington), Vol. 3, Pt 1, pp. 21–187.

GILBERT, G. K. (1909) The convexity of hilltops, *Journal of Geology*, Vol. 17, pp. 344–51.

GILBERT, G. K. (1914) The transportation of debris by running water, *U.S. Geological Survey, Professional Paper 86*, 263 pp.

GILBERT, G. K. (1928) Studies of Basin Range structure, *U.S. Geological Survey, Professional Paper 153*, 92 pp.

GOULD, B. A. (1859) *Reply to the Statement of the Trustees of the Dudley Observatory* (Albany), 366 pp.

GOULD, B. A. (1866) On a new and brilliant variable star, *American Journal of Science*, 2nd Ser., Vol. 42, pp. 80–3.

GOULD, C. N. (1959) *Covered Wagon Geologist* (University of Oklahoma Press, Norman), 282 pp.

GREEN, A. H. (1876) *Geology for Students and General Readers: Part I. Physical Geology* (Daldy, Isbister and Co., London), 552 pp. (2nd edn 1882).

GRIFFIN, P. F. (1952) *The contribution of Richard Elwood Dodge to Educational Geography* (Ph.D. Thesis, Columbia University).

GRUND, A. (1904) Die Karsthydrographie: Studien aus Westbosnien, *Penck's Geographische Abhandlungen*, Vol. 7 (3), pp. 103–200.

GRUND, A. (1910) Beiträge zur Morphologie des dinarischen Gebirges, *Penck's Geographische Abhandlungen*, Vol. 9, pp. 1–236.

GULLIVER, F. P. (1899) Shoreline topography, *Proceedings of the American Academy of Arts and Sciences*, Vol. 34, pp. 151–258.

HALLOWELL, A. D. (1884) *James and Lucretia Mott: Life and letters* (Houghton Mifflin, Boston and New York), 566 pp.

HALLOWELL, W. P. (1893) *Record of a Branch of the Hallowell Family* (Hallowell, Philadelphia), 246 pp.

HETTNER, A. (1921) *Die Oberflächenformen des Festlandes* (Teubner, Leipzig and Berlin), 250 pp. (2nd edn 1928, translated into English as *The Surface Features of the Earth*, by P. Tilley, Macmillan, London, 1972, 193 pp.)

HOBBS, W. H. (1907) Nathaniel Southgate Shaler, *Transactions of the Wisconsin Academy of Science, Arts and Letters*, Vol. 15, Pt 2, pp. 925–8.

HOTCHKIN, S. F. (1892) *The York Road, Old and New* (Binder and Kelly, Philadelphia), 516 pp.

HUNTINGTON, E. (1912) William Morris Davis, Geographer, *Bulletin of the Geographical Society of Philadelphia*, Vol. 10, pp. 224–34.

HUNTINGTON, E. (1914) The solar hypothesis of climatic changes, *Bulletin of the Geological Society of America*, Vol. 25, pp. 477–590.

HUXLEY, T. H. (1877) *Physiography* (Macmillan, London), 384 pp.

JAMES, H. (1930) *Charles W. Eliot: President of Harvard University 1869–1909*, 2 vols (Constable, London).

JAMES, P. E. (1967) On the origin and persistence of error in geography, *Annals of the Association of American Geographers*, Vol. 57, pp. 1–24.

JOHNSON, A. M. (1951) *William Morris Davis (1815–1891): A story of a nineteenth century American* (Privately printed, Washington, D.C.), 57 pp.

JOHNSON, D. W. (1919) *Shore Processes and Shoreline Development* (Wiley, New York), 584 pp.

JOHNSON, D. W. (1929) The geographic prospect, *Annals of the Association of American Geographers*, Vol. 19, pp. 167–231.

JOHNSON, D. W. (1931) *Stream Sculpture on the Atlantic Slope* (Columbia University Press, New York), 142 pp.

JOHNSON, D. W. (1932A) Rock fans of arid regions, *American Journal of Science*, Vol. 23, pp. 389–416.

JOHNSON, D. W. (1932B) Rock planes of arid regions; *Geographical Review*, Vol. 22, pp. 656–65.

JOHNSON, D. W. (1932C) Principles of marine level correlation, *Geographical Review*, Vol. 22, pp. 294–8.

JOHNSON, D. W. (1934) William Morris Davis, *Science*, Vol. 79, pp. 445–9.

JOHNSON, D. W. (1938–42) Studies in scientific method, *Journal of Geomorphology*, Vol. 1, pp. 64–6 and 147–52; Vol. 2, pp. 366–72; Vol. 3, pp. 59–64, 156–62, 256–62 and 353–5; Vol. 4, pp. 145–9 and 328–32; Vol. 5, pp. 73–7 and 171–3.

JUDSON, S. (1960) William Morris Davis: An appraisal, *Zeitschrift für Geomorphologie*, Band 4, pp. 193–201.

JUDSON, S. (1971) William Morris Davis, *Dictionary of Scientific Biography*, Vol. 3 (Charles Scribner's Sons, New York), pp. 592–6.

KANEV, D. D. (1957) William Morris Davis (in Bulgarian); *Geografia, Godina 7 – Knizhka*, Vol. 10, pp. 20–1.

KEYES, C. (1934) William Morris Davis and physiography, *The Pan-American Geologist*, Vol. 62, pp. 1–16.

KING, C. (1878) Systematic Geology, In *Report of the U.S. Geological Exploration of the 40th Parallel*, Vol. 1 (Washington), 803 pp.

KING, L. C. (1953) Canons of landscape evolution,· *Bulletin of the Geological Society of America*, Vol. 64, pp. 721–52.

KNADLER, G. A. (1958) *Isaiah Bowman: Backgrounds to his contributions to thought* (D.Ed. Thesis, Indiana University).

KNOPF, A. (1918) A geologic reconnaissance of the Inyo Range and the eastern slope of the southern Sierra Nevada, *U.S. Geological Survey, Professional Paper 110*, 130 pp.

KRUG-GENTHE, M. (1903) Die Geographie in den Vereinigten Staaten, *Geographische Zeitschrift*, Vol. 9, pp. 626–37 and 666–85. (Contains one of the first clear discussions in German of the Davis cycle.)

LAWSON, A. C. (1897) The post-Pleistocene diastrophism of the coast of southern California, *Univ. of California Dept. of Geology, Bulletin*, Vol. 1, pp. 115–60.

LAWSON, A. C. (1915) Epigene profiles of the desert, *Univ. of California, Publications in Geology, Bulletin 9*, pp. 23–48.

LAWSON, A. C. (1925) The Cypress Plain, *Univ. of California Dept. of Geology, Bulletin*, Vol. 15 (6), pp. 153–8.

LE CONTE, J. (1880) The old river-beds of California, *American Journal of Science*, Vol. 19, pp. 176–90.

LE CONTE, J. (1886) A post-Tertiary elevation of the Sierra Nevada shown by the river beds, *American Journal of Science*, Vol. 32, pp. 167–81.

LEE, W. T. (1925) Carlsbad Cavern, New Mexico, *Scientific Monthly*, Vol. 21, pp. 186–90.

LEIGHLY, J. (1955) What has happened to physical geography?, *Annals of the Association of American Geographers*, Vol. 45, pp. 309–18.

LOBECK, A. K. (1944) Douglas Johnson, *Annals of the Association of American Geographers*, Vol. 34, pp. 216–22.

LOUDERBACK, G. D. (1904) Basin range structure of the Humboldt Region, *Bulletin of the Geological Society of America*, Vol. 15, pp. 289–346.

LÖWL, F. (1882) Die Entstehung der Durchbruchstäler, *Petermanns Geographische Mitteilungen*, Vol. 28, pp. 405–16.

MACKIN, J. H. (1948) The concept of the graded river, *Bulletin of the Geological Society of America*, Vol. 59, pp. 463–512.

MARTIN, G. J. (1968) *Mark Jefferson: Geographer* (Eastern Michigan University Press, Ypsilanti, Michigan), 370 pp.

MARTIN, G. J. (1972) Robert LeMoyne Barrett, 1871–1969: Last of the founding members of the Association of American Geographers, *Professional Geographer*, Vol. 24 (1), pp. 29–31.

MARTIN, G. J. (Forthcoming) *The Life and Thought of Ellsworth Huntington.*

MARTIN, L. (1950) William Morris Davis: Investigator, Teacher and Leader in geomorphology, *Annals of the Association of American Geographers*, Vol. 40, pp. 172–80.

MATSON, G. C. (1909) Water resources of the Bluegrass Region, Kentucky, *U.S. Geological Survey, Water Supply Paper 233*, 223 pp.

MCGEE, W J (1888) Three formations of the middle Atlantic slope, *American Journal of Science*, Vol. 35, pp. 120–43, 328–30, 367–88 and 448–66.

MCGEE, W J (1897) Sheetflood erosion, *Bulletin of the Geological Society of America*, Vol. 8, pp. 87–112.

MEINZER, O. E. (1923) The occurrence of ground water in the United States, *U.S. Geological Survey, Water Supply Paper 489*, 321 pp.

MERRILL, G. P. (1920) Contributions to a history of American State Geological and Natural History Surveys; *Smithsonian Institution, U.S. National Museum, Bulletin 109*, 549 pp.

MILL, H. R. (1934) William Morris Davis, *Geographical Journal*, Vol. 84, pp. 93–5.

MILL, H. R. (1951) *An Autobiography* (Longmans, London), 224 pp.

NIELSON, W. A. (Ed.) (1926) *Charles W. Eliot: The Man and his beliefs*, 2 vols (Harpers, New York).

OESTREICH, K. (1934) William Morris Davis, *Petermanns Geographische Mitteilungen*, Year 78, p. 136.

PAIGE, S. (1912) Rock-cut surfaces in the desert ranges, *Journal of Geology*, Vol. 20, pp. 442–50.

PASSARGE, S. (1904) *Die Kalahari* (Berlin), 289 pp.

PASSARGE, S. (1919) *Die Grundlagen der Landschaftskunde ... Band I: Beschreibende Landschafteskunde* (Friedrichsen, Hamburg), 210 pp.

PASSARGE, S. (1920) *Die Grundlagen der Landschaftskunde ... Band III: Die Oberflächengestaltung der Erde* (Friedrichsen, Hamburg), 558 pp.

PENCK, A. (1894) *Morphologie der Erdoberfläche*, 2 vols (Engelhorn, Stuttgart), 486 pp and 696 pp.

PENCK, A. and BRÜCKNER, E. (1901–9) *Die Alpen im Eiszeitalter*, 3 vols (Chr. Herm. Tauchnitz, Leipzig), 1199 pp.

PENCK, A. (1913) Die Formen der Landoberfläche und Verschiebungen der Klimagürtel, *Sitzungsber. Preuss. Akad. Wissensch.*, Jan.–June, pp. 77–97.

PENCK, A. (1914) The shifting of climatic belts, *Scottish Geographical Magazine*, Vol. 30, pp. 281–93.

PENCK, A. (1919) Die Gipfelflur der Alpen, *Sitzungsber. Preuss. Akad. Wissensch., Math.-Phys. Kl.* XVII, pp. 256–68. (The summit level of the Alps.)

PENCK, A. (1926) Reply to the review by I. Bowman, *Geographical Review*, Vol. 16, pp. 350–1.

PENCK, W. (1920A) Der Südrand der Puna de Atacama (Nordwestargentinien): Ein Beitrag zur Kenntnis des andinen Gebirgstypus und zu der Frage der Gebirgsbildung, *Abhandl. Math.-Phys. Kl. sächs. Akad. Wissensch., Leipzig, Band 37*, No. 1, 420 pp.

PENCK, W. (1920B) Wesen und Grundlagen der morphologischen Analyse, *Bericht Math.-Phys. Kl. sächs. Akad. Wissensch., Leipzig, Band 72*, pp. 65–102.

PENCK, W. (1924) *Die morphologische Analyse: Ein Kapitel der physikalischen Geologie*; Geographische Abhandlungen. 2. Reihe, Heft 2 (Stuttgart), 283 pp.

PENCK, W. (1925) Die Piedmontflächen des südlichen Schwarzwaldes, *Zeitschr. Gessellch. Erdk. Berlin*, pp. 83–108. (The piedmont flats of the southern Black Forest.) Translation by M. Simons, 1961.

PENCK, W. (1953) *Morphological Analysis of Landforms*, Trans. by Czech, H. and Boswell, K. C. (Macmillan, London), 429 pp.

PLATT, R. S. (1957) A note on Rollin D. Salisbury, *Annals of the Association of American Geographers*, Vol. 47, p. 276.

POWELL, J. W. (1876) *Report on the Geology of the Eastern Portion of the Uinta Mountains* (Washington), 218 pp.

PUMPELLY, R. (1918) *My Reminiscences*, 2 vols (Holt, New York).

PUMPELLY, R. (1920) *Travels and Adventures of Raphael Pumpelly*, Ed. by O. S. Rice (Holt, New York), 367 pp.

PUTNAM, W. C. (1937) The marine erosional cycle for a steeply sloping shoreline of emergence, *Journal of Geology*, Vol. 43, pp. 844–50.

RAND, J. C. (1890) *One of a Thousand: A series of biographical sketches of one thousand representative men resident in the Commonwealth of Massachusetts (1888–9)* (Boston), 707 pp.

RICHTER, E. (1907) Beitrage für landeskunde Bosniens und der Herzegowina, *Wiss. Mitt. aus Bosnien und der Herzegowina*, Vol. 10, pp. 383–545.

RIGDON, V. E. (1933) *The Contributions of Williams Morris Davis to Geography in America* (Ph.D. Dissertation, University of Nebraska).

RUSSELL, I. C. (1895) *The Lakes of North America* (Ginn, Boston), 125 pp.

SAARMANN, G. (1951) William Morris Davis, *Petermanns Geographische Mitteilungen*, Year 95, pp. 193–5.

SALISBURY, R. D. (1924) *Physiography*, 3rd edn (John Murray, London), 676 pp.

SAWICKI, L. (1909) Ein beitrage zum geographischen Zyklus in Karst, *Geographische Zeitschrift*, Vol. 15, pp. 187–204 and 259–81.

SAYLES, R. W. (1931) Bermuda during the Ice Age, *Proceedings of the American Academy of Arts and Sciences*, Vol. 66, pp. 382–467.

SCHUCHERT, C. (1910) Paleogeography of North America, *Bulletin of the Geological Society of America*, Vol. 20, pp. 427–606.

SCHUMM, S. A. (1963) The disparity between present rates of denudation and orogeny, *U.S. Geological Survey, Professional Paper 454-H*, 13 pp.

SIMONS, M. (1962) The morphological analysis of landforms: A new review of the work of Walther Penck, *Transactions of the Institute of British Geographers, No. 31*, pp. 1–14.

SMEDLEY, R. C. (1883) *History of the Underground Railroad in Chester and neighboring Counties of Pennsylvania* (Lancaster, Pa.), 407 pp.

SPURR, J. E. (1901) Origin and structure of the basin ranges, *Bulletin of the Geological Society of America*, Vol. 12, pp. 217–70.

STILL, W. (1872) *The Underground Rail Road* (Philadelphia), 780 pp.

STODDART, D. R. (1962) Coral Islands by Charles Darwin, *Atoll Research Bulletin*, No. 88, 20 pp.

STODDART, D. R. (1966) Darwin's Impact on Geography, *Annals of the Association of American Geographers*, Vol. 56, pp. 683–98.

STODDART, D. R. (1969) Ecology and morphology of recent coral reefs, *Biological Reviews*, Vol. 44, pp. 433–98.

STRAHLER, A. N. (1950) Davis' concepts of slope development viewed in the light of recent quantitative investigations, *Annals of the Association of American Geographers*, Vol. 40, pp. 209–13.

SUESS, E. (1883–1908) *Das antlitz der Erde*, 3 vols (Tempsky, Vienna). Vol. 2, 1888, translated by E. de Margerie, 1900 (Armand Colin, Paris), and by H. B. C. and W. J. Sollas, 1906 (Oxford).

SWINNERTON, A. C. (1932) Origin of limestone caverns, *Bulletin of the Geological Society of America*, Vol. 43, pp. 663–94.

Symposium (1940) Walther Penck's contribution to geomorphology, *Annals of the Association of American Geographers*, Vol. 30, pp. 219–80.

TARR, R. S. (1898) The peneplain, *American Geologist*, Vol. 21, pp. 351–70.

TEITZE, E. (1878) Einige Beonerkungen über die Bildung von Querthälern, *Jahrbuch Geologischen Reichsanstalt*, Vol. 28, pp. 581–610.

THORNBURY, W. D. (1954) *Principles of Geomorphology*, 1st edn (Wiley, New York), 618 pp.

TILLEY, P. (1968) Early challenges to Davis' concept of the cycle of erosion; *Professional Geographer*, Vol. 20, pp. 265–9.

THOMAS, J. (1870), *Universal Pronouncing Dictionary of Biography and Mythology*, 2 vols (Lippincott, Philadelphia and London).

TUAN, Y. F. (1957) The misleading antithesis of Penckian and Davisian concepts of slope retreat in waning development, *Proceedings of the Indiana Academy of Science*, Vol. 67, pp. 212–14.

TUAN, Y. F. (1959) Pediments in Southeastern Arizona, *University of California Publications in Geography*, Vol. 13, 163 pp.

VILLARD, O. W. (1910) *John Brown 1800–1859* (Constable, London), 738 pp.

VIPONT, E. (1954) *The Story of Quakerism 1652–1952* (The Bannisdale Press, London), 312 pp.

VISHER, S. S. (1961) The Association of American Geographers in its early years (Abst.), *Annals of the Association of American Geographers*, Vol. 51, p. 426.

VON ENGELN, O. D. (1942) *Geomorphology* (Macmillan, New York), 655 pp.

WALCOTT, C. D. (1891) Correlation Papers: Cambrian, *U.S. Geological Survey, Bulletin 81*, 447 pp.

WALTHER, J. (1900) *Das Gesetz der Wüstenbildung in Gegenwart und Vorzeit* (Quelle and Meyer, Leipzig), 342 pp.

WELLER, J. M. (1927) The Geology of Edmonson County, *Kentucky Geological Survey*, Ser. 6, Vol. 28, 246 pp.

WHITBECK, R. H. (1910) The present trend of geography in the United States, *Geographical Journal*, Vol. 35, pp. 419–25.

Who's Who in America (1932–3) William Morris Davis (Marquis, Chicago), Vol. 17, p. 669.

WILLIAMS, H. S. (1897) On the southern Devonian formations, *American Journal of Science*, Fourth Series, Vol. 3, pp. 393–403.

WOODFORD, A. O. (1926) The visit of W. M. Davis to Pomona College, *Pomona College Quarterly Magazine*, March 1926 (Reprinted in the *Harvard Alumni Bulletin*, Vol. 28, No. 29, p. 864).

WOOLDRIDGE, S. W. and LINTON, D. L. (1939) Structure, surface and drainage in south-east England; *Transactions of the Institute of British Geographers, No. 10*, 120 pp.

WOOLDRIDGE, S. W. (1955) 'Geographical Essays' by W. M. Davis; *Geographical Journal*, Vol. 121, pp. 89–90.

WRIGHT, J. K. (1963) Wild geographers I have known, *Professional Geographer*, Vol. 15 (4), pp. 1–4.

WRIGHT, J. K. (1966) *Human Nature in Geography* (Harvard University Press), 362 pp.

WRIGLEY, G. M. (1951) Isaiah Bowman, *Geographical Review*, Vol. 41, pp. 7–65.

Indexes

Subject and Place Index
Index of Persons

Subject and Place Index

In the study of landforms principle and place are symbiotic and cannot safely be divorced. But place-names have posed a problem because Davis was a great traveller and loved the minutiae of itineraries. In the following index, places and areas on which a considerable amount of information is provided are given separate entries. Places and areas on which a small amount of information is provided are listed under the major administrative or political unit in which they are situated as this grouping seemed likely to be most helpful to the reader. However a few places, such as Numadzu and Bra, on which no useful information is given except the name, are omitted for the sake of brevity.

Here, and in the *Index of Persons*, references to figures are added in italic at the end of the entries.

D. M. Beckinsale

Index of Persons

References to photographs and drawings are given in italics at the end of the entries.

The informative comments on some of the Davis and Mott family are quoted from an abridged record of family traits that Davis sent to the National Academy of Sciences at their request in 1932.

T - #0123 - 071024 - C0 - 234/156/48 - PB - 9780415567954 - Gloss Lamination